CAMBRIDGE MONOGRAPHS
ON MATHEMATICAL PHYSICS

QUANTUM FIELDS IN CURVED SPACE

QUANTUM FIELDS IN CURVED SPACE

N. D. BIRRELL

Logica Pty Ltd, Australia

AND

P. C. W. DAVIES

Professor of Theoretical Physics, University of Newcastle upon Tyne

CAMBRIDGE UNIVERSITY PRESS
Cambridge, New York, Melbourne, Madrid, Cape Town, Singapore,
São Paulo, Delhi, Dubai, Tokyo, Mexico City

Cambridge University Press
The Edinburgh Building, Cambridge CB2 8RU, UK

Published in the United States of America by
Cambridge University Press, New York

www.cambridge.org
Information on this title: www.cambridge.org/9780521278584

First published 1982
First paperback edition (with corrections) 1984
Reprinted 1989, 1992, 1994

A catalogue record for this publication is available from the British Library

Library of Congress catalogue card number: 81-3851

ISBN 978-0-521-23385-9 Hardback
ISBN 978-0-521-27858-4 Paperback

Contents

Preface to the paperback edition

Since the book first went to press, there have been several important advances in this subject area. The topic of interacting fields in curved space has been greatly developed, especially in connection with the phenomenon of symmetry breaking and restoration in the very early universe, where both high temperatures and spacetime curvature are significant. A direct consequence of this work has been the formulation of the so-called inflationary universe scenario, in which the universe undergoes a de Sitter phase in the very early stages. This work has focussed attention once more on quantum field theory in de Sitter space, and on the calculation of $\langle \phi^2 \rangle$. A comprehensive review of the inflationary scenario is given in *The Very Early Universe*, edited by G.W. Gibbons, S.W. Hawking and S.T.C. Siklos (Cambridge University Press, 1983).

Further results of a technical nature have recently been obtained concerning a number of the topics considered in this book. Mention should be made of the work of M.S. Fawcett, who has finally calculated the quantum stress tensor for a Schwarzschild black hole (*Commun. Math. Phys.*, **81** (1983), 103), and of W.G. Unruh & R.M. Wald, who have clarified the thermodynamic properties of black holes by appealing to the effects of accelerated mirrors close to the event horizon (*Phys. Rev. D*, **25** (1982), 942; **27** (1983), 2271). Interest has also arisen over field theories in higher-dimensional spacetimes, in which Casimir and other vacuum effects become important. For a review, see E. Witten, *Nucl. Phys. B*, **186** (1981), 412. Finally, much further work has been done on the properties of particle detectors (see, for example, the paper by K.J. Hinton in *J. Phys. A: Gen. Phys.*, **16** (1983), 1937).

We are grateful to K.J. Hinton, J. Pfautsch, S.D. Unwin and W.R. Walker for assistance in revising the text.

Note added at 1986 reprinting

We would like to thank Professor H. Minn for providing corrections to the original printing.

Preface

The subject of quantum field theory in curved spacetime, as
mation to an as yet inaccessible theory of quantum gravity,
tremendously in importance during the last decade. In this bo
attempted to collect and unify the vast number of papers
contributed to the rapid development of this area. The book al
some original material, especially in connection with partic
models and adiabatic states.

The treatment is intended to be both pedagogical and ar
assume no previous acquaintance with the subject, but the rea
preferably be familiar with basic quantum field theory at th
Bjorken & Drell (1965) and with general relativity at the level of
(1972) or Misner, Thorne & Wheeler (1973). The theory is develo
basics, and many technical expressions are listed for the first tin
place. The reader's attention is drawn to the list of convent
abbreviations on page ix, and the extensive references and bibli

In preparing this book we have drawn upon the material of a v
number of authors. In adapting certain published material (includ
of the authors) we have gratuitously made what we conside
corrections, occasionally without explicitly warning the reader that
of that material differs from the original publications.

The bulk of the text was written while we worked together
Department of Mathematics, Kings College, London. We are
indebted to many colleagues there and elsewhere for assistance.
thanks are extended to T.S. Bunch, S.M. Christensen, N.A. Dought
Dowker, M.J. Duff, L.H. Ford, S.A. Fulling, C.J. Isham, G. Kenne
Parker and R.M. Wald for critical reading of sections of manuscri

Finally we should like to thank Mrs J. Bunn for typing the manu
and the Science Research Council for financial support.

Conventions and abbreviations

Our notation for quantum field theory mainly follows that of Bjorken & Drell (1965). The sign conventions for the metric and curvature tensors are $(- - -)$ in the terminology of Misner, Thorne & Wheeler (1973). That is, the metric signature is $(+ - - -)$; $R^{\alpha}{}_{\beta\gamma\delta} = \partial_\delta \Gamma^{\alpha}{}_{\beta\gamma} - \ldots$; $R_{\mu\nu} = R^{\alpha}{}_{\mu\alpha\nu}$. Formulae can be changed from our notation to the often used Misner, Thorne & Wheeler $(+ + +)$ conventions by changing the signs of $g_{\mu\nu}$, $\Box \equiv g^{\mu\nu}\nabla_\mu \nabla_\nu$, $R^{\alpha}{}_{\beta\gamma\delta}$, $R_{\mu\nu}$, $T_\mu{}^\nu$ but leaving $R_{\alpha\beta\gamma\delta}$, $R_\mu{}^\nu$, R and $T_{\mu\nu}$ unchanged. For the majority of the book we use units in which $\hbar = c = G = 1$.

The following special symbols and abbreviations are used throughout:

$*$	complex conjugate
$\dagger$ or h.c.	Hermitian conjugate
$\bar{\ }$	Dirac adjoint
$\dfrac{\partial}{\partial x^\mu}$ or ∂_μ or $,\mu$	partial derivative
∇_μ or $;\mu$	covariant derivative
Re (Im)	real (imaginary) part
tr	trace
ln	natural logarithm
k_{B}	Boltzmann's constant
γ	Euler's constant
$[A,B]$	$AB - BA$
$\{A,B\}$	$AB + BA$
$a_{(\mu,\nu)}$	$\frac{1}{2}(a_{\mu,\nu} + a_{\nu,\mu})$
$\simeq$	approximately equal to
$\sim$	order of magnitude estimate
$\approx$	asymptotically approximate to
$\equiv$	defined to be equal to
$: \, :$	normal ordering

1
Introduction

The last decade has witnessed remarkable progress in the construction of a unified theory of the forces of nature. The electromagnetic and weak interactions have received a unified description with the Weinberg–Salam theory (Weinberg 1967, Salam 1968), while attempts to incorporate the strong interaction as described by quantum chromodynamics into a wider gauge theory seem to be achieving success with the so-called grand unified theories (Georgi & Glashow 1974, for a review see Cline & Mills 1978).

The odd one out in this successive unification is gravity. Not only does gravity stand apart from the other three forces of nature, it stubbornly resists attempts to provide it with a quantum framework. The quantization of the gravitational field has been pursued with great ingenuity and vigour over the past forty years (for reviews see Isham 1975, 1979*a*, 1981) but a completely satisfactory quantum theory of gravity remains elusive. Perhaps the most hopeful current approaches are the supergravity theories, in which the graviton is regarded as only one member of a multiplet of gauge particles including both fermions and bosons (Freedman, van Nieuwenhuizen & Ferrara 1976, Deser & Zumino 1976; for a review see van Nieuwenhuizen & Freedman 1979).

In the absence of a viable theory of quantum gravity, can one say anything at all about the influence of the gravitational field on quantum phenomena? In the early days of quantum theory, many calculations were undertaken in which the electromagnetic field was considered as a classical background field, interacting with quantized matter. Such a semiclassical approximation readily yields some results that are in complete accordance with the full theory of quantum electrodynamics (see, for example, Schiff 1949, chapter 11). One may therefore hope that a similar regime exists for quantum aspects of gravity, in which the gravitational field is retained as a classical background, while the matter fields are quantized in the usual way. Adopting Einstein's general theory of relativity as a description of gravity, one is led to the subject of quantum field theory in a curved background spacetime, which is the subject of this book.

It was originally pointed out by Planck (1899) that the universal constants G, $\hbar$ and c could be combined to give a new fundamental unit of length, the Planck length $(G\hbar/c^3)^{\frac{1}{2}} = 1.616 \times 10^{-33}$ cm, and time, the Planck time $(G\hbar/c^5)^{\frac{1}{2}} = 5.39 \times 10^{-44}$ s. If the gravitational field is treated as a small perturbation, and attempts are made to quantize it along the lines of quantum electrodynamics (Q.E.D.), then the square of the Planck length appears in the role of coupling constant. Unlike Q.E.D., however, in which the coupling constant $e^2/\hbar c$ is dimensionless (and small) the Planck length has dimensions. Effects can become large when the length and time scales of the quantum processes of interest fall below the Planck value. When this happens, the higher orders of perturbation theory become comparable with the lowest order, and the whole concept of a small perturbation expansion breaks down.

The Planck values therefore mark the frontier at which a full theory of quantum gravity, preferably non-perturbative in character, must be invoked. Nevertheless, one might hope that when the distances and times involved are much larger than the Planck values, the quantum effects of the gravitational field will be negligible. As the Planck length is so small (twenty powers of ten below the size of an atomic nucleus) this appears to leave much scope for a semiclassical theory.

There is, however, a problem with this naive reasoning. According to the equivalence principle, which lies at the very foundation of metric theories of gravity, all forms of matter and energy couple equally strongly to gravity. This includes, of course, gravitational energy; crudely speaking, gravity gravitates. In quantum language, we may say that the graviton is just as much subject to an external gravitational field as, say, a photon. Consequently, whenever a classical background gravitational field produces important effects involving (real or virtual) photons, one must allow for equally important effects involving gravitons. It follows that quantum gravity will enter in a non-trivial way *at all* scales of distance and time, whenever interesting quantum field effects occur. Thus, the basic nonlinearity of gravity frustrates all attempts to ignore quantum gravity (Duff 1981).

In spite of this complication, it may still be possible to proceed with a semiclassical description. In ordinary classical relativity one frequently wishes to discuss the propagation of gravitational waves in a curved background spacetime. Although one is dealing with the vacuum Einstein equation, the small disturbance that represents the gravitational wave can be separated off from the background spacetime:

$$g_{\mu\nu} = g^{c}_{\mu\nu} + \bar{g}_{\mu\nu} \qquad (1.1)$$

where $\bar{g}_{\mu\nu}$ represents the wave and $g^{c}_{\mu\nu}$ the background spacetime. The waves can then be treated as a null fluid much like any other, and their contribution to the left-hand side of Einstein's equations can be cast in a form that enables them to be taken over to the right and treated as a part of the source, i.e., as a part of $T_{\mu\nu}$.

In this spirit, it seems reasonable to suppose that the 'graviton' field, representing linearized perturbations on the background spacetime, can be included along with all the other quantum fields as part of the matter rather than the geometry. So long as one remains well clear of Planck dimensions, such a linearized approximation should work. However, until a full theory of quantum gravity is available, this approximation will be open to question. (Whether or not one should call a theory that makes provision for gravitons 'semiclassical' is, of course, a matter for debate.) It is analogous to treating photon emission by an atom immersed in a background electric or magnetic field.

DeWitt (1967*a*,*b*) has used (1.1) as the starting point for an approach to the complete quantization of gravity – the so called 'background field' method. In this approach $g^{c}_{\mu\nu}$ is the classical metric of some background spacetime, and $\bar{g}_{\mu\nu}$ is taken to be a quantum field propagating in this background. The Einstein action and the matter field actions can be expanded in powers of $\bar{g}$ about g^{c}, and Feynman rules derived. The lowest order, one-loop, quantum processes are shown in the Feynman diagrams in fig. 1. The wavy line represents the propagator for the field $\bar{g}_{\mu\nu}$ propagating in the background $g^{c}_{\mu\nu}$, while the uniform line represents the corresponding propagator for a matter field. Since neither fig. 1*a* nor 1*b* contains any graviton vertices, they both represent expressions which are independent of the gravitational coupling constant G, and so are always of comparable magnitude. Thus, at the one-loop level, the quantization of the

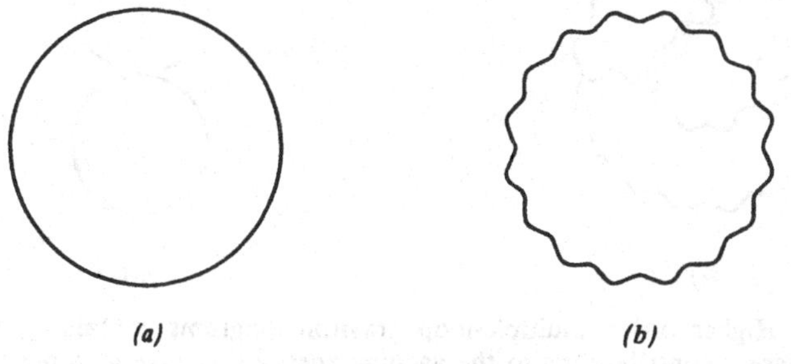

Fig. 1. Lowest order (one-loop) contribution to the vacuum energy from matter fields (*a*) and linearized gravitons (*b*) propagating in a prescribed background gravitational field.

gravitational field in the background $g^{c}_{\mu\nu}$ is equally as important as the quantization of the matter fields.

Physically, the single closed loops represent an infinite vacuum or zero-point energy, that in the case of a flat background spacetime (i.e., conventional quantum field theory) is artificially removed by subtraction, or by so-called normal ordering (see §2.4). When the background is curved, however, a more elaborate procedure is necessary involving the dynamics of the gravitational field. This is the device known as *renormalization*, and is familiar from Q.E.D., which is also plagued by divergences.

In the latter case, the divergences are removed by renormalization of particle masses, charges and wavefunctions. Only a finite number of quantities needs to be renormalized, a feature that qualifies Q.E.D. for the status of a 'renormalizable theory'. This 'renormalizability' depends crucially on the fact that the coupling constant $e^2/\hbar c$ is dimensionless. In contrast, G has units of $(\text{length})^2$ (in natural units $\hbar = c = 1$), which gives rise, via a simple power counting argument, to an unending sequence of new divergences at each order. Higher order terms in the expansion of the

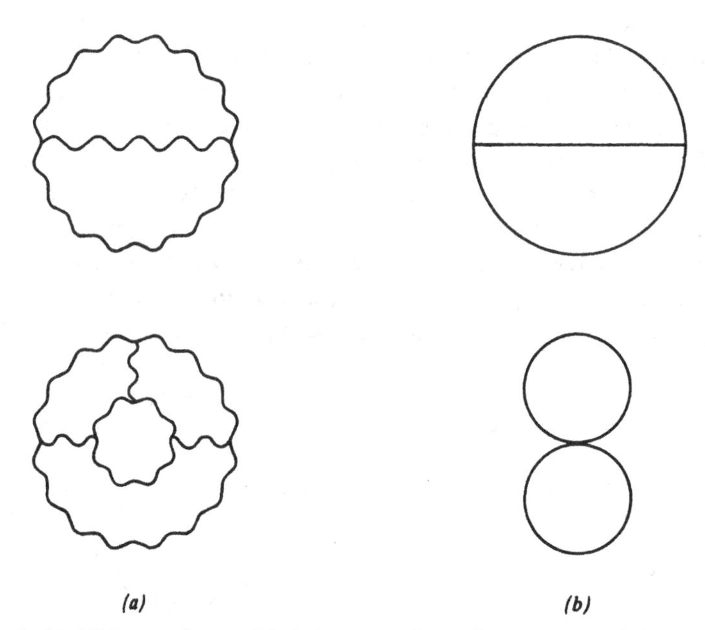

Fig. 2. (*a*) Higher order, multiple-loop graviton diagrams containing vertices. These represent contributions to the vacuum energy that diverge more strongly than that from fig. 1(*b*), and render quantum gravity non-renormalizable. (*b*) When non-gravitational interactions are introduced, multiple loops can also occur in the matter field diagrams.

gravitational action in powers of $\bar{g}$ produce graviton Feynman diagrams with multiple loops (see fig. 2*a*). There is a simple relationship between the degree of divergence and the number of such loops in a diagram (see, for example, Duff 1975), such that with increasing numbers of loops one encounters more and more virulent divergences. This fact renders quantum gravity 'unrenormalizable' – with each new order more new physical quantities have to be invented to absorb the infinities. It is for this reason that the quantization of the gravitational field has not been satisfactorily accomplished.

If one truncates the expansion of the combined theory of gravity and matter at some particular number of loops, then the finite number of divergent quantities can be removed by renormalization of a finite number of physical quantities. For example, as we shall see in chapter 6, at the one-loop level, renormalization of G, the cosmological constant Λ, and the coupling constants of two new geometrical tensors suffices to render the theory finite. Thus, in a sense, this truncated theory could be considered renormalizable.

It should be noted that if one attempts to carry out quantization by expanding around flat spacetime, taking $g^{c}_{\mu\nu} = \eta_{\mu\nu}$ (the Minkowski space metric) in (1.1), then the renormalization cannot be carried out in a generally covariant way (see, for example, Duff 1975). This, combined with other difficulties (Christensen & Duff 1980), obliges one to consider fields propagating in a background with arbitrary metric $g^{c}_{\mu\nu}$.

In this book we consider, for the most part, the quantum theory of gravity plus matter truncated at the one-loop level. For free matter fields, there are no higher loop processes anyway and fig. 1*a* gives the exact contribution of these fields (to the effective action defined in chapter 6). Fig. 1*b* gives only the contribution from the gravitons which is of zeroth order in G. As the loop expansion is an expansion in $\hbar$ (Nambu 1966) the theory truncated at the one-loop level contains all terms of the complete theory to order $\hbar$, and is in that sense the first order quantum correction to general relativity.

If self- or mutually-interacting matter fields are included in the theory then the one-loop contribution of the matter fields shown in fig. 1*a* is no longer exact, as there now exist multiple-loop diagrams involving vertices (fig. 2*b*). As it is the effects of interactions between matter fields ('particles') that are most often observed in the laboratory, such multiple-loop matter field diagrams should be included in our discussions. However, if we wish to work to a consistent order in $\hbar$, we should then have to include *graviton* diagrams also with arbitrary numbers of loops and so be confronted with the non-renormalizability of gravity. However, each extra graviton loop in a connected Feynman diagram in general introduces a factor of G, while each matter field loop introduces a factor of the relevant coupling constant (such as e^2). If l is a typical length or time scale for the system under consideration, then provided $l^{-2}G \ll e^2$, the effect of additional graviton loops will be insignificant compared with that of additional matter loops. Thus, even in the case of interacting matter fields, there is a large regime in which quantum gravity can be limited to the one-loop level with some justification.

Given that we possess at least a reasonable approximation to a theory of gravitational effects on quantum fields, how important are the processes described thereby? Crudely speaking, non-trivial gravitational effects occur in quantum field modes for which the wavelength λ is comparable with some characteristic length scale of the background spacetime. Thus, near a black hole of radius r, the quantum field modes with wavelength $\gtrsim r$ are seriously disturbed by the presence of the hole. Similarly, if the gravitational field changes on a timescale t, then quantum field modes with frequency $\lesssim t^{-1}$ are seriously disturbed. Thus, if one regards 10^{-13} cm and 10^{-23} s as the length and time scales characteristic of important quantum processes, one finds that only in the vicinity of microscopic black holes or in the earliest epochs of the big bang can important effects be expected.

The weakness of gravity therefore effectively precludes phenomena that can be studied in the laboratory and, unless microscopic black holes are much more numerous than present estimates suggest, it precludes any possibility at all of direct observational verification. Quantum field theory in curved spacetime must, it seems, rest entirely upon theoretical considerations.

The paucity of experimental checks renders all the more significant the results of Hawking (1975). His study of quantum black holes and the discovery of their thermal emission is a cornerstone of the theory developed in this book. Hawking's result is compelling for two reasons. First, it appears to be very fundamental, and has been derived in several different ways. Second, it establishes a strong connection between black holes and thermodynamics that was suspected before the application of quantum theory to black holes. It is therefore tempting to suppose that the Hawking effect has exposed a small corner of a broad new area of fundamental physics in which gravity, quantum field theory and thermodynamics are closely interwoven. If this is the case, then their synthesis would almost certainly lead to important new advances in physics, including some with observational consequences.

The situation can be compared with the early days of kinetic theory. The atomic hypothesis was not really open to direct experimental verification in the mid-nineteenth century due to the smallness of atomic effects. Nevertheless, the fully developed theory was capable of reaching beyond the atomic domain and predicting new phenomena in gas dynamics that could be checked. Similarly, one hopes that quantum gravity would, if it became properly understood, intrude into other, more accessible, areas of physics.

Investigation of the effects of gravity on quantum fields dates at least since the work of Schrödinger (1932). After the Second World War there was a surge of interest in quantizing the gravitational field, but direct investigation of particle creation effects in a background gravitational field really began in earnest with the work of Parker in the late 1960s followed by the investigations of Zel'dovich and coworkers. These early investigations dwelt on the cosmological consequences of particle creation. They were hampered by the lack of systematic techniques for studying the stress–energy–momentum tensor, $T_{\mu\nu}$, of the gravitationally disturbed quantum fields. The stress-tensor is important for two reasons. It can be used to assess the importance of quantum effects on the dynamics of the gravitational field itself (i.e., the back-reaction problem). Also, it is frequently a more useful probe of the physical situation than a particle count. In regions of strong gravity, vacuum polarization effects, akin to those in Q.E.D., can lead to important phenomena even in the absence of actual particle creation.

In the mid-seventies, a great deal of effort was expended in developing rigorous techniques for computing $\langle T_{\mu\nu} \rangle$. Being formally infinite, these techniques involve renormalization, so some of them were borrowed from ordinary (Minkowski space) quantum field theory and from the full quantum gravity theory, while others were developed specially.

An essential feature of these techniques is that they all yield a covariantly conserved (i.e. divergenceless) $\langle T_{\mu\nu} \rangle$, which is therefore a suitable candidate for the right-hand side of a semiclassical Einstein equation. It might seem that if particle pairs are created from the vacuum then $\langle T_{\mu\nu} \rangle$ should not be conserved. Indeed, Hawking (1970) showed that a conserved $\langle T_{\mu\nu} \rangle$, subject to the dominant energy condition (basically that energy and pressure should always remain positive – see Hawking & Ellis 1973) was incompatible with particle creation. However, one of the unusual features of curved space quantum field theory is that spacetime curvature can induce negative stress–energy–momentum in the vacuum, thereby violating the dominant energy condition, and circumventing Hawking's result (Zel'dovich & Pitaevsky 1971). Thus particle creation is compatible with a conserved $\langle T_{\mu\nu} \rangle$. The possibility of violating the energy conditions also opens the way to the avoidance of spacetime singularities (see §7.4).

With Hawking's discovery of thermal black hole emission, combined with these improving techniques for computing $\langle T_{\mu\nu} \rangle$, the subject of quantum field theory in curved spacetime, or one-loop quantum gravity, enjoyed a period of very rapid expansion, with several hundred papers appearing in the literature. While some of the interpretative issues have been contentious (most notably the physical significance to be attached to particles versus $\langle T_{\mu\nu} \rangle$, the effect of spacetime singularities, the existence of so-called conformal anomalies, and the criteria to be used in the specification of appropriate quantum states) there is now broad agreement on most of the technical results. In the chapters that follow the treatment will inevitably reflect some of the interpretative opinions of the authors, but we have tried as far as possible to adhere to what we understand to be the 'majority view'. We have tried to avoid making vague statements of a physical nature about 'particles' and 'energy', where possible stating our conclusions in operational terms – what a hypothetical observer moving in such-and-such a way would actually measure with a particular piece of apparatus.

The organization of material is conventional. In chapter 2 we review the basic concepts of ordinary Minkowski space quantum field theory and establish our notation. Chapter 3 generalizes these ideas to curved spacetime and introduces the notion of particle creation by gravitational fields. We also give what is to date the most complete and careful description of the concepts of adiabatic states, and of particle detectors. Applications to flat and curved spacetime follow in chapters 4 and 5, with several explicit examples worked in detail.

A central feature of the book, chapter 6, gives an exhaustive account of regularization and renormalization techniques. Several methods have been published, and we have endeavoured not only to discuss them all (with examples) but to unify them as much as possible. Armed with these results, the reader can tackle chapter 7, that applies the methods of chapter 6 to a number of important examples.

The full range of techniques developed in the earlier chapters are then deployed for a detailed discussion of black holes, which occupies the whole of chapter 8. Considerations of space compelled us to limit many of the important and fascinating physical implications to a catalogue of references. We have been unable, for example, to dwell at length on the thermodynamic aspects of black holes and their possible extension to more general gravitational fields; nor have we been able to give a detailed account of time asymmetry, Poincaré cycles, black holes versus white holes and the implications of quantum spacetime singularities. In addition, nearly all the exciting work on the astrophysical and cosmological consequences of quantum black holes has been omitted.

The final chapter attempts to go beyond what may be considered as the 'first round' of results in the subject of quantum field theory in curved spacetime. Here we briefly outline the generalization of the theory that will be necessary to accommodate non-gravitational interactions. Special interest centres on issues such as the effect of self- and mutual-field interactions on particle creation and vacuum stress, and whether a theory that is renormalizable in Minkowski space, such as Q.E.D. or $\lambda\phi^4$ theory, remains renormalizable in the presence of a non-trivial topology and/or spacetime curvature.

It is inevitable in a book of this sort that there will be some ragged edges. The subject is still rapidly evolving, and many gaps remain to be filled. Nevertheless, the last year or so has witnessed a period of consolidation and reflection, so that we have been able to present a reasonably coherent and self-contained account.

2
Quantum field theory in Minkowski space

In this chapter we shall summarize the essential features of ordinary Minkowski space quantum field theory, with which we assume the reader has a working knowledge. A great deal of the formalism can be extended to curved spacetime and non-trivial topologies with little or no modification. In the later chapters we shall follow the treatment given here.

Most of the detailed analysis will refer to a scalar field, but the main results will be listed for higher spins also. This restriction will enable the important features of curved space quantum field theory to emerge with the minimum of mathematical complexity.

Much of the chapter will be familiar from textbooks such as Bjorken & Drell (1965), but the reader should take special note of the results on the expectation value of the stress–energy–momentum tensor and vacuum divergence (§2.4), as these will play a central role in what follows. Special importance also attaches to Green functions, treated in detail in §2.7. The reader may be unfamiliar with thermal Green functions and metric Euclideanization. As these will be essential for an understanding of the quantum black hole system, an outline of this topic is given here.

Finally, although we shall not develop a lot of our formalism using the Feynman path-integral technique, we do make use of the basic structure of the path integral in the work on renormalization in chapter 6, and again on interacting fields in curved space in chapter 9. While it is not necessary for the reader to master the path-integral formulation, the basic outline given in §2.8 may be found helpful.

2.1 Scalar field

Consider a scalar field $\phi(t,\mathbf{x})$ defined at all points $(t,\mathbf{x})$ of an n-dimensional Minkowski spacetime, satisfying the field equation

$$(\Box + m^2)\phi = 0 \tag{2.1}$$

where $\Box \equiv \eta^{\mu\nu}\partial_\mu\partial_\nu$ and $\eta^{\mu\nu}$ is the Minkowskian metric tensor. The quantity m is to be interpreted as the mass of the field quanta when the theory is quantized. In what follows we shall frequently abbreviate the spacetime point $(t,\mathbf{x}) = (x^0,\mathbf{x})$ as x.

Equation (2.1) may be obtained from the Lagrangian density

$$\mathscr{L}(x) = \tfrac{1}{2}(\eta^{\alpha\beta}\phi_{,\alpha}\phi_{,\beta} - m^2\phi^2) \tag{2.2}$$

by constructing the action

$$S = \int \mathscr{L}(x)\mathrm{d}^n x \tag{2.3}$$

and demanding that for variations with respect to ϕ

$$\delta S = 0. \tag{2.4}$$

One set of solutions of (2.1) is

$$u_{\mathbf{k}}(t,\mathbf{x}) \propto \mathrm{e}^{\mathrm{i}\mathbf{k}\cdot\mathbf{x} - \mathrm{i}\omega t} \tag{2.5}$$

where

$$\omega \equiv (k^2 + m^2)^{\frac{1}{2}} \tag{2.6}$$

$$k \equiv |\mathbf{k}| = \left(\sum_{i=1}^{n-1} k_i^2\right)^{\frac{1}{2}} \tag{2.7}$$

and the Cartesian components of $\mathbf{k}$ can take the values

$$-\infty < k_i < \infty, \quad i = 1, \dots, n-1.$$

The modes (2.5) are said to be positive frequency with respect to t, being eigenfunctions of the operator $\partial/\partial t$:

$$\frac{\partial}{\partial t}u_{\mathbf{k}}(t,\mathbf{x}) = -\mathrm{i}\omega u_{\mathbf{k}}(t,\mathbf{x}), \quad \text{with } \omega > 0. \tag{2.8}$$

Define the scalar product

$$\begin{aligned}(\phi_1,\phi_2) &= -\mathrm{i}\int\{\phi_1(x)\partial_t\phi_2^*(x) - [\partial_t\phi_1(x)]\phi_2^*(x)\}\mathrm{d}^{n-1}x \\ &= -\mathrm{i}\int_t \phi_1(x)\overleftrightarrow{\partial_t}\phi_2^*(x)\mathrm{d}^{n-1}x,\end{aligned} \tag{2.9}$$

where t denotes a spacelike hyperplane of simultaneity at instant t. Then the $u_{\mathbf{k}}$ modes (2.5) are orthogonal

$$(u_{\mathbf{k}}, u_{\mathbf{k}'}) = 0, \quad \mathbf{k} \neq \mathbf{k}'. \tag{2.10}$$

If we choose

$$u_{\mathbf{k}} = [2\omega(2\pi)^{n-1}]^{-\frac{1}{2}} e^{i\mathbf{k}\cdot\mathbf{x} - i\omega t} \tag{2.11}$$

then the $u_{\mathbf{k}}$ functions are normalized in the scalar product (2.9):

$$(u_{\mathbf{k}}, u_{\mathbf{k}'}) = \delta^{n-1}(\mathbf{k} - \mathbf{k}'). \tag{2.12}$$

For many purposes it is more convenient to restrict the solutions $u_{\mathbf{k}}$ to the interior of a spacelike $(n-1)$-torus of side L (i.e., choose periodic boundary conditions). Then

$$u_{\mathbf{k}} = (2L^{n-1}\omega)^{-\frac{1}{2}} e^{i\mathbf{k}\cdot\mathbf{x} - i\omega t} \tag{2.13}$$

where

$$k_i = 2\pi j_i/L, \quad j_i = 0, \pm 1, \pm 2, \ldots, \quad i = 1, \ldots, n-1.$$

Thus

$$(u_{\mathbf{k}}, u_{\mathbf{k}'}) = \delta_{\mathbf{k}\mathbf{k}'}. \tag{2.14}$$

To convert from continuum to discrete (box) normalization one should replace each $\int d^{n-1}k$ by

$$(2\pi/L)^{n-1} \prod_{i=1}^{n-1} \sum_{j_i} \equiv (2\pi/L)^{n-1} \sum_{\mathbf{k}}.$$

2.2 Quantization

The system is quantized in the canonical quantization scheme by treating the field ϕ as an operator, and imposing the following equal time commutation relations

$$\left.\begin{aligned} [\phi(t,\mathbf{x}), \phi(t,\mathbf{x}')] &= 0 \\ [\pi(t,\mathbf{x}), \pi(t,\mathbf{x}')] &= 0 \\ [\phi(t,\mathbf{x}), \pi(t,\mathbf{x}')] &= i\delta^{n-1}(\mathbf{x} - \mathbf{x}') \end{aligned}\right\} \tag{2.15}$$

where π is the canonically conjugate variable to ϕ defined by

$$\pi = \frac{\partial \mathscr{L}}{\partial(\partial_t \phi)} = \partial_t \phi. \tag{2.16}$$

The field modes (2.11) or (2.13) and their respective complex conjugates form a complete orthonormal basis with scalar product (2.9), so ϕ may be expanded as

$$\phi(t,\mathbf{x}) = \sum_{\mathbf{k}} [a_{\mathbf{k}} u_{\mathbf{k}}(t,\mathbf{x}) + a_{\mathbf{k}}^{\dagger} u_{\mathbf{k}}^{*}(t,\mathbf{x})]. \tag{2.17}$$

The equal time commutation relations for ϕ and π are then equivalent to

$$\left.\begin{aligned} [a_{\mathbf{k}}, a_{\mathbf{k}'}] &= 0 \\ [a_{\mathbf{k}}^{\dagger}, a_{\mathbf{k}'}^{\dagger}] &= 0 \\ [a_{\mathbf{k}}, a_{\mathbf{k}'}^{\dagger}] &= \delta_{\mathbf{k}\mathbf{k}'}. \end{aligned}\right\} \tag{2.18}$$

In the Heisenberg picture, the quantum states span a Hilbert space. A convenient basis in this Hilbert space is the so-called Fock representation. The normalized basis ket vectors, denoted $|\,\rangle$, can be constructed from the vector $|0\rangle$, called the vacuum, or no-particle state, the physical significance of which will be discussed shortly. The state $|0\rangle$ has the property that it is annihilated by all the $a_{\mathbf{k}}$ operators:

$$a_{\mathbf{k}}|0\rangle = 0, \quad \forall \mathbf{k}. \tag{2.19}$$

The state obtained by operating on $|0\rangle$ with $a_{\mathbf{k}}^{\dagger}$ is called a one-particle state, and is denoted by $|1_{\mathbf{k}}\rangle$

$$|1_{\mathbf{k}}\rangle = a_{\mathbf{k}}^{\dagger}|0\rangle. \tag{2.20}$$

Similarly one may construct many-particle states

$$|1_{\mathbf{k}_1}, 1_{\mathbf{k}_2}, \ldots, 1_{\mathbf{k}_j}\rangle = a_{\mathbf{k}_1}^{\dagger} a_{\mathbf{k}_2}^{\dagger} \ldots a_{\mathbf{k}_j}^{\dagger}|0\rangle, \tag{2.21}$$

if all $\mathbf{k}_1$, $\mathbf{k}_2, \ldots, \mathbf{k}_j$ are distinct. If any $a_{\mathbf{k}}^{\dagger}$ are repeated, then

$$|{}^{1}n_{\mathbf{k}_1}, {}^{2}n_{\mathbf{k}_2}, \ldots, {}^{j}n_{\mathbf{k}_j}\rangle = ({}^{1}n!\,{}^{2}n! \ldots {}^{j}n!)^{-\frac{1}{2}} (a_{\mathbf{k}_1}^{\dagger})^{{}^{1}n} (a_{\mathbf{k}_2}^{\dagger})^{{}^{2}n} \ldots (a_{\mathbf{k}_j}^{\dagger})^{{}^{j}n}|0\rangle, \tag{2.22}$$

the $n!$ terms being necessary to accommodate the Bose statistics of identical scalar particles. Also

$$a_{\mathbf{k}}^{\dagger}|n_{\mathbf{k}}\rangle = (n+1)^{\frac{1}{2}}|(n+1)_{\mathbf{k}}\rangle \tag{2.23}$$

$$a_{\mathbf{k}}|n_{\mathbf{k}}\rangle = n^{\frac{1}{2}}|(n-1)_{\mathbf{k}}\rangle. \tag{2.24}$$

The basis vectors are normalized according to

$$\langle {}^{1}n_{\mathbf{k}_1}, {}^{2}n_{\mathbf{k}_2}, \ldots, {}^{r}n_{\mathbf{k}_r} | {}^{1}m_{\mathbf{k}'_1}, {}^{2}m_{\mathbf{k}'_2}, \ldots, {}^{s}m_{\mathbf{k}'_s}\rangle$$

$$= \delta_{rs} \sum_{\alpha} \delta_{{}^{1}n\,{}^{\alpha(1)}m} \ldots \delta_{{}^{r}n\,{}^{\alpha(s)}m} \delta_{\mathbf{k}_1\mathbf{k}'_{\alpha(1)}} \ldots \delta_{\mathbf{k}_r\mathbf{k}'_{\alpha(s)}} \tag{2.25}$$

where the sum is over all permutations α of the integers $1 \ldots s$.

2.3 Energy–momentum

To explore the significance of these Fock states, it is instructive to examine the Hamiltonian and momentum operators for the field. These quantities are obtained from the stress–energy–momentum tensor, $T_{\mu\nu}$, henceforth abbreviated as stress-tensor. $T_{\mu\nu}$ may be constructed in a standard manner (see (3.189)) to be

$$T_{\alpha\beta} = \phi_{,\alpha}\phi_{,\beta} - \tfrac{1}{2}\eta_{\alpha\beta}\eta^{\lambda\delta}\phi_{,\lambda}\phi_{,\delta} + \tfrac{1}{2}m^2\phi^2\eta_{\alpha\beta} \tag{2.26}$$

from which one obtains for the Hamiltonian density

$$T_{tt} = \tfrac{1}{2}\left[(\partial_t\phi)^2 + \sum_{i=1}^{n-1}(\partial_i\phi)^2 + m^2\phi^2\right] \tag{2.27}$$

and for the momentum density

$$T_{ti} = \partial_t\phi\partial_i\phi, \quad i = 1, \ldots, n-1, \tag{2.28}$$

in terms of Minkowski coordinates.

Substituting ϕ from (2.17) into (2.27) and (2.28), and integrating over all space, yields

$$H \equiv \int_t T_{tt}\mathrm{d}^{n-1}x = \tfrac{1}{2}\sum_{\mathbf{k}}(a_{\mathbf{k}}^{\dagger}a_{\mathbf{k}} + a_{\mathbf{k}}a_{\mathbf{k}}^{\dagger})\omega \tag{2.29}$$

$$P_i \equiv \int_t T_{ti}\mathrm{d}^{n-1}x = \sum_{\mathbf{k}} a_{\mathbf{k}}^{\dagger}a_{\mathbf{k}}k_i \tag{2.30}$$

for the Hamiltonian and momentum component operators, respectively. Using the commutation relations (2.18), equation (2.29) may be recast in a more suggestive form

$$H = \sum_{\mathbf{k}}(a_{\mathbf{k}}^{\dagger}a_{\mathbf{k}} + \tfrac{1}{2})\omega. \tag{2.31}$$

Clearly, both H and P_i commute with the operators

$$\left.\begin{aligned} N_{\mathbf{k}} &\equiv a_{\mathbf{k}}^{\dagger}a_{\mathbf{k}} \\ \text{and} \qquad & \\ N &\equiv \sum_{\mathbf{k}} N_{\mathbf{k}} \end{aligned}\right\} \tag{2.32}$$

$$[N, H] = [N, P_i] = 0. \tag{2.33}$$

The significance of N is revealed by taking its expectation values for the

Fock states. From (2.19) and (2.24) one obtains

$$\langle 0|N_{\mathbf{k}}|0\rangle = 0, \quad \forall \mathbf{k} \tag{2.34}$$

$$\langle {}^1n_{\mathbf{k}_1}, {}^2n_{\mathbf{k}_2}, \ldots, {}^jn_{\mathbf{k}_j}|N_{\mathbf{k}_i}|{}^1n_{\mathbf{k}_1}, {}^2n_{\mathbf{k}_2}, \ldots, {}^jn_{\mathbf{k}_j}\rangle = {}^in. \tag{2.35}$$

Thus, the expectation value of the operator $N_{\mathbf{k}_i}$ is the integer ${}^in$, that is, the entry in the ket vector under the label $\mathbf{k}_i$. Similarly, if $N_{\mathbf{k}_i}$ in (2.35) is summed over all i

$$\langle |N| \rangle = \sum_i {}^in. \tag{2.36}$$

This simple relationship between $N_{\mathbf{k}}$ and n suggests the name 'number operator for the mode $\mathbf{k}$' for $N_{\mathbf{k}}$ and 'total number operator' for N. Because of relations (2.33), eigenstates of N are also eigenstates of H and $\mathbf{P}$. For each increment of one in the number ${}^in$, $\langle |H| \rangle$ and $\langle |\mathbf{P}| \rangle$ increase by ω_i and $\mathbf{k}_i$ respectively. We can therefore interpret the ${}^in$ as labelling the *number of quanta*, each of energy ω_i and momentum $\mathbf{k}_i$, in the mode labelled by $\mathbf{k}_i$. Thus, the state $|{}^1n_{\mathbf{k}_1}, {}^2n_{\mathbf{k}_2}, \ldots, {}^jn_{\mathbf{k}_j}\rangle$ is a state containing ${}^1n$ quanta in the mode with momentum $\mathbf{k}_1$, ${}^2n$ quanta in the mode with momentum $\mathbf{k}_2$ and so on.

Returning to (2.23) and (2.24), a useful physical interpretation is now available for the operators $a_{\mathbf{k}}$ and $a_{\mathbf{k}}^\dagger$. The former reduces the number of quanta in mode $\mathbf{k}$ by one, while the latter increases this number by one. Thus $a_{\mathbf{k}}$ is referred to as an *annihilation operator* and $a_{\mathbf{k}}^\dagger$ as a *creation operator*, for quanta in the mode $\mathbf{k}$.

2.4 Vacuum energy divergence

Special interest attaches to the state $|0\rangle$. This is the no-particle, or *vacuum state*. It carries zero momentum

$$\langle 0|\mathbf{P}|0\rangle = 0, \tag{2.37}$$

a result which follows immediately from (2.30). We should also expect it to carry zero energy, as no field quanta are present. However inspection of (2.31) reveals a term $\sum_{\mathbf{k}} \frac{1}{2}\omega$, so

$$\langle 0|H|0\rangle = \langle 0|0\rangle \sum_{\mathbf{k}} \tfrac{1}{2}\omega = \sum_{\mathbf{k}} \tfrac{1}{2}\omega \tag{2.38}$$

where we have used the normalization condition $\langle 0|0\rangle = 1$.

Not only is the right-hand side of (2.38) nonzero; it is actually infinite

$$\sum_{\mathbf{k}} \tfrac{1}{2}\omega = \tfrac{1}{2}(L/2\pi)^{n-1}\int \omega \mathrm{d}^{n-1}k$$
$$= (L^2/4\pi)^{(n-1)/2}\frac{1}{\Gamma((n-1)/2)}\int_0^\infty (k^2+m^2)^{\frac{1}{2}}k^{n-2}\mathrm{d}k \quad (2.39)$$

which diverges like k^n for large k. This divergence can be usefully analysed by performing the integral in (2.39) with n continued away from integral values to obtain

$$-L^{n-1}2^{-n-1}\pi^{-n/2}m^n\Gamma(-n/2).$$

The Γ-function contains poles at all even integral values of $n \geq 0$. This method of temporarily making divergent quantities finite by continuing the dimension of the spacetime away from integer values forms the basis of dimensional regularization (see chapter 6).

The fact that (2.39) is divergent apparently indicates that the vacuum contains an infinite density of energy. The trouble comes from the $\frac{1}{2}\omega$ zero-point energy associated with each simple harmonic oscillator mode of the scalar field. As ω has no upper bound the zero-point energy can be arbitrarily large. This is a problem which will plague the subject of quantum fields in curved spacetime throughout. However, in flat spacetime, it is easily circumvented. Energy as such is not measurable in non-gravitational physics, so we can rescale – or *renormalize* – the zero point of energy, even by an infinite amount, without affecting observable quantities. This may be accomplished by simply throwing away the $\frac{1}{2}\sum_{\mathbf{k}}\omega$ term in (2.31) or, more elegantly, by defining a *normal ordering* operation, denoted by : :, in which one demands that wherever a product of creation and annihilation operators appears, it is understood that all annihilation operators stand to the right of the creation operators. Thus, returning to the form (2.29), the normal ordering operation demands

$$:a_{\mathbf{k}}a_{\mathbf{k}}^{\dagger}: = a_{\mathbf{k}}^{\dagger}a_{\mathbf{k}} \quad (2.40)$$

whence

$$:H: = \sum_{\mathbf{k}} a_{\mathbf{k}}^{\dagger}a_{\mathbf{k}}\omega \quad (2.41)$$

and the troublesome $\frac{1}{2}\omega$ term has disappeared.

Finally, let us return to $T_{\mu\nu}$ and expression (2.26). Using the expansion

(2.17) for ϕ in terms of the modes $u_{\mathbf{k}}$, we have

$$\phi_{,\alpha}\phi_{,\beta} = \sum_{\mathbf{k}}\sum_{\mathbf{k}'}(a_{\mathbf{k}}\partial_\alpha u_{\mathbf{k}} + a_{\mathbf{k}}^\dagger \partial_\alpha u_{\mathbf{k}}^*)(a_{\mathbf{k}'}\partial_\beta u_{\mathbf{k}'} + a_{\mathbf{k}'}^\dagger \partial_\beta u_{\mathbf{k}'}^*).$$

From (2.19) and the associated result

$$\langle 0|a_{\mathbf{k}}^\dagger = 0 \tag{2.42}$$

together with

$$\langle 0|a_{\mathbf{k}}a_{\mathbf{k}'}^\dagger|0\rangle = \delta_{\mathbf{k}\mathbf{k}'}$$

one obtains

$$\langle 0|\phi_{,\alpha}\phi_{,\beta}|0\rangle = \sum_{\mathbf{k}} u_{\mathbf{k},\alpha}u_{\mathbf{k},\beta}^*.$$

In general

$$\langle 0|T_{\alpha\beta}|0\rangle = \sum_{\mathbf{k}} T_{\alpha\beta}[u_{\mathbf{k}}, u_{\mathbf{k}}^*], \tag{2.43}$$

where $T_{\alpha\beta}[\phi, \phi]$ denotes the bilinear expression (2.26) for $T_{\alpha\beta}$. Similarly

$$\langle {}^1n_{\mathbf{k}_1}, {}^2n_{\mathbf{k}_2}, \ldots, |T_{\alpha\beta}|{}^1n_{\mathbf{k}_1}, {}^2n_{\mathbf{k}_2}, \ldots\rangle = \sum_{\mathbf{k}} T_{\alpha\beta}[u_{\mathbf{k}}, u_{\mathbf{k}}^*] + 2\sum_i {}^i n T_{\alpha\beta}[u_{\mathbf{k}_i}, u_{\mathbf{k}_i}^*]. \tag{2.44}$$

2.5 Dirac spinor field

So far attention has been restricted to the scalar (spin-zero) field. The quantization of higher-spin fields proceeds in close analogy. In particular the spin $\frac{1}{2}$ fermion field ψ possesses the Lagrangian density

$$\mathscr{L} = \tfrac{1}{2}\mathrm{i}(\bar{\psi}\gamma^\alpha\psi_{,\alpha} - \bar{\psi}_{,\alpha}\gamma^\alpha\psi) - m\bar{\psi}\psi \tag{2.45}$$

where $\bar{\psi}$ is the Dirac adjoint of ψ (i.e., $\psi^\dagger\gamma^0$), and γ^μ are Dirac matrices that satisfy the anticommutation relations

$$\{\gamma^\alpha, \gamma^\beta\} = 2\eta^{\alpha\beta}. \tag{2.46}$$

(See, for example, Delbourgo & Prasad 1974 for the properties of γ-matrices in n-dimensions.)

Variation of $\bar{\psi}$ in the action $S = \int \mathscr{L}\,\mathrm{d}^n x$ leads to the Dirac equation

$$\mathrm{i}\gamma^\alpha\psi_{,\alpha} - m\psi = 0 \tag{2.47}$$

for a particle of mass m. The field ψ carries spinor labels (ψ^a) which have been suppressed in the above expressions. Similarly we suppress labels on the Dirac matrices (γ_{ab}).

A complete set of mode solutions of the Dirac equation is given (in discrete normalization) by

$$\left.\begin{aligned} u_{\mathbf{k},s}(t,\mathbf{x}) &= N\,u(\mathbf{k},s)\mathrm{e}^{\mathrm{i}\mathbf{k}\cdot\mathbf{x}-\mathrm{i}\omega t} \\ v_{\mathbf{k},s}(t,\mathbf{x}) &= N\,v(\mathbf{k},s)\mathrm{e}^{-\mathrm{i}\mathbf{k}\cdot\mathbf{x}+\mathrm{i}\omega t}, \end{aligned}\right\} \tag{2.48}$$

where

$$N = \begin{cases} (m/\omega L^{n-1})^{\frac{1}{2}}, & m \neq 0 \\ (2\omega L^{n-1})^{-\frac{1}{2}}, & m = 0. \end{cases} \tag{2.49}$$

The familiar constant positive and negative energy spinors $u(\mathbf{k},s)$, $v(\mathbf{k},s)$ (see, for example, Bjorken & Drell 1965, chapters 3 and 10) exist in two independent spin states and are normalized according to

$$u^{\dagger}(\mathbf{k},s)u(\mathbf{k},s') = v^{\dagger}(\mathbf{k},s)v(\mathbf{k},s') = \begin{cases} (\omega/m)\delta_{ss'}, & m \neq 0 \\ 2\omega\delta_{ss'}, & m = 0. \end{cases} \tag{2.50}$$

We can thus expand the field ψ as

$$\psi(t,\mathbf{x}) = \sum_{\pm s}\sum_{\mathbf{k}}[b_{\mathbf{k}}(s)u_{\mathbf{k},s}(t,\mathbf{x}) + d^{\dagger}_{\mathbf{k}}(s)v_{\mathbf{k},s}(t,\mathbf{x})] \tag{2.51}$$

which is normalized with respect to the inner product

$$(\psi,\phi) = \int_t \mathrm{d}^{n-1}x\,\bar{\psi}(t,\mathbf{x})\gamma_0\phi(t,\mathbf{x}). \tag{2.52}$$

The operators $b_{\mathbf{k}}(s)$, $d_{\mathbf{k}}(s)$, $b^{\dagger}_{\mathbf{k}}(s)$, $d^{\dagger}_{\mathbf{k}}(s)$ all anticommute except in the following cases:

$$\{b_{\mathbf{k}}(s), b^{\dagger}_{\mathbf{k}'}(s')\} = \{d_{\mathbf{k}}(s), d^{\dagger}_{\mathbf{k}'}(s')\} = \delta_{ss'}\delta_{\mathbf{k}\mathbf{k}'}. \tag{2.53}$$

By constructing Hamiltonian, momentum and angular momentum operators and considering their expectation values in the Fock basis, one finds that $b^{\dagger}_{\mathbf{k}}(s)$ is the creation operator for quanta in a mode of momentum $\mathbf{k}$, energy ω and spin s, while $d^{\dagger}_{\mathbf{k}}(s)$ annihilates quanta in a mode of momentum $(-\mathbf{k})$, energy $(-\omega)$, spin s. Physically $b^{\dagger}_{\mathbf{k}}(s)$ and $d^{\dagger}_{\mathbf{k}}(s)$ represent creation operators for electrons and positrons respectively, while $b_{\mathbf{k}}(s)$, $d_{\mathbf{k}}(s)$ are the corresponding annihilation operators.

More detailed examination of the stress-tensor for the Dirac field is postponed until its treatment in curved spacetime in §3.8.

2.6 Electromagnetic field

The electromagnetic (massless, spin 1) field is described by the Lagrangian density

$$\mathscr{L} = -\tfrac{1}{4}F_{\alpha\beta}F^{\alpha\beta} \tag{2.54}$$

where

$$F_{\alpha\beta} = A_{\alpha,\beta} - A_{\beta,\alpha} \tag{2.55}$$

is the Maxwell field strength tensor. Variation of the action $S = \int \mathscr{L}\, d^n x$ then yields

$$F^{\alpha\beta}{}_{,\beta} = 0, \tag{2.56}$$

which along with the identity

$$F_{\alpha\beta,\gamma} + F_{\beta\gamma,\alpha} + F_{\gamma\alpha,\beta} = 0 \tag{2.57}$$

constitutes Maxwell's equations.

The field strength tensor (2.55) and hence the Lagrangian (2.54) are invariant under local gauge transformations

$$A_\alpha \to A_\alpha^\Lambda = A_\alpha + \partial_\alpha \Lambda(x) \tag{2.58}$$

where $\Lambda(x)$ is an arbitrary differentiable scalar function. This gauge invariance prevents the straightforward quantization of the theory and must be broken, usually by adding to the Lagrangian density a gauge-fixing term

$$\mathscr{L}_G = -\tfrac{1}{2}\zeta^{-1}(A^\alpha{}_{,\alpha})^2, \tag{2.59}$$

where ζ is a parameter determining the choice of gauge; $\zeta = 1$ being the Feynman gauge and $\zeta \to 0$ the Landau gauge.

The field equation resulting from the inclusion of the gauge-breaking term is

$$[\eta_{\alpha\beta}\Box - (1 - \zeta^{-1})\partial_\alpha \partial_\beta]A^\beta = 0. \tag{2.60}$$

In the Feynman gauge these equations reduce to

$$\Box A_\alpha = 0 \tag{2.61}$$

with solution

$$A^\alpha(t, \mathbf{x}) = \sum_{\mathbf{k}\lambda} [a_{\mathbf{k}\lambda} u^\alpha_{\mathbf{k}\lambda}(t, \mathbf{x}) + a^\dagger_{\mathbf{k}\lambda} u^{\alpha}_{\mathbf{k}\lambda}{}^*(t, \mathbf{x})], \tag{2.62}$$

where the plane wave modes $u^{\alpha}_{\mathbf{k}\lambda}$ are given by

$$u^{\alpha}_{\mathbf{k}\lambda}(t, \mathbf{x}) = (2L^{n-1}\omega)^{-\frac{1}{2}} e^{\alpha}_{\mathbf{k}\lambda} e^{i\mathbf{k}\cdot\mathbf{x} - i\omega t}, \tag{2.63}$$

with $e^{\alpha}_{\mathbf{k}\lambda}$, $\lambda = 1, 2, 3, 4$ labelling independent polarization vectors associated with mode $\mathbf{k}$. These vectors can be chosen to form an orthonormal system with

$$\eta_{\alpha\beta} e^{\alpha}_{\mathbf{k}\lambda} e^{\beta}_{\mathbf{k}\lambda'} = \eta_{\lambda\lambda'}. \tag{2.64}$$

The quantization of the fields now proceeds in much the same way as for the scalar field; however, because the physical photon has only two independent (transverse) degrees of freedom, while the Lorentz covariant choice of gauge used here has kept all four potentials A^{α} on an equal footing, the results so obtained do not immediately make physical sense. To obtain a physical interpretation of the quantized theory, the so-called Gupta–Bleuler formalism can be used (Gupta 1950, Bleuler 1950; see also the textbook treatments of Jauch & Rohrlich 1955, Bogolubov & Shirkov 1959 or Schweber 1961). We shall not adopt this formalism here, but shall instead treat the quantization of the electromagnetic field by the path-integral approach, which introduces a new feature, i.e., the so-called ghost fields, that assume considerable importance in the curved spacetime treatment. Before turning to this, however, the subject of Green functions must be considered.

2.7 Green functions

Vacuum expectation values of various products of free field operators can be identified with various Green functions of the wave equation. Treating first the scalar field, of particular importance are the expectation values of the commutator and anticommutator of the fields, denoted respectively by

$$iG(x, x') = \langle 0|[\phi(x), \phi(x')]|0\rangle \tag{2.65}$$

$$G^{(1)}(x, x') = \langle 0|\{\phi(x), \phi(x')\}|0\rangle. \tag{2.66}$$

G is known as the Pauli–Jordan or Schwinger function while $G^{(1)}$ is sometimes called Hadamard's elementary function. These Green functions can be split into their positive and negative frequency parts as

$$\left.\begin{aligned} iG(x, x') &= G^{+}(x, x') - G^{-}(x, x') \\ G^{(1)}(x, x') &= G^{+}(x, x') + G^{-}(x, x') \end{aligned}\right\} \tag{2.67}$$

where $G^{\pm}$, known as the Wightman functions, are given by

$$\left.\begin{aligned} G^{+}(x, x') &= \langle 0|\phi(x)\phi(x')|0\rangle \\ G^{-}(x, x') &= \langle 0|\phi(x')\phi(x)|0\rangle . \end{aligned}\right\} \tag{2.68}$$

The Feynman propagator G_{F} is defined as the time-ordered product of fields

$$\begin{aligned} \mathrm{i}G_{\mathrm{F}}(x, x') &= \langle 0|T(\phi(x)\phi(x'))|0\rangle \\ &= \theta(t-t')G^{+}(x, x') + \theta(t'-t)G^{-}(x, x') \end{aligned} \tag{2.69}$$

where

$$\theta(t) = \begin{cases} 1, & t > 0 \\ 0, & t < 0. \end{cases}$$

Finally, retarded and advanced Green functions are defined respectively by

$$\left.\begin{aligned} G_{\mathrm{R}}(x, x') &= -\theta(t-t')G(x, x') \\ G_{\mathrm{A}}(x, x') &= \theta(t'-t)G(x, x') \end{aligned}\right\} \tag{2.70}$$

and their average is denoted as

$$\bar{G}(x, x') = \tfrac{1}{2}[G_{\mathrm{R}}(x, x') + G_{\mathrm{A}}(x, x')], \tag{2.71}$$

which is related to G_{F} by

$$G_{\mathrm{F}}(x, x') = -\bar{G}(x, x') - \tfrac{1}{2}\mathrm{i}G^{(1)}(x, x'). \tag{2.72}$$

Using the field equation (2.1) it is clear that G, $G^{(1)}$, $G^{\pm}$ all satisfy the homogeneous equation

$$(\Box_x + m^2)\mathcal{G}(x, x') = 0. \tag{2.73}$$

Also, using $\partial_t\theta(t-t') = \delta(t-t')$ and the equal time commutators (2.15) one obtains the following equations for $G_{\mathrm{F}}, G_{\mathrm{R}}$ and G_{A}:

$$(\Box_x + m^2)G_{\mathrm{F}}(x, x') = -\delta^n(x-x') \tag{2.74}$$

$$(\Box_x + m^2)G_{\mathrm{R,A}}(x, x') = \delta^n(x-x'). \tag{2.75}$$

The Green functions $G_{\mathrm{F,R,A}}$ describe the propagation of field disturbances subject to certain boundary conditions.

Integral representations for the Green functions can be obtained by substituting the mode decomposition (2.17) for ϕ into the definitions of the Green functions as vacuum expectation values. One finds that all the Green

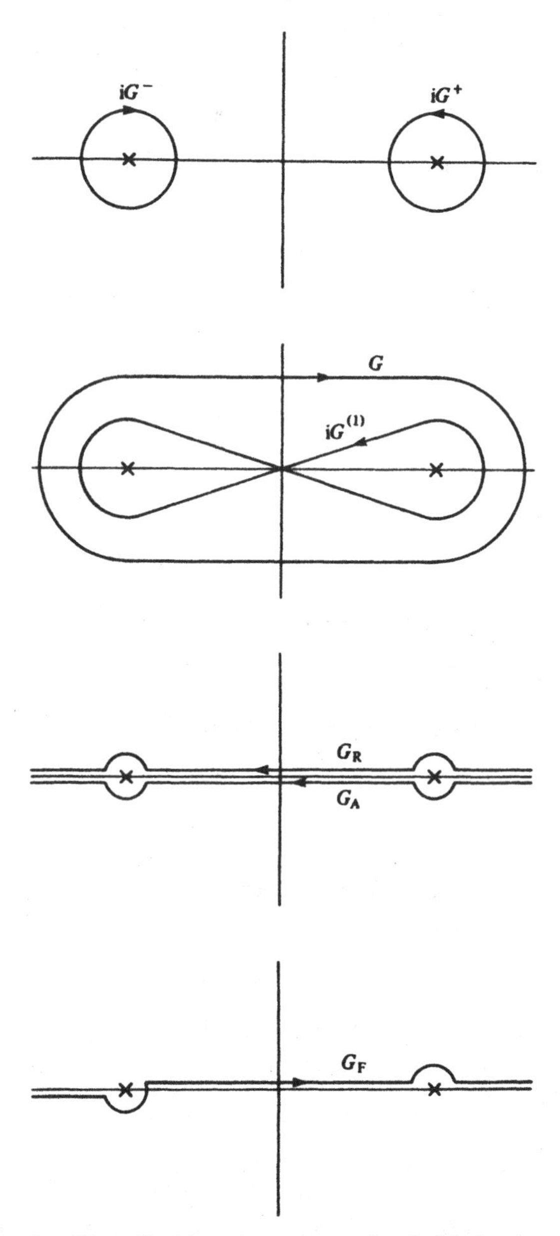

Fig. 3. The various Green functions are each associated with the above contours in the complex k^0 plane. The poles on the real axis at $k^0 = \pm(|\mathbf{k}|^2 + m^2)^{\frac{1}{2}}$ are marked with crosses. The open contours should be envisaged as closed by infinite semicircles in the upper- or lower-half planes.

functions can be represented as

$$\mathcal{G}(x,x')=(2\pi)^{-n}\int\frac{\exp[\mathrm{i}\mathbf{k}\cdot(\mathbf{x}-\mathbf{x}')-\mathrm{i}k^0(t-t')]}{(k^0)^2-|\mathbf{k}|^2-m^2}\mathrm{d}^n k. \tag{2.76}$$

The integral has poles at $k^0=\pm(|\mathbf{k}|^2+m^2)^{\frac{1}{2}}$. Considered as a contour integral, the k^0 integration may be performed by deforming the contour around the poles. The way in which this deformation is performed (see fig. 3) depends on the boundary conditions on the field and determines which of the various Green functions is obtained from (2.76).

For example, the integral corresponding to G_F yields

$$G_\mathrm{F}(x,x')=\frac{-\mathrm{i}\pi}{(4\pi\mathrm{i})^{n/2}}\left(\frac{2m^2}{-\sigma+\mathrm{i}\varepsilon}\right)^{(n-2)/4}H^{(2)}_{\frac{1}{2}n-1}\{[2m^2(\sigma-\mathrm{i}\varepsilon)]^{\frac{1}{2}}\} \tag{2.77}$$

where $\sigma=\frac{1}{2}(x-x')^2=\frac{1}{2}\eta_{\alpha\beta}(x^\alpha-x'^\alpha)(x^\beta-x'^\beta)$ and $H^{(2)}$ is a Hankel function of the second kind. The $-\mathrm{i}\varepsilon$ is added to σ to indicate that G_F is really the boundary value of a function which is analytic in the lower-half σ plane.

In the massless limit the Green functions are customarily denoted by D rather than G. In this limit the Feynman propagator in four dimensions reduces to

$$D_\mathrm{F}(x,x')=(\mathrm{i}/8\pi^2\sigma)-(1/8\pi)\delta(\sigma) \tag{2.78}$$

while Hadamard's elementary function becomes

$$D^{(1)}(x,x')=-1/4\pi^2\sigma. \tag{2.79}$$

For the fields of spin $\frac{1}{2}$ and 1 we note here only a few of the Green functions. For Dirac spinors one defines

$$\mathrm{i}S_\mathrm{F}(x,x')=\langle 0|T(\psi(x)\bar{\psi}(x'))|0\rangle \tag{2.80}$$

$$S^{(1)}(x,x')=\langle 0|[\psi(x),\bar{\psi}(x')]|0\rangle \tag{2.81}$$

which satisfy

$$(\mathrm{i}\gamma^\alpha\partial_\alpha-m)S_\mathrm{F}(x,x')=\delta^n(x-x'), \tag{2.82}$$

$$(\mathrm{i}\gamma^\alpha\partial_\alpha-m)S^{(1)}(x,x')=0 \tag{2.83}$$

and can be written in terms of G_F and $G^{(1)}$ as

$$S_\mathrm{F}(x,x')=(\mathrm{i}\gamma^\alpha\partial_\alpha+m)G_\mathrm{F}(x,x') \tag{2.84}$$

$$S^{(1)}(x,x')=-(\mathrm{i}\gamma^\alpha\partial_\alpha+m)G^{(1)}(x,x'). \tag{2.85}$$

Note that, for example, S_F is a matrix in the spinor indices of the fields in

(2.80), which when written in full reads

$$iS_F(x,x')_{ab} = \langle 0|\psi_a(x)\psi_c^\dagger(x')|0\rangle(\gamma^0)^c{}_b\theta(t-t')$$
$$-\langle 0|\psi_c^\dagger(x')\psi_a(x)|0\rangle(\gamma^0)^c{}_b\theta(t'-t).$$

The Feynman propagator for the electromagnetic field is defined by

$$iD_{F\alpha\beta}(x,x') = \langle 0|T(A_\alpha(x)A_\beta(x'))|0\rangle \tag{2.86}$$

which is obviously gauge dependent. Using (2.60) one finds that

$$[\eta_{\alpha\lambda}\Box_x - (1-\zeta^{-1})\partial_\alpha^x\partial_\lambda^x]D_F^{\lambda\beta}(x,x') = \delta_\alpha^\beta\delta^n(x-x') \tag{2.87}$$

which gives an integral representation for the propagator:

$$D_{F\alpha\beta}(x,x')$$
$$= (2\pi)^{-n}\int\frac{[-\eta_{\alpha\beta}+(1-\zeta)k_\alpha k_\beta/k^2]}{(k^0)^2-|\mathbf{k}|^2}\exp[\mathrm{i}\mathbf{k}\cdot(\mathbf{x}-\mathbf{x}')-\mathrm{i}k^0(t-t')]\mathrm{d}^nk. \tag{2.88}$$

The contour for the k^0 integral in (2.88) is the same as for G_F (see fig. 3). In particular, in the Feynman gauge ($\zeta = 1$)

$$D_{F\alpha\beta}(x,x') = -\eta_{\alpha\beta}D_F(x,x'). \tag{2.89}$$

Note also that if the gauge-breaking term is removed by letting $\zeta\to\infty$ then (2.88) becomes infinite. That is, unless the gauge-breaking term is present the differential operator on the left-hand side of (2.87) is not invertible.

Examination of the integral representations of the Feynman Green functions, such as (2.76) and (2.88), reveals a useful mathematical property that is frequently exploited in practical calculations. Inspection of the contour for G_F in the complex k^0 plane shows that the topological relation between the contour (assumed closed by an infinite semicircle) and the poles remains unchanged if it is rotated anticlockwise through 90^0 to lie along the imaginary k^0 axis from $-\mathrm{i}\infty$ to $\mathrm{i}\infty$. If the integration variable k^0 is now changed to $\kappa = -\mathrm{i}k^0$ and the variables $\tau = -\mathrm{i}t$ and $\tau' = -\mathrm{i}t'$ replace t and t', then the contour of integration will once again run along the real axis but will no longer intersect the poles.

For example, in the scalar case one obtains

$$G_F(t,\mathbf{x};t',\mathbf{x}') = -\mathrm{i}G_E(\mathrm{i}\tau,\mathbf{x};\mathrm{i}\tau',\mathbf{x}') \tag{2.90}$$

where

$$G_E(\tau,\mathbf{x};\tau',\mathbf{x}') = (2\pi)^{-n}\int_{-\infty}^{\infty}\frac{\exp[\mathrm{i}\mathbf{k}\cdot(\mathbf{x}-\mathbf{x}')+\mathrm{i}\kappa(\tau-\tau')]}{\kappa^2+|\mathbf{k}|^2+m^2}\mathrm{d}\kappa\,\mathrm{d}^{n-1}k. \tag{2.91}$$

G_E is the 'Euclidean' Green function which satisfies

$$(\Box_x - m^2)G_E(x, x') = -\delta^n(x - x'). \tag{2.92}$$

In (2.92), $\Box$ is the elliptic operator

$$\frac{\partial^2}{\partial \tau^2} + \frac{\partial^2}{\partial (x^1)^2} + \ldots + \frac{\partial^2}{\partial (x^{n-1})^2},$$

which is the d'Alembertian on an n-dimensional *Euclidean* space, rather than on Minkowski space. This corresponds to considering the properties of the field ϕ in Euclidean space.

The advantage of Euclidean field theory is that the elliptic operator has a unique, well-defined inverse, because the poles in the integral representation (2.91) lie on the imaginary rather than the real axis. Hence, it is often mathematically convenient to work in Euclidean space, and 'rotate' back to pseudo-Euclidean spacetime, using (2.90), at the end of the calculation. The boundary conditions for the Feynman propagator are automatically imposed by this procedure. (Note that none of the other contours in fig. 3 can be so rotated without intersecting the poles.) For a more detailed discussion (in the context of curved spacetimes) see Candelas & Raine (1977*a*) and Wald (1979*b*).

The Green functions introduced so far have all been calculated as expectation values of products of field operators in a pure state, namely the vacuum state. These Green functions are suitable for describing systems at zero temperature. However a system at nonzero temperature is not described by a pure state but one that is statistically distributed over all such states. The Green functions for systems at nonzero temperature are thus given by the average over all pure states of the expectation value of the products of field operators in those pure states (Kadanoff & Baym 1962, Abrikosov, Gorkov & Dzyaloskinskii 1963, Mattuck 1967, Fetter & Walecka 1971).

Suppose that $|\psi_i\rangle$ is a pure state, being an eigenstate of the Hamiltonian (2.41) with energy eigenvalue E_i. Then it will also be an eigenstate of the total number operator N of (2.32) with number eigenvalue n_i say. Since both the number of particles and the energy are variable, an equilibrium system at temperature T is described by a grand canonical ensemble of states. The probability that the system will be in the state $|\psi_i\rangle$ is given by

$$\rho_i = e^{-\beta(E_i - \mu n_i)}/Z \tag{2.93}$$

where

$$\beta = 1/k_B T \tag{2.94}$$

k_B being Boltzmann's constant, μ the chemical potential,

$$Z = \sum_j e^{-\beta(E_j - \mu n_j)} = e^{-\beta\Omega} \tag{2.95}$$

the grand partition function, and Ω the thermodynamic potential. The ensemble average at a temperature $T = (k_B\beta)^{-1}$ of any operator A is thus

$$\langle A \rangle_\beta = \sum_i \rho_i \langle \psi_i | A | \psi_i \rangle. \tag{2.96}$$

Introducing the quantum *density operator* defined by

$$\rho = \exp[\beta(\Omega + \mu N - H)] \tag{2.97}$$

then

$$\rho_i = \langle \psi_i | \rho | \psi_i \rangle. \tag{2.98}$$

The requirement of unit total probability becomes

$$\operatorname{tr} \rho \equiv \sum_i \langle \psi_i | \rho | \psi_i \rangle = 1, \tag{2.99}$$

and (2.96) reduces to

$$\langle A \rangle_\beta = \operatorname{tr} \rho A. \tag{2.100}$$

Now we can define nonzero temperature Green functions (also called thermal or temperature Green functions) simply by replacing the vacuum expectation values in the definitions of zero-temperature Green functions by the ensemble average $\langle\ \rangle_\beta$. For example, from (2.68) we define in the case of scalar fields

$$\left.\begin{aligned} G_\beta^+(x, x') &= \langle \phi(x)\phi(x') \rangle_\beta \\ G_\beta^-(x, x') &= \langle \phi(x')\phi(x) \rangle_\beta. \end{aligned}\right\} \tag{2.101}$$

Assuming for now that the chemical potential vanishes we have the following important property of these thermal Green functions:

$$G_\beta^\pm(t, \mathbf{x}; t', \mathbf{x}') = G_\beta^\mp(t + i\beta, \mathbf{x}; t', \mathbf{x}'). \tag{2.102}$$

This relation is obtained from the Heisenberg equations of motion

$$\phi(t, \mathbf{x}) = e^{iH(t - t_0)}\phi(t_0, \mathbf{x})e^{-iH(t - t_0)} \tag{2.103}$$

as follows:

$$\begin{aligned} G_\beta^+(t, \mathbf{x}; t', \mathbf{x}') &= \operatorname{tr}[e^{-\beta H}\phi(t, \mathbf{x})\phi(t', \mathbf{x}')]/\operatorname{tr}(e^{-\beta H}) \\ &= \operatorname{tr}[e^{-\beta H}\phi(t, \mathbf{x})e^{\beta H}e^{-\beta H}\phi(t', \mathbf{x}')]/\operatorname{tr}(e^{-\beta H}) \end{aligned}$$

$$= \operatorname{tr}[\phi(t+\mathrm{i}\beta,\mathbf{x})\mathrm{e}^{-\beta H}\phi(t',\mathbf{x}')]/\operatorname{tr}(\mathrm{e}^{-\beta H})$$

$$= \operatorname{tr}[\mathrm{e}^{-\beta H}\phi(t',\mathbf{x}')\phi(t+\mathrm{i}\beta,\mathbf{x})]/\operatorname{tr}(\mathrm{e}^{-\beta H})$$

$$= G_\beta^-(t+\mathrm{i}\beta,\mathbf{x};t',\mathbf{x}') \qquad (2.104a)$$

and similarly for G_β^-. In arriving at the above result we have used the property $\operatorname{tr} AB = \operatorname{tr} BA$. From (2.102) similar properties for the other Green functions can be obtained; for example, from (2.67) it follows that

$$G_\beta^{(1)}(t,\mathbf{x};t',\mathbf{x}') = G_\beta^{(1)}(t+\mathrm{i}\beta,\mathbf{x};t',\mathbf{x}'). \qquad (2.104b)$$

Had the chemical potential been retained it would merely have introduced the factor $\mathrm{e}^{\beta\mu}$ on the right-hand side. Note, however, that

$$\mathrm{i}G_\beta(x,x') = \mathrm{i}G(x,x') = [\phi(x),\phi(x')], \qquad (2.105)$$

because the commutator of a free scalar field is a c-number (this follows from (2.18)) and thus its statistical and vacuum expectation values are equal. This will not in general be the case in interacting theories where the (unequal time) commutator can be an operator.

Using (2.105) and relations such as (2.104) for the other Green functions we can write integral representations for the thermal Green functions. Starting with G_β we can use (2.105) to write its Fourier transform as

$$\mathrm{i}G(x,x') = \mathrm{i}G_\beta(x,x') = (1/2\pi)\int_{-\infty}^{\infty} \mathrm{d}\omega\, c(\omega;x,x')\mathrm{e}^{-\mathrm{i}\omega(t-t')} \qquad (2.106)$$

in which $c(\omega;x,x')$ can easily be calculated from (2.76) with the appropriate contour:

$$c(\omega;x,x') = (2\pi)^{1-n}\int \mathrm{d}^{n-1}k\,\delta(\omega^2-|\mathbf{k}|^2-m^2)[\theta(\omega)-\theta(-\omega)]\mathrm{e}^{\mathrm{i}\mathbf{k}\cdot(\mathbf{x}-\mathbf{x}')}. \qquad (2.107)$$

If we also write the Fourier transform of $G_\beta^\pm$ as

$$G_\beta^\pm(x,x') = (1/2\pi)\int_{-\infty}^{\infty} \mathrm{d}\omega\, g^\pm(\omega)\mathrm{e}^{-\mathrm{i}\omega(t-t')} \qquad (2.108)$$

then, from (2.67),

$$c(\omega) = g^+(\omega) - g^-(\omega). \qquad (2.109)$$

The relation (2.104a) implies that

$$g^+(\omega) = \mathrm{e}^{\beta\omega}g^-(\omega)$$

which, when used in conjunction with (2.109) yields

$$g^{\pm}(\omega) = \pm c(\omega)(1 - e^{\mp\beta\omega})^{-1}$$

giving the integral representations

$$G_{\beta}^{\pm}(x, x') = \pm \int_{-\infty}^{\infty} \frac{d\omega}{2\pi} \frac{c(\omega)}{1 - e^{\mp\beta\omega}} e^{-i\omega(t - t')}. \tag{2.110}$$

From this equation the representations of other Green functions can be obtained. In particular, by explicitly evaluating the integrals, making use of expansions of the factors $(1 - e^{\mp\beta\omega})^{-1}$ in powers of $e^{\mp\beta\omega}$, one finds that

$$G_{\beta}^{(1)}(t, \mathbf{x}; t', \mathbf{x}') = \sum_{k=-\infty}^{\infty} G^{(1)}(t + ik\beta, \mathbf{x}; t', \mathbf{x}'). \tag{2.111}$$

That is, the thermal Green function can be written as an infinite imaginary-time image sum of the corresponding zero-temperature Green function.

In the spin $\frac{1}{2}$ case, because the fields anticommute rather than commute one finds that $S_{\beta}^{(1)}$, the nonzero temperature version of (2.81), satisfies an antiperiodicity condition

$$S_{\beta}^{(1)}(t, \mathbf{x}; t', \mathbf{x}') = -S_{\beta}^{(1)}(t + i\beta, \mathbf{x}, t', \mathbf{x}') \tag{2.112}$$

rather than the periodicity condition of (2.104). One then obtains an image sum similar to (2.111), but reflecting the antiperiodicity by including a factor $(-1)^k$:

$$S_{\beta}^{(1)}(t, \mathbf{x}; t', \mathbf{x}') = \sum_{k=-\infty}^{\infty} (-1)^k S^{(1)}(t + ik\beta, \mathbf{x}; t', \mathbf{x}'). \tag{2.113}$$

The treatment of commuting, spin 1 fields is similar to the scalar case (see, for example, Brown & Maclay 1969).

2.8 Path-integral quantization

The canonical quantization scheme outlined in §2.1 is only one of several approaches to quantum field theory. One could, for example, have started with the covariant commutation relations

$$[\phi(x), \phi(x')] = iG(x, x') \tag{2.114}$$

(recall (2.65), noting that G is a c-number), in place of the canonical commutation relations (2.15). The scheme (2.114) has the advantage of being closer to the spirit of general relativity because it does not single out a particular time t. It leads immediately to the same commutation relations (2.18) for the creation and annihilation operators. In a globally hyperbolic spacetime (see §3.1) the covariant and canonical approaches are equivalent.

Another quantization technique, especially well suited to the rigorous treatment of the functional analysis that is encountered in quantum field theory, is the C^* algebra approach of Segal (1967) (for a review of the relevance of this approach to curved spacetime, see Isham 1978*a*).

Finally, the path-integral quantization of Feynman (Feynman & Hibbs 1965) is a powerful approach to quantum gravity and to the quantization of interacting fields, with their attendant problems of renormalization. As we shall have occasion to use this technique in later chapters, we shall give a short review in this section. More detailed treatments can be found in Rzewuski (1969), Abers & Lee (1973), Taylor (1976), Frampton (1977), Nash (1978) and Itzykson & Zuber (1980).

The basic object of the theory is the functional integral for a field ϕ with action S

$$Z[J] = \langle \text{out}, 0|0, \text{in}\rangle = \int \mathscr{D}[\phi] \exp\{\mathrm{i}S[\phi] + \mathrm{i}\int \mathrm{d}^n x\, J(x)\phi(x)\} \quad (2.115)$$

taken over the space of functions ϕ with an appropriate measure. The quantity Z is known as the generating functional for the theory and gives the transition amplitude from the initial vacuum $|0, \text{in}\rangle$ to the final vacuum $|0, \text{out}\rangle$ in the presence of a source $J(x)$ which is producing particles. When the source is switched off the two vacua reduce to the usual source-free Minkowski space vacuum $|0\rangle$ and one has

$$Z[0] = \langle 0|0\rangle \quad (2.116)$$

(usually normalized to unity).

It follows by functional differentiation of Z with respect to J that

$$\mathrm{i}^j\langle 0|T(\phi(x_1)\ldots\phi(x_j))|0\rangle_c = \left(\frac{\delta^j \ln Z}{\delta J(x_1)\ldots\delta J(x_j)}\right)_{J=0} \quad (2.117)$$

giving the connected, time-ordered Green functions of the theory (c denoting that only connected Feynman diagrams are included in perturbation theory).

As an example, we consider the case of the free scalar field with

$$S[\phi] = \int \mathrm{d}^n x [\mathscr{L}_0(x) + \tfrac{1}{2}\mathrm{i}\varepsilon\phi^2(x)]$$

where $\mathscr{L}_0$ is the free field Lagrangian (2.2) and the infinitesimal factor (which is related to the boundary conditions on ϕ) can be used to make the functional integral convergent. Substituting (2.2) for $\mathscr{L}_0$ and integrating by parts, the action becomes

$$S[\phi] = \int \mathrm{d}^n x[-\tfrac{1}{2}\phi(\Box + m^2 - \mathrm{i}\varepsilon)\phi] \tag{2.118}$$

where we have discarded a boundary term. Using (2.118) the exponent in (2.115) can be written suggestively in the form

$$-\tfrac{1}{2}\int \mathrm{d}^n x\, \mathrm{d}^n y\, \phi(x) K_{xy} \phi(y) + \int J(x)\phi(x)\mathrm{d}^n x \tag{2.119}$$

where the symmetric operator

$$K_{xy} = (\Box_x + m^2 - \mathrm{i}\varepsilon)\delta^n(x-y) \tag{2.120}$$

can formally be treated as a symmetric matrix K with continuous indices x, y, having the properties

$$\int \mathrm{d}^n y\, K^{\frac{1}{2}}_{xy} K^{\frac{1}{2}}_{yz} = K_{xz} \tag{2.121}$$

$$\int \mathrm{d}^n y\, K^{\frac{1}{2}}_{xy} K^{-\frac{1}{2}}_{yz} = \delta^n(x-z) \tag{2.122}$$

$$K^{-1}_{xy} = -G_{\mathrm{F}}(x,y), \tag{2.123}$$

the last result following by inverting the definition of the Feynman propagator (2.74). These properties become well defined within the context of the functional integral $\mathscr{D}[\phi]$.

Changing the integration variable from ϕ to

$$\phi'(x) = \int \mathrm{d}^n y\, K^{\frac{1}{2}}_{xy} \phi(y), \tag{2.124}$$

the quadratic form (2.119) may be recast, using (2.121)–(2.123), as

$$-\tfrac{1}{2}\int \mathrm{d}^n x \left[\phi'(x) - \int \mathrm{d}^n y\, J(y) K^{-\frac{1}{2}}_{yx}\right]^2 - \tfrac{1}{2}\int \mathrm{d}^n x\, \mathrm{d}^n y\, J(x) G_{\mathrm{F}}(x,y) J(y). \tag{2.125}$$

When (2.125) is substituted into the exponent in the functional integral, the second term is independent of ϕ and can be removed from the integral, while the first term yields an integral of the Gaussian type, and can be performed to give a numerical factor.

Thus

$$Z(J) \propto (\det K^{\frac{1}{2}})^{-1} \exp\left[-\tfrac{1}{2}\mathrm{i}\int \mathrm{d}^n x\, \mathrm{d}^n y\, J(x) G_{\mathrm{F}}(x, y) J(y)\right] \quad (2.126)$$

where

$$(\det K^{\frac{1}{2}})^{-1} = [\det(-G_{\mathrm{F}})]^{\frac{1}{2}} = \exp[\tfrac{1}{2}\operatorname{tr}\ln(-G_{\mathrm{F}})] \quad (2.127)$$

is the Jacobian arising from the change of variable (2.124).

It follows by inspection of (2.126) that, for example

$$\left(\frac{\delta^2 \ln Z}{\delta J(x)\delta J(y)}\right)_{J=0} = -\langle 0|T(\phi(x)\phi(y))|0\rangle = -\mathrm{i}G_{\mathrm{F}}(x, y)$$

in agreement with the definition (2.69).

For the spin $\frac{1}{2}$ field ψ, the generating functional Z is taken to be

$$Z(\eta, \bar{\eta}) = \int \mathscr{D}[\psi]\mathscr{D}[\bar{\psi}] \exp\left\{\mathrm{i}\int \mathrm{d}^n x[\mathscr{L}_0(x) + \bar{\eta}(x)\psi(x) + \eta(x)\bar{\psi}(x)]\right\} \quad (2.128)$$

where $\eta, \bar{\eta}$ are anticommuting external currents and $\mathscr{L}_0$ is given by (2.45). In place of (2.126) one obtains

$$Z(\eta, \bar{\eta}) \propto (\det S_{\mathrm{F}})^{-1} \exp\left[-\mathrm{i}\int \mathrm{d}^n x\, \bar{\eta}(x) S_{\mathrm{F}}(x, y)\eta(y)\right]. \quad (2.129)$$

The case of the electromagnetic field runs into complications associated with the gauge symmetry. To see this, one examines the analogue of (2.115)

$$Z(J) = \int \mathscr{D}[A_\alpha] \exp\left\{\mathrm{i}\int \mathrm{d}^n x[\mathscr{L}_0(x) + J^\beta(x) A_\beta(x)]\right\} \quad (2.130)$$

with $\mathscr{L}_0(x)$ given by (2.54). The action in the exponent of (2.130) can be written

$$\int \mathrm{d}^n x\, \mathscr{L}_0(x) = -\tfrac{1}{4}\int F_{\alpha\beta}F^{\alpha\beta}\mathrm{d}^n x = -\tfrac{1}{2}\int \mathrm{d}^n x\, \mathrm{d}^n y\, A_\alpha(x) K^{\alpha\beta}_{xy} A_\beta(y), \quad (2.131)$$

$$K^{\alpha\beta}_{xy} = (\eta^{\alpha\beta}\Box_x - \partial^\alpha_x \partial^\beta_x)\delta^n(x - y). \quad (2.132)$$

Inspection of (2.87) and (2.88) in the limit $\zeta \to \infty$ shows that $K^{-1} = D_{\mathrm{F}}(\zeta \to \infty)$ is singular. Thus, as we shall see below, quantization of the electromagnetic field based on the above naive description cannot proceed. This difficulty was mentioned in connection with gauge invariance on page 19, where a gauge-breaking term (2.59) was introduced into the Lagrangian (this term is removed by taking the $\zeta \to \infty$ limit) which yields an invertible wave operator, and Green function (2.88), for finite ζ.

Because $\mathscr{L}_0$ is manifestly gauge-invariant, it is independent of the longitudinal and timelike components of A_α, i.e., K projects out the transverse field components. Hence, any variation of the longitudinal and timelike components of A_α will leave $\mathscr{L}_0$ unchanged. More generally, variation of A_α related to the original A_α by the gauge transformation (2.58) will leave $\mathscr{L}_0$ unchanged.

If one envisages the space of all functions A_α, then under the action of the gauge transformation (2.58), a point (i.e., function A_α) in the space will be mapped into points along a line, representing other functions related to the original A_α by the continuous gauge transformation. This line is called the *orbit* of the gauge group associated with (2.58). Problems with the path integral occur because variations of A_μ in $\mathscr{D}[A_\alpha]$ that lie *along* these orbits do not induce any variation in $\mathscr{L}_0(x)$. To see this, note that the functional integration in (2.130) is over an infinite volume of function space, so in order for the integral to converge it is necessary for $\mathscr{L}_0(x) \to \infty$ as $A_\alpha \to \infty$ to produce a declining exponential. However, when $A_\alpha \to \infty$ along the gauge orbits, $\mathscr{L}_0$ remains constant (and hence finite) and fails to produce the required convergence. As the volume of the 'orbit subspace' is also infinite, the functional integral is undefined, unless a way can be found to renormalize it by dividing out this infinite volume.

One way to cure this ill was pointed out by Fadeev & Popov (1967), following earlier work of Feynman (1963) and DeWitt (1965, 1967*b*). Variations in A_α are required to be restricted to functions belonging to distinct orbits. This can be achieved by choosing a 'hypersurface' in the space of A_α that intersects each orbit only once. Then rather than integrating over the whole space, one integrates only over the hypersurface.

The equation for such a hypersurface may be written

$$F[A_\alpha] = 0, \tag{2.133}$$

so it might be supposed that one need merely insert $\delta[F(A_\alpha)]$ into the integrand of (2.130). However, to ensure a gauge-invariant result, more care is needed. Two neighbouring hypersurfaces will be related by

$$F(A_\alpha^\Lambda(x)) = F(A_\alpha(x)) + \int \mathrm{d}^n y M_{xy} \Lambda(y) + \mathrm{O}(\Lambda^2) \tag{2.134}$$

where Λ parametrizes the gauge transformation, $F(A_\alpha) = [F(A_\alpha^\Lambda)]_{\Lambda=0}$, and M depends on the choice of gauge. For example, in the Landau gauge

$F(A_\alpha)=A^\alpha{}_{,\alpha}$, so

$$F(A_\alpha^\Lambda)=A^\alpha{}_{,\alpha}+\Box\Lambda=F(A_\alpha)+\Box\Lambda \tag{2.135}$$

from which it follows that

$$M_{xy}=\Box\delta^n(x-y). \tag{2.136}$$

It is convenient to define the quantity $\Delta_F(A_\alpha)$ by

$$\Delta_F^{-1}(A_\alpha)=\int\mathscr{D}[\Lambda]\delta[F(A_\alpha^\Lambda)] \tag{2.137}$$

which is easily shown to be gauge invariant. When restricted to the hypersurface defined by (2.133), the first term on the right of (2.134) vanishes so, symbolically, $F(A_\alpha^\Lambda)=M\Lambda$. The integral in (2.137) can then be performed immediately, by changing the variable from Λ to $M\Lambda$:

$$\Delta_F^{-1}(A_\alpha)=(\det M)^{-1}=\exp(-\operatorname{tr}\ln M) \tag{2.138}$$

$(\det M)^{-1}$ being the Jacobean arising from this change of variable.

Returning to the afflicted path integral (2.130) (with J set to zero) we may insert, using (2.137), the unit operator $\Delta_F(A_\alpha)\int\mathscr{D}[\Lambda]\delta[F(A_\alpha^\Lambda)]$ into the integrand without changing anything. Exploiting the gauge invariance of Δ_F we may change its argument to A_α^Λ, and then with a change of integration variable the path integral may be written

$$\int\mathscr{D}[A_\alpha]\int\mathscr{D}[\Lambda]\Delta_F(A_\alpha)\delta[F(A_\alpha)]\exp\left[\mathrm{i}\int\mathscr{L}_0(x)\mathrm{d}^n x\right] \tag{2.139}$$

from which it is obvious that the integrand of the $\mathscr{D}[\Lambda]$ integral is independent of Λ. Hence we may factor it out. Although infinite, this integral is independent of the fields A_α, so it may simply be divided out from (2.139) to obtain a new definition of Z that does not suffer from singularity problems:

$$Z=\int\mathscr{D}[A_\alpha]\delta[F(A_\alpha)]\exp\left\{\mathrm{i}\int\mathrm{d}^n x[\mathscr{L}_0(x)+J^\alpha(x)A_\alpha(x)-\mathrm{i}\operatorname{tr}\ln M]\right\}. \tag{2.140}$$

In arriving at (2.140) we have used the fact that the factor $\delta[F(A_\alpha)]$ in the integrand of (2.139) restricts Δ_F to the hypersurface (2.133), enabling us to use the result (2.138).

The restriction of the integration to the hypersurface $F(A_\alpha)=0$ does not, therefore, merely introduce a factor $\delta[F(A_\alpha)]$ into the integrand, but also introduces an extra term into the field action. This additional contribution can be regarded as due to an additional fictitious field. In fact, it is not difficult to show that in the Landau gauge

$$\int \mathscr{D}[c]\mathscr{D}[c^*]\exp\left(\mathrm{i}\int \eta^{\alpha\beta}\partial_\alpha c \partial_\beta c^*\right) = \exp(-\operatorname{tr}\ln M) \tag{2.141}$$

where c and c^* are massless scalar fields, but satisfying anticommutation relations, for which reason they are called 'Fadeev–Popov ghost fields'. As they do not couple to the vector fields A_α, the ghost fields are frequently ignored in flat spacetime quantum field theory. In curved spacetime, however, they play an important role.

The vector part of the path integral may be further reduced in the Landau gauge by also expressing $\delta[F(A_\alpha)] = \delta(A^\alpha{}_{,\alpha})$ as an exponential:

$$\delta(A^\alpha{}_{,\alpha}) = \lim_{\zeta\to 0} \exp\left[-\frac{\mathrm{i}}{2\zeta}\int (A^\alpha{}_{,\alpha})^2 \mathrm{d}^n x\right]. \tag{2.142}$$

Comparison with (2.59) reveals that the exponent here is simply $\mathrm{i}\mathscr{L}_G$ in the Landau ($\zeta\to 0$) gauge. Hence the vector generating functional is

$$\int \mathscr{D}[A_\alpha]\exp\left[\mathrm{i}\int \mathrm{d}^n x(\mathscr{L}_0 + \mathscr{L}_G)\right]. \tag{2.143}$$

This functional integral may be evaluated along the same lines as the scalar case and is found to be proportional to

$$\exp\left[-\tfrac{1}{2}\mathrm{i}\int \mathrm{d}^n x\, \mathrm{d}^n y\, J_\alpha(x) D_F^{\alpha\beta}(x-y) J_\beta(y)\right] \tag{2.144}$$

where $D_F^{\alpha\beta}$ is given from (2.88)

Although the total generating functional, (2.144) multiplied by (2.141), has been derived in the Landau gauge $\zeta\to 0$, it remains true for any ζ.

We end this section with some brief remarks about the convergence of the path integral (2.115). In general, because the action S is real, the exponent in the integrand is pure imaginary. Hence the integral is generally not properly defined when taken over the whole function space. In the foregoing this was ameliorated by the careful use of the iε factor. As remarked in §2.7, the employment of iε to define the Feynman Green functions is equivalent to passing to an imaginary time description, in which the field is defined in Euclidean space rather than Minkowski space. If the vacuum expectation values of the time-ordered field products, such as (2.117), are analytic in the complex t plane, then one can construct a field quantization in Euclidean space, in which the path integrals assume a well-defined, strongly convergent Gaussian form (at least for a wide class of Lagrangians), and recover the Minkowski space theory at the end by 'rotating' back from it to t. This technique is frequently used in practice. When passing to curved spacetime, however, it may happen that no 'Euclideanized' (i.e., positive definite metric) spacetime exists that corresponds to the original pseudo-Riemannian spacetime.

3

Quantum field theory in curved spacetime

The basic formalism of quantum field theory is generalized to curved spacetime in this chapter, in a straightforward way. The discussion is preceded by a very brief summary of pseudo-Riemannian geometry. The treatment is in no way intended to be complete, and we refer the reader to Weinberg (1972), Hawking & Ellis (1973), or Misner, Thorne & Wheeler (1973) for further details. Readers unfamiliar with conformal transformations and Penrose conformal diagrams are advised to read §3.1 carefully, however.

The basic generalization of the particle concept to curved spacetime is readily accomplished. What is not so easy is the physical interpretation of the formalism so developed. There has, in fact, been a certain amount of controversy over the meaning – and meaningfulness – of the particle concept when a background gravitational field is present. In some cases, such as for static spacetimes, the concept seems well defined, while in others (e.g. spacetimes that admit closed timelike world lines or do not everywhere possess Cauchy surfaces) the notion of particle can seem hopelessly obscure. We restrict consideration to 'well-behaved' spacetimes, and do not embark upon a philosophical discourse about the meaning of particles. Instead we relate the formalism directly to what an actual particle detector might be expected to register in the particular quantum state of interest. It is in this concrete operational sense that we define particles in curved spacetime. Although this approach has been studied before, we give the most developed treatment of particle detectors so far.

Building upon an explicit example of particle creation by a changing background gravitational field, we present a detailed and in-depth analysis of adiabaticity – the definition of particles in a quasi-static spacetime. It is here that one expects to make contact with standard quantum field theory, for we know that to be a good approximation in the (relatively) slowly expanding universe that we inhabit. The treatment reveals that the high frequency behaviour of the field is independent of the quantum state or the global structure of the spacetime, and depends purely on the local geometry. This turns out to be of crucial significance for the regularization and renormalization programme dealt with in later chapters.

We give what is intended to be a complete treatment of the concept of adiabatic states in §3.5. As a result, this section requires careful reading. What is essential is the application of the adiabatic limit to the Feynman propagator, dealt with in §3.6. This culminates in the so-called DeWitt–Schwinger representation of G_F, given by the expansion (3.141). Its rôle in renormalization theory is so crucial that the importance of this expansion cannot be overemphasized. We recommend that §3.5 be read briefly in the first instance, with attention concentrated on the physical remarks, and re-read in depth when the subsequent applications have been examined.

The section on the conformal vacuum, §3.7, is much easier to follow, and relatively important for later applications, especially to concrete calculations of cosmological particle creation and quantum vacuum stress.

A final section on higher-spin fields in curved spacetime is merely a summary of standard formalism, though the expressions for the stress–energy–momentum tensors (3.190)–(3.195) will be frequently used.

3.1 Spacetime structure

We assume spacetime to be a C^∞ n-dimensional, globally hyperbolic, pseudo-Riemannian manifold (for more detailed discussion see Hawking & Ellis 1973). These conditions may be more restrictive than necessary to construct a viable quantum field theory (for example, see Avis, Isham & Storey 1978 for quantum field theory in a non-globally hyperbolic spacetime). The differentiability conditions ensure the existence of differential equations and the global hyperbolicity ensures the existence of Cauchy hypersurfaces.

The pseudo-Riemannian metric $g_{\mu\nu}$ associated with the line element

$$ds^2 = g_{\mu\nu}(x)dx^\mu dx^\nu, \quad \mu, \nu = 0, 1, \ldots, (n-1)$$

has signature $n-2$. Several coordinate patches with associated $g_{\mu\nu}$ may be needed to cover the entire manifold. We define the determinant

$$g \equiv |\det g_{\mu\nu}|.$$

We shall frequently make use of Penrose conformal diagrams (Penrose 1964) for depicting the causal structure of spacetime. This is a device that enables the whole of an infinite spacetime to be represented as a finite diagram (compact manifold), by applying a *conformal transformation* to the metric structure. Conformal transformations, which shrink or stretch the manifold, must be distinguished from coordinate transformations $x^\mu \rightarrow x'^\mu$

which merely relabel the coordinates in some patch, leaving the geometry itself unchanged. A conformal transformation of the metric may be described by

$$g_{\mu\nu}(x) \rightarrow \bar{g}_{\mu\nu}(x) = \Omega^2(x) g_{\mu\nu}(x) \tag{3.1}$$

for some continuous, non-vanishing, finite, real function $\Omega(x)$.

From such a transformation of the metric, the conformal transformation properties of various other quantities can be derived. For example, the Christoffel symbol, Ricci tensor and Ricci scalar transform respectively as

$$\Gamma^{\rho}{}_{\mu\nu} \rightarrow \bar{\Gamma}^{\rho}{}_{\mu\nu} = \Gamma^{\rho}{}_{\mu\nu} + \Omega^{-1}(\delta^{\rho}_{\mu}\Omega_{;\nu} + \delta^{\rho}_{\nu}\Omega_{;\mu} - g_{\mu\nu}g^{\rho\alpha}\Omega_{;\alpha}) \tag{3.2}$$

$$R^{\nu}{}_{\mu} \rightarrow \bar{R}^{\nu}{}_{\mu} = \Omega^{-2}R^{\nu}{}_{\mu} - (n-2)\Omega^{-1}(\Omega^{-1})_{;\mu\rho}g^{\rho\nu} + (n-2)^{-1}\Omega^{-n}(\Omega^{n-2})_{;\rho\sigma}g^{\rho\sigma}\delta^{\nu}_{\mu} \tag{3.3}$$

$$R \rightarrow \bar{R} = \Omega^{-2}R + 2(n-1)\Omega^{-3}\Omega_{;\mu\nu}g^{\mu\nu} + (n-1)(n-4)\Omega^{-4}\Omega_{;\mu}\Omega_{;\nu}g^{\mu\nu} \tag{3.4}$$

from which one obtains the following useful transformation:

$$[\Box + \tfrac{1}{4}(n-2)R/(n-1)]\phi \rightarrow [\bar{\Box} + \tfrac{1}{4}(n-2)\bar{R}/(n-1)]\bar{\phi} = \Omega^{-(n+2)/2}[\Box + \tfrac{1}{4}(n-2)R/(n-1)]\phi \tag{3.5}$$

where

$$\Box\phi = g^{\mu\nu}\nabla_{\mu}\nabla_{\nu}\phi = (-g)^{-\frac{1}{2}}\partial_{\mu}[(-g)^{\frac{1}{2}}g^{\mu\nu}\partial_{\nu}\phi] \tag{3.6}$$

$$\bar{\phi}(x) \equiv \Omega^{(2-n)/2}(x)\phi(x). \tag{3.7}$$

As a simple illustration of a Penrose diagram, consider two-dimensional Minkowski space, which has the line element

$$ds^2 = dt^2 - dx^2. \tag{3.8}$$

We frequently work in terms of null coordinates, u, v, defined by

$$\left.\begin{aligned} u &= t - x \\ v &= t + x \end{aligned}\right\} \tag{3.9}$$

in which the line element (3.8) becomes

$$ds^2 = du\,dv \tag{3.10}$$

so that

$$g_{\mu\nu} = \tfrac{1}{2}\begin{bmatrix} 0 & 1 \\ 1 & 0 \end{bmatrix}. \tag{3.11}$$

Suppose we perform the *coordinate transformation*

$$\left.\begin{aligned} u' &= 2\tan^{-1}u \\ v' &= 2\tan^{-1}v \end{aligned}\right\} \tag{3.12}$$

where

$$-\pi \leq u',\ v' \leq \pi. \tag{3.13}$$

Then from (3.10)

$$\mathrm{d}s^2 = \tfrac{1}{4}\sec^2\tfrac{1}{2}u' \sec^2\tfrac{1}{2}v' \mathrm{d}u'\mathrm{d}v' \tag{3.14}$$

so that

$$g_{\mu\nu}(u', v') = \tfrac{1}{8}\sec^2\tfrac{1}{2}u' \sec^2\tfrac{1}{2}v' \begin{bmatrix} 0 & 1 \\ 1 & 0 \end{bmatrix}. \tag{3.15}$$

If we now perform a *conformal transformation* with

$$\Omega^2(x) = (\tfrac{1}{4}\sec^2\tfrac{1}{2}u' \sec^2\tfrac{1}{2}v')^{-1}$$

then

$$g_{\mu\nu}(u', v') \to \bar{g}_{\mu\nu}(u', v') = \tfrac{1}{2}\begin{bmatrix} 0 & 1 \\ 1 & 0 \end{bmatrix} \tag{3.16}$$

and the conformally-related line element is given by

$$\mathrm{d}\bar{s}^2 = \mathrm{d}u'\mathrm{d}v'. \tag{3.17}$$

This has the same form as the original Minkowski space line element (3.10), but only covers the compact region (3.13) depicted in fig. 4. The effect of the conformal transformation (3.16) has been to shrink in infinity to the boundary lines on the diagram.

The boundary of the figure contains several features of interest. First note that all null rays remain at 45° in the Penrose diagram: conformal transformations leave the null cones invariant. Thus any causal analysis may proceed with the null rays drawn as is usual in Minkowski space. Clearly all null rays will terminate on the diagonal boundary lines labelled $\mathscr{I}^+$ and $\mathscr{I}^-$ and called future and past null infinity respectively. Asymptotically timelike lines converge on points marked i^+ (future timelike infinity) and i^- (past timelike infinity). Similarly, asymptotically spacelike lines converge on i^0 (spacelike infinity).

The analysis here applies also to four-dimensional Minkowski space if each point on the diagram is considered as a 2-sphere, except for points on

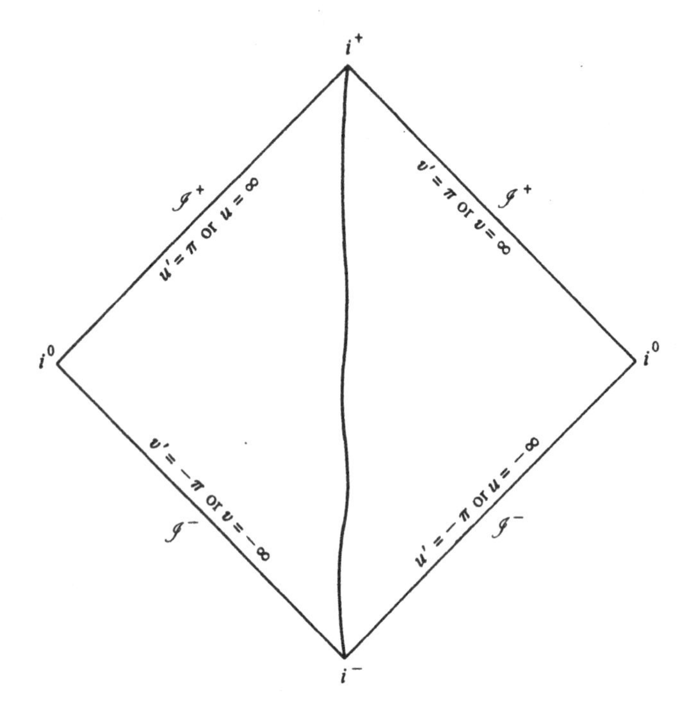

Fig. 4. Penrose conformal diagram of Minkowski space. The compact region $-\pi \leq u', v' \leq \pi$ is conformal to the whole of Minkowski space $(-\infty \leq u, v \leq \infty)$. Null rays $u, v =$ constant remain at 45°. The world line of an asymptotically timelike observer is shown.

the vertical symmetry axis and i^0, which represent spacetime points. Thus $\mathscr{I}^+$ and $\mathscr{I}^-$ are really null 3-surfaces.

A timelike line that is asymptotically null may occur, such as if a particle undergoes uniform acceleration for all time, approaching the speed of light as $t \to \infty$. The world line for such a particle is drawn in fig. 5. The null asymptote has the property that events that lie above it are forever causally inaccessible to the accelerated particle. That is, those events cannot communicate with the particle (though the converse may not be true). The null asymptote is therefore an *event horizon* for the accelerated particle, though not for an unaccelerated particle.

As another illustration of Penrose diagrams, we shall consider four-dimensional Schwarzschild spacetime, described by the line element

$$ds^2 = (1 - 2M/r)dt^2 - (1 - 2M/r)^{-1}dr^2 - r^2(d\theta^2 + \sin^2\theta \, d\phi^2). \quad (3.18)$$

This spacetime is the unique spherically symmetric vacuum solution of

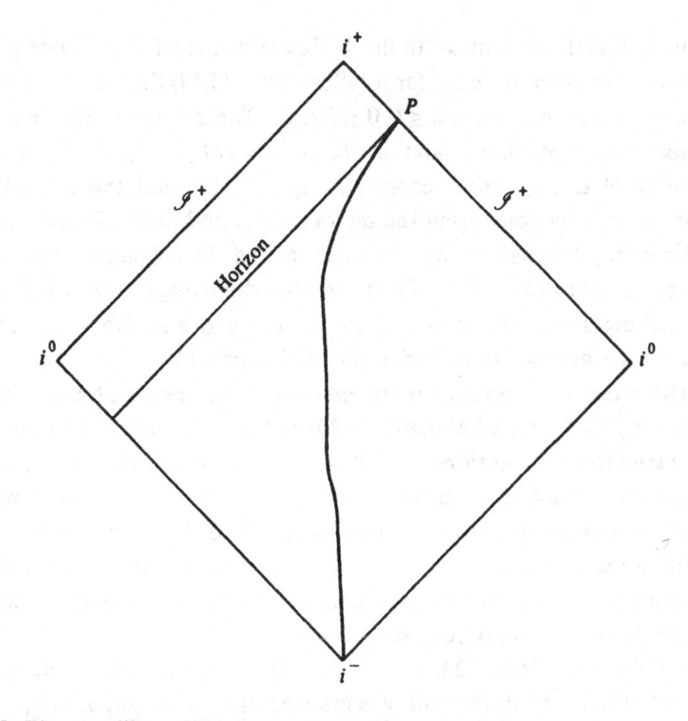

Fig. 5. The timelike world line represents an observer who accelerates continuously to the right, so that he asymptotically approaches the speed of light. This world line does not, therefore, reach i^+, but intersects $\mathscr{I}^+$ at P. The backward null ray through P thus acts as an event horizon, because events above it can never be witnessed by the observer: all null rays from such events intersect $\mathscr{I}^+$ between P and i^+.

Einstein's equation and is often used to represent the empty space region surrounding a spherical star or collapsing body of mass M.

Transforming to *Kruskal* coordinates defined (Kruskal 1960) by

$$\left.\begin{aligned} \bar{u} &= -4M\,\mathrm{e}^{-u/4M} \\ \bar{v} &= 4M\,\mathrm{e}^{v/4M} \end{aligned}\right\} \tag{3.19}$$

where $u = t - r^*, v = t + r^*$ and $r^* = r + 2M\ln|(r/2M) - 1|$, (3.18) becomes

$$\mathrm{d}s^2 = (2M/r)\mathrm{e}^{-r/2M}\mathrm{d}\bar{u}\,\mathrm{d}\bar{v} - r^2(\mathrm{d}\theta^2 + \sin^2\theta\,\mathrm{d}\phi^2). \tag{3.20}$$

The $\mathrm{d}\bar{u}\,\mathrm{d}\bar{v}$ part of this metric is conformal to two-dimensional Minkowski space described by (3.10), which in turn may be compactified by the coordinate and conformal transformations (3.12) and (3.16). The resulting diagram is therefore identical to fig. 4. However, the left-hand edge of the diamond is not $\mathscr{I}$ in this case, for it follows from (3.19) that $\bar{u}, \bar{v}$ are defined only in the quadrant $-\infty < \bar{u} \leq 0, 0 \leq \bar{v} < \infty$. The left-hand edge therefore comprises segments $\bar{u} = 0$ and $\bar{v} = 0$, or $r = 2M$, $t = \pm\infty$. This is a singularity of the u, v(or t, r) coordinate system, but not the $\bar{u}, \bar{v}$ system, as may be seen by comparing the metrics (3.18) and (3.20). Therefore, the spacetime may be analytically extended beyond the left-hand edge, using the Kruskal coordinates $\bar{u}, \bar{v}$ defined over the whole plane $-\infty < \bar{u}, \bar{v} < \infty$. The resulting Penrose diagram is shown in fig. 6 and the spacetime is known as the maximally extended Kruskal manifold.

The horizontal zig-zag lines represent $r = 0$ in the past and future. This is a singularity in the Kruskal metric (3.20) and, in fact, in the spacetime, as may be seen from inspection of the Riemann tensor. The manifold cannot be analytically continued across these edges. There is another asymptotically Minkowskian spacetime region marked II, to the left of the diagram, containing the $\mathscr{I}$ surfaces $\bar{u} = +\infty, \bar{v} = -\infty$. This half of the manifold is geometrically identical to the right-hand diamond marked I, except that the time direction is formally reversed $(t \to -t)$.

The null ray $\bar{u} = 0$ $(r = 2M, t = +\infty)$ is the latest retarded null ray to reach $\mathscr{I}^+$: for $\bar{u} > 0$, all the null rays intersect the future singularity, $r = 0$. The ray $\bar{u} = 0$ (representing an outward directed 2-surface in the full four-dimensional picture) is therefore an event horizon for observers restricted to region I. Similarly $\bar{v} = 0$ $(r = 2M, t = -\infty)$ is an event horizon for observers restricted to region II. The empty regions III and IV are thus black holes for these respective sets of observers (and white holes for their opposite numbers). The world lines of observers who avoid the black and white holes converge on i^+ and i^- in their respective regions. The existence of an event horizon prevents any communication between regions I and II.

The maximally extended Kruskal spacetime is everywhere (except at $r = 0$) a solution of the vacuum Einstein equation. In the real world, where a black hole is more likely to form from the implosion of a star, the vacuum equations only apply to the region outside the star and so only a fragment of fig. 6 will be relevant. Nevertheless, as we shall see in chapter 8, the difference as far as observers in region I are concerned is negligible at late times $(t \to \infty)$, after the implosion has occurred.

Frequently our analysis will be restricted to spacetimes with special geometrical symmetries. These can be described using Killing vectors ξ^μ, which are solutions of Killing's equation

$$\mathscr{L}_\xi g_{\mu\nu}(x) = 0 \tag{3.21}$$

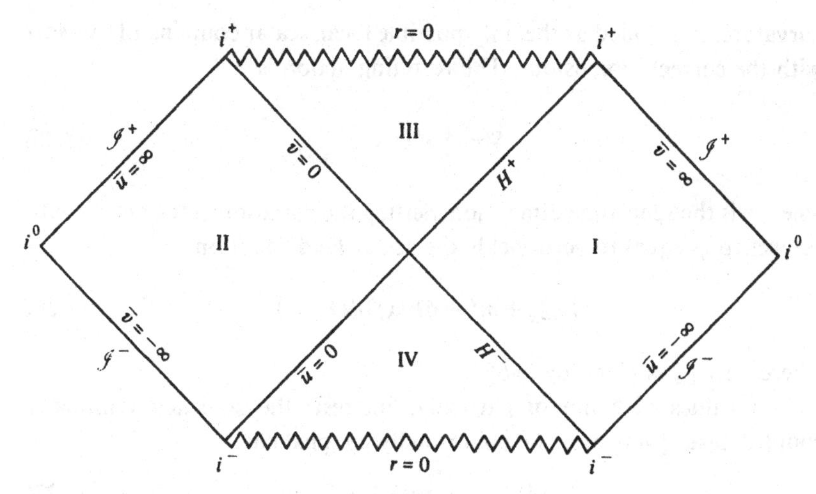

Fig. 6. Penrose diagram of the maximally extended Kruskal manifold. The physical singularity is marked by a wavy line. The future and past horizons, labelled $H^{\pm}$ respectively, are represented by the null rays $\bar{u} = 0$, $\bar{v} = 0$. Observers who do not fall into the black hole (or emerge from the white hole) are restricted to either of the diamond shaped regions marked I and II.

where $\mathscr{L}_{\xi}$ is the Lie derivative along the vector field ξ^{μ}. Equation (3.21) can be written as

$$\xi_{\mu;\nu} + \xi_{\nu;\mu} = 0. \tag{3.22}$$

We shall also be interested in the symmetries associated with conformal flatness, when the spacetime is conformal to Minkowski space. The geometry then admits a *conformal* Killing vector field, which satisfies the conformal generalization of (3.21):

$$\mathscr{L}_{\xi} g_{\mu\nu}(x) = \lambda(x) g_{\mu\nu}(x) \tag{3.23}$$

where $\lambda(x)$ is some (non-singular, non-vanishing) scalar function.

3.2 Scalar field quantization

Formally, field quantization in curved spacetime proceeds in close analogy to the Minkowski space case. We start with Lagrangian density (see §3.8)

$$\mathscr{L}(x) = \tfrac{1}{2}[-g(x)]^{\frac{1}{2}}\{g^{\mu\nu}(x)\phi(x)_{,\mu}\phi(x)_{,\nu} - [m^2 + \xi R(x)]\phi^2(x)\} \tag{3.24}$$

where $\phi(x)$ is the scalar field and m the mass of the field quanta. The coupling between the scalar field and the gravitational field represented by the term $\xi R\phi^2$, where ξ is a numerical factor and $R(x)$ is the Ricci scalar curvature, is included as the only possible local, scalar coupling of this sort with the correct dimensions. The resulting action is

$$S = \int \mathscr{L}(x)\mathrm{d}^n x \tag{3.25}$$

where n is the spacetime dimension. Setting the variation of the action with respect to ϕ equal to zero yields the scalar field equation

$$[\Box_x + m^2 + \xi R(x)]\phi(x) = 0 \tag{3.26}$$

where now $\Box$ is given by (3.6).

Two values of ξ are of particular interest: the so-called minimally coupled case, $\xi = 0$, and the conformally coupled case

$$\xi = \tfrac{1}{4}[(n-2)/(n-1)] \equiv \xi(n). \tag{3.27}$$

In this latter case, if $m = 0$ the action and hence the field equations are invariant under conformal transformations (3.1) if the field is assumed to transform as in (3.7). Indeed, from (3.5) it is clear that if

$$[\Box + \tfrac{1}{4}(n-2)R/(n-1)]\phi = 0,$$

then also

$$[\bar{\Box} + \tfrac{1}{4}(n-2)\bar{R}/(n-1)]\bar{\phi} = 0.$$

The scalar product (2.9) is generalized to

$$(\phi_1, \phi_2) = -\mathrm{i}\int_\Sigma \phi_1(x)\overleftrightarrow{\partial_\mu}\phi_2^*(x)[-g_\Sigma(x)]^{\frac{1}{2}}\mathrm{d}\Sigma^\mu \tag{3.28}$$

where $\mathrm{d}\Sigma^\mu = n^\mu\,\mathrm{d}\Sigma$, with n^μ a future-directed unit vector orthogonal to the spacelike hypersurface Σ and $\mathrm{d}\Sigma$ is the volume element in Σ. The hypersurface Σ is taken to be a Cauchy surface in the (globally hyperbolic) spacetime and one can show, using Gauss' theorem (see Hawking & Ellis 1973, §2.8), that the value of (ϕ_1, ϕ_2) is independent of Σ.

There exists a complete set of mode solutions $u_i(x)$ of (3.26) which are orthonormal in the product (3.28): i.e., satisfying

$$(u_i, u_j) = \delta_{ij}, \quad (u_i^*, u_j^*) = -\delta_{ij}, \quad (u_i, u_j^*) = 0. \tag{3.29}$$

The index i schematically represents the set of quantities necessary to label the modes. The field ϕ may be expanded as in (2.17):

$$\phi(x) = \sum_i [a_i u_i(x) + a_i^\dagger u_i^*(x)]. \tag{3.30}$$

The covariant quantization of the theory is implemented by adopting the commutation relations (2.18):

$$[a_i, a_j^\dagger] = \delta_{ij}, \text{ etc.} \tag{3.31}$$

The construction of a vacuum state, Fock space, etc. can then proceed exactly as described for the Minkowski space case in §2.1. This time, however, there is an inherent ambiguity in the formalism (Fulling 1973). In Minkowski space there is a natural set of modes, namely (2.11), that are closely associated with the natural rectangular coordinate system (t, x, y, z). In turn, these natural coordinates are associated with the Poincaré group, the action of which leaves the Minkowski line element unchanged. Specifically, the vector $\partial/\partial t$ is a Killing vector of Minkowski space, orthogonal to the spacelike hypersurfaces $t =$ constant, and the modes (2.11) are eigenfunctions of this Killing vector with eigenvalues $-i\omega$ for $\omega > 0$ (positive frequency). The vacuum is invariant under the action of the Poincaré group.

In curved spacetime the Poincaré group is no longer a symmetry group of the spacetime (see in this connection Urbantke 1969). Indeed, in general there will be no Killing vectors at all with which to define positive frequency modes. In some special classes of spacetime there may be symmetry under certain restricted transformations, for example rotations or translations, or the de Sitter group. In these cases, there may exist 'natural' coordinates associated with the Killing vectors – analogues of rectangular coordinates in Minkowski space. But even if such coordinates do exist, we shall see that they do not enjoy the same central physical status in quantum field theory as their Minkowski space counterparts. In general, though, no such privileged coordinates are available and no natural mode decomposition of ϕ based on the separation of the wave equation (3.26) in these privileged coordinates will present itself. Indeed, the whole spirit of general relativity, expressed through the principle of general covariance, is that coordinate systems are physically irrelevant.

Consider, therefore, a second complete orthonormal set of modes $\bar{u}_j(x)$. The field ϕ may be expanded in this set also

$$\phi(x) = \sum_j [\bar{a}_j \bar{u}_j(x) + \bar{a}_j^\dagger \bar{u}_j^*(x)]. \tag{3.32}$$

This decomposition of ϕ defines a new vacuum state $|\bar{0}\rangle$:

$$\bar{a}_j |\bar{0}\rangle = 0, \quad \forall j \tag{3.33}$$

and a new Fock space.

As both sets are complete, the new modes $\bar{u}_j$ can be expanded in terms of the old:

$$\bar{u}_j = \sum_i (\alpha_{ji} u_i + \beta_{ji} u_i^*). \tag{3.34}$$

Conversely

$$u_i = \sum_j (\alpha_{ji}^* \bar{u}_j - \beta_{ji} \bar{u}_j^*). \tag{3.35}$$

These relations are known as Bogolubov transformations (Bogolubov 1958). The matrices α_{ij}, β_{ij} are called Bogolubov coefficients, and by using (3.34) and (3.29) they can be evaluated as

$$\alpha_{ij} = (\bar{u}_i, u_j), \quad \beta_{ij} = -(\bar{u}_i, u_j^*). \tag{3.36}$$

Equating the expansions (3.30) and (3.32) and making use of (3.34), (3.35) and the orthonormality of the modes, (3.29), one obtains

$$a_i = \sum_j (\alpha_{ji} \bar{a}_j + \beta_{ji}^* \bar{a}_j^\dagger) \tag{3.37}$$

and

$$\bar{a}_j = \sum_i (\alpha_{ji}^* a_i - \beta_{ji}^* a_i^\dagger). \tag{3.38}$$

The Bogolubov coefficients possess the following properties

$$\sum_k (\alpha_{ik} \alpha_{jk}^* - \beta_{ik} \beta_{jk}^*) = \delta_{ij}, \tag{3.39}$$

$$\sum_k (\alpha_{ik} \beta_{jk} - \beta_{ik} \alpha_{jk}) = 0. \tag{3.40}$$

It follows immediately from (3.37) that the two Fock spaces based on the two choices of modes u_i and $\bar{u}_j$ are different so long as $\beta_{ji} \neq 0$. For example $|\bar{0}\rangle$ will not be annihilated by a_i:

$$a_i|\bar{0}\rangle = \sum_j \beta_{ji}^* |\bar{1}_j\rangle \neq 0 \tag{3.41}$$

in contrast to (3.33). In fact, the expectation value of the operator $N_i = a_i^\dagger a_i$ for the number of u_i-mode particles in the state $|\bar{0}\rangle$ is

$$\langle \bar{0}|N_i|\bar{0}\rangle = \sum_j |\beta_{ji}|^2, \tag{3.42}$$

which is to say that the vacuum of the $\bar{u}_j$ modes contains $\Sigma_j |\beta_{ji}|^2$ particles in the u_i mode.

Note that if u_j are positive frequency modes with respect to some timelike Killing vector field ξ, satisfying

$$\mathscr{L}_\xi u_j = -\mathrm{i}\omega u_j, \quad \omega > 0 \tag{3.43}$$

(cf. (2.8) which can be written $\mathscr{L}\partial_t u_k = -\mathrm{i}\omega u_k$), and $\bar{u}_k$ are a linear combination of u_j alone (not u_j^*), i.e., containing only positive frequencies with respect to ξ, then $\beta_{jk} = 0$. In that case $\bar{a}_k|0\rangle = 0$ as well as $a_j|0\rangle = 0$. Thus, the two sets of modes u_j and $\bar{u}_k$ share a common vacuum state. If any $\beta_{jk} \neq 0$, the $\bar{u}_k$ will contain a mixture of positive-(u_j) and negative-(u_j^*) frequency modes, and particles will be present.

More generally the Fock space based on $|0\rangle$ can be related to that based on $|\bar{0}\rangle$ using the completeness of the Fock space basis elements:

$$|{}^1n_{i_1}, {}^2n_{i_2}, \ldots\rangle$$

$$= \sum_{k=0}^{\infty} \frac{1}{k!} \sum_{j_1 \ldots j_k} |\bar{1}_{j_1}, \bar{1}_{j_2}, \ldots, \bar{1}_{j_k}\rangle\langle\bar{1}_{j_1}, \bar{1}_{j_2}, \ldots, \bar{1}_{j_k}|{}^1n_{i_1}, {}^2n_{i_2}, \ldots\rangle. \tag{3.44}$$

In the notation used here we have, for example,

$$|{}^1n_{i_1}\rangle = |1_{i_1}, 1_{i_1}, \ldots, 1_{i_1}\rangle/({}^1n_i!)^{\frac{1}{2}}$$

where '1_{i_1}' is repeated ${}^1n_i$ times. The matrix element $\langle\bar{1}_{j_1}, \bar{1}_{j_2}, \ldots, \bar{1}_{j_k}|{}^1n_{i_1}, {}^2n_{i_2}, \ldots\rangle$ can be thought of as the transition amplitude or *S*-matrix element for a transition from the state $|{}^1n_{i_1}, {}^2n_{i_2}, \ldots\rangle$ to $|\bar{1}_{j_1}, \bar{1}_{j_2}, \ldots, \bar{1}_{j_k}\rangle$. These *S*-matrix elements can be written in terms of the Bogolubov coefficients. In particular, for the vacuum to many-particle amplitudes one finds (DeWitt 1975)

$$\langle\bar{0}|1_{j_1}, 1_{j_2}, \ldots, 1_{j_k}\rangle = \begin{cases} \mathrm{i}^{k/2}\langle\bar{0}|0\rangle \sum_p \Lambda_{p_1p_2}\cdots\Lambda_{p_{k-1}p_k} & k \text{ even} \\ 0 & k \text{ odd} \end{cases} \tag{3.45}$$

$$\langle\bar{1}_{j_1}, \bar{1}_{j_2}, \ldots, \bar{1}_{j_k}|0\rangle = \begin{cases} \mathrm{i}^{k/2}\langle\bar{0}|0\rangle \sum_p V_{p_1p_2}\cdots V_{p_{k-1}p_k} & k \text{ even} \\ 0 & k \text{ odd} \end{cases} \tag{3.46}$$

where p represents all distinct permutations of $\{j_1 \ldots j_k\}$ and

$$\left.\begin{aligned} \Lambda_{ij} &= -\mathrm{i}\sum_k \beta_{kj}\alpha_{ik}^{-1} \\ V_{ij} &= \mathrm{i}\sum_k \beta_{jk}^*\alpha_{ki}^{-1}. \end{aligned}\right\} \tag{3.47}$$

The many-particle to many-particle amplitudes may be found in Birrell & Taylor (1980). We defer consideration of the vacuum-to-vacuum amplitude until chapter 6.

In addition to the particle states and Bogolubov coefficients, we shall need various Green functions introduced in §2.7. The same notation and definitions will be used for the Green functions as in §2.7, except that now the field $\phi(x)$ satisfies the curved spacetime equation (3.26), and care must be taken in the specification of the vacuum state $|0\rangle$. Using (3.26), curved spacetime generalizations of the Green function equations (2.73)–(2.75) can be obtained. For example, for the Feynman propagator

$$\mathrm{i}G_{\mathrm{F}}(x, x') = \langle 0|T(\phi(x)\phi(x'))|0\rangle \tag{3.48}$$

one obtains

$$[\Box_x + m^2 + \xi R(x)]G_{\mathrm{F}}(x, x') = -[-g(x)]^{-\frac{1}{2}}\delta^n(x - x'), \tag{3.49}$$

in place of (2.74). It is important to note that (3.49) by itself does not specify the state $|0\rangle$ appearing in (3.48), nor does it ensure that the solution has the properties of a time-ordered product. To fix the state and impose the time ordering, boundary conditions must be imposed on the solution of (3.49). In Minkowski space these boundary conditions take the form of the choice of contour used in (2.76). In curved spacetime the specification of boundary conditions will not be so simple, and will depend on the global features of the particular case under consideration. (For example, radiation which is retarded at its source will not remain so elsewhere because of 'backscattering' off the spacetime curvature.) Detailed discussions of wave propagation in curved spacetime, and the properties of Green functions, have been given for example, by DeWitt & Brehme (1960), and Friedlander (1975).

3.3 Meaning of the particle concept: particle detectors

The question naturally arises as to which set of modes furnishes the 'best' description of a physical vacuum, i.e., corresponds most closely to our actual experience of 'no particles'. It turns out that this question cannot be answered as stated, because it is necessary to specify also the details of the quantum measurement process that is used to detect the presence of quanta. In particular the state of motion of the measuring device can affect whether or not particles are observed to be present. For example, a free-falling detector will not always register the same particle density as a non-inertial, accelerating detector. In fact, this is even true in Minkowski space: an accelerated detector will register quanta even in the vacuum state defined by (2.19).

The special feature of Minkowski space is not that there is a unique vacuum (there is not), but that the conventional vacuum state as defined in terms of the modes (2.11) is the agreed vacuum for *all inertial* measuring devices, throughout the spacetime. This is because the vacuum defined by (2.19) is invariant under the Poincaré group and so are the set of inertial observers in Minkowski space.

One of the lessons learned from the development of this subject has been the realization that the particle concept does not generally have universal significance. Particles may register their presence on some detectors but not others, so there is an essential observer-dependent quality about them. One is still free to assert the presence of particles, but without specifying the state of motion of the detector, the concept is not very useful, even in Minkowski space.

Part of the reason for the nebulousness of the particle concept is its *global* nature. The modes are defined on the whole of spacetime (or at least a large patch) so that a particular observer's specification of the field mode decomposition, and hence the number operator describing the response of a particle detector carried by him, will depend, for example, on the observer's entire past history. To obtain a more objective probe of the state of a field one must construct locally-defined quantities, such as $\langle\psi|T_{\mu\nu}(x)|\psi\rangle$, which assumes a particular value at the point x of spacetime. The stress-tensor is objective in the sense that, for a fixed state $|\psi\rangle$, the results of different measuring devices can be related in the familiar fashion by the usual tensor transformation. For example, if $\langle\psi|T_{\mu\nu}(x)|\psi\rangle = 0$ for one observer, it will vanish for all observers. This is in contrast to the particle concept, where one observer may detect no particles while another may, as we shall see, disagree.

In many problems of interest the spacetime can be treated as asymptotically Minkowskian in the remote past and/or future. Under these circumstances the choice of the 'natural' Minkowskian vacuum defined by (2.19) has a well-understood physical meaning, i.e., the absence of particles according to *all inertial* observers in the asymptotic region – usually taken to be the commonly accepted idea of a vacuum. We refer to the remote past and future as the *in* and *out* regions respectively. This terminology is borrowed from Minkowskian quantum field theory where it is assumed that as $t \to \pm\infty$ all the field interactions approach zero. The analogous situation here would be to suppose that in the in and out regions spacetime admits natural particle states and a privileged quantum vacuum. This can either be Minkowski space, or some other spacetime of high symmetry such as the Einstein static universe. Whether a particular spacetime constitutes a suitable in or out region may also depend on the quantum field of interest. In the case of massless, conformally coupled fields, a conformally flat spacetime, even if not static, may still be a good candidate (see §3.7).

Since we work in the Heisenberg picture, if we choose the state of the quantum field in the in region to be the vacuum state, then it will remain in that state during its subsequent evolution. However, as will soon be demonstrated, at later times, outside the in region, freely falling particle detectors may still register particles in this 'vacuum' state. In particular, if there is also an out region, then the in vacuum may not coincide with the out vacuum. In that case a natural (e.g., inertial) class of observers in the out region will detect the presence of particles. We can therefore say that particles have been 'created' by the time-dependent external gravitational field. This is an especially useful description if the in and out regions are Minkowskian so that all inertial observers in the out region register the presence of quanta. Analogous processes of particle creation by external electromagnetic fields are well known (see for example, Gitman 1977, where references to earlier works are given). The possibility of similar particle production due to spacetime curvature was discussed over forty years ago by Schrödinger (1939), while other early work is due to DeWitt (1953), Takahashi & Umezawa (1957) and Imamura (1960). The first thorough treatment of particle production by an external gravitational field was given by Parker (1966, 1968, 1969), and Sexl & Urbantke (1967, 1969).

To illustrate these considerations we shall treat a model of a particle detector due to Unruh (1976) and DeWitt (1979). It consists of an idealized point particle with internal energy levels labelled by the energy E, coupled via a monopole interaction with a scalar field ϕ. We work in four-dimensional Minkowski space.

Suppose the particle detector moves along the world line described by the functions $x^\mu(\tau)$, where τ is the detector's proper time. The detector–field interaction is described by the interaction Lagrangian $cm(\tau)\phi[x(\tau)]$, where c is a small coupling constant and m is the detector's monopole moment operator. Suppose the field ϕ is in the vacuum state $|0_M\rangle$ defined by (2.19), where the subscript M stands for 'Minkowski vacuum'. For a general trajectory, the detector will not remain in its ground state E_0, but will undergo a transition to an excited state $E > E_0$, while the field will make a transition to an excited state $|\psi\rangle$. For sufficiently small c the amplitude for this transition may be given by first order perturbation theory (see §9.1) as

$$ic\langle E, \psi| \int_{-\infty}^{\infty} m(\tau)\phi[x(\tau)]\,\mathrm{d}\tau\,|0_M, E_0\rangle$$

(The limits of integration may be confined to a smaller interval provided that the detector coupling is switched off adiabatically outside that interval.)

Using the equation for the time evolution of $m(\tau)$

$$m(\tau) = e^{iH_0\tau} m(0) e^{-iH_0\tau},$$

where $H_0|E\rangle = E|E\rangle$, the above transition amplitude factorizes to give

$$ic\langle E|m(0)|E_0\rangle \int_{-\infty}^{\infty} e^{i(E-E_0)\tau} \langle \psi|\phi(x)|0_M\rangle \, d\tau. \tag{3.50}$$

If ϕ is expanded in terms of standard Minkowski plane wave modes (2.17) it is clear that, to this order of perturbation theory, transitions can only occur to the state $|\psi\rangle = |1_{\mathbf{k}}\rangle$ containing one quantum of frequency $\omega = (|\mathbf{k}|^2 + m^2)^{\frac{1}{2}}$, for some $\mathbf{k}$. Then (in the continuum normalization (2.11))

$$\langle 1_{\mathbf{k}}|\phi(x)|0_M\rangle = \int d^3k' (16\pi^3\omega')^{-\frac{1}{2}} \langle 1_{\mathbf{k}}|a^{\dagger}_{\mathbf{k}'}|0_M\rangle e^{-i\mathbf{k}'\cdot\mathbf{x}+i\omega' t}$$

$$= (16\pi^3\omega)^{-\frac{1}{2}} e^{-i\mathbf{k}\cdot\mathbf{x}+i\omega t} \tag{3.51}$$

We must now take into account that $\mathbf{x}$ in (3.51) is not an independent variable but is determined by the detector's trajectory. Suppose it follows an inertial world line, i.e.,

$$\mathbf{x} = \mathbf{x}_0 + \mathbf{v}t = \mathbf{x}_0 + \mathbf{v}\tau(1 - v^2)^{-\frac{1}{2}} \tag{3.52}$$

where $\mathbf{x}_0 =$ constant, $\mathbf{v} =$ constant, $|\mathbf{v}| < 1$, then the integral in (3.50) (with $\psi = 1_{\mathbf{k}}$) becomes

$$(16\pi^3\omega)^{-\frac{1}{2}} e^{-i\mathbf{k}\cdot\mathbf{x}_0} \int_{-\infty}^{\infty} e^{i(E-E_0)\tau} e^{i\tau(\omega-\mathbf{k}\cdot\mathbf{v})(1-v^2)^{-\frac{1}{2}}} d\tau$$

$$= (4\pi\omega)^{-\frac{1}{2}} e^{-i\mathbf{k}\cdot\mathbf{x}_0} \delta(E - E_0 + (\omega - \mathbf{k}\cdot\mathbf{v})(1 - v^2)^{-\frac{1}{2}}). \tag{3.53}$$

However as $\mathbf{k}\cdot\mathbf{v} \le |\mathbf{k}||\mathbf{v}| < \omega$ and $E > E_0$, the argument of the δ-function is always > 0 and the transition amplitude vanishes. The transition is forbidden on energy conservation grounds – a direct consequence of Poincaré invariance.

If, on the other hand, instead of (3.52) we had chosen a more complicated trajectory, the integral in (3.50) would not have yielded a δ-function and the result would be nonzero. In such a case it is of interest to calculate the transition probability to *all* possible E and ψ, obtained by squaring the modulus of (3.50), and summing over E and the complete set ψ, to obtain

$$c^2 \sum_E |\langle E|m(0)|E_0\rangle|^2 \mathscr{F}(E-E_0), \tag{3.54}$$

where

$$\mathscr{F}(E) = \int_{-\infty}^{\infty} \mathrm{d}\tau \int_{-\infty}^{\infty} \mathrm{d}\tau' \mathrm{e}^{-\mathrm{i}E(\tau-\tau')} G^+(x(\tau), x(\tau')). \tag{3.55}$$

The detector *response function* $\mathscr{F}(E)$, is independent of the details of the detector, and is determined by the positive frequency Wightman Green function G^+ defined by (2.68). It represents the bath of 'particles' that the detector effectively experiences as a result of its motion. The remaining factor in (3.54) represents the *selectivity* of the detector to this bath, and clearly depends on the internal structure of the detector itself.

In the cases of detector trajectories in Minkowski space for which

$$G^+(x(\tau), x(\tau')) = g(\Delta\tau) \tag{3.56}$$

$$\Delta\tau \equiv \tau - \tau' \tag{3.57}$$

for some function g, the system is invariant under time translations in the reference frame of the detector ($\tau \to \tau + \text{constant}$). This means that the detector is in equilibrium with the ϕ field, so that the number of quanta absorbed by the detector per unit τ is constant. If this rate is nonzero, the transition probability will diverge, as the transition amplitude (3.54) is computed for an infinite proper time interval. This can be seen immediately from (3.55), because the Wightman function will be a function of $\tau - \tau'$ only. Thus the double integration reduces to a Fourier transform of the two-point function multiplied by an infinite time integral.

This sort of circumstance frequently arises in quantum theory, and may be dealt with by adiabatically switching off the coupling as $\tau \to \pm\infty$, or considering instead the transition probability per unit proper time:

$$c^2 \sum_E |\langle E|m(0)|E_0\rangle|^2 \int_{-\infty}^{\infty} \mathrm{d}(\Delta\tau) \mathrm{e}^{-\mathrm{i}(E-E_0)\Delta\tau} G^+(\Delta\tau). \tag{3.58}$$

To simplify the following examples we now restrict attention to a massless scalar field ϕ. Then the positive frequency Wightman function is easily evaluated from (2.76), using the appropriate contour in fig. 3, to be

$$D^+(x, x') = -1/4\pi^2[(t-t'-\mathrm{i}\varepsilon)^2 - |\mathbf{x}-\mathbf{x}'|^2], \tag{3.59}$$

where the small imaginary part $i\varepsilon$, $\varepsilon > 0$, can be interpreted using the equation

$$1/(x \mp i\varepsilon) = (P/x) \pm i\pi\delta(x). \tag{3.60}$$

In the case of the inertial trajectory (3.52), equation (3.59) becomes

$$D^+(\Delta\tau) = -1/4\pi^2(\Delta\tau - i\varepsilon)^2, \tag{3.61}$$

(where we have absorbed a positive factor $(1 - v^2)^{-\frac{1}{2}}$ into ε) and the integral in (3.58) can be calculated as a contour integral, closing the contour in an infinite semicircle in the lower-half $\Delta\tau$ plane, since $E - E_0 > 0$. However the pole in the integrand is at $\Delta\tau = i\varepsilon$, which is in the upper-half plane, so the result is zero, as expected. No particles are detected.

As another example of this special equilibrium case, consider that the detector moves along a hyperbolic trajectory in the (t, z) plane:

$$x = y = 0, \quad z = (t^2 + \alpha^2)^{\frac{1}{2}} \quad \alpha = \text{constant}. \tag{3.62}$$

This represents a detector that accelerates uniformly with acceleration α^{-1} in the frame of the detector (see, for example, Rindler (1969)). The detector's proper time τ is related to t by

$$t = \alpha \sinh(\tau/\alpha) \tag{3.63}$$

so, from (3.59), we find

$$D^+(\Delta\tau) = -\left[16\pi^2\alpha^2 \sinh^2\left(\frac{\tau - \tau'}{2\alpha} - \frac{i\varepsilon}{\alpha}\right)\right]^{-1} \tag{3.64}$$

where we have absorbed a positive function of τ, τ' into ε. Using the identity

$$\operatorname{cosec}^2 \pi x = \pi^{-2} \sum_{k=-\infty}^{\infty} (x - k)^{-2} \tag{3.65}$$

we can write (3.64) as

$$D^+(\Delta\tau) = -(4\pi^2)^{-1} \sum_{k=-\infty}^{\infty} (\Delta\tau - 2i\varepsilon + 2\pi i\alpha k)^{-2}. \tag{3.66}$$

Substituting this into (3.58) and performing the Fourier transform with the help of a contour integral yields

$$\frac{c^2}{2\pi} \sum_E \frac{(E - E_0)|\langle E|m(0)|E_0\rangle|^2}{e^{2\pi(E - E_0)\alpha} - 1}. \tag{3.67}$$

The appearance of the Planck factor $[e^{2\pi(E - E_0)\alpha} - 1]^{-1}$ in (3.67) indicates that the equilibrium between the accelerated detector and the ϕ field in the state $|0_M\rangle$ is the same as that which would have been achieved had the detector remained unaccelerated, but immersed in a bath of thermal radiation at the temperature

$$T = 1/2\pi\alpha k_B = \text{acceleration}/2\pi k_B \qquad (3.68)$$

where k_B is Boltzmann's constant.

The same conclusion can also be drawn from an examination of the *thermal* Green function for an *inertial* detector (Dowker 1977). This we have already derived for the case of $G_\beta^{(1)}$ in (2.111) (a similar derivation for G_β^+ is more complicated). Putting $\mathbf{x} = \mathbf{x}' = 0$, $t = \tau$, $t' = \tau'$ in (2.111) and using (2.79) gives, for a massless field,

$$D_\beta^{(1)}(\Delta\tau) = -(2\pi^2)^{-1} \sum_{k=-\infty}^{\infty} (\Delta\tau + ik\beta)^{-2}, \qquad (3.69)$$

where the principal value is implied.

Let us compare this result with the $D^{(1)}$ function for the accelerated observer moving through a quantum field in the Minkowski *vacuum* state. It can be calculated from D^+ by making use of (2.67) in the form

$$D^{(1)}(\Delta\tau) = D^+(\Delta\tau) + D^+(-\Delta\tau).$$

Applying this identity to (3.66), noting (3.60), one observes that

$$D^{(1)}(\Delta\tau) = D_\beta^{(1)}(\Delta\tau)$$

where $\beta = 1/k_B T$ and T is given by (3.68). We conclude that the vacuum Green function for a uniformly accelerated detector is the same as the thermal Green function for an inertial detector.

What does this mean physically? It is often stated that a uniformly accelerated observer will 'see' thermal radiation (Davies 1975, Unruh 1976) even though the field ϕ is in the vacuum state $|0_M\rangle$ and, as far as inertial observers are concerned, no particles are detected whatever. Certainly the accelerated detector absorbs energy and makes transitions to excited states, just as if it were bathed in thermal radiation. However, as implied in §2.4, $\langle 0_M|{:}T_{\mu\nu}{:}|0_M\rangle = 0$. Transforming to the accelerated frame using the usual tensor transformation law gives $\langle 0_M|{:}T'_{\mu\nu}{:}|0_M\rangle = 0$ also, so both accelerated and unaccelerated observers agree that the stress–energy–momentum of the ϕ field vanishes. This has led to the descriptions 'quasi' or 'fictitious' particles for the quanta that excite the accelerated detector, but really the phenomenon is more an indication that the traditional quantum particle concept is applicable only in very restrictive circumstances.

If the state $|0_M\rangle$ cannot supply the energy to excite the detector, how can we reconcile its excitation with the principle of energy conservation? Moreover, the transition that elevates the detector from energy level E_0 to E is accompanied by the appearance of a quantum in the ϕ field ($|0_M\rangle \rightarrow |1_{\mathbf{k}}\rangle$). This means that *both* the detector and the field gain energy.

The explanation comes from a consideration of the agency that brings about the acceleration of the detector in the first place. As the detector accelerates, its coupling to the ϕ field causes the *emission* of quanta, which produces a resistance against the accelerating force. The work done by the external force to overcome this resistance supplies the missing energy that feeds into the field via the quanta emitted from the detector, and also into the detector which simultaneously makes upward transitions. But as far as the detector is concerned, the net effect is the absorption of thermally distributed quanta.

Returning to the question of a suitable definition of a quantum vacuum state and Fock space could a case be made that such definitions based on the accelerated system are equally valid contenders as the traditional Minkowski-based constructs? In response it may be objected that basing one's treatment of these concepts on the considerations of accelerated observers is a fraud, because inertial observers occupy a special status in most physical theory. Hence, as far as Minkowski space is concerned, the vacuum $|0_M\rangle$ is a strong candidate for the 'correct' or 'physical' vacuum – the experiences of the accelerated observers being 'distorted' by the effects of their non-uniform motion. The trouble is that when gravitational fields are present, inertial observers become free-falling observers, and in general no two free-falling detectors will agree on a choice of vacuum. Only in exceptional cases of high symmetry will a set of detector trajectories exist that all register no particles, and this set may not even be free falling (e.g. the observers who accelerate to remain stationary close to a spherical star – see §8.4).

It is a simple matter to generalize the experiences of the detector to curved spacetime by replacing G^+ by its curved spacetime counterpart (see page 48) and the vacuum state $|0_M\rangle$ by some more general vacuum. Examples will be given in §§3.6 and 3.7.

It is also interesting to examine the case when the quantum field is not in a vacuum state but a many-particle state (2.22) (or its curved space counterpart). Then G^+ is replaced by

$$\langle {}^1n_{\mathbf{k}_1}, {}^2n_{\mathbf{k}_2}, \ldots, {}^jn_{\mathbf{k}_j}|\phi(x)\phi(x')|{}^1n_{\mathbf{k}_1}, {}^2n_{\mathbf{k}_2}, \ldots, {}^jn_{\mathbf{k}_j}\rangle$$

$$= G^+(x, x') + \sum_i {}^in u_{\mathbf{k}_i}(x)u^*_{\mathbf{k}_i}(x') + \sum_i {}^in u^*_{\mathbf{k}_i}(x)u_{\mathbf{k}_i}(x'). \qquad (3.70)$$

Passing to the continuum limit, the right-hand side of (3.70) is replaced by

$$G^+(x,x')+\int d^{n-1}k\, n_{\mathbf{k}} u_{\mathbf{k}}(x)u_{\mathbf{k}}^*(x')+\int d^{n-1}k\, n_{\mathbf{k}} u_{\mathbf{k}}^*(x)u_{\mathbf{k}}(x') \quad (3.71)$$

where $n_{\mathbf{k}}$ is the number density of quanta in k-space.

For an inertial detector moving along the trajectory (3.52) in n-dimensional Minkowski space, only the last term of (3.71) yields a contribution to the detector response function (3.55):

$$\frac{\mathscr{F}(E)}{T}=(2\pi)^{1-n}\int_{-\infty}^{\infty} d(\Delta\tau)e^{-iE\Delta\tau}\int\frac{d^{n-1}k}{2\omega}\exp\left[i(\omega-\mathbf{k}\cdot\mathbf{v})\Delta\tau(1-v^2)^{-\frac{1}{2}}\right]n_{\mathbf{k}} \quad (3.72)$$

where T is the total duration for which the detector is switched on. If $\mathbf{v}=0$, the $\Delta\tau$ integration may be performed to yield $2\pi\delta(E-\omega)$. If in addition the quanta are distributed isotropically, then $n_{\mathbf{k}}=n_k$ and the $d^{n-1}k$ integral may be performed to yield

$$\frac{\mathscr{F}(E)}{T}=\frac{2^{2-n}\pi^{(3-n)/2}}{\Gamma((n-1)/2)}(E^2-m^2)^{(n-3)/2}n_{(E^2-m^2)^{\frac{1}{2}}}\theta(E-m). \quad (3.73)$$

Substituting (3.73) into (3.54), one notices that the presence of the function $\theta(E-E_0-m)$ shows that the absorption of a single quantum of mass m by the detector will not occur unless the energy level spacing $E-E_0$ in the detector is at least equal to the particle rest energy m. Further examination of (3.73) reveals that the transition response rate of the detector to the bath of quanta is proportional to the number of quanta in the mode of interest, as would be expected on physical grounds. The energy dependence however, is related in a complicated way to the particle number spectrum, and will be additionally complicated by the selectivity of the detector (energy dependence of the matrix element in (3.54)).

For $\mathbf{v}\neq 0$, the angular integration in (3.72) must be performed first. The τ integration then yields the difference between two θ-functions (rather than a δ-function) which delimits the range of the dk integration. In the massless four-dimensional case one obtains

$$\frac{\mathscr{F}(E)}{T}=\frac{1}{4\pi}\left(\frac{1-v^2}{v^2}\right)^{\frac{1}{2}}\int_{E^-}^{E^+}n_k dk, \quad (3.74)$$

where $E^{\pm}=E[(1\pm v)/(1\mp v)]^{\frac{1}{2}}$. This expression is readily understood physically. As the detector moves through the isotropic bath of radiation, a particular transition with energy $E-E_0$ will not select quanta from only one mode, but a whole range, varying from blueshifted lower energy modes in the forward direction to redshifted higher energy modes in the backward direction. The factors $[(1+v)/(1-v)]^{\frac{1}{2}}$ and $[(1-v)/(1+v)]^{\frac{1}{2}}$ are recognized as the usual Doppler blue- and redshift factors respectively, and one observes that the response function is proportional to the total number of quanta within the energy range made accessible by the Doppler shifting.

Given the vagueness of the particle concept in more general situations, one might wonder what meaning should be attached to the Bogolubov coefficients α and β introduced in the last section, and in particular what is meant by the expectation value for the number of quanta in a mode i, given by (3.42).

In general there is *no* simple relation between $\langle N_i \rangle$ and the particle number as measured by a detector, even if it is freely falling. However, in one special case a simple relation does exist. Consider a spacetime that is asymptotically static in the remote past and future, and construct the vacuum state in the in region associated with the standard plane wave modes of the form (2.11). This is the usual, physical, no-particle state, and an inertial detector in the in region would certainly register no quanta.

Of course, if the spacetime is only asymptotically static, the upper limits of the integrals in (3.55) are no longer strictly speaking infinity. We make the assumption that the monopole detector interaction is adiabatically switched off before the detector enters the non-static region. Similarly in the out region, the detector will be adiabatically switched on after the spacetime motion has ceased. This enables us to continue to use $\pm\infty$ in the limits as though the spacetime were always static. As already remarked, this type of assumption – of restricting an interaction to a finite duration using adiabatic switching – is routine in quantum theory, and so long as the duration of interaction is very much greater than $(E-E_0)^{-1}$, the spurious excitation of the detector by the switching process itself will be negligible.

In the out region the in modes will no longer in general be plane waves of fixed frequency, like (2.11). What, therefore, will a particle detector make of the *in* vacuum state in the *out* region? Clearly, an inertial detector in the out region will not register quanta if the quantum field is in the *out* vacuum state associated with modes that take the form (2.11) in the *out* region. However, it will, as we shall see, generally register the presence of some quanta if the field is in the *in* vacuum state.

To find out what the detector does register, one must analyse (in connection with (3.55)) the Wightman function G^+ (constructed using the in vacuum) evaluated in the out region. This will generally be a complicated function of the position variables $\mathbf{x}, \mathbf{x}'$. However, in the simple case of a homogeneous universe, such as the asymptotically static, spatially flat, Robertson–Walker model, G^+ will be invariant under spatial translations and rotations. We shall restrict discussion to this simple case.

If the in modes are denoted by $u_{\mathbf{k}}^{\text{in}}$ and the in vacuum by $|0, \text{in}\rangle$ then we must consider

$$G_{\text{in}}^+(x, x') = \langle \text{in}, 0|\phi(x)\phi(x')|0, \text{in}\rangle = \int u_{\mathbf{k}}^{\text{in}}(x)u_{\mathbf{k}}^{\text{in}*}(x')\mathrm{d}^{n-1}k \quad (3.75)$$

where x and x' are both situated in the out region. Denoting the out modes (i.e., the modes that reduce to the standard plane wave form (2.11) in the out region) by $u_{\mathbf{k}}^{\text{out}}$, we may use (3.34) to expand $u_{\mathbf{k}}^{\text{in}}$ in terms of $u_{\mathbf{k}}^{\text{out}}$. Because of the spatial homogeneity, both sets of modes, $u_{\mathbf{k}}^{\text{out}}$ and $u_{\mathbf{k}}^{\text{in}}$, will remain as plane waves (i.e., $\propto \mathrm{e}^{\mathrm{i}\mathbf{k}\cdot\mathbf{x}}$) throughout, but with altered time dependence. The scalar products $(u_{\mathbf{k}}^{\text{in}}, u_{\mathbf{k}'}^{\text{out}})$ and $(u_{\mathbf{k}}^{\text{in}}, u_{\mathbf{k}'}^{\text{out}*})$ will therefore be proportional to $\delta_{\mathbf{k}\mathbf{k}'}$ and $\delta_{-\mathbf{k}\mathbf{k}'}$, respectively (see (3.28)). It follows from (3.36) that the Bogolubov transformation will be both diagonai and isotropic, i.e., will have the form

$$\alpha_{\mathbf{k}\mathbf{k}'} = \alpha_k \delta_{\mathbf{k}\mathbf{k}'} \quad (3.76)$$

$$\beta_{\mathbf{k}\mathbf{k}'} = \beta_k \delta_{-\mathbf{k}\mathbf{k}'} \quad (3.77)$$

where $k = |\mathbf{k}|$, so that

$$u_{\mathbf{k}}^{\text{in}}(x) = \alpha_k u_{\mathbf{k}}^{\text{out}}(x) + \beta_k u_{-\mathbf{k}}^{\text{out}*}(x). \quad (3.78)$$

Substituting (3.78) into (3.75) yields

$$G_{\text{in}}^+ = \int \mathrm{d}^{n-1}k[|\alpha_k|^2 u_{\mathbf{k}}^{\text{out}}(x)u_{\mathbf{k}}^{\text{out}*}(x') + \alpha_k\beta_k^* u_{\mathbf{k}}^{\text{out}}(x)u_{-\mathbf{k}}^{\text{out}}(x')$$

$$+ \beta_k\alpha_k^* u_{-\mathbf{k}}^{\text{out}*}(x)u_{\mathbf{k}}^{\text{out}*}(x') + |\beta_k|^2 u_{-\mathbf{k}}^{\text{out}*}(x)u_{-\mathbf{k}}^{\text{out}}(x')]. \quad (3.79)$$

Inspection of (3.79) shows that, because of the $\mathrm{e}^{\mathrm{i}\mathbf{k}\cdot\mathbf{x}}$ dependence of $u_{\mathbf{k}}$, G^+ depends on position only through $\mathbf{x} - \mathbf{x}'$, i.e., it is invariant under spatial translations as expected. It follows immediately from this that when $\mathbf{x}$ and $\mathbf{x}'$ are restricted to the inertial detector trajectories (3.52), the position dependence disappears from G^+. The detector response (3.55) is therefore location-independent, as expected on grounds of homogeneity.

In the out region, the $u_{\mathbf{k}}^{\text{out}}$ modes are simply standard plane waves, so the τ and τ' integrations in (3.55) are immediate, and yield δ-functions. The only non-vanishing term to survive from (3.79) is the final one.

The detector response function per unit time is

$$\frac{\mathscr{F}(E)}{T} = \frac{2^{2-n}\pi^{(3-n)/2}}{\Gamma((n-1)/2)} \int_0^\infty \frac{\mathrm{d}k\, k^{n-2}}{(k^2+m^2)^{\frac{1}{2}}} |\beta_k|^2 \delta(E - (k^2+m^2)^{\frac{1}{2}})$$

$$= \frac{2^{2-n}\pi^{(3-n)/2}}{\Gamma((n-1)/2)} (E^2 - m^2)^{(n-3)/2} |\beta_{(E^2-m^2)^{\frac{1}{2}}}|^2 \theta(E-m) \qquad (3.80)$$

where for simplicity we have restricted the detector trajectory to the $\mathbf{v} = 0$ case, and chosen the Robertson–Walker scale factor to be unity in the out region. This response function is identical to that associated with an isotropic bath of quanta with $|\beta_k|^2$ particles in mode k in a permanently static spacetime, as revealed by comparison of (3.80) with (3.73).

The simple form of this result implies the interpretation of $|\beta_k|^2$ as the number of quanta created in mode k by the cosmological motion, and is completely convincing. Of course, the use of a detector is not the only criterion by which the physical reasonableness of this interpretation can be based. Evaluation of $\langle T_{\mu\nu} \rangle$ would also reveal, in the out region, a bath of energy identical to that of $|\beta_k|^2$ particles in mode k in a permanently static spacetime.

3.4 Cosmological particle creation: a simple example

To see how, in practice, particle creation can occur in a spacetime with Minkowskian in and out regions, we shall consider a simple example. A suitable spacetime is a two-dimensional Robertson–Walker universe with line element

$$\mathrm{d}s^2 = \mathrm{d}t^2 - a^2(t)\mathrm{d}x^2 \qquad (3.81)$$

where the spatial sections expand (or contract) uniformly as described by the single scalar function $a(t)$. Introducing the new time parameter η (the so-called conformal time) defined by $\mathrm{d}\eta = \mathrm{d}t/a$, whence

$$t = \int^t \mathrm{d}t' = \int^\eta a(\eta')\mathrm{d}\eta', \qquad (3.82)$$

(3.81) may be recast as

$$\mathrm{d}s^2 = a^2(\eta)(\mathrm{d}\eta^2 - \mathrm{d}x^2) = C(\eta)(\mathrm{d}\eta^2 - \mathrm{d}x^2), \qquad (3.83)$$

where we have defined the 'conformal scale factor' $C(\eta) = a^2(\eta)$. This form of the line element is manifestly conformal to Minkowski space (see §3.1).

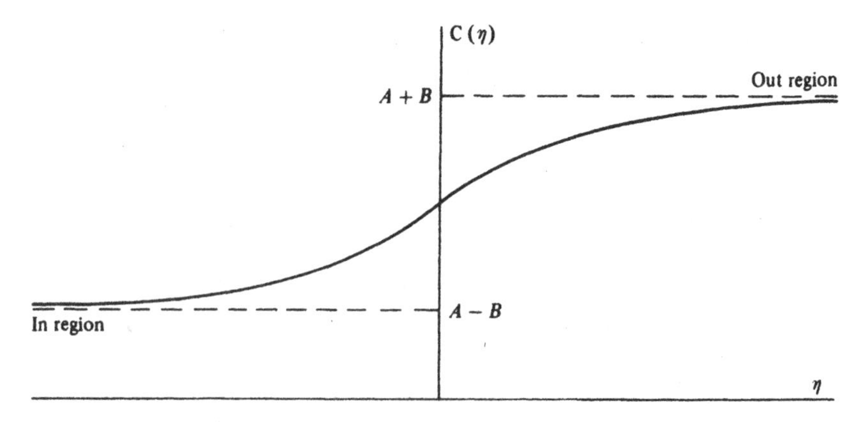

Fig. 7. The conformal scale factor $C(\eta) = A + B \tanh \rho\eta$ represents an asymptotically static universe that undergoes a period of smooth expansion.

Suppose that

$$C(\eta) = A + B \tanh \rho\eta, \quad A, B, \rho \text{ constants}, \tag{3.84}$$

then in the far past and future the spacetime becomes Minkowskian since

$$C(\eta) \to A \pm B, \quad \eta \to \pm\infty$$

(see fig. 7). We consider the production of massive, minimally coupled scalar particles in this spacetime; an investigation first carried out by Bernard & Duncan (1977). Note that in two dimensions minimal and conformal coupling are equivalent (see (3.27)).

Since $C(\eta)$ is not a function of x (the spatial coordinate) spatial translation invariance is still a symmetry in this spacetime, so we can separate the variables in the scalar mode functions appearing in (3.30):

$$u_k(\eta, x) = (2\pi)^{-\frac{1}{2}} e^{ikx} \chi_k(\eta). \tag{3.85}$$

Substituting (3.85) in place of ϕ into the scalar field equation (3.26), with $\xi = 0$ and the metric given by (3.83), one obtains an ordinary differential equation for $\chi_k(\eta)$:

$$\frac{d^2}{d\eta^2}\chi_k(\eta) + (k^2 + C(\eta)m^2)\chi_k(\eta) = 0. \tag{3.86}$$

This equation can be solved in terms of hypergeometric functions. The normalized modes which behave like the positive frequency Minkowski space modes (2.11) in the remote past ($\eta, t \to -\infty$) are

$$u_k^{\text{in}}(\eta, x) = (4\pi\omega_{\text{in}})^{-\frac{1}{2}} \exp\{ikx - \mathrm{i}\omega_+\eta - (\mathrm{i}\omega_-/\rho)\ln[2\cosh(\rho\eta)]\}$$
$$\times\, {}_2F_1(1 + (\mathrm{i}\omega_-/\rho), \mathrm{i}\omega_-/\rho; 1 - (\mathrm{i}\omega_{\text{in}}/\rho); \tfrac{1}{2}(1 + \tanh\rho\eta))$$
$$\xrightarrow[\eta \to -\infty]{} (4\pi\omega_{\text{in}})^{-\frac{1}{2}} e^{ikx - \mathrm{i}\omega_{\text{in}}\eta}, \tag{3.87}$$

where

$$\left.\begin{aligned} \omega_{\text{in}} &= [k^2 + m^2(A - B)]^{\frac{1}{2}} \\ \omega_{\text{out}} &= [k^2 + m^2(A + B)]^{\frac{1}{2}} \\ \omega_\pm &= \tfrac{1}{2}(\omega_{\text{out}} \pm \omega_{\text{in}}). \end{aligned}\right\} \tag{3.88}$$

On the other hand, the modes which behave like positive frequency Minkowski modes in the out region as $\eta \to +\infty$ are found to be

$$u_k^{\text{out}}(\eta, x) = (4\pi\omega_{\text{out}})^{-\frac{1}{2}} \exp\{ikx - \mathrm{i}\omega_+\eta - (\mathrm{i}\omega_-/\rho)\ln[2\cosh(\rho\eta)]\}$$
$$\times\, {}_2F_1(1 + (\mathrm{i}\omega_-/\rho), \mathrm{i}\omega_-/\rho; 1 + (\mathrm{i}\omega_{\text{out}}/\rho); \tfrac{1}{2}(1 - \tanh\rho\eta))$$
$$\xrightarrow[\eta \to +\infty]{} (4\pi\omega_{\text{out}})^{-\frac{1}{2}} e^{ikx - \mathrm{i}\omega_{\text{out}}\eta}. \tag{3.89}$$

Clearly u_k^{in} and u_k^{out} are not equal, which means that the β Bogolubov coefficient in (3.34) must be non-vanishing. To see this explicitly we can use the linear transformation properties of hypergeometric functions (see, for example, Abramowitz & Stegun 1965, Equations (15.3.6), (15.3.3)) to write u_k^{in} in terms of u_k^{out} as (cf. (3.78))

$$u_k^{\text{in}}(\eta, x) = \alpha_k u_k^{\text{out}}(\eta, x) + \beta_k u_{-k}^{\text{out}*}(\eta, x) \tag{3.90}$$

where

$$\alpha_k = \left(\frac{\omega_{\text{out}}}{\omega_{\text{in}}}\right)^{\frac{1}{2}} \frac{\Gamma(1 - (\mathrm{i}\omega_{\text{in}}/\rho))\Gamma(-\mathrm{i}\omega_{\text{out}}/\rho)}{\Gamma(-\mathrm{i}\omega_+/\rho)\Gamma(1 - (\mathrm{i}\omega_+/\rho))} \tag{3.91}$$

$$\beta_k = \left(\frac{\omega_{\text{out}}}{\omega_{\text{in}}}\right)^{\frac{1}{2}} \frac{\Gamma(1 - (\mathrm{i}\omega_{\text{in}}/\rho))\Gamma(\mathrm{i}\omega_{\text{out}}/\rho)}{\Gamma(\mathrm{i}\omega_-/\rho)\Gamma(1 + (\mathrm{i}\omega_-/\rho))}. \tag{3.92}$$

Comparison of (3.90) with (3.34) reveals that the Bogolubov coefficients are given by (cf. (3.76), (3.77))

$$\alpha_{kk'} = \alpha_k \delta_{kk'}, \quad \beta_{kk'} = \beta_k \delta_{-kk'}. \tag{3.93}$$

From (3.91) and (3.92) one obtains

$$|\alpha_k|^2 = \frac{\sinh^2(\pi\omega_+/\rho)}{\sinh(\pi\omega_{\text{in}}/\rho)\sinh(\pi\omega_{\text{out}}/\rho)} \tag{3.94}$$

$$|\beta_k|^2 = \frac{\sinh^2(\pi\omega_-/\rho)}{\sinh(\pi\omega_{\text{in}}/\rho)\sinh(\pi\omega_{\text{out}}/\rho)} \tag{3.95}$$

from which the normalization condition

$$|\alpha_k|^2 - |\beta_k|^2 = 1 \tag{3.96}$$

follows immediately (see (3.39)).

Consider the case that the quantum field resides in the state $|0, \text{in}\rangle$, defined by (3.33) in terms of the in region modes u_k^{in} (which we associate with the modes $\bar{u}_k$ of §3.2). In the remote past, where the spacetime is Minkowskian, all inertial particle detectors will register an absence of particles, so that unaccelerated observers there would identify the quantum state with a physical vacuum.

In the out region ($\eta \to +\infty$), the spacetime is also Minkowskian and the quantum field is also in the state $|0, \text{in}\rangle$ (as we are working in the Heisenberg picture), but in contrast to the situation in the in region, $|0, \text{in}\rangle$ is not regarded by inertial observers in the out region as the physical vacuum, this rôle being reserved for the state $|0, \text{out}\rangle$, defined in terms of modes u_k^{out}. Indeed, unaccelerated particle detectors there will register the presence of quanta (see (3.80)). In the mode k, the expected number of detected quanta is given by (3.95). We can therefore describe this quantum development as the creation of particles into the mode k as a consequence of the cosmic expansion.

3.5 Adiabatic vacuum

The results of the previous section have a readily visualizable physical description. In the massless limit, $\omega_- \to 0$, and the right-hand side of (3.95) vanishes: no particle production occurs. This is an example of a conformally trivial situation, i.e., a conformally invariant field propagating in a spacetime that is conformal to Minkowski space (see §3.1). Particle production takes place only when the conformal symmetry is broken by the presence of a mass, which provides a length scale for the theory. The production process can be regarded as caused by the coupling of the spacetime expansion to the quantum field via the mass. The changing 'gravitational field' feeds energy into the perturbed scalar field modes.

When attributing the production of field quanta to the changing gravitational field, it seems natural to describe the particles themselves as being produced *during* the period of expansion. Certainly no particles are produced in the asymptotically static regions. Moreover, if (3.95) represents a final density of particles which were not present in the in region, one would naively expect that a measurement taking place at an intermediate time, during the period of expansion, would reveal a particle density somewhere between zero and the value given by (3.95).

Unfortunately, these naive ideas do not stand up to scrutiny. As explained in §3.3, when spacetime is curved, no natural definition of particles is generally available. In spite of this, because of the special symmetry of the Robertson–Walker spacetime, one can identify a privileged class of observers, i.e., the comoving observers, who see the universe as expanding precisely isotropically. One might then wish to identify particles in the expansion region with the excitation of comoving particle detectors. This aspect will be studied in the next section.

Even if, for reasons of symmetry, a particular definition of particle is achieved as suggested above, the particle number will not be constant, a fact which makes its measurement inherently uncertain. If the average particle creation rate over an interval Δt is A, then to achieve a precise measurement of particle number, one must choose Δt such that $|A|\Delta t \ll 1$. However, there is also an uncertainty $(m\Delta t)^{-1}$ in the number of particles due to the Heisenberg energy–time uncertainty relation. Thus the total uncertainty in the particle number over a time interval Δt is (Parker 1969)

$$\Delta N \gtrsim (m\Delta t)^{-1} + |A|\Delta t,$$

which has a minimum value $2(|A|/m)^{\frac{1}{2}}$ when $\Delta t = (m|A|)^{-\frac{1}{2}}$. So long as $|A| \neq 0$ or $m \neq \infty$, this inherent uncertainty in N is non-vanishing.

In spite of this, we know from the success of standard Minkowski space quantum field theory that there must exist some sort of approximation to the curved space theory for which the particle number is 'almost meaningful'; we do, after all, inhabit an expanding universe. The above physical argument suggests that if the creation rate of particles is low, or the mass of the particles is high, then the notion of a well-defined particle number becomes a useful concept.

How can these ideas be made more precise?

The density and rate of particle production will obviously depend on the vigour of the expansion motion. In the limit of very weak expansion we would expect the creation rate to fall smoothly to zero, thereby recovering the Minkowski space theory. Inspection of (3.95) reveals that $|\beta_k|^2 \propto B^2 \to 0$

as the total *amount* of expansion approaches zero, and the Minkowski space limit is achieved. However, the particle creation falls much more sharply to zero if the expansion *rate* approaches zero. The rate is here parametrized by ρ, and for $\rho \to 0$ one obtains an exponential decline

$$|\beta_k|^2 \to e^{-2\pi\omega_{in}/\rho} \to 0. \tag{3.97}$$

The relevant 'slowness' parameter is ρ/ω_{in} which becomes small if $\rho \ll k$ or m. This condition is easy to understand physically. One expects the expansion motion to excite modes of the field for which $\omega \lesssim$ the expansion rate. For ω much greater than this, the particle production is exponentially suppressed. Thus, the high k modes are only excited very inefficiently. Similarly, the production of high-mass particles is exponentially small because of the large amount of energy which must emerge from the changing gravitational field to supply the particles' rest mass.

The fact that the creation of high k or m quanta is so strongly suppressed implies that, in the out region ($\eta \to \infty$) inertial particle detectors will register quanta in these energetic modes only extremely infrequently. The field which began vacuous in all modes at $\eta \to -\infty$, ends up 'almost' vacuous in the high energy modes. The slower the rate of expansion during the intermediate phase, the greater the probability that a given mode remains empty of quanta. As the probability declines exponentially in energy at high k and m, the approximation of neglecting any quanta that may have been created as a result of the expansion improves rapidly with energy.

If a high energy 'in' mode remains vacuous to high probability in the out region, it seems obvious that it must also be vacuous in the intermediate region, during the period of expansion. This idea can only be made meaningful, however, by specifying the motion of the particle detector. In the next section we shall show that a *comoving* detector will indeed almost certainly fail to register quanta in the high energy modes during the intermediate phase.

Although the above remarks are well illustrated by the example of the previous section, they apply completely generally to any Robertson–Walker spacetime with a smooth (C^∞) scale factor $C(\eta)$. In particular, the rapid decline in quanta as m or $k \to \infty$ is a completely general feature.

In the case of a Robertson–Walker model universe with static in and out regions, the situation is clear. If either the in or out vacuum states are chosen as the state of the quantum field, then a comoving particle detector will, over its entire world line, almost certainly fail to detect quanta in the high energy modes. So long as the mode frequency is much greater than the expansion rate, the probability of no detector response will remain very close to unity. However, for the lower modes, there will be quanta registered, signalling a breakdown of the meaningful approximation to a vacuum state. (Similar remarks apply if the field is chosen to be in a many-particle in or out state.) Moreover, because the in and out vacuum states are equally good in this respect, any linear combination of them will also lead to the above physical detector characteristics.

If there are no static in or out regions, an approximate definition of particles cannot be based on the above construction. Instead, a method must be found of selecting those exact mode solutions of the field equation that come in some sense 'closest' to the Minkowski space limit. Physically this might be envisaged as a construction that 'least disturbs' the field by the expansion, i.e., results in a definition of particles for which there is *minimal* particle production by the changing geometry. Such a construction has been given by Parker (1966, 1968, 1969, 1971, 1972) and has been subsequently developed by Parker & Fulling (1974), Fulling, Parker & Hu (1974), Fulling (1979), Bunch, Christensen & Fulling (1978), Birrell (1978), Hu (1978, 1979) and Bunch (1980*a*).

To arrive at a precise mathematical description of the above ideas will evidently entail some sort of high-mass expansion of the field modes. We give here a treatment valid for conformally coupled scalar fields in spatially flat Robertson–Walker spacetimes. Generalizations of this situation are given in the references cited above.

The line element for the spacetime is

$$ds^2 = C(\eta)[d\eta^2 - \sum_i (dx^i)^2] \tag{3.98}$$

where we take $C(\eta)$ to be a C^∞ function of the conformal time η. Because of the homogeneity of the spatial sections, the mode solutions of the wave equation are separable

$$u_{\mathbf{k}} = (2\pi)^{(1-n)/2} C^{(2-n)/4}(\eta) e^{i\mathbf{k}\cdot\mathbf{x}} \chi_k(\eta) \tag{3.99}$$

where $k = |\mathbf{k}|$. For a conformally coupled field χ_k satisfies the equation

$$\frac{d^2}{d\eta^2}\chi_k(\eta) + \omega_k^{\,2}(\eta)\chi_k(\eta) = 0 \tag{3.100}$$

where

$$\omega_k^{\,2}(\eta) = k^2 + C(\eta)m^2. \tag{3.101}$$

Equation (3.100) is reminiscent of the classical equation of motion for a harmonic oscillator with time-dependent frequency; for example, a simple pendulum whose length is slowly shortened, thereby decreasing the period. This problem was important in the formulation of quantum theory, because it appears that the energy E of one quantum of oscillation ($h\nu$) is insufficient to make up a whole quantum as the frequency ν rises. However, Einstein showed that so long as the pendulum length is decreased infinitely slowly, E/ν is an adiabatic invariant, and the number of quanta is conserved, irrespective of how great is the change in pendulum length (see, for example, Chandrasekhar 1958).

In the cosmological problem, we shall find that, similarly, the number of quanta (i.e., particle number) is an adiabatic invariant, irrespective of the total *quantity* of cosmological expansion, so long as the expansion *rate* is infinitely slow.

Equation (3.100) possesses the formal WKB-type solutions

$$\chi_k = (2W_k)^{-\frac{1}{2}}\exp\left[-\mathrm{i}\int^{\eta} W_k(\eta')\mathrm{d}\eta'\right] \tag{3.102}$$

where W_k satisfies the nonlinear equation

$$W_k^{\,2}(\eta) = \omega_k^{\,2}(\eta) - \frac{1}{2}\left(\frac{\ddot{W}_k}{W_k} - \frac{3}{2}\frac{\dot{W}_k^{\,2}}{W_k^{\,2}}\right). \tag{3.103}$$

There is, of course, an arbitrary phase factor implicit in (3.102) that can be specified by giving a lower limit to the integral.

If the spacetime is slowly varying, then the derivative terms in (3.103) will be small compared to $\omega_k^{\,2}$, so a zeroth order approximation is to substitute

$$W_k^{(0)}(\eta) \equiv \omega_k(\eta) \tag{3.104}$$

into the integrand of (3.102). This solution obviously reduces to the standard Minkowski space modes as $C(\eta) \to$ constant.

Solutions to (3.103) may be approximated by iteration, using $W_k^{(0)}$ as the lowest order (Bunch 1980*a* – this method is equivalent to that of Liouville 1837, Chakroborty 1973). To clarify the 'slowness' property, it is helpful to introduce a parameter T known as the adiabatic parameter. If η is temporarily replaced by η/T (we take $T = 1$ at the end of the calculation) then the adiabatic limit of slow expansion may be examined by investigating what happens if $T \to \infty$.

With this device (3.100) may be rewritten

$$\frac{\mathrm{d}^2\chi(\eta_1)}{\mathrm{d}\eta_1^{\,2}} + T^2\omega_k^{\,2}(\eta_1)\chi_k(\eta_1) = 0 \tag{3.105}$$

where $\eta_1 = \eta/T$. Evidently

$$\frac{d}{d\eta}C(\eta/T) = \frac{1}{T}\frac{d}{d\eta_1}C(\eta_1) \tag{3.106}$$

so that in the limit $T \to \infty$, $C(\eta_1)$ and all its derivatives with respect to η vary infinitely slowly. Thus we can reproduce the effects of a slowly varying $C(\eta)$ by treating instead a large T approximation.

When an expansion in inverse powers of T is performed, the term of order T^{-n} will be called the nth adiabatic order. From (3.106) it is clear that the adiabatic order is in that case equivalent to the number of derivatives of C. It follows on dimensional grounds that if a quantity has dimensions m^d, a term of adiabatic order A in its expansion will contain $A - d$ powers of m^{-1} and k^{-1}.

The next iteration of (3.103) yields

$$(W_k^{(2)})^2 = \omega_k^2 - \frac{1}{2}\left(\frac{\ddot{\omega}_k}{\omega_k} - \frac{3\dot{\omega}_k^2}{2\omega_k^2}\right) \tag{3.107}$$

which involves two derivatives of ω_k, hence C, so is of second adiabatic order in the slowness approximation. The Ath iterate yields a term of $2A$th adiabatic order. (When inserting $W_k^{(2)}$ in (3.102) to calculate the expansion to order two, the square root of (3.107) need be expanded only to terms of adiabatic order two.) In what follows we shall denote the Ath order adiabatic approximation to χ_k by $\chi_k^{(A)}$, and the associated modes (3.99) by $u_k^{(A)}$.

Suppose that instead of using the exact solution χ_k given by (3.102) one used instead the zeroth order adiabatic approximation obtained by replacing W_k by $W_k^{(0)}$. In an in region, where the universe is static, both expressions yield the usual Minkowski solutions, with constant frequency. As the universe expands, however, the exact and approximate expressions will begin to differ, but only by terms of adiabatic order higher than zero, and this remains true however much the universe expands.

As an illustration, consider the case

$$C(\eta) = 1 + e^{a\eta}, \quad a = \text{constant}, \quad \eta < 0, \tag{3.108}$$

which has an asymptotically static in region, followed by a period of expansion. The exact solution, which reduces to the usual positive frequency Minkowski space solution in the remote past is (Birrell 1979*a*)

$$\chi_k = \frac{\Gamma(1 - (2i\omega_k^-/a))}{(2\omega_k^-)^{\frac{1}{2}}}\left(\frac{m}{a}\right)^{2i\omega_k^-/a} J_{-2i\omega_k^-/a}(e^{a\eta/2}) \tag{3.109}$$

where J is a Bessel function and $\omega_k^- = \omega_k(-\infty) = (k^2 + m^2)^{\frac{1}{2}}$. On the other hand, the zeroth order adiabatic solution (which also reduces to the standard Minkowski space positive frequency modes in the in region) is

$$\chi_k^{(0)} = 2^{-\frac{1}{2}}(k^2 + m^2 + m^2 e^{a\eta})^{-1/4} \exp\left[-\mathrm{i}\int (k^2 + m^2 + m^2 e^{a\eta})^{\frac{1}{2}} \mathrm{d}\eta\right]$$

$$= 2^{-\frac{1}{2}}(k^2 + m^2 + m^2 e^{a\eta})^{-1/4} \exp\left\{-\frac{2\mathrm{i}}{a}\left[(k^2 + m^2 + m^2 e^{a\eta})^{\frac{1}{2}} - (k^2 + m^2)^{\frac{1}{2}} \tanh^{-1}\left(\frac{k^2 + m^2}{k^2 + m^2 + m^2 e^{a\eta}}\right)^{\frac{1}{2}}\right]\right\}. \quad (3.110)$$

Clearly the exact and zeroth order adiabatic solutions agree in the remote past where the expansion is infinitely slow. Also, by using the standard asymptotic expansion of (3.109), for large m, it is easily verified that, to within an irrelevant arbitrary phase factor (associated with the indefinite lower limit of the integral in (3.102)), (3.109) reduces to (3.110) plus terms of adiabatic order greater than zero, regardless of the value of η.

In general there exists the following relation

$$u_{\mathbf{k}} = \alpha_k^{(A)}(\eta) u_{\mathbf{k}}^{(A)} + \beta_k^{(A)}(\eta) u_{\mathbf{k}}^{(A)*}, \quad (3.111)$$

defining an *exact* field mode (i.e., an exact solution of the field equation) in terms of the adiabatic approximation $u_{\mathbf{k}}^{(A)}$. Clearly $\alpha_k^{(A)}$ and $\beta_k^{(A)}$ must be constant to order A, because $u_{\mathbf{k}}^{(A)}$ and $u_{-\mathbf{k}}^{(A)}$ are solutions of the field equation to this order. Suppose that we make the particular choice

$$\begin{aligned} \alpha_k^{(A)}(\eta_0) &= 1 + O(T^{-(A+1)}) \\ \beta_k^{(A)}(\eta_0) &= 0 + O(T^{-(A+1)}) \end{aligned} \quad (3.112)$$

for some fixed time η_0. Then it follows that α and β are given by (3.112) for all time. The modes $u_{\mathbf{k}}$ defined by (3.111), (3.112) are said to be 'adiabatic positive frequency modes' to adiabatic order A. It is important to note that, this appellation notwithstanding, the modes $u_{\mathbf{k}}$ are not adiabatic approximations – they are exact. However, they are not defined uniquely by the prescription (3.111) and (3.112). There exist an infinite number of sets of such modes corresponding to different choices of η_0.

Special importance attaches to the exact modes $u_{\mathbf{k}}^{\mathrm{in}}$ that reduce to the standard plane wave positive frequency exponential solutions in a static in region. Another set, $u_{\mathbf{k}}^{\mathrm{out}}$, reduce to standard form in a static out region. In these static regions, all terms of adiabatic order greater than zero in

$u_{\mathbf{k}}^{(A)}$ vanish, so that such exact modes are of adiabatic positive frequency to infinite order. Thus, from the discussion associated with (3.112), the β Bogolubov coefficient connecting the two sets of modes $u_{\mathbf{k}}^{\text{in}}$ and $u_{\mathbf{k}}^{\text{out}}$, must fall off faster than any inverse power of T in the adiabatic limit $T \to \infty$. This was indeed found to be the case in the example of the previous section. It implies that the particle number associated with the quantization of either of these sets of modes is an adiabatic invariant during the cosmic expansion, in direct analogy to the problem of the pendulum with varying length.

If instead of using a set of exact field modes that reduce to standard form in, say, the in region, one used instead exact solutions that match on to Ath order *approximate* adiabatic modes at some later moment η_0, as in (3.111), (i.e., $u_{\mathbf{k}}$ and $\partial u_{\mathbf{k}}/\partial t$ are equated to adiabatic order A to $u_{\mathbf{k}}^{(A)}$ and $\partial u_{\mathbf{k}}^{(A)}/\partial t$ at $\eta = \eta_0$), then in the in region, these matched exact solutions will no longer reduce to standard form, and will in general be a linear combination of positive and negative frequency plane wave modes. A vacuum state constructed by quantizing these 'distorted' modes will not, therefore, be the same as the usual physical vacuum in the in region. That is, an inertial particle detector would register a bath of quanta when the field is in this 'distorted' vacuum. Nevertheless, the number spectrum of these quanta will generally fall off at large energy like $k^{-(A+1)}$ (or $m^{-(A+1)}$), reflecting the fact that the field modes have been matched on to approximate modes that differ from the standard in modes only by terms of adiabatic order $A + 1$ and higher. Thus, the 'distorted' vacuum may be regarded as an *adiabatic approximation* to the in vacuum. This adiabatic vacuum will with high probability leave the high energy modes vacuous, so that a comoving detector will, with high probability, register zero quanta in these modes. The probability that any given mode is empty of quanta for all time will not, however, be as high as in the case of the in or out vacuum states, with their associated more rapid decline at high energy (faster than any power of k or m).

Two features of the adiabatic vacuum concept should be properly understood. First, as remarked, the adiabatic vacuum is not some sort of approximate state based on approximate field modes. The adiabatic-approximate modes are themselves, of course, merely approximate solutions (to order A) of the field equation, but they are only used as a mathematical template against which to match exact mode solutions at some $\eta = \eta_0$. It is the *exact* modes themselves that become quantized. The associated vacuum state (i.e., the Ath order adiabatic vacuum) of these exact modes is a perfectly good contender for a vacuum state. True, it may not represent the experiences of a *comoving* detector as well as, say, the $(A + 1)$th adiabatic, or the in, vacuum states, but as far as the quantum field theory is concerned, it is respectable enough.

Secondly, there is no *unique* Ath order adiabatic vacuum, as the matching procedure may take place at any η_0. The associated exact modes will differ, for all time, only by terms of higher adiabatic order, so all are candidates for quantization and for the construction of an Ath order adiabatic vacuum. All such vacuum states will have comparable high energy behaviour, but will generally differ in the structure of the low energy modes.

Although an adiabatic vacuum is less specific than a vacuum state associated with a static in or out region, its representation of physical particles (in the sense of the experiences of a comoving particle detector) is nevertheless the best that is available if the spacetime *has no static* in and out regions. Current cosmological theory suggests that this is the case in the real universe.

As an illustration of the use of the adiabatic approximation in an asymptotically *non-static* four-dimensional cosmological model, consider the scale factor

$$C(\eta) = a^2 + b^2\eta^2, \quad -\infty < \eta < \infty \tag{3.113}$$

where a and b are constant (Audretsch & Schäfer 1978*a*). In the asymptotic regions, $\eta \to \pm\infty$, the model approaches the radiation-dominated Friedmann cosmology with

$$a(t) \equiv C^{\frac{1}{2}}(t) \propto t^{\frac{1}{2}}. \tag{3.114}$$

The time symmetry inherent in (3.113) indicates that the space contracts to a minimum scale factor at $\eta = 0$, 'bounces', and re-expands (see fig. 8).

Although not asymptotically static, the spacetime is nevertheless slowly varying as $\eta \to \pm\infty$, since

$$\frac{\mathrm{d}^l}{\mathrm{d}\eta^l}(\dot{C}/C) \to 0, \quad l \geq 0. \tag{3.115}$$

Thus, in the limit, the expansion rate vanishes and the adiabatic approximation becomes exact, so all adiabatic orders yield identical results as $\eta \to \pm\infty$.

More generally, the zeroth order adiabatic approximation becomes good whenever

$$\omega_k(\eta) = (k^2 + m^2a^2 + m^2b^2\eta^2)^{\frac{1}{2}}$$

becomes large compared with the derivatives on the left-hand side of (3.115). To see when this occurs, one studies (3.105) in which the factor

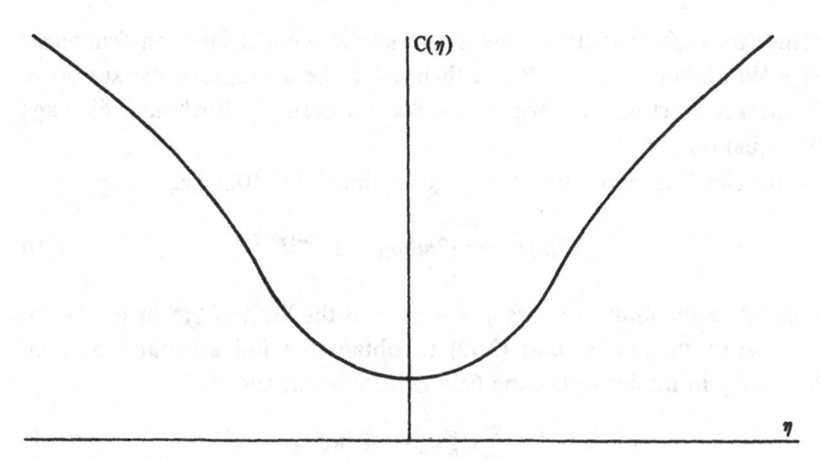

Fig. 8. The conformal scale factor $C(\eta) = a^2 + b^2\eta^2$ represents a universe that contracts to a minimum 'size' then re-expands symmetrically. At large η it behaves like a radiation-dominated Friedmann model.

$T^2\omega_k^2(\eta_1)$ can in this case be written as

$$T^2\omega_k^2(\eta_1) = mbT^2\lambda + m^2b^2\eta_1^2T^4$$

where

$$\lambda \equiv (ma^2/b) + (k^2/mb). \tag{3.116}$$

As the derivative $\dot{C}/C$ behaves like T^{-1}, one can now see that, for fixed $\eta_1 = \eta/T$, the zeroth order adiabatic approximation will be a good one for large λ, or large η, or large mb, or for any number of these quantities becoming large together.

Let us consider the limit of large λ with mb and η fixed. Substituting

$$W_k^{(0)} = \omega_k(\eta) = (mb\lambda)^{\frac{1}{2}} + O(T^{-2})$$

into (3.102) yields

$$\chi_k^{(0)}(\eta) \xrightarrow{\lambda\to\infty} (2mb\lambda)^{-\frac{1}{4}} \exp[-i(mb\lambda)^{\frac{1}{2}}\eta], \quad \eta \text{ fixed}. \tag{3.117}$$

For comparison, one may now construct the exact solutions of (3.100) which reduce to (3.117) plus terms of $O(T^{-1})$(i.e., $O(\lambda^{-\frac{1}{2}})$) in the limit of large λ:

$$\chi_k^{\text{in}}(\eta) = (2mb)^{-1/4}e^{-\pi\lambda/8}D_{-(1-i\lambda)/2}[(i-1)(mb)^{\frac{1}{2}}\eta], \quad \text{for } \eta < 0 \tag{3.118}$$

and

$$\chi_k^{\text{out}}(\eta) = \chi_k^{\text{in}}(-\eta)^*, \quad \text{for } \eta > 0. \tag{3.119}$$

(This can be checked by writing the parabolic cylinder function D in terms of a Whittaker function W and then using the asymptotic expansion of Whittaker functions for large index. See, for example, Bucholz 1969, page 99, equation (19*a*).)

One can also check that the large $|\eta|$ limit of (3.102), i.e.,

$$\chi_k^{(0)}(\eta) \xrightarrow{\eta \to \pm \infty} (2mb|\eta|)^{-\frac{1}{2}} e^{\mp imb\eta^2/2} \tag{3.120}$$

is equal to the limit of χ_k^{in} as $\eta \to -\infty$, and the limit of χ_k^{out} as $\eta \to +\infty$.

Substituting (3.118) into (3.99) to obtain the full adiabatic positive frequency in modes $u_{\mathbf{k}}^{\text{in}}(x)$, the field ϕ may be expanded

$$\phi = \sum_{\mathbf{k}} (a_{\mathbf{k}}^{\text{in}} u_{\mathbf{k}}^{\text{in}} + a_{\mathbf{k}}^{\text{in}\dagger} u_{\mathbf{k}}^{\text{in}*}) \tag{3.121}$$

and the associated adiabatic vacuum state $|0_{\text{in}}^A\rangle$ defined by

$$a_{\mathbf{k}}^{\text{in}}|0_{\text{in}}^A\rangle = 0. \tag{3.122}$$

Similarly one may use (3.119) to define another adiabatic vacuum state of the same adiabatic order

$$a_{\mathbf{k}}^{\text{out}}|0_{\text{out}}^A\rangle = 0. \tag{3.123}$$

Because χ_k^{in} is matched to $\chi_k^{(0)}$ at $\eta = -\infty$, where all higher order adiabatic corrections vanish by virtue of (3.115), it is matched to $\chi_k^{(A)}$ for arbitrary A at $\eta = -\infty$. Thus the vacuum $|0_{\text{in}}^A\rangle$ defines an adiabatic in vacuum of infinite adiabatic order. Similarly $|0_{\text{out}}^A\rangle$ defines an adiabatic out vacuum of infinite adiabatic order. To see this explicitly, one may evaluate the Bogolubov transformation between $u_{\mathbf{k}}^{\text{in}}$ and $u_{\mathbf{k}}^{\text{out}}$:

$$u_k^{\text{in}} = \frac{\mathrm{i}(2\pi)^{\frac{1}{2}} \, e^{-\pi\lambda/4}}{\Gamma(\frac{1}{2}(1 - \mathrm{i}\lambda))} u_{\mathbf{k}}^{\text{out}} - \mathrm{i}e^{-\pi\lambda/2} u_{\mathbf{k}}^{\text{out}*} \tag{3.124}$$

(see, for example, Gradshteyn & Ryzhik 1965, equation (9.248(3))). Thus, if the quantum state is chosen to be $|0_{\text{in}}^A\rangle$, a comoving particle detector in the out region will detect a spectrum

$$|\beta_{\mathbf{k}}|^2 = \exp\{-\pi[(k^2/mb) + (ma^2/b)]\} \tag{3.125}$$

which does indeed fall off faster than any inverse power of k or m at high energies. It also falls faster than any positive power of b (the 'slowness' parameter) as $b \to 0$. Notice also that (3.125) remains finite as $a \to 0$, which corresponds to a model that intersects a singularity at $\eta = 0$.

As pointed out by Audretsch & Schäfer (1978*a*), the spectrum (3.125) is the same as that for a non-relativistic thermal gas of particles with momentum $kC^{-\frac{1}{2}}(\eta)$ at a chemical potential $-\frac{1}{2}ma^2 C^{-1}(\eta)$ and temperature $b/(2\pi Ck_B)$, where k_B is, as before, Boltzmann's constant.

An alternative formulation of particle creation by a gravitational field, called Hamiltonian diagonalization, attempts to define particle states at each instant in analogy with the Minkowskian definition (see, for example, Imamura 1960, Grib & Mamaev 1969, 1971, Berger 1975, Castagnino, Verbeure & Weder 1974, 1975, Grib, Mamaev & Mostepanenko 1976, 1980*a*, *b*). Generally this procedure predicts vastly more creation than the methods employed here, and does not lead to the above-noted rapid decline in particle number at large k. Schäfer & Dehnen (1977), for example, predict the present particle creation rate to be 16 particles per km^3 per year – very many orders of magnitude in excess of what would be expected on physical grounds. Hamiltonian diagonalization has been strongly criticised, most notably by Fulling (1979) (see also Raine & Winlove 1975).

Analyses of the conceptual foundations of the particle concept in curved spacetime have been given by Ashtekar & Magnon (1975*a*, *b*), Hájíček (1976), Volovich, Zagrebnov & Frolov (1977), and Martellini, Sodano & Vitiello (1978).

3.6 Adiabatic expansion of Green functions

In the previous section it was shown that the high frequency behaviour of a massive scalar field is relatively insensitive to the long term time-dependence of the background Robertson–Walker spacetime. This is because the high frequency components of the field only probe the geometry in the immediate vicinity of the spacetime point of interest, and in this restricted neighbourhood the metric only changes by a small amount. In contrast the long wavelength, low frequency modes probe the entire manifold, and their structure is sensitive to the geometry and hence the particular adiabatic construction.

In some applications, e.g., the regularization of ultraviolet divergences, only the high frequency field behaviour is of interest. It is then only necessary to deal with high frequency approximations. Because the high frequencies probe only the short distances, one is led to examine short distance approximations.

Special interest attaches to the short distance behaviour of the Green functions, such as $G_F(x, x')$ in the limit $x \to x'$. We follow here a treatment of Bunch & Parker (1979) who obtained an adiabatic expansion of G_F.

Introducing Riemann normal coordinates y^μ for the point x, with origin at the point x'(e.g., Kreyszig 1968, Petrov 1969), one may expand

$$g_{\mu\nu}(x) = \eta_{\mu\nu} + \tfrac{1}{3}R_{\mu\alpha\nu\beta}y^\alpha y^\beta - \tfrac{1}{6}R_{\mu\alpha\nu\beta;\gamma}y^\alpha y^\beta y^\gamma + [\tfrac{1}{20}R_{\mu\alpha\nu\beta;\gamma\delta} + \tfrac{2}{45}R_{\alpha\mu\beta\lambda}R^\lambda{}_{\gamma\nu\delta}]y^\alpha y^\beta y^\gamma y^\delta + \ldots$$

where $\eta_{\mu\nu}$ is the Minkowski metric tensor, and the coefficients are all evaluated at $y = 0$.

Defining

$$\mathscr{G}_{\mathrm{F}}(x, x') = (-g(x))^{\frac{1}{4}}G_{\mathrm{F}}(x, x') \tag{3.126}$$

and its Fourier transform by

$$\mathscr{G}_{\mathrm{F}}(x, x') = (2\pi)^{-n}\int \mathrm{d}^n k\, \mathrm{e}^{-iky}\mathscr{G}_{\mathrm{F}}(k) \tag{3.127}$$

where $ky = \eta^{\alpha\beta}k_\alpha y_\beta$, one can work in a sort of localized momentum space. Expanding (3.49) in normal coordinates and converting to k-space, $\mathscr{G}_{\mathrm{F}}(k)$ can readily be solved by iteration to any adiabatic order. The result to adiabatic order four (i.e., four derivatives of the metric) is

$$\mathscr{G}_{\mathrm{F}}(k) \approx (k^2 - m^2)^{-1} - (\tfrac{1}{6} - \xi)R(k^2 - m^2)^{-2} + \tfrac{1}{2}\mathrm{i}(\tfrac{1}{6} - \xi)R_{;\alpha}\partial^\alpha(k^2 - m^2)^{-2} - \tfrac{1}{3}a_{\alpha\beta}\partial^\alpha\partial^\beta(k^2 - m^2)^{-2} + [(\tfrac{1}{6} - \xi)^2R^2 + \tfrac{2}{3}a^\lambda{}_\lambda](k^2 - m^2)^{-3} \tag{3.128}$$

where $\partial_\alpha = \partial/\partial k^\alpha$,

$$a_{\alpha\beta} = \tfrac{1}{2}(\xi - \tfrac{1}{6})R_{;\alpha\beta} + \tfrac{1}{120}R_{;\alpha\beta} - \tfrac{1}{40}R_{\alpha\beta;\lambda}{}^\lambda - \tfrac{1}{30}R_\alpha{}^\lambda R_{\lambda\beta} + \tfrac{1}{60}R^\kappa{}_\alpha{}^\lambda{}_\beta R_{\kappa\lambda} + \tfrac{1}{60}R^{\lambda\mu\kappa}{}_\alpha R_{\lambda\mu\kappa\beta}, \tag{3.129}$$

and we are using the symbol $\approx$ to indicate that this is an asymptotic expansion. One ensures that (3.127) represents a time-ordered product (as in (3.48)) by performing the k^0 integral along the appropriate (Feynman) contour in fig. 3. This is equivalent to replacing m^2 by $m^2 - \mathrm{i}\varepsilon$. Similarly, the adiabatic expansions of other Green functions can be obtained by using the other contours in fig. 3.

Substituting (3.128) into (3.127) yields

$$\mathscr{G}_{\mathrm{F}}(x, x') \approx \int \frac{\mathrm{d}^n k}{(2\pi)^n}\mathrm{e}^{-iky}\left[a_0(x, x') + a_1(x, x')\left(-\frac{\partial}{\partial m^2}\right) + a_2(x, x')\left(\frac{\partial}{\partial m^2}\right)^2\right](k^2 - m^2)^{-1} \tag{3.130}$$

where

$$a_0(x, x') \equiv 1 \tag{3.131}$$

and, to adiabatic order 4,

$$a_1(x, x') = (\tfrac{1}{6} - \xi)R - \tfrac{1}{2}(\tfrac{1}{6} - \xi)R_{;\alpha}y^\alpha - \tfrac{1}{3}a_{\alpha\beta}y^\alpha y^\beta \tag{3.132}$$

$$a_2(x, x') = \tfrac{1}{2}(\tfrac{1}{6} - \xi)^2 R^2 + \tfrac{1}{3}a^\lambda{}_\lambda \tag{3.133}$$

with all geometric quantities on the right-hand side of (3.132) and (3.133) evaluated at x'.

If one uses the integral representation

$$(k^2 - m^2 + \mathrm{i}\varepsilon)^{-1} = -\mathrm{i}\int_0^\infty \mathrm{d}s\, \mathrm{e}^{\mathrm{i}s(k^2 - m^2 + \mathrm{i}\varepsilon)} \tag{3.134}$$

in (3.130), then the $\mathrm{d}^n k$ integration may be interchanged with the ds integration, and performed explicitly to yield (dropping the iε)

$$\mathscr{G}_{\mathrm{F}}(x, x') = -\mathrm{i}(4\pi)^{-n/2}\int_0^\infty \mathrm{i}\,\mathrm{d}s\,(\mathrm{i}s)^{-n/2}\exp[-\mathrm{i}m^2 s + (\sigma/2\mathrm{i}s)]F(x, x'; \mathrm{i}s). \tag{3.135}$$

In (3.135) the function σ is defined by

$$\sigma(x, x') = \tfrac{1}{2}y_\alpha y^\alpha, \tag{3.136}$$

which is one-half of the square of the proper distance between x and x', while the function F has the following asymptotic adiabatic expansion

$$F(x, x'; \mathrm{i}s) \approx a_0(x, x') + a_1(x, x')\mathrm{i}s + a_2(x, x')(\mathrm{i}s)^2 + \ldots \tag{3.137}$$

Using (3.126), equation (3.135) gives a representation of $G_{\mathrm{F}}(x, x')$ originally derived by DeWitt (1965, 1975), following the work of Schwinger (1951*a*) (and hence labelled DS, see also Fock 1937, Nambu 1950)

$$G_{\mathrm{F}}^{\mathrm{DS}}(x, x') = -\mathrm{i}\Delta^{\frac{1}{2}}(x, x')(4\pi)^{-n/2}\int_0^\infty \mathrm{i}\,\mathrm{d}s\,(\mathrm{i}s)^{-n/2}$$

$$\exp[-\mathrm{i}m^2 s + (\sigma/2\mathrm{i}s)]F(x, x'; \mathrm{i}s) \tag{3.138}$$

where Δ is the Van Vleck determinant (Van Vleck 1928)

$$\Delta(x, x') = -\det[\partial_\mu\partial_\nu\sigma(x, x')][g(x)g(x')]^{-\frac{1}{2}} \tag{3.139}$$

In the normal coordinates about x' that we are currently using, Δ reduces to $[-g(x)]^{-\frac{1}{2}}$. In the treatment of DeWitt, the extension of the asymptotic expansion (3.137) of F to all adiabatic orders is written as

$$F(x,x';\mathrm{i}s) \approx \sum_{j=0}^{\infty} a_j(x,x')(\mathrm{i}s)^j \tag{3.140}$$

with $a_0(x,x')=1$, the other a_j being given by recursion relations which enable their adiabatic expansions to be obtained (Christensen 1976). It is important to note that the so-called DeWitt–Schwinger 'proper time' representation, (3.138), is intended to be an exact representation of the Feynman propagator. The expansions (3.137) and (3.140) are, however, only asymptotic approximations in the limit of large adiabatic parameter T.

If (3.140) is substituted into (3.138) the integral can be performed to give the adiabatic expansion of the Feynman propagator in coordinate space:

$$G_{\mathrm{F}}^{\mathrm{DS}}(x,x') \approx \frac{-\mathrm{i}\pi\Delta^{\frac{1}{2}}(x,x')}{(4\pi\mathrm{i})^{n/2}} \sum_{j=0}^{\infty} a_j(x,x')\left(-\frac{\partial}{\partial m^2}\right)^j \times\left[\left(\frac{2m^2}{-\sigma}\right)^{(n-2)/4} H^{(2)}_{(n-2)/2}\big((2m^2\sigma)^{\frac{1}{2}}\big)\right] \tag{3.141}$$

in which, strictly, a small imaginary part $\mathrm{i}\varepsilon$ should be subtracted from σ.

Since we have not imposed global boundary conditions on the Green function solution of (3.49), the expansion (3.141) does not determine the particular vacuum state in (3.48). In particular, the '$\mathrm{i}\varepsilon$' in the expansion of G_{F} only ensures that (3.141) represents the expectation value, in some set of states, of a time-ordered product of fields. Under some circumstances the use of '$\mathrm{i}\varepsilon$' in the exact representation (3.138) may give additional information concerning the global nature of the states – see, for example, the discussion on page 48. Expressed differently, the same high frequency behaviour of (3.48) results from almost all choices of vacuum state, a fact which will turn out to be of considerable importance.

In a Robertson–Walker spacetime, the Feynman Green function calculated as an expectation value in an adiabatic vacuum $|0^A\rangle$

$$\mathrm{i}G_{\mathrm{F}}^{A}(x,x') = \langle 0^A|T(\phi(x)\phi(x'))|0^A\rangle, \tag{3.142}$$

should have an expansion of the form (3.138) and (3.140). The vacuum $|0^A\rangle$ is defined in terms of a set of adiabatic positive frequency modes $u_{\mathbf{k}}$ given by (3.111), (3.112) for some η_0, and so one can write (3.142) as

$$\mathrm{i}G_{\mathrm{F}}^{A}(x,x') = \theta(x^0 - x^{0\prime})\int \mathrm{d}^{n-1}k\, u_{\mathbf{k}}(x)u_{\mathbf{k}}^*(x') + \theta(x^{0\prime} - x^0)\int \mathrm{d}^{n-1}k\, u_{\mathbf{k}}^*(x)u_{\mathbf{k}}(x'). \tag{3.143}$$

To obtain an expansion of (3.143) to adiabatic order A, it is only necessary to use the expansion of $u_{\mathbf{k}}$ to this order; i.e., to calculate (3.143) with $u_{\mathbf{k}}$ replaced by $u_{\mathbf{k}}^{A}$. Doing this, one can explicitly verify that the expansions of (3.142) and (3.143) agree to order A (Birrell 1978, Bunch, Christensen & Fulling 1978, Bunch & Parker 1979). In particular, one can check that the 'iε' is necessary to guarantee the time ordering in (3.142) and (3.143).

We are now in a position to discuss the response of a particle detector to an Ath order adiabatic vacuum. To do this, one must substitute

$$G_{A}^{+}(x(\tau), x(\tau')) \equiv \langle 0^{A}|\phi(x(\tau))\phi(x(\tau'))|0^{A}\rangle \tag{3.144}$$

into (3.55), to obtain the detector response function along the world line $x(\tau)$. If we only wish to consider the contribution to the probability of terms in this exact two-point function up to adiabatic order A, then, for any spacetime, we can calculate these terms using the adiabatic expansion $u_{\mathbf{k}}^{(A)}$ for the modes. As in the case of G_{F}, the expansion of G_{A}^{+} will be equal to the expansion to order A of the DeWitt–Schwinger representation G_{DS}^{+}. The expansion of G_{DS}^{+} can be calculated in momentum space as

$$\mathrm{i}G_{\mathrm{DS}}^{+}(x, x') = [-g(x)]^{-1/4}\mathscr{G}^{+}(x, x')$$

where, to adiabatic order four, $\mathscr{G}^{+}$ is given by the right-hand side of (3.130) with the k^0 integral performed along the contour shown in fig. 9. Using the expansion for $[-g(x)]^{-1/4}$ in normal coordinates about x' one observes that G^{+} (to arbitrary adiabatic order) will be a sum of terms with the general

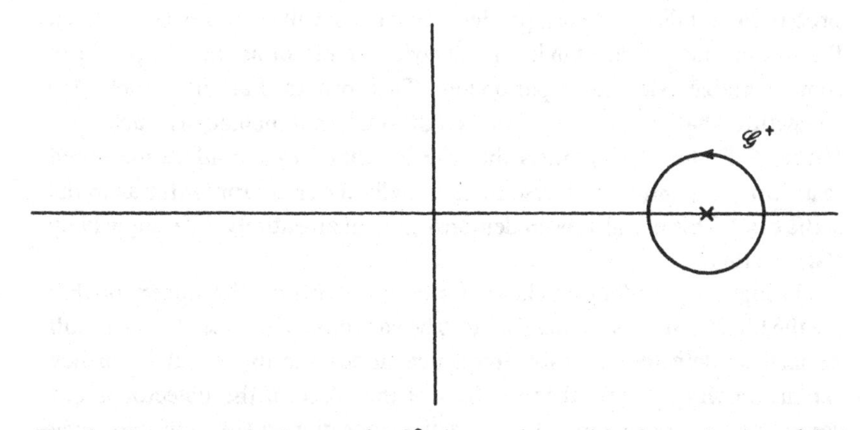

Fig. 9. The contour in the complex k^0 plane to be used in the evaluation of the integral giving $\mathscr{G}^{+}$. The cross indicates the pole at $k^0 = (|\mathbf{k}|^2 + m^2)^{\frac{1}{2}}$.

structure

$$\int \frac{\mathrm{d}^n k}{(2\pi)^n} \frac{\mathrm{e}^{-iky}}{(k^2 - m^2)^p} S_{\mu\nu\ldots\lambda}(x') y^\mu y^\nu \ldots y^\lambda \tag{3.145}$$

where S is a geometrical tensor and, as before, y^μ are the normal coordinates about x' of the point x. The k^0 integral can be performed around the contour shown in fig. 9 to give a linear combination of terms having the form

$$\int \frac{\mathrm{d}^{n-1} k}{(2\pi)^{n-1}} \frac{\mathrm{e}^{\mathrm{i}\mathbf{k}\cdot\mathbf{y} - \mathrm{i}\omega y^0}}{(2\omega)^r} S_{\mu\nu\ldots\lambda}(x')(y^0)^q y^\mu y^\nu \ldots y^\lambda \tag{3.146}$$

where r and q are integers.

We now restrict attention to detectors that are comoving, i.e., with world lines $x(\tau)$ such that $x^i =$ constant and $\tau = t$ (the Robertson–Walker cosmic time). Using the properties of normal coordinates, $G^+_{\mathrm{DS}}(x(\tau), x(\tau'))$ will be given by a linear combination of terms like (3.146) with $\mathbf{y} = 0, y^0 = t$. Substitution of such a term into (3.55) yields integrals of the form

$$\int_{-\infty}^{\infty} \mathrm{e}^{-\mathrm{i}(\omega + E - E_0)\tau} (\tau)^l \mathrm{d}\tau, \quad (l \text{ an integer}).$$

Because $\omega + E - E_0 > 0$, all such integrals vanish as they are derivatives of $\delta(\omega + E - E_0)$.

In conclusion, we may say that an adiabatic vacuum of order A is a state for which a comoving particle detector remains unexcited with a probability that differs from unity only by terms of order $A + 1$. That is, the probability for detection of a particle of energy ω will fall to zero faster than the Ath inverse power of ω in the ultraviolet limit, or as the Ath power of some suitable 'slowness' parameter. This proves Parker's (1969, §E) conjecture that adiabatic particles (although not named as such – see Parker & Fulling 1974) satisfy the criterion that they should 'be measured in a slowly expanding universe by essentially the same apparatus as in the static case'. This has also been demonstrated in an entirely different way by Parker (1966).

The higher the order of A chosen for the construction, the more probably are the high frequency modes found to be vacuous. Of course, there will still be particles detected in the low frequency modes; the above high frequency expansion will not reveal the details of this. Also, if the detector is not comoving, there will generally be particle detection in the high frequency modes that falls to zero slower than T^{-A}.

3.7. Conformal vacuum

Despite the appealing properties of adiabatic vacuum states, the fact remains that in a curved spacetime a particular set of mode solutions of the field equation and the corresponding vacuum and many-particle states do not *in general* have direct physical significance. In particular, a 'vacuum' state may not necessarily be measured as devoid of quanta, even by a freely-falling detector. Nevertheless, if there exist geometrical symmetries in the spacetime of interest, it may be that a particular set of modes and particle states emerge as in some sense 'natural'.

One case of special interest to which we shall frequently return is that of *conformal* triviality, i.e., a conformally invariant field propagating in a conformally flat spacetime (a spacetime conformal to Minkowski space). The symmetry of such spacetimes is manifested by the existence of a conformal Killing vector satisfying (3.23). Examples of conformally flat spacetimes are all two-dimensional spacetimes and the spatially flat Robertson–Walker cosmological models. Their metric tensors may always be cast in the form

$$g_{\mu\nu}(x) = \Omega^2(x)\eta_{\mu\nu} \tag{3.147}$$

where $\eta_{\mu\nu}$ is the Minkowski metric tensor.

The conformally invariant scalar wave equation requires $m = 0$ and the choice of ξ given by (3.27):

$$[\Box + \tfrac{1}{4}(n-2)R/(n-1)]\phi = 0. \tag{3.148}$$

Under the conformal transformation

$$g_{\mu\nu} \to \Omega^{-2} g_{\mu\nu} = \eta_{\mu\nu} \tag{3.149}$$

one obtains from (3.1), (3.5) and (3.148)

$$\bar{\Box}\bar{\phi} \equiv \eta^{\mu\nu}\partial_\mu\partial_\nu(\Omega^{(n-2)/2}\phi) = 0, \tag{3.150}$$

as $\bar{R} = 0$ in Minkowski space. The power $(n-2)/2$ of Ω is known as the conformal weight of the scalar field. Equation (3.150) possesses the familiar Minkowski space mode solutions for $\bar{\phi}$ (cf. (2.11))

$$\bar{u}_{\mathbf{k}}(x) = [2\omega(2\pi)^{n-1}]^{-\frac{1}{2}} e^{-ik\cdot x}, \quad k^0 = \omega. \tag{3.151}$$

These modes are positive frequency with respect to the timelike conformal Killing vector ∂_η. That is

$$\mathscr{L}_{\partial_\eta}\bar{u}_{\mathbf{k}}(x) = -i\omega\bar{u}_{\mathbf{k}}(x), \quad \omega > 0. \tag{3.152}$$

Noting from (3.150) that $\phi = \Omega^{(2-n)/2}\,\bar{\phi}$, the mode decomposition (3.30) for ϕ can in this case be written

$$\phi(x) = \Omega^{(2-n)/2}(x)\sum_{\mathbf{k}}\left[a_{\mathbf{k}}\bar{u}_{\mathbf{k}}(x) + a_{\mathbf{k}}^{\dagger}\bar{u}_{\mathbf{k}}^{*}(x)\right] \tag{3.153}$$

with $\bar{u}_{\mathbf{k}}$ given by (3.151). The vacuum state associated with these modes, defined by $a_{\mathbf{k}}|0\rangle = 0$, is called the *conformal vacuum*. Similar states exist for the massless spin $\frac{1}{2}$ field, and the electromagnetic fields (see §3.8).

The Green function equation (3.49) may be treated similarly. Using (3.5),

$$[\Box_x + \tfrac{1}{4}(n-2)R(x)/(n-1)]D_F(x,x') = -[-g(x)]^{-\frac{1}{2}}\delta^n(x-x')$$

becomes

$$\Omega^{-(n+2)/2}(x)\eta^{\mu\nu}\partial_\mu\partial_\nu(\Omega^{(n-2)/2}(x)D_F(x,x')) = -\Omega^{-n}(x)\delta^n(x-x').$$

Thus

$$\begin{aligned}\eta^{\mu\nu}\partial_\mu\partial_\nu(\Omega^{(n-2)/2}(x)D_F(x,x')) &= -\Omega^{(2-n)/2}(x)\delta^n(x-x')\\ &= -\Omega^{(2-n)/2}(x')\delta^n(x-x')\end{aligned}$$

which implies

$$D_F(x,x') = \Omega^{(2-n)/2}(x)\tilde{D}_F(x,x')\Omega^{(2-n)/2}(x') \tag{3.154}$$

where $\tilde{D}_F(x,x')$ is the massless version of the Minkowski space Feynman Green function satisfying (2.74). This relation can also be derived using the definition (3.48) of the Green function in terms of the field. Similar relations hold for the other scalar Green functions. For massless spin $\frac{1}{2}$ and spin 1 fields, analogous relations hold, with the power of Ω determined by the conformal weight appropriate to the field (see §3.8).

We shall now use these results to analyse the behaviour of a comoving particle detector in a Robertson–Walker universe described by the line element (3.98), to see how it responds to the conformal vacuum state. We work in four dimensions.

In Robertson–Walker spacetime the comoving geodesics $\mathbf{x} = \text{constant}$ map under the conformal transformation (3.149) into the geodesics $\mathbf{x} =$ constant in Minkowski space. However the proper time τ along these geodesics, which coincides with cosmic time t, does not coincide with the conformal time parameter $x^0 = \eta$ of (3.98) (see also(3.82)). Along the comoving geodesics the Green function $D^+(x,x')$ reduces to

$$D^+(\eta,\eta') = -\frac{C^{-\frac{1}{2}}(\eta)C^{-\frac{1}{2}}(\eta')}{4\pi^2(\eta-\eta'-\mathrm{i}\varepsilon)^2} \tag{3.155}$$

where we have used (3.59) and (3.154) with G_F replaced by D^+ and $\Omega^2 = C$. Substituting this into the response function (3.55) and using (3.82) to change from integrals with respect to τ to integrals with respect to η gives

$$\mathscr{F}(E) = -\frac{1}{4\pi^2}\int d\eta \int d\eta' \frac{\exp\left[-iE\int_{\eta'}^{\eta} C^{\frac{1}{2}}(\eta'')d\eta''\right]}{(\eta - \eta' - i\varepsilon)^2} \tag{3.156}$$

which will *not* in general vanish, owing to the presence of the $C^{\frac{1}{2}}$ factor in the exponent. Explicit examples of this are given in §§5.3, 5.4.

Thus, even a comoving particle detector – the nearest analogue in Robertson–Walker space to an inertial observer in Minkowski space – will in general register the presence of particles in the conformal vacuum. On the other hand, inspection of (3.152) reveals that the modes which are positive frequency with respect to the conformal vacuum at one time remain so for all time. Thus, if a field satisfying a conformally invariant wave equation in a conformally flat spacetime is in the conformal vacuum at one time it will remain so for all time, and there will be no particle production (Parker 1966, 1968, 1969, 1971, 1973; see also page 189). This will be the case for neutrinos (if they are massless) and photons, though not for gravitons (Grishchuk 1974, 1975). The absence of particle production is to be interpreted in the following sense. If the cosmological expansion is allowed to cease (smoothly), then an inertial particle detector adiabatically switched on after the expansion has ceased will register no particles. The positive response of the detector that is switched on during the expansion might therefore be described as spurious, but it is really only another illustration that in regions of nonzero spacetime curvature the particle concept loses much of its intuitive meaning.

The conformal vacua of a slightly more general class of spacetimes will be discussed in §5.2 (see also Candelas & Dowker 1979).

3.8 Fields of arbitrary spin in curved spacetime

In Minkowski space field theory, the spin of a field can be classified according to the field's properties under infinitesimal Lorentz transformations

$$x^\alpha \to \bar{x}^\alpha = \Lambda^\alpha{}_\beta x^\beta = (\delta^\alpha{}_\beta + \omega^\alpha{}_\beta)x^\beta \tag{3.157}$$

$$\omega_{\alpha\beta} = -\omega_{\beta\alpha} \tag{3.158}$$

$$|\omega^\alpha{}_\beta| \ll 1.$$

(For the time being we restrict attention to four-dimensional Minkowski space.)

A general multicomponent field $T^{\gamma\delta\cdots\lambda}(x)$ transforms to

$$[D(\Lambda)]^{\gamma'\delta'\ldots\lambda'}{}_{\gamma\delta\ldots\lambda}T^{\gamma\delta\ldots\lambda} \tag{3.159}$$

where

$$D(\Lambda) = 1 + \tfrac{1}{2}\omega^{\alpha\beta}\Sigma_{\alpha\beta}. \tag{3.160}$$

In order that Lorentz transformations form a group, the antisymmetric $\Sigma_{\alpha\beta}$ (called the group generators) are constrained to satisfy the commutation identities

$$[\Sigma_{\alpha\beta}, \Sigma_{\gamma\delta}] = \eta_{\gamma\beta}\Sigma_{\alpha\delta} - \eta_{\alpha\gamma}\Sigma_{\beta\delta} + \eta_{\delta\beta}\Sigma_{\gamma\alpha} - \eta_{\delta\alpha}\Sigma_{\gamma\beta}, \tag{3.161}$$

or, equivalently,

$$\left.\begin{aligned} \mathbf{a}\times\mathbf{a} &= \mathrm{i}\mathbf{a} \\ \mathbf{b}\times\mathbf{b} &= \mathrm{i}\mathbf{b} \\ [a_i, b_i] &= 0 \end{aligned}\right\} \tag{3.162}$$

where

$$\left.\begin{aligned} a_i &= \tfrac{1}{2}(-\mathrm{i}\varepsilon_i{}^{kl}\Sigma_{kl} + \Sigma_{i0}) \\ b_j &= \tfrac{1}{2}(-\mathrm{i}\varepsilon_j{}^{kl}\Sigma_{kl} - \Sigma_{j0}). \end{aligned}\right\} \tag{3.163}$$

Equations (3.162) are recognized as the commutation relations for two independent angular momentum operators **a** and **b**. Each can be represented by an infinite matrix, which can be reduced to submatrices labelled by an integer or half integer A or B, where

$$\mathbf{a}^2 = A(A+1), \quad \mathbf{b}^2 = B(B+1), \tag{3.164}$$

acting in a vector space with $2A + 1$ or $2B + 1$ dimensions. These separately irreducible representations are combined as a direct product to form representations of the Lorentz group, labelled now by the two integers or half integers (A, B), acting in a vector space of $(2A + 1)(2B + 1)$ dimensions.

For example, a vector field T^γ transforms to $\Lambda^{\gamma'}{}_\gamma T^\gamma$, so that, using (3.157) and (3.160),

$$[\Sigma_{\alpha\beta}]^\gamma{}_\delta = \delta_\alpha{}^\gamma\eta_{\beta\delta} - \delta_\beta{}^\gamma\eta_{\alpha\delta} \tag{3.165}$$

which yields, from (3.164), $A = \frac{1}{2}, B = \frac{1}{2}$. Thus, the vector field can be classified with the $(\frac{1}{2},\frac{1}{2})$ irreducible representation of the Lorentz group. This field is therefore a spin $\frac{1}{2}+\frac{1}{2}=$ spin 1 field.

Similarly, a scalar field clearly has $\Sigma_{\alpha\beta}=0$, thus being identified with the $(0,0)$ irreducible representation, i.e., spin 0. A second rank tensor field transforms to $\Lambda^{\gamma'}{}_{\gamma}\Lambda^{\delta'}{}_{\delta}T^{\gamma\delta}$, so the $D(\Lambda)$ is simply a product of two vector field $D(\Lambda)$. The product representation $(\frac{1}{2},\frac{1}{2})\otimes(\frac{1}{2},\frac{1}{2})$ reduces to four irreducible representations: $(1,1)$, $(1,0)$, $(0,1)$ and $(0,0)$. The tensor field therefore contains components with spins 2, 1 and 0.

Finally, for the Dirac spinor field one must choose

$$\Sigma_{\alpha\beta}=\tfrac{1}{4}[\gamma_\alpha,\gamma_\beta] \tag{3.166}$$

where γ are Dirac matrices, associated with the $(\frac{1}{2},0)$ and $(0,\frac{1}{2})$, i.e., spin $\frac{1}{2}$, irreducible representations.

We wish to generalize these considerations to curved spacetime without losing the connection with the Lorentz group. This can be achieved by employing the so-called tetrad, or vierbein formalism (see, for example, Synge 1960, Weinberg 1972), which can easily be extended to n dimensions (n-beins). The essence of this approach is to erect normal coordinates y^α_X at each spacetime point X. In terms of y^α_X the metric at X is then simply $\eta_{\alpha\beta}$. In terms of a more general coordinate system, however, the metric tensor will be more complicated, but is related to $\eta_{\alpha\beta}$ by

$$g_{\mu\nu}(x)=V^\alpha_\mu(x)V^\beta_\nu(x)\eta_{\alpha\beta} \tag{3.167}$$

where

$$V^\alpha_\mu(X)=\left(\frac{\partial y^\alpha_X}{\partial x^\mu}\right)_{x=X}, \quad \alpha=0,1,2,3$$

is called a vierbein. Note that the label α refers to the local inertial frame associated with the normal coordinates y^α_X at X, while μ is associated with the general coordinate system x^μ. We adopt the convention in this section that labels from the beginning of the Greek alphabet refer to the former, and those from the end refer to the latter.

As the general coordinate system is arbitrary, we can consider the effect of changing the x^μ while leaving the y^α_X fixed. Then V^α_μ transforms as a covariant vector

$$V^\alpha_\mu\to\frac{\partial x^\nu}{\partial x'^\mu}V^\alpha_\nu. \tag{3.168}$$

Clearly, it is also possible to Lorentz transform the y^α_X arbitrarily at each point X:

$$y^{\alpha}{}_X \to y'^{\alpha}{}_X = \Lambda^{\alpha}{}_{\beta}(X) y_X{}^{\beta}. \tag{3.169}$$

In this case, V^{α}_{μ} transforms as a Lorentz contravariant vector

$$V^{\alpha}{}_{\mu}(X) \to \Lambda^{\alpha}{}_{\beta}(X) V^{\beta}{}_{\mu}(X), \tag{3.170}$$

which obviously leaves the metric (3.167) invariant.

If a generally covariant vector A_{μ} is contracted into $V_{\alpha}{}^{\mu}$, the resulting object

$$A_{\alpha} = V_{\alpha}{}^{\mu} A_{\mu}$$

transforms as a collection of four scalars under general coordinate transformations, while under the local Lorentz transformations (3.169) it behaves as a vector. Thus, by use of vierbeins, one can convert general tensors into local, Lorentz-transforming tensors, shifting the additional spacetime dependence into the vierbeins.

If expression (3.159) is written schematically as $D(\Lambda)\psi$, where ψ is a tensor field, then the derivative of ψ, $\partial_{\alpha}\psi$, will also be a tensor field in Minkowski space. Under Lorentz transformations, $\partial_{\alpha}\psi$ will become $\Lambda_{\alpha}{}^{\beta} D(\Lambda) \partial_{\beta}\psi$. When passing to curved spacetime, we wish to generalize the derivative ∂_{α} to a covariant derivative ∇_{α}, but retaining this simple transformation property for arbitrary local Lorentz transformations at each spacetime point:

$$\nabla_{\alpha}\psi \to \Lambda_{\alpha}{}^{\beta}(x) D(\Lambda(x)) \nabla_{\beta}\psi(x). \tag{3.171}$$

This may be achieved by defining

$$\nabla_{\alpha} = V_{\alpha}{}^{\mu}(\partial_{\mu} + \Gamma_{\mu}) \tag{3.172}$$

where the connection

$$\Gamma_{\mu}(x) = \tfrac{1}{2}\Sigma^{\alpha\beta} V_{\alpha}{}^{\nu}(x) \Big(\nabla_{\mu} V_{\beta\nu}(x) \Big), \tag{3.173}$$

$\Sigma^{\alpha\beta}$ being the generator of the Lorentz group associated with the particular representation $D(\Lambda)$ under which ψ transforms, and $V_{\beta\nu} = g_{\mu\nu} V_{\beta}{}^{\mu}$.

The utility of the property (3.171) is that any function of ψ and $\nabla_{\alpha}\psi$ that is a scalar under Lorentz transformations in Minkowski space, remains a scalar under local changes in the vierbein, as well as under general coordinate transformations. Thus, the Lagrangian of the field may be generalized to curved spacetime by replacing all derivatives ∂_{α} by ∇_{α} and contracting all vectors, tensors, etc. into n-beins ($A_{\alpha} \to V_{\alpha}{}^{\mu} A_{\mu}$, etc.).

For example, for a scalar field ϕ, the spin is zero and $\Sigma_{\alpha\beta} = 0$, so (3.172) reveals that $\nabla_\alpha = \partial_\alpha$. The Lagrangian (2.2) therefore becomes

$$\mathscr{L}(x) = \tfrac{1}{2}(-g)^{\frac{1}{2}}(\eta^{\alpha\beta} V_\alpha{}^\mu \partial_\mu \phi V^\nu{}_\beta \partial_\nu \phi - m^2\phi^2). \tag{3.174}$$

The factor $[-g(x)]^{\frac{1}{2}} = \det(V^\alpha{}_\mu)$ has been included to make $\mathscr{L}(x)$ a scalar density, and hence to make the action

$$S = \int \mathscr{L}(x) \mathrm{d}^4 x \tag{3.175}$$

a scalar. Using (3.167), equation (3.174) reduces to the expression quoted in (3.24), with $\xi = 0$. The term ξR, which vanishes in flat spacetime, is the only additional geometric scalar that can be added to the Lagrangian.

The Lagrangian density for a spin $\frac{1}{2}$ field in Minkowski space is given by (2.45). In curved spacetime this becomes, using (3.166) for the Σs in (3.173),

$$\begin{aligned} \mathscr{L}(x) &= \det V\{\tfrac{1}{2}\mathrm{i}[\bar{\psi}\gamma^\alpha V_\alpha{}^\mu \nabla_\mu \psi - V_\alpha{}^\mu(\nabla_\mu \bar{\psi})\gamma^\alpha \psi] - m\bar{\psi}\psi\} \\ &= \det V\{\tfrac{1}{2}\mathrm{i}[\bar{\psi}\gamma^\mu \nabla_\mu \psi - (\nabla_\mu \bar{\psi})\gamma^\mu \psi] - m\bar{\psi}\psi\} \end{aligned} \tag{3.176}$$

where $\gamma^\mu = V_\alpha{}^\mu \gamma^\alpha$ are the curved space counterparts of the Dirac γ matrices and which, from (3.167), clearly satisfy

$$\{\gamma^\mu, \gamma^\nu\} = 2g^{\mu\nu}, \tag{3.177}$$

i.e., the curved space generalization of (2.46).

The Lagrangian (3.176), which can also be used in n dimensions, is conformally invariant in the massless limit so long as ψ transforms as

$$\psi \to \Omega^{(1-n)/2}(x)\psi \tag{3.178}$$

under the transformations (3.1).

Variation of the action S with respect to $\bar{\psi}$ yields the covariant Dirac equation

$$\mathrm{i}\gamma^\mu \nabla_\mu \psi - m\psi = 0 \tag{3.179}$$

(Fock 1929, Fock & Ivanenko 1929, Bargmann 1932, Schrödinger 1932).

Next, the electromagnetic field $F_{\mu\nu}$ is a spin 1 field. The Minkowski space expression (2.54), together with the gauge-breaking and ghost Lagrangians, must be generalized by replacing A_α by $V_\alpha{}^\mu A_\mu$ and ∂_α by ∇_α with Γ_μ given by (3.173). In this case, $\Sigma_{\alpha\beta}$ is given by (3.165) for the $(\frac{1}{2}, \frac{1}{2})$ representation of the Lorentz group. The result is, for the Maxwell term

$$\mathscr{L}(x) = -\tfrac{1}{4}(-g)^{\frac{1}{2}} F_{\mu\nu} F^{\mu\nu} \tag{3.180}$$

where

$$F_{\mu\nu} = A_{\mu;\nu} - A_{\nu;\mu} = A_{\mu,\nu} - A_{\nu,\mu} \tag{3.181}$$

(the connection terms cancelling), and for the gauge-breaking term

$$\mathscr{L}_{\mathrm{G}} = -\tfrac{1}{2}\zeta^{-1}(A^{\mu}{}_{;\mu})^2. \tag{3.182}$$

The ghost term is dealt with under the scalar case, and results in the Lagrangian density (see (2.141))

$$\mathscr{L}_{\text{ghost}} = g^{\mu\nu}\partial_\mu c \partial_\nu c^*. \tag{3.183}$$

Variation of $S = \int(\mathscr{L} + \mathscr{L}_{\mathrm{G}})\,\mathrm{d}^4x$ with respect to A_μ yields

$$F_{\mu\nu;}{}^{\nu} + \zeta^{-1}(A^{\nu}{}_{;\nu})_{;\mu} = 0 \tag{3.184}$$

or

$$A_{\mu;\nu}{}^{\nu} + R_{\mu}{}^{\rho}A_{\rho} - (1-\zeta^{-1})A_{\nu;}{}^{\nu}{}_{\mu} = 0. \tag{3.185}$$

It is important to note that this theory is only conformally invariant in four spacetime dimensions.

Detailed discussions of spin 1 fields in curved spacetime have been given, in particular, by Schrödinger (1939), Parker (1972) and Mashoon (1973).

The method used above for transforming spin 0, $\frac{1}{2}$ and 1 theories to curved spacetime can equally well be used for theories of arbitrary spin fields. Such theories have been investigated in flat spacetime by, for example, Dirac (1936), Fierz (1939), Fierz & Pauli (1939), Wichmann (1962), Barut, Muzinich & Williams (1963), Weinberg (1964*a*, *b*, *c*) and Dowker (1967*a*, *b*) (see also the review of Mohan 1968).

The consequences of arbitrary spin theories in curved spacetimes have, for example, been studied by Belifante (1940), Hatalkar (1954), Duan' (1956), Penrose (1965), Dowker & Dowker (1966*a*, *b*), Dowker (1972), Grensing (1977), Christensen & Duff (1978*a*, 1979) and Birrell (1979*b*). We shall not pursue this interesting topic any further here.

As in the case of scalar fields, one can write down curved spacetime generalizations of the Green functions and Green function equations for fields of nonzero spin. For example, the Feynman propagators for spins $\frac{1}{2}$ and 1 still have the form (2.80) and (2.86) respectively, but now satisfy the equations (cf. (2.82), (2.87))

$$[\mathrm{i}\gamma^{\mu}(x)\nabla_{\mu}^{x} - m]S_{\mathrm{F}}(x, x') = [-g(x)]^{-\frac{1}{2}}\delta^{n}(x - x'), \tag{3.186}$$

and

$$[g_{\mu\rho}(x)\Box_x + R_{\mu\rho}(x) - (1 - \zeta^{-1})\nabla^x_\mu \nabla^x_\rho] D_F^{\rho\nu}(x, x')$$
$$= [-g(x)]^{-\frac{1}{2}} \delta^\nu_\mu \delta^n(x - x'), \tag{3.187}$$

which are derived using (3.179) and (3.185) respectively. In the spin $\frac{1}{2}$ case, the curved spacetime generalization of (2.84) is

$$S_F(x, x') = [i\gamma^\mu(x)\nabla^x_\mu + m] G_F(x, x') \tag{3.188}$$

where G_F is now a bi-spinor satisfying (3.49) with $\xi = \frac{1}{4}$ and $\Box_x = g^{\mu\nu}(x)\nabla^x_\mu \nabla^x_\nu$, as can be seen by substitution of (3.188) into (3.186).

Note that as in the scalar case the differential equations for the propagators do not determine the vacuum states appearing in the curved spacetime versions of (2.80) or (2.86), without the imposition of boundary conditions. These boundary conditions will depend on the particular physical system under study. Also, as in the scalar case one can obtain adiabatic (DeWitt–Schwinger) expansions for the propagator. Such an expansion has been given for spin $\frac{1}{2}$ by Bunch & Parker (1979) using the momentum space construction of the previous section, while the results for spin $\frac{1}{2}$ and 1 have been obtained using the DeWitt–Schwinger proper time method by Christensen (1978) (see also DeWitt & Brehme 1960, DeWitt 1965, Adler, Lieberman & Ng 1977). Particular terms (those having relevance to the conformal anomaly – see §6.3) in the adiabatic expansion of the propagator for high-spin fields have been obtained by Christensen & Duff (1978*a*, 1979) and Birrell (1979*b*).

We finally note that the stress-tensor for a field theory of arbitrary spin in curved spacetime can be obtained by variation of the action with respect to the metric:

$$T_{\mu\nu}(x) = \frac{2}{[-g(x)]^{\frac{1}{2}}} \frac{\delta S}{\delta g^{\mu\nu}(x)} = \frac{V_{\alpha\mu}(x)}{\det[V(x)]} \frac{\delta S}{\delta V^\nu_\alpha(x)} \tag{3.189}$$

(the factor of $(-g)^{-\frac{1}{2}}$ being inserted to give a tensor rather than a tensor density). In particular, the stress-tensors for spin $0, \frac{1}{2}$, and 1 are found to be

$$T_{\mu\nu}(s=0) = (1 - 2\xi)\phi_{;\mu}\phi_{;\nu} + (2\xi - \tfrac{1}{2}) g_{\mu\nu} g^{\rho\sigma}\phi_{;\rho}\phi_{;\sigma} - 2\xi\phi_{;\mu\nu}\phi$$
$$+ \frac{2}{n}\xi g_{\mu\nu}\phi\Box\phi - \xi\left[R_{\mu\nu} - \tfrac{1}{2}Rg_{\mu\nu} + \frac{2(n-1)}{n}\xi R g_{\mu\nu}\right]\phi^2$$
$$+ 2\left[\tfrac{1}{4} - \left(1 - \frac{1}{n}\right)\xi\right] m^2 g_{\mu\nu}\phi^2 \tag{3.190}$$

$$T_{\mu\nu}(s=\tfrac{1}{2}) = \tfrac{1}{2}\mathrm{i}[\bar{\psi}\gamma_{(\mu}\nabla_{\nu)}\psi - (\nabla_{(\mu}\bar{\psi})\gamma_{\nu)}\psi] \tag{3.191}$$

$$T_{\mu\nu}(s=1) = T^{\mathrm{Maxwell}}_{\mu\nu} + T^{\mathrm{G}}_{\mu\nu} + T^{\mathrm{ghost}}_{\mu\nu} \tag{3.192}$$

$$T^{\mathrm{Maxwell}}_{\mu\nu} = \tfrac{1}{4}g_{\mu\nu}F^{\rho\sigma}F_{\rho\sigma} - F_{\mu}{}^{\rho}F_{\rho\nu} \tag{3.193}$$

$$T^{\mathrm{G}}_{\mu\nu} = \zeta^{-1}\{A_{\mu}A^{\rho}{}_{;\rho\nu} + A^{\rho}{}_{;\rho\mu}A_{\nu} - g_{\mu\nu}[A^{\rho}A^{\sigma}{}_{;\sigma\rho} + \tfrac{1}{2}(A^{\rho}{}_{;\rho})^{2}]\} \tag{3.194}$$

$$T^{\mathrm{ghost}}_{\mu\nu} = -c^{*}_{;\mu}c_{;\nu} - c^{*}_{;\nu}c_{;\mu} - g_{\mu\nu}g^{\rho\sigma}c^{*}_{;\rho}c_{;\sigma}. \tag{3.195}$$

These equations are obtained using (3.189) with the Lagrangian densities (3.24) and (3.176) for $s=0$ and $s=\frac{1}{2}$ respectively, while the spin 1 result is obtained using the sum of the Lagrangian densities (3.180), (3.182) and (3.183). In obtaining the spin 0 and $\frac{1}{2}$ results in the symmetrized form given, the relevant field equations have been applied, and in the cases of spin $\frac{1}{2}$ and 1 integration by parts has been used. We also note the following variational formulae which are useful for the derivation of these results:

$$\left.\begin{aligned}
\delta g^{\mu\nu} &= -g^{\mu\rho}g^{\nu\sigma}\delta g_{\rho\sigma} \\
\delta(-g)^{\frac{1}{2}} &= \tfrac{1}{2}(-g)^{\frac{1}{2}}g^{\mu\nu}\delta g_{\mu\nu} \\
\delta R &= -R^{\mu\nu}\delta g_{\mu\nu} + g^{\rho\sigma}g^{\mu\nu}(\delta g_{\rho\sigma;\mu\nu} + \delta g_{\rho\mu;\sigma\nu}) \\
\delta g_{\mu\nu} &= -[g_{\mu\rho}V^{\alpha}{}_{\nu} + g_{\nu\rho}V^{\alpha}{}_{\mu})\delta V_{\alpha}{}^{\rho}.
\end{aligned}\right\} \tag{3.196}$$

4

Flat spacetime examples

Having invested so much effort in mastering curved space quantum field theory, the reader may be dismayed to return to the topic of flat spacetime. Flat spacetime does not, however, imply Minkowski space quantum field theory.

We consider three main topics in which the general curved spacetime formalism must be applied to achieve sensible results, even though the geometry is flat. This enables some non-trivial geometrical effects to be explored within the considerable simplification afforded by a flat geometry. In particular, we are able to discuss $\langle T_{\mu\nu} \rangle$ in some special cases without employing the full theory of curved space regularization and renormalization to be developed in chapter 6.

The first case examines the effects of a non-trivial topology. We do not treat particle creation at this stage, but limit the discussion to $\langle T_{\mu\nu} \rangle$, which is nonzero even for the vacuum. This topic is one of the few in our subject which makes contact with laboratory physics, for the disturbance to the electromagnetic vacuum induced by the presence of two parallel conducting plates is actually observable. The force of attraction that appears is called the Casimir effect, and has been extensively discussed in the literature.

The treatment of boundary surfaces leads naturally to a very simple, yet extremely illuminating, system that is well worth studying in detail. This is the case of the 'moving mirror', in which a boundary at which the quantum field is constrained moves about. The ensuing creation of particles provides a heuristic model of more complicated systems to be treated later. In particular, the evaporating black hole enjoys a very close analogue in a certain type of accelerating mirror arrangement. The reader is especially recommended to follow the details leading to the evaluation of the Bogolubov transformation (4.60), as these are identical to the black hole case, and so are not repeated in chapter 8.

The third topic in this chapter is also closely related to the black hole case, but is of considerable intrinsic interest. It concerns the experiences of an observer (particle detector) that accelerates uniformly through the

Minkowski vacuum state. The result – that the accelerating observer perceives a bath of radiation with an apparently thermal spectrum – is an intriguing one that casts light on the relationship between event horizons, quantum field theory and entropy. Once again, attention is especially directed to the discussion of the Bogolubov transformation on pages 114–116. Although somewhat technical, the treatment (particularly of the analyticity properties of the field modes) is directly relevant to the black hole system, and will repay careful study.

4.1 Cylindrical two-dimensional spacetime

The simplest generalization of Minkowski space quantum field theory is the introduction of non-trivial topological structures in a locally flat spacetime. The easiest such generalization is the $R^1 \times S^1$ two-dimensional spacetime with compactified (closed) spatial sections. This spacetime has the two-dimensional Minkowski space line element (3.8) or (3.10), but the spatial points x and $x + L$ are identified, where L is the periodicity length ('circumference of the universe'). This spacetime is shown in fig. 10.

The effect of the space closure is to restrict the field modes (2.11) to a discrete set (cf. (2.13)):

$$u_k = (2L\omega)^{-\frac{1}{2}} e^{i(kx - \omega t)} \tag{4.1}$$

where $k = 2\pi n/L$, $n = 0, \pm 1, \pm 2, \pm 3, \ldots$ Restricting attention to the massless case, $\omega = |k|$, the modes labelled by positive values of n have the form

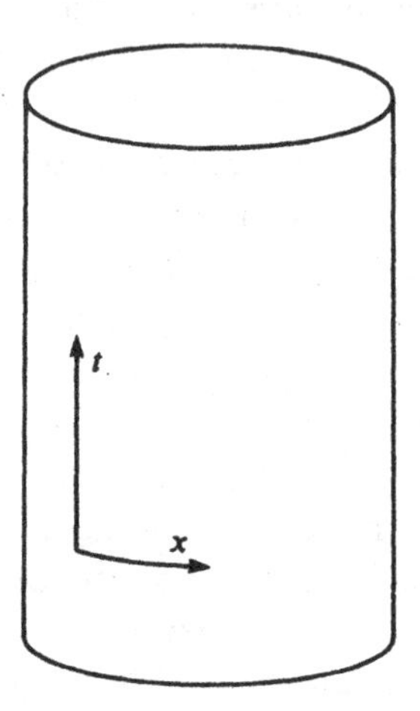

Fig. 10. Two-dimensional spacetime with compact spatial sections ($R^1 \times S^1$). The circumference of the cylinder is L.

$\exp[ik(x-t)]$ and represent waves that move from left to right, while negative values of n give $\exp[ik(x+t)]$, which represent left-moving waves.

In arriving at (4.1) we have imposed periodic boundary conditions on u_k (i.e., $u_k(t,x) = u_k(t, x+nL)$). One can also consider imposing antiperiodic boundary conditions $u_k(t,x) = (-1)^n u_k(t, x+nL)$, in which case the modes are given by (4.1) but with $k = 2\pi(n+\frac{1}{2})/L$, $n = 0, \pm 1, \pm 2, \pm 3, \ldots$ In the latter case the scalar field is to be regarded as a section through a non-product bundle and is sometimes referred to as a *twisted* field (see Isham 1978*b*).

Because the field modes are forced into a discrete set the field energy will be disturbed. In this two-dimensional case one finds from (2.26) the Cartesian components of the stress-tensor operator to be

$$T_{tt} = T_{xx} = \frac{1}{2}\left(\frac{\partial \phi}{\partial t}\right)^2 + \frac{1}{2}\left(\frac{\partial \phi}{\partial x}\right)^2 \tag{4.2}$$

$$T_{tx} = T_{xt} = \frac{\partial \phi}{\partial t}\frac{\partial \phi}{\partial x}. \tag{4.3}$$

We shall evaluate $\langle 0_L | T_{\mu\nu} | 0_L \rangle$, where $|0_L\rangle$ is the vacuum associated with the discrete modes (4.1). This state clearly has the property: $|0_L\rangle \to |0\rangle$ as $L \to \infty$, $|0\rangle$ being the usual Minkowski space vacuum. (See Fulling 1973 for a discussion of the inequivalence of $|0_L\rangle$ and $|0_{L'}\rangle$ for $L \neq L'$.)

Using (2.43) with the modes (4.1) (and hence the vacuum $|0_L\rangle$) one obtains

$$\langle 0_L | T_{tt} | 0_L \rangle = (1/2L) \sum_{n=-\infty}^{\infty} |k| = (2\pi/L^2) \sum_{n=0}^{\infty} n \tag{4.4}$$

which is clearly infinite. This was expected, as the $R^1 \times S^1$ system suffers from the same ultraviolet divergence properties as Minkowski space. The compactified spatial sections can modify the long wavelength modes, but the ultraviolet behaviour is unchanged.

In the case of the Minkowski space vacuum energy calculation given in §2.4, the ultraviolet divergence was removed by normal ordering with respect to the creation and annihilation operators of the Fock space associated with the modes (2.11). In the case of a general state $|\psi\rangle$ in this Fock space normal ordering reduces to

$$\langle \psi | :T_{\alpha\beta}: | \psi \rangle = \langle \psi | T_{\alpha\beta} | \psi \rangle - \langle 0 | T_{\alpha\beta} | 0 \rangle, \tag{4.5}$$

which, in particular, guarantees $\langle 0 | :T_{\alpha\beta}: | 0 \rangle = 0$. If we consider Minkowski space as the covering space of $R^1 \times S^1$, then $|0_L\rangle$ can be considered as a state in the above Fock space (Fulling 1973) and one can remove the divergence in $\langle 0_L|T_{\alpha\beta}|0_L\rangle$ by applying (4.5). (For a rigorous discussion of this step see Kay 1979.) In particular

$$\langle 0_L|:T_{tt}:|0_L\rangle \equiv \langle 0_L|T_{tt}|0_L\rangle - \langle 0|T_{tt}|0\rangle$$
$$= \langle 0_L|T_{tt}|0_L\rangle - \lim_{L'\to\infty} \langle 0_{L'}|T_{tt}|0_{L'}\rangle. \quad (4.6)$$

Because both terms on the right-hand side of (4.6) are individually divergent, they cannot be subtracted without careful analysis. We defer the detailed discussion of this issue until chapter 6. Here we follow the simple procedure of introducing the cut-off factor $e^{-\alpha|k|}$ into the divergent sums of the type (4.4), and let $\alpha \to 0$ at the end of the calculation. Although *ad hoc*, this step is justified by the rigorous treatment given later.

With the cut-off factor, the sum (4.4) is finite and is readily performed:

$$\langle 0_L|T_{tt}|0_L\rangle = (2\pi/L^2)\sum_{n=0}^{\infty} n e^{-2\pi\alpha n/L} = (2\pi/L^2)e^{2\pi\alpha/L}(e^{2\pi\alpha/L}-1)^{-2}, \quad (4.7)$$

which may be expanded about $\alpha = 0$

$$\langle 0_L|T_{tt}|0_L\rangle = (1/2\pi\alpha^2) - (\pi/6L^2) + O(\alpha^3)$$

with a similar expansion for $\langle 0_{L'}|T_{tt}|0_{L'}\rangle$. Thus,

$$\lim_{L'\to\infty} \langle 0_{L'}|T_{tt}|0_{L'}\rangle = 1/2\pi\alpha^2.$$

Substituting these results into (4.6) and taking $\alpha \to 0$ one finds

$$\langle 0_L|:T_{tt}:|0_L\rangle = -\pi/6L^2.$$

This procedure can also be used to show that $\langle 0_L|:T_{tx}:|0_L\rangle = 0$.

Although $\langle T_{\alpha\beta}\rangle$ diverges when evaluated for both states $|0\rangle$ and $|0_L\rangle$ the difference between the two results is finite. Thus, if we require that $\langle 0|:T_{\alpha\beta}:|0\rangle = 0$, then the state $|0_L\rangle$ contains a finite, negative energy density

$$\rho = \langle 0_L|:T_{tt}:|0_L\rangle = -\pi/6L^2 \quad (4.8)$$

and pressure

$$p = \langle 0_L|:T_{xx}:|0_L\rangle = -\pi/6L^2. \quad (4.9)$$

The cloud of negative vacuum energy is distributed uniformly throughout the $R^1 \times S^1$ universe with total energy $-\pi/6L$.

In the case of twisted fields the vacuum energy can be calculated

(Isham 1978*b*) along similar lines. Instead of (4.7) one now has

$$\langle 0_L|T_{tt}|0_L\rangle = (\pi/2L^2)\sum_{n=-\infty}^{\infty}|2n+1|e^{-\pi\alpha|2n+1|/L}$$

$$= (\pi/2L^2)\left[\sum_{n=-\infty}^{\infty}|n|e^{-\pi\alpha|n|/L} - \sum_{n=-\infty}^{\infty}2|n|e^{-2\pi\alpha|n|/L}\right]$$

$$= 2[2\pi/(2L)^2]\sum_{n=0}^{\infty}ne^{-2\pi\alpha n/(2L)} - (2\pi/L^2)\sum_{n=0}^{\infty}ne^{-2\pi\alpha n/L}, \quad (4.10)$$

the second line resulting from the fact that the sum over odd n is the same as the sum over all n minus the sum over even n. The final term of (4.10) is precisely minus the sum in (4.7), while the preceding term is twice the sum in (4.7) with L replaced by $2L$. One can therefore read off the finite part of ρ from (4.8) as

$$\rho = -2\pi/6(2L)^2 + \pi/6L^2 = \pi/12L^2. \quad (4.11)$$

The vacuum energy of the twisted scalar field is $-\frac{1}{2}$ of that for untwisted fields.

Finally, for a massless spin $\frac{1}{2}$ field, use of (2.51) and (3.191) specialized to flat spacetime yields (Davies & Unruh 1977)

$$\langle 0_L|T_{\mu\nu}|0_L\rangle = \tfrac{1}{2}i\sum_{\pm s}\sum_k[\bar{v}_{k,s}\gamma_{(\mu}\partial_{\nu)}v_{k,s} - \partial_{(\nu}\bar{v}_{k,s}\gamma_{\mu)}v_{k,s}], \quad (4.12)$$

from which, using (2.48) and (2.50), one obtains

$$\langle 0_L|T_{tt}|0_L\rangle = -(1/2L)\sum_{\pm s}\sum_k v^\dagger(k,s)v(k,s)$$

$$= -(2/L)\sum_k|k| = \begin{cases} -(8\pi/L^2)\sum_{n=0}^{\infty}n & \text{untwisted} \\ -(2\pi/L^2)\sum_{n=-\infty}^{\infty}|2n+1| & \text{twisted} \end{cases} \quad (4.13)$$

In both cases the result is simply minus four times the corresponding scalar field result.

Twisted spinor fields can be introduced in precisely the same way as twisted scalar fields (Isham 1978*c*, Avis & Isham 1979*a*, *b*) and their vacuum energy has been calculated in various topologies by DeWitt, Hart & Isham (1979) and Banach & Dowker (1979). Furthermore, Avis & Isham (1979*a*, *b*) have argued that, in most spacetimes with non-trivial topology, if the vacuum generating functional $Z(0,0)$ (see (2.129)) is to be invariant under

Lorentz transformations (3.170) of the vierbein, then one *must* include both twisted and untwisted spinors in one's calculations. Thus, twisted fields should not be considered as a mathematical curiosity, but rather as being equally as important as untwisted fields.

4.2 Use of Green functions

Rather than working directly with the stress-tensor $T_{\mu\nu}$ as given by (4.2) and (4.3), and using an arbitrary cut-off function in the mode integrals, the results of the previous section can be obtained more elegantly by working instead with the Green functions defined in §2.7.

It is also convenient to work in the null coordinates u and v defined in (3.9). In terms of u and v the stress-tensor components $T_{\mu\nu}(u, v)$ for a massless scalar field in two dimensions are simply

$$T_{uu} = (\partial_u \phi)^2 \tag{4.14}$$

$$T_{vv} = (\partial_v \phi)^2 \tag{4.15}$$

$$T_{uv} = T_{vu} = \tfrac{1}{2}\partial_u \phi \partial_v \phi. \tag{4.16}$$

Also

$$T_{tt} = T_{uu} + T_{vv} + 2T_{uv} \tag{4.17}$$

$$T_{xx} = T_{uu} + T_{vv} - 2T_{uv} \tag{4.18}$$

$$T_{tx} = T_{xt} = T_{vv} - T_{uu}. \tag{4.19}$$

It follows from (4.14) and (2.66) that

$$\langle 0_L | T_{uu}(u, v) | 0_L \rangle = \lim_{v'', v' \to v} \lim_{u'', u' \to u} \partial_{u''} \partial_{u'} \tfrac{1}{2} D_L^{(1)}(u'', v''; u', v') \tag{4.20}$$

where we have affixed the subscript L to $D^{(1)}$ to indicate that it is calculated in the $R^1 \times S^1$ vacuum $|0_L\rangle$ of the previous section. (Recall that a Green function is denoted by D if it is that of a massless field.) We have also chosen to write (4.20) in a form which is symmetric under interchange of (u'', v'') with (u', v'). In addition, one may treat a thermal state $|\beta_L\rangle$ at temperature $T = (k_B \beta)^{-1}$ by using the thermal Green function $D_{L,\beta}^{(1)}$ (see (2.111), generalizing it to $R^1 \times S^1$) in place of $D_L^{(1)}$:

$$\langle \beta_L | T_{uu} | \beta_L \rangle = \lim_{v'', v' \to v} \lim_{u'', u' \to u} \partial_{u''} \partial_{u'} \tfrac{1}{2} D_{L,\beta}^{(1)}(u'', v''; u', v'), \tag{4.21}$$

where (4.21) reduces to (4.20) as $\beta \to \infty$ $(T \to 0)$.

We first construct the zero-temperature Green function using the modes

(4.1). We have

$$D_L^{(1)}(u'',v'';u',v') = \langle 0_L|\{\phi(u'',v''),\phi(u',v')\}|0_L\rangle$$
$$= \sum_{n=-\infty}^{\infty} [u_k(u'',v'')u_k^*(u',v') + \text{c.c.}]$$
$$= (1/2\pi)\sum_{n=1}^{\infty} n^{-1}(e^{(-2\pi n i/L)\Delta u} + e^{(-2\pi n i/L)\Delta v}) + \text{c.c.} \tag{4.22}$$

where $\Delta u = u'' - u'$, $\Delta v = v'' - v'$. In arriving at (4.22) we have discarded the infinite term that arises from $n = 0$ in the above summation. This infrared (zero frequency) divergence is a feature of two-dimensional spacetime massless quantum field theory. The discarded term would in any case have disappeared had the differentiation in (4.21) been performed before passing to the massless limit.

Because the exponents in (4.22) are purely imaginary, the summations are not absolutely convergent. The Green function $D_L^{(1)}$ must be defined carefully in the distributional sense, and some attention to this topic has been given by Bunch, Christensen & Fulling (1978). Here we shall simply insert into the exponent an infinitesimal real part of the appropriate sign to make the summations converge absolutely. This will always be done in the similar cases which follow.

Performing the summations in (4.22) yields

$$D_L^{(1)}(u'',v'';u',v') = -(1/4\pi)\ln[16\sin^2(\pi\Delta u/L)\sin^2(\pi\Delta v/L)]. \tag{4.23}$$

If instead we consider a twisted scalar field, then $D_L^{(1)}$ is given by the second line of (4.22) with n replaced by $n + \frac{1}{2}$. (Note that there is no infrared divergent term in this case.) One obtains

$$D_L^{(1)}(u'',v'';u',v') = -(1/4\pi)\ln[\tan^2(\pi\Delta u/2L)\tan^2(\pi\Delta v/2L)]. \tag{4.24}$$

The thermal Green function is now obtained from $D_L^{(1)}$ by taking the infinite image sum in (2.111):

$$D_{L,\beta}^{(1)} = -(1/4\pi)\sum_{m=-\infty}^{\infty} \ln\{16\sin^2[\pi(\Delta u + im\beta)/L]\sin^2[\pi(\Delta v + im\beta)/L]\} \quad \text{(untwisted)} \tag{4.25}$$

$$D_{L,\beta}^{(1)} = -(1/4\pi)\sum_{m=-\infty}^{\infty} \ln\{\tan^2[\pi(\Delta u + im\beta)/2L]\tan^2[\pi(\Delta v + im\beta)/2L]\} \quad \text{(twisted).} \tag{4.26}$$

The $m = 0$ term is the zero-temperature Green function $D_L^{(1)}$.

The stress-tensor now follows using (4.21):

$$\begin{aligned}\langle \beta_L | T_{uu} | \beta_L \rangle &= \lim_{\Delta u \to 0} -(\pi/4L^2)\operatorname{cosec}^2(\pi\Delta u/L) \\ &\quad + (\pi/2L^2) \sum_{m=1}^{\infty} \operatorname{cosech}^2(\pi m\beta/L), \quad \text{(untwisted)} \\ &= \lim_{\Delta u \to 0} (\pi/16L^2)[1 - \operatorname{cosec}^2(\pi\Delta u/2L)] \\ &\quad - (\pi/8L^2) \sum_{m=1}^{\infty} [\operatorname{sech}^2(\pi m\beta/2L) \\ &\quad + \operatorname{cosech}^2(\pi m\beta/2L)] \quad \text{(twisted).}\end{aligned}$$

In the zero-temperature limit, $\beta \to \infty$, only the first (m-independent) terms survive. As expected, these diverge like $(\Delta u)^{-2}$ in the limit $\Delta u \to 0$. This is the infinite vacuum energy. To renormalize, we subtract

$$\langle 0 | T_{uu} | 0 \rangle = \lim_{L \to \infty} \langle 0_L | T_{uu} | 0_L \rangle = -1/4\pi\Delta u^2.$$

We may then take the limit $\Delta u \to 0$ and arrive at the finite result

$$-(\pi/12L^2) + (\pi/2L^2) \sum_{m=1}^{\infty} \operatorname{cosech}^2(\pi m\beta/L) \qquad \text{(untwisted)}$$

$$(\pi/24L^2) + (\pi/8L^2) \sum_{m=1}^{\infty} [\operatorname{sech}^2(\pi m\beta/2L) + \operatorname{cosech}^2(\pi m\beta/2L)] \quad \text{(twisted)} \tag{4.27}$$

Because of the symmetry of $D^{(1)}_{L,\beta}$ under interchange of u and v it readily follows that

$$\langle \beta_L | T_{vv} | \beta_L \rangle = \langle \beta_L | T_{uu} | \beta_L \rangle. \tag{4.28}$$

Also, the fact that $D^{(1)}_{L,\beta}$ can be written as a v-independent function plus a u-independent function gives from (4.16)

$$\langle \beta_L | T_{uv} | \beta_L \rangle = \langle \beta_L | T_{vu} | \beta_L \rangle = 0. \tag{4.29}$$

Then, from (4.17), the energy densities are simply twice the quantities in (4.27), which agree in the zero-temperature limit ($\beta \to \infty$) with (4.8) and (4.11) respectively.

4.3 Boundary effects

So far we have restricted attention to manifolds without boundaries. Even if spacetime itself is unbounded, the quantum field may still be constrained by the presence of material boundaries. For example, an electromagnetic field will be modified in the presence of conducting surfaces. This possibility offers a valuable opportunity to test in the laboratory some of the geometrical effects that underlie curved space quantum field theory.

In the previous section it was pointed out that the vacuum expectation value of the stress-tensor $T_{\mu\nu}$ formally diverges, even in flat spacetime. If the topology is non-trivial, then the difference in vacuum stress between the non-trivial and trivial spacetimes is finite and nonzero. Similarly, conducting surfaces alter the topology of the field configuration and can lead to a nonzero vacuum stress.

To investigate these possibilities we first consider the simple case of an infinite plane in unbounded four-dimensional Minkowski space (DeWitt 1975, 1979), and a massless scalar field constrained to vanish at the plane's surface (Dirichlet boundary conditions). The field modes will no longer have the form (2.5) because the field reflects from the boundary, which we take to lie along the plane $x_3 = 0$. Instead one must work with modes of the form

$$\sin |k_3| x_3 e^{ik_1 x_1 + ik_2 x_2 - i\omega t} \tag{4.30}$$

which vanish at $x_3 = 0$. The corresponding vacuum state will also be different.

The Green function will no longer be given by (2.79). Its form may be found using the method of images

$$D_{\mathrm{B}}^{(1)}(x, x') = \frac{1}{2\pi^2}\Bigg(\frac{1}{(x_1 - x_1')^2 + (x_2 - x_2')^2 + (x_3 - x_3')^2 - (t - t')^2} - \frac{1}{(x_1 - x_1')^2 + (x_2 - x_2')^2 + (x_3 + x_3')^2 - (t - t')^2}\Bigg) \tag{4.31}$$

which vanishes, by construction, for $x_3 = 0$ and $x_3' = 0$. The first term of this expression is identical to that of unbounded Minkowski space, and diverges quadratically as expected when $x \to x'$; when differentiated to form $\langle 0|T_{\mu\nu}|0\rangle$, as explained in the previous section, (4.31) will yield a quartically divergent expression. The effect of the presence of the boundary can be computed by subtracting (2.79), the Green function for unbounded Minkowski space, from (4.31); i.e., discarding the first term. The remaining term is finite in the limit $x \to x'$. This must now be substituted into the four-dimensional analogue of (4.20). For example, from (2.27) one obtains ($m = 0$)

$$\langle 0|T_{tt}|0\rangle_{\mathrm{B}} = \lim_{\substack{t',x'_1,x'_2,x'_3 \to t,x_1,x_2,x_3 \\ t'',x''_1,x''_2,x''_3 \to t,x_1,x_2,x_3}} \tfrac{1}{4}(\partial_{t''}\partial_{t'} + \partial_{x''_1}\partial_{x'_1} + \partial_{x''_2}\partial_{x'_2}$$

$$+ \partial_{x''_3}\partial_{x'_3})[D_{\mathrm{B}}^{(1)}(x'', x') - D^{(1)}(x'', x')]$$

$$= -1/16\pi^2 x_3^{\,4}. \tag{4.32}$$

Similarly $\langle 0|T_{ii}|0\rangle_{\mathrm{B}} = +(16\pi^2 x_3^4)^{-1}$, all other components being zero.

Far from the boundary ($x_3 \to \infty$) the vacuum stress vanishes as expected. However, the presence of the boundary clearly modifies the vacuum stress in its vicinity. Moreover, this stress actually diverges as the surface is approached ($x_3 \to 0$). Integrating over all space yields an infinite vacuum energy per unit area of the boundary surface, even though we have already subtracted from $\langle T_{\mu\nu}\rangle$ the infinite vacuum energy of unbounded Minkowski space.

The origin of this infinite surface energy is not hard to find. The Green function (4.31) is deliberately constructed to vanish at $x_3 = 0$, even when $x \to x'$. On the other hand, the function $D^{(1)}(x, x')$ given by (2.79) clearly diverges on the boundary when $x \to x'$. So when the Green function is rendered finite by subtracting the latter from the former, the difference will also diverge on the boundary. Inspection of (4.31) shows that the second term taken on its own does indeed diverge at $x_3 = 0$ as $x \to x'$. Thus, the simple device of subtracting the infinite vacuum effects associated with unbounded Minkowski space to remove the quartically divergent vacuum stress works at all spacetime points except those on the boundary, where the quartic divergence persists, in the form of an x_3^{-4} term.

It is clear from this argument that an infinite surface energy will arise quite generally when the field is constrained to vanish on a boundary of some arbitrary shape. However, although $D_{\mathrm{B}}^{(1)} - D^{(1)}$ will diverge on the boundary, it does not necessarily follow that $\langle T_{\mu\nu}\rangle$ will do so, as the latter quantity is constructed from the former by a complicated formula.

To investigate this, consider the general form that $\langle T_{\mu\nu}\rangle$ might have in the vicinity of a single plane boundary at $x_3 = 0$ (DeWitt 1979). On grounds of symmetry, this tensor can only be built out of $\eta_{\mu\nu}$ and $\hat{x}_3{}^\mu \hat{x}_3{}^\nu$ where $\hat{x}_3{}^\mu$ is the unit normal vector to the boundary. Moreover, it can only be a function of x_3. Consequently

$$\langle T^{\mu\nu}\rangle = f(x_3)\eta^{\mu\nu} + g(x_3)\hat{x}_3{}^\mu \hat{x}_3{}^\nu. \tag{4.33}$$

If the covariant conservation condition is imposed (see §6.3)

$$\partial_\mu \langle T^{\mu\nu}\rangle = 0 \tag{4.34}$$

it is concluded that f and g can differ only by a constant. The right-hand side of (4.33) therefore must have the form

$$g(x_3)(\eta^{\mu\nu} + \hat{x}_3{}^{\mu}\hat{x}_3{}^{\nu}) + \alpha\eta^{\mu\nu}$$

where α is a constant. The trace of this quantity is $3g(x_3) + 4\alpha$. If the stress-tensor is required to be traceless, this forces $g(x_3) = -4\alpha/3 = \text{constant}$. As the renormalized vacuum stress must approach zero far from the boundary, the constant must be zero in this case. Hence the renormalized vacuum stress will vanish for a field with a traceless stress-tensor. This is the case for the electromagnetic and neutrino fields, but not for the massless scalar field described by $T_{\mu\nu}$ as given by (2.26).

In §3.2 it was pointed out that in curved space, the scalar wave equation can acquire an additional term ξR. The corresponding stress-tensor operator is given by (3.190). Even in the flat space limit, (3.190) does not reduce to (2.26) if $\xi \neq 0$. (This curious fact is less surprising if it is remembered that $T_{\mu\nu}$ is obtained by varying the metric $g_{\mu\nu}$ in the field Lagrangian. Thus, even if one takes the limit $g_{\mu\nu} \to \eta_{\mu\nu}$ at the end of the calculation, one retains contributions to $T_{\mu\nu}$ associated with the $\xi R\phi^2$ term in (3.24).) The choice $\xi = \frac{1}{6}$ corresponds to a conformally invariant scalar field equation, which implies a traceless $T_{\mu\nu}$:

$$T_{\mu\nu}(\xi = \tfrac{1}{6}) = \tfrac{2}{3}\phi_{,\mu}\phi_{,\nu} - \tfrac{1}{6}\eta_{\mu\nu}\eta^{\sigma\rho}\phi_{,\sigma}\phi_{,\rho} - \tfrac{1}{3}\phi\phi_{;\mu\nu} + \tfrac{1}{12}\eta_{\mu\nu}\phi\Box\phi, \quad (4.35)$$

$$T^{\mu}{}_{\mu} = 0. \quad (4.36)$$

Sometimes (4.35) is called the 'new improved stress-tensor' (Chernikov & Tagirov 1968, Callan, Coleman & Jackiw 1970). For such a scalar field $\langle 0|T_{\mu\nu}|0\rangle = 0$ near a plane boundary.

Although the above argument implies that the infinite surface stress on a plane boundary will not be present for a conformally invariant field, the reasoning depended crucially on the symmetries associated with a flat surface. If the boundary is curved, the divergent surface energy reappears. It may be shown that, in general, as the surface is approached (Deutsch & Candelas 1979, Kennedy, Critchley & Dowker 1980)

$$\langle T_{\mu\nu}\rangle \propto \varepsilon^{-3}\chi_{\mu\nu} + O(\varepsilon^{-2})$$

for a conformally invariant field, where ε is the (small) distance from the surface and $\chi_{\mu\nu}$ the second fundamental form of the boundary. We shall return to this topic, and the physical significance of the infinite surface energy, in §6.6.

Restricting attention for now to plane boundaries and conformally invariant fields, it is easy to generalize the problem to the case where more than one boundary is present. Casimir (1948) considered the vacuum energy associated with the electromagnetic field in the region between two parallel reflecting planes. The restrictions of conservation and tracelessness require that

$$\langle T^{\mu\nu}\rangle = A(\tfrac{1}{4}\eta^{\mu\nu} + \hat{x}_3{}^{\mu}\hat{x}_3{}^{\nu})$$

where A is constant and the planes are again orthogonal to $\hat{x}_3$. On dimensional grounds $A \propto a^{-4}$ where a is the separation distance between the two planes. Calculation (Casimir 1948) shows that

$$A = -\pi^2/180a^4 \tag{4.37}$$

in the vacuum case.

The reason why A is nonzero is because the boundary planes constrain the electromagnetic field modes in the x_3 direction to form a discrete set. In this respect, the field suffers a topological distortion similar to the case of the $R^1 \times S^1$ model considered in §4.1, and the constant A may be calculated by following the method of that section. Alternatively, the $D_{\mathrm{B}}^{(1)}$ Green function may be computed as an infinite image sum (infinite reflections between the planes).

In the *scalar* field case

$$D_{\mathrm{B}}^{(1)}(x,x') = \frac{1}{2\pi^2}\sum_{n=-\infty}^{\infty}\left(\frac{1}{(x_1-x_1')^2+(x_2-x_2')^2+(x_3-x_3'-an)^2-(t-t')^2} - \frac{1}{(x_1-x_1')^2+(x_2-x_2')^2+(x_3+x_3'-an)^2-(t-t')^2}\right) \tag{4.38}$$

which vanishes at x_3 or $x_3' = 0$ and x_3 or $x_3' = a$, as required. The infinite vacuum divergence may be removed by discarding the $n = 0$ term. Use of (4.35) then yields

$$\langle 0|T_{\mu\nu}|0\rangle_{\mathrm{B}} = \frac{-\pi^2}{1440a^4}\begin{bmatrix} 1 & 0 & 0 & 0 \\ 0 & -1 & 0 & 0 \\ 0 & 0 & -1 & 0 \\ 0 & 0 & 0 & 3 \end{bmatrix} \tag{4.39}$$

in Cartesian coordinates. The coefficient of (4.39) is one-half the value of the electromagnetic case on account of the fact that the scalar field has only half as many modes (i.e., one rather than two polarization states).

If the quantum field is not in a vacuum state, but at finite temperature T, then we may replace $D_B^{(1)}$ in (4.38) by a thermal Green function, i.e., replace $t-t'$ by $t-t'-im\beta$ (where $\beta = 1/k_B T$) and sum over all m from $-\infty$ to $+\infty$. The corresponding energy density between the planes is

$$-kTa^{-3}f'(\xi)+a^{-4}f(\xi)$$

where

$$f(\xi) = -\frac{1}{8\pi^2}\sum_{m,n=1}^{\infty}\frac{(2\xi)^4}{[n^2+4\xi^2m^2]^2}$$

and $\xi = k_B Ta$. In the limit of small Ta, the temperature correction to the energy density is

$$(kT)^3\zeta(3)/2\pi a$$

where ζ is a zeta-function, while for large Ta it is

$$\tfrac{1}{30}\pi^2(kT)^4.$$

The electromagnetic values are twice these; a detailed treatment has been given by Brown & Maclay (1969).

The presence of a vacuum stress between parallel boundary planes implies that there will be a force of attraction between two electrically neutral conducting surfaces. Casimir's calculation (see (4.37)) indicates a force per unit area of the surfaces given by

$$F = -\frac{\partial E}{\partial a} = \frac{-\pi^2}{240a^4}.$$

Forces arising in a manner similar to this have been detected experimentally (see, for example, Sparnaay 1958 and Tabor & Winterton 1969). The 'Casimir effect' has since been the subject of much study, extremely detailed analyses having been given by Lifschitz (1955) and Balian & Duplantier (1977, 1978).

Another plane boundary problem has been investigated by Dowker & Kennedy (1978), and Deutsch & Candelas (1979). They find for the renormalized conformal scalar field vacuum stress-tensor in the wedge between two inclined planes

$$\langle T_{\mu\nu}\rangle = \frac{1}{1440r^4\alpha^2}\left(\frac{\pi^2}{\alpha^2}-\frac{\alpha^2}{\pi^2}\right)\begin{bmatrix}1 & 0 & 0 & 0\\ 0 & -1 & 0 & 0\\ 0 & 0 & 3 & 0\\ 0 & 0 & 0 & -1\end{bmatrix}$$

where α is the angle between the planes, r is the distance from the axis of intersection and cylindrical polar coordinates are used for the spacelike part of the tensor.

The electromagnetic case yields (Deutsch & Candelas 1979)

$$\frac{1}{720\pi^2 r^4}\left(\frac{\pi^2}{\alpha^2}+11\right)\left(\frac{\pi^2}{\alpha^2}-1\right)\begin{bmatrix}1 & 0 & 0 & 0\\ 0 & -1 & 0 & 0\\ 0 & 0 & 3 & 0\\ 0 & 0 & 0 & -1\end{bmatrix}.$$

In both cases the finite answer may be obtained by the usual subtraction of the infinite vacuum stress associated with unbounded Minkowski space.

4.4 Moving mirrors

Until now, the flat spacetime examples have all involved static situations, i.e., the disturbance of the quantum state induces vacuum energy and stress, but no particles are created. If the parallel plates discussed in the preceding section in connection with the Casimir effect were moved rapidly around, the irreversible production of entropy would occur and new quanta would appear between the plates. Even the motion of a single reflecting boundary (mirror) can create particles.

We shall treat this problem in detail for a two-dimensional spacetime where the conformal triviality can be exploited. Explicit examples for four dimensions are rather sparse (Candelas & Raine 1976, 1977*b*). We follow the treatment of Fulling & Davies (1976) and Davies & Fulling (1977*a*).

Suppose the mirror (which in two dimensions degenerates to a reflecting point) moves along the trajectory

$$x = z(t), \quad |\dot{z}(t)| < 1, \tag{4.40}$$

$$z(t) = 0, \quad t < 0.$$

A massless scalar field ϕ, satisfying the field equation

$$\Box\phi = \frac{\partial^2\phi}{\partial u\partial v} = 0 \tag{4.41}$$

with reflection boundary condition

$$\phi(t, z(t)) = 0 \tag{4.42}$$

has a set of mode solutions

$$u_k^{\text{in}}(u, v) = \mathrm{i}(4\pi\omega)^{-\frac{1}{2}}(\mathrm{e}^{-\mathrm{i}\omega v} - \mathrm{e}^{-\mathrm{i}\omega(2\tau_u - u)}) \tag{4.43}$$

where $\omega = |k|$ and τ_u is determined implicitly by the trajectory (4.40) through

$$\tau_u - z(\tau_u) = u. \tag{4.44}$$

The solutions (4.43) apply to the right of the mirror. The incoming (left-moving) waves $e^{-i\omega v}$ correspond to standard exponential modes all the way from $\mathscr{I}^-$ (see fig. 11) to the mirror surface, but the right-moving (reflected) waves are complicated because of the Doppler shift suffered during the reflection from the moving mirror. This asymmetry between u and v is chosen to correspond to the usual retarded boundary condition (no incoming field quanta from $\mathscr{I}^-$).

In the region to the right of the mirror the field ϕ, constrained by (4.42), can be expanded in terms of the modes (4.43) as

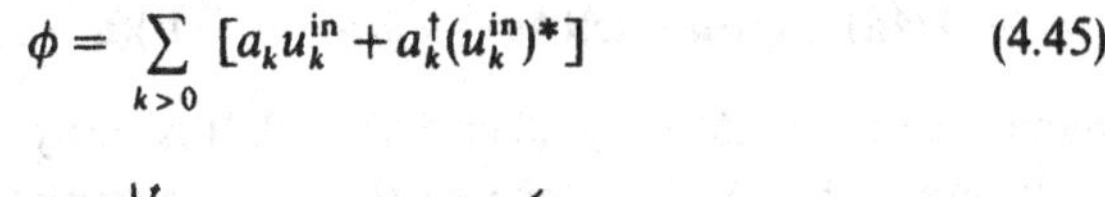

$$\phi = \sum_{k>0} [a_k u_k^{\text{in}} + a_k^\dagger (u_k^{\text{in}})^*] \tag{4.45}$$

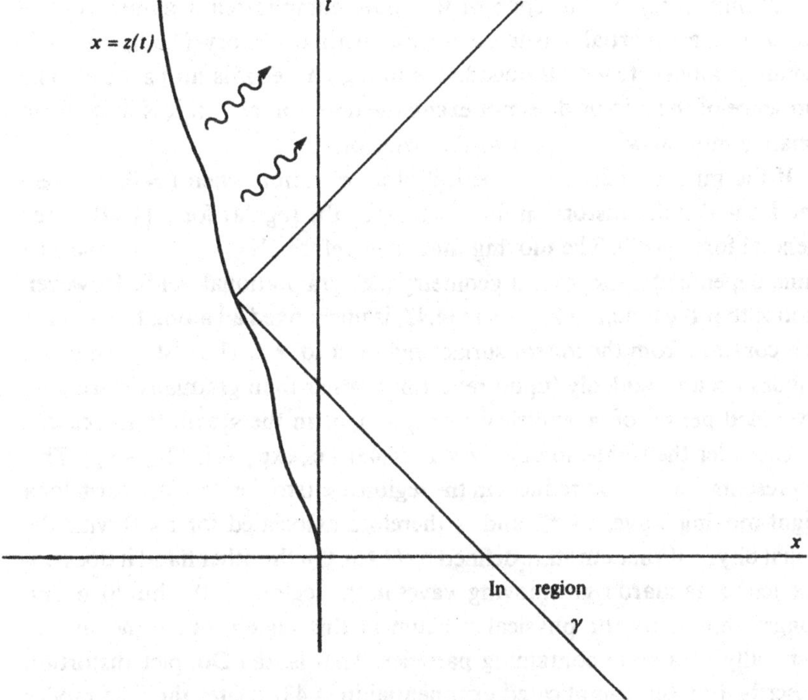

Fig. 11. Radiation by a moving mirror. The scalar field is constrained to vanish on the boundary $x = z(t)$. Null rays incoming from $\mathscr{I}^-$, such as γ, reflect from the mirror to $\mathscr{I}^+$. Wave modes reflected after $t = 0$ suffer a Doppler shift in frequency due to the 'mirror' motion.

and an 'in' vacuum defined by

$$a_k|0, \text{in}\rangle = 0. \tag{4.46}$$

We have labelled this state and the modes (4.43) by 'in' since in the in region, $t \leq 0$, the mirror is at rest at $z = 0$ in the (u, v) frame. Then $\tau_u = u$ and (4.43) reduces to

$$u_k^{\text{in}} = \text{i}(4\pi\omega)^{-\frac{1}{2}}(\text{e}^{-\text{i}\omega v} - \text{e}^{-\text{i}\omega u}) = (\pi\omega)^{-\frac{1}{2}} \sin \omega x\, \text{e}^{-\text{i}\omega t}, \quad t \leq 0, \tag{4.47}$$

which is a positive frequency mode with respect to Minkowski time t.

The state $|0, \text{in}\rangle$ may be considered to be void of particles for $t < 0$. Indeed, the Wightman function in the in region is

$$\begin{aligned}
&\langle \text{in}, 0|\phi(x)\phi(x')|0, \text{in}\rangle \\
&= -(1/4\pi)\ln[(\Delta u - \text{i}\varepsilon)(\Delta v - \text{i}\varepsilon)/(v - u' - \text{i}\varepsilon)(u - v' - \text{i}\varepsilon)]
\end{aligned} \tag{4.48}$$

where $\Delta u = u - u', \Delta v = v - v'$ and $t, t' < 0$. It is easily verified by substituting (4.48) into (3.55) that in spite of the more complicated argument of the logarithm an inertial particle detector, with trajectory (3.52), which is adiabatically switched off outside the in region, records no particles. The presence of the mirror does not excite the detector, even if it is in uniform relative motion with respect to the detector.

If the mirror undergoes a period of acceleration when $t > 0$, the field modes will suffer distortion for $u > 0$, from the regular form (4.47) to the general form (4.43). The moving mirror therefore plays the same rôle as a time-dependent background geometry (i.e., gravitational field). However, notice that the function $2\tau_u - u$ in (4.43) is unchanged all along the null ray $u =$ constant from the mirror surface right out to $\mathscr{I}^+$. The distortion of the modes occurs suddenly (upon reflection) rather than gradually during an extended period of geometrical disruption, as in the gravitational case.

Consider the right-moving piece of (4.43), i.e., $\exp[-\text{i}\omega(2\tau_u - u)]$. This represents a wave that reduces in the region $u < 0$ to the standard form for a right-moving wave, $\text{e}^{-\text{i}\omega u}$, and is therefore associated for $t < 0$ with the usual physical vacuum state defined by (4.46). On the other hand it does not reduce to standard right-moving waves in the region $u > 0$. Thus $|0, \text{in}\rangle$ no longer represents the physical vacuum in this region, but rather it will generally be a state containing particles. That is, the Doppler distortion described by the complicated exponential in (4.43) excites the field modes and causes particles to appear. Physically this is described by saying that the moving mirror creates particles, which stream away to the right along the null rays $u =$ constant.

To confirm the physical reality of these quanta, one can examine the experiences of an inertial particle detector in the region $u > 0$. The Wightman function in the in vacuum is

$$D^+(u, v; u', v')$$
$$= -(1/4\pi)\ln[(p(u) - p(u') - i\varepsilon)(v - v' - i\varepsilon)/(v - p(u') - i\varepsilon)(p(u) - v' - i\varepsilon)] \tag{4.49}$$

where we have defined

$$p(u) = 2\tau_u - u. \tag{4.50}$$

For a general $x(t)$, hence $p(u)$, the Fourier transform of (4.49) will be nonzero, so when D^+ is substituted in (3.55) it predicts a nonzero response from the detector.

As an illustration, consider mirror trajectories with the asymptotic form

$$z(t) \to -t - Ae^{-2\kappa t} + B \text{ as } t \to \infty \tag{4.51}$$

where A, B, κ are constants > 0. The behaviour at earlier times will be irrelevant as we shall only consider the particle flux in the asymptotic region as $t \to \infty$. However, one example of a trajectory with asymptotic form (4.51), which joins smoothly (C^1) onto a static trajectory for $t < 0$, is $z(t) = -\ln(\cosh \kappa t)$.

The class (4.51) of trajectories is of special geometrical interest because only the null rays with $v < B$ can reflect from the mirror. All rays with $v > B$ pass undisturbed to the left-hand portion of $\mathscr{I}^+$ (see fig. 12). The ray $v = B$ therefore acts as a sort of horizon, and equispaced lines of constant u, when traced back, through reflection, to $\mathscr{I}^-$ crowd up along $v = B$. In chapter 8 we shall see that an identical situation occurs in the collapse of a star to form a black hole.

From (4.44) and (4.51) one obtains

$$p(u) \equiv 2\tau_u - u \to B - Ae^{-\kappa(u+B)} \text{ as } u \to \infty. \tag{4.52}$$

We assume that the detector moves along the trajectory

$$x = x_0 + wt, \quad w \text{ constant}. \tag{4.53}$$

The expression (4.49) for D^+ may be written as a sum of four terms by factorizing the argument of the logarithm. When $p(u)$ is chosen to be (4.52) and the resulting D^+ is used in (3.55) for the detector response function (assuming the detector to be switched off adiabatically at early times), three of the terms give vanishing contributions. The remaining term, involving

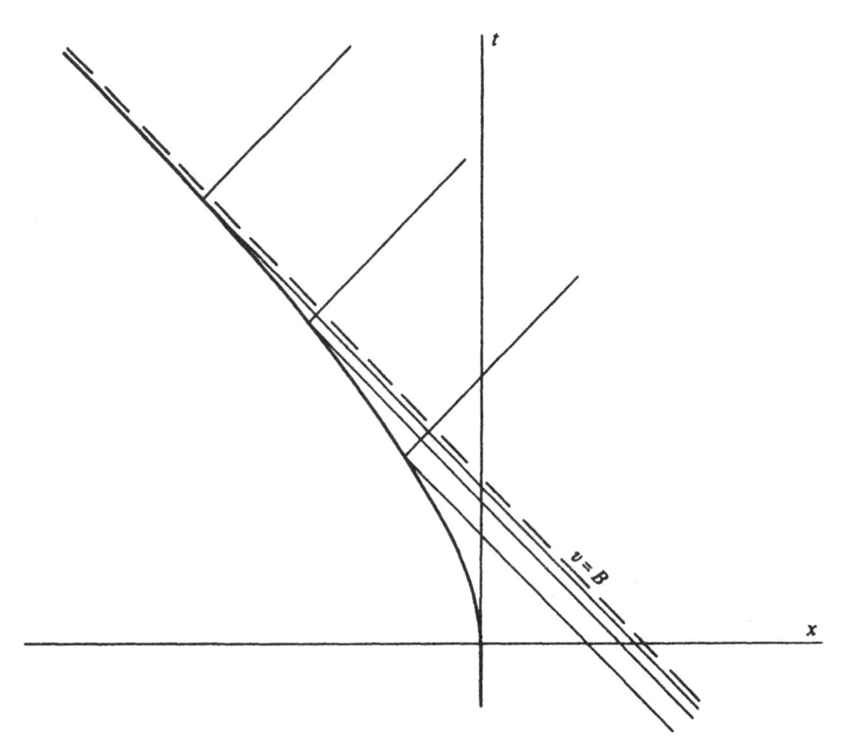

Fig. 12. The mirror trajectory, $z(t) = -t - Ae^{-2\kappa t} + B$, is asymptotic to the null ray, $v = B$. Late time surfaces of constant phase from $\mathscr{I}^+$ (equispaced null rays $u = \text{constant}$) crowd up along $v = B$ when reflected back off the mirror. Advanced (left-moving) rays later than $v = B$ do not intersect the mirror at all. The radiation excited from the vacuum by this mirror trajectory has a thermal spectrum at late times.

$p(u') - p(u)$, yields

$\mathscr{F}(E)$/unit time

$$= -(1/4\pi)\int_{-\infty}^{\infty} e^{-iE\Delta\tau}\ln\{\sinh[\tfrac{1}{2}\kappa((1-w)/(1+w))^{\frac{1}{2}}\Delta\tau - i\varepsilon]\}\mathrm{d}\Delta\tau. \tag{4.54}$$

Using the identities

$$\sinh x = x\prod_{m=1}^{\infty}(m\pi - ix)(m\pi + ix)/(m\pi)^2$$

and

$$\int_{\sigma}^{\infty}\frac{e^{-i\omega x}}{\omega(e^{\beta\omega}-1)}\mathrm{d}\omega \underset{\sigma\to 0}{\rightarrow} -\ln\left[\prod_{m=1}^{\infty}\sigma e^{\gamma}(\beta m + ix)\right]$$

where γ is Euler's constant, the right-hand side of (4.54) may be reduced to

$$\frac{1}{2\pi}\int_{-\infty}^{\infty} e^{-iE\Delta\tau} \int_{0}^{\infty} \frac{\cos\{\frac{1}{2}\omega\kappa\Delta\tau[(1-w)/(1+w)]^{\frac{1}{2}}\}}{\omega(e^{\pi\omega}-1)} d\omega d\Delta\tau.$$

Interchanging the order of integration gives a sum of two δ-functions, only one of which contributes to the ω integral giving

$$\mathscr{F}(E)/\text{unit time} = \frac{1}{E(e^{E/k_B T}-1)}, \tag{4.55}$$

where

$$k_B T = (\kappa/2\pi)[(1-w)/(1+w)]^{\frac{1}{2}}. \tag{4.56}$$

Inspection of this result shows that the detector responds to a flux of particles from the mirror that is constant in time and has the spectrum of thermal radiation travelling in the u-direction with a temperature given by (4.56). The factor $[(1-w)/(1+w)]^{\frac{1}{2}}$ is the Doppler shift due to the motion of the detector at velocity w relative to the radiation, resulting in a red- or blue-shifting of the temperature. If the detector is at rest relative to the privileged frame defined by the mirror in its initially static phase (i.e., $w = 0$) then

$$k_B T = \kappa/2\pi. \tag{4.57}$$

The thermal nature of the flux of radiation from a mirror moving along the trajectory (4.51) can also be deduced by explicitly evaluating the Bogolubov transformation between the in and the out modes. The modes u_k^{in} given by (4.43) assume a simple form on $\mathscr{I}^-$, but after reflection from the mirror they are generally complicated functions on $\mathscr{I}^+$. Conversely one can define modes u_k^{out} that are simple complex exponentials on $\mathscr{I}^+$, but complicated functions in the in region, i.e., on $\mathscr{I}^-$. The Bogolubov transformation between u_k^{in} and u_k^{out} can be evaluated on any spacelike surface; in the case of the asymptotically null trajectory (4.51) it is convenient to evaluate the transformation at $t = 0$, in the in region. The modes u_k^{in} then assume their simple form (4.47), but u_k^{out} will be complicated.

To determine the form of u_k^{out} at $t = 0$ we note that standard right-moving exponential waves $e^{-i\omega u}$ on $\mathscr{I}^+$ will, when traced back in time and reflected off the mirror into the in region, assume the form of complicated left-moving waves, $e^{-i\omega f(v)}$. Portions of the waves $e^{-i\omega u}$ at late time u will correspond to left-moving waves on $\mathscr{I}^-$ that crowd up along the null asymptote $v = B$. Thus $f(v)$ will be a rapidly varying function of v in this region. By symmetry, the function f will be the inverse of the function p defined by (4.52); in this case

$$f(v) \sim -\kappa^{-1} \ln[(B-v)/A] - B, \quad v < B, \tag{4.58}$$

in the region $v \to B$. We are not interested here in the left-moving modes in the region $v > B$, as these merely continue undisturbed to the left as far as $\mathscr{I}^+$, and do not contribute to the production of the thermal radiation from the surface of the mirror. Hence we put $u_k^{\text{out}} = 0$ for $v \geq B$.

We therefore wish to evaluate the Bogolubov transformation between u_k^{in} and that portion of u_k^{out} that corresponds to right-moving waves $e^{-i\omega u}$ travelling towards $\mathscr{I}^+$, i.e., the portion of $e^{-i\omega f(v)}$ in the region $v < B$. We use the scalar products (3.36) (with $\bar{u} = u^{\text{in}}$ and $u = u^{\text{out}}$), integrating along $t = 0$ as far as $v = B$ only. After an integration by parts, one obtains

$$\left.\begin{matrix}\alpha_{\omega'\omega} \\ \beta_{\omega'\omega}\end{matrix}\right\} = \pm(2\pi)^{-1} \mathrm{i}(\omega'/\omega)^{\frac{1}{2}} \int_0^B \mathrm{e}^{\pm \mathrm{i}\omega f(x) - \mathrm{i}\omega' x} \mathrm{d}x. \tag{4.59}$$

Because $f(x)$ is rapidly varying near $x = B$, most of the contribution to the integral comes from this region, so we have discarded the boundary term in (4.59) at $x = 0$. We may also approximate (4.59) by using the asymptotic form (4.58) for all x. The resulting Bogolubov coefficients will then be good approximations in describing the radiation flux at late time (large u) in the out region. The integration may now be performed explicitly in terms of incomplete Γ-functions. However, the rapidly varying waveform near $v = B$ represents very high frequencies ω', so in the above approximation we may let $\omega' \to \infty$ and recover an ordinary Γ-function:

$$\left.\begin{matrix}\alpha_{\omega'\omega} \\ \beta_{\omega'\omega}\end{matrix}\right\} = \mp(4\pi^2\omega\omega')^{-\frac{1}{2}} \mathrm{e}^{\pm\pi\omega/2\kappa} \mathrm{e}^{\pm \mathrm{i}\omega D - \mathrm{i}\omega' B} (\omega')^{\pm \mathrm{i}\omega/\kappa} \Gamma(1 \mp \mathrm{i}\omega/\kappa) \tag{4.60}$$

where $D = \kappa^{-1} \ln A - B$, from which it follows that

$$|\beta_{\omega'\omega}|^2 = \frac{1}{2\pi\kappa\omega'}\left(\frac{1}{\mathrm{e}^{\omega/k_B T} - 1}\right) \tag{4.61}$$

with $k_B T$ given by (4.57). This result is in agreement with the spectrum displayed in (4.55).

Note that

$$\langle 0, \mathrm{in}|N_\omega|\mathrm{in}, 0\rangle = \int_0^\infty |\beta_{\omega'\omega}|^2 \mathrm{d}\omega' \tag{4.62}$$

diverges logarithmically. This is because if the mirror continues to accelerate for all time then the steady flux of radiation will accumulate an infinite number of quanta per mode. The result may be converted into a number of quanta per $d\omega$ per unit time by constructing finite wave packets rather than plane wave modes. This topic will be discussed further in chapter 8.

As another illustration of the computation of a Bogolubov transformation in the moving mirror system we shall investigate the hyperbolic trajectory of a mirror which recedes with uniform acceleration. We take the trajectory to be

$$\begin{aligned} z(t) &= B - (B^2 + t^2)^{\frac{1}{2}}, \quad t > 0 \\ &= 0, \qquad\qquad\qquad t < 0 \end{aligned} \tag{4.63}$$

where B is a constant. This is qualitatively similar to the case shown in fig. 12, also having a null asymptote along $v = B$. Again the out modes, when traced back to $t = 0$, have the form $e^{-i\omega f(v)}$ for $v < B$. Thus (4.59) still gives the Bogolubov coefficients, but with $f(x)$ appropriate to the trajectory (4.63). One finds $f(x) = Bx/(B - x)$, from which it follows that

$$\alpha_{\omega'\omega} = i(B/\pi)e^{-i(\omega+\omega')B}K_1(2iB(\omega\omega')^{\frac{1}{2}}) \tag{4.64}$$

$$\beta_{\omega'\omega} = (B/\pi)e^{i(\omega-\omega')B}K_1(2B(\omega\omega')^{\frac{1}{2}}). \tag{4.65}$$

where K_1 is a modified Bessel function. We shall return to this example in §7.1.

It is also possible to treat in detail particle production in the region between two moving mirrors (Moore 1970, Fulling & Davies 1976).

4.5 Quantum field theory in Rindler space

Two examples have now been given in which a detector responds to radiation with a thermal spectrum. In §3.3 it was shown that a uniformly accelerating detector perceives the usual Minkowski vacuum state to be a thermal bath of radiation, whereas in the previous section we saw how an inertial detector responds to a thermal flux of radiation streaming away from a mirror that recedes along the non-uniformly accelerating trajectory (4.51). These two situations are physically rather distinct, and it is usual to regard the thermal radiation in the latter case as in some sense more physical than in the former. This point of view is supported by a computation of $\langle T_{\mu\nu} \rangle$ in the two cases (see §7.1). In spite of their rather different physical status, the two examples are closely related geometrically. The conformal transformation that 'straightens' out the trajectory (4.51) simultaneously converts the straight (i.e., inertial) detector trajectory to a hyperbola, i.e., that of a uniformly accelerating detector (Davies & Fulling 1977*a*).

The thermal radiation detected by a uniformly accelerated system may be deduced by an entirely different analysis that throws important light on the intimate association between the quantum particle concept and the causal and topological structure of spacetime that will prove useful when we come to discuss quantum black holes. (For further details on this section see, for example, Fulling 1973, 1977, Davies 1975, Unruh 1976, Candelas & Deutsch 1977, Dowker 1977, Troost & van Dam 1977 and Horibe 1979.)

Consider two-dimensional Minkowski space with metric (3.8) or (3.10) i.e.,

$$ds^2 = d\bar{u}\, d\bar{v} = dt^2 - dx^2. \tag{4.66}$$

(A bar has been appended to the Minkowski space null coordinates $\bar{u}$, $\bar{v}$ for ease of comparison with later results.) Under the following coordinate transformation

$$t = a^{-1}e^{a\xi} \sinh a\eta \tag{4.67}$$

$$x = a^{-1}e^{a\xi} \cosh a\eta, \tag{4.68}$$

$a =$ constant > 0 and $-\infty < \eta$, $\xi < \infty$, or equivalently

$$\bar{u} = -a^{-1}e^{-au} \tag{4.69}$$

$$\bar{v} = a^{-1}e^{av} \tag{4.70}$$

where $u = \eta - \xi$, $v = \eta + \xi$, (4.66) becomes

$$ds^2 = e^{2a\xi}du\, dv = e^{2a\xi}(d\eta^2 - d\xi^2). \tag{4.71}$$

The coordinates (η, ξ) cover only a quadrant of Minkowski space, namely the wedge $x > |t|$ shown in fig. 13. Lines of constant η are straight $(x \propto t)$ while lines of constant ξ are hyperbolae

$$x^2 - t^2 = a^{-2}e^{2a\xi} = \text{constant}. \tag{4.72}$$

They therefore represent the world lines of uniformly accelerated observers treated in §3.3. Comparison of (3.62) with (4.72) shows that

$$ae^{-a\xi} = \alpha^{-1} = \text{proper acceleration}. \tag{4.73}$$

Thus, lines of large positive ξ (far from $x = t = 0$) represent weakly accelerated observers, while the hyperbolae that closely approach $x = t = 0$ have large negative ξ and hence a high proper acceleration. All the

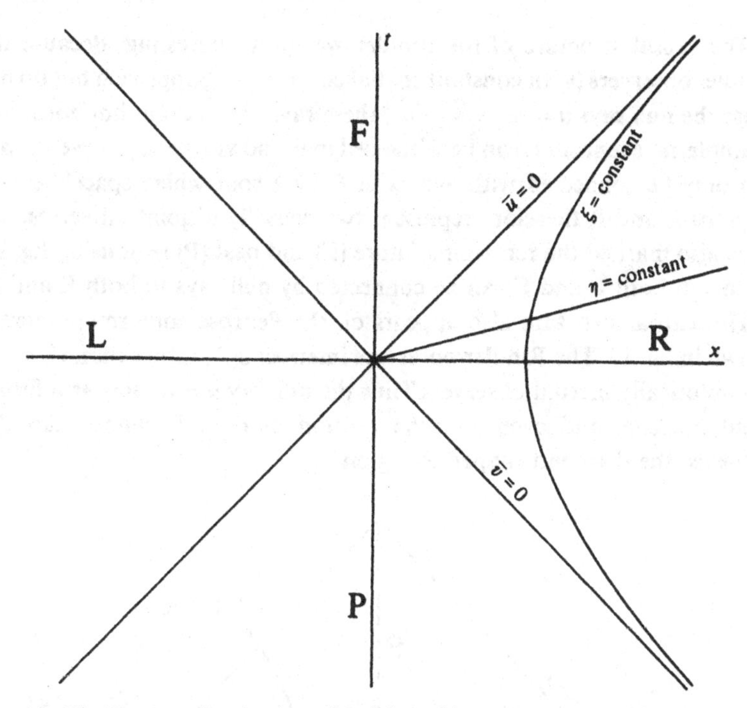

Fig. 13. Rindler coordinatization of Minkowski space. In R and L, time coordinates $\eta=$ constant are straight lines through the origin, space coordinates $\xi=$ constant are hyperbolae (corresponding to the world lines of uniformly accelerated observers) with null asymptotes $\bar{u}=0, \bar{v}=0$, which act as event horizons. The four regions R, L, F and P must be covered by separate coordinate patches. Rindler coordinates are non-analytic across $\bar{u}=0$ and $\bar{v}=0$.

hyperbolae are asymptotic to the null rays $\bar{u}=0, \bar{v}=0$ (or $u=\infty, v=-\infty$), which means that the accelerated observers approach the speed of light as $\eta \to \pm\infty$. These observers' proper time τ is related to ξ and η by

$$\tau = e^{a\xi}\eta. \tag{4.74}$$

The system (η, ξ) is known as the Rindler coordinate system (Rindler 1966), and the portion $x > |t|$ of Minkowski space is called the Rindler wedge. Uniformly accelerated observers are sometimes referred to as Rindler observers.

A second Rindler wedge $x < |t|$ may be obtained by reflecting the first in the t- and then the x-axis. This is achieved by changing the signs of the right-hand sides of the transformation equations (4.67)–(4.70). We label the left- and right-hand wedges by L and R respectively. Note that the sign reversals in L mean that, crudely speaking, the direction of time there is reflected, i.e., increasing t corresponds to decreasing η.

The causal structure of the Rindler wedge is interesting. Because the Rindler observers (with constant spatial coordinate ξ) approach but do not cross the null rays $u = \infty$, $v = -\infty$, these rays act as event horizons. For example, no events in L can be witnessed in R and vice versa, as events in L can only be connected with events in R by a somewhere spacelike line. Regions L and R therefore represent two causally disjoint universes. We have also marked the remaining future (F) and past (P) regions on fig. 13. Events in both P and F can be connected by null rays to both L and R.

This causal structure also appears on the Penrose conformal diagram shown in fig. 14. The Rindler observers intersect $\mathscr{I}^{\pm}$, rather than $i^{\pm}$ as do asymptotically inertial observers. Thus the null ray $u = \infty$ acts as a future event horizon, and events in the portion marked F cannot causally influence the diamond shaped R region.

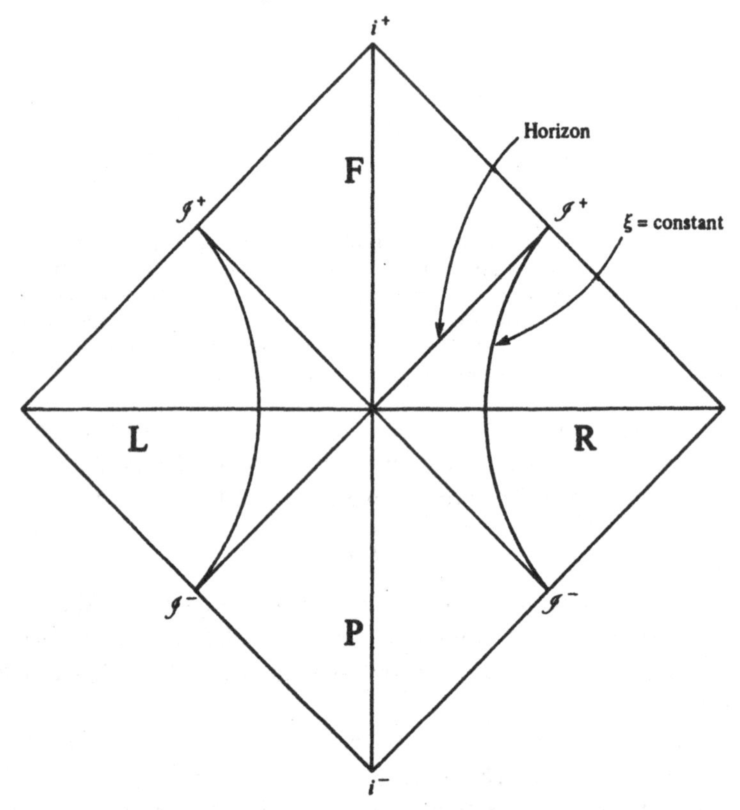

Fig. 14. Conformal diagram of Rindler system. The regions R, L, F and P are represented by diamond shaped regions. The timelike lines $\xi = \text{constant}$ intersect $\mathscr{I}^{\pm}$, not $i^{\pm}$. Clearly, events in F cannot be witnessed in R, so the null ray $\bar{u} = 0$ ($u = \infty$) acts as an event horizon.

Consider the quantization of a massless scalar field ϕ in two-dimensional Minkowski spacetime. The wave equation

$$\Box\phi \equiv \left(\frac{\partial^2}{\partial t^2} - \frac{\partial^2}{\partial x^2}\right)\phi \equiv \frac{\partial^2 \phi}{\partial \bar{u}\partial \bar{v}} = 0 \tag{4.75}$$

possesses standard orthonormal mode solutions

$$\bar{u}_k = (4\pi\omega)^{-\frac{1}{2}} e^{ikx - i\omega t} \tag{4.76}$$

(i.e., (2.11) with $n = 2$) where $\omega = |k| > 0$ and $-\infty < k < \infty$. These modes are positive frequency with respect to the timelike Killing vector ∂_t, in that they satisfy

$$\mathscr{L}_{\partial_t} \bar{u}_k = -i\omega \bar{u}_k. \tag{4.77}$$

The modes with $k > 0$ consist of right-moving waves

$$(4\pi\omega)^{-\frac{1}{2}} e^{-i\omega\bar{u}} \tag{4.78}$$

along the rays $\bar{u} =$ constant, while for $k < 0$ one has left-moving waves along $\bar{v} =$ constant

$$(4\pi\omega)^{-\frac{1}{2}} e^{-i\omega\bar{v}}. \tag{4.79}$$

The Minkowski vacuum state $|0_M\rangle$ and associated Fock space are constructed by expanding ϕ in terms of $\bar{u}_k$ as explained in §2.2.

In the Rindler regions R and L one may adopt an alternative quantization prescription, based not on the modes $\bar{u}_k$, but their Rindler counterparts u_k. The metric (4.71) is conformal to the whole of Minkowski space, for under the conformal transformation $g_{\mu\nu} \to e^{-2a\xi} g_{\mu\nu}$, (4.71) reduces to $d\eta^2 - d\xi^2$ with $-\infty < \eta, \xi < \infty$. Because the wave equation is conformally invariant, we can write it in Rindler coordinates as (see (3.150))

$$e^{2a\xi}\Box\phi = \left(\frac{\partial^2}{\partial \eta^2} - \frac{\partial^2}{\partial \xi^2}\right)\phi = \frac{\partial^2 \phi}{\partial u \partial v} = 0, \tag{4.80}$$

for which there exist mode solutions

$$u_k = (4\pi\omega)^{-\frac{1}{2}} e^{ik\xi \pm i\omega\eta} \tag{4.81}$$

$$\omega = |k| > 0, \quad -\infty < k < \infty.$$

The upper sign in (4.81) applies in region L, the lower sign in region R. The presence of the sign change can either be regarded as due to the 'time reversal' in L, or due to the fact that a right-moving wave in R moves towards increasing values of ξ, while in L it moves towards decreasing values of ξ. In any case, the modes (4.81) are those which satisfy the normalization condition (3.29). They are positive frequency with respect to the timelike Killing vector $+\partial_\eta$ in R and $-\partial_\eta$ in L, satisfying

$$\mathscr{L}_{\pm\partial_\eta} u_k = -\mathrm{i}\omega u_k \tag{4.82}$$

(in R and L respectively) in place of (4.77).

The fact that (4.81) has the same functional form as (4.76) is a consequence of the conformal triviality of the system (i.e., Rindler space is conformal to Minkowski space and the wave equation (2.1) is conformally invariant).

Define

$$\left.\begin{aligned} {}^{\mathrm{R}}u_k &= (4\pi\omega)^{-\frac{1}{2}} \mathrm{e}^{ik\xi - \mathrm{i}\omega\eta}, && \text{in R} \\ &= 0, && \text{in L} \end{aligned}\right\} \tag{4.83}$$

$$\left.\begin{aligned} {}^{\mathrm{L}}u_k &= (4\pi\omega)^{-\frac{1}{2}} \mathrm{e}^{ik\xi + \mathrm{i}\omega\eta}, && \text{in L} \\ &= 0, && \text{in R.} \end{aligned}\right\} \tag{4.84}$$

The set (4.83) is complete in the Rindler region R, while (4.84) is complete in L, but neither set separately is complete on all of Minkowski space. However, both sets together are so complete, and lines η = constant taken across both R and L are Cauchy surfaces for the whole spacetime. Therefore, the modes (4.83) and (4.84) can also be analytically continued (Boulware 1975*a*, *b*) into regions F and P (a becomes imaginary). Thus, these Rindler modes are every bit as good a basis for quantizing the ϕ field as the Minkowski space basis (4.76).

The field may be expanded in either set

$$\phi = \sum_{k=-\infty}^{\infty} (a_k \bar{u}_k + a_k^\dagger \bar{u}_k^*) \tag{4.85}$$

(cf. (2.17)), or

$$\phi = \sum_{k=-\infty}^{\infty} (b_k^{(1)} {}^{\mathrm{L}}u_k + b_k^{(1)\dagger} {}^{\mathrm{L}}u_k^* + b_k^{(2)} {}^{\mathrm{R}}u_k + b_k^{(2)\dagger} {}^{\mathrm{R}}u_k^*), \tag{4.86}$$

yielding two alternative Fock spaces, and two vacuum states, $|0_{\mathrm{M}}\rangle$ or $|0_{\mathrm{R}}\rangle$ (the subscripts M and R standing for 'Minkowski' and 'Rindler' respectively) defined by

$$a_k|0_{\mathrm{M}}\rangle = 0 \tag{4.87}$$

or

$$b_k^{(1)}|0_{\mathrm{R}}\rangle = b_k^{(2)}|0_{\mathrm{R}}\rangle = 0. \tag{4.88}$$

The fact that these vacuum states are not equivalent is obvious by inspection of the structure of the Rindler modes (4.81). Because of the sign change in the exponent at $\bar{u} = \bar{v} = 0$ (the 'crossover point' between L and R), the functions ${}^{\mathrm{R}}u_k$ do not go over smoothly to ${}^{\mathrm{L}}u_k$ as one passes from R to L. This means that as one passes from $\bar{u} < 0$ to $\bar{u} > 0$ (or $\bar{v} < 0$ to $\bar{v} > 0$) the right- (or left-) moving modes are non-analytic at this point. In contrast, the positive frequency Minkowski modes (4.78) and (4.79) are analytic not only on the real $\bar{u}$ (or $\bar{v}$) axis, but they are also analytic and bounded in the entire lower half of the complex $\bar{u}$ (or $\bar{v}$) planes. This analyticity property remains true of *any* pure positive frequency functions, i.e., any linear superposition of these positive frequency Minkowski modes. Hence the Rindler modes, by virtue of their non-analyticity at $\bar{u} = \bar{v} = 0$, cannot be a combination of pure positive frequency Minkowski modes, but must also contain negative frequencies. It will be recalled from §3.2 that the mixing of positive and negative frequencies implies that the vacuum states cannot be the same, i.e., the vacuum associated with one set of modes contains particles associated with the other set of modes.

To determine what Rindler particles are present in the Minkowski vacuum, we must determine the Bogolubov transformation between the two sets of modes. This may be achieved using (3.36), which are essentially Fourier transforms of the Rindler modes. An alternative, more elegant, method due to Unruh (1976), is to note that although ${}^{\mathrm{L}}u_k$ and ${}^{\mathrm{R}}u_k$ are non-analytic, the two (un-normalized) combinations

$$ {}^{\mathrm{R}}u_k + \mathrm{e}^{-\pi\omega/a}\, {}^{\mathrm{L}}u^*_{-k} \tag{4.89} $$

and

$$ {}^{\mathrm{R}}u^*_{-k} + \mathrm{e}^{\pi\omega/a}\, {}^{\mathrm{L}}u_k \tag{4.90} $$

are analytic and bounded, both for all real $\bar{u}$, $\bar{v}$, and everywhere in the lower half complex $\bar{u}$ and $\bar{v}$ planes.

To see this, one can write (4.89) and (4.90) in Minkowski coordinates, where they are proportional to

$$ \begin{aligned} &\bar{u}^{\mathrm{i}\omega/a}, \quad k > 0 \\ &\bar{v}^{-\mathrm{i}\omega/a}, \quad k < 0 \end{aligned} \tag{4.91} $$

and

$$ \begin{aligned} &\bar{v}^{\mathrm{i}\omega/a}, \quad k > 0 \\ &\bar{u}^{-\mathrm{i}\omega/a}, \quad k < 0 \end{aligned} \tag{4.92} $$

($\omega = |k|$) respectively, for all $\bar{u}$, $\bar{v}$ in the range $-\infty$ to ∞ (i.e., in both L and

R); they are clearly analytic across $\bar{u} = \bar{v} = 0$. They are also analytic in the lower-half complex $\bar{u}$ and $\bar{v}$ planes if the branch cut of the complex powers in (4.91) and (4.92) is taken to lie in the upper-half plane (i.e., $\ln(-1) = -i\pi$, which fixes the signs of the exponents $e^{\pm\pi\omega/a}$ in (4.89) and (4.90)).

Because the modes (4.89) and (4.90) share the positive frequency analyticity properties of the Minkowski modes $\bar{u}_k$, they must also share a common vacuum state $|0_M\rangle$ (see remark on page 47). Thus, instead of (4.85) we can expand ϕ in terms of (4.89) and (4.90) as

$$\phi = \sum_{k=-\infty}^{\infty} [2\sinh(\pi\omega/a)]^{-\frac{1}{2}}[d_k^{(1)}(e^{\pi\omega/2a}\,{}^{R}u_k + e^{-\pi\omega/2a}\,{}^{L}u^*_{-k}) + d_k^{(2)}(e^{-\pi\omega/2a}\,{}^{R}u^*_{-k} + e^{\pi\omega/2a}\,{}^{L}u_k)] + \text{h.c.} \tag{4.93}$$

where now

$$d_k^{(1)}|0_M\rangle = d_k^{(2)}|0_M\rangle = 0, \tag{4.94}$$

and we have also introduced a normalization factor. The operators $b_k^{(1,2)}$ can be related to $d_k^{(1,2)}$ by taking the inner products $(\phi, {}^{R}u_k)$, $(\phi, {}^{L}u_k)$, first with ϕ given by (4.86) and then by (4.93). One obtains

$$b_k^{(1)} = [2\sinh(\pi\omega/a)]^{-\frac{1}{2}}[e^{\pi\omega/2a}d_k^{(2)} + e^{-\pi\omega/2a}d_{-k}^{(1)\dagger}] \tag{4.95}$$

$$b_k^{(2)} = [2\sinh(\pi\omega/a)]^{-\frac{1}{2}}[e^{\pi\omega/2a}d_k^{(1)} + e^{-\pi\omega/2a}d_{-k}^{(2)\dagger}]. \tag{4.96}$$

These Bogolubov transformations provide the required relation between the states $|0_R\rangle$ and $|0_M\rangle$.

Now consider an accelerating Rindler observer at $\xi = \text{constant}$. From (4.74) one sees that such an observer's proper time is proportional to η. One therefore expects that the vacuum for Rindler observers will be $|0_R\rangle$ as this is the state associated with modes which are positive frequency with respect to η. Thus, according to (4.86), a Rindler observer in L (respectively R) will detect particles counted by the number operator $b_k^{(1)\dagger}b_k^{(1)}$ (respectively $b_k^{(2)\dagger}b_k^{(2)}$). If the field is in the state $|0_M\rangle$ (that is, it is devoid of usual Minkowski space particles) then, using (4.95) and (4.96) it may be deduced that a Rindler observer will detect

$$\langle 0_M|b_k^{(1,2)\dagger}b_k^{(1,2)}|0_M\rangle = e^{-\pi\omega/a}/[2\sinh(\pi\omega/a)] = (e^{2\pi\omega/a} - 1)^{-1} \tag{4.97}$$

particles in mode k. This is precisely the Planck spectrum for radiation at temperature $T_0 = a/2\pi k_B$. The temperature T as seen by the accelerated observer is given by the Tolman relation

$$T = (g_{00})^{-\frac{1}{2}}T_0 \tag{4.98}$$

(see, for example, Tolman 1934, Landau & Lifshitz, 1958, §27, or Balazs 1958). Here g_{00} is obtained from the metric (4.71), in terms of which the accelerated observer has constant spatial coordinate. Using (4.73) and (4.98) one obtains $T = 1/(2\pi\alpha k_B)$, which is in exact agreement with (3.68) obtained for an accelerated detector in Minkowski space. The particles detected by such an observer are often called *Rindler particles*. The quantum field theory of accelerated observers and associated thermal effects have been considered in great generality by Sanchez (1979). Possible physical effects have been examined by Barshay & Troost (1978) and Hosoya (1979).

Here we predict thermal particles seen by an accelerating observer in flat space, but under a conformal transformation we could obtain a thermal bath seen by an inertial observer in curved space. This is the so-called Hawking effect (see chapter 8).

5
Curved spacetime examples

This chapter is devoted to a direct application of the curved spacetime quantum field theory developed in chapter 3. We treat particle creation by time-dependent gravitational fields by examing a variety of expanding and contracting cosmological models. Most of the models are special cases of the Robertson–Walker homogeneous isotropic spacetimes, chosen either for their simplicity, or special interest in illuminating certain aspects of the formalism.

All the main cases that have appeared in the literature are collected here. The Milne universe (technically flat spacetime) and de Sitter space are especially useful for illustrating the role of adiabaticity in assessing the physical reasonableness of a quantum state. De Sitter space also enjoys the advantage of being the only time-dependent cosmological model for which both the particle creation effects and the vacuum stress (deferred until §6.4) have been explicitly evaluated by all known techniques.

A small but important section, §5.5, presents a classification scheme that relates the vacuum states in conformally-related spacetimes. This topic too has a 'thermal' aspect to it. It will turn out to be of relevance for the computation of $\langle T_{\mu\nu} \rangle$ in Robertson–Walker spacetimes in chapter 6 and chapter 7.

The final section is an attempt to go beyond the simple Robertson–Walker models and treat the subject of anisotropy in cosmology. This is an issue of central importance in modern cosmological theory, because the observed high degree of isotropy in the universe is without adequate explanation. Here, once again, our theory makes contact with the real world, albeit in a rather model-dependent and modest way. It seems probable that quantum particle creation played a large part in determining the constituents and condition of the primeval cosmological fluid. Moreover, the back-reaction of quantum effects on the gravitational dynamics, to be discussed in chapters 6 and 7, certainly played a role in the dissipation of primeval anisotropy. The resulting entropy, in the form of quantum particles, remains with us today as an observational constraint on the quantum theory of the very early universe.

5.1 Robertson–Walker spacetimes

Most of the examples to be considered in this chapter will involve fields propagating in particular Robertson–Walker spacetimes. Such spacetimes are especially important in providing cosmological models which are in good agreement with observation. Robertson–Walker spacetimes with flat spatial sections were discussed in §§3.4–3.6. We start by summarizing the essential formulae, and then extend them to Robertson–Walker spacetimes with hyperbolic and spherical spatial sections. Attention is restricted to scalar fields, the extension to fields of higher spins following along the lines of §3.8.

The line element for n-dimensional Robertson–Walker spacetimes with flat spatial sections is

$$\mathrm{d}s^2 = C(\eta)\left[\mathrm{d}\eta^2 - \sum_{i=1}^{n-1} (\mathrm{d}x^i)^2\right]. \tag{5.1}$$

The mode decomposition (3.30) for the field ϕ is

$$\phi(x) = \int \mathrm{d}^{n-1}k[a_{\mathbf{k}}u_{\mathbf{k}}(x) + a_{\mathbf{k}}^{\dagger}u_{\mathbf{k}}^{*}(x)]. \tag{5.2}$$

The modes $u_{\mathbf{k}}$ can be written in a separated form as

$$u_{\mathbf{k}}(x) = (2\pi)^{(1-n)/2}\mathrm{e}^{\mathrm{i}\mathbf{k}\cdot\mathbf{x}}C^{(2-n)/4}(\eta)\chi_k(\eta) \tag{5.3}$$

where $k = |\mathbf{k}| = (\sum_{i=1}^{n-1} k_i^2)^{\frac{1}{2}}$ and χ_k satisfies

$$\frac{\mathrm{d}^2\chi_k}{\mathrm{d}\eta^2} + \{k^2 + C(\eta)[m^2 + (\xi - \xi(n))R(\eta)]\}\chi_k = 0 \tag{5.4}$$

with

$$\xi(n) \equiv \tfrac{1}{4}[(n-2)/(n-1)]. \tag{5.5}$$

The normalization condition (3.29) reduces to a condition on the Wronskian of the solutions χ_k:

$$\chi_k\partial_\eta\chi_k^* - \chi_k^*\partial_\eta\chi_k = \mathrm{i}. \tag{5.6}$$

These equations can easily be extended to Robertson–Walker spacetimes with curved spatial sections. For simplicity we confine our attention to four-dimensional spacetimes; equations which allow the generalization to arbitrary dimensions can be found in Bander & Itzykson (1966). The general Robertson–Walker line element is

$$ds^2 = dt^2 - a^2(t) \sum_{i,j=1}^{3} h_{ij} dx^i dx^j \tag{5.7}$$

where

$$\begin{aligned}\sum_{i,j=1}^{3} h_{ij} dx^i dx^j &= (1 - Kr^2)^{-1} dr^2 + r^2(d\theta^2 + \sin^2\theta \, d\phi^2) \\ &= d\chi^2 + f^2(\chi)(d\theta^2 + \sin^2\theta \, d\phi^2)\end{aligned} \tag{5.8}$$

and

$$f(\chi) = r = \begin{cases} \sin \chi, & 0 \le \chi \le 2\pi, \quad K = +1 \\ \chi, & 0 \le \chi < \infty, \quad K = 0 \\ \sinh \chi, & 0 \le \chi < \infty, \quad K = -1. \end{cases} \tag{5.9}$$

Equation (5.8) gives the line element on the spatial sections, which are hyperbolic, flat or closed depending on whether $K = -1, 0, 1$ respectively. Writing $C(\eta) = a^2(t)$, with the conformal time parameter η given by

$$\eta = \int^t a^{-1}(t') dt', \tag{5.10}$$

the line element (5.7) can be recast in the form

$$ds^2 = C(\eta)\left(d\eta^2 - \sum_{i,j=1}^{3} h_{ij} dx^i dx^j \right). \tag{5.11}$$

Defining $\Upsilon = (1 - Kr^2)^{-1}$ and $D = \dot{C}/C$, where a dot denotes differentiation with respect to η, the nonzero Christoffel symbols for the metric (5.11) are (indices 0, 1, 2, 3 corresponding to η, r, θ, ϕ respectively)

$$\begin{aligned} &\Gamma^0_{00} = \Gamma^1_{01} = \Gamma^2_{02} = \Gamma^3_{03} = \tfrac{1}{2}D, \quad \Gamma^0_{11} = \tfrac{1}{2}D\Upsilon, \quad \Gamma^0_{22} = \tfrac{1}{2}Dr^2, \\ &\Gamma^0_{33} = \tfrac{1}{2}Dr^2 \sin^2\theta, \quad \Gamma^1_{11} = Kr\Upsilon \quad \Gamma^1_{22} = -r\Upsilon^{-1}, \\ &\Gamma^1_{33} = -r\Upsilon^{-1}\sin^2\theta, \quad \Gamma^2_{12} = \Gamma^3_{13} = 1/r, \\ &\Gamma^2_{33} = -\sin\theta\cos\theta, \quad \Gamma^3_{23} = \cot\theta. \end{aligned} \tag{5.12}$$

From (5.12) the nonzero components of the Ricci tensor, and hence the Ricci scalar, can be computed:

$$\begin{aligned} &R_{00} = \tfrac{3}{2}\dot{D}, \quad R_{11} = -\tfrac{1}{2}(\dot{D} + D^2)\Upsilon - 2K\Upsilon, \\ &R_{22} = \Upsilon^{-1} r^2 R_{11}, \quad R_{33} = \sin^2\theta \, R_{22}, \end{aligned} \tag{5.13}$$

$$R = C^{-1}[3\dot{D} + \tfrac{3}{2}D^2 + 6K]. \tag{5.14}$$

The scalar field ϕ satisfies (3.26) with R given by (5.14). Its mode decomposition can now be written as

$$\phi(x) = \int d\tilde{\mu}(k)[a_{\mathbf{k}}u_{\mathbf{k}}(x) + a_{\mathbf{k}}^{\dagger}u_{\mathbf{k}}^{*}(x)] \tag{5.15}$$

where the measure $\tilde{\mu}(k)$ will be defined shortly. The modes are separated as

$$u_{\mathbf{k}}(x) = C^{-\frac{1}{2}}(\eta)\mathscr{Y}_{\mathbf{k}}(\mathbf{x})\chi_k(\eta) \tag{5.16}$$

with $\mathbf{x} = (r, \theta, \phi)$ or (χ, θ, ϕ) and $\mathscr{Y}_{\mathbf{k}}(\mathbf{x})$ a solution of

$$\Delta^{(3)}\mathscr{Y}_{\mathbf{k}}(\mathbf{x}) = -(k^2 - K)\mathscr{Y}_{\mathbf{k}}(\mathbf{x}). \tag{5.17}$$

In (5.17) $\Delta^{(3)}$ is the Laplacian associated with the spatial metric h_{ij}:

$$\Delta^{(3)}\mathscr{Y}_{\mathbf{k}} \equiv h^{-\frac{1}{2}}\partial_i\left(h^{\frac{1}{2}}\, h^{ij}\partial_j\mathscr{Y}_{\mathbf{k}}\right), \tag{5.18}$$

$h = \det(h_{ij})$. With this separation χ_k satisfies (5.4) with $n = 4$, $(\xi(4) = \frac{1}{6})$. Further, if the functions $\mathscr{Y}_{\mathbf{k}}$ are normalized such that

$$\int d^3x\, h^{\frac{1}{2}}\, \mathscr{Y}_{\mathbf{k}}(\mathbf{x})\mathscr{Y}_{\mathbf{k}'}^{*}(\mathbf{x}) = \delta(\mathbf{k}, \mathbf{k}') \tag{5.19}$$

where $\delta(\mathbf{k}, \mathbf{k}')$ is the δ-function with respect to the measure $\tilde{\mu}$:

$$\int d\tilde{\mu}(k') f(\mathbf{k}')\delta(\mathbf{k}, \mathbf{k}') = f(\mathbf{k}), \tag{5.20}$$

then the normalization condition (3.29) once again reduces to (5.6).

The eigenfunctions $\mathscr{Y}_{\mathbf{k}}$ of the three-dimensional Laplacian are (Parker & Fulling 1974)

$$\mathscr{Y}_{\mathbf{k}}(\mathbf{x}) = \begin{cases} (2\pi)^{-\frac{3}{2}} e^{i\mathbf{k}\cdot\mathbf{x}}, & \mathbf{k} = (k_1, k_2, k_3), \quad (K = 0) \\ \Pi_{kJ}^{(\pm)}(\chi)Y_J^M(\theta, \phi), & \mathbf{k} = (k, J, M), \quad (K = \pm 1) \end{cases} \tag{5.21}$$

where

$$-\infty < k_i < \infty; \quad k = |\mathbf{k}|, \quad (K = 0)$$

$$M = -J, -J+1, \ldots, J; \begin{cases} J = 0, 1, \ldots, k-1; & k = 1, 2, \ldots, \quad (K = 1) \\ J = 0, 1, \ldots; & 0 < k < \infty, \quad (K = -1) \end{cases} \tag{5.22}$$

The Y_J^M are spherical harmonics. The functions $\Pi^{(-)}$ are defined by (see, for example, Bander & Itzykson (1966), or Dolginov & Toptygin (1959)):

$$\Pi_{kJ}^{(-)}(\chi) = [\tfrac{1}{2}\pi k^2(k^2+1)\ldots(k^2+J^2)]^{-\frac{1}{2}} \sinh^J \chi \left(\frac{\mathrm{d}}{\mathrm{d}\cosh\chi}\right)^{1+J} \cos k\chi. \tag{5.23}$$

One can obtain the functions $\Pi_{kJ}^{(+)}(\chi)$ from $\Pi_{kJ}^{(-)}(\chi)$ by replacing k by $-ik$ and χ by $-i\chi$ in the latter (see, for example, Lifshitz & Khalatnikov (1963)).

With these definitions, the measure $\tilde{\mu}(k)$ is defined by

$$\int \mathrm{d}\tilde{\mu}(k) = \left\{ \begin{array}{ll} \int \mathrm{d}^3 k, & (K=0) \\ \sum\limits_{k,J,M}, & (K=1) \\ \int_0^\infty \mathrm{d}k \sum\limits_{J,M}, & (K=-1) \end{array} \right\} \tag{5.24}$$

Finally, we note that it is also possible to write the spatial part of the mode decomposition in terms of Gegenbauer polynomials rather than the $\Pi^{(+)}$ functions (see, for example, Ford 1976).

5.2 Static Robertson–Walker spacetimes

The simplest cases in which the preceding analysis can be applied are the static spacetimes with $C(\eta) = c = a^2 =$ constant. The static spacetime with flat spatial sections ($K = 0$) is, of course, Minkowski space, which we shall not consider again. Rather, we discuss the closed ($K = 1$) Einstein universe (Einstein 1917) and the static spacetime with hyperbolic spatial sections ($K = -1$).

In these cases the scalar curvature (5.14) reduces to

$$R = 6K/c, \tag{5.25}$$

which permits normalized solutions of (5.4) in four dimensions to be written down immediately:

$$\chi_k(\eta) = (2\omega_k)^{-\frac{1}{2}} e^{-i\omega_k \eta} \tag{5.26}$$

where

$$\begin{aligned} \omega_k^2 &= k^2 + \mu^2 \\ &= k^2 + cm^2 + (6\xi - 1)K. \end{aligned} \tag{5.27}$$

Note from (5.10) that $\eta = t/a$, so that the solutions (5.26) are positive frequency with respect to the Killing vectors ∂_η and ∂_t. Since the spacetimes are static and admit these global, timelike Killing vectors, the definition of particles here is no more ambiguous than the definition of particles in Minkowski space. In particular, using (5.16) and (5.26) to construct positive frequency modes

$$u_{\mathbf{k}}(\mathbf{x}) = (2c\omega_k)^{-\frac{1}{2}}\mathscr{Y}_{\mathbf{k}}(\mathbf{x})e^{-i\omega_k\eta} \tag{5.28}$$

for the expansion (5.15), a vacuum state $|0\rangle$ is defined by

$$a_{\mathbf{k}}|0\rangle = 0. \tag{5.29}$$

One expects this vacuum to be as physically reasonable as the standard Minkowski vacuum, a conjecture which can be verified by an investigation of particle detectors. The Wightman function in the Einstein universe has been calculated by Dowker (1971) (see also Critchley 1976, Dowker & Critchley 1977*a* and Bunch & Davies 1977*a*) who finds

$$G^+(x, x') = \frac{i\mu}{8\pi c \sin(\Delta\chi)} \times \sum_{n=-\infty}^{\infty} \frac{(\Delta\chi + 2\pi n)H_1^{(2)}\{\mu[(\Delta\eta - i\varepsilon)^2 - (\Delta\chi + 2n\pi)^2]^{\frac{1}{2}}\}}{[(\Delta\eta - i\varepsilon)^2 - (\Delta\chi + 2\pi n)^2]^{\frac{1}{2}}}, \quad (K = +1) \tag{5.30}$$

where H is a Hankel function. We have defined as usual

$$\Delta\chi = \chi - \chi', \quad \Delta\eta = \eta - \eta' \tag{5.31}$$

and, without loss of generality, have used the isotropy of the spacetime to orient the axes such that $\phi = \phi'$, $\theta = \theta'$. The form of the Wightman function is that of a sum of Minkowski space positive frequency Wightman functions (cf. (2.77)). Using this fact it is not difficult to show using (3.55) that a particle detector at rest in the (χ, η) frame registers no particles.

We note for later use that in the massless, conformally coupled limit ($m \to 0$, $\xi \to \frac{1}{6}$, hence $\mu \to 0$) (5.30) reduces to

$$D^+(x, x') = \frac{1}{8\pi^2 c[\cos(\Delta\eta - i\varepsilon) - \cos(\Delta\chi)]}, \quad (K = +1). \tag{5.32}$$

The massless, conformally coupled Wightman function for a scalar field in a static, hyperbolic spacetime has been calculated by Bunch (1978*a*), who finds

$$D^+(x, x') = \frac{\Delta\chi}{4\pi^2 c \sinh(\Delta\chi)[\Delta\chi^2 - (\Delta\eta - i\varepsilon)^2]}, \quad (K = -1) \tag{5.33}$$

while for a massive field and arbitrary ξ one obtains

$$G^+(x,x')$$
$$= -\frac{i\mu}{8\pi c \sinh(\Delta\chi)} \frac{\Delta\chi H_1^{(2)}\{\mu[(\Delta\eta - i\varepsilon)^2 - (\Delta\chi)^2]^{\frac{1}{2}}\}}{[(\Delta\eta - i\varepsilon)^2 - (\Delta\chi)^2]^{\frac{1}{2}}}, \quad (K = -1). \tag{5.34}$$

Once again the similarity to the Minkowski space Wightman function guarantees that a detector at rest in the (χ, η) frame will register no particles.

If the spacetime is not static then in general solutions to (5.4) are very difficult to find and it is only in special cases or using approximation methods that progress can be made. However, for a massless, conformally coupled field, the conformal invariance of the field equation immediately allows one to write down the modes; they are given by (5.28) with c replaced by $C(\eta)$. This is analogous to the spatially flat case discussed in §3.7, and the modes so obtained are, as in §3.7, positive frequency with respect to the conformal Killing vector ∂_η. From these modes the Green functions in the conformal vacuum are readily obtained. For example, for $K = \pm 1$, G^+ in the conformal vacuum is obtained by replacing c by $C^{\frac{1}{2}}(\eta)C^{\frac{1}{2}}(\eta')$ in (5.32) and (5.33) respectively.

5.3 The Milne universe

The case of Robertson–Walker spacetimes with $a(t) = t$ is particularly interesting, not only because the conformally non-trivial $(m \neq 0)$ field equation can be solved exactly, but because it sheds light upon the nature of the relationship between the conformal and the adiabatic vacua (Fulling, Parker & Hu, 1974, Bunch 1977, 1978*a*, Davies & Fulling 1977*b*, Bunch, Christensen & Fulling 1978).

The four-dimensional hyperbolic spacetime with $a(t) = t$ (known as the Milne universe; Milne 1932) and its two-dimensional counterpart, are in fact merely unconventional coordinatizations of flat spacetime, analogous to the Rindler system (see §4.5). Nevertheless, as in the Rindler case, the Milne model leads to non-trivial quantum effects, and can also cast light on the uniformly expanding cosmological models with $K = 0$ and $+1$ (which are not merely flat spacetime in disguise).

The two-dimensional Milne universe has line element

$$\begin{aligned} ds^2 &= dt^2 - a^2t^2dx^2 \\ &= e^{2a\eta}(d\eta^2 - dx^2) \end{aligned} \tag{5.35}$$

where $|t| = a^{-1}e^{a\eta}$ and a is a constant. Under the coordinate transformation

$$y^0 = a^{-1}e^{a\eta}\cosh ax, \quad y^1 = a^{-1}e^{a\eta}\sinh ax \tag{5.36}$$

(5.35) reduces to

$$ds^2 = (dy^0)^2 - (dy^1)^2, \quad 0 < y^0 < \infty \quad -\infty < y^1 < \infty \tag{5.37}$$

(see fig. 15).

The massive wave equation with $\xi = 0$ and $C = e^{2a\eta}$ is readily solved in terms of either Bessel or Hankel functions, from which two complete sets of normalized modes $\{\bar{u}_\mathbf{k}, \bar{u}^*_\mathbf{k}\}$ and $\{u_\mathbf{k}, u^*_\mathbf{k}\}$ can be constructed using (5.3) with

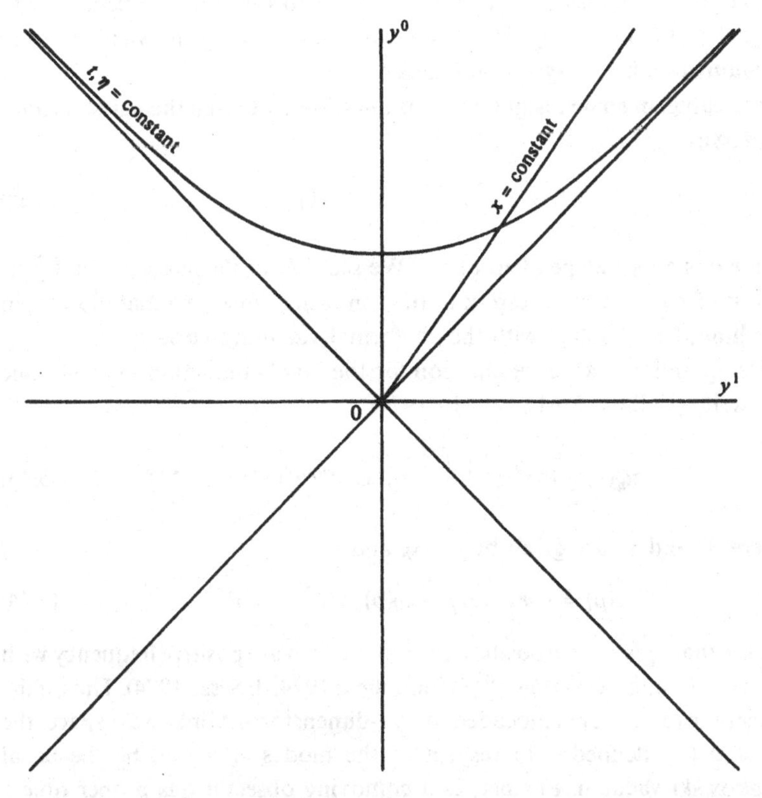

Fig. 15. Milne universe. The t, x coordinates cover the wedge of Minkowski space corresponding to region F in fig. 13. This region is therefore like the Rindler wedge on its side. The lines x = constant radiating from the origin represent the world lines of observers who perceive a universe expanding from a big bang origin at 0.

$$\bar{\chi}_k = [(2a/\pi)\sinh(\pi k/a)]^{-\frac{1}{2}} J_{-ik/a}(mt) \tag{5.38}$$

$$\chi_k = \tfrac{1}{2}(\pi/a)^{\frac{1}{2}} e^{\pi k/2a} H^{(2)}_{ik/a}(mt), \quad (k > 0). \tag{5.39}$$

These two complete sets of modes are related by a Bogolubov transformation:

$$\bar{\chi}_k = \alpha_k \chi_k + \beta_k \chi_k^* \tag{5.40}$$

with

$$\alpha_k = [e^{\pi k/a}/2\sinh(\pi k/a)]^{\frac{1}{2}}, \quad \beta_k = [e^{-\pi k/a}/2\sinh(\pi k/a)]^{\frac{1}{2}}, \tag{5.41}$$

from which the similarity with the Rindler case is manifest (compare (5.41) with (4.96)).

It is clear from (5.40) that the state $|\bar{0}\rangle$ defined by (3.32) and (3.33) with respect to the modes $\bar{u}_{\mathbf{k}}$ will be inequivalent to the state $|0\rangle$ defined with respect to the modes $u_{\mathbf{k}}$. This leads us to ask once again, whether either vacuum state is in any way privileged.

We can gain some insight into this question by taking the massless limit of (5.38):

$$\bar{\chi}_k \xrightarrow[m \to 0]{} e^{i\theta} e^{-ik\eta}/(2k)^{\frac{1}{2}}, \tag{5.42}$$

where θ is an η-independent phase. We see that in the massless limit $\bar{\chi}_k$ is positive frequency with respect to the conformal time η, so that $|\bar{0}\rangle$ can, in this limit, be identified with the conformal vacuum (see §3.7).

Using an integral representation for the Hankel function in (5.39), one can write (Fulling, Parker & Hu 1974)

$$u_{\mathbf{k}}(x) = (8\pi^2 a)^{-\frac{1}{2}} \int_{-\infty}^{\infty} d\rho \; e^{-i\omega(\rho)y^0} e^{ip(\rho)y^1} e^{-ik\rho/a} \tag{5.43}$$

where y^0 and y^1 are given by (5.36), and

$$p(\rho) = -m\sinh\rho, \quad \omega(\rho) = (p^2 + m^2)^{\frac{1}{2}}. \tag{5.44}$$

We see that $u_{\mathbf{k}}$ is a superposition of modes which are positive frequency with respect to Minkowski time y^0 (Sommerfield 1974, di Sessa 1974). Thus, if the spacetime (5.35) were embedded in two-dimensional Minkowski space, the vacuum $|0\rangle$ defined with respect to the modes $u_{\mathbf{k}}$ would be the usual Minkowski vacuum. Further, as a comoving observer has proper time $t \propto y^0$, such an observer will see no particles in this vacuum.

Another useful fact about the modes $u_{\mathbf{k}}$ can be obtained by comparing them with the positive frequency adiabatic modes based on (3.102) (which, for conformal coupling, also holds in two dimensions). To zeroth order

$$W_k^{(0)} = \omega_k = (k^2 + m^2 e^{2a\eta})^{\frac{1}{2}},$$

which when substituted into (3.105) yields a factor

$$T^2 \omega_k^2(\eta_1) = T^2 k^2 + T^2 m^2 e^{2a\eta_1 T}.$$

Clearly ω_k becomes large compared with the derivatives of $\dot{C}/C = O(T^{-1})$ (for fixed $\eta_1 = \eta/T$) as k, m or η become large either individually or together.

In the limit of large η

$$W_k \to \omega_k \simeq m e^{a\eta}$$

which when substituted in (3.102) gives

$$\chi_k^{(A)} \to (2m e^{a\eta})^{-\frac{1}{2}} \exp(-im e^{a\eta}/a), \quad \eta \to \infty. \tag{5.45}$$

(A indicates that this limit holds to any adiabatic order. All higher order corrections vanish as $\eta \to \infty$.) On the other hand, using the asymptotic expansion of Hankel functions, one can easily show that (up to a constant phase factor) (5.39) reduces to (5.45) in the limit of large η (i.e., large $|t|$). One can also verify that in the limit of large k, or large k and m, (5.39) reduces to the corresponding limit of (3.102).

Thus the modes $u_{\mathbf{k}}$ are positive frequency with respect to the adiabatic definition given in §3.5, and the vacuum $|0\rangle$ is an adiabatic vacuum of infinite adiabatic order. So in this case the adiabatic definition of positive frequency agrees with the definition inherited from Minkowski space in a natural way. Further, not only does an inertial particle detector register no particles to any finite adiabatic order in this vacuum, it registers strictly no particles at all.

Since an inertial detector registers no particles in the adiabatic vacuum $|0\rangle$, it is clear from (5.40) that it must detect particles in the vacuum $|\bar{0}\rangle$. Confining attention to the massless case, this implies that a comoving detector will register particles in the Milne conformal vacuum as indeed it would in almost any other conformal vacuum (see §3.7).

The presence of particles in the conformal vacuum can be seen in another way: by calculating the conformal expectation value of the stress-tensor, $\langle \bar{0}|T_{\mu\nu}|\bar{0}\rangle$. To do this rigorously (Davies & Fulling 1977*b*) one must use the regularization and renormalization techniques described in the next chapter, and applied to this example in §7.1. However, even without regularization and renormalization, it is possible to calculate the finite difference between the expectation values of $T_{\mu\nu}$ in the $|0\rangle$ and $|\bar{0}\rangle$ states respectively. In the massless limit one finds (Bunch 1977, Bunch,

Christensen & Fulling 1978)

$$\langle \bar{0}|T_{tt}|\bar{0}\rangle - \langle 0|T_{tt}|0\rangle = -(1/\pi a^2 t^2)\int_0^\infty k|\beta_k|^2 \mathrm{d}k$$

$$= -(1/\pi)\int_0^\infty q(\mathrm{e}^{2\pi q t} - 1)^{-1}\mathrm{d}q = -1/24\pi t^2, \qquad (5.46)$$

where we have used (5.41).

Next one can argue that because $|0\rangle$ is equivalent to the usual Minkowski space vacuum, the renormalized (or normal ordered) value of $\langle 0|T_{tt}|0\rangle$ must be zero, so that (5.46) gives the renormalized value of $\langle \bar{0}|T_{tt}|\bar{0}\rangle$. (These ideas will be made more precise in the next chapter.) Thus the conformal vacuum contains a negative energy density of radiation with a Planck spectrum at temperature $(2\pi t)^{-1}$. The relation between the two vacuum states is therefore closely similar to that between the Minkowski and Rindler definitions, even to the extent of the thermal association.

This explicit example nicely illustrates the physical differences between the adiabatic and conformal vacuum states. The difference is perhaps not surprising when it is remembered that the former is constructed from an analysis of the field behaviour at large mass, whereas the latter exploits the conformal symmetry of the massless case.

The situation is similar in the four-dimensional ($K = -1$) Milne universe. Expressions (5.38) and (5.39) still provide solutions of the field equation in the conformally coupled ($\xi = \frac{1}{6}$) case. The states $|0\rangle$ and $|\bar{0}\rangle$ are still adiabatic and (in the massless limit) conformal vacuum states, respectively. In fact, similar results hold even in the $K = 0$ and $+1$ models with $a(t) = t$, even though one can no longer appeal to the Minkowskian covering space to give special significance to $|0\rangle$.

The $K = 0$ model serves to illustrate a curious phenomenon concerning the massless limit of the adiabatic vacuum. Equation (5.4) reduces in this case to

$$\frac{\mathrm{d}^2\chi_k}{\mathrm{d}\eta^2} + [k^2 + (6\xi - 1)a^2]\chi_k = 0. \qquad (5.47)$$

The curvature term, $(6\xi - 1)a^2$, enters in the same way as would a mass. However, for $\xi < \frac{1}{6}$ it is negative (corresponding to a tachyonic mass), a state of affairs that is usually taken to imply that the vacuum state is unstable. If a self-interaction term such as $\lambda\phi^4$ is added to the field Lagrangian, spontaneous symmetry breaking can occur (Goldstone 1961). This feature also arises in other cosmological models (see §5.4) and has been exploited in the work of Frolov, Grib & Mostepanenko (1977, 1978).

The fact that particle states become ill-defined for $\xi < \frac{1}{6}$ in the above example may also be deduced by examining the adiabatically constructed massive modes in the $K = 0$ model:

$$\chi_k = \tfrac{1}{2}(\pi/a)^{\frac{1}{2}} e^{-i\pi\nu/2} H_\nu^{(2)}(mt) \tag{5.48}$$

where

$$\nu = [1 - 6\xi - (k/a)^2]^{\frac{1}{2}}.$$

The solutions (5.48), which reduce to (5.45) for large $|t|$, can be used to construct an adiabatic vacuum in the usual way. However, in the limit of small mass χ_k reduces to

$$\tfrac{1}{2}i(\pi/a)^{\frac{1}{2}} \operatorname{cosec}(\pi\nu)\left(\frac{e^{-i\pi\nu/2}(\frac{1}{2}ma^{-1})^{-\nu}e^{-\nu\eta}}{\Gamma(1-\nu)} - \frac{e^{i\pi\nu/2}(\frac{1}{2}ma^{-1})^{\nu}e^{\nu\eta}}{\Gamma(1+\nu)}\right)$$

which diverges as $m \to 0$ if $\xi < \frac{1}{6}$, $k^2/a^2 < 1 - 6\xi$.

Particle creation in this model has been considered in detail by Chitre & Hartle (1977) using analytic continuation methods (see §8.5).

5.4 De Sitter space

De Sitter space (de Sitter 1917*a*, *b*) is the curved spacetime which has been most studied by quantum field theorists (see bibliography and references cited below). The reason for this special attention stems from the fact that de Sitter space is the unique maximally symmetric curved spacetime (see, for example, Weinberg 1972). It enjoys the same degree of symmetry as Minkowski space (ten Killing vectors), which greatly facilitates technical computations as far as quantum field theory is concerned. Even so, the presence of curvature, and non-trivial global properties, introduce new aspects to the quantization of fields in de Sitter space. We confine our attention in this section to scalar fields, although higher-spin fields have been treated in many of the papers cited; in particular, fields of arbitrary spin on de Sitter space have been discussed by Dowker & Critchley (1976*a*, *b*), Grensing (1977) and Birrell (1979*b*). See also §6.4.

Four-dimensional de Sitter space is most easily represented as the hyperboloid

$$z_0{}^2 - z_1{}^2 - z_2{}^2 - z_3{}^2 - z_4{}^2 = -\alpha^2 \tag{5.49}$$

embedded in five-dimensional Minkowski space with metric

$$ds^2 = dz_0{}^2 - dz_1{}^2 - dz_2{}^2 - dz_3{}^2 - dz_4{}^2 \tag{5.50}$$

(see, for example, Schrödinger 1956, Hawking & Ellis 1973 and, in particular, Fulling 1972, where the connection between the geometry of de Sitter space and quantum field theory is discussed).

From the form of (5.49) it is clear that the symmetry group of de Sitter space is the ten parameter group $SO(1,4)$ of homogeneous 'Lorentz transformations' in the five-dimensional embedding space (see for example, Gürsey 1964) known as the de Sitter group. Just as the Poincaré group plays a central role in the quantization of fields in Minkowski space, so the de Sitter group of symmetries on de Sitter space is fundamental to the discussion of quantization to be given here.

As usual, the first step in scalar field quantization is the solution of the field equation (3.26). To accomplish this, a coordinatization of de Sitter space must first be specified. From the discussions of the Rindler and Milne spacetimes already given it should be clear that the choice of a particular coordinate system can lead to what appears as an especially natural choice (or choices) of 'vacuum' state. For example, in the Rindler case, the (Rindler) vacuum defined by positive frequency modes (4.81) is the most natural for the coordinatization (4.71), while for the coordinatization (4.66), the Minkowski vacuum is the more natural choice. We are therefore led to consider in turn the three most widely used coordinatizations of the de Sitter space hyperboloid and to discuss the properties of the resulting solutions of the field equation.

Consider first the coordinates $(t, \mathbf{x})$ defined by

$$\left.\begin{aligned} z_0 &= \alpha \sinh(t/\alpha) + \tfrac{1}{2}\alpha^{-1} e^{t/\alpha} |\mathbf{x}|^2 \\ z_4 &= \alpha \cosh(t/\alpha) - \tfrac{1}{2}\alpha^{-1} e^{t/\alpha} |\mathbf{x}|^2 \\ z_i &= e^{t/\alpha} x_i, \quad i = 1, 2, 3, \quad -\infty < t, x_i < \infty. \end{aligned}\right\} \tag{5.51}$$

which cover the half of the de Sitter manifold with $z_0 + z_4 > 0$ (see fig. 16). In these coordinates the line element (5.50) becomes

$$ds^2 = dt^2 - e^{2t/\alpha} \sum_{i=1}^{3} (dx^i)^2, \tag{5.52}$$

which is the line element for the steady-state universe of Bondi & Gold (1948) and Hoyle (1948). In terms of the conformal time

$$\eta = -\alpha e^{-t/\alpha}, \quad -\infty < \eta < 0 \tag{5.53}$$

the line element (5.52) becomes

$$ds^2 = (\alpha^2/\eta^2)[d\eta^2 - \sum_{i=1}^{3} (dx^i)^2], \tag{5.54}$$

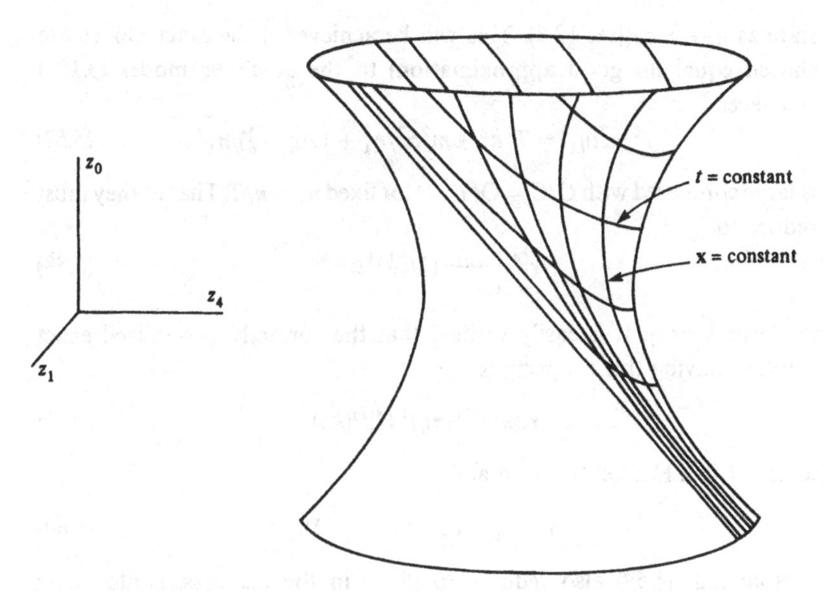

Fig. 16. De Sitter space represented as a hyperboloid embedded in five-dimensional flat spacetime (two dimensions are suppressed in the figure). The coordinates shown cover only half the space, called the steady-state universe, and are bounded by the null rays $z_0 + z_4 = 0$ ($t = -\infty$). The 3-spaces t = constant are flat, and expand with increasing t.

revealing that this portion of de Sitter space is conformal to a portion of Minkowski space. The remaining half of de Sitter space can be coordinatized by reversing the signs of the right-hand sides in (5.51), or by allowing the conformal time η to range over all real numbers $-\infty < \eta < \infty$.

Since (5.54) has the form of a spatially flat Robertson–Walker line element (5.1) with $C(\eta) = (\alpha/\eta)^2$, we can work within the framework of §5.1. The Ricci scalar is calculated from (5.14) to be

$$R = 12\alpha^{-2}, \tag{5.55}$$

which, when substituted into (5.4) gives an equation for the η-dependent factor χ_k of the field modes. This equation can be solved in terms of Bessel or Hankel functions. The particular linear combination used to define the 'positive frequency' modes will, of course, determine the choice of vacuum.

Since

$$\frac{\mathrm{d}^l}{\mathrm{d}\eta^l}\left(\frac{\dot{C}}{C}\right) \xrightarrow[\eta \to \pm\infty]{} 0 \tag{5.56}$$

we expect an adiabatic vacuum to be a reasonable definition of a no-particle state as $\eta \to \pm\infty$ (see §3.5). This will be achieved if the exact modes are chosen equal (to good approximation) to the adiabatic modes (3.102) whenever

$$T^2\omega_k^2(\eta_1) = T^2k^2 + m^2\alpha^2/\eta_1^2 + 12(\xi - \tfrac{1}{6})/\eta_1{}^2 \tag{5.57}$$

is large compared with $\dot{C}/C = O(T^{-1})$ for fixed $\eta_1 = \eta/T$. That is, they must reduce to

$$\chi_k^{(A)} \xrightarrow[k,\eta\to\infty]{} (1/2k)^{\frac{1}{2}}\,\mathrm{e}^{-ik\eta} \tag{5.58}$$

for large k or η. It is easily verified that the correctly normalized exact solution having this property is

$$\chi_k(\eta) = \tfrac{1}{2}(\pi\eta)^{\frac{1}{2}} H_\nu^{(2)}(k\eta) \tag{5.59}$$

where H is a Hankel function and

$$\nu^2 = \tfrac{9}{4} - 12(m^2R^{-1} + \xi). \tag{5.60}$$

Note that (5.59) also reduces to (5.58) in the massless, conformally coupled ($\xi = \frac{1}{6}$) limit. It follows that, unlike in the Milne universe, here the conformal vacuum coincides with the massless limit of the adiabatic vacuum.

Since (5.59) has the asymptotic form of the adiabatic modes (5.58) as $\eta \to +\infty$ as well as when $\eta \to -\infty$, there is no particle production, even though the scale factor passes through a coordinate singularity at $\eta = 0$. This is expected of a vacuum which is invariant under the de Sitter group. However, just as Poincaré invariance of the Minkowski space vacuum does not guarantee its uniqueness, neither does de Sitter invariance define a unique vacuum in de Sitter space (Chernikov & Tagirov 1968). Thus it has been necessary to invoke the adiabatic prescription for defining positive frequency to choose between the various de Sitter invariant contenders for a vacuum state.

To see how well the chosen 'vacuum' accords with the 'no-particle' state of a comoving detector, one must obtain the positive frequency Wightman function G^+ for this vacuum. This has been calculated as a mode sum using (5.59) by several authors (for example, Schomblond & Spindel 1976, Critchley 1976, Bunch 1977, Bunch & Davies 1978*a*), who find

$$G^+(x, x') = (16\pi\alpha^2)^{-1}(\tfrac{1}{4} - \nu^2)\sec \pi\nu \\ \times F\left(\tfrac{3}{2} + \nu, \tfrac{3}{2} - \nu; 2; 1 + \frac{(\Delta\eta - i\varepsilon)^2 - |\Delta\mathbf{x}|^2}{4\eta\eta'}\right) \tag{5.61}$$

where F is a hypergeometric function. In the massless, conformally coupled limit, (5.61) reduces to

$$D^{+}(x, x') = \frac{-\eta\eta'}{4\pi^2\alpha^2[(\Delta\eta - i\varepsilon)^2 - |\Delta\mathbf{x}|^2]} \tag{5.62}$$

which bears the expected conformal relation to the Minkowski space Wightman function (3.59) (see §3.7).

Substituting (5.62) into (3.55), and setting $\Delta\mathbf{x} = 0$, we can calculate the response function of a comoving observer with proper time t moving in the conformal (equivalently adiabatic) vacuum. Confining attention to the 'steady-state' part of de Sitter space defined by (5.53) we find

$$\begin{aligned}\mathscr{F}(E) &= \frac{1}{16\pi^2\alpha^2}\int_{-\infty}^{\infty} d\left(\frac{t+t'}{2}\right)\int_{-\infty}^{\infty} d\Delta t \frac{e^{-iE\Delta t}}{\sin^2[(i\Delta t/2\alpha) + \varepsilon]} \\ &= \frac{1}{16\pi^4\alpha^2}\int_{-\infty}^{\infty} d\left(\frac{t+t'}{2}\right)\sum_{n=-\infty}^{\infty}\int_{-\infty}^{\infty} d\Delta t \frac{e^{-iE\Delta t}}{[(i\Delta t/2\alpha\pi) + \varepsilon - n]^2}.\end{aligned}$$

Performing a contour integral for Δt, closing the contour in the lower-half complex Δt plane ($E > 0$), one obtains

$$\mathscr{F}(E)/(\text{unit time}) = (E/2\pi)(e^{2\alpha\pi E} - 1)^{-1}, \tag{5.63}$$

which is a thermal spectrum with temperature $T = 1/(2\pi\alpha k_B)$. Thus, although no particles are created, a comoving observer inhabiting this portion of de Sitter space will perceive a thermal bath of radiation. This is another example of the general phenomenon, discussed in §3.7, of the nonzero response of a comoving detector to a conformal vacuum.

The result obtained here is not in conflict with the fact that the conformal vacuum is also an adiabatic vacuum of arbitrary order, for it was shown in §3.6 that the probability for a comoving detector to detect a particle of energy E in such a vacuum should fall to zero faster than any inverse power of E in the limit of large E (see page 78). This is certainly the case with (5.63). Similarly it falls to zero faster than any inverse power of the adiabatic parameter α, which gives a measure of the magnitude of the curvature of the spacetime ($R = 12/\alpha^2$).

A second coordinate system (t, χ, θ, ϕ) frequently employed is defined by

$$\left.\begin{aligned} z_0 &= \alpha \sinh(t/\alpha), \\ z_1 &= \alpha \cosh(t/\alpha)\cos\chi, \\ z_2 &= \alpha \cosh(t/\alpha)\sin\chi\cos\theta, \\ z_3 &= \alpha \cosh(t/\alpha)\sin\chi\sin\theta\cos\phi, \\ z_4 &= \alpha \cosh(t/\alpha)\sin\chi\sin\theta\sin\phi, \end{aligned}\right\} \tag{5.64}$$

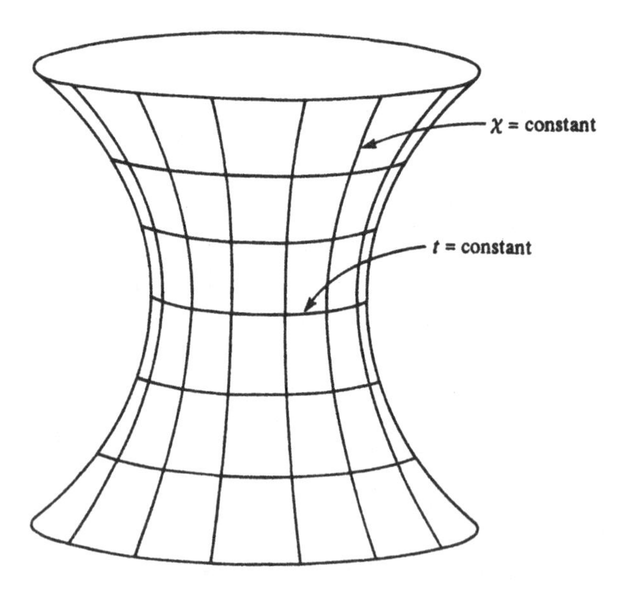

Fig. 17. De Sitter space represented as a contracting and re-expanding compact space with metric (5.65).

in which the line element (5.50) becomes

$$ds^2 = dt^2 - \alpha^2 \cosh^2(t/\alpha)[d\chi^2 + \sin^2\chi(d\theta^2 + \sin^2\theta\, d\phi^2)]. \quad (5.65)$$

If $-\infty < t < \infty$, $0 \le \chi \le \pi$, $0 \le \theta \le \pi$, $0 \le \phi \le 2\pi$ then the coordinates cover the whole de Sitter manifold. The metric (5.65) is that of a $K = +1$ (closed) Robertson–Walker spacetime (see fig. 17).

Introducing the conformal time

$$\eta = 2\tan^{-1}(e^{t/\alpha}), \quad 0 \le \eta < \pi, \quad (5.66)$$

the line element (5.65) can be cast into a form which is manifestly conformal to the Einstein universe:

$$ds^2 = \alpha^2 \sin^{-2}\eta[d\eta^2 - d\chi^2 - \sin^2\chi(d\theta^2 + \sin^2\theta\, d\phi^2)]. \quad (5.67)$$

Using this line element, (5.4) reduces to

$$\frac{d^2\chi_k}{d\eta^2} + \{k^2 + \operatorname{cosec}^2\eta[m^2\alpha^2 + 12(\xi - \tfrac{1}{6})]\}\chi_k = 0, \quad (5.68)$$

which possesses solutions that can be written in terms of Legendre functions (Tagirov 1973, Critchley 1976):

$$\chi_k = \sin^{\frac{1}{2}}\eta[AP^{\nu}_{k-\frac{1}{2}}(-\cos\eta) + BQ^{\nu}_{k-\frac{1}{2}}(-\cos\eta)] \quad (5.69)$$

where, from (5.22), $k = 1, 2, 3, \ldots$ and ν is given by (5.60). The choice of constants A, B determine the vacuum. Unfortunately, because

$$\dot{C}/C = -2\cot\eta \rightarrow \pm\infty \text{ as } t \rightarrow \pm\infty,$$

the spacetime is not slowly expanding in these limits, and we cannot define physically reasonable adiabatic in and out vacua.

This difficulty is highlighted if one applies a Liouville transformation (Liouville 1837, see also Fulling 1979) to (5.68). In this case such a transformation can be effected by changing back from variable η to t and defining

$$\chi_1(t) = \sin^{-\frac{1}{2}}(\eta)\chi(\eta),$$

whence (5.68) becomes

$$\frac{d^2\chi_1}{dt^2} + \alpha^{-2}[\sin^2\eta(k^2 - \tfrac{1}{2}) - \tfrac{1}{4}\cos^2\eta + m^2\alpha^2 + 12(\xi - \tfrac{1}{6})]\chi_1 = 0. \quad (5.70)$$

In the distant past and future, $t \rightarrow \pm\infty$, and this equation reduces to

$$\frac{d^2\chi_1}{dt^2} - (\nu/\alpha)^2\chi_1 = 0$$

with a solution

$$\chi_1 \propto e^{-\nu t/\alpha} \quad (5.71)$$

which is positive frequency with respect to t (for $\nu^2 < 0, \operatorname{Im}\nu > 0$). Some authors (e.g., Gutzwiller 1956, Critchley 1976, Rumpf 1976*a*, *b*; see also Dowker & Critchley 1976*a*) have defined vacuum states by taking positive frequency solutions of (5.68) or (5.70) to be those which behave like (5.71) for large $|t|$. With such a definition they have found that the in and out vacua are not equivalent and that the Bogolubov transformation connecting them is constant (independent of frequency), implying an infinite amount of particle production. (Fulling, 1972, seems to have been the first person to remark on the inappropriate nature of this vacuum. See also Schäfer 1978.) This pathological result is explained by the fact that the regions $t \rightarrow \pm\infty$, and hence the above defined vacuum states, are not adiabatic in this case. However it follows from §3.5 that if one requires the Bogolubov coefficient β_k to fall off for large k sufficiently fast, so as to guarantee a finite total particle density, then it is necessary to use an adiabatic vacuum of order two or more.

Although adiabatic in and out regions do not exist, it is still possible to define adiabatic vacua as being those which are vacuous in the high momentum modes. Recall that the adiabatic modes (3.102) become good approximations to *exact* adiabatic positive frequency modes when

$$T^2\omega_k^2(\eta_1) = T^2k^2 + T^2\operatorname{cosec}^2(\eta_1 T)[m^2\alpha^2 + 12(\xi - \tfrac{1}{6})] \qquad (5.72)$$

becomes large with respect to the derivatives of $\dot{C}/C$ for fixed $\eta_1 = \eta/T$. This will clearly be the case for large k, m or α (for fixed η). It is not, however, the case for large $|t|$, i.e. $\eta \to 0$ or π, whence $T \to 0, \pi/\eta_1$. For example, as $\eta \to \pi$, $T \to \pi/\eta_1$, (5.72) becomes infinite at the same rate as $\mathrm{d}^l(\dot{C}/C)/\mathrm{d}\eta^l$ for $l = 1$, and more slowly for $l > 1$.

In the limit of large k (3.102) yields, to zeroth order

$$\chi_k^{(0)}(\eta) \xrightarrow[k\to\infty]{} \mathrm{e}^{-\mathrm{i}k\eta}/(2|k|)^{\frac{1}{2}}. \qquad (5.73)$$

Choosing constants A, B such that the correctly normalized solution (5.69) reduces to (5.73) for large k, one obtains (Chernikov & Tagirov 1968, Tagirov 1973)

$$\chi_k(\eta) = \sin^{\frac{1}{2}}\eta[\pi\Gamma(k + \tfrac{1}{2} - \nu)/4\Gamma(k + \tfrac{1}{2} + \nu)]^{\frac{1}{2}}\mathrm{e}^{+\mathrm{i}\nu\pi/2}$$
$$\times [P_{k-\frac{1}{2}}^{\nu}(-\cos\eta) - (2\mathrm{i}/\pi)Q_{k-\frac{1}{2}}^{\nu}(-\cos\eta)]. \qquad (5.74)$$

Since the solution (5.74) reduces to (5.73) in the limit of large k, regardless of the value of η, it defines an adiabatic vacuum for all time. Thus, as in the previous coordinatization of de Sitter space there is no 'particle' production; (5.74) defines a stable adiabatic vacuum. Further, in the massless conformally coupled limit the modes (5.74) reduce to the right-hand side of (5.73), so once again the conformal vacuum and the massless limit of the adiabatic vacuum coincide.

One can now use (5.74) to evaluate the positive frequency Wightman function for the adiabatic vacuum (Tagirov 1973). It turns out to be precisely the same function of the coordinates z_μ of the embedding space as (5.61). Thus the vacuum defined in the coordinatization (5.54) is equivalent to that defined in the coordinatization (5.67). Moreover, it turns out that a particle detector at rest in this coordinate system will perceive this vacuum state as a bath of thermal radiation with temperature $T = 1/(2\pi\alpha k_{\mathrm{B}})$, exactly as in the steady-state case (see (5.63)). This result was originally discovered by Gibbons & Hawking (1977*a*), who used it to argue for an observer-dependent formulation of the particle concept in quantum gravitational physics.

The final coordinates that we shall study is the static system defined by

$$\left.\begin{aligned} z_0 &= (\alpha^2 - r^2)^{\frac{1}{2}} \sinh(t/\alpha) \\ z_1 &= (\alpha^2 - r^2)^{\frac{1}{2}} \cosh(t/\alpha) \\ z_2 &= r\sin\theta\cos\phi \\ z_3 &= r\sin\theta\sin\phi \\ z_4 &= r\cos\theta, \quad 0 \leq r < \alpha \end{aligned}\right\} \tag{5.75}$$

which only covers the half of the de Sitter manifold with $z_0 + z_1 > 0$ (see fig. 18). In these coordinates the line element (5.50) becomes

$$ds^2 = [1 - (r^2/\alpha^2)]dt^2 - [1 - (r^2/\alpha^2)]^{-1}dr^2 - r^2(d\theta^2 + \sin^2\theta\, d\phi^2). \tag{5.76}$$

The line element (5.76) possesses a coordinate singularity at $r = \alpha$, which is the event horizon for an observer situated at $r = 0$, following the trajectory of the Killing vector ∂_t. The structure of (5.76) is reminiscent of the Schwarzschild line element (3.18), and one may develop similar field quantizations in the two cases (Gibbons & Hawking 1977*a*, Lapedes 1978*b* and Lohiya & Panchapakesan 1978, 1979; see also Hájíček 1977).

With this formulation of quantum field theory in de Sitter space, the natural choice of field modes and vacuum state associated with the coordinate system (5.75) enjoys a different status from the previous two formulations that we have discussed as the vacuum state is not invariant

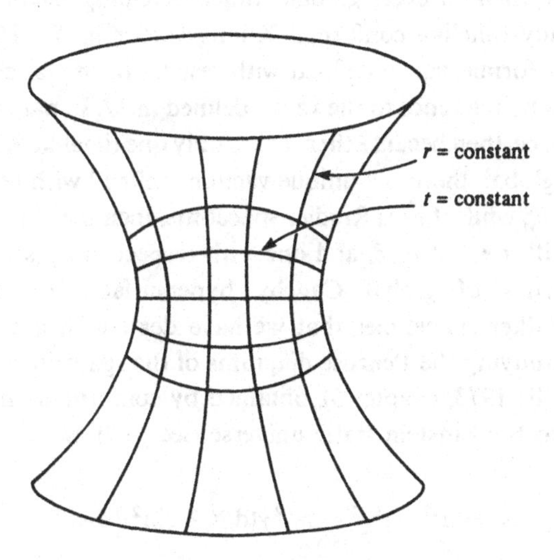

Fig. 18. De Sitter space as a static spacetime with metric (5.76).

under the de Sitter group (see Chernikov & Tagirov 1968). In fact, in §6.3 we shall show that the expectation value of the quantum field stress-tensor in this vacuum state diverges at the event horizon $r = \alpha$. As the location of the horizon is dependent on the origin of radial coordinates, each point in de Sitter space (i.e., each observer) must be associated with a different choice of vacuum state. Hence, this prescription leads to a vacuum that is not even translation invariant, and which comoving observers apparently perceive as a bath of thermal radiation. (See also the discussion of Schrödinger, 1956, §1.4, on the relation between the observer at $r = 0$ and the properties of the static coordinate system.)

5.5 Classification of conformal vacua

Before turning to a more general class of spacetimes, we can consolidate much of the information of the previous examples using a classification scheme proposed by Candelas & Dowker (1979). This approach will also prove to be very useful for discussing stress-tensors in various spacetimes.

The idea behind the scheme is to classify the conformal vacuum states associated with various spacetimes, and it proceeds as follows: Suppose that M is a curved spacetime which is conformally mapped into the flat spacetime $\tilde{M}$ (which need not necessarily be unbounded Minkowski space), and further suppose that Σ is a global Cauchy hypersurface of M which is mapped under the conformal transformation to a global Cauchy hypersurface $\tilde{\Sigma}$ of $\tilde{M}$, then for every globally timelike Killing vector in $\tilde{M}$, there exists a globally timelike conformal Killing vector in M. Thus, we can classify the conformal vacua defined with respect to the latter conformal Killing vectors by reference to the vacua defined in $\tilde{M}$. In particular if $\tilde{M}$ is Minkowski space, then because there exists only one timelike Killing vector field which is global, there is a unique vacuum defined with respect to the Killing vector ∂_t, while if $\tilde{M}$ is Rindler spacetime, then there are two vacua, one defined with respect to ∂_t and one with respect to ∂_η (see (4.82)).

The properties of global Cauchy hypersurfaces of the various Robertson–Walker spacetimes that we have dealt with are most easily discussed by studying the Penrose diagrams of the spacetimes (see §3.1 or Hawking & Ellis 1973, chapter 5), obtained by conformally mapping the spacetimes into the Einstein static universe (see §5.2), which has the line element

$$ds^2 = dt^2 - d\chi^2 - \sin^2\chi(d\theta^2 + \sin^2\theta\, d\phi^2). \tag{5.77}$$

Suppressing the coordinates θ, ϕ, the Einstein universe can be represented

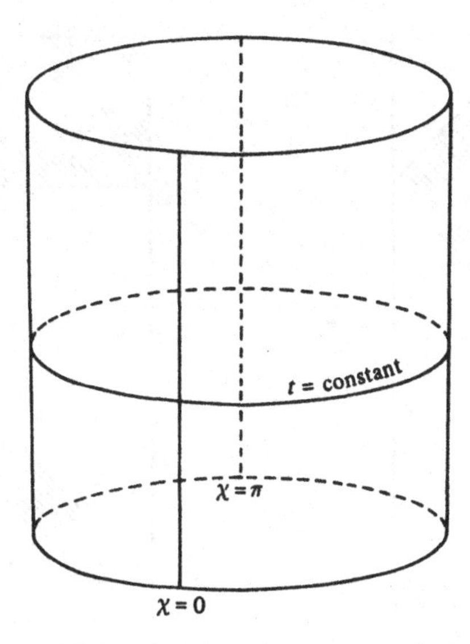

Fig. 19. Representation of the Einstein universe as a cylinder. The coordinates θ and ϕ are suppressed, t runs vertically from $-\infty$ to $+\infty$ and χ runs around the cylinder.

as the cylinder shown in fig. 19. Unwrapping this cylinder gives the region $-\infty < t < \infty$, $0 \leq \chi < \pi$ of the t, χ plane. The conformal images of other spacetimes on this region constitute the Penrose diagrams that we wish to study.

The Robertson–Walker cases are shown in fig. 20.

In the $K = 1, 0, -1$ Robertson–Walker spacetimes, figs. 20*b*, *d*, *f* respectively, the exact range of the conformal time will depend on the form of $C(\eta)$ appearing in (5.11). We have restricted the range to $\eta > 0$, but if this restriction is lifted and the spacetime is extended through what is usually a singularity at $\eta = 0$, then the bottom halves of the diagrams are filled in by reflection in the line $t = 0$. Similarly, if the steady-state universe in the form (5.54) is extended by allowing the range $-\infty < \eta < \infty$, then the top half of fig. 20*c* is filled in by reflection in $t = 0$.

The most important fact revealed by these diagrams is that the spacetimes represented by figs. 20*a*–*d* all have the surface $t = 0, 0 \leq \chi < \pi$ as the common conformal image of a global Cauchy hypersurface, while the spacetimes in figs. 20*e*, *f* have $t = \pi/2$, $0 \leq \chi < \pi/2$ as a common surface.

From this we can immediately classify the spacetimes of figs. 20 into two classes: (i) figs. 20*a*–*d*, only having conformal vacua defined with respect to the conformal Killing vector inherited by conformal transformation from the timelike Killing vector ∂_t of Minkowski space (t being Minkowski time);

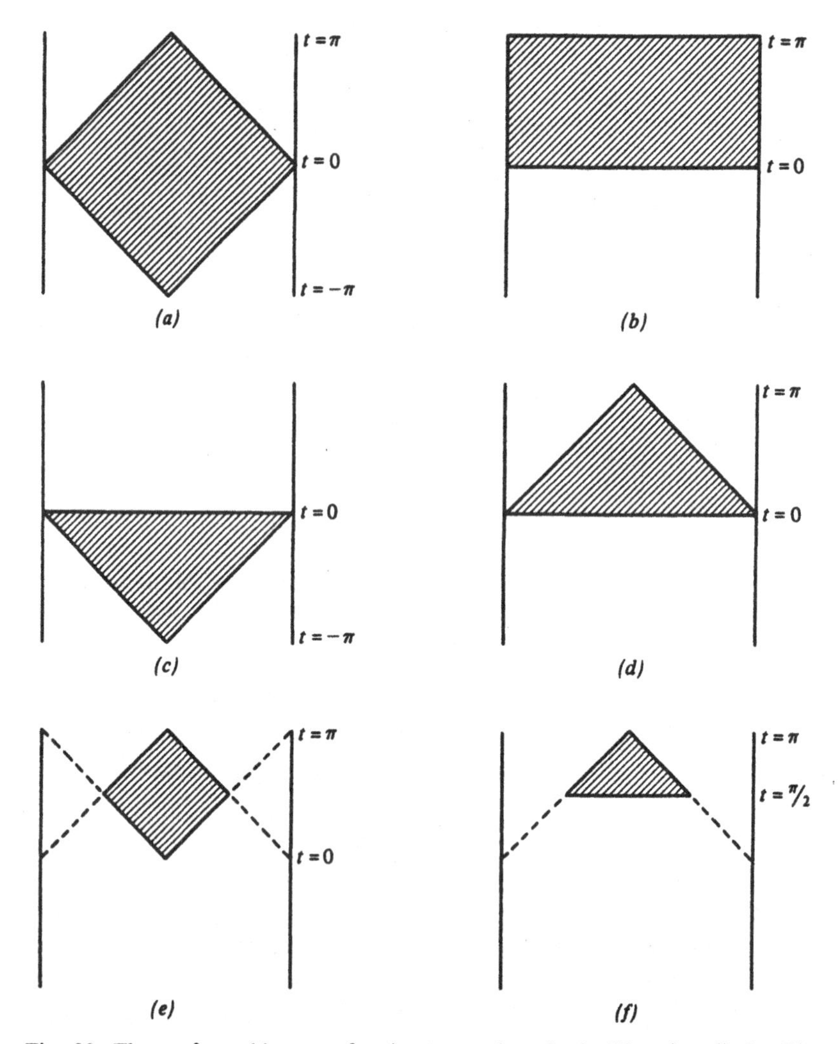

Fig. 20. The conformal images of various spacetimes in the Einstein cylinder. The coordinate t is the time coordinate in the Einstein cylinder (fig. 19). (*a*) Minkowski space, (*b*) de Sitter space and $K = 1$ Robertson – Walker spacetime, (*c*) steady-state universe, (*d*) $K = 0$ Robertson – Walker spacetime, (*e*) the Rindler wedge, static de Sitter space, the open Einstein universe and the Milne universe, (*f*) $K = -1$ Robertson – Walker spacetime. (After Candelas & Dowker 1979.)

(ii) figs. 20*e*, *f*, having two possible conformal vacua defined with respect to the conformal Killing vector inherited from the timelike Killing vector ∂_t and ∂_η of Rindler spacetime.

Since both the Minkowski and Rindler spacetimes are flat, the classification of conformal vacua has been reduced to purely topological considerations. Different topologies can be distinguished by the value of the topological invariant called the Euler–Poincaré characteristic (see, Gilkey 1975), which has been used by Christensen & Duff (1978*b*) to characterize the (Euclideanized) Minkowski and Rindler vacua. It takes the value 1 for Minkowski space (topology R^4) and 0 for Rindler spacetime (topology $R^3 \times S^1$).

What do these topological characterizations mean physically? In §4.5 we showed that the Minkowski vacuum contains a thermal spectrum of Rindler particles at temperature $a/2\pi k_B$. One can also demonstrate this by showing that the Green functions in the Minkowski vacuum are Rindler thermal Green functions (see §2.7). In a similar way one can relate the vacua of other spacetimes in fig. 20 which have similar geometries but different topologies. For example, static de Sitter space (5.76) and de Sitter space (5.65) have the same curvature, but static de Sitter space is a member of the Rindler class, fig. 20*e*, while de Sitter space is a member of the Minkowski class, fig. 20*b*. From this one can deduce that the de Sitter conformal vacuum is given by the thermalization of the static de Sitter ∂_η conformal vacuum at a temperature $1/2\pi\alpha k_B$. Similarly the conformal vacuum of the Einstein universe is the thermalization at imaginary temperature $T = 1/2\pi i k_B$ of the ∂_η conformal vacuum of the open Einstein universe (see Candelas & Dowker 1979 for further discussion).

These results are summarized in fig. 21, where the vertical arrows denote conformal transformations and the horizontal lines denote thermalization at the temperature shown. In addition to this, from our discussion of

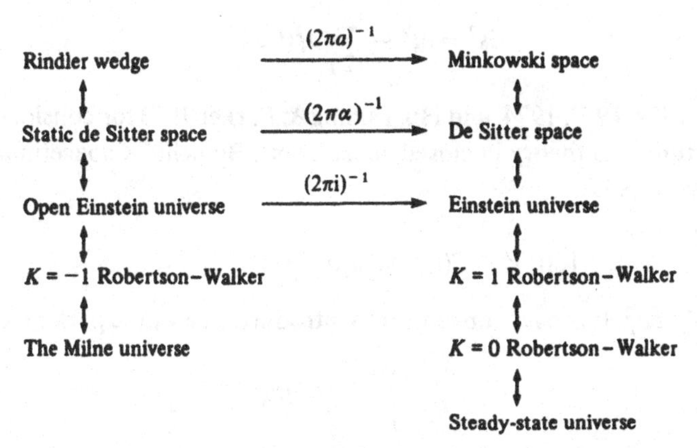

Fig. 21. Thermal and conformal relationships between various spacetimes. The vertical arrows indicate that the spacetimes are related by conformal transformations, while horizontal arrows represent relationship by thermalization at the temperature given. (After Candelas & Dowker 1979).

particle detectors in the steady-state universe (see (5.63)), we know that the conformal vacuum in the steady-state universe can be obtained from the Minkowski vacuum by a thermalization at temperature $1/2\pi\alpha k_B$. That is, the vertical conformal relation in the diagram between the Minkowski and steady-state vacua is also a thermalization relation in this case. One can verify that a similar relation holds between the Einstein and the de Sitter (line element (5.65)) conformal vacua.

5.6 Bianchi I spacetimes and perturbation theory

So far, all of our examples have involved isotropic spacetimes. These are especially interesting because the universe today is observed to be to a high degree of approximation isotropic. The origin of this isotropy has long been a puzzle to cosmologists (see, for example, the review of MacCallum 1979). Amongst other speculations, it has been suggested (Zel'dovich 1970) that the universe may have begun anisotropic, but rapidly isotropized as a result of quantum effects in the primeval phase of the big bang. The investigation of this suggestion is technically difficult, because it involves the back-reaction of quantum field effects, induced by spacetime curvature on the dynamics of the gravitational field. We shall briefly return to this aspect at the end of chapter 7 after the stress–energy–momentum tensor has been discussed. Here we restrict attention to the quantum field theory.

Consider the four-dimensional (spatially flat) Bianchi type I spacetimes (Bianchi 1918) with line element

$$ds^2 = dt^2 - \sum_{i=1}^{3} a_i^2(t)dx_i^2. \tag{5.78}$$

(See Hu 1972, 1973, 1974, and Hu, Fulling & Parker 1973 for consideration of quantum field theory in closed, anisotropic, Bianchi IX spacetimes. See also Berger 1975.)

Define

$$[a(t)]^2 = C(t) \equiv (a_1 a_2 a_3)^{\frac{2}{3}} = (C_1 C_2 C_3)^{\frac{1}{3}}, \tag{5.79}$$

where $C_i \equiv a_i^2$. It proves convenient to introduce a new time parameter η by

$$\eta = \int^t a^{-1}(t')dt', \tag{5.80}$$

which reduces to the conformal time (5.10) in the isotropic limit $a_1 = a_2 = a_3 = a$. If we also define

$$d_i = \dot{C}_i / C_i \tag{5.81}$$

where a dot, as usual, denotes differentiation with respect to η, and

$$D \equiv \tfrac{1}{3} \sum_{i=1}^{3} d_i = \dot{C}/C, \tag{5.82}$$

$$Q \equiv \tfrac{1}{72} \sum_{i<j} (d_i - d_j)^2, \tag{5.83}$$

then the nonzero Christoffel symbols for the metric (5.78) are

$$\Gamma^{\eta}{}_{\eta\eta} = \tfrac{1}{2}D, \quad \Gamma^{\eta}{}_{ii} = \tfrac{1}{2}d_i C_i/C, \quad \Gamma^{i}{}_{i\eta} = \Gamma^{i}{}_{\eta i} = \tfrac{1}{2}d_i. \tag{5.84}$$

From these one may obtain the nonzero components of the Ricci tensor (see, for example, Fulling, Parker & Hu 1974):

$$R_{\eta\eta} = \tfrac{3}{2}\dot{D} + 6Q, \quad R_{ii} = -\tfrac{1}{2}C^{-1}C_i(\dot{d}_i + d_i D), \tag{5.85}$$

and the Ricci scalar:

$$R = C^{-1}[3\dot{D} + \tfrac{3}{2}D^2 + 6Q]. \tag{5.86}$$

Further, defining

$$\left.\begin{aligned} S &= \tfrac{1}{144} \sum_{\substack{i<j \\ k \neq i,j}} (d_k - D)(d_i - d_j)^2 \\ U &= \tfrac{1}{72} \sum_{i<j} (\dot{d}_i - \dot{d}_j)^2 \end{aligned}\right\} \tag{5.87}$$

the square of the Weyl tensor becomes

$$C^{\alpha\beta\gamma\delta}C_{\alpha\beta\gamma\delta} = 4C^{-2}(3U - 4\dot{S} + 12Q^2), \tag{5.88}$$

which gives a measure of the anisotropy of the spacetime. Because it is not conformally flat we are unable to define conformal vacuum states for field theories with conformally invariant Lagrangians, as we did in the case of the Robertson–Walker spacetimes. Adiabatic vacua still exist though (Fulling, Parker & Hu, 1974, Hu 1978, Fulling 1979).

The mode decomposition (3.30) with measure appropriate to flat spatial sections, is

$$\phi(x) = \int d^3k[a_{\mathbf{k}}u_{\mathbf{k}}(x) + a^{\dagger}_{\mathbf{k}}u^{*}_{\mathbf{k}}(x)]. \tag{5.89}$$

Because spatial translation invariance is still a symmetry of Bianchi I models, the modes can be separated (cf. (5.3)):

$$u_{\mathbf{k}}(x) = (2\pi)^{-\frac{3}{2}} e^{i\mathbf{k}\cdot\mathbf{x}} C^{-\frac{1}{4}}(\eta)\chi_{\mathbf{k}}(\eta). \tag{5.90}$$

Substituting this into (3.26) one obtains an equation for $\chi_{\mathbf{k}}$ (cf. (5.4)):

$$\frac{d^2\chi_{\mathbf{k}}}{d\eta^2} + \left\{C(\eta)\left(\sum_{i=1}^{3}\frac{k_i^2}{C_i(\eta)} + m^2 + (\xi - \tfrac{1}{6})R(\eta)\right) + Q(\eta)\right\}\chi_{\mathbf{k}} = 0. \tag{5.91}$$

From (3.28), the normalization condition becomes (cf. (5.6))

$$\chi_{\mathbf{k}}\partial_\eta\chi_{\mathbf{k}}^* - \chi_{\mathbf{k}}^*\partial_\eta\chi_{\mathbf{k}} = i. \tag{5.92}$$

As in §3.5, a positive frequency generalized WKB-type solution to (5.91) may be obtained from

$$\chi_{\mathbf{k}} = (2W_{\mathbf{k}})^{-\frac{1}{2}}\exp\left[-i\int^{\eta} W_{\mathbf{k}}(\eta')d\eta'\right], \tag{5.93}$$

where $W_{\mathbf{k}}$ is constructed to any adiabatic order A (to give $W_{\mathbf{k}}^{(A)}$, $\chi_{\mathbf{k}}^{(A)}$) by iteration from

$$W_{\mathbf{k}}^{(0)} = \omega_{\mathbf{k}} = \left\{C(\eta)\left(\sum_{i=1}^{3}\frac{k_i^2}{C_i} + m^2\right)\right\}^{\frac{1}{2}}. \tag{5.94}$$

One may now go on to define adiabatic vacuum states using the procedure already described for the Robertson–Walker models, i.e., by matching exact positive frequency solutions of (5.91) to the above approximate solutions at some instant of time.

Unfortunately it is extremely difficult to find exact solutions of (5.91). One simple case can, however, be found:

$$a_1(t) = t, \quad a_2 = a_3 = 1, \quad 0 < t < \infty, -\infty < x^i < \infty. \tag{5.95}$$

This spacetime is similar to the Milne universe of §5.3, as may be seen by employing the coordinate transformation

$$y^0 = t\cosh x^1, \quad y^1 = t\sinh x^1, \quad y^2 = x^2, \quad y^3 = x^3 \tag{5.96}$$

under which the line element (5.78) with (5.95) reduces to that for Minkowski space, with the restriction $y^0 > |y^1|$.

In this very simple case, (5.91) can be solved in terms of Bessel or Hankel functions (Fulling, Parker & Hu 1974, Nariai 1976, 1977*a*, *b*, *c*). Inspection then shows that the solution given in terms of the Hankel function $H^{(2)}$ is positive frequency with respect to the adiabatic definition. As in the Milne case, it also turns out to be positive frequency with respect to the usual definition in the covering Minkowski space.

In view of the difficulty in finding exact solutions of (5.91), it is natural to investigate approximation methods for solving it. One such method, which was originally exploited by Zel'dovich & Starobinsky (1971, 1977) and has subsequently been developed by Birrell & Davies (1980*a*), is the solution of (5.91) for small anisotropic perturbations about a Robertson–Walker spacetime (see also §7.3).

We give below an outline of their method. The line element (5.78) is first cast in the form

$$ds^2 = C(\eta)\{d\eta^2 - \sum_{i=1}^{3} [1 + h_i(\eta)](dx^i)^2\} \tag{5.97}$$

and attention restricted to cases in which

$$\max |h_i(\eta)| \ll 1 \tag{5.98}$$

and, for simplicity of the ensuing discussion,

$$\sum_{i=1}^{3} h_i(\eta) = 0. \tag{5.99}$$

We can now expand (5.91) to first order in h_i, to obtain

$$\frac{d^2\chi_{\mathbf{k}}}{d\eta^2} + \left[k^2 + m^2C(\eta) + (\xi - \tfrac{1}{6})C(\eta)R_1(\eta) - \sum_{i=1}^{3} h_i(\eta)k_i^2\right]\chi_{\mathbf{k}} = 0, \tag{5.100}$$

where R_1 is the Ricci scalar for the isotropic spacetime obtained by setting $h_i = 0$:

$$R_1 = C^{-1}[3\dot{D} + \tfrac{3}{2}D^2]. \tag{5.101}$$

Imposing the conditions

$$\left.\begin{array}{ll} \text{(i)} & h_i(\eta) \to 0 \text{ as } \eta \to \pm\infty \\ \text{(ii)} & C(\eta)R(\eta) \to 0 \text{ as } \eta \to \pm\infty \text{ if } \xi \neq \tfrac{1}{6} \\ \text{(iii)} & C(\eta) \to C(\infty) = C(-\infty) < \infty \text{ as } \eta \to \pm\infty \text{ if } m \neq 0, \end{array}\right\} \tag{5.102}$$

then the normalized positive frequency solution of (5.100) as $\eta \to -\infty$ is immediately given as

$$\chi_{\mathbf{k}}^{\text{in}}(\eta) = (2\omega)^{-\frac{1}{2}} e^{-i\omega\eta}, \tag{5.103}$$

with

$$\omega^2 = k^2 + m^2C(\infty). \tag{5.104}$$

A method of relaxing the conditions (ii) and (iii) above using Liouville transformations has been given by Birrell (1979*a*, *c*).

With the initial condition (5.102), (5.100) can be written as an integral equation

$$\chi_{\mathbf{k}}(\eta) = \chi_{\mathbf{k}}^{\text{in}}(\eta) + \omega^{-1} \int_{-\infty}^{\eta} V_{\mathbf{k}}(\eta') \sin[\omega(\eta - \eta')] \chi_{\mathbf{k}}(\eta') \mathrm{d}\eta'. \quad (5.105)$$

where

$$V_{\mathbf{k}}(\eta) = \sum_i h_i(\eta) k_i^2 + m^2[C(\infty) - C(\eta)] - (\xi - \tfrac{1}{6}) C(\eta) R_1(\eta). \quad (5.106)$$

In the late time region (5.105) possesses the solution

$$\chi_{\mathbf{k}}^{\text{out}}(\eta) = \alpha_{\mathbf{k}} \chi_{\mathbf{k}}^{\text{in}}(\eta) + \beta_{\mathbf{k}} \chi_{\mathbf{k}}^{\text{in}*}(\eta), \quad (5.107)$$

where the Bogolubov coefficients are given by

$$\alpha_{\mathbf{k}} = 1 + \mathrm{i} \int_{-\infty}^{\infty} \chi_{\mathbf{k}}^{\text{in}*}(\eta) V_{\mathbf{k}}(\eta) \chi_{\mathbf{k}}(\eta) \mathrm{d}\eta \quad (5.108)$$

$$\beta_{\mathbf{k}} = -\mathrm{i} \int_{-\infty}^{\infty} \chi_{\mathbf{k}}^{\text{in}}(\eta) V_{\mathbf{k}}(\eta) \chi_{\mathbf{k}}(\eta) \mathrm{d}\eta. \quad (5.109)$$

If we treat $V_{\mathbf{k}}(\eta)$ as 'small', then we can solve (5.105) by iteration. To lowest order in $V_{\mathbf{k}}$, one has $\chi_{\mathbf{k}}(\eta) = \chi_{\mathbf{k}}^{\text{in}}(\eta)$, which, when substituted in (5.108) and (5.109) gives the Bogolubov coefficients to first order in $V_{\mathbf{k}}$:

$$\alpha_{\mathbf{k}} = 1 + (\mathrm{i}/2\omega) \int_{-\infty}^{\infty} V_{\mathbf{k}}(\eta) \mathrm{d}\eta \quad (5.110)$$

$$\beta_{\mathbf{k}} = -(\mathrm{i}/2\omega) \int_{-\infty}^{\infty} \mathrm{e}^{-2\mathrm{i}\omega\eta} V_{\mathbf{k}}(\eta) \mathrm{d}\eta. \quad (5.111)$$

The formalism can also be used for cosmological models which expand from the singularity at $\eta = 0$, so long as $V_{\mathbf{k}}(\eta)$ vanishes as $\eta \to 0$. One simply replaces the lower limits of η integrations by zero.

The technique is really one of perturbation expansion about conformal triviality, which is broken by the anisotropy. We shall develop the generalized case, however, in which the presence of a (small) mass or (small) non-conformal coupling also contributes to the conformal symmetry breaking. Thus, for $m = 0$, $\xi = \frac{1}{6}$, the condition of small $V_{\mathbf{k}}$ corresponds simply to the small anisotropy condition (5.98). If $m \neq 0$, $\xi \neq \frac{1}{6}$, then the small $V_{\mathbf{k}}$ approximation will be valid provided m is small and $\xi \simeq \frac{1}{6}$. If $h_i \equiv 0$ these latter conditions imply a small deviation from conformal invariance in a Robertson–Walker spacetime with scale factor $a(\eta) = C^{\frac{1}{2}}(\eta)$.

If the quantum state chosen corresponds to the in vacuum, then in the out region ($\eta \to \infty$) the number density (per unit proper volume) is (cf (3.42))

$$n = (2\pi a)^{-3} \int |\beta_{\mathbf{k}}|^2 \mathrm{d}^3 k \tag{5.112}$$

and the energy density is

$$\rho = (2\pi)^{-3} a^{-4} \int |\beta_{\mathbf{k}}|^2 \omega \, \mathrm{d}^3 k. \tag{5.113}$$

Substituting (5.111) and (5.106) into (5.112), and performing the momentum integral (Birrell & Davies 1980*a*) one obtains for the second order (in $V_{\mathbf{k}}$) approximation to the number density

$$\begin{aligned} n = {} & (960\pi a^3)^{-1} \int_{-\infty}^{\infty} [C^2(\eta) C^{\alpha\beta\gamma\delta} C_{\alpha\beta\gamma\delta}(\eta) + 60 V^2(\eta)] \mathrm{d}\eta \\ & + \bar{m}(960 a^3)^{-1} \int_{-\infty}^{\infty} \mathrm{d}\eta_1 \int_{-\infty}^{\infty} \mathrm{d}\eta_2 \Big\{ F(\bar{m}(\eta_1 - \eta_2)) \Big[-60\, V(\eta_1) V(\eta_2) \\ & + (8\bar{m}^4 - 6\bar{m}^2 \partial_{\eta_1} \partial_{\eta_2} + \tfrac{3}{2} \partial^2_{\eta_1} \partial^2_{\eta_2}) \sum_i h_i(\eta_1) h_i(\eta_2) \Big] \\ & - J_1(2\bar{m}(\eta_1 - \eta_2)) \Big[60 V(\eta_1) V(\eta_2) + \tfrac{1}{2} \sum_i \ddot{h}_i(\eta_1) \ddot{h}_i(\eta_2) \Big] \Big\} \end{aligned} \tag{5.114}$$

where

$$\bar{m}^2 \equiv C(\infty) m^2, \tag{5.115}$$

$$V(\eta) \equiv m^2 [C(\infty) - C(\eta)] - (\xi - \tfrac{1}{6}) C(\eta) R_{\mathrm{I}}(\eta), \tag{5.116}$$

$$F(x) \equiv -(1/\pi) + x[J_0(2x) \mathbf{H}_{-1}(2x) + \mathbf{H}_0(2x) J_1(2x)], \tag{5.117}$$

$\mathbf{H}$ denoting Struve functions and J Bessel functions. We have also used the fact that to second order in h_i, (5.88) reduces to $C^{\alpha\beta\gamma\delta} C_{\alpha\beta\gamma\delta} = \tfrac{1}{2} C^{-2} \sum_i (\ddot{h}_i)^2$. In the massless limit (5.114) reduces to

$$n = (960\pi a^3)^{-1} \int_{-\infty}^{\infty} C^2(\eta) [C^{\alpha\beta\gamma\delta} C_{\alpha\beta\gamma\delta} + 60(\xi - \tfrac{1}{6})^2 R_{\mathrm{I}}^2] \mathrm{d}\eta, \tag{5.118}$$

which is purely geometrical. In arriving at (5.114) and (5.118) we have, in addition to (5.102), imposed the condition

$$\dot{h}_i(\eta), \ddot{h}_i(\eta) \to 0 \quad \text{as } \eta \to \pm\infty. \tag{5.119}$$

If we further demand that

$$\dddot{h}_i(\eta) \to 0 \quad \text{as}\, \eta \to \pm\infty, \tag{5.120}$$

then the second order approximation to the energy density is calculated from (5.113) in a similar way:

$$\rho = (3840\pi^2 a^4)^{-1} \int_{-\infty}^{\infty} \mathrm{d}\eta_1 \int_{-\infty}^{\infty} \mathrm{d}\eta_2 \operatorname{Re} K_0(2\mathrm{i}\bar{m}(\eta_1 - \eta_2))(\partial_{\eta_1}\partial_{\eta_2} - 4\bar{m}^2)$$
$$\times \left[120V(\eta_1)V(\eta_2) + (\partial_{\eta_1}\partial_{\eta_2} - 4\bar{m}^2)^2 \sum_i h_i(\eta_1)h_i(\eta_2) \right], \tag{5.121}$$

where K_0 is a modified Bessel function.

In the massless limit (5.121) reduces to

$$\rho = -(3840\pi^2 a^4)^{-1} \int_{-\infty}^{\infty} \mathrm{d}\eta_1 \int_{-\infty}^{\infty} \mathrm{d}\eta_2 \ln[2\mathrm{i}\mu(\eta_1 - \eta_2)]$$
$$\times \left\{ 120\dot{V}(\eta_1)\dot{V}(\eta_2) + \sum_i \dddot{h}_i(\eta_1)\dddot{h}_i(\eta_2) \right\}. \tag{5.122}$$

In this expression μ is an arbitrary mass which can be changed without altering the value of ρ, because

$$\int_{-\infty}^{\infty} \dot{V}(\eta)\mathrm{d}\eta = \int_{-\infty}^{\infty} \dddot{h}_i(\eta)\mathrm{d}\eta = 0.$$

Unfortunately there are very few choices of h_i or V for which the integrals in (5.114) or (5.121) can be evaluated in terms of known functions, though these expressions are still of great value for numerical calculations. A related method of numerically (and sometimes analytically) calculating *exact* results has been given by Birrell (1979*a*, *c*).

A wider class of problems can be solved in closed form by returning to (5.111) to perform the η integration before calculating n or ρ from (5.112) and (5.113). As an example, consider the spacetime with line element (5.97) and

$$h_i(\eta) = \mathrm{e}^{-\alpha\eta^2} \cos(\beta\eta^2 + \delta_i) \tag{5.123}$$

where α, β, δ_i are constants, and (5.99) is satisfied by demanding that the δ_i differ from one another by $2\pi/3$.

From (5.106) and (5.111) the anisotropic contribution to the Bogolubov coefficient $\beta_{\mathbf{k}}$ is found to be

$$\beta_{\mathbf{k}} = -\frac{\mathrm{i}\pi^{\frac{1}{2}}}{2\omega} \sum_i k_i^2 \operatorname{Re}\left(\frac{\mathrm{e}^{-\omega 2/(\alpha + \mathrm{i}\beta)}}{(\alpha + \mathrm{i}\beta)^{\frac{1}{2}}} \mathrm{e}^{-\mathrm{i}\delta_i} \right). \tag{5.124}$$

Substituting this into (5.113) gives the following contribution to the energy density in the out region:

$$\rho = \frac{\bar{m}^2}{1536\pi^{\frac{1}{2}}a^4} \frac{(\alpha^2+\beta^2)^{\frac{3}{2}}}{\alpha^2} \mathrm{e}^{-3\alpha\bar{m}^2/(\alpha^2+\beta^2)} W_{-\frac{3}{2},\frac{3}{2}}\left(\frac{2\alpha\bar{m}^2}{\alpha^2+\beta^2}\right), \tag{5.125}$$

where W is a Whittaker function. In the limit $\xi = \frac{1}{6}$, $m = 0$, this is the only contribution to the energy density and it reduces to

$$\rho = \frac{1}{2880\pi} \frac{(\alpha^2+\beta^2)^{\frac{3}{2}}}{\alpha^3 a^4}. \tag{5.126}$$

Since only condition (i) of (5.102) applies in this limit, the energy density given by (5.126) is valid for any choice of scale factor $a(\eta) = C^{\frac{1}{2}}(\eta)$, allowing various cosmological hypotheses to be tested (Birrell & Davies 1980*a*, see also §7.4).

6
Stress-tensor renormalization

In previous chapters the production of quanta by a changing gravitational field was studied in detail. It was pointed out that only in exceptional circumstances does the particle concept in curved space quantum field theory correspond closely to the intuitive physical picture of a subatomic particle. In general, no natural definition of particle exists, and particle detectors will respond in a variety of ways that bear no simple relation to the usual conception of the quantity of matter present.

For some purposes it is advantageous to study the expectation values of other observables. Part of the problem with the particle concept concerns the fact that it is defined globally, in terms of field modes, and so is sensitive to the large scale structure of spacetime. In contrast, physical detectors are at least quasi-local in nature. It therefore seems worthwhile to investigate physical quantities that are defined locally, i.e., at a spacetime point, rather than globally. One such object of interest is the stress–energy–momentum (or stress) tensor, $T_{\mu\nu}(x)$, at the point x. In addition to describing part of the physical structure of the quantum field at x, the stress-tensor also acts as the source of gravity in Einstein's field equation. It therefore plays an important part in any attempt to model a self-consistent dynamics involving the gravitational field coupled to the quantum field. For many investigators, especially astrophysicists, it is this back-reaction of the quantum processes on the background geometry that is of primary interest.

In this chapter, we shall present the formalism necessary for the calculation of $\langle T_{\mu\nu} \rangle$ for a variety of quantum fields. It is a subject that involves some unusual subtleties, and requires delicate handling. Much of the chapter is devoted to a careful discussion of several procedures, known as regularization techniques, for computing a finite, renormalized $\langle T_{\mu\nu} \rangle$ from an apparently meaningless infinite quantity. While the reader is likely to encounter any one of these procedures in the published literature, with some notable exceptions, the end results are independent of which particular method of regularization is employed. We present all the major techniques, both for completeness and to demonstrate their mutual consistency and interrelation. It is not necessary, however, to master each individual method to understand the later sections, where we apply the results to some background spacetimes of special interest.

In much of our treatment we work with the action rather than $T_{\mu\nu}$, because this permits a more elegant approach, especially when combined with complex analytic methods. However, the treatment is very formal, and often involves the manipulation of quantities that are not obviously well-defined for the problem under study. The reader may well feel uneasy with such unfamiliar quantities as $\ln G_F$, and prefer the more 'nuts and bolts' approach involving $\langle T_{\mu\nu} \rangle$ directly, especially when combined with the non-analytic regularization techniques.

Along with chapter 3, this chapter is fundamental to the subject of the book. It is considerably longer and in some ways more technical than the other chapters. On a first reading we recommend that attention be mainly directed to the dimensional continuation method of regularization, and the topic of conformal anomalies. Then, after gaining familiarity with the application of dimensional regularization to renormalization, and having studied some specific examples, such as de Sitter space in §6.4, the reader will want to study the other techniques. In particular, point-splitting and adiabatic regularization offer the best starting point for those wishing to embark upon actual calculations involving explicit models. This chapter contains all the basic information necessary to begin such calculations.

The final section of the chapter can be read profitably even without a full understanding of regularization and renormalization theory.

Early discussion of regularization in curved space quantum field theory can be found in the work of Utiyama & DeWitt (1962), and Halpern (1967). Some physical conjectures were made by Sakharov (1967) and followed up by Grib & Mamaev (1969).

A somewhat different formalism from that described here has been given by Kibble & Randjbar-Daemi (1980) and Randjbar-Daemi, Kay & Kibble (1980) within the framework of Kibble's (1978) nonlinear generalization of quantum mechanics.

6.1 The fundamental problem

In §2.4 it was pointed out that the expectation value of H, even in the Minkowski vacuum state, is infinite. Again in §§4.1–4.3 we found that $\langle 0|T_{\mu\nu}|0\rangle$ was ultraviolet divergent. This behaviour is symptomatic of the problems that afflict any attempt to evaluate the expectation values of an operator that is quadratic in the field strength. For example, $\langle 0|\phi^2(x)|0\rangle$ may be obtained from the limit as $x' \to x$ of $G^{(1)}(x, x')$ (see (2.66)). But inspection of the DeWitt–Schwinger expansion (3.141) reveals that, quite generally, G_F (and hence $G^{(1)}$) diverges like $\sigma^{(2-n)/2}$ (or $\ln \sigma$ for $n=2$) as $\sigma \to 0$.

In Minkowski space quantum field theory, the divergence is simply discarded, for example by the use of normal ordering (see (2.41)). When the topology is non-trivial, but the geometry still flat, as in §§4.1.–4.2, one may employ an ultraviolet regulator function $e^{-\alpha|k|}$, to cut off the ultraviolet divergence, and then take the difference between $\langle T_{\mu\nu} \rangle$ in the topology of interest and its (cut-off) value in Minkowski space, letting $\alpha \to 0$ at the end of the calculation. Alternatively, in the Green function approach, the unbounded Minkowski space expression for $G^{(1)}(x, x')$ is first subtracted from the $G^{(1)}$ function evaluated in the topology of interest, and only after this manoeuvre is the limit $x \to x'$ taken.

There are two reasons why these simple devices cannot be trusted when spacetime is curved. The first concerns the rôle of $T_{\mu\nu}$ in gravity theory. In non-gravitational physics, only energy differences are observable, so that an infinite vacuum energy causes little embarrassment – one simply renormalizes the zero point by an infinite amount. When gravity is taken into account, however, this is not satisfactory. Energy is a source of gravity, and will bring about the very spacetime curvature whose effects we are trying to study. It cannot simply be thrown away; we are not free to rescale the zero point of energy. Instead, a more elaborate renormalization scheme is required involving the dynamics of the gravitational field.

A second reason that we run into trouble can be illustrated by a simple example. Consider the spatially flat Robertson–Walker spacetime described by the scale factor

$$a(t) = (1 - A^2 t^2)^{\frac{1}{2}}, \quad A \text{ constant} \tag{6.1}$$

which has physical singularities at $t = \pm A^{-1}$.

In four dimensions the massless scalar field satisfying the wave equation

$$\Box \phi = 0 \tag{6.2}$$

is not conformally coupled, but mode solutions are easily found (Bunch & Davies 1978*b*)

$$u_{\mathbf{k}} = (16\pi^3)^{-\frac{1}{2}} C^{-\frac{1}{2}}(\eta)(k^2 + A^2)^{-\frac{1}{4}} \exp\left[i\mathbf{k}\cdot\mathbf{x} - i(k^2 + A^2)^{\frac{1}{2}}\eta\right] \tag{6.3}$$

where $C(\eta) = a^2(\eta) = \cos^2 A\eta$ and η is the conformal time.

A Fock space may be constructed from the modes (6.3), and the vacuum state $|0\rangle$ used to evaluate $\langle 0| T_{\mu\nu} |0\rangle$. From (3.190) we find for the minimally coupled case $\xi = 0$

$$T_{\mu\nu} = \phi_{,\mu}\phi_{,\nu} - \tfrac{1}{2}g_{\mu\nu}g^{\sigma\rho}\phi_{,\sigma}\phi_{,\rho} \tag{6.4}$$

which is of the same form as the flat space expression (2.26) with $m = 0$. Using the mode formula (2.43) (which is valid in curved space), one obtains for the energy density

$$\langle 0|T_0{}^0|0\rangle = (1/32\pi^3C^2)\int d^3k[(k^2 + A^2)^{\frac{1}{2}} + (k^2 + \tfrac{1}{4}D^2)(k^2 + A^2)^{-\frac{1}{2}}] \tag{6.5}$$

where $D(\eta) = C^{-1}\partial C/\partial\eta$.

The integral diverges quartically as expected. If an ultraviolet cut-off factor $\exp[-\alpha(k^2 + A^2)^{\frac{1}{2}}]$ is introduced into (6.5), the integrals may be performed in terms of MacDonald functions. Expanding in powers of α, one readily obtains

$$\rho a^4 = (32\pi^2)^{-1}[48/\alpha^4 + (D^2 - 8A^2)/\alpha^2 + A^2(\tfrac{1}{2}D^2 - A^2)\ln\alpha] + O(\alpha^0) \tag{6.6}$$

where we have written the energy density as ρ and C^2 as a^4. The left-hand side of (6.6) represents the energy of massless radiation in a volume a^3, which redshifts as the universe expands.

To recover the Minkowski space result from (6.6) one may put $a = 1$ and $D = A = 0$. Only the first term on the right-hand side of (6.6) remains. It follows that in curved spacetime even if this term is discarded, ρa^4 still diverges in the limit $\alpha \to 0$. That is, the difference in zero-point energy between the Robertson–Walker universe and Minkowski space is still infinite. We cannot cure the divergence in $\langle 0|T_{00}|0\rangle$ simply by discarding a Minkowski-type term.

Inspection of (6.6) reveals that in addition to the expected quartic divergence, we have to contend with quadratic and logarithmic divergent terms also. Evidently the control of infinities in $\langle T_{\mu\nu}\rangle$ will involve a considerably more elaborate procedure than in the case of flat spacetime.

How are we to make physical sense out of a divergent result? To obtain a finite answer will obviously entail the subtraction of infinite quantities, but this can be done in an infinite variety of ways. Some additional criteria must be imposed in order that a unique answer be obtained. Divergences also plague Minkowski space quantum field theory, especially when field interactions are permitted. In quantum electrodynamics, infinite subtractions may be carried out systematically to yield finite results that are in good agreement with experiment, provided the subtractions are performed covariantly (see, for example, Schwinger 1951*a*, Valatin 1954*a*, *b*, *c*). Therefore we should endeavour to maintain general coordinate invariance when handling the divergences of $\langle T_{\mu\nu}\rangle$. In addition to maintaining general covariance, one might also require $\langle T_{\mu\nu}\rangle$ to possess a certain number of 'physically reasonable' properties. If enough such restrictions are imposed on $\langle T_{\mu\nu}\rangle$, then the subtraction procedure might be defined uniquely. In §6.6 we shall see that such an approach can indeed be implemented.

An alternative strategy is to treat the computation of $\langle T_{\mu\nu}\rangle$ as part of a wider dynamical theory involving gravity. In the semiclassical theory being considered in this book, the gravitational field is treated classically, while the matter fields (including the graviton to one-loop level – see chapter 1) are treated quantum mechanically. This is reminiscent of the successful semiclassical theory of electrodynamics, where the classical electromagnetic field is coupled to the *expectation value* of the electric current operator. By analogy, we seek a theory based on Einstein's field equation

$$R_{\mu\nu} - \tfrac{1}{2}Rg_{\mu\nu} + \Lambda g_{\mu\nu} = -8\pi G T_{\mu\nu}, \tag{6.7}$$

but with the stress-tensor source regarded now as a quantum expectation value:

$$R_{\mu\nu} - \tfrac{1}{2}Rg_{\mu\nu} + \Lambda_{\mathrm{B}} g_{\mu\nu} = -8\pi G_{\mathrm{B}}\langle T_{\mu\nu}\rangle. \tag{6.8}$$

The reason for the subscripts B on the cosmological constant Λ and Newton's constant G, will be made clear in due course. In this chapter we shall explicitly incorporate factors of G, which elsewhere are set equal to one.

The classical Einstein equation (6.7) can be derived from the action

$$S = S_{\mathrm{g}} + S_{\mathrm{m}} \tag{6.9}$$

by the condition

$$\frac{2}{(-g)^{\frac{1}{2}}}\frac{\delta S}{\delta g^{\mu\nu}} = 0. \tag{6.10}$$

The first term on the right of (6.9) is the gravitational action

$$S_{\mathrm{g}} = \int L_{\mathrm{g}}(-g)^{\frac{1}{2}}\,\mathrm{d}^n x = \int (-g)^{\frac{1}{2}}(16\pi G_{\mathrm{B}})^{-1}(R - 2\Lambda)\mathrm{d}^n x \tag{6.11}$$

for which $2(-g)^{-\frac{1}{2}}\delta S_{g}/\delta g^{\mu\nu}$ yields the left-hand side of (6.7). The second term in (6.9) is the classical matter action, for which (cf. (3.189))

$$\frac{2}{(-g)^{\frac{1}{2}}}\frac{\delta S_{\mathrm{m}}}{\delta g^{\mu\nu}} = T_{\mu\nu}, \tag{6.12}$$

yields the right-hand side of (6.7).

How can this procedure work in the semiclassical case (6.8)? We seek a quantity W, called the *effective action* for the quantum matter fields, which, when functionally differentiated, yields $\langle T_{\mu\nu}\rangle$:

$$\frac{2}{(-g)^{\frac{1}{2}}}\frac{\delta W}{\delta g^{\mu\nu}} = \langle T_{\mu\nu}\rangle, \tag{6.13}$$

where the precise meaning of $\langle\;\rangle$ used here on $T_{\mu\nu}$ will be elucidated below.

To discover the structure of W, let us return to first principles, recalling the path-integral quantization procedure outlined in §2.8. Our notation will imply a treatment for the scalar field, but the formal manipulations are identical for fields of higher spins. In §2.8, the generating functional

$$Z[J] = \int \mathscr{D}[\phi]\exp\left\{\mathrm{i}S_{\mathrm{m}}[\phi] + \mathrm{i}\int J(x)\phi(x)\mathrm{d}^n x\right\} \tag{6.14}$$

was interpreted physically as the vacuum persistence amplitude $\langle \text{out}, 0|0, \text{in}\rangle$. The presence of the external current J can cause the initial vacuum state $|0, \text{in}\rangle$ to be unstable, i.e., it can bring about the production of particles. In flat space, in the limit $J = 0$, no particles are produced, and we have the normalization condition

$$Z[0] \equiv \langle \text{out}, 0|0, \text{in}\rangle_{J=0} = \langle 0|0\rangle = 1. \tag{6.15}$$

However, when spacetime is curved, we have seen that, in general, $|0, \text{out}\rangle \neq |0, \text{in}\rangle$, even in the absence of source currents J. Hence (6.15) will no longer apply.

Path-integral quantization still works in curved spacetime; one simply treats S_{m} in (6.14) as the curved spacetime matter action, and $J(x)$ as a current density (a scalar density in the case of scalar fields). One can thus set $J = 0$ in (6.14) and examine the variation of $Z[0]$:

$$\begin{aligned}\delta Z[0] &= \mathrm{i}\int \mathscr{D}[\phi]\delta S_{\mathrm{m}} e^{\mathrm{i}S_{\mathrm{m}}[\phi]} \\ &= \mathrm{i}\langle \text{out}, 0|\delta S_{\mathrm{m}}|0, \text{in}\rangle,\end{aligned} \tag{6.16}$$

which is a statement of Schwinger's variational principle (Schwinger 1951*b*). From (6.16) and (6.12) one immediately obtains

$$\frac{2}{(-g)^{\frac{1}{2}}}\frac{\delta Z[0]}{\delta g^{\mu\nu}} = \mathrm{i}\langle \text{out}, 0|T_{\mu\nu}|0, \text{in}\rangle. \tag{6.17}$$

Noting that the matter action S_{m} appears exponentiated in (6.14), we may therefore identify

$$Z[0] = e^{iW}, \tag{6.18}$$

whence,

$$W = -i \ln \langle \text{out}, 0|0, \text{in} \rangle \tag{6.19}$$

and, from (6.17),

$$\frac{2}{(-g)^{\frac{1}{2}}} \frac{\delta W}{\delta g^{\mu\nu}} = \frac{\langle \text{out}, 0| T_{\mu\nu} |0, \text{in} \rangle}{\langle \text{out}, 0|0, \text{in} \rangle} \tag{6.20}$$

The functional $Z[0]$ is evaluated in much the same way as in flat spacetime (see page 30). The main differences arise from (i) the replacement of the measure $d^n x$ by the covariant measure $d^n x[-g(x)]^{\frac{1}{2}}$; (ii) the replacement of the identity $\delta^n(x-y)$ by $\delta^n(x-y)[-g(y)]^{-\frac{1}{2}}$ for which

$$\int d^n x[-g(x)]^{\frac{1}{2}} \delta^n(x-y)[-g(y)]^{-\frac{1}{2}} = 1;$$

(iii) the replacement of (2.120) by

$$K_{xy} = (\Box_x + m^2 - i\varepsilon + \xi R)\delta^n(x-y)[-g(y)]^{-\frac{1}{2}} \tag{6.21}$$

which, using

$$\int d^n y[-g(y)]^{\frac{1}{2}} K_{xy} K_{yz}^{-1} = \delta(x-z)[-g(z)]^{-\frac{1}{2}} \tag{6.22}$$

and (3.49), implies

$$K_{xz}^{-1} = -G_F(x, z) \tag{6.23}$$

(cf. (2.123)). Following through steps similar to those on page 30 one finds (as in flat spacetime)

$$Z[0] \propto [\det(-G_F)]^{\frac{1}{2}} \tag{6.24}$$

where the proportionality constant is metric-independent and can be ignored. Thus we arrive at

$$W = -i \ln Z[0] = -\tfrac{1}{2} i \,\text{tr}\, [\ln(-G_F)]. \tag{6.25}$$

In (6.25) G_F is to be interpreted as an operator which acts on a space of vectors $|x\rangle$, normalized by

$$\langle x|x' \rangle = \delta^n(x-x')[-g(x)]^{-\frac{1}{2}}, \tag{6.26}$$

in such a way that

$$G_F(x, x') = \langle x|G_F|x'\rangle. \tag{6.27}$$

The trace of an operator M which acts in this space, is defined by

$$\operatorname{tr} M = \int d^n x[-g(x)]^{\frac{1}{2}} M_{xx} = \int d^n x[-g(x)]^{\frac{1}{2}} \langle x|M|x\rangle. \tag{6.28}$$

To make sense of the formal expression (6.23) we must use a representation for the Feynman Green function G_F. We shall use the DeWitt–Schwinger representation given by the proper time integral (3.138). Writing the operator equivalent of (6.23) as

$$G_F = -K^{-1} = -i\int_0^\infty e^{-iKs} ds, \tag{6.29}$$

we have from (3.138) the useful result

$$\langle x|e^{-iKs}|x'\rangle = i(4\pi)^{-n/2}\Delta^{\frac{1}{2}}(x, x')e^{-im^2 s + \sigma/2is} F(x, x'; is)(is)^{-n/2}. \tag{6.30}$$

Now, assuming K to have a small negative imaginary part,

$$\int_\Lambda^\infty e^{-iKs}(is)^{-1} i ds = -\operatorname{Ei}(-i\Lambda K) \tag{6.31}$$

where Ei is the exponential integral function, having for small values of its argument the expansion

$$\operatorname{Ei}(x) = \gamma + \ln(-x) + O(x), \tag{6.32}$$

γ being Euler's constant. Substituting (6.32) into (6.31) and letting $\Lambda \to 0$ yields

$$\ln(-G_F) = -\ln(K) = \int_0^\infty e^{-iKs}(is)^{-1} i ds, \tag{6.33}$$

which is correct up to the addition of a metric-independent (infinite) constant that can be ignored in what follows. Thus, in the DeWitt–Schwinger representation (6.30) (or (3.138))

$$\langle x|\ln(-G_F^{DS})|x'\rangle = -\int_{m^2}^\infty G_F^{DS}(x, x') dm^2, \tag{6.34}$$

where the integral with respect to m^2 brings down the extra power of $(is)^{-1}$ that appears in (6.33).

Returning to the expression (6.25) for W, (6.28) and (6.34) yield

$$W = \tfrac{1}{2}\mathrm{i}\int \mathrm{d}^n x[-g(x)]^{\frac{1}{2}} \lim_{x'\to x}\int_{m^2}^{\infty} \mathrm{d}m^2 G_{\mathrm{F}}^{\mathrm{DS}}(x, x'). \tag{6.35}$$

Interchanging the order of integration and taking the limit $x'\to x$ one obtains

$$W = \tfrac{1}{2}\mathrm{i}\int_{m^2}^{\infty} \mathrm{d}m^2 \int \mathrm{d}^n x[-g(x)]^{\frac{1}{2}} G_{\mathrm{F}}^{\mathrm{DS}}(x, x),$$

and the $\mathrm{d}^n x$ integral is seen to be precisely the expression for the one-loop Feynman diagram fig. 1*a* (see chapter 9 for Feynman rules). Thus W is known as the *one-loop effective action*. In the case of fermion effective actions, there would be a remaining trace over spinorial indices.

From (6.35) we may define an *effective Lagrangian density* $\mathscr{L}_{\mathrm{eff}}$, by

$$W = \int \mathscr{L}_{\mathrm{eff}}(x)\mathrm{d}^n x \equiv \int [-g(x)]^{\frac{1}{2}} L_{\mathrm{eff}}(x)\mathrm{d}^n x \tag{6.36}$$

whence

$$L_{\mathrm{eff}}(x) = [-g(x)]^{-\frac{1}{2}}\mathscr{L}_{\mathrm{eff}}(x) = \tfrac{1}{2}\mathrm{i}\lim_{x'\to x}\int_{m^2}^{\infty} \mathrm{d}m^2 G_{\mathrm{F}}^{\mathrm{DS}}(x, x'). \tag{6.37}$$

Inspection of (3.137) and (3.138) shows that L_{eff} diverges at the lower end of the s integral because the $\sigma/2s$ damping factor in the exponent vanishes in the limit $x'\to x$. (Convergence at the upper end is guaranteed by the $-\mathrm{i}\varepsilon$ that is implicitly added to m^2 in the DeWitt–Schwinger representation of G_{F}.) In four dimensions, the potentially divergent terms in the DeWitt–Schwinger expansion of L_{eff} are

$$L_{\mathrm{div}} = -\lim_{x'\to x}\frac{\Delta^{\frac{1}{2}}(x, x')}{32\pi^2}\int_0^{\infty}\frac{\mathrm{d}s}{s^3}\mathrm{e}^{-\mathrm{i}(m^2 s - \sigma/2s)}[a_0(x, x')$$

$$+ a_1(x, x')\mathrm{i}s + a_2(x, x')(\mathrm{i}s)^2] \tag{6.38}$$

where the coefficients a_0, a_1 and a_2 are given by (3.131)–(3.133). The remaining terms in this asymptotic expansion, involving a_3 and higher, are finite in the limit $x'\to x$.

The divergences in L_{eff} are, of course, the same as those that afflict $\langle T_{\mu\nu}\rangle$. Inspection of (3.131)–(3.133) in the limit $x'\to x$ reveals that the term in square brackets in (6.38) is entirely geometrical, i.e., it is built out of local tensors, $R_{\mu\nu\sigma\tau}$, and its contractions. This feature is readily understood. The divergences arise because of the ultraviolet behaviour of the field modes. These short wavelengths, however, only probe the local geometry in the neighbourhood of x – they are not sensitive to the global features of the spacetime, such as the topology or past behaviour. They are also independent of the quantum state employed, as we shall see below.

Because L_{div} is purely geometrical, it is better regarded as a contribution to the *gravitational* rather than the quantum matter Lagrangian. Although it arises from the action of the quantum matter field, it behaves as a quantity constructed solely from the gravitational field (the metric). Of course this will not be true of the remaining, finite portions of L_{eff}, which include the long wavelength part. This can probe the large scale structure of the manifold, and is also sensitive to the quantum state.

6.2 Renormalization in the effective action

Let us determine the precise form of the geometrical L_{div} terms, to compare them with the conventional gravitational Lagrangian L_g that appears in (6.11). This is a delicate matter because (6.38) is, of course, infinite. What we require is to display the divergent terms in the form $\infty \times$ geometrical object. This can be done in a variety of ways.

For example, in n dimensions, the asymptotic (adiabatic) expansion of L_{eff} is

$$L_{\text{eff}} \approx \lim_{x' \to x} \frac{\Delta^{\frac{1}{2}}(x, x')}{2(4\pi)^{n/2}} \sum_{j=0}^{\infty} a_j(x, x') \int_0^{\infty} (\mathrm{i}s)^{j-1-n/2} \mathrm{e}^{-\mathrm{i}(m^2 s - \sigma/2s)} \mathrm{i} \, \mathrm{d}s \tag{6.39}$$

of which the first $\frac{1}{2}n + 1$ terms are divergent as $\sigma \to 0$. If n is treated as a variable which can be analytically continued throughout the complex plane, then we may take the $x' \to x$ limit

$$L_{\text{eff}} \approx \tfrac{1}{2}(4\pi)^{-n/2} \sum_{j=0}^{\infty} a_j(x) \int_0^{\infty} (\mathrm{i}s)^{j-1-n/2} \mathrm{e}^{-\mathrm{i}m^2 s} \mathrm{i} \, \mathrm{d}s \tag{6.40}$$

$$= \tfrac{1}{2}(4\pi)^{-n/2} \sum_{j=0}^{\infty} a_j(x)(m^2)^{n/2-j} \Gamma(j - n/2) \tag{6.41}$$

where $a_j(x) \equiv a_j(x, x)$.

In what follows we shall wish to retain the units of L_{eff} as $(\text{length})^{-4}$, even when $n \neq 4$. It is therefore necessary to introduce an arbitrary mass scale μ and to rewrite (6.41) as

$$L_{\text{eff}} \approx \tfrac{1}{2}(4\pi)^{-n/2}(m/\mu)^{n-4} \sum_{j=0}^{\infty} a_j(x) m^{4-2j} \Gamma(j - n/2). \tag{6.42}$$

As $n \to 4$, the first three terms of (6.42) diverge because of poles in the Γ-functions:

$$\left.\begin{aligned}\Gamma\left(-\frac{n}{2}\right) &= \frac{4}{n(n-2)}\left(\frac{2}{4-n}-\gamma\right)+O(n-4)\\ \Gamma\left(1-\frac{n}{2}\right) &= \frac{2}{2-n}\left(\frac{2}{4-n}-\gamma\right)+O(n-4)\\ \Gamma\left(2-\frac{n}{2}\right) &= \frac{2}{4-n}-\gamma+O(n-4).\end{aligned}\right\} \tag{6.43}$$

Calling these first three terms L_{div}, we have (see Bunch 1979)

$$L_{\mathrm{div}} = -(4\pi)^{-n/2}\left\{\frac{1}{n-4}+\frac{1}{2}\left[(\gamma+\ln\left(\frac{m^2}{\mu^2}\right)\right]\right\}\left(\frac{4m^4a_0}{n(n-2)}-\frac{2m^2a_1}{n-2}+a_2\right) \tag{6.44}$$

where we have used the expansion

$$(m/\mu)^{n-4} = 1+\tfrac{1}{2}(n-4)\ln(m^2/\mu^2)+O((n-4)^2) \tag{6.45}$$

and dropped terms in (6.44) that vanish when $n \to 4$.

The functions a_0, a_1 and a_2 are given by taking the coincidence limits of (3.131)–(3.133) and (3.129):

$$a_0(x) = 1 \tag{6.46}$$

$$a_1(x) = (\tfrac{1}{6}-\xi)R \tag{6.47}$$

$$a_2(x) = \frac{1}{180}R_{\alpha\beta\gamma\delta}R^{\alpha\beta\gamma\delta} - \frac{1}{180}R^{\alpha\beta}R_{\alpha\beta} - \tfrac{1}{6}(\tfrac{1}{5}-\xi)\Box R + \tfrac{1}{2}(\tfrac{1}{6}-\xi)^2R^2. \tag{6.48}$$

It is now evident that L_{div} as given by (6.44) is a purely geometrical expression. At this stage we recall that L_{eff} is only part of the total Lagrangian. There is also a gravitational part. Because L_{div} is purely geometrical, we can try to absorb it into the gravitational Lagrangian.

Using (6.11), the total gravitational Lagrangian density now becomes $(-g)^{\frac{1}{2}}$ multiplied by

$$-\left(A+\frac{\Lambda_B}{8\pi G_B}\right)+\left(B+\frac{1}{16\pi G_B}\right)R-\frac{a_2(x)}{(4\pi)^{n/2}}\left\{\frac{1}{n-4}+\frac{1}{2}\left[\gamma+\ln\left(\frac{m^2}{\mu^2}\right)\right]\right\} \tag{6.49}$$

where

$$A = \frac{4m^4}{(4\pi)^{n/2}n(n-2)}\left\{\frac{1}{n-4}+\frac{1}{2}\left[\gamma+\ln\left(\frac{m^2}{\mu^2}\right)\right]\right\}$$

and

$$B = \frac{2m^2(\frac{1}{6} - \xi)}{(4\pi)^{n/2}(n-2)}\left\{\frac{1}{n-4} + \frac{1}{2}\left[\gamma + \ln\left(\frac{m^2}{\mu^2}\right)\right]\right\}.$$

The first term in (6.49) is a constant. The contribution from L_{div}, i.e., the term A, is indistinguishable physically from Λ_B. This is the part of the gravitational Lagrangian that gives rise to the so-called cosmological term $\Lambda g_{\mu\nu}$ in the gravitational field equations (6.8). Hence, the effect of the scalar quantum field is to change, or renormalize, the cosmological constant from Λ_B to

$$\Lambda \equiv \Lambda_B + \frac{32\pi m^4 G_B}{(4\pi)^{n/2} n(n-2)}\left\{\frac{1}{n-4} + \tfrac{1}{2}\left[\gamma + \ln\left(\frac{m^2}{\mu^2}\right)\right]\right\}. \tag{6.50}$$

Because a physical observation will only yield the renormalized value, Λ, we need not ask about the value of Λ_B, nor worry about the fact that, when we finally let $n \to 4$, the term in curly brackets in (6.50) diverges: we never see this term in isolation. This technique of absorbing an infinite quantity into a renormalized physical quantity is familiar in quantum field theory. For example, in quantum electrodynamics, an electron is 'dressed' in a cloud of virtual photons that contribute (infinitely) to the total mass of the electron. We only measure the renormalized electron mass (which is, of course, finite). The electron is inseparable from its photon cloud, so we never see the 'bare' mass. Adopting this terminology, we may say that the 'bare' cosmological constant, Λ_B, is never observed either. The appellation 'bare' prompts the use of the B subscript.

Turning to the second term in (6.49), we see that L_{div} also renormalizes Newton's gravitational constant, by changing G_B to

$$G = G_B/(1 + 16\pi G_B B). \tag{6.51}$$

The final term in (6.49) is not to be found in the usual Einstein Lagrangian. The factor $a_2(x)$ is of adiabatic order four, being of fourth order in derivatives of the metric (see (6.48)), and so represents a higher order correction to the general theory of relativity, which only contains terms with up to second derivatives of the metric. When this extra term is inserted into S_g, the left-hand side of the field equation becomes modified to

$$R_{\mu\nu} - \tfrac{1}{2}Rg_{\mu\nu} + \Lambda g_{\mu\nu} + \alpha\,{}^{(1)}H_{\mu\nu} + \beta\,{}^{(2)}H_{\mu\nu} + \gamma H_{\mu\nu} \tag{6.52}$$

where

$${}^{(1)}H_{\mu\nu} \equiv \frac{1}{(-g)^{\frac{1}{2}}}\frac{\delta}{\delta g^{\mu\nu}}\int(-g)^{\frac{1}{2}}R^2\,\mathrm{d}^n x = 2R_{;\mu\nu} - 2g_{\mu\nu}\Box R - \tfrac{1}{2}g_{\mu\nu}R^2 + 2RR_{\mu\nu} \tag{6.53}$$

$$
\begin{aligned}
{}^{(2)}H_{\mu\nu} &\equiv \frac{1}{(-g)^{\frac{1}{2}}}\frac{\delta}{\delta g^{\mu\nu}}\int(-g)^{\frac{1}{2}}R^{\alpha\beta}R_{\alpha\beta}\mathrm{d}^n x \\
&= R_{;\mu\nu} - \tfrac{1}{2}g_{\mu\nu}\Box R - \Box R_{\mu\nu} - \tfrac{1}{2}g_{\mu\nu}R^{\alpha\beta}R_{\alpha\beta} + 2R^{\alpha\beta}R_{\alpha\beta\mu\nu} \\
&= 2R_{\mu\ ;\nu\alpha}^{\ \alpha} - \Box R_{\mu\nu} - \tfrac{1}{2}g_{\mu\nu}\Box R + 2R_{\mu}^{\ \alpha}R_{\alpha\nu} - \tfrac{1}{2}g_{\mu\nu}R^{\alpha\beta}R_{\alpha\beta}
\end{aligned} \tag{6.54}
$$

$$
\begin{aligned}
H_{\mu\nu} &\equiv \frac{1}{(-g)^{\frac{1}{2}}}\frac{\delta}{\delta g^{\mu\nu}}\int(-g)^{\frac{1}{2}}R^{\alpha\beta\gamma\delta}R_{\alpha\beta\gamma\delta}\mathrm{d}^n x \\
&= -\tfrac{1}{2}g_{\mu\nu}R^{\alpha\beta\gamma\delta}R_{\alpha\beta\gamma\delta} + 2R_{\mu\alpha\beta\gamma}R_{\nu}^{\ \alpha\beta\gamma} - 4\Box R_{\mu\nu} + 2R_{;\mu\nu} \\
&\quad - 4R_{\mu\alpha}R^{\alpha}{}_{\nu} + 4R^{\alpha\beta}R_{\alpha\mu\beta\nu}.
\end{aligned} \tag{6.55}
$$

Note that in the special case $n = 4$, the generalized Gauss–Bonnet theorem (Chern 1955, 1962) states that

$$
\int \mathrm{d}^4x[-g(x)]^{\frac{1}{2}}(R_{\alpha\beta\gamma\delta}R^{\alpha\beta\gamma\delta} + R^2 - 4R_{\alpha\beta}R^{\alpha\beta}) \tag{6.56}
$$

is a topological invariant (called the Euler number), so that its metric variation will vanish identically. It then follows from (6.53)–(6.55) that

$$
H_{\mu\nu} = -{}^{(1)}H_{\mu\nu} + 4\,{}^{(2)}H_{\mu\nu}. \tag{6.57}
$$

The coefficients α, β and γ in (6.52) all contain $1/(n-4)$ and so diverge as n approaches the physical dimension 4. We must therefore introduce terms of adiabatic order 4 into the original gravitational Lagrangian with bare coefficients a_B, b_B, c_B into which the divergent terms involving α, β, γ can be absorbed to yield renormalized coefficients a, b, c. Because of (6.57), at the physical dimension $n = 4$, only two of these coefficients are independent, so we may choose $c = 0$. The values of a and b can only be determined by experiment. In order to avoid conflict with observation, it is necessary to assume that both a and b are very small numerically (see, for example, Stelle 1977, 1978, Horowitz & Wald 1978). In principle there is no reason why these renormalized quantities may not be set equal to zero, thus recovering Einstein's theory. Quantum field theory merely indicates that terms involving higher derivatives of the metric are *a priori* expected.

The device of allowing n to be continued analytically away from the physical dimension enables us to render the formally divergent L_{div} temporarily finite, so that its component pieces may be manipulated meaningfully and absorbed into the gravitational Lagrangian. The technique of altering a formally divergent expression for such purposes is called *regularization*. The approach used above is known as dimensional regularization and was first used in interacting quantum field theories in Minkowski space (Ashmore 1972, Bollini & Giambiagi 1972, 't Hooft & Veltman 1972). At the end of the calculation, when renormalization of the bare constants has been achieved, the regularization may be relaxed, e.g., $n \to 4$.

Once the terms L_{div} have been removed from L_{eff}, the remainder is finite and will be called the renormalized effective Lagrangian:

$$L_{\text{ren}} \equiv L_{\text{eff}} - L_{\text{div}}. \tag{6.58}$$

In four dimensions the asymptotic expansion of L_{ren} will consist of all terms with $j \geq 3$ in (6.40). Putting $x' = x$ and $n = 4$ we can write this asymptotic expansion as

$$L_{\text{ren}} \approx \frac{1}{32\pi^2} \int_0^\infty \sum_{j=3}^{\infty} a_j(x)(is)^{j-3} e^{-im^2 s} i ds \tag{6.59}$$

which may be integrated three times by parts to yield (see (3.140))

$$-\frac{1}{64\pi^2} \int_0^\infty \ln(is) \frac{\partial^3}{\partial(is)^3} [F(x, x; is) e^{-ism^2}] d(is)$$

$$+\frac{1}{64\pi^2} \int_0^\infty \ln(is) \frac{\partial^3}{\partial(is)^3} \{[(a_0 + a_1(is) + a_2(is)^2] e^{-ism^2}\} i ds. \tag{6.60}$$

The latter, finite, term simply renormalizes Λ, G, a, b and c by finite amounts, i.e., it is of the same form as L_{div}, involving constant $\times\, a_j$, $j = 1, 2, 3$. Obviously the renormalized effective Lagrangian will always be ambiguous up to terms of this type – finite renormalization terms – so we can drop the second term in (6.60).

For the same reason we need not worry about the choice of the mass scale μ introduced in (6.42). Rescaling μ changes L_{div} by a finite amount, but only by altering the coefficients of the geometrical terms a_0, a_1 and a_2. In practice, one would choose a fixed value of μ and use the results of one's calculations with this value of μ to calibrate the instruments used to measure the constants Λ, G, a and b. Once these constants have been measured, further calculations using this same value of μ and the measured values of the constants can be used to make predictions about the outcome of experiments using the previously calibrated instruments. If the value of μ is changed one must either recalibrate one's instruments or else change the values of Λ, G, a and b. The effect of either of these changes will leave invariant the predictions made about the outcome of experiments. Such an analysis of the rescaling of μ in interacting field theories in Minkowski space leads to a renormalization group equation ('t Hooft 1973, see also §9.2). Therefore we may write in place of (6.60)

$$L_{\text{ren}} = -\frac{1}{64\pi^2}\int_0^\infty \ln(\text{i}s)\frac{\partial^3}{\partial(\text{i}s)^3}[F(x, x; \text{i}s)\text{e}^{-\text{i}sm^2}]\text{i}\,\text{d}s \tag{6.61}$$

where it is understood that any finite multiple of a_0, a_1 or a_2 may be added to this expression. Having been derived from an asymptotic expansion for F, (6.61) cannot be regarded as the complete Lagrangian associated with the physical, renormalized $\langle T_{\mu\nu}\rangle$, the construction of which is deferred until §6.4. We display L_{ren} in the above form for completeness only. It is worth noting, however, that in principle the complete renormalized Lagrangian could be computed from (6.61) if the exact expression for F were available.

Besides dimensional regularization, other regularization techniques are available. Consider an eigenfunction expansion of the operator K^{-1}:

$$K^{-1} = -G_{\text{F}} = \sum_m \frac{|m\rangle\langle m|}{\lambda_m} \tag{6.62}$$

where

$$K|m\rangle = \lambda_m|m\rangle, \tag{6.63}$$

$$\langle n|m\rangle = \delta_{nm},$$

and

$$\sum_m |m\rangle\langle m| = 1. \tag{6.64}$$

Then

$$K^\nu|m\rangle = \lambda_m^\nu|m\rangle \tag{6.65}$$

so

$$(-G_{\text{F}})^\nu = \sum_m \lambda_m^{-\nu}|m\rangle\langle m| \tag{6.66}$$

whence

$$\text{tr}(-G_{\text{F}})^\nu = \int \text{d}^4x[-g(x)]^{\frac{1}{2}}\sum_m \lambda_m^{-\nu}\langle x|m\rangle\langle m|x\rangle = \sum_m \lambda_m^{-\nu}, \tag{6.67}$$

where we have used the completeness relation

$$\int \text{d}^4x[-g(x)]^{\frac{1}{2}}|x\rangle\langle x| = 1. \tag{6.68}$$

The right-hand side of (6.67) is reminiscent of Riemann's ζ-function, $\sum_{m=1}^{\infty} m^{-\nu}$, and may be used to define a *generalized ζ-function* (Dowker & Critchley 1976*a*, *b*, 1977*b*, Hawking 1977*a*, Gibbons 1977*a*):

$$\operatorname{tr}(-G_F)^{\nu} = \sum_m \lambda_m^{-\nu} \equiv \zeta(\nu). \tag{6.69}$$

We now wish to obtain the effective action (6.25) in terms of the generalized ζ-function. First we note that the argument of the logarithm in (6.25) should strictly be dimensionless. As $G_F(x, x')$ has dimensions $(\text{mass})^{n-2}$ and $|x\rangle$, from (6.26), has dimensions of $(\text{mass})^{n/2}$, we see from (6.27) that the operator G_F has dimensions of $(\text{mass})^{-2}$. We can thus make the argument of the logarithm in (6.25) dimensionless for all n by inserting a factor μ^2, writing

$$W = -\tfrac{1}{2}\mathrm{i}\operatorname{tr}[\ln(-\mu^2 G_F)]. \tag{6.70}$$

We have, in fact, already adjusted (6.24) by just such a constant in going to (6.25). The parameter μ plays a similar role in ζ-function regularization to the μ introduced in (6.42) as part of dimensional regularization. As before, changes in μ result only in finite changes to the inherently ambiguous coefficients of the renormalization terms.

The effective action (6.70) can be written in terms of the generalized ζ-function as

$$\begin{aligned} W &= -\tfrac{1}{2}\mathrm{i} \lim_{\nu\to 0} \operatorname{tr}\frac{\mathrm{d}}{\mathrm{d}\nu}(-\mu^2 G_F)^{\nu} \\ &= \lim_{\nu\to 0}\{-\tfrac{1}{2}\mathrm{i}\mu^{2\nu}[\zeta'(\nu) + \zeta(\nu)\ln\mu^2]\}. \end{aligned} \tag{6.71}$$

The introduction of the generalized ζ-function must be understood as a purely formal operation. In general (6.69) will not converge for all values of ν. However, it may be defined by analytic continuation from regions where it does converge. We then find that $\zeta(0)$ and $\zeta'(0)$ are finite, allowing us to write

$$W = -\tfrac{1}{2}\mathrm{i}[\zeta'(0) + \zeta(0)\ln\mu^2]. \tag{6.72}$$

Thus, by this formal analytic process, the divergences in the effective action W have been eliminated. It is perhaps worth comparing this result with the corresponding result for the usual Riemann ζ-function: The series $\sum_{m=1}^{\infty} m^{-\nu}$ is clearly divergent if ν is set equal to zero, yet Riemann's ζ-function, defined by this series for $\nu > 1$, when analytically continued to $\nu = 0$, gives the finite value $-\tfrac{1}{2}$.

To show that $\zeta(0)$ and $\zeta'(0)$ appearing in (6.72) are finite, we shall evaluate them explicitly in terms of the DeWitt–Schwinger proper time representation. We first note that

$$\int_0^\infty (\mathrm{i}s)^{\nu-1}\mathrm{e}^{-\mathrm{i}Ks}\mathrm{d}(\mathrm{i}s) = K^{-\nu}\Gamma(\nu), \tag{6.73}$$

from which

$$(-G_\mathrm{F})^\nu = K^{-\nu} = [\Gamma(\nu)]^{-1}\int_0^\infty (\mathrm{i}s)^{\nu-1}\mathrm{e}^{-\mathrm{i}Ks}\mathrm{i}\,\mathrm{d}s. \tag{6.74}$$

Now using (6.69) and (6.30) one obtains in four dimensions

$$\zeta(\nu) = \mathrm{i}[\Gamma(\nu)]^{-1}(4\pi)^{-2}\int \mathrm{d}^4x[-g(x)]^{\frac{1}{2}}\int_0^\infty (\mathrm{i}s)^{\nu-3}\mathrm{e}^{-\mathrm{i}m^2s}F(x,x;\mathrm{i}s)\mathrm{i}\,\mathrm{d}s. \tag{6.75}$$

If Re $\nu > 2$, we can perform three integrations by parts. Recalling that m^2 is understood to mean $m^2 - \mathrm{i}\varepsilon$, one finds that the boundary terms vanish, to give

$$\zeta(\nu) = -\frac{\mathrm{i}(4\pi)^{-2}}{\Gamma(\nu+1)(\nu-1)(\nu-2)}\int \mathrm{d}^4x[-g(x)]^{\frac{1}{2}}\int_0^\infty (\mathrm{i}s)^\nu\frac{\partial^3}{\partial(\mathrm{i}s)^3}$$
$$\times [F(x,x;\mathrm{i}s)\mathrm{e}^{-\mathrm{i}sm^2}]\mathrm{i}\,\mathrm{d}s, \tag{6.76}$$

from which one obtains

$$\zeta(0) = \mathrm{i}(4\pi)^{-2}\int \mathrm{d}^4x[-g(x)]^{\frac{1}{2}}[\tfrac{1}{2}m^4 - m^2a_1(x) + a_2(x)], \tag{6.77}$$

where we have used (3.140) to evaluate

$$\frac{\partial^2}{\partial(\mathrm{i}s)^2}[F(x,x;\mathrm{i}s)\mathrm{e}^{-\mathrm{i}sm^2}]_{s=0}.$$

Since $\zeta(0)$ only contains finite renormalization terms, it is now clear that changes of μ in (6.72) will only result in finite changes of the renormalized coefficients in the complete action. Thus, for our present purposes we can ignore the $\zeta(0)$ term in (6.72).

Differentiation of (6.76) with respect to ν yields

$$\zeta'(0) = \tfrac{1}{2}\mathrm{i}(4\pi)^{-2}\Big\{(\gamma - \tfrac{3}{2})\int \mathrm{d}^4x[-g(x)]^{\frac{1}{2}}[\tfrac{1}{2}m^4 - m^2a_1(x) + a_2(x)]$$
$$- \int \mathrm{d}^4x[-g(x)]^{\frac{1}{2}}\int_0^\infty \ln(\mathrm{i}s)\frac{\partial^3}{\partial(\mathrm{i}s)^3}[F(x,x;\mathrm{i}s)\mathrm{e}^{-\mathrm{i}sm^2}]\mathrm{i}\,\mathrm{d}s\Big\}, \tag{6.78}$$

which is finite. The first integral in this expression only contains finite renormalization terms which once again can be ignored. Inserting the remaining term into (6.72) and using the definition (6.36), one obtains

$$L_{\text{ren}} = -\frac{1}{64\pi^2}\int_0^\infty \ln(\text{i}s)\frac{\partial^3}{\partial(\text{i}s)^3}[F(x,x;\text{i}s)\text{e}^{-\text{i}sm^2}]\text{i}\,\text{d}s, \qquad (6.79)$$

$$= L_{\text{eff}} + \text{finite renormalization terms}$$

which agrees with the result (6.61) obtained by dimensional regularization.

Note that the ζ-function technique described here does not require explicit infinite renormalization of coupling constants in the gravitational Lagrangian. (This is not the case, however, for the method adopted by Dowker & Critchley 1976*a*, *b*, 1977*b*.) The analytic continuation method converts a manifestly infinite series into a finite result (e.g., $\sum_{m=1}^{\infty} m^{-\nu}$ with $\nu = 0$ becomes $\frac{1}{2}$). Clearly, an infinite term has been tacitly discarded in this formal procedure.

It might therefore be supposed that variations of the technique would produce different answers. For example, if instead of (6.71), the logarithm in (6.70) were represented as follows

$$\lim_{\nu\to 0}[\nu^{-1}(-\mu^2 G_{\text{F}})^{\nu} - \nu^{-1}]$$

then we would have

$$W = -\tfrac{1}{2}\text{i}\lim_{\nu\to 0}\nu^{-1}\zeta(0) - \tfrac{1}{2}\text{i}[\zeta'(0) + \zeta(0)\ln\mu^2] + c \qquad (6.80)$$

where c is an infinite, metric-independent constant that can be safely discarded. Equation (6.80) differs from (6.72) only in the first term which, on inspection of (6.77), is seen to represent an infinite renormalization of the gravitational action. Thus, whether or not an infinite renormalization is carried out, the *renormalized* effective action is the same.

There is still an element of doubt about whether the renormalized effective action contains some ambiguity. Can one be sure that these renormalization procedures, involving questionable formal manipulations of divergent quantities, will always give the 'correct' answer? This issue can only be resolved by first deciding what physical criteria a 'correct' answer should satisfy, a subject to be investigated in §6.6. Secondly, a rigorous mathematical foundation for the formal manipulations given here is necessary. Such a treatment of flat space quantum field theory has been given by, for example, Taylor (1960, 1963) and Caianello (1973).

Another frequently employed regularization technique is called point-splitting, or point-separation. The basic idea is to return to (6.37), keeping x'

and x separated by an infinitesimal distance in a non-null direction. So long as $\sigma \neq 0$, L_{eff} remains finite, and the integral in (6.38) can be performed in terms of Hankel functions. (Alternatively, the first three terms of the sum in (3.138) can be substituted into (6.37) to give the same result). Expanding the Hankel functions in an asymptotic expansion in powers of σ and retaining only terms which do not vanish as $\sigma \to 0$, one obtains (Christensen 1978), in the four-dimensional case

$$\begin{aligned} L_{\text{div}} = \lim_{x' \to x} (1/8\pi^2)\Delta^{\frac{1}{2}}(x,x')\{a_0(x,x')[\sigma^{-2} + \tfrac{1}{2}m^2\sigma^{-1} \\ - \tfrac{1}{4}m^4 \times (\gamma + \tfrac{1}{2}\ln|\tfrac{1}{2}m^2\sigma|) + \tfrac{3}{16}m^4] \\ - a_1(x,x')[\tfrac{1}{2}\sigma^{-1} - \tfrac{1}{2}m^2(\gamma + \tfrac{1}{2}\ln|\tfrac{1}{2}m^2\sigma|) + \tfrac{1}{4}m^2] \\ - \tfrac{1}{2}a_2(x,x')[\gamma + \tfrac{1}{2}\ln|\tfrac{1}{2}m^2\sigma|]\}. \end{aligned} \tag{6.81}$$

The quantities $a_0(x,x')$, $a_1(x,x')$, $a_2(x,x')$ and $\Delta(x,x')$ may now be expanded in powers of σ. The leading term in the resulting expansion of (6.81) diverges like σ^{-2}. The coefficients in the expansion are all geometrical, involving the Riemann tensor and its derivatives, but a glance at (3.132) and (3.133) shows that when $x' \neq x$, these coefficients will contain vector quantities arising from the y^α factors. From (3.136), or general theory (DeWitt 1965), these vectors can be written in terms of σ as

$$y^\mu = \sigma^{;\mu} \equiv \sigma^\mu, \tag{6.82}$$

whence (3.136) reads

$$\sigma = \tfrac{1}{2}\sigma_\mu\sigma^\mu. \tag{6.83}$$

It is convenient to parametrize the strength of the divergences by one quantity ε, proportional to the geodesic distance between x' and x, and parametrize separately the (non-null) direction of splitting by a unit vector t^μ (see fig. 22). This is done by writing

$$\sigma^\mu = 2\varepsilon t^\mu \tag{6.84}$$

where

$$t^\mu t_\mu \equiv \Sigma = \pm 1, \tag{6.85}$$

depending on whether t^μ is timelike or spacelike respectively. Then (6.83) becomes

$$\sigma = 2\varepsilon^2\Sigma. \tag{6.86}$$

(The factor of 2 in (6.86) is introduced for later convenience.) In this

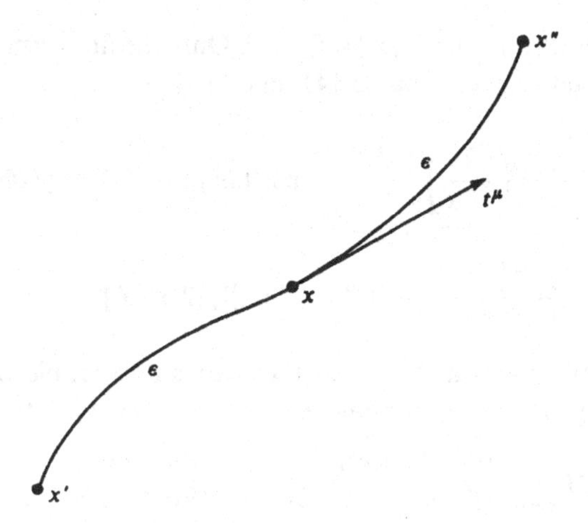

Fig. 22. The points x'', x' lie along a non-null geodesic through x, parametrized by t^μ, each a proper distance ε from x.

parametrization, the leading divergence in L_{div} behaves like ε^{-4}, and a typical direction-dependent coefficient will have the form

$$R_{\mu\nu}R_{\rho\sigma}t^\mu t^\nu t^\rho t^\sigma.$$

(Further details will be given in §6.4.) Such terms cannot be absorbed by renormalization, as the vector t^μ is a purely artificial construct and not part of the dynamics.

To rid the theory of direction-dependent terms, one may average over all directions using a suitable measure (Adler, Lieberman & Ng 1977). The troublesome terms then reduce to combinations of R, R^2, $R_{\alpha\beta}R^{\alpha\beta}$ and $R_{\alpha\beta\gamma\delta}R^{\alpha\beta\gamma\delta}$, which may be removed in the usual way by renormalization of Λ, G, a, b and c in the gravitational action. The finite remainder, $L_{\text{ren}} = L_{\text{eff}} - L_{\text{div}}$, will obviously reproduce (6.59) and (6.61), and so agrees with both dimensional and ζ-function regularization. If we are interested in the field equations rather than the action, then no renormalization of c is necessary. This is because the point-splitting technique works throughout the computation in four dimensions, for which (6.56) is a topological invariant and hence does not contribute to the field equations when substituted into the functional derivative (6.10). It may therefore be used to eliminate one of the three constants a, b and c from the field equations.

It is instructive to examine further the close relation between the point separation and dimensional techniques of regularization. Both start from the DeWitt–Schwinger representation (6.39) in which the s integration can be performed (recall (3.141)). Consider the simplest case of all: Minkowski space. Then $a_0(x,x')=1, a_j(x,x')=0, j>0$. Only the first term of the series contributes and one has from (3.141) and (6.37)

$$L_{\text{eff}}=\frac{\pi}{2(4\pi\mathrm{i})^{n/2}}\lim_{\sigma\to 0}\left(\frac{2}{-\sigma}\right)^{(n-2)/4}\int_{m^2}^{\infty}\mathrm{d}\bar{m}^2(\bar{m}^2)^{(n-2)/4}H^{(2)}_{\frac{1}{2}n-1}[(2\bar{m}^2\sigma)^{\frac{1}{2}}]$$

$$=\lim_{\sigma\to 0}\frac{(-1)^{-(n+1)/2}\pi}{2(4\pi)^{n/2}}(2m^2)^{n/4}\sigma^{-n/4}H^{(2)}_{n/2}[(2m^2\sigma)^{\frac{1}{2}}]. \tag{6.87}$$

If dimensional regularization is used, then we want to be able to set $\sigma=0$ in (6.87) and expand the result about $n=4$. From the asymptotic forms

$$H^{(2)}_\nu(z)\underset{z\to 0}{\longrightarrow}\begin{cases}(\mathrm{i}/\pi)\Gamma(\nu)(\frac{1}{2}z)^{-\nu}, & \operatorname{Re}\nu>0 \qquad (6.88a)\\ (\mathrm{i}/\pi)\mathrm{e}^{-\pi\nu\mathrm{i}}\Gamma(-\nu)(\frac{1}{2}z)^{\nu}, & \operatorname{Re}\nu<0 \qquad (6.88b)\end{cases}$$

it is clear that, for (6.87) to remain finite as $\sigma\to 0$, one must continue n to have negative real part, in which case

$$L_{\text{eff}}\underset{\sigma\to 0}{\longrightarrow}\frac{(-1)^{-n}}{2(4\pi)^{n/2}}m^n\Gamma\left(-\frac{n}{2}\right)\underset{n\to 4}{\longrightarrow}\frac{-m^4}{32\pi^2(n-4)}.$$

If, on the other hand, point-separation is used, then we want to be able to set $n=4$ at the outset and use σ as the regulator. In this case (6.88a) must be used, and one obtains

$$L_{\text{eff}}\underset{\sigma\to 0}{\longrightarrow}\frac{(-1)^{-n/2}2^{\frac{1}{2}n-1}}{(4\pi)^{n/2}}\Gamma\left(\frac{n}{2}\right)\sigma^{-n/2}\underset{n\to 4}{\longrightarrow}\frac{1}{8\pi^2\sigma^2}.$$

Note that in the massless case the divergent term vanishes completely in dimensional regularization and no renormalization at all is necessary. In contrast, in the point-separation expression, all m-dependence disappears so that L_{eff} still diverges like σ^{-2} in the massless limit. Renormalization is therefore necessary in this method.

In the case of two-dimensional spacetimes, inspection of (6.41) shows that for $n=2$ only the $j=0$ and 1 terms in the DeWitt–Schwinger series are potentially divergent. These terms may be removed by renormalizing Λ and G in the effective action as before. It should be noted that

$$\int \mathrm{d}^2x[-g(x)]^{\frac{1}{2}}R(x)$$

is a topological invariant in two dimensions, so that the variation of this term in the effective action gives no contribution to the two-dimensional 'Einstein equation'. Put another way, the two-dimensional Einstein tensor,

$R_{\mu\nu} - \frac{1}{2}Rg_{\mu\nu}$, vanishes identically. We must, however, include a term involving R in the effective action if we wish to remove all of the divergences by renormalization. The renormalized effective Lagrangian is (cf. (6.61))

$$L_{\mathrm{ren}} = -\frac{1}{8\pi}\int_0^\infty \ln(\mathrm{i}s)\frac{\partial^2}{\partial(\mathrm{i}s)^2}[F(x,x;\mathrm{i}s)\mathrm{e}^{-\mathrm{i}sm^2}]\mathrm{d}(\mathrm{i}s). \tag{6.89}$$

In the case of higher-spin fields the above treatment goes through almost without change. For the spinor fields, the effective action can be written in terms of the bi-spinor G_{F} defined by (3.188):

$$W_{(\frac{1}{2})} = \tfrac{1}{2}\mathrm{i}\,\mathrm{tr}\,[\ln(-G_{\mathrm{F}})], \tag{6.90}$$

where the trace is now taken over spinor indices as well. The difference in sign between (6.90) and (6.25) is a direct result of the anticommuting nature of the spinor fields as opposed to the commuting properties of the scalar field. The case of the electromagnetic field is complicated by the presence of ghost fields (see §2.8). From the path-integral approach one finds

$$W_{\mathrm{EM}} = -\tfrac{1}{2}\mathrm{i}\,\mathrm{tr}\,[\ln(D_{\mathrm{F}})] + W_{\mathrm{ghost}} \tag{6.91}$$

where D_{F} is defined by (3.187) and W_{ghost} is equal to -2 times the minimally coupled scalar effective action (the factor -2 being due to the fact that there are two, anticommuting, scalar ghost fields).

To evaluate the effective actions (6.90), (6.91) in terms of the DeWitt–Schwinger proper time expansions, one requires the appropriate coefficients a_i. These have been given by DeWitt (1965) and Christensen (1978), who also discusses in detail the point-separation renormalization of the effective actions. (In the electromagnetic case it is necessary to temporarily add a mass term to the Lagrangian for the DeWitt–Schwinger method to be useful.) We record here only the x, x' coincidence limits of the coefficients which occur in the divergent terms in four dimensions. For the spin $\frac{1}{2}$ case, the coefficients in the expansion of G_{F} in (3.188) and (6.90) are spinors:

$$\begin{aligned}
a_0(x) &= \mathbb{1}\\
a_1(x) &= -\tfrac{1}{12}R(x)\mathbb{1}\\
a_2(x) &= (\tfrac{1}{288}R^2 + \tfrac{1}{120}\Box R - \tfrac{1}{180}R^{\mu\nu}R_{\mu\nu} + \tfrac{1}{180}R^{\mu\nu\rho\sigma}R_{\mu\nu\rho\sigma})\mathbb{1}\\
&\quad + \tfrac{1}{48}\Sigma_{\mu\nu}\Sigma_{\rho\sigma}R^{\mu\nu\xi\lambda}R^{\rho\sigma}{}_{\xi\lambda},
\end{aligned} \tag{6.92}$$

where $\mathbb{1}$ is the unit spinor and

$$\Sigma_{\mu\nu} = \tfrac{1}{4}[\gamma_\mu, \gamma_\nu] = V_\mu{}^\alpha V_\nu{}^\beta \Sigma_{\alpha\beta},$$

γ_μ being the curved space gamma matrices defined on page 85, and $\Sigma_{\alpha\beta}$ being given by (3.166). For later use we record here the result of taking the trace over spinor indices of (6.92):

$$\operatorname{tr} a_0(x) = \operatorname{tr} \mathbb{1} \equiv \jmath$$

$$\operatorname{tr} a_1(x) = -\tfrac{1}{12}\jmath R(x)$$

$$\operatorname{tr} a_2(x) = \tfrac{1}{720}\jmath[\tfrac{5}{2}R^2 + 6\Box R - \tfrac{7}{2}R^{\mu\nu\rho\sigma}R_{\mu\nu\rho\sigma} - 4R^{\mu\nu}R_{\mu\nu}], \qquad (6.93)$$

where $\jmath$ is the number of spinor components (i.e., the dimension of the gamma matrices used). The coefficients in the expansions of $D_{F\mu\nu}$ of (3.187) and (6.91) are tensors, and are given in the Feynman gauge $\zeta = 1$ by

$$a_{0\mu\nu}(x) = g_{\mu\nu}(x)$$

$$a_{1\mu\nu}(x) = \tfrac{1}{6}Rg_{\mu\nu} - R_{\mu\nu}$$

$$a_{2\mu\nu}(x) = -\tfrac{1}{6}RR_{\mu\nu} + \tfrac{1}{6}\Box R_{\mu\nu} + \tfrac{1}{2}R_{\mu\rho}R^{\rho}{}_{\nu} - \tfrac{1}{12}R^{\lambda\sigma\rho}{}_{\mu}R_{\lambda\sigma\rho\nu}$$
$$+ (\tfrac{1}{72}R^2 - \tfrac{1}{30}\Box R - \tfrac{1}{180}R^{\rho\sigma}R_{\rho\sigma} + \tfrac{1}{180}R^{\rho\sigma\lambda\omega}R_{\rho\sigma\lambda\omega})g_{\mu\nu}. \qquad (6.94)$$

To summarize the conclusions of this rather technical section: The technique of renormalization enables us to work with an action for the coupled gravitational–quantum matter fields in the form

$$S = S_g + W$$

and to transfer divergent pieces of W into a suitably general S_g, absorbing the infinities into renormalized coupling constants. Thus

$$S = (S_g)_{\text{ren}} + W_{\text{ren}}$$

where $(S_g)_{\text{ren}}$ contains the renormalized, physical constants, and W_{ren} is now finite. Inserting S into (6.10) yields the semiclassical equation (in four dimensions)

$$R_{\mu\nu} - \tfrac{1}{2}Rg_{\mu\nu} + \Lambda g_{\mu\nu} + a^{(1)}H_{\mu\nu} + b^{(2)}H_{\mu\nu} = -8\pi G\frac{\langle \text{out}, 0|T_{\mu\nu}|0, \text{in}\rangle_{\text{ren}}}{\langle \text{out}, 0|0, \text{in}\rangle}, \qquad (6.95)$$

where the right-hand side is now finite and Λ, a, b and G must be determined by measurement.

The renormalization did not make explicit use of $|0, \text{out}\rangle$ or $|0, \text{in}\rangle$. In particular, it is not necessary to assume that asymptotic in and out regions exist, where the vacuum concept is simple or corresponds closely to the concept of a physical vacuum. The in and out vacuum states enter here purely formally. The appearance of $\langle \text{out}, 0|T_{\mu\nu}|0, \text{in}\rangle$, rather than some other arrangement of vacuum states, is related to the boundary conditions implicit in the DeWitt–Schwinger representation of the Feynman propagator G_F. This is controlled by the use of the iε factor (see page 76). Frequently we are more interested in the quantity $\langle \text{in}, 0|T_{\mu\nu}|0, \text{in}\rangle$ (or perhaps $\langle \text{out}, 0|T_{\mu\nu}|0, \text{out}\rangle$). The effective action which yields these vacuum expectation values can be obtained by using (6.25) with G_F written in terms of the appropriate vacuum expectation value (e.g., for $\langle \text{in}, 0|T_{\mu\nu}|0, \text{in}\rangle$ use $G_F(x, x') = -\mathrm{i}\langle \text{in}, 0|T(\phi(x)\phi(x'))|0, \text{in}\rangle$). Fortunately we do not need to embark upon a fresh discussion of renormalization for each case. The divergences arise from the short distance (high momentum) behaviour of the propagator (see §3.6), which is independent of global or state-dependent effects. Therefore one expects the divergences in all three above forms for $\langle T_{\mu\nu}\rangle$ to be the same (DeWitt 1975). This expectation is confirmed by using (3.45), (3.46) to write $|0, \text{out}\rangle$ in terms of many-particle in states. The quantity $\langle 1_{j_1}, 1_{j_2}, 1_{j_3}, \ldots, 1_{j_k}, \text{in}|T_{\mu\nu}|0, \text{in}\rangle$ may be evaluated by expanding $T_{\mu\nu}$ in creation and annihilation operators, and allowing them to act on $|0, \text{in}\rangle$. After some work one arrives at the result

$$\langle \text{in}, 0|T_{\mu\nu}|0, \text{in}\rangle = \frac{\langle \text{out}, 0|T_{\mu\nu}|0, \text{in}\rangle}{\langle \text{out}, 0|0, \text{in}\rangle} - \mathrm{i}\sum_{i,j}\Lambda_{ij}T_{\mu\nu}(u^*_{\text{in},i}, u^*_{\text{in},j}) \tag{6.96}$$

where $u_{\text{in},i}$ are modes in the in region, and the notation in the final term is explained on page 17. A similar expression holds for $\langle \text{out}, 0|T_{\mu\nu}|0, \text{out}\rangle$ with Λ_{ij} replaced by V_{ij} from (3.46) and u_{in} replaced by u_{out}. The final term in (6.96) is finite. Hence the divergences present in $\langle \text{in}, 0|T_{\mu\nu}|0, \text{in}\rangle$ are the same as those present in $\langle \text{out}, 0|T_{\mu\nu}|0, \text{in}\rangle/\langle \text{out}, 0|0, \text{in}\rangle$.

6.3 Conformal anomalies and the massless case

Special interest attaches to field theories in which the classical action S is invariant under conformal transformations (see §3.1)

$$g_{\mu\nu}(x) \to \Omega^2(x)g_{\mu\nu}(x) = \bar{g}_{\mu\nu}(x). \tag{6.97}$$

From the definition of functional differentiation one has

$$S[\bar{g}_{\mu\nu}] = S[g_{\mu\nu}] + \int \frac{\delta S[\bar{g}_{\mu\nu}]}{\delta \bar{g}^{\rho\sigma}(x)}\delta \bar{g}^{\rho\sigma}(x)\,\mathrm{d}^n x, \tag{6.98}$$

which, using $\delta\bar{g}^{\mu\nu}(x) = -2\bar{g}^{\mu\nu}(x)\Omega^{-1}(x)\delta\Omega(x)$, and (6.12), gives

$$S[\bar{g}_{\mu\nu}] = S[g_{\mu\nu}] - \int [-\bar{g}(x)]^{\frac{1}{2}} T_\rho{}^\rho[\bar{g}_{\mu\nu}(x)]\Omega^{-1}(x)\delta\Omega(x)\mathrm{d}^n x. \tag{6.99}$$

From this equation one immediately obtains

$$T_\rho{}^\rho[g_{\mu\nu}(x)] = -\frac{\Omega(x)}{[-g(x)]^{\frac{1}{2}}} \frac{\delta S[\bar{g}_{\mu\nu}]}{\delta\Omega(x)}\bigg|_{\Omega=1}, \tag{6.100}$$

and it is clear that if the classical action is invariant under the conformal transformations (6.97), then the classical stress-tensor is traceless. This is easily verified explicitly using (3.190), (3.191) and (3.193) for the massless scalar field with $\xi = \xi(n)$, the massless spin $\frac{1}{2}$ field and the four-dimensional electromagnetic field respectively. Because conformal transformations are essentially a rescaling of lengths at each spacetime point x, the presence of a mass and hence a fixed length scale in the theory will always break the conformal invariance. Therefore we are led to the massless limit of the regularization and renormalization procedures used in the previous section. This involves some delicate issues.

Although all the higher order ($j > 2$) terms in the DeWitt–Schwinger expansion of the effective Lagrangian (6.42) are infrared divergent at $n = 4$ as $m \to 0$, we can still use this expansion to yield the ultraviolet divergent terms arising from $j = 0$, 1, and 2 in the four-dimensional case. We may put $m = 0$ immediately in the $j = 0$ and 1 terms in the expansion, because they are of positive power for $n \sim 4$. These terms therefore vanish. The only non-vanishing potentially ultraviolet divergent term is therefore $j = 2$:

$$\tfrac{1}{2}(4\pi)^{-n/2}(m/\mu)^{n-4}a_2(x)\,\Gamma(2 - n/2), \tag{6.101}$$

which must be handled carefully.

Substituting for $a_2(x)$ with $\xi = \xi(n)$ from (6.48), and rearranging terms, we may write the divergent term in the effective action arising from (6.101) as follows

$$W_{\text{div}} = \tfrac{1}{2}(4\pi)^{-n/2}(m/\mu)^{n-4}\Gamma(2 - n/2)\int \mathrm{d}^n x[-g(x)]^{\frac{1}{2}}a_2(x) \tag{6.102}$$

$$= \tfrac{1}{2}(4\pi)^{-n/2}(m/\mu)^{n-4}\Gamma(2 - n/2)\int \mathrm{d}^n x[-g(x)]^{\frac{1}{2}}[\alpha F(x) + \beta G(x)] + O(n-4) \tag{6.103}$$

where

$$F = R^{\alpha\beta\gamma\delta}R_{\alpha\beta\gamma\delta} - 2R^{\alpha\beta}R_{\alpha\beta} + \tfrac{1}{3}R^2 \tag{6.104}$$

and

$$G = R^{\alpha\beta\gamma\delta}R_{\alpha\beta\gamma\delta} - 4R^{\alpha\beta}R_{\alpha\beta} + R^2, \tag{6.105}$$

while the coefficients are

$$\alpha = \tfrac{1}{120}, \qquad \beta = -\tfrac{1}{360}. \tag{6.106}$$

In obtaining (6.103) we have dropped the $\Box R$ and R^2 terms from $a_2(x)$, the first because it is a total divergence, and so will not contribute to the action, the second because its coefficient is proportional to $(n-4)^2$ when the conformal coupling $\xi = \xi(n) = \frac{1}{4}(n-2)/(n-1)$ is inserted. In the limit $n \to 4$ this coefficient beats the $(n-4)^{-1}$ singularity from the Γ function in (6.43), causing this term to vanish.

The reason for decomposing a_2 into F and G is that in four dimensions (and only four) F is the square of the Weyl tensor $C_{\alpha\beta\gamma\delta}C^{\alpha\beta\gamma\delta}$. Moreover $\int(-g)^{\frac{1}{2}}G\,\mathrm{d}^4x$ is a topological invariant (see(6.56)). Both of these quantities remain invariant under conformal transformations. It follows that, at $n = 4$, W_{div} in the massless conformally coupled limit is invariant under conformal transformations.

However, we must not relax the regularization and pass to $n = 4$ before computing the physical quantities of interest, and away from $n = 4$ W_{div} is *not* conformally invariant (though W is). We shall find that a vestige of this conformal breakdown survives in the physical quantities even when we put $n = 4$ at the end of the calculation.

To see this, one may use the identities (Duff 1977)

$$\frac{2}{(-g)^{\frac{1}{2}}}g^{\mu\nu}\frac{\delta}{\delta g^{\mu\nu}}\int(-g)^{\frac{1}{2}}F\,\mathrm{d}^nx = -(n-4)(F - \tfrac{2}{3}\Box R) \tag{6.107}$$

$$\frac{2}{(-g)^{\frac{1}{2}}}g^{\mu\nu}\frac{\delta}{\delta g^{\mu\nu}}\int(-g)^{\frac{1}{2}}G\,\mathrm{d}^nx = -(n-4)G, \tag{6.108}$$

which enable the contribution of W_{div} to the trace of the stress-tensor to be evaluated immediately:

$$\langle T_\mu{}^\mu\rangle_{\mathrm{div}} = \frac{2}{(-g)^{\frac{1}{2}}}g^{\mu\nu}\frac{\delta W_{\mathrm{div}}}{\delta g^{\mu\nu}} = \tfrac{1}{2}(4\pi)^{-n/2}(m/\mu)^{n-4}(4-n)\Gamma(2-n/2)$$
$$\times[\alpha(F - \tfrac{2}{3}\Box R) + \beta G] + O(n-4). \tag{6.109}$$

From (6.43) we see that the $(n-4)$ factors arising from (6.107) and (6.108) cancel the $(n-4)^{-1}$ divergence from $\Gamma(2-n/2)$ to yield, as $n \to 4$.

$$\langle T_\mu{}^\mu\rangle_{\mathrm{div}} = (1/16\pi^2)[\alpha(F - \tfrac{2}{3}\Box R) + \beta G]. \tag{6.110}$$

Since this result is independent of m/μ, which has been retained in (6.109) essentially as an infrared cut-off, we can finally set $m = 0$, without changing the finite result (6.110). Note that because W_{div} is local, and independent of the state, so is $\langle T_\mu{}^\mu\rangle_{\mathrm{div}}$. It depends only on the geometry at x.

Now because W is conformally invariant in the massless, conformally coupled limit, the expectation value of the trace of the *total* stress-tensor is zero:

$$\langle T_\mu{}^\mu\rangle\bigg|_{m=0,\xi=1/6} = -\frac{\Omega(x)}{[-g(x)]^{\frac{1}{2}}}\frac{\delta W[\bar{g}_{\mu\nu}]}{\delta\Omega(x)}\bigg|_{m=0,\xi=1/6,\Omega=1} = 0. \quad (6.111)$$

It follows that if the divergent portion $\langle T_{\mu\nu}\rangle_{\mathrm{div}}$ has acquired the trace (6.110), then the finite, renormalized residue, $\langle T_{\mu\nu}\rangle_{\mathrm{ren}}$, must also have a trace, i.e., the negative of (6.110):

$$\langle T_\mu{}^\mu\rangle_{\mathrm{ren}} = -(1/16\pi^2)[\alpha(F - \tfrac{2}{3}\Box R) + \beta G] \quad (6.112)$$

$$= -a_2/16\pi^2 \quad (6.113)$$

$$= -(1/2880\pi^2)[R_{\alpha\beta\gamma\delta}R^{\alpha\beta\gamma\delta} - R_{\alpha\beta}R^{\alpha\beta} - \Box R] \quad (6.114)$$

$$= -(1/2880\pi^2)[C_{\alpha\beta\gamma\delta}C^{\alpha\beta\gamma\delta} + R_{\alpha\beta}R^{\alpha\beta} - \tfrac{1}{3}R^2 - \Box R] \quad (6.115)$$

The trace (6.112)–(6.115) has appeared in the theory even though the classical stress-tensor is traceless, and even though both W and W_{div} remain conformally invariant in four dimensions. It has arisen because the non-conformal nature of W_{div} (though not W) away from $n = 4$ leaves a finite imprint at $n = 4$ due to the $(n-4)^{-1}$ divergent nature of W_{div}. This result is known as a conformal, or trace, anomaly. A similar symmetry breaking, known as the axial vector anomaly, had previously been discovered (Adler 1969) in quantum electrodynamics.

The existence of an anomalous $\Box R$ term in the trace was originally found by Capper & Duff (1974, 1975). Following this, Deser, Duff & Isham (1976) demonstrated that the other terms might also arise, and in several subsequent investigations (see bibliography) their coefficients were established. Some of the early work was marred by mistakes (see Bunch 1979) and controversy. Fortunately, it is now known that all the regularization schemes predict the same conformal anomaly for the scalar field.

The appearance of the anomalous trace is closely associated with the scaling behaviour of the effective action and hence with the renormalization group (see chapter 9). To investigate this point we recall the technique of ζ-function regularization. Following along lines similar to those on page 80, one finds that under the conformal transformation

(6.97), the Feynman propagator for a massless, conformally coupled scalar field transforms to $\Omega^{-(n-2)/2}(x)G_F(x,x')\Omega^{-(n-2)/2}(x')$. Inspection of (6.26) reveals that $|x\rangle$ transforms to $\Omega^{-n/2}(x)|x\rangle$, so we see that the operator G_F, defined by (6.27) transforms to $\Omega^2 G_F$. Using this result in (6.70), one obtains

$$W[\bar{g}_{\mu\nu}] = -\tfrac{1}{2}\mathrm{i}\,\mathrm{tr}\,\{\ln[-(\mu\Omega)^2 G_F]\}, \tag{6.116}$$

in which Ω appears only in the combination $\mu\Omega$. Therefore, applying (6.100) to W, we may write

$$\langle T_\mu{}^\mu(x)\rangle = -\frac{\Omega(x)}{(-g)^{\frac{1}{2}}}\frac{\delta W[\bar{g}_{\mu\nu}]}{\delta\Omega(x)}\bigg|_{\Omega=1} = -\frac{\mu(x)}{(-g)^{\frac{1}{2}}}\frac{\delta W[\mu]}{\delta\mu(x)}\bigg|_{\mu=1}, \tag{6.117}$$

in which we are temporarily regarding μ as a function of x, and W to be a functional of μ. With this in mind we now employ (6.72) and (6.77) to write the massless limit of W as

$$W = -\tfrac{1}{2}\mathrm{i}\zeta'(0) + (1/32\pi^2)\int \mathrm{d}^4x[-g(x)]^{\frac{1}{2}}\,a_2(x)\ln[\mu^2(x)], \tag{6.118}$$

which when substituted in (6.117) yields

$$\langle T_\mu{}^\mu(x)\rangle_{\mathrm{ren}} = -a_2(x)/16\pi^2. \tag{6.119}$$

Evidently the analytic continuation (in ν) used in the ζ-function method not only renders W finite, as was shown in the previous section, but also breaks the conformal invariance, producing the trace (6.119). This expression for the trace is in exact agreement with (6.113) obtained using dimensional regularization.

The calculation of the trace anomaly in spacetime of other than four dimensions is easily achieved (see, for example, Christensen 1978, Dowker & Kennedy 1978). First note that when n is odd, L_{eff}, given by (6.42), is finite. Hence one may conclude that there is no anomaly in odd-dimensional spacetimes.

For n even, and equal to n_0, only the first $1+n_0/2$ terms in (6.42) are ultraviolet divergent, i.e., possess poles in their respective Γ functions at $n=n_0$. Of these, all but the $a_{n_0/2}$ term vanish at $n=n_0$ when $m\to 0$. The latter term contains the factor m^{n-n_0} and does not vanish. This term produces the anomalous trace

$$\langle T_\mu{}^\mu\rangle_{\mathrm{ren}} = -a_{n_0/2}/(4\pi)^{n_0/2}, \tag{6.120}$$

a result which can be confirmed using ζ-function regularization.

One interesting case is $n_0 = 2$. We then have

$$\langle T_\mu{}^\mu \rangle_{\text{ren}} = -a_1/4\pi = -R/24\pi, \tag{6.121}$$

where we have used (6.47) with $\xi = 0$ (corresponding to conformal coupling in two dimensions).

It may be wondered whether the anomalous trace can be removed by adding suitable counterterms to the effective Lagrangian. There are no local geometrical quantities that, when varied in the effective action, yield a contribution to $\langle T_{\mu\nu} \rangle_{\text{ren}}$ whose trace will cancel the entire anomaly. However, it could be removed if one were to entertain a dynamical theory containing more complicated actions (Brown & Dutton 1978, Fradkin & Vilkovisky 1978). It is also possible to remove the $\Box R$ term from (6.114) or (6.115) by using the identity

$$\frac{2}{(-g)^{\frac{1}{2}}} g^{\mu\nu} \frac{\delta}{\delta g^{\mu\nu}} \int (-g)^{\frac{1}{2}} R^2 \mathrm{d}^4 x = -12 \Box R \tag{6.122}$$

(see (6.53)). By adding an R^2 term to L_{eff}(recall from (6.104)–(6.106) that $\alpha F + \beta G$ does not contain an R^2 term) the coefficient of $\Box R$ in the anomalous trace may be varied at will. It could, if desired, be made equal to zero. Of course, the introduction of an R^2 term in L_{eff} means that W_{eff} is no longer invariant under conformal transformations. Whether or not this is reasonable can only be answered by experiment. There seems to be no compelling theoretical reason why the conformal symmetry of the dynamics should be broken by hand, so we shall retain the $\Box R$ term in the anomaly in what follows (see, however, Horowitz & Wald 1978).

We now turn to the trace anomaly for higher-spin fields. So long as we require W to be conformally invariant, then the anomalous trace can only be of the form (6.112) for some α and β. However, if W is not conformally invariant (e.g., massless scalar field with $\xi \neq \xi(n)$), then the anomalous trace may contain additional $\Box R$ and R^2 terms. It will also, of course, contain an ordinary, i.e., non-anomalous, component that will depend on the quantum state. The latter contribution will not generally be of a geometrical, or even local form (see chapter 7 for some explicit examples). The anomalous portion of the trace will be the state-independent, local, geometrical piece.

As remarked above, the appearance of a conformal anomaly can be understood in the case when dimensional regularization is used as the fact that, although W is conformally invariant in n dimensions, W_{div} is only conformally invariant in four dimensions. When the dimensionality is analytically continued to n, the breakdown of conformal invariance away from $n = 4$ survives in $\langle T_\mu{}^\mu \rangle_{\text{div}}$ (and hence $\langle T_\mu{}^\mu \rangle_{\text{ren}}$) at $n = 4$. One may have to contend with theories in which not only W_{div}, but W itself is only conformally invariant at some particular dimension. Two examples of this are the scalar case, if one insists on fixing ξ independently of n, and the electromagnetic field, which is only conformally invariant in four dimensions. In these cases, dimensional regularization yields different results from the other regularization schemes. However, the difference is only in the coefficient of the $\Box R$ anomaly, which we have seen can anyway be changed by the addition of an R^2 counterterm to L_{eff}, and so must be determined experimentally.

The generalization of (6.119) to fields of arbitrary spin has been given by Christensen & Duff (1978*a*):

$$\langle T_\mu{}^\mu\rangle_{\text{ren}} = -\frac{(-1)^{2A+2B}}{16\pi^2}\operatorname{tr} a_2(A,B) \tag{6.123}$$

where (A, B) labels the representation of the Lorentz group under which the field concerned transforms (see (3.164)). This trace can be written in terms of four parameters as

$$\langle T_\mu{}^\mu\rangle_{\text{ren}} = (2880\pi^2)^{-1}\{aC_{\alpha\beta\gamma\delta}C^{\alpha\beta\gamma\delta} + b(R_{\alpha\beta}R^{\alpha\beta} - \tfrac{1}{3}R^2) + c\Box R + \mathrm{d}R^2\}. \tag{6.124}$$

The coefficients a, b, c and d can all be expressed as simple polynomials in A and B. The general formulae are given by Christensen & Duff (1979) for the particular higher-spin field equations used by them. We list in table 1 some of the more important results for fields with spins ≤ 2.

Note that table 1 only gives the anomalous contribution. If the field is not conformally invariant, then there will be additional, non-anomalous

Table 1. *The coefficients appearing in the trace anomaly equation* (6.124) *for fields of various spins. The results for spin* $\frac{1}{2}$ *assume two component spinors, the results for four components being obtained by multiplication by two. The crosses indicate where consistency conditions for higher-spin fields require the corresponding geometrical object in the anomaly to vanish* (*Christensen & Duff* 1980).

(A,B)	a	b	c	d
$(0,0)$	-1	-1	$(6-30\xi)$	$-90(\xi-\frac{1}{6})^2$
$(\frac{1}{2},0)$	$-\frac{7}{4}$	$-\frac{11}{2}$	3	0
$(\frac{1}{2},\frac{1}{2})$	11	-64	-6	-5
$(1,0)$	-33	27	12	$-\frac{5}{2}$
$(1,\frac{1}{2})$	$\frac{291}{4}$	$\times$	$\times$	$\frac{61}{8}$
$(1,1)$	-189	$\times$	$\times$	$-\frac{747}{4}$

contributions, as already remarked. In particular, in the scalar case $\xi \neq \frac{1}{6}$, one has the non-anomalous contribution

$$6(\xi - \tfrac{1}{6})\{\langle \phi_{;\mu}\phi^{;\mu}\rangle + \xi R\langle \phi^2\rangle\}, \tag{6.125}$$

which depends on the quantum state chosen. The scalar, $(\frac{1}{2}, 0)$ spin $\frac{1}{2}$, and $(\frac{1}{2}, \frac{1}{2})$ spin 1 results in the table can be obtained from (6.48), (6.93) and (6.94) respectively.

In general, a higher-spin physical field of interest will not correspond to a single representation (A, B) of the Lorentz group, but will be a linear combination of several such representations. For example, the electromagnetic field contains scalar ghost contributions, so to obtain the electromagnetic anomaly, one must subtract from the $(\frac{1}{2}, \frac{1}{2})$ anomaly twice the $(0, 0)$ anomaly. In table 2 we list these various physical combinations for the massless fields with spins ≤ 2. The results have been computed using ζ-function regularization. As remarked, use of dimensional regularization alters the coefficient of $\Box R$ in the electromagnetic case. Brown & Cassidy (1977*b*) find $c = 12$.

There is a useful consistency condition on the coefficients a, b, c and d that arises when dimensional regularization is applied to a theory which is conformally invariant in n-dimensions (Duff 1977). A comparison of (6.112) and (6.124) reveals that

$$a = -180(\alpha + \beta), \quad b = 360\beta, \quad c = 120\alpha, \quad d = 0 \tag{6.126}$$

from which one obtains the constraints

$$2a + b + 3c = 0, \quad d = 0. \tag{6.127}$$

Clearly only two of the four coefficients a, b, c and d remain to be determined.

The trace anomaly is especially important in the special case that the

Table 2. *The coefficients in the trace anomaly for the physical massless fields of spin* ≤ 2. *In the calculation of the spin 1 and spin 2 results* (0,0) *represents the minimally coupled scalar field* ($\xi = 0$). *The crosses have the same meaning as in table* 1.

Spin	(A,B)	a	b	c	d
0	$(0,0)$	-1	-1	$(6-30\xi)$	$-90(\xi-\frac{1}{6})^2$
$\frac{1}{2}$	$(\frac{1}{2},0)$	$-\frac{7}{4}$	$-\frac{11}{2}$	3	0
1	$(\frac{1}{2},\frac{1}{2})-2(0,0)$	13	-62	-18	0
$\frac{3}{2}$	$(1,\frac{1}{2})-2(\frac{1}{2},0)$	$\frac{233}{4}$	×	×	$\frac{61}{8}$
2	$(1,1,)+(0,0)-2(\frac{1}{2},\frac{1}{2})$	-212	×	×	$-\frac{717}{4}$

background spacetime is conformally flat. If the quantum field is also conformally invariant, then we have a conformally trivial situation (see §3.7). In this case, it turns out that the anomalous trace determines the entire stress-tensor once the quantum state has been specified (Brown & Cassidy 1977*a*, Bunch & Davies 1977*b*, Davies 1977*a*, *b*).

To see this we use (6.99) with S replaced by W_{ren}:

$$W_{\text{ren}}[\bar{g}_{\mu\nu}] = W_{\text{ren}}[g_{\mu\nu}] - \int [-\bar{g}(x)]^{\frac{1}{2}} \langle T_\rho{}^\rho[\bar{g}_{\mu\nu}(x)] \rangle_{\text{ren}} \Omega^{-1}(x) \delta\Omega(x) \mathrm{d}^n x. \tag{6.128}$$

Now using (6.13) and the fact that

$$\bar{g}^{\nu\sigma} \frac{\delta}{\delta \bar{g}^{\mu\sigma}} = g^{\nu\sigma} \frac{\delta}{\delta g^{\mu\sigma}},$$

one obtains

$$\begin{aligned}\langle T_\mu{}^\nu[\bar{g}_{\kappa\lambda}(x)] \rangle_{\text{ren}} &= (g/\bar{g})^{\frac{1}{2}} \langle T_\mu{}^\nu[g_{\kappa\lambda}(x)] \rangle_{\text{ren}} \\ &\quad - \frac{2}{[-\bar{g}(x)]^{\frac{1}{2}}} \bar{g}^{\nu\sigma}(x) \frac{\delta}{\delta \bar{g}^{\mu\sigma}} \int [-\bar{g}(x')]^{\frac{1}{2}} \\ &\quad \times \langle T_\rho{}^\rho[\bar{g}_{\kappa\lambda}(x')] \rangle_{\text{ren}} \Omega^{-1}(x') \delta\Omega(x') \mathrm{d}^n x'. \end{aligned} \tag{6.129}$$

For conformally invariant field theories, the trace in the integral on the right-hand side of this equation is purely anomalous, which we know to be local and state-independent. Thus, independently of the specification of the state on the left-hand side one may perform the variational integration (Brown & Cassidy 1977*a*, DeWitt 1979). Alternatively, within the framework of dimensional regularization, one may proceed as follows: From the discussion associated with (6.111)–(6.115), we have

$$\begin{aligned}\langle T_\rho{}^\rho[\bar{g}_{\kappa\lambda}(x)] \rangle_{\text{ren}} &= -\langle T_\rho{}^\rho[\bar{g}_{\kappa\lambda}(x)] \rangle_{\text{div}} \\ &= \frac{\Omega(x)}{[-\bar{g}(x)]^{\frac{1}{2}}} \frac{\delta W_{\text{div}}[\bar{g}_{\kappa\lambda}]}{\delta\Omega(x)}, \end{aligned} \tag{6.130}$$

which, when substituted in (6.129), permits the integration to be carried out immediately, giving

$$\begin{aligned}\langle T_\mu{}^\nu[\bar{g}_{\kappa\lambda}(x)] \rangle_{\text{ren}} &= (g/\bar{g})^{\frac{1}{2}} \langle T_\mu{}^\nu[g_{\kappa\lambda}(x)] \rangle_{\text{ren}} \\ &\quad - \frac{2}{[-\bar{g}(x)]^{\frac{1}{2}}} \bar{g}^{\nu\sigma}(x) \frac{\delta}{\delta \bar{g}^{\mu\sigma}(x)} W_{\text{div}}[\bar{g}_{\kappa\lambda}] \\ &\quad + \frac{2}{[-\bar{g}(x)]^{\frac{1}{2}}} g^{\nu\sigma} \frac{\delta}{\delta g^{\mu\sigma}(x)} W_{\text{div}}[g_{\kappa\lambda}]. \end{aligned} \tag{6.131}$$

We first apply this result to the two-dimensional case. Using (6.41) and (6.47) with $\xi = \xi(2) = 0$, and the fact that

$$\Gamma(1 - n/2) = 2/(2 - n) + O(1),$$

one has

$$W_{\text{div}}[g_{\kappa\lambda}] = -[1/4\pi(n-2)] \int [-g(x')]^{\frac{1}{2}} a_1 [g_{\kappa\lambda}(x')] \mathrm{d}^n x'$$

$$= -[1/24\pi(n-2)] \int [-g(x')]^{\frac{1}{2}} R(x') \mathrm{d}^n x', \qquad (6.132)$$

in which terms of order $n - 2$ have been dropped. Substituting (6.132) and its counterpart for $\bar{g}_{\kappa\lambda}$ into (6.131), one obtains

$$\langle T_\mu{}^\nu[\bar{g}_{\kappa\lambda}(x)]\rangle_{\text{ren}} = (g/\bar{g})^{\frac{1}{2}} \langle T_\mu{}^\nu[g_{\kappa\lambda}(x)]\rangle_{\text{ren}}$$
$$+ [1/12\pi(n-2)][(\bar{R}_\mu{}^\nu - \tfrac{1}{2}\delta_\mu{}^\nu \bar{R})$$
$$-(R_\mu{}^\nu - \tfrac{1}{2}\delta_\mu{}^\nu R)], \qquad (6.133)$$

which, using (3.3) and (3.4) for $\bar{R}_\mu{}^\nu$ and $\bar{R}$, gives

$$\langle T_\mu{}^\nu[\bar{g}_{\kappa\lambda}(x)]\rangle_{\text{ren}} = (g/\bar{g})^{\frac{1}{2}} \langle T_\mu{}^\nu[g_{\kappa\lambda}(x)]\rangle_{\text{ren}}$$
$$+ (1/12\pi)[(\Omega^{-3}\Omega_{;\rho\mu} - 2\Omega^{-4}\Omega_{;\rho}\Omega_{;\mu})g^{\rho\nu}$$
$$+ \delta_\mu{}^\nu g^{\rho\sigma}(\tfrac{3}{2}\Omega^{-4}\Omega_{;\rho}\Omega_{;\sigma} - \Omega^{-3}\Omega_{;\rho\sigma})]. \qquad (6.134)$$

It is not difficult to show using (6.93) that (6.132) and hence (6.134) also holds in the case of two-component spin $\frac{1}{2}$ spinor fields.

All two-dimensional spacetimes are conformally flat:

$$g_{\mu\nu} = C(x)\eta_{\mu\nu}.$$

Thus (6.134) with $g_{\kappa\lambda} = \eta_{\kappa\lambda}$, and $\Omega = C^{\frac{1}{2}}$, enables one to write the expectation value of the stress-tensor in any two-dimensional curved spacetime in terms of its expectation value in flat spacetime. This result takes a particularly simple form in null coordinates (3.9) in which

$$\mathrm{d}s^2 = C(u, v)\mathrm{d}u\,\mathrm{d}v, \qquad (6.135)$$

when (6.134) gives

$$\langle T_\mu{}^\nu[g_{\kappa\lambda}(x)]\rangle_{\text{ren}} = (-g)^{-\frac{1}{2}} \langle T_\mu{}^\nu[\eta_{\kappa\lambda}(x)]\rangle_{\text{ren}} + \theta_\mu{}^\nu$$
$$-(1/48\pi)R\delta_\mu{}^\nu, \qquad (6.136)$$

(Davies 1977*b*) where

$$\left.\begin{aligned}\theta_{uu} &= -(1/12\pi)C^{\frac{1}{2}}\partial_u^2 C^{-\frac{1}{2}}\\ \theta_{vv} &= -(1/12\pi)C^{\frac{1}{2}}\partial_v^2 C^{-\frac{1}{2}}\\ \theta_{uv} &= \theta_{vu} = 0.\end{aligned}\right\} \tag{6.137}$$

If the state used in evaluating the expectation value in flat spacetime is a vacuum state, then the state appearing in the curved spacetime expectation value is a conformal vacuum. As discussed in §5.5, whether or not the flat spacetime vacuum is the usual Minkowski space vacuum depends on whether the curved spacetime is conformal to all, or only a part of Minkowski space. If it is conformal to all of Minkowski space, then the usual vacuum is indeed used, and the first term on the right-hand side of (6.136) vanishes. Otherwise this term will give a nonzero contribution. We shall pursue this point further below.

We now turn to the four-dimensional conformally trivial case. From (6.103)

$$W_{\text{div}} = -[1/16\pi^2(n-4)]\int \mathrm{d}^n x[-g(x)]^{\frac{1}{2}}[\alpha F(x) + \beta G(x)] + O(1). \tag{6.138}$$

Substituting (6.138) in (6.131) with $\bar{g}_{\kappa\lambda}$ replaced by $g_{\kappa\lambda}$, and $g_{\kappa\lambda}$ replaced by $\bar{g}_{\kappa\lambda}$, a flat spacetime metric (e.g., $\eta_{\kappa\lambda}$), and performing the functional differentiations (Brown & Cassidy 1977*b*, Bunch 1979) one obtains

$$\langle T_\mu{}^\nu[g_{\kappa\lambda}]\rangle_{\text{ren}} = (\bar{g}/g)^{\frac{1}{2}}\langle T_\mu{}^\nu[\bar{g}_{\kappa\lambda}]\rangle_{\text{ren}} - (1/16\pi^2)\{\tfrac{1}{9}\alpha\,{}^{(1)}H_\mu{}^\nu + 2\beta\,{}^{(3)}H_\mu{}^\nu\}, \tag{6.139}$$

where ${}^{(1)}H_{\mu\nu}$ is given by (6.53), and

$$\begin{aligned}{}^{(3)}H_{\mu\nu} &\equiv \tfrac{1}{12}R^2 g_{\mu\nu} - R^{\rho\sigma}R_{\rho\mu\sigma\nu}\\ &= R_\mu{}^\rho R_{\rho\nu} - \tfrac{2}{3}RR_{\mu\nu} - \tfrac{1}{2}R_{\rho\sigma}R^{\rho\sigma}g_{\mu\nu} + \tfrac{1}{4}R^2 g_{\mu\nu}.\end{aligned} \tag{6.140}$$

In the scalar case, α and β are given by (6.106) and

$$\langle T_\mu{}^\nu[g_{\kappa\lambda}]\rangle_{\text{ren}} = (\bar{g}/g)^{\frac{1}{2}}\langle T_\mu{}^\nu[\bar{g}_{\kappa\lambda}]\rangle_{\text{ren}} - (1/2880\pi^2)[\tfrac{1}{6}\,{}^{(1)}H_\mu{}^\nu - {}^{(3)}H_\mu{}^\nu]. \tag{6.141}$$

For other conformally invariant field theories, α and β can be obtained from the anomaly coefficients a, b, c, d of (6.124), by using (6.126).

The result (6.139) has been derived in a less formal manner by Bunch &

Davies (1977*b*). As their method appeals more to the underlying physics of the situation we shall briefly outline here the main steps involved.

We first note that because $\langle T_{\mu\nu}\rangle_{\text{ren}}$ is derived from an effective action W_{ren}, it must automatically be covariantly conserved

$$\langle T_\mu{}^\nu\rangle_{\text{ren};\nu} = 0. \tag{6.142}$$

Next, we note that if the quantum state is a conformal vacuum, or any of the associated excited states, then $\langle T_\mu{}^\nu\rangle_{\text{ren}}$ must be a purely *local* tensor, i.e., it depends only on the geometry at x. This follows because G_{F} is purely local (it may be constructed by simple conformal scaling from Minkowski space where it is manifestly local; see, for example, (3.154)). Moreover, the differentiation of G_{F} required to produce $\langle T_\mu{}^\nu\rangle_{\text{ren}}$ and the renormalization procedure, are also both local (and covariant) processes.

Mindful of the above two restrictions we proceed to find the most general local conserved tensor with the required units of (length)$^{-4}$. From the relation

$$\begin{aligned} R_{\alpha\beta\gamma\delta} = \tfrac{1}{2}(g_{\alpha\gamma}R_{\beta\delta} - g_{\alpha\delta}R_{\beta\gamma} - g_{\beta\gamma}R_{\alpha\delta} + g_{\beta\delta}R_{\alpha\gamma} \\ + \tfrac{1}{6}R(g_{\alpha\delta}g_{\beta\gamma} - g_{\alpha\gamma}g_{\beta\delta}) + C_{\alpha\beta\gamma\delta}, \end{aligned} \tag{6.143}$$

it is clear that in a conformally flat spacetime, where $C_{\alpha\beta\gamma\delta} = 0$, the Riemann tensor is determined entirely in terms of the Ricci tensor and its contractions. Taking into account further degeneracy due to the conformal symmetry (Davies, Fulling, Christensen & Bunch 1977) one arrives at six independent geometrical quantities with the correct units:

$$R_\mu{}^\alpha R_{\alpha\nu}, RR_{\mu\nu}, R_{;\mu\nu}, R_{\alpha\beta}R^{\alpha\beta}g_{\mu\nu}, R^2 g_{\mu\nu}, g_{\mu\nu}\Box R. \tag{6.144}$$

We require linear combinations of these tensors that will be covariantly conserved. Two local conserved tensors are already known: ${}^{(1)}H_{\mu\nu}$ and ${}^{(2)}H_{\mu\nu}$. In conformally flat spacetimes, however, they are not independent:

$${}^{(2)}H_{\mu\nu} = \tfrac{1}{3}{}^{(1)}H_{\mu\nu}. \tag{6.145}$$

One readily verifies, using (6.142) that the only other local conserved tensor is ${}^{(3)}H_{\mu\nu}$ defined by (6.140). Although purely geometrical the latter tensor is only accidentally conserved in conformally flat spacetimes, i.e., it cannot be derived from variation of a geometrical term in the action, nor is it the limit of a tensor that is conserved in a non-conformally flat spacetime (Ginzburg, Kirzhnits & Lyubushin 1971).

In addition to these two geometrical tensors, there may be another conserved tensor that is local, but non-geometrical, i.e., a tensor determined entirely by the local geometry, but which cannot be expressed in terms of $R_{\alpha\beta}$ or R. Call this extra tensor ${}^{(4)}H_{\mu\nu}$. Thus

$$\langle T_\mu{}^\nu \rangle_{\text{ren}} = A\,{}^{(1)}H_\mu{}^\nu + B\,{}^{(3)}H_\mu{}^\nu + {}^{(4)}H_\mu{}^\nu \tag{6.146}$$

for some constants A and B. Taking the trace of (6.146) yields

$$\langle T_\mu{}^\mu \rangle_{\text{ren}} = -6A\,\Box R - B(R_{\alpha\beta}R^{\alpha\beta} - \tfrac{1}{3}R^2) + {}^{(4)}H_\mu{}^\mu, \tag{6.147}$$

which when compared with (6.112) (remembering $F = C_{\alpha\beta\gamma\delta}C^{\alpha\beta\gamma\delta} = 0$) gives

$$A = -\alpha/144\pi^2, \quad B = -\beta/8\pi^2, \quad {}^{(4)}H_\mu{}^\mu = 0. \tag{6.148}$$

The anomalous trace therefore determines $\langle T_{\mu\nu} \rangle_{\text{ren}}$ up to the local, conserved, traceless tensor ${}^{(4)}H_{\mu\nu}$, the value of which will in any case depend on the choice of quantum state. Comparison of (6.139) with (6.148) reveals exact agreement, with ${}^{(4)}H_{\mu\nu}$ identified as the flat spacetime 'boundary' term $(\bar{g}/g)^{\frac{1}{2}} \langle T_\mu{}^\nu[\bar{g}_{\kappa\lambda}] \rangle_{\text{ren}}$.

We now apply the form (6.139) to some specific examples, namely, the Robertson–Walker spacetimes discussed in §5.1. Being homogeneous and isotropic, any uniform distribution of matter (including the conformal vacuum state) in this class of spacetimes will share their geometrical symmetries. This restricts $\langle T_{\mu\nu} \rangle_{\text{ren}}$ in the following way

$$\left.\begin{aligned} &\langle T_1{}^1 \rangle = \langle T_2{}^2 \rangle = \langle T_3{}^3 \rangle \\ &\langle T_\mu{}^\nu \rangle_{\text{ren}} = 0, \quad \mu \neq \nu. \end{aligned}\right\} \tag{6.149}$$

Working in the coordinate system in which the metric takes the form (5.11) one can use (5.13) to obtain

$$\left.\begin{aligned} {}^{(1)}H_{00} &= C^{-1}[-9\ddot{D}D + \tfrac{9}{2}\dot{D}^2 + \tfrac{27}{8}D^4 + 9KD^2 - 18K^2] \\ {}^{(1)}H_{11} &= C^{-1}\Upsilon[6\dddot{D} - 3\ddot{D}D + \tfrac{3}{2}\dot{D}^2 - 9\dot{D}D^2 + \tfrac{9}{8}D^4 \\ &\quad - 12K\dot{D} + 3KD^2 - 6K^2] \end{aligned}\right\} \tag{6.150}$$

$$\left.\begin{aligned} {}^{(3)}H_{00} &= C^{-1}[\tfrac{3}{16}D^4 + \tfrac{3}{2}KD^2 + 3K^2] \\ {}^{(3)}H_{11} &= C^{-1}\Upsilon[-\tfrac{1}{2}\dot{D}D^2 + \tfrac{1}{16}D^4 - 2K\dot{D} + \tfrac{1}{2}KD^2 + K^2]. \end{aligned}\right\} \tag{6.151}$$

In the case of the spacetimes shown in figs. 20*a–d* the conformal vacuum is based on the Minkowski space vacuum, so the first term on the right-hand side of (6.139) vanishes, and the conformal vacuum expectation value of the stress-tensor is given in terms of (6.150) and (6.151) alone. In particular, for the Einstein universe, for which $C = a^2 = $ constant, $D = 0$ (see

§5.2), one obtains

$$\langle T_\mu{}^\nu \rangle_{\text{Einstein}} = \frac{p(s)}{2\pi^2 a^4} \operatorname{diag}(1, -\tfrac{1}{3}, -\tfrac{1}{3}, -\tfrac{1}{3}), \tag{6.152}$$

where $p(s)$ is a spin-dependent coefficient which takes the values

$$p(0) = \tfrac{1}{240}, \quad p(\tfrac{1}{2}) = \tfrac{17}{960}, \quad p(1) = \tfrac{11}{120}.$$

(See Kennedy & Unwin 1980 for the case of an Einstein universe with points identified.) In the closed de Sitter spacetime (5.65), or the steady-state universe (5.52), one readily finds

$$\langle T_\mu{}^\nu \rangle_{\substack{\text{steady}\\ \text{state}}} = \frac{q(s)}{960\pi^2 \alpha^4} \delta_\mu{}^\nu, \tag{6.153}$$

where

$$q(0) = 1, \quad q(\tfrac{1}{2}) = \tfrac{11}{2}, \quad q(1) = 62.$$

For the spacetimes shown in figs. 20*e*, *f*, the conformal vacuum may be based on the Minkowskian or Rindler vacua (see §4.5). In the latter case the first term on the right-hand side of (6.139) no longer vanishes. The four-dimensional Rindler line element can be written as

$$ds^2 = \zeta^2 d\eta^2 - d\zeta^2 - dy^2 - dz^2, \quad (0 < \zeta < \infty) \tag{6.154}$$

which is the four-dimensional extension of (4.71), with $\zeta = a^{-1} e^{a\xi}$ and the η of (4.71) being replaced by $a^{-1}\eta$. Under the coordinate transformation

$$\begin{aligned} \zeta &= \Upsilon^{\frac{1}{2}}(1 - r\Upsilon^{\frac{1}{2}} \cos\theta)^{-1}, \\ y &= r\Upsilon^{\frac{1}{2}} \sin\theta \cos\phi (1 - r\Upsilon^{\frac{1}{2}} \cos\theta)^{-1}, \\ z &= r\Upsilon^{\frac{1}{2}} \sin\theta \sin\phi (1 - r\Upsilon^{\frac{1}{2}} \cos\theta)^{-1}, \end{aligned} \tag{6.155}$$

with Υ having its $K = -1$ value, $(1 + r^2)^{-1}$, (6.154) takes the form

$$ds^2 = \zeta^2 [d\eta^2 - \Upsilon dr^2 - r^2(d\theta^2 + \sin^2\theta\, d\phi^2)], \tag{6.156}$$

which clearly shows the conformal relation between the Rindler and $K = -1$ Robertson–Walker spacetimes. The Rindler vacuum expectation value of the stress-tensor has been evaluated by Candelas & Deutsch (1977, 1978), who obtain the result in the interesting form

$$\langle T_\mu{}^\nu[\eta_{\lambda\kappa}] \rangle_{\text{Rindler}} = \frac{h(s)}{2\pi^2 \zeta^4} \int_0^\infty \frac{d\nu\, \nu(\nu^2 + s^2)}{e^{2\pi\nu} - (-1)^{2s}} \operatorname{diag}(-1, \tfrac{1}{3}, \tfrac{1}{3}, \tfrac{1}{3}) \tag{6.157}$$

where $h(s)$ is the number of helicity states of the spin s field.

It is now a straightforward matter to obtain $\langle T_\mu{}^\nu \rangle_{\text{ren}}$ in the open static universe (§5.2). The geometrical contributions, ${}^{(1)}H_\mu{}^\nu$, ${}^{(3)}H_\mu{}^\nu$ will be the same as for the Einstein universe, so

$$\langle T_\mu{}^\nu \rangle'_{\substack{\text{static}\\ \text{open}}} = (\zeta^4/a^4)\langle T_\mu{}^\nu \rangle_{\text{Rindler}} + \langle T_\mu{}^\nu \rangle_{\text{Einstein}} = 0. \quad (6.158)$$

The latter equality follows from the fact that $p(s)$ in (6.152) can be represented as

$$p(s) = h(s) \int_0^\infty \frac{\mathrm{d}\nu \; \nu(\nu^2 + s^2)}{\mathrm{e}^{2\pi\nu} - (-1)^{2s}}. \quad (6.159)$$

Alternatively, one can independently obtain the result (6.158) (Bunch 1978*a*; see §7.2) and so arrive at (6.157). The thermal connection involved in these various vacuum states, represented schematically in fig. 20, is immediately evident from the Planckian-type integrals in (6.157) and (6.159).

Finally, the static form of de Sitter spacetime (5.76) provides another interesting example conformally related to the Rindler spacetime. The metric (5.76) written in the form

$$\begin{aligned} \mathrm{d}s^2 = {} & [1 - (r^2/\alpha^2)]\{\mathrm{d}t^2 - [1 - (r^2/\alpha^2)]^{-2}\mathrm{d}r^2 \\ & - r^2[1 - (r^2/\alpha^2)]^{-1}(\mathrm{d}\theta^2 + \sin^2\theta \, \mathrm{d}\phi^2)\}, \end{aligned}$$

transforms, under the change of variable $r/\alpha = r'[1 + (r')^2]^{-\frac{1}{2}}$, to

$$\mathrm{d}s^2 = (\alpha^2 - r^2)[(\mathrm{d}t^2/\alpha^2) - \Upsilon \mathrm{d}r'^2 - r'^2(\mathrm{d}\theta^2 + \sin^2\theta \, \mathrm{d}\phi^2)] \quad (6.160)$$

with $\Upsilon = [1 + (r')^2]^{-1}$, which is manifestly conformal to (6.156) if $\eta = t/\alpha$. The contribution from the geometrical terms of (6.139) is independent of coordinates, and so will be the same as the steady-state case (6.153). Thus, one immediately obtains for the stress-tensor in the static de Sitter (Rindler) conformal vacuum

$$\begin{aligned} \langle T_\mu{}^\nu \rangle_{\substack{\text{static}\\ \text{de Sitter}}} &= \zeta^4(\alpha^2 - r^2)^{-2}\langle T_\mu{}^\nu \rangle_{\text{Rindler}} + \langle T_\mu{}^\nu \rangle_{\substack{\text{steady}\\ \text{state}}} \\ &= \frac{-p(s)}{2\pi^2}(\alpha^2 - r^2)^{-2} \operatorname{diag}(1, -\tfrac{1}{3}, -\tfrac{1}{3}, -\tfrac{1}{3}) \\ &\quad + \frac{q(s)}{960\pi^2\alpha^4}\delta_\mu{}^\nu, \end{aligned} \quad (6.161)$$

which clearly diverges at the horizon ($r = \alpha$). This is the undesirable feature of the 'observer-dependent' de Sitter vacuum mentioned at the end of §5.4.

A useful relation between the trace anomaly and the complete stress-tensor in Robertson–Walker spacetimes has been given by Parker (1979). Robertson–Walker spacetimes possess a conformal Killing vector field ξ^μ (see §3.1)

$$\xi^\mu = a(t)\delta^\mu{}_t \tag{6.162}$$

where $a = C^{\frac{1}{2}}$, and the λ of (3.23) is here given by

$$\lambda = 2a'(t) \tag{6.163}$$

the prime denoting differentiation with respect to t. From the conservation equation (6.142), and (3.23), we conclude

$$\begin{aligned}[\langle T^{\mu\nu}\rangle_{\text{ren}}\xi_\nu]_{;\mu} &= \langle T^{\mu\nu}\rangle_{\text{ren}}\xi_{(\nu;\mu)} \\ &= \tfrac{1}{2}\lambda\langle T^{\mu\nu}\rangle_{\text{ren}}g_{\mu\nu}.\end{aligned} \tag{6.164}$$

Integrating (6.164) over a 4-volume bounded by constant time hypersurfaces t_1 and t_2 gives

$$\begin{aligned}\tfrac{1}{2}\int d^4x(-g)^{\frac{1}{2}}\lambda\langle T_\mu{}^\mu\rangle_{\text{ren}} &= \int_{t_2} d^3x(-g)^{\frac{1}{2}}\langle T^{t\nu}\rangle_{\text{ren}}\xi_\nu \\ &\quad - \int_{t_1} d^3x(-g)^{\frac{1}{2}}\langle T^{t\nu}\rangle_{\text{ren}}\xi_\nu\end{aligned} \tag{6.165}$$

where we have used the divergence theorem. Using the fact that $\langle T_{\mu\nu}\rangle$ is a function of t alone, (6.165) reduces to

$$\rho(t_2)a^4(t_2) = \rho(t_1)a^4(t_1) - \int_{t_1}^{t_2} a^3a'\langle T_\mu{}^\mu\rangle_{\text{ren}}dt, \tag{6.166}$$

where $\rho = \langle T^{tt}\rangle_{\text{ren}}$ is the proper energy density as measured by a comoving observer, and we have used (6.162) and (6.163).

If (6.166) is applied to ${}^{(4)}H^{\mu\nu}$ alone, then its tracelessness implies

$${}^{(4)}H_1{}^1 = -\tfrac{1}{3}{}^{(4)}H_t{}^t = -\tfrac{1}{3}\rho(t) \propto a^{-4}(t) \tag{6.167}$$

which expresses the fact that ${}^{(4)}H_{\mu\nu}$ describes massless radiation with classical behaviour, that redshifts in the usual way as the universe expands: the total energy in a comoving volume, ${}^{(4)}H_{tt}a^3$, decreases like a^{-1}. The conservation equation (6.142) reduces in this case to the simple form

$$p\,da^3 + d(\rho a^3) = 0 \tag{6.168}$$

where the pressure $p = -{}^{(4)}H_1{}^1$.

If (6.166) is applied to the entire stress-tensor $\langle T_{\mu\nu}\rangle_{\text{ren}}$ given by (6.146), then we must take into account the non-vanishing trace (6.147). Using the formulae (5.13).

$$R_{\alpha\beta}R^{\alpha\beta} = C^{-2}[3\dot{D}^2 + \tfrac{3}{2}\dot{D}D^2 + \tfrac{3}{4}D^4 + 6K\dot{D} + 6KD^2 + 12K^2] \tag{6.169}$$

$$\Box R = C^{-2}[3\dddot{D} - \tfrac{9}{2}\dot{D}D^2 - 6K\dot{D}], \tag{6.170}$$

it is easily verified that the integrand of (6.166) is an exact time derivative. Integrating the trace term yields

$$3[B(a'^4 + 2Ka'^2) + 12A(-a^2a'a''' - aa'^2a'' + \tfrac{1}{2}a^2a''^2 + \tfrac{3}{2}a'^4 + Ka'^2)], \tag{6.171}$$

where A, B are given in (6.148).

In the case of an asymptotically static spacetime, (6.171) vanishes and (6.166) then yields

$$\rho(\text{in})a^4(\text{in}) = \rho(\text{out})a^4(\text{out}) \tag{6.172}$$

which expresses the fact that any energy present in the field originally (e.g., due to space curvature or the presence of quanta) merely redshifts like classical radiation into the out region. No augmentation or reduction of the field energy occurs as a result of the expansion. This is not surprising, perhaps, in a conformally trivial situation. Indeed, we already know from §3.7 that no quanta are created from the conformal vacuum by the expansion motion. Equation (6.172) is another expression of this.

6.4 Computing the renormalized stress-tensor

In the previous sections of this chapter it has been demonstrated how the formally divergent quantity $\langle T_{\mu\nu}\rangle$ can be rendered finite by renormalization of coupling constants in the gravitational action, i.e., by adding infinite counterterms that are purely geometrical and can be regarded as part of the gravitational dynamics. In practice, of course, interest centres not on the divergences themselves, but on the finite remainder, since this is supposed to be the physically relevant portion. In particular, $\langle T_{\mu\nu}\rangle_{\text{ren}}$ is the quantity that is intended to reside on the right-hand side of the generalized Einstein equation (6.95). Whilst some conformally trivial systems allow $\langle T_{\mu\nu}\rangle_{\text{ren}}$ to be computed entirely from a knowledge of the trace anomaly (see §6.3) in general such short cuts are not available, and laborious 'brute force' techniques are necessary. In this section we shall outline how such techniques can be employed.

First it has to be decided what quantity one wishes to compute. The standard development of the DeWitt–Schwinger representation yields

$$\frac{\langle \text{out}, 0 | T_{\mu\nu} | 0, \text{in} \rangle}{\langle \text{out}, 0 | 0, \text{in} \rangle}, \tag{6.173}$$

but this is not, obviously, a true expectation value. That role is reserved for a quantity of the sort $\langle \psi | T_{\mu\nu} | \psi \rangle$. Working in the Heisenberg picture, it is natural to specify the quantum state in the in region, e.g., $\langle \text{in}, 0 | T_{\mu\nu}(x) | 0, \text{in} \rangle$, though x may refer to any point in the spacetime. As shown at the end of §6.2, the divergences, hence renormalization, of the various vacuum expressions are identical. Only the finite remainder differs.

In principle, it is possible to compute $\langle \psi | T_{\mu\nu} | \psi \rangle$ using the finite renormalized effective Lagrangian L_{ren} combined with a computation of the Λ coefficients in (6.96). Besides being extremely complicated, however, the DeWitt–Schwinger representation is not trustworthy at long wavelengths if computed by an asymptotic expansion of the form used to obtain (6.61). Instead, it is better to work with $\langle \psi | T_{\mu\nu} | \psi \rangle$ from scratch, using the DeWitt–Schwinger representation solely for the purpose of renormalization.

Because $\langle T_{\mu\nu} \rangle_{\text{ren}}$ is determined by the low-energy, long-wavelength portion of $\langle T_{\mu\nu} \rangle$, it will be sensitive to the global structure of the spacetime manifold, and the quantum state chosen. It will not, therefore be a geometrical, or even a local, object in general. A variety of techniques is available for computing $\langle T_{\mu\nu} \rangle_{\text{ren}}$, depending on the regularization mechanism adopted.

Although our formal discussion of renormalization was based on the action functional, in a practical calculation it is not possible to follow this route. This is because in order to carry out the functional differentiation of W_{ren} with respect to $g_{\mu\nu}$ to form $\langle T_{\mu\nu} \rangle_{\text{ren}}$, it is generally necessary to know W_{ren} for all geometries $g_{\mu\nu}$. This is impossibly difficult. There is however one exception, namely the conformally trivial case, where the entire stress-tensor is determined by the conformal scaling behaviour alone. Apart from this case, it is necessary to work directly with $\langle T_{\mu\nu} \rangle$.

The number of explicit cases where analytic regularization techniques can be profitably employed is rather limited, owing to the paucity of spacetimes in which the relevant field equation can be solved in n dimensions, or divergent quantities cast into recognizable forms of generalized ζ-functions.

One interesting case in which dimensional regularization can be used to advantage, due to the high degree of geometrical symmetry, is de Sitter space. If a vacuum state is chosen that is invariant under the de Sitter group (see §5.4) then, as the only (maximally) form invariant, rank two tensor under the de Sitter group is $g_{\mu\nu}$ (see, for example, Weinberg 1972, §13.4), we must have

$$\langle T_{\mu\nu} \rangle = T g_{\mu\nu}/n \tag{6.174}$$

where T is the trace of the stress-tensor

$$T = \langle T_\mu{}^\mu \rangle. \tag{6.175}$$

For a massive scalar field, one obtains from (3.190) using the field equation (3.26)

$$T_\mu{}^\mu = m^2 \phi^2 + (n-1)[\xi - \xi(n)] \Box \phi^2 \tag{6.176}$$

whence, from (2.66),

$$T = \langle T_\mu{}^\mu \rangle = \tfrac{1}{2} m^2 G^{(1)}(x,x) + \tfrac{1}{2}(n-1)[\xi - \xi(n)] \Box G^{(1)}(x,x). \tag{6.177}$$

Now $G^{(1)}(x,x)$ diverges at $n=4$, but can be finite for $n \neq 4$. It has been obtained for all n by Candelas & Raine (1975)($\xi = 0$ only) and Dowker & Critchley (1976*b*). The result is

$$G^{(1)}(x,x) = \frac{2\alpha^2}{(4\pi\alpha^2)^{n/2}} \frac{\Gamma(\nu(n) - \frac{1}{2} + n/2)\Gamma(-\nu(n) - \frac{1}{2} + n/2)}{\Gamma(\frac{1}{2} + \nu(n))\Gamma(\frac{1}{2} - \nu(n))} \Gamma(1 - n/2) \tag{6.178}$$

where α is the radius of the de Sitter universe (see (5.49)) and

$$[\nu(n)]^2 = \tfrac{1}{4}(n-1)^2 - m^2\alpha^2 - \xi n(n-1) \tag{6.179}$$

cf. (5.60). Note that as $G^{(1)}$ is independent of x, only the first term in (6.177) will contribute to the trace:

$$T = \tfrac{1}{2} m^2 G^{(1)}(x,x). \tag{6.180}$$

The pole at $n=4$ is manifest in (6.178). It must be removed by expanding $G^{(1)}$ about $n=4$, and subtracting from it the adiabatic expansion of $G^{(1)}_{\mathrm{DS}}$, truncated at an appropriate adiabatic order A, also expanded about $n=4$. We shall denote the truncated DeWitt–Schwinger expansion by ${}^{(A)}G^{(1)}_{\mathrm{DS}}$ in what follows. The rationale behind this step is that the low order adiabatic terms in the expansion of $G^{(1)}_{\mathrm{DS}}$ are the ones which, through (6.37), can be used to form L_{div}. Subtraction of such terms from $G^{(1)}$ is then equivalent to renormalization of L_{eff} via (6.58).

To determine the adiabatic order A at which the expansion of $G^{(1)}_{\mathrm{DS}}$ should be truncated, we note that fourth order adiabatic terms appear in L_{div} (in the form of $a_2(x)$). Hence we must retain in $G^{(1)}_{\mathrm{DS}}$ those terms which, when used to construct L_{div}, would yield terms up to fourth adiabatic order. In this respect one must remember that differentiation of a term will increase its adiabatic order. However, in the de Sitter space example, one sees from (6.177) and (6.180) that no derivatives are involved in computing $\langle T_{\mu\nu}\rangle$ from $G^{(1)}$. As the adiabatic order of $\langle T_{\mu\nu}\rangle$ is the same as that of L_{div}, this implies that the expansion of $G^{(1)}_{\mathrm{DS}}$ should be truncated at order 4 (in n dimensions, truncation is at adiabatic order n). Explicitly, using (3.135) and (3.137), and following the same steps as led to (6.41), one obtains

$$
\begin{aligned}
{}^{(4)}G^{(1)}_{\mathrm{DS}} = 2m^{n-4}(4\pi)^{-n/2}\{&a_0(x)m^2\Gamma(1-n/2) + a_1(x)\Gamma(2-n/2) \\
&+ a_2(x)m^{-2}\Gamma(3-n/2)\},
\end{aligned} \tag{6.181}
$$

where, in n-dimensional de Sitter space

$$
a_0(x) = 1, \quad a_1(x) = (\tfrac{1}{6} - \xi)n(n-1)\alpha^{-2}
$$

and a_2, which is needed only in four dimensions, is given by

$$
a_2(x) = [2(1-6\xi)^2 - \tfrac{1}{15}]\alpha^{-4}, \quad (n = 4).
$$

Subtracting (6.181) from (6.178), and expanding the result about $n = 4$ gives

$$
\begin{aligned}
&G^{(1)}(x,x) - {}^{(4)}G^{(1)}_{\mathrm{DS}}(x,x) \\
&\quad = (1/8\pi^2\alpha^2)\{(m^2\alpha^2 + 12\xi - 2)[\psi(\tfrac{3}{2}+\nu) + \psi(\tfrac{3}{2}-\nu) - \ln(m^2\alpha^2) - 1] \\
&\qquad + m^2\alpha^2 - \tfrac{2}{3} - (\alpha/m)^2 a_2\} + O(n-4).
\end{aligned} \tag{6.182}
$$

Now using (6.180) and (6.174), the renormalized stress-tensor at $n = 4$ can be obtained (Dowker & Critchley 1976*b*):

$$
\begin{aligned}
\langle T_{\mu\nu}\rangle_{\mathrm{ren}} = (g_{\mu\nu}/64\pi^2)\{&m^2[m^2 + (\xi - \tfrac{1}{6})R][\psi(\tfrac{3}{2}+\nu) \\
&+ \psi(\tfrac{3}{2}-\nu) - \ln(12m^2R^{-1})] - m^2(\xi - \tfrac{1}{6})R \\
&- \tfrac{1}{18}m^2R - \tfrac{1}{2}(\xi - \tfrac{1}{6})^2R^2 + \tfrac{1}{2160}R^2\},
\end{aligned} \tag{6.183}
$$

where we have used $R = 12\alpha^{-2}$. Note that in the massless, conformally coupled ($\xi - \frac{1}{6}$) limit, the trace of (6.183) as given by the renormalized version of (6.180) results entirely from the final term in (6.182):

$$
T_{\mathrm{ren}} = -a_2/16\pi^2,
$$

in agreement with (6.120).

Next we turn to an example of the use of ζ-function regularization (Dowker & Banach 1978). In §5.2 we obtained normal mode solutions of the conformally coupled massless scalar wave equation in the Einstein static universe (see (5.28)). Because the spatial sections are compact, the *total* renormalized vacuum energy due to the space curvature and non-trivial topology will be finite. Hence we may compute this quantity rather than the energy density, and then use the fact that the spatial homogeneity implies a uniform energy density, to compute $\langle 0|T_0{}^0|0\rangle$ by simply dividing out the total volume $(2\pi^2a^3)$ of space.

The total energy is given by

$$E = \int d^3x\, h^{\frac{1}{2}} \langle 0|T_0{}^0|0\rangle, \tag{6.184}$$

which, for a conformally coupled, massless scalar field in the Einstein universe, is

$$E = \tfrac{1}{2}\int d\mu(k)(k/a), \tag{6.185}$$

where we have used (3.190), (5.15), (5.28) and (5.19). This expression is analogous to the Minkowski space result (2.38); $\frac{1}{2}(k/a)$ being the proper energy eigenvalue of modes labelled by k. Using (5.24) we have

$$E = \tfrac{1}{2}\sum_{k=1}^{\infty}\sum_{J=0}^{k-1}\sum_{M=-J}^{J}(k/a)$$

$$= \tfrac{1}{2}\sum_{k=1}^{\infty} k^2(k/a), \tag{6.186}$$

showing that the contribution to the energy from modes labelled by k has degeneracy k^2. The sum in (6.186) diverges like k^4. To regularize it we replace the energy eigenvalue $k/2a$ by $(k/2a)^{-s}$. Then

$$E = \lim_{s\to -1}\sum_{k=1}^{\infty} k^2(k/2a)^{-s}. \tag{6.187}$$

Now the sum in (6.187) is proportional to a Riemann ζ-function, namely

$$(2a)^s\zeta(s-2). \tag{6.188}$$

But this quantity, when analytically continued back to $s = -1$, is *finite*. So from (6.187)

$$E = \zeta(-3)/2a = 1/240a. \tag{6.189}$$

Dividing E by the proper volume of space, the energy density is simply

$$\rho = 1/480\pi^2a^4, \tag{6.190}$$

and because the trace anomaly vanishes in the Einstein universe, the complete renormalized stress-tensor may be deduced from (6.190):

$$\langle 0|T_\mu{}^\nu|0\rangle_{\text{ren}} = \frac{1}{480\pi^2 a^4}\,\text{diag}(1, -\tfrac{1}{3}, -\tfrac{1}{3}, -\tfrac{1}{3}). \tag{6.191}$$

This result was first obtained by Ford (1976) using a mode sum cut-off and is in agreement with (6.152) (see also Dowker & Critchley 1976*c*).

As a final example of the analytic techniques, consider the 'Casimir' energy density in flat $R^1 \times S^1$ spacetime due to a massless conformally coupled scalar field (Isham 1978*b*). The vacuum energy density is equal to (see §4.1)

$$(2\pi/L^2) \sum_{n=0}^{\infty} n \quad \text{or} \quad (\pi/L^2) \sum_{n=-\infty}^{\infty} |n+\tfrac{1}{2}| \tag{6.192}$$

for untwisted and twisted fields respectively. Replacing n and $|n+\frac{1}{2}|$ by n^{-s} and $|n+\frac{1}{2}|^{-s}$, (6.192) may be re-expressed as ζ-functions analytically continued back to $s = -1$:

$$(2\pi/L^2)\zeta(-1) = -\pi/6L^2 \tag{6.193}$$

and

$$(\pi/L^2)[\zeta(-1,\tfrac{1}{2}) + \zeta(-1,-\tfrac{1}{2}) + \tfrac{1}{2}] = \pi/12L^2, \tag{6.194}$$

respectively, where

$$\zeta(s,q) \equiv \sum_{n=0}^{\infty} (q+n)^{-s}, \quad \text{Re}\, s > 1 \tag{6.195}$$

and $\zeta(-1,q)$ is given, for example, by Gradshteyn & Ryzhik (1965), (9.531). Equations (6.193) and (6.194) are in agreement with (4.8) and (4.11) obtained using cut-off methods.

Notice that in neither of the latter two examples has it been necessary to actually *renormalize* anything. The analytic continuation has simply discarded the appropriate divergent terms to yield a finite answer. Whether or not the continuation always contrives to discard exactly the right terms (i.e., those terms that, in another regularization scheme, would be absorbed into renormalization of coupling constants) has not been proved, though it is generally believed to be true.

From the point of view of practical computation, possibly the most efficient regularization technique is point-splitting. Once again, it is not in general possible to compute $\langle T_{\mu\nu}\rangle_{\text{ren}}$ by functionally differentiating a renormalized effective action. Instead, one works directly with $\langle T_{\mu\nu}\rangle$, or preferably with $G^{(1)}(x,x')$ (which is simpler). Renormalization may be carried out on $\langle T_{\mu\nu}\rangle$, by subtracting from it terms up to adiabatic order n in $\langle T_{\mu\nu}\rangle_{\mathrm{DS}}$, which is formed by differentiation of $G^{(1)}_{\mathrm{DS}}(x,x')$. These are the terms which arise from L_{div}, and their subtraction is equivalent to renormalization of constants in the generalized Einstein action. Alternatively, one can form $\langle T_{\mu\nu}\rangle_{\mathrm{ren}}$ by operating on

$$G^{(1)}_{\mathrm{ren}}(x,x') = G^{(1)}(x,x') - {}^{(n)}G^{(1)}_{\mathrm{DS}}(x,x'), \tag{6.196}$$

with a differential operator obtained from (3.190) in a manner to be described shortly. In ${}^{(n)}G^{(1)}_{\mathrm{DS}}$, only those terms which make a contribution to the stress-tensor of adiabatic order n or less are retained. These will be terms in $G^{(1)}_{\mathrm{DS}}$ which are of order n or less, although care must be exercised because differentiation can increase the adiabatic order of a quantity. Thus, the procedure for computing the renormalized stress-tensor using point-splitting can be summarized as follows:

(1) Solve the field equation for a complete set of normal modes from which particle states may be defined.
(2) Construct $G^{(1)}(x,x')$ as a mode sum.
(3) Form $G^{(1)}_{\mathrm{ren}}$ according to (6.196), truncating the expansion of $G^{(1)}_{\mathrm{DS}}$ at order n.
(4) Operate on $G^{(1)}_{\mathrm{ren}}$ to form $\langle 0|T_{\mu\nu}(x,x')|0\rangle_{\mathrm{ren}}$, discarding any terms of adiabatic order greater than n which have appeared from differentiation of terms in ${}^{(n)}G^{(1)}_{\mathrm{DS}}$.
(5) Let $x'\to x$ and display the finite result $\langle 0|T_{\mu\nu}(x)|0\rangle_{\mathrm{ren}}$.

The state $|0\rangle$ will, of course, be dependent on the definition of 'positive frequency' modes at step (1).

The differentiation of $G^{(1)}(x,x')$ is generally a complicated procedure. Formally one has

$$\langle T_{\mu\nu}(x)\rangle = \lim_{x'\to x} \mathscr{D}_{\mu\nu}(x,x')G^{(1)}(x,x'). \tag{6.197}$$

The Green function $G^{(1)}(x,x')$ is not a scalar function of x, but a *bi-scalar* of the two spacetime points x and x'; that is, it transforms like a scalar at each point. (Higher spin fields involve bi-spinors, bi-vectors, etc.) Consequently the differential operator $\mathscr{D}_{\mu\nu}(x,x')$ is a non-local operator. For example, the first term of (3.190) gives rise to the expectation value

$$(1-2\xi)\langle 0|\nabla_\mu\phi(x)\nabla_\nu\phi(x)|0\rangle, \tag{6.198}$$

which, using the point-splitting technique, is treated as

$$\lim_{x'\to x} \tfrac{1}{2}(1-2\xi)[\nabla_\mu\nabla_{\nu'} + \nabla_{\mu'}\nabla_\nu]\tfrac{1}{2}G^{(1)}(x,x'), \tag{6.199}$$

where the prime on a derivative indicates that it acts at x' rather than at x.

This means that the resulting object is not a tensor, but a bi-vector. (Expression (6.199) has been written in the most symmetrical way possible, although this is not essential.)

To construct a tensor from a bi-vector, and to maintain general covariance, it is necessary to parallel transport the derivative vector (spinors etc.) back to the same spacetime point, which could be the midpoint between x, x', one of the end points (e.g. x), or perhaps somewhere else. Differences between the parallel-transported and non-transported results will arise, even when the points x, x' are made to coincide, from a σ^{-1} factor in the expansion of $G^{(1)}(x, x')$ multiplying a σ-order transport correction. Fortunately, these complicated corrections have been worked out once and for all (for a general spacetime) by Christensen (1976, 1978), Davies, Fulling, Christensen & Bunch (1977), Bunch (1977) and Adler, Lieberman & Ng (1977, 1978), so we shall not go into the details here.

It should be noted, however, that if $G^{(1)}$ is renormalized first (i.e., before differentiation to form $\langle T_{\mu\nu} \rangle$) according to (6.196), then all σ^{-1} terms are in any case removed, so any transport corrections are of order σ and vanish when we let $\sigma \to 0$ at the end of the calculation. Thus, only if one insists on first constructing an unrenormalized stress-tensor will the effects of parallel transport need to be taken into account in a practical calculation.

We consider the symmetric case where one constructs $G^{(1)}(x'', x')$, the two points x'', x' lying at (small) equal proper distance from the spacetime point x of interest, on either side, along a non-null geodesic through x (see fig. 22). Symmetrization, while not essential, does simplify some of the expressions. At the end of the calculation we let both x'' and x' approach x. The (small) proper distance from x to x' is denoted ε and the direction of the geodesic is parametrized by the tangent vector t^μ at x. (If the separated points x'', x' remain in a normal neighbourhood of x, then this geodesic will be unique.)

To make use of this formalism, the first step is to convert $G^{(1)}$ into a function of ε and t^μ rather than x'' and x'. This involves solving the equation for the geodesic joining x'', x and x' in terms of ε and t^μ as a power series in ε up to order ε^{n+1}. The results are available in two important cases: two spacetime dimensions and four-dimensional Robertson–Walker spacetimes. For the two-dimensional case (Davies & Fulling 1977*b*)

$$u(\varepsilon) = u + \varepsilon t^u - \tfrac{1}{2} C^{-1} C_{,u} (t^u)^2 \varepsilon^2$$
$$+ \tfrac{1}{6} C [C^{-3}(3C_{,u}^2 - C C_{,uu}) t^u - \tfrac{1}{4} R t^v] (t^u)^2 \varepsilon^3 + \ldots, \quad (6.200)$$

where we use null coordinates u and v and the metric (6.135). A similar equation holds for $v(\varepsilon)$, with u and v interchanged. It is understood that C and its derivatives are all evaluated at the central point $x = (u, v)$. The end points x', x'' are given by $x' = (u(\varepsilon), v(\varepsilon)), x'' = (u(-\varepsilon), v(-\varepsilon))$.

For the Robertson–Walker spacetime with line element (5.7), one can choose $\theta'' = \theta' = \theta, \phi'' = \phi' = \phi$ without loss of generality, because the spacetime is isotropic. Then one requires only $\eta(\varepsilon)$ and $r(\varepsilon)$, which are given by (Bunch 1977, Bunch & Davies 1977*a*)

$$\left.\begin{aligned} \eta(\varepsilon) &= \eta + \varepsilon t^\eta + \frac{1}{2!}\varepsilon^2 t_2^\eta + \frac{1}{3!}\varepsilon^3 t_3^\eta + \frac{1}{4!}\varepsilon^4 t_4^\eta + \ldots \\ r(\varepsilon) &= r + \varepsilon t^r + \frac{1}{2!}\varepsilon^2 t_2^r + \frac{1}{3!}\varepsilon^3 t_3^r + \frac{1}{4!}\varepsilon^4 t_4^r + \ldots \end{aligned}\right\} \tag{6.201}$$

where

$$\begin{aligned}
t_2^\eta &= -\tfrac{1}{2}D(t^\eta)^2 - \tfrac{1}{2}D\Upsilon(t^r)^2 \\
t_2^r &= -Dt^\eta t^r - Kr\Upsilon(t^r)^2 \\
t_3^\eta &= (-\tfrac{1}{2}\dot{D} + \tfrac{1}{2}D^2)(t^\eta)^3 + (-\tfrac{1}{2}\dot{D} + \tfrac{3}{2}D^2)\Upsilon t^\eta(t^r)^2 \\
t_3^r &= (-\dot{D} + \tfrac{3}{2}D^2)(t^\eta)^2 t^r + 3DKr\Upsilon t^\eta(t^r)^2 + (\tfrac{1}{2}D^2 - K)\Upsilon(t^r)^3 \\
t_4^\eta &= (-\tfrac{1}{2}\ddot{D} + \tfrac{7}{4}\dot{D}D - \tfrac{3}{4}D^3)(t^\eta)^4 + (-\tfrac{1}{2}\ddot{D} + 5\dot{D}D - \tfrac{9}{2}D^3)\Upsilon(t^\eta t^r)^2 \\
&\quad + (\tfrac{1}{4}\dot{D}D - \tfrac{3}{4}D^3)\Upsilon^2(t^r)^4 \\
t_4^r &= (-\ddot{D} + 5\dot{D}D - 3D^3)(t^\eta)^3 t^r + (4\dot{D}Kr - 9D^2Kr)\Upsilon(t^\eta t^r)^2 \\
&\quad + (2\dot{D}D - 3D^3 + 6DK)\Upsilon t^\eta(t^r)^3 + (-2D^2Kr + K^2r)\,\Upsilon^2(t^r)^4 \\
t_5^\eta &= (-\tfrac{1}{2}\dddot{D} + \tfrac{11}{4}\ddot{D}D + \tfrac{7}{4}\dot{D}^2 - \tfrac{23}{4}\dot{D}D^2 + \tfrac{3}{2}D^4)(t^\eta)^5 \\
&\quad + (-\tfrac{1}{2}\dddot{D} + \tfrac{15}{2}\ddot{D}D + 5\dot{D}^2 - 32\dot{D}D^2 + 15D^4)\Upsilon(t^\eta)^3(t^r)^2 \\
&\quad + (\tfrac{3}{4}\ddot{D}D + \tfrac{1}{4}\dot{D}^2 - \tfrac{33}{4}\dot{D}D^2 + \tfrac{15}{2}D^4)\Upsilon^2 t^\eta(t^r)^4 \\
t_5^r &= (-\dddot{D} + \tfrac{15}{2}\ddot{D}D + 5\dot{D}^2 - \tfrac{43}{2}\dot{D}D^2 + \tfrac{15}{2}D^4)(t^\eta)^4 t^r \\
&\quad + (5\ddot{D}Kr - 35\dot{D}DKr + 30D^3Kr)\Upsilon(t^\eta)^3(t^r)^2 \\
&\quad + (\tfrac{7}{2}\ddot{D}D + 2\dot{D}^2 - \tfrac{47}{2}\dot{D}D^2 + 15D^4 + 10\dot{D}K - 30D^2K)\Upsilon(t^\eta)^2(t^r)^3 \\
&\quad + (-10\dot{D}DKr + 20D^3Kr - 10DK^2r)\Upsilon^2 t^\eta(t^r)^4 \\
&\quad + (-\dot{D}D^2 + \tfrac{3}{2}D^4 - 5D^2K + K^2)\Upsilon^2(t^r)^5.
\end{aligned} \tag{6.202}$$

The notation here is the same as in (5.12) and we have written $t^\eta = t^0, t^r = t^1$ etc.

Once $G^{(1)}$ has been cast into ε, t^μ notation, a useful check on the algebra is to ensure that the divergent terms agree with those of $G^{(1)}_{\mathrm{DS}}(x, x')$. The leading terms in $G^{(1)}_{\mathrm{DS}}$ for the scalar case are (Bunch, Christensen & Fulling 1978, Bunch 1977)

$$\begin{aligned} G^{(1)}_{\mathrm{DS}}(\varepsilon, t^\mu) = &-(1/\pi)\{[1-(m^2+\xi R)\varepsilon^2\,\Sigma](\gamma+\tfrac{1}{2}\ln|m^2\varepsilon^2|) \\ &+\tfrac{1}{2}(\xi-\tfrac{1}{6})m^{-2}R+\varepsilon^2\Sigma[m^2+\tfrac{1}{2}(\xi-\tfrac{1}{6})R]\} \\ &+\mathrm{O}(\varepsilon^2T^{-2})+\mathrm{O}(\varepsilon^4)+\mathrm{O}(T^{-4}), \end{aligned} \tag{6.203}$$

in two dimensions, and (Bunch 1977, Bunch & Davies 1978*b*)

$$\begin{aligned} G^{(1)}_{\mathrm{DS}}(\varepsilon, t^\mu) = &-\frac{1}{8\pi^2\varepsilon^2\Sigma}+\frac{m^2+(\xi-\frac{1}{6})R}{4\pi^2}(\tfrac{1}{2}\ln|m^2\varepsilon^2|+\gamma) \\ &+\frac{1}{24\pi^2}R_{\alpha\beta}\frac{t^\alpha t^\beta}{\Sigma}-\frac{m^2}{8\pi^2}+\frac{\varepsilon^2\Sigma}{24\pi^2}(\tfrac{1}{2}\ln|m^2\varepsilon^2|+\gamma)\bigg\{-3m^4 \\ &-2m^2R_{\alpha\beta}\frac{t^\alpha t^\beta}{\Sigma}-6(\xi-\tfrac{1}{6})m^2R+(\xi-\tfrac{1}{6})\bigg[-2RR_{\alpha\beta}\frac{t^\alpha t^\beta}{\Sigma} \\ &+R_{;\alpha\beta}\frac{t^\alpha t^\beta}{\Sigma}-\Box R-3(\xi-\tfrac{1}{6})R^2\bigg]\bigg\} \\ &+\frac{\varepsilon^2\Sigma}{1440\pi^2}\bigg[225m^4+360m^2(\xi-\tfrac{1}{6})R+60m^2R_{\alpha\beta}\frac{t^\alpha t^\beta}{\Sigma} \\ &+6R_{\alpha\beta;\gamma\delta}\frac{t^\alpha t^\beta t^\gamma t^\delta}{\Sigma^2}-14R_{\alpha\beta}R_{\gamma\delta}\frac{t^\alpha t^\beta t^\gamma t^\delta}{\Sigma^2} \\ &+4R_{\alpha\rho}R^\rho{}_\beta\frac{t^\alpha t^\beta}{\Sigma}-\tfrac{4}{3}RR_{\alpha\beta}\frac{t^\alpha t^\beta}{\Sigma}-\Box R-R^{\alpha\beta}R_{\alpha\beta} \\ &+\tfrac{1}{3}R^2+30(\xi-\tfrac{1}{6})\Box R+90(\xi-\tfrac{1}{6})^2R^2\bigg] \\ &+\frac{1}{1440\pi^2m^2}[R^{\alpha\beta}R_{\alpha\beta}-\tfrac{1}{3}R^2-\Box R+30(\xi-\tfrac{1}{6})\Box R \\ &+90(\xi-\tfrac{1}{6})^2R^2]+\mathrm{O}(\varepsilon^4)+\mathrm{O}(T^{-6}) \end{aligned} \tag{6.204}$$

in conformally flat four-dimensional spacetimes. In these expressions, t^μ is normalized according to (6.85), and, as previously, $\mathrm{O}(T^{-p})$ indicates terms of adiabatic order p (i.e., p derivatives of the metric).

Renormalization of $G^{(1)}(x'', x')$ is now immediate. Writing it as a function of ε, t^μ, one simply subtracts from it all of (6.203) for two dimensions or

(6.204) for four dimensions. However, one must not yet take the limit $\varepsilon \to 0$, even though $G^{(1)}_{\text{ren}}(x'', x')$ remains finite as $x'', x' \to x$. First it is necessary to differentiate to form $\langle 0 | T_{\mu\nu}(x'', x') | 0 \rangle_{\text{ren}}$, discarding any terms of higher adiabatic order than n that result from differentiation of terms from (6.203) or (6.204).

For the general form

$$G^{(1)} = c + \varepsilon^2 \Sigma [e_{\alpha\beta}(t^\alpha t^\beta / \Sigma) + f] + \varepsilon^2 \Sigma [q_{\alpha\beta}(t^\alpha t^\beta / \Sigma) + r](\gamma + \tfrac{1}{2} \ln |\alpha^2 \varepsilon^2|) \tag{6.205}$$

where α and γ are constants, and $c, f, r, e_{\alpha\beta}$ and $q_{\alpha\beta}$ can be functions of x, the resulting stress-tensor is (Bunch 1977; see also Davies, Fulling, Christensen & Bunch 1977 and Bunch & Davies 1978*b*) in the coincidence limit

$$\begin{aligned}\langle T_{\mu\nu} \rangle = {} & (\tfrac{1}{2} - \xi) T^{(1)}_{\mu\nu} + \left[\left(\frac{n-1}{n} \right) (\xi - \xi(n)) - \frac{1}{n} (\tfrac{1}{2} - \xi) \right] g_{\mu\nu} T^{(1)\sigma}_{\ \ \sigma} \\ & + \xi T^{(2)}_{\mu\nu} - \frac{1}{n} g_{\mu\nu} \xi T^{(2)\sigma}_{\ \ \sigma} \\ & - \tfrac{1}{2} c \xi \left[R_{\mu\nu} - \tfrac{1}{2} R g_{\mu\nu} - \left(1 - \frac{n}{2} \right) \frac{1}{n} R g_{\mu\nu} \right. \\ & \left. + \frac{2(n-1)}{n} (\xi - \xi(n)) R g_{\mu\nu} \right] \\ & + \left[\frac{1}{2n} - \left(\frac{n-1}{n} \right) (\xi - \xi(n)) \right] m^2 g_{\mu\nu} c \end{aligned} \tag{6.206}$$

where

$$\begin{aligned} T^{(1)}_{\mu\nu} \equiv {} & \langle 0 | \{ \phi_{;\mu}, \phi_{;\nu} \} | 0 \rangle \\ = {} & -\tfrac{1}{2} (q_{\mu\nu} + r g_{\mu\nu}) [\gamma + \tfrac{1}{2} \ln |\alpha^2 m^{-2}|] \\ & - \tfrac{1}{4} q_{\alpha\beta} \frac{t^\alpha t^\beta}{\Sigma} \left(g_{\mu\nu} - \frac{2 t_\mu t_\nu}{\Sigma} \right) - q_{(\mu\alpha} t^\alpha t_{\nu)} \Sigma^{-1} \\ & - \tfrac{1}{4} r \left(g_{\mu\nu} + \frac{2 t_\mu t_\nu}{\Sigma} \right) + \tfrac{1}{4} c_{;\mu\nu} - \tfrac{1}{2} e_{\mu\nu} - \tfrac{1}{2} f g_{\mu\nu} \end{aligned} \tag{6.207}$$

and

$$T^{(2)}_{\mu\nu} = - \langle 0 | \{ \phi, \phi_{;\mu\nu} \} | 0 \rangle = T^{(1)}_{\mu\nu} - \tfrac{1}{2} c_{;\mu\nu}. \tag{6.208}$$

In the conformally coupled case, $\xi = \xi(n)$, (6.206) manifestly has the trace $\frac{1}{2} m^2 c$. Note that terms of order ε^2 in $G^{(1)}$ make a finite contribution, when differentiated, to $\langle T_{\mu\nu} \rangle$.

Inspection of (6.206) shows that the only terms in (6.203) and (6.204) which will produce a contribution to $\langle T_{\mu\nu}\rangle_{\text{ren}}$ of adiabatic order greater than n are those of order m^{-2}. The only contribution from these terms which is of sufficiently low adiabatic order not to be discarded by the above prescription comes from their substitution as c in the final term of (6.206). Thus the only contribution of the terms of order m^{-2} to $\langle T_{\mu\nu}\rangle_{\text{ren}}$ are of order m^0, and there is no difficulty in taking the massless limit of $\langle T_{\mu\nu}\rangle_{\text{ren}}$ if this is desired. The mass must be kept nonzero until after differentiation to avoid infrared divergences at intermediate steps of the calculation.

As an illustration of this rather cumbersome procedure let us evaluate $\langle T_{\mu\nu}\rangle_{\text{ren}}$ for a massless, conformally coupled scalar field in two-dimensional spacetime in the conformal vacuum state. (Several other examples will be given in the next chapter.) This conformally trivial case has already been solved in §6.3 (see (6.136)). So as to provide a case in which the first term on the right-hand side of (6.136) is nonzero, we shall include compactification of the spatial sections. The Green functions are given by (4.23) and (4.24) for the twisted and untwisted fields respectively:

$$D_{\text{L}}^{(1)}(x'',x') = \begin{cases} -(1/4\pi)\ln\left[16\sin^2(\pi\Delta u/L)\sin^2(\pi\Delta v/L)\right] & \text{untwisted} \\ -(1/4\pi)\ln\left[\tan^2(\pi\Delta u/2L)\tan^2(\pi\Delta v/2L)\right] & \text{twisted} \end{cases}$$

where $\Delta u = u(\varepsilon) - u(-\varepsilon)$, $\Delta v = v(\varepsilon) - v(-\varepsilon)$. Although when used in §4.2. these Green functions referred to flat spacetime, they remain the same in the present example (see (3.154)) because (i) the spacetime is conformally flat and (ii) the conformal weight for the scalar field in two dimensions is zero.

Using (6.200) one immediately obtains power series expansions in ε for $D_{\text{L}}^{(1)}(x'',x')$:

$$\begin{aligned} \text{constant} &- (1/2\pi)\ln|\varepsilon^2 C^{-1}| + (\varepsilon^2/2\pi)\{\tfrac{1}{12}R\Sigma + \tfrac{2}{3}(\alpha\pi^2/L^2)[(t^u)^2 + (t^v)^2] \\ &+ \tfrac{1}{6}C^{-2}(CC_{,uu} - 3C_{,u}^2)(t^u)^2 \\ &+ \tfrac{1}{6}C^{-2}(CC_{,vv} - 3C_{,v}^2)(t^v)^2\} + O(\varepsilon^4), \end{aligned}$$

where $\alpha = 1$ for the untwisted field and $-\frac{1}{2}$ for the twisted field, and we have used the normalization condition (6.85), which in the metric (6.135) has the form

$$Ct^u t^v = \Sigma. \tag{6.209}$$

To renormalize, we now subtract (6.203), setting $\xi = 0$ (conformal coupling), and $m = 0$ except in the $m^{-2}R$ term (see remarks above). The remainder, aside from the $m^{-2}R$ term, is

$$\begin{aligned}&\text{constant} + (1/2\pi)\ln C + \varepsilon^2\Sigma\{(\pi\alpha/3L^2)\Sigma^{-1}[(t^u)^2 + (t^v)^2]\\&\quad + (1/12\pi)\Sigma^{-1}C^{-2}(CC_{,uu} - 3C_{,u}^2)(t^u)^2\\&\quad + (1/12\pi)\Sigma^{-1}C^{-2}(CC_{,vv} - 3C_{,v}^2)(t^v)^2\\&\quad - (1/24\pi)R\} + O(\varepsilon^2).\end{aligned}$$

One may identify scalar functions c and f and a traceless tensor $e_{\alpha\beta}$ where

$$e_{uu} = (\pi\alpha/3L^2) + \tfrac{1}{12\pi}C^{-2}(CC_{,uu} - 3C_{,u}^2)$$

$$e_{vv} = (\pi\alpha/3L^2) + \tfrac{1}{12\pi}C^{-2}(CC_{,vv} - 3C_{,v}^2)$$

$$e_{uv} = e_{vu} = 0$$

to be inserted in the general formula (6.205). All other terms are zero.

Reading off the relevant terms from (6.206)–(6.208), with $n = 2$, one obtains

$$\langle 0|T_{uu}|0\rangle_{\text{ren}} = -\frac{\pi\alpha}{12L^2} + \frac{1}{24\pi}\left[\frac{C_{,uu}}{C} - \frac{3C_{,u}^2}{2C^2}\right] \tag{6.210}$$

$$\langle 0|T_{vv}|0\rangle_{\text{ren}} = -\frac{\pi\alpha}{12L^2} + \frac{1}{24\pi}\left[\frac{C_{,vv}}{C} - \frac{3}{2}\frac{C_{,v}^2}{C^2}\right]. \tag{6.211}$$

The $-R/12\pi m^2$ term does not contribute to these stress-tensor components when inserted as c in the final term in (6.206) (since $g_{uu} = g_{vv} = 0$) but it does give rise to components

$$\begin{aligned}\langle 0|T_{uv}|0\rangle_{\text{ren}} &= \langle 0|T_{vu}|0\rangle_{\text{ren}} = \tfrac{1}{4}C\langle 0|T_\alpha^{\alpha}|0\rangle_{\text{ren}}\\&= -RC/96\pi.\end{aligned} \tag{6.212}$$

Clearly the m^{-2} term gives rise to the trace anomaly. These results are in complete agreement with (6.136), when the zero temperature (leading) term of (4.27)–(4.29) is used for $\langle T_\mu^{\nu}[\eta_{\kappa\lambda}]\rangle_{\text{ren}}$.

In the above we have discussed how $\langle T_{\mu\nu}\rangle_{\text{ren}}$ can be obtained by forming $\langle T_{\mu\nu}\rangle$, suitably regularized, and subtracting from it $\langle T_{\mu\nu}\rangle_{\text{DS}}$, formed from terms in the DeWitt–Schwinger expansion. In practice this was achieved by first constructing $G^{(1)}$, suitably regularized, subtracting from it terms in $G_{\text{DS}}^{(1)}$ and then differentiating to form the stress-tensor. The terms that were subtracted were chosen to be equivalent to renormalization of coefficients in the generalized Einstein *action*, which resulted in the subtraction of all terms in $\langle T_{\mu\nu}\rangle_{\text{DS}}$ up to adiabatic order n (in n dimensions).

It may appear perverse to compute a renormalized $\langle T_{\mu\nu}\rangle$ by consideration of $G^{(1)}$ and L_{eff}, rather than the gravitational field equations themselves, in which $\langle T_{\mu\nu}\rangle$ appears on the right-hand side. One could, after all, attempt to absorb the divergent pieces of $\langle T_{\mu\nu}\rangle$ directly into geometrical terms on the left-hand side of the generalized Einstein equation, given by (6.95), by renormalizing the constants G, Λ, a and b. There is, however, a subtlety involved in proceeding this way, the illumination of which is useful for understanding the origin of the trace anomaly.

First consider dimensional regularization. Repeating the considerations of §6.2 for an arbitrary field having

$$\left.\begin{aligned} &\operatorname{tr} a_0 = s \\ &\operatorname{tr} a_1 = zR \\ &\operatorname{tr} a_2 = wR^{\alpha\beta\gamma\delta}R_{\alpha\beta\gamma\delta} + xR^{\alpha\beta}R_{\alpha\beta} + yR^2 \end{aligned}\right\} \tag{6.213}$$

one obtains

$$\begin{aligned} \langle T_{\mu\nu}\rangle_{\text{div}} = &-\frac{1}{(4\pi)^{-n/2}}\left[\frac{1}{n-4}+\frac{1}{2}\left(\gamma+\ln\left(\frac{m^2}{\mu^2}\right)\right)\right] \\ &\times\left\{-\frac{4sm^4}{n(n-2)}g_{\mu\nu}-\frac{4m^2 z}{(n-2)}(R_{\mu\nu}-\tfrac{1}{2}g_{\mu\nu}R)\right. \\ &\left.+\,2wH_{\mu\nu}+2x\,{}^{(2)}H_{\mu\nu}+2y\,{}^{(1)}H_{\mu\nu}\right\}. \end{aligned} \tag{6.214}$$

These terms may be removed from $\langle T_{\mu\nu}\rangle$ by absorbing them into terms in generalized Einstein equation in n dimensions, the left-hand side of which is given by (6.52). Comparison of (6.213) with (6.123) and (6.124) gives w, x, y in terms of the anomaly coefficients a, b, c, d for a field transforming under the (A, B) representation of the Lorentz group:

$$\left.\begin{aligned} w &= -(180)^{-1}(-1)^{2A+2B}a \\ x &= -(180)^{-1}(-1)^{2A+2B}(b-2a) \\ y &= -(180)^{-1}(-1)^{2A+2B}(d-\tfrac{1}{3}b+\tfrac{1}{3}a). \end{aligned}\right\} \tag{6.215}$$

It is convenient to rewrite (6.214) as

$$\langle T_{\mu\nu}\rangle_{\text{div}} = -\frac{1}{(4\pi)^{n/2}}\left[\frac{1}{n-4}+\frac{1}{2}\left(\gamma+\ln\left(\frac{m^2}{\mu^2}\right)\right)\right]$$
$$\times\left\{-\frac{4sm^4}{n(n-2)}g_{\mu\nu} - wC^{\alpha\beta\gamma\delta}C_{\alpha\beta\gamma\delta}g_{\mu\nu} + 4wC_{\mu\alpha\beta\gamma}C^{\alpha\beta\gamma}{}_{\nu}\right.$$
$$-\frac{4zm^2}{(n-2)}(R_{\mu\nu}-\tfrac{1}{2}g_{\mu\nu}R)+2(4w+x)A_{\mu\nu}$$
$$\left.+{}^{(1)}H_{\mu\nu}[y+\tfrac{1}{3}(w+x)]\right\} - (4\pi)^{-n/2}(4w+x)$$
$$\times\left\{-\frac{{}^{(1)}H_{\mu\nu}}{6(n-1)}+2{}^{(3)}H_{\mu\nu}\left[\frac{(n-2)}{(n-3)}-\frac{w}{4w+x}\right]\right\},$$

(6.216)

(Bunch 1979) where the *traceless tensor* $A_{\mu\nu}$ is defined by

$$A_{\mu\nu} = -\frac{n}{4(n-1)}{}^{(1)}H_{\mu\nu} + {}^{(2)}H_{\mu\nu} - \frac{(n-2)(n-4)}{4(n-3)}{}^{(3)}H_{\mu\nu} \tag{6.217}$$

and ${}^{(3)}H_{\mu\nu}$ is the generalization to n-dimensional, non-conformally flat spacetimes of (6.140):

$$\begin{aligned}{}^{(3)}H_{\mu\nu} = {} & \frac{4(n-3)}{(n-2)^2}R^{\rho}_{\mu}R_{\rho\nu} - \frac{2n(n-3)}{(n-1)(n-2)^2}RR_{\mu\nu} \\ & -\frac{2(n-3)}{(n-2)^2}R_{\rho\sigma}R^{\rho\sigma}g_{\mu\nu} \\ & +\frac{(n+2)(n-3)}{2(n-1)(n-2)^2}R^2g_{\mu\nu} + \frac{4}{(n-2)}C_{\rho\mu\sigma\nu}R^{\rho\sigma}.\end{aligned} \tag{6.218}$$

In obtaining (6.217) we have also used the identity

$$\begin{aligned}C_{\alpha\beta\gamma\delta} = {} & R_{\alpha\beta\gamma\delta} - (n-2)^{-1}(g_{\alpha\delta}R_{\beta\gamma} + g_{\beta\gamma}R_{\alpha\delta} - g_{\alpha\gamma}R_{\beta\delta} - g_{\beta\delta}R_{\alpha\gamma}) \\ & + (n-1)^{-1}(n-2)^{-1}(g_{\alpha\gamma}g_{\beta\delta} - g_{\alpha\delta}g_{\beta\gamma})R.\end{aligned} \tag{6.219}$$

In the conformally invariant case (6.216) simplifies because we may use the relations (6.127). This yields $w + x = -3y$, causing the divergent term proportional to ${}^{(1)}H_{\mu\nu}$ in (6.216) to vanish. If we also pass to the massless limit, the only surviving terms that diverge as $n \to 4$ are the $A_{\mu\nu}$ and Weyl tensor terms. The former is traceless, while the trace of the latter is $O(n-4)$.

This combines with the factor $(n-4)^{-1}$ to yield a finite trace. Adding it to the traces of the (finite) ${}^{(1)}H_{\mu\nu}$ and ${}^{(3)}H_{\mu\nu}$ terms in (6.216) yields the negative of the familiar trace anomaly. It follows that when $\langle T_{\mu\nu}\rangle_{\text{div}}$ is subtracted from the traceless $\langle T_{\mu\nu}\rangle$ to yield $\langle T_{\mu\nu}\rangle_{\text{ren}}$, the latter will acquire an anomalous trace. There is thus no obstacle to dimensionally renormalizing $\langle T_{\mu\nu}\rangle$ in the gravitational field equations themselves.

The situation is different, however, if one tries to follow this route using point-splitting regularization. In four dimensions the left-hand side of the generalized Einstein equation is

$$R_{\mu\nu} - \tfrac{1}{2}Rg_{\mu\nu} + \Lambda g_{\mu\nu} + \alpha\,{}^{(1)}H_{\mu\nu} + \beta\,{}^{(2)}H_{\mu\nu}. \tag{6.220}$$

The tensor $H_{\mu\nu}$ is absent in (6.220) because in four dimensions the relation (6.57) can be employed. Using (6.81) to form the point-split regularized effective action, and functionally differentiating to form $\langle T_{\mu\nu}\rangle_{\text{div}}$, one finds that all the divergent terms can indeed be removed by renormalization of coefficients in the generalized Einstein equation with the left-hand side (6.220). There is, however, a problem. In the massless, conformally coupled limit, one obtains (Christensen 1978)

$$\begin{aligned}\langle T_{\mu\nu}\rangle_{\text{div}} = &-(1/480\pi^2)({}^{(2)}H_{\mu\nu} - \tfrac{1}{3}\,{}^{(1)}H_{\mu\nu})(\gamma + \tfrac{1}{2}\ln|\tfrac{1}{2}\mu^2\sigma|)\\ &+ (\text{geometrical terms that remain finite as } \sigma \to 0),\end{aligned} \tag{6.221}$$

where we have once again averaged over splitting vector directions and μ is an arbitrary infrared cut-off parameter.

The first term in (6.221) may be removed by renormalization of α and β in (6.220), but one finds that the remaining (finite) terms may not. However, from (6.53) and (6.54), the trace of the divergent terms in (6.221) is proportional to $\Box R$ and so its removal cannot give the full trace anomaly (6.113)–(6.115). One finds that the finite geometrical terms in (6.221), which cannot be removed by renormalization, are precisely those which make up the remainder of the anomaly. In contrast, if the *action* is renormalized, then the whole of W_{div}, including the parts which give the finite terms in (6.221), is removed from W, leading to the usual trace anomaly. Thus the resulting $\langle T_{\mu\nu}\rangle_{\text{ren}}$ depends crucially on whether renormalization is carried out before or after differentiation to obtain the stress-tensor.

Closer examination shows why this is so. The renormalized gravitational action is

$$S_{\text{grav,ren}} = \int (-g)^{\frac{1}{2}}\{(16\pi G)^{-1}(R - 2\Lambda) + aR^2 + bR^{\alpha\beta}R_{\alpha\beta} + cR^{\alpha\beta\gamma\delta}R_{\alpha\beta\gamma\delta}\}\,\mathrm{d}^4x.$$

However, a, b, and c (which should not be confused with the coefficients in (6.124)) are not strictly constants in the point-splitting regularization scheme. They contain σ-dependent divergences, such as σ^{-2} and $\ln \sigma$, that arise from (6.81), and σ is both a function of x and a functional of the metric. If the regularization is relaxed so that the points come together and these terms diverge, then this x-dependence is removed. That is what happens when renormalization of the action is undertaken. On the other hand, if renormalization of the gravitational field equation is required then, as with dimensional regularization, the regularization can only be relaxed after the renormalization has been effected, otherwise terms will be lost. Thus, it is necessary to functionally differentiate W with $\sigma \neq 0$. Consequently W is a non-local functional of the geometry, because σ is a bi-scalar $\sigma(x, x')$. Additional finite terms come from the differentiation of the σ-dependence, these being precisely the geometrical terms in (6.221) (in the massless, conformally coupled scalar case) and supplying the missing terms in the anomalous trace.

To see this explicitly, consider the portion of the massless effective action that comes from the final term of L_{div} in (6.81):

$$-(1/32\pi^2)\int(-g)^{\frac{1}{2}}\, a_2(x, x')\ln|\tfrac{1}{2}m^2\sigma(x, x')|\,\mathrm{d}^4x. \tag{6.222}$$

Here m is kept nonzero to act as an infrared cut-off. If the action is renormalized, and $\sigma \to 0$, this term is removed from W, a crucial step that yields an anomalous trace proportional to $a_2(x, x)$. However, if we functionally differentiate first, we obtain a finite contribution to $\langle T_{\mu\nu}\rangle$ from σ:

$$(1/64\pi^2)a_2(x)g_{\mu\nu}, \tag{6.223}$$

where we have used (Christensen 1978)

$$\frac{\delta\sigma(x, x')}{\delta g^{\mu\nu}} = -\tfrac{1}{2}\sigma_\mu\sigma_\nu + \mathrm{O}(\sigma^3{}_\mu)$$

and averaged over the σ_μ. If it is argued that, being finite, and not having the form of any of the tensors in (6.220), this term should be retained in $\langle T_{\mu\nu}\rangle_{\text{ren}}$, then $\langle T_{\mu\nu}\rangle_{\text{ren}}$ will remain traceless. If, however, it is regarded as part of the divergences (having come from the functional differentiation of a divergent term in L_{div}), then it must simply be discarded, yielding the correct trace $\langle T_{\mu\nu}\rangle_{\text{ren}}$. This is the procedure advocated in the steps on page 195, since the term (6.223) is of adiabatic order four and, by steps (3) and (4), is to be subtracted.

Notice that if one is prepared to contemplate a non-local action even *after* regularization has been removed, then it is possible, as the above calculation indicates, to add any multiple of $a_2(x)$ to the trace $\langle T_{\mu\nu}\rangle_{\text{ren}}$. As remarked, Brown & Dutton (1978) have argued that a non-local action should be used to remove the conformal anomaly entirely.

6.5 Other regularization methods

Although point-splitting and analytic methods are the most developed regularization schemes, a number of other methods have been proposed. In Robertson–Walker spacetimes these methods are all related to the so-called adiabatic regularization scheme of Parker & Fulling (1974), which we shall consider first.

Adiabatic regularization is in fact a misnomer for a *subtraction* scheme which can encompass point-splitting, dimensional regularization or any other regularization method. Instead of subtracting the adiabatic expansion of $G^{(1)}_{\text{DS}}$ to form a renormalized Green function, as in the previous section, adiabatic regularization works with subtractions based on the adiabatic expansion of the *modes*. The possibility of doing this arises from the relation, discussed in §3.6,

$$^{(A)}G^{(1)}_{\text{DS}}(x'', x') = \langle 0^A | \{\phi(x''), \phi(x')\} | 0^A \rangle |^{(A)}, \tag{6.224}$$

where the left-hand side is the expansion to adiabatic order A of $G^{(1)}_{\text{DS}}$, and the right-hand side is the expansion to order A of the expectation value of a field in an Ath order adiabatic vacuum. Since the right-hand side is only required to order A, it is possible to replace the exact field ϕ by its expansion to order A, $\phi^{(A)}$, which can be formed as a mode expansion in terms of the modes $u^A_{\mathbf{k}}$ of §3.5. Thus

$$\begin{aligned} ^{(A)}G^{(1)}_{\text{DS}}(x'',x') &= \langle 0^A | \{\phi^{(A)}(x''), \phi^{(A)}(x')\} | 0^A \rangle |^{(A)} \\ &= \int \mathrm{d}\tilde{\mu}(k) \{u^{(A)}_{\mathbf{k}}(x'')u^{(A)*}_{\mathbf{k}}(x') + u^{(A)*}_{\mathbf{k}}(x'')u^{(A)}_{\mathbf{k}}(x')\}|^{(A)}, \end{aligned} \tag{6.225}$$

in which the measure is given by (5.24). In the case of $K = \pm 1$ Robertson–Walker spacetimes with spatial curvature, the adiabatic formalism goes through with only obvious changes from the $K = 0$ case discussed in §3.5. The symbol $|^{(A)}$ in (6.225) indicates that the cross terms from the field products that are of adiabatic order greater than A are to be discarded.

As remarked above, point-splitting regularization and renormalization consists of subtracting what is really an adiabatic expansion, either from $\mathscr{L}_{\text{eff}}$, or $\langle T_{\mu\nu}\rangle$, or from $G^{(1)}$. In the latter case, the left-hand side of (6.225), with $A = n$, is expressed in the form of a DeWitt–Schwinger series, and subtracted from $G^{(1)}(x'', x')$. However, one could work instead with the right-hand side, i.e., the adiabatic mode sum, and subtract this from $G^{(1)}$ instead. The advantage of working directly with adiabatic modes (rather than an adiabatic expansion of an exact $G^{(1)}$) is that in some cases it may be possible to arrange for explicit cancellation of divergences *before* the mode integral is even performed, in which case potentially divergent terms never appear in the calculation; only finite integrals are involved. Thus, no actual regularization at all will be necessary. The points x'' and x' can be brought together *ab initio*. (If $\langle T_{\mu\nu}\rangle$ is required then differentiation of $G^{(1)}(x'', x')$ must be performed before x'', $x' \to x$.)

To illustrate this so called 'adiabatic regularization' technique, consider a massless, conformally coupled scalar field in two-dimensional Robertson–Walker spacetime with R^2 topology (Bunch 1977, 1978*b*). Following the treatment of §§4.1, 4.2, using the conformal vacuum, one obtains

$$\langle 0|T_{uu}|0\rangle = \langle 0|T_{vv}|0\rangle = \int \frac{\partial u_k}{\partial u}\frac{\partial u_k^*}{\partial u}\mathrm{d}k = \frac{1}{4\pi}\int_0^\infty k\,\mathrm{d}k \tag{6.226}$$

$$\langle 0|T_{uv}|0\rangle = \langle 0|T_{vu}|0\rangle = 0, \tag{6.227}$$

in which we have already put the points together in anticipation of ultimately having a finite mode integral to perform.

In two dimensions the adiabatic expansion of the modes is given by (see e.g. (3.86))

$$u_k^{(A)}(x) = (2\pi)^{-\frac{1}{2}}\mathrm{e}^{\mathrm{i}kx}\chi_k^{(A)}(\eta), \tag{6.228}$$

where $\chi_k^{(A)}$ is given by (3.102) expanded to order A. Using (6.228) in (6.225), and differentiating to find the stress-tensor, one has

$$\langle 0^A|T_{uu}|0^A\rangle\Big|^{(A)} = \frac{1}{16\pi}\int_{-\infty}^{\infty}\frac{\mathrm{d}k}{W_k^{(A)}}\left[(W_k^{(A)}+k)^2+\tfrac{1}{4}\left(\frac{\dot{W}_k^{(A)}}{W_k^{(A)}}\right)^2\right]^{(A)} \tag{6.229}$$

$$\langle 0^A|T_{vv}|0^A\rangle\Big|^{(A)} = \frac{1}{16\pi}\int_{-\infty}^{\infty}\frac{\mathrm{d}k}{W_k^{(A)}}\left[(W_k^{(A)}-k)^2+\tfrac{1}{4}\left(\frac{\dot{W}_k^{(A)}}{W_k^{(A)}}\right)^2\right]^{(A)} \tag{6.230}$$

$$\langle 0^A|T_{uv}|0^A\rangle\Big|^{(A)} = \langle 0^A|T_{vu}|0^A\rangle\Big|^{(A)} = \frac{Cm^2}{16\pi}\int_{-\infty}^{\infty}\frac{\mathrm{d}k}{W_k^{(A)}}\Big|^{(A)} \tag{6.231}$$

According to our adiabatic subtraction prescription (which is motivated by the desire to achieve equivalence with a result obtained by renormalization of the effective Lagrangian), we subtract (6.229)–(6.231) with $A = 2$, from the respective quantities in (6.226) and (6.227), obtaining $W_k^{(2)}$ by iteration as described on page 67. One finds

$$W_k^{(2)} = \omega_k - \frac{\ddot{C}m^2}{8\omega_k^3} + \frac{5\dot{C}^2m^4}{32\omega_k^5}, \tag{6.232}$$

which, when substituted in (6.229) and (6.230) yields

$$\langle 0^{A=2}|T_{uu}|0^{A=2}\rangle|^{(2)} = \langle 0^{A=2}|T_{vv}|0^{A=2}\rangle|^{(2)}$$

$$= \frac{1}{8\pi}\int_0^\infty \left[\omega_k + \frac{k^2}{\omega_k} - \frac{1}{8}\frac{\ddot{C}m^2}{\omega_k^3} + \frac{1}{8}\frac{\ddot{C}k^2m^2}{\omega_k^5} + \frac{7}{32}\frac{\dot{C}^2m^4}{\omega_k^5} - \frac{5}{32}\frac{\dot{C}^2k^2m^4}{\omega_k^7}\right]\mathrm{d}k. \tag{6.233}$$

Subtracting this from (6.226) one obtains

$$\langle T_{uu}\rangle_{\text{ren}} = \langle T_{vv}\rangle_{\text{ren}} = -\frac{1}{8\pi}\int_0^\infty \frac{(\omega_k - k)^2}{\omega_k}\mathrm{d}k$$

$$+\frac{1}{64\pi}\int_0^\infty \left(\frac{\ddot{C}m^2}{\omega_k^3} - \frac{\ddot{C}k^2m^2}{\omega_k^5} - \frac{7}{4}\frac{\dot{C}^2m^4}{\omega_k^5} + \frac{5}{4}\frac{\dot{C}^2k^2m^4}{\omega_k^7}\right)\mathrm{d}k, \tag{6.234}$$

in which the limit $m \to 0$ is to be taken at the end. As expected, both integrals are *finite*. The first vanishes as $m \to 0$, while the second integral yields

$$\langle T_{uu}\rangle_{\text{ren}} = \langle T_{vv}\rangle_{\text{ren}} = \frac{1}{96\pi}\left(\frac{\ddot{C}}{C} - \frac{3\dot{C}^2}{2C^2}\right) = \theta_{uu} = \theta_{vv}, \tag{6.235}$$

with $\theta_{\mu\nu}$ given by (6.137).

Similarly, one obtains

$$\langle T_{uv}\rangle_{\text{ren}} = \langle T_{vu}\rangle_{\text{ren}} = -\frac{C\ddot{C}m^4}{64\pi}\int_0^\infty \frac{\mathrm{d}k}{\omega_k^5} + \frac{5C\dot{C}^2m^6}{256\pi}\int_0^\infty \frac{\mathrm{d}k}{\omega_k^7}$$

$$= -\frac{1}{96\pi}\left(\frac{\ddot{C}}{C} - \frac{\dot{C}^2}{C^2}\right) = -\frac{R}{48\pi}g_{\mu\nu}. \tag{6.236}$$

Hence,

$$\langle T_{\mu\nu}\rangle_{\text{ren}} = \theta_{\mu\nu} - (48\pi)^{-1}Rg_{\mu\nu}, \tag{6.237}$$

which is a special case of (6.136) and (6.210)–(6.212) that were obtained by different methods. The analogous four-dimensional calculation has also been performed by Bunch (1977, 1978*b*), giving agreement with (6.141).

In most cases the exact mode solutions will be too complicated to allow cancellation of potentially divergent terms before the mode sum is performed. In such cases a true regularization scheme must be used. If point-splitting or dimensional regularization is employed, then the method is precisely the same as we have used in previous sections. However, the relation (6.225) has the advantage of allowing other regularizations to be used, with the knowledge that the final result will be identical to that obtained by renormalization of constants in the generalized Einstein equation. The regularization is merely employed as a convenience in the manipulation of a quantity which is known already to be finite. In particular, one can use regularizations which are not manifestly covariant, but which are calculationally more convenient than the covariant schemes. For example, Birrell (1978) has used an exponential cut-off $e^{-\alpha k}$ to greatly facilitate calculations, while Bunch (1980*b*) has used a (non-covariant) form of dimensional regularization to examine the divergences in $\langle 0^A|T_{\mu\nu}|0^A\rangle$. Alternatively, Hu (1979) has exploited the form of (6.225) as a mode sum to formally parametrize $\langle 0^A|T_{\mu\nu}|0^A\rangle$ in terms of a single simple logarithmically divergent integral, a method that he has also used in Bianchi I spacetimes (Hu 1978) using the formulation of Parker, Fulling & Hu (1974).

Finally, the adiabatic regularization scheme is particularly useful for numerical calculations (Birrell 1978), where one may proceed as follows: Differentiating (6.225) to form the stress-tensor, and letting the points come together, one has (e.g., (6.229)–(6.231))

$$\langle 0^A|T_{\mu\nu}(x)|0^A\rangle|^{(A)} = \int \mathrm{d}\tilde{\mu}(k) T^{(A)}_{\mu\nu}(\mathbf{k};x) \tag{6.238}$$

where $T^{(A)}_{\mu\nu}(\mathbf{k},x)$ is given in terms of the modes $u^{(A)}_{\mathbf{k}}(x)$. Similarly, the expectation value of the exact stress-tensor in the spacetime of interest can be written as (e.g., (6.226), (6.227))

$$\langle T_{\mu\nu}(x)\rangle = \int \mathrm{d}\tilde{\mu}(k) T_{\mu\nu}(\mathbf{k};x), \tag{6.239}$$

where $T_{\mu\nu}(\mathbf{k};x)$ is given in terms of the exact modes $u_{\mathbf{k}}(x)$. Then the renormalized expectation value is given by

$$\begin{aligned}\langle T_{\mu\nu}\rangle_{\mathrm{ren}} &= \langle T_{\mu\nu}(x)\rangle - \langle 0^A|T_{\mu\nu}(x)|0^A\rangle|^{(A)} \\ &= \int \mathrm{d}\tilde{\mu}(k)[T_{\mu\nu}(\mathbf{k},x) - T^{(A)}_{\mu\nu}(\mathbf{k},x)]\end{aligned} \tag{6.240}$$

where $A = n$ in an n-dimensional spacetime. The portion $T_{\mu\nu}(\mathbf{k},x)$ can be calculated numerically without difficulty given the modes and their derivatives (either from analytic expressions or from a numerical solution of the field equation), while $T^{(A)}_{\mu\nu}$ with $A = 2$ or 4 is trivially evaluated. Since the potentially divergent terms in (6.240) must cancel by construction, the integral is finite, and can be computed using straightforward quadrature methods. Numerical calculation is impossible using point-splitting or dimensional regularization because of the inability to take numerically the limit $x'' \to x'$ or $n \to n_0$ (the physical dimension) respectively.

Closely related to adiabatic regularization are the n-wave and Pauli–Villars regularization schemes. The n-wave regularization method was introduced by Zel'dovich & Starobinsky (1971) as a variant of the Pauli–Villars scheme, and has been shown by Parker & Fulling (1974) to be equivalent to adiabatic regularization. Here, we shall only discuss the Pauli–Villars method proper.

This method has long been used in Minkowski space interacting quantum field theory as a regularization technique (Pauli & Villars 1949). The basic idea is to augment the physical field Lagrangian of interest with contributions from additional fictitious fields, and arrange the parameters of the theory such that the divergent terms from the fictitious fields exactly cancel those of the physical field thus artificially rendering the theory finite. Several fictitious fields may be necessary to remove all of the divergences. To avoid the appearance of fictitious particles being created (and unitarity breaking down) and to remove the regularization, the masses of these field quanta are allowed to go to infinity at the end of the calculation. The infinities that would reappear in this limit are removed by renormalization, as in the other regularization schemes. In this case the potentially divergent terms are functions of the fictitious particle masses. By allowing some of the fictitious scalar fields to anticommute, rather than commute (or in the case of spinor fields, commute rather than anticommute), they can be arranged to contribute negatively to the stress-tensor, and so cancel the divergences of the opposite sign from the physical field.

This technique has been employed by Vilenkin (1978) to obtain the trace anomaly for the conformally coupled scalar field, and by Bernard & Duncan (1977) who give a detailed treatment of stress-tensor regularization. The idea of using a high-mass approximation to isolate and subtract the purely geometrical divergent terms in $\langle T_{\mu\nu} \rangle$ is essentially the same as both the adiabatic approach and the use of the DeWitt–Schwinger expansion in powers m^{-1}. Thus, Pauli–Villars regularization is basically the same as these other techniques in its approach to renormalization (Bunch, Christensen & Fulling 1978).

The DeWitt–Schwinger expansion (6.39) in the coincidence limit

$x = x'(\sigma = 0)$ yields

$$L_{\text{eff}} = \tfrac{1}{2}\mathrm{i}(4\pi)^{-n/2} \sum_{j=0}^{\infty} a_j(x) m^{n-2j} \int_0^{\infty} (\mathrm{i}\bar{s})^{j-1-n/2} \mathrm{e}^{-\mathrm{i}\bar{s}} \mathrm{d}\bar{s} \qquad (6.241)$$

where we have changed variables: $\bar{s} = m^2 s$. The first $n/2$ terms are ultraviolet divergent. If we add to L_{eff} the contributions from the fictitious scalar regulator fields in an appropriate combination, these divergences can be formally cancelled. For example, in two dimensions, three additional fields are required: a commuting field with mass $(2M^2 - m^2)^{\frac{1}{2}}$ and two anticommuting fields, each with mass M. (At the end of the calculation we let $M \to \infty$.) The coefficient of the leading order (quadratic) divergent integral in (6.241) is then

$$a_0 m^2 + a_0(2M^2 - m^2) - 2a_0 M^2 = 0$$

while that of the next order (logarithmic) divergence is likewise

$$a_0 + a_0 - 2a_0 = 0.$$

In four dimensions, five extra scalar fields are necessary: for example, two anticommuting fields each with mass $(M^2 + m^2)^{\frac{1}{2}}$, one more with mass $(4M^2 + m^2)^{\frac{1}{2}}$ and two commuting fields with mass $(3M^2 + m^2)^{\frac{1}{2}}$. It is easy to check that this arrangement exactly cancels the a_0, a_1 and a_2 terms of (6.241).

Although the above procedure formally eliminates the ultraviolet divergent terms of (6.241), the change of variables $\bar{s} = m^2 s$ is not strictly legitimate for terms with $j = 0, 1$, which are divergent. Indeed, had we retained the s variable, the complete integral, including contributions from regulator fields, for $n = 2$, would be

$$\int_0^{\infty} (\mathrm{i}s)^{j-2}[\mathrm{e}^{-\mathrm{i}m^2 s} + \mathrm{e}^{-\mathrm{i}(2M^2 - m^2)s} - 2\mathrm{e}^{-\mathrm{i}M^2 s}]\mathrm{d}s$$

$$= -4 \int_0^{\infty} (\mathrm{i}s)^{j-2} \mathrm{e}^{-\mathrm{i}M^2 s} \sin^2[\tfrac{1}{2}(M^2 - m^2)s]\mathrm{d}s. \qquad (6.242)$$

The integral is easily evaluated. For $j > 1$ it vanishes when $M \to \infty$. For $j = 0$, 1 it diverges in the limit $M \to \infty$. These divergences were missed by the change of variables to $\bar{s}$. They are the divergences which we expect to appear when the regularization is removed and can be eliminated by renormalization of the generalized Einstein action (or the gravitational field equations) in the usual way. At the end of this section we shall give a practical computation where these M-dependent divergences will be explicitly displayed.

First, however, it is instructive to see in detail how the trace anomaly arises in the Pauli–Villars method. The trace of the classical expression for $T_{\mu\nu}$ in the case of conformal coupling ($\xi = \xi(n)$) is $m^2\phi^2$, which clearly vanishes for $m = 0$. However $\langle T_\mu{}^\mu \rangle$ does not vanish because one must include in this quantity contributions from the regulator fields for which $M \to \infty$ rather than zero. Thus, in two dimensions

$$\langle 0|T_\mu^\mu|0\rangle = M^2 G^{(1)}_{2M^2}(x,x) - M^2 G^{(1)}_{M^2}(x,x) \tag{6.243}$$

where we have put $m = 0$ and used $G^{(1)}(x,x) = 2\langle 0|\phi^2(x)|0\rangle$. Using the DeWitt–Schwinger expansion for $G^{(1)}$

$$\langle 0|T_\mu^\mu|0\rangle = (M^2/2\pi) \sum_{j=0}^{\infty} a_j(x) \int_0^\infty (\mathrm{i}s)^{j-1} [\mathrm{e}^{-2\mathrm{i}M^2 s} - \mathrm{e}^{-\mathrm{i}M^2 s}]\mathrm{i}\,\mathrm{d}s. \tag{6.244}$$

Treating the integral in a similar way to (6.242), one sees that only the a_0 term is divergent as $M \to \infty$. This will be removed in the renormalization of Λ in (6.220). Evaluating the other integrals, one finds that all terms vanish as $M \to \infty$ except that involving a_1, which yields

$$\langle 0|T_\mu^\mu|0\rangle_{\text{ren}} \xrightarrow[M\to\infty]{} -a_1(x)/4\pi, \tag{6.245}$$

in agreement with (6.121).

In four dimensions, one similarly finds that the only surviving term after renormalization and relaxation of the regularization ($M \to \infty$) comes from the term proportional to m^{-2} in $G^{(1)}$, leaving $-a_2/16\pi^2$, the familiar result. In odd spacetime dimensions, there is no term proportional to m^{-2} in $G^{(1)}_{\text{DS}}$, so the trace anomaly vanishes.

To illustrate the use of the Pauli–Villars method in a practical calculation, we shall follow the example given by Bernard & Duncan (1977). This example has already been discussed in §3.4 in connection with particle creation.

Consider a massive, conformally coupled scalar field in two-dimensional Robertson–Walker spacetime with metric given by (3.83) and (3.84). Near the region $\eta \to -\infty$, the deviation from flat spacetime may be considered small. If we choose $A = 1 + b/2, B = b/2$, then the conformal factor

$$\begin{aligned} C(\eta) &= 1 + \tfrac{1}{2}b(1 + \tanh \rho\eta) \\ &\simeq 1 + b\mathrm{e}^{2\rho\eta} \end{aligned} \tag{6.246}$$

for $\rho\eta \to -\infty$. The quantity ρ^{-1} may be envisaged as equivalent to the adiabatic parameter T of §3.5.

To first order in $be^{2\rho\eta}$, the modes (3.87) reduce to

$$u_k^{\text{in}} \approx (4\pi\omega_{\text{in}})^{-\frac{1}{2}} e^{ikx - i\omega_{\text{in}}\eta} \left[1 - \frac{m^2 be^{2\rho\eta}}{4\rho^2(1 - i\omega_{\text{in}}/\rho)} \right] \tag{6.247}$$

where now $\omega_{\text{in}}^2 = k^2 + m^2$.

From (3.190), putting $n = 2$, $\xi = 0$

$$T_{\mu\nu} = \phi_{;\mu}\phi_{;\nu} - \tfrac{1}{2} g_{\mu\nu} g^{\sigma\rho} \phi_{;\sigma}\phi_{;\rho} + \tfrac{1}{2} m^2 g_{\mu\nu} \phi^2. \tag{6.248}$$

Then using (2.43) (which remains valid in curved spacetime) with modes (6.247), one readily obtains, to first order,

$$\langle \text{in}, 0 | T_{\eta\eta} | 0, \text{in} \rangle = \frac{1}{4\pi} \int_{-\infty}^{\infty} (k^2 + m^2)^{\frac{1}{2}} dk + \frac{m^2 be^{2\rho\eta}}{8\pi} \int_{-\infty}^{\infty} \frac{dk}{(k^2 + m^2)^{\frac{1}{2}}} \tag{6.249}$$

$$\langle \text{in}, 0 | T_{xx} | 0, \text{in} \rangle = \frac{1}{4\pi} \int_{-\infty}^{\infty} (k^2 + m^2)^{\frac{1}{2}} dk - \frac{m^2}{4\pi} \int_{-\infty}^{\infty} \frac{dk}{(k^2 + m^2)^{\frac{1}{2}}}$$
$$- \frac{m^2 be^{2\rho\eta}}{8\pi} \int_{-\infty}^{\infty} \frac{dk}{(k^2 + m^2)^{\frac{1}{2}}} + \frac{m^4 be^{2\rho\eta}}{8\pi} \int_{-\infty}^{\infty} \frac{dk}{(\rho^2 + k^2 + m^2)(k^2 + m^2)^{\frac{1}{2}}}. \tag{6.250}$$

These expressions contain the usual quadratic and logarithmic divergences that are to be cancelled by the regulator fields. When these are added to the above expressions one has

$$\int_{-\infty}^{\infty} [(k^2 + m^2)^{\frac{1}{2}} + (k^2 + 2M^2 - m^2)^{\frac{1}{2}} - 2(k^2 + M^2)^{\frac{1}{2}}] dk$$
$$= \tfrac{1}{2} m^2 \ln[(2M^2 - m^2)/m^2] + M^2 \ln[M^2/(2M^2 - m^2)] \tag{6.251}$$

and

$$\int_{-\infty}^{\infty} [m^2(k^2 + m^2)^{-\frac{1}{2}} + (2M^2 - m^2)(k^2 + 2M^2 - m^2)^{-\frac{1}{2}}$$
$$- 2M^2(k^2 + M^2)^{-\frac{1}{2}}] dk$$
$$= m^2 \ln[(2M^2 - m^2)/m^2] + 2M^2 \ln[M^2/(2M^2 - m^2)], \tag{6.252}$$

for the previously quadratic and logarithmically divergent terms, respectively. Substituting (6.251) and (6.252) into (6.249) and (6.250) (plus their regulator counterparts) yields, to the order considered here

$$\langle \text{in}, 0 | T_{\eta\eta} | 0, \text{in} \rangle = (C/8\pi)\{m^2 \ln[(2M^2 - m^2)/m^2]$$
$$+ 2M^2 \ln[M^2/(2M^2 - m^2)]\} \tag{6.253}$$

$$\langle \text{in}, 0|T_{xx}|0, \text{in}\rangle = -(C/8\pi)\{m^2 \ln[(2M^2 - m^2)/m^2] + 2M^2 \ln[M^2/(2M^2 - m^2)]\} + f(m^2) + f(2M^2 - m^2) - 2f(M^2) \tag{6.254}$$

where

$$f(m^2) = \frac{m^4 b e^{2\rho\eta}}{8\pi} \int_{-\infty}^{\infty} \frac{dk}{(k^2 + \rho^2 + m^2)(k^2 + m^2)^{\frac{1}{2}}} = \frac{m^4 b e^{2\rho\eta}}{8\pi\rho(m^2 + \rho^2)^{\frac{1}{2}}} \ln\left[\frac{(m^2 + \rho^2)^{\frac{1}{2}} + \rho}{(m^2 + \rho^2)^{\frac{1}{2}} - \rho}\right]. \tag{6.255}$$

Noting also that $\langle T_{\eta x}\rangle = \langle T_{x\eta}\rangle = 0$, these expressions can be written

$$\langle \text{in}, 0|T_{\mu\nu}|0, \text{in}\rangle = (g_{\mu\nu}/8\pi)\{m^2 \ln[(2M^2 - m^2)/m^2] + 2M^2 \ln[M^2/(2M^2 - m^2)]\} + \eta_{\mu 1}\eta_{\nu 1}[f(m^2) + f(2M^2 - m^2) - 2f(M^2)]. \tag{6.256}$$

The first term on the right-hand side of (6.256) can be absorbed in renormalization of the cosmological constant Λ in the Einstein equation (it diverges as $M \to \infty$). There is no renormalization associated with a_1 in the effective Lagrangian, because this gives rise to a multiple of the Einstein tensor in the Einstein equation and this vanishes identically in two dimensions. The remaining term in (6.256) is finite as $M \to \infty$, and, taking the limit, is easily evaluated. One has

$$\langle \text{in}, 0 \,|\, T_{\mu\nu}|0, \text{in}\rangle_{\text{ren}} = -\eta_{\mu 1}\eta_{\nu 1} \frac{b e^{2\rho\eta}}{4\pi}\left[m^2 - \tfrac{2}{3}\rho^2 - \frac{m^4}{2\rho(m^2 + \rho^2)^{\frac{1}{2}}} \ln\left(\frac{(m^2 + \rho^2)^{\frac{1}{2}} + \rho}{(m^2 + \rho^2)^{\frac{1}{2}} - \rho}\right)\right]. \tag{6.257}$$

Note that, for $m = 0$, the trace of (6.257) gives, to lowest order in $be^{2\rho\eta}$, the correct conformal anomaly:

$$\langle \text{in}, 0|T_\mu{}^\mu|0, \text{in}\rangle_{\text{ren}} = -be^{2\rho\eta}\rho^2/6\pi \simeq -R/24\pi. \tag{6.258}$$

One disadvantage of the use of Pauli–Villars regularization is that one must always be able to work with massive fields (even if in some approximation). In many calculations, the massless field equation is often considerably simpler than its massive counterpart.

6.6 Physical significance of the stress-tensor

The enormous amount of effort that has been invested in developing techniques for computing $\langle T_{\mu\nu}\rangle_{\text{ren}}$ calls for a searching appraisal of the physical basis that underlies this quantity. In the previous sections it has been shown how, by a variety of complicated mathematical devices, one may extract from a meaningless formal quantity, $\langle T_{\mu\nu} \rangle$, a residue that, at least in the cases explicitly investigated, must be considered a serious contender for the quantity to be placed on the right-hand side of the gravitational field equation.

Not withstanding the physical reasonableness of the answers obtained, the construction of $\langle T_{\mu\nu} \rangle_{\text{ren}}$ by infinite renormalization of gravitational coupling constants is open to criticism. The DeWitt–Schwinger proper time integral, the inverse mass expansion of G_{F}, and many of the more formal manipulations of §§6.1 and 6.2 (such as those involving tr ln G_{F}) are at best ill-defined for hyperbolic operators, and may not even exist in some cases. Few rigorous results have been proved.

In addition to these mathematical problems, there is the question of whether the semiclassical theory makes sense at all. How would one measure $\langle T_{\mu\nu} \rangle$? When is the semiclassical approximation valid? If $\langle T_{\mu\nu} \rangle$ can only be measured gravitationally (i.e., through its entry into the field equation) can one consistently neglect higher order (i.e., two-loop graviton, etc.) contributions? What is the relation between higher order corrections and the heuristic expectation that $\langle T^2_{\mu\nu} \rangle$ should be $\ll \langle T_{\mu\nu} \rangle$ if the latter quantity is to accurately approximate some average distribution of quantum stress–energy–momentum? When, if ever, will the 'back-reaction' (i.e. gravitational dynamics modified by gravitationally induced $\langle T_{\mu\nu} \rangle$) be approximately determined by $\langle T_{\mu\nu} \rangle_{\text{ren}}$ computed at the one-loop level? Misgivings about these issues have been expressed by a number of authors (for example Duff 1981). Many of them might be resolved if a full theory of quantum gravity were available, to which one could claim that the semiclassical theory is some sort of approximation.

One approach to the physical significance of $\langle T_{\mu\nu} \rangle$ is to abandon renormalization altogether, and to ask, simply, that if the semiclassical theory is to make physical sense, what criteria might one wish $\langle T_{\mu\nu} \rangle$ to satisfy? If these criteria are too restrictive, no such object may exist; too loose, and it may not be unique.

The approach of attempting to define a unique, existing $\langle T_{\mu\nu} \rangle$ purely by imposing physical criteria ('axioms') was instigated by Christensen (1975) and has been pursued with great success by Wald (1977, 1978*a*, *b*), whose work has considerably strengthened the results of the renormalization programme. In the weaker system of axioms, Wald proposes that any physically meaningful $\langle T_{\mu\nu} \rangle$ ought at least to satisfy four eminently reasonable conditions:

(1) covariant conservation
(2) causality
(3) standard results for 'off-diagonal' elements
(4) standard results in Minkowski space.

The first condition is simply relation (6.142) which is necessary if $\langle T_{\mu\nu}\rangle$ is to appear on the right-hand side of the gravitational field equations, as the left-hand side is divergenceless. The causality axiom is rather subtle. The precise statement is:

'For a fixed 'in' state, $\langle T_{\mu\nu}\rangle$ at a point p in spacetime depends only on the spacetime geometry to the causal past of p'. By this it is meant that changes in the metric structure of the spacetime outside the past null cone through p ought not to effect $\langle T_{\mu\nu}\rangle$, so long as the state of the quantum field in the remote past is unaltered. (One can conceive of alterations outside the past null cone that nevertheless modify the Fock space based on the 'in' modes. These are not permitted.) A time-reversed statement then applies to fixed 'out' states and changes in the geometry outside the future null cone.

Condition (3) is simply the observation that as $\langle\Phi|T_{\mu\nu}|\Psi\rangle$ is in any case finite for orthogonal states, $\langle\Phi|\Psi\rangle=0$, the value of this quantity ought to be the usual (i.e., formal) one. By condition (4), we mean that the normal ordering procedure in Minkowski space (see §2.4) should be valid.

It is now straightforward to prove a remarkable result: if $\langle T_{\mu\nu}\rangle$ satisfies the first three of the four conditions above, then it is unique to within a local conserved tensor.

The proof runs as follows (see Wald, 1977, for a detailed exposition). If $T_{\mu\nu}$ and $\tilde{T}_{\mu\nu}$ are two (renormalized) stress-tensor operators satisfying conditions (1)–(3), then our aim is to show that the expectation value of

$$U_{\mu\nu}\equiv T_{\mu\nu}-\tilde{T}_{\mu\nu} \tag{6.259}$$

is a local, conserved tensor.

First note that by condition (3), the matrix elements of $U_{\mu\nu}$ between orthogonal states must vanish (because $\langle\Phi|T_{\mu\nu}|\Psi\rangle=\langle\Phi|\tilde{T}_{\mu\nu}|\Psi\rangle$). Furthermore, putting $|\Pi_\pm\rangle=2^{-\frac{1}{2}}(|\Psi\rangle\pm|\Phi\rangle)$, then

$$\langle\Pi_+|U_{\mu\nu}|\Pi_-\rangle=0$$

so

$$\langle\Psi|U_{\mu\nu}|\Psi\rangle-\langle\Phi|U_{\mu\nu}|\Phi\rangle=0,\quad \forall\Psi,\Phi, \tag{6.260}$$

i.e., all the diagonal elements (expectation values) are equal. Thus $U_{\mu\nu}$ is a multiple of the identity operator:

$$U_{\mu\nu}=u_{\mu\nu}I, \tag{6.261}$$

where $u_{\mu\nu}$ is an ordinary c-number tensor field.

We can now see that $u_{\mu\nu}$ must be a local tensor, for if we take the expectation value of $U_{\mu\nu}$ in some normalized 'in' state, we have

$$\langle \text{in} | U_{\mu\nu}(p) | \text{in} \rangle = u_{\mu\nu}(p). \tag{6.262}$$

But this would also be our answer if we took a normalized 'out' state

$$\langle \text{out} | U_{\mu\nu}(p) | \text{out} \rangle = u_{\mu\nu}(p). \tag{6.263}$$

As condition (2) requires that $u_{\mu\nu}(p)$ in (6.262) can only depend on the geometry in the causal past of p, while $u_{\mu\nu}(p)$ in (6.263) is similarly restricted by the geometry in the causal future, these two objects can only be equal if they depend solely on the geometry in the intersection of the past and future null cones, i.e., at p. Hence $u_{\mu\nu}(p)$ is a local tensor at p.

Finally, by condition (1), $u_{\mu\nu}$ must be conserved:

$$u^{\mu\nu}{}_{;\nu} = 0. \tag{6.264}$$

Thus, $\langle T_{\mu\nu} \rangle$ is unique to within a local, conserved tensor. However, any conserved tensor that is a function solely of the local geometry more properly belongs to the left-hand side of the gravitational field equations anyway, i.e., it is more reasonable to regard it as part of the gravitational dynamics than of the quantum field.

Even in renormalization theory, we are always free to take a local conserved tensor (e.g., ${}^{(1)}H_{\mu\nu}$) and place it on the right-hand side of the field equations if we wish. A physical measurement can resolve this ambiguity by determining the coefficient of such a term. Alternatively, one might attempt to advance arguments why no such term ought to be present on physical grounds (i.e., require that its coefficient be zero). For example, Wald (1977) suggested a fifth condition based on stability criteria for the gravitational dynamics, which, together with condition (4), uniquely fixes $u_{\mu\nu} = 0$. This fifth condition is now considered to be of dubious value as we shall discuss shortly.

One would like to know whether or not the renormalization prescriptions described in the earlier sections of this chapter satisfy Wald's conditions. If they do, then we have good reason to believe the results, whether the removal of the divergences is regarded as a legitimate renormalization of gravitational coupling constants, or merely as an *ad hoc* ansatz. This is especially important in the case of point-splitting, because one is unable to renormalize $\langle T_{\mu\nu} \rangle$ in the gravitational field equations anyway, it being necessary (guided by renormalization in the effective action) to discard a term in order to achieve the full conformal anomaly (see discussion on page 205). Moreover, the procedure of averaging over directions of the splitting vector t_μ prior to renormalization is open to the objection that no unique measure on the space of directions exists; one merely chooses a 'natural' one.

The point-splitting prescription starts with the DeWitt–Schwinger expansion of $G^{(1)}(x, x')$. If this expansion exists, then, in four dimensions, one can see by inspection of (3.141) that it will have the general form

$$S(x, x') = (U/\sigma) + V \ln \sigma + W \qquad (6.265)$$

where $U = U(x, x')$ and

$$V(x, x') = \sum_{l=0}^{\infty} V_l(x, x')\sigma^l, \quad W(x, x') = \sum_{l=0}^{\infty} W_l(x, x')\sigma^l. \qquad (6.266)$$

Expressed in this form, a Green function is called a Hadamard elementary solution, after the extensive work by Hadamard (1923) on the singularity structure of second order elliptic and hyperbolic equations. The question of when $G^{(1)}(x, x') = \langle 0|\{\phi(x), \phi(x')\}|0\rangle$ has the form of a Hadamard elementary solution (i.e., in what spacetimes and for what states $|0\rangle$) is still open, but it is generally believed to possess this form for a wide class of spacetimes and boundary conditions in which we are interested. In particular, Fulling, Sweeny & Wald (1978) have proved that if $G^{(1)}$ has the Hadamard singularity structure in an open neighbourhood of a Cauchy surface, then it has this form everywhere. This has as a corollary the result that, for a spacetime which is flat in the past of some Cauchy surface, $G^{(1)}$ calculated in the in vacuum will have the Hadamard form everywhere because it manifestly has that form in ordinary flat space quantum field theory.

The coefficients in (6.266) may be found by substituting (6.265) into the field equation and solving recursively (DeWitt & Brehme 1960, Garabedian 1964, Adler, Lieberman & Ng 1977). This procedure uniquely determines V. Moreover, W is uniquely determined once $W_0(x, x')$ is specified. Its specification may be regarded as the imposition of a boundary condition on the field, and it uniquely characterizes $S(x, x')$. An overall normalization then fixes U.

It is convenient to change the form of the Hadamard solution slightly and write it as

$$S(x, x') = [2/(4\pi)^2]\Delta^{\frac{1}{2}}[-(2/\sigma) + v \ln \sigma + w], \qquad (6.267)$$

where $\Delta(x, x')$ is defined by (3.139) and series analogous to (6.266) exist for v and w. When (6.267) is expanded in inverse powers of mass, it reproduces the DeWitt–Schwinger Green function expansion $G^{(1)}_{\mathrm{DS}}$, given in terms of the expansion (3.141), which was used as the basis of our renormalization programme. In four dimensions, this involved subtracting from $G^{(1)}(x, x')$ the first three terms (terms up to adiabatic order four) of the inverse mass expansion of $G_{\mathrm{DS}}^{(1)}$. However, to investigate consistency with the Wald axioms, it is preferable to work, not with a truncated inverse mass expansion, but with the exact Hadamard form (6.267). This is especially true in the massless limit, where the w terms of the DeWitt–Schwinger expansion diverge (being an expansion in m^{-1}). Shortly we shall discover the relation between renormalization based on $S(x, x')$ as compared with $G_{\mathrm{DS}}^{(1)}(x, x')$.

The Green function $G^{(1)}(x, x')$ is the basic object from which $\langle 0|T_{\mu\nu}|0\rangle$ is constructed by differentiation. We assume that $G^{(1)}$ has the Hadamard form in a practical calculation. Now $G^{(1)}(x, x')$ is manifestly symmetric in x, x' and hence satisfies the field equation in both of these variables. In the case of the Hadamard solution (6.267), one can prove (Fulling, Sweeny & Wald 1978) that Δ and v are symmetric for all Hadamard solutions. However, the same conclusion cannot be drawn for w, as this non-singular part is only specified uniquely once $w_0(x, x')$ is given, and can vary from solution to solution, depending on boundary conditions. Thus, although $w(x, x')$ is symmetric for $G^{(1)}(x, x')$ by construction, it will not be so in general. We may therefore conclude that $S(x, x')$ will not, in general, satisfy the field equation in x', even though it will, by definition, satisfy it in x.

We wish to discover a way of removing the pole terms from $G^{(1)}(x, x')$ in such a fashion as to respect the Wald axioms and hence achieve an (almost) unique answer when the residue is differentiated to give $\langle T_{\mu\nu}(x)\rangle$. One likely procedure is to subtract from $G^{(1)}(x, x')$ a Hadamard solution $S(x, x')$ (Wald 1977, 1978*a*, Adler, Lieberman & Ng 1977). To make the procedure unambiguous, however, one must decide on the boundary conditions for S, i.e., specify $w_0(x, x')$. Since for massless fields in Minkowski space $D^{(1)}$ is given by (2.79), i.e., $w = 0$ (w = constant in the massive case), for consistency with condition (4) we assume that $w_0(x, x') = 0$ in S. From our previous result we know that any other w_0 which is consistent with all four conditions, can yield a result which differs at most by a local conserved tensor from that obtained using $w_0 = 0$.

With this choice, $w_0 = 0$, we have

$$G^{\mathrm{B}} = G^{(1)} - S \tag{6.268}$$

where G^{B} (B standing for 'boundary-condition-dependent part') is a smooth, uniquely determined function of x and x'. From it, one may construct $\langle T_{\mu\nu}^{\mathrm{B}}\rangle$ by differentiation (see §6.4 for details of this step).

The result clearly satisfies Wald's condition (2), because the two pieces,

$G^{(1)}$ and S, from which it is constructed, are both causal, $G^{(1)}$ because the ϕ field from which it is computed propagates causally, and S because it is a purely local object, so does not depend on the geometry outside the null cones. When $x' \to x$, G^{B} depends only on the geometry on and inside the null cone through x.

Condition (3) can also easily be shown to be satisfied. From (6.260) we can express the matrix elements of $T^{\mathrm{B}}_{\mu\nu}$ between orthogonal states in the form

$$\begin{aligned}\langle \Pi_+ | T^{\mathrm{B}}_{\mu\nu} | \Pi_- \rangle &= \tfrac{1}{2}\langle \Psi | T^{\mathrm{B}}_{\mu\nu} | \Psi \rangle - \tfrac{1}{2}\langle \Phi | T^{\mathrm{B}}_{\mu\nu} | \Phi \rangle \\ &= \tfrac{1}{2} \lim_{x' \to x} \mathscr{D}_{\mu\nu}(G^{\mathrm{B}}_{\Psi} - G^{\mathrm{B}}_{\Phi}) \\ &= \tfrac{1}{2} \lim_{x' \to x} \mathscr{D}_{\mu\nu}(G_{\Psi} - G_{\Phi}) \end{aligned} \tag{6.269}$$

where $\mathscr{D}_{\mu\nu}$ is the operator discussed on page 195 and G_{Ψ}, etc., denote $\langle \Psi | \{\phi(x), \phi(x')\} | \Psi \rangle$, etc. The final equality in (6.269) follows from the fact that the same $S(x, x')$ is subtracted from both G_{Ψ} and G_{Φ}, by the 'renormalization' ansatz (6.268). According to condition (3), (6.269) is supposed to be the usual (formal, 'unrenormalized') expression, which it clearly is, since it is independent of S.

Unfortunately, the ansatz (6.268) does not satisfy Wald's condition (1). As remarked above, $S(x, x')$ will not in general satisfy the field equation in both x and x'. For the boundary condition $w_0 = 0$, it does not, in fact, satisfy this equation in x'. Calculation shows that, in the massless case

$$\nabla^{\nu}\langle T^{\mathrm{B}}_{\mu\nu} \rangle = \tfrac{1}{4} \lim_{x' \to x} \nabla_{\mu}[\Box_{x'} + \tfrac{1}{6}R(x')]G^{\mathrm{B}}(x, x'). \tag{6.270}$$

so that $\langle T^{\mathrm{B}}_{\mu\nu} \rangle$ fails to be conserved to the extent that it fails to satisfy the field equation in x'. Wald (1978*a*) evaluates the right-hand side of (6.270) and obtains

$$\nabla_{\mu} a_2(x)/64\pi^2 \tag{6.271}$$

where $a_2(x)$ is the usual coefficient in the DeWitt–Schwinger expansion. Thus, to construct a conserved $\langle T_{\mu\nu} \rangle$ that is still consistent with the conditions (2)–(4) one must take

$$\langle T_{\mu\nu}(x) \rangle = \langle T^{\mathrm{B}}_{\mu\nu}(x) \rangle - a_2(x) g_{\mu\nu}(x)/64\pi^2. \tag{6.272}$$

Whereas $g^{\mu\nu}\langle T^{\mathrm{B}}_{\mu\nu} \rangle = 0$ by construction,

$$g^{\mu\nu}\langle T_{\mu\nu} \rangle = -a_2(x)/16\pi^2 \tag{6.273}$$

in agreement with the conformal anomaly (6.119). As our result is unique up to a local conserved tensor, and as there exists no such tensor with the trace (6.273), we may conclude that the conformal anomaly is an inevitable consequence of a local, semiclassical theory. Of course one does have the freedom to adjust the $\Box R$ part of the anomaly by adding multiples of the local conserved tensors ${}^{(1)}H_{\mu\nu}$ and ${}^{(2)}H_{\mu\nu}$.

Wald originally suggested as a fifth condition, based on stability arguments, that $\langle T_{\mu\nu}\rangle$ should contain no terms of adiabatic order greater than three (such as terms giving rise to $\Box R$ in the trace). However, one can show that the stress-tensor will in general contain non-local terms involving higher adiabatic order contributions which cannot be removed by addition of (local) counterterms proportional to ${}^{(1)}H_{\mu\nu}$ and ${}^{(2)}H_{\mu\nu}$. Horowitz (1980) has suggested that, for massless fields, the presence of such non-local, higher derivative terms will lead to instability about flat spacetime. The full implication of such results for the semiclassical theory is still under investigation.

What is the relationship between $S(x, x')$ as used here to render $G^{(1)}$ finite, and the DeWitt–Schwinger $G^{(1)}_{\mathrm{DS}}(x, x')$ used in the previous sections to renormalize $G^{(1)}$? First note that (in four dimensions) we are only interested in terms in the expansion of $G^{(1)}_{\mathrm{DS}}$ up to adiabatic order four, and up to order σ. Terms of higher order in σ do not contribute to $\langle T_{\mu\nu}\rangle$ as $\sigma \to 0$. From (3.141) these terms are

$$\begin{aligned} G^{(1)}(x, x') = (\Delta^{\frac{1}{2}}/4\pi^2)\{&a_0[-(1/\sigma) + m^2L(1 - \tfrac{1}{4}m^2\sigma) - \tfrac{1}{2}m^2 + \tfrac{5}{16}m^2\sigma] \\ &- a_1[L(1 - \tfrac{1}{2}m^2\sigma) + \tfrac{1}{2}m^2\sigma] \\ &- a_2\sigma[\tfrac{1}{2}L - \tfrac{1}{4}] + (1/2m^2)a_2 + O(T^{-6}) + O(\sigma^2)\}, \end{aligned} \qquad (6.274)$$

where $L \equiv \gamma + \tfrac{1}{2}\ln|\tfrac{1}{2}m^2\sigma|$, γ being Euler's constant.

Clearly the divergent and logarithmic terms in (6.274) have the Hadamard form (6.267), and, by the uniqueness property, must agree with the corresponding terms in $S(x, x')$. Moreover, because $G^{(1)}_{\mathrm{DS}}$ satisfies the field equation at each adiabatic order and at each power of σ, the determination of the coefficients of the terms $a_0m^2, a_0m^4\sigma, a_1m^2\sigma$ and $a_2\sigma$ are independent of the m^{-2} terms and higher. Hence they will be the same as in the Hadamard $S(x, x')$. This is confirmed by explicit calculation (Adler, Lieberman & Ng 1977). Thus the difference between $G^{(1)}_{\mathrm{DS}}$ and S is the 'anomaly term', i.e., the ultimate term in (6.274), which as $x' \to x$ is $a_2/(8\pi^2m^2)$. In a practical calculation, one renormalizes $\langle T_{\mu\nu}\rangle$ by subtracting (6.274) from $G^{(1)}$, and after differentiation dropping all terms resulting from (6.274) of adiabatic order greater than four. As discussed on page 205 this procedure is equivalent to subtracting the final term in (6.274) only in its contribution made via the term $2\{\frac{1}{4}-[1-(1/n)]\xi\}m^2g_{\mu\nu}\phi^2$ in (3.190). Thus, in the four-dimensional, conformally-coupled, massless limit, the final term in (6.274) makes the single contribution $-a_2g_{\mu\nu}/(64\pi^2)$ to $\langle T_{\mu\nu}\rangle$, and so this is the only difference resulting in $\langle T_{\mu\nu}\rangle$ from the subtraction of $G^{(1)}_{\mathrm{DS}}$ rather than S. But in the method in which S is subtracted, precisely this difference term is added on in (6.272) to maintain covariant conservation. (The necessity of including this term can be seen from the fact that $G^{(1)}$ satisfies the field equation in x, order by order in the adiabatic parameter. The conservation of $\langle T_{\mu\nu}\rangle$ depends on this property. Thus, when computing a conserved $\langle T_{\mu\nu}\rangle$, one must work to a consistent adiabatic order. The term $m^2\phi^2$, because it contains no derivatives, must involve two higher adiabatic orders in the expansion (6.274) than do the other terms (containing two derivatives).) We may therefore conclude that the point-splitting renormalization prescription given in §§6.2 and 6.4 is equivalent to the Wald subtraction method based on the Hadamard solution, which we know to yield an (almost) unique answer (see also Bunch, Christensen & Fulling 1978 for the two-dimensional case). Thus, the point-splitting renormalization method and the other renormalization techniques that we have shown are closely related to it, can be used with the confidence that the results they produce are consistent with the physically very reasonable conditions (1)–(4), and are (to within local conserved tensors) uniquely determined. This implies that the resulting $\langle T_{\mu\nu}\rangle_{\mathrm{ren}}$ must be the physically correct answer if the semiclassical theory is to make sense at all.

Finally, we turn to another aspect of the physical interpretation of $\langle T_{\mu\nu}\rangle$ concerning boundaries. It is usually supposed spacetime is unbounded. However, material surfaces can act as effective boundaries, e.g., the presence of conductors constrains the electromagnetic field to vanish, at least in some approximation (ignoring a 'skin depth'). The general mathemetical analysis (Minakshisundaram & Pleijel 1949, McKean & Singer 1967, Greiner 1971, Stewartson & Waechter 1971, Waechter 1972, Gilkey 1975, Kennedy 1978) of elliptic operators in Riemannian space and, by association, of the wave operator in pseudo-Riemannian spacetime, incorporates the effects of a boundary ∂M to a manifold M in the DeWitt–Schwinger series or its elliptic equivalent. One finds that when a boundary is present, the effective action given by (6.36) and (6.40) is augmented by a surface effective action (see, for example, Kennedy, Critchley & Dowker 1980)

$$W_{\mathrm{S}} = \int_{\partial M} \mathrm{d}^{n-1}x(\pm h)^{\frac{1}{2}} L_{\mathrm{S}}(x), \tag{6.275}$$

where h is the determinant of the metric h_{ij} induced on the boundary by the spacetime metric $g_{\mu\nu}$ ($\pm$ according to whether ∂M is timelike or spacelike respectively) and the asymptotic expansion of L_S (comparable to (6.40)) is

$$L_S \approx \tfrac{1}{2}(4\pi)^{-n/2} \sum_{j=0}^{\infty} b_{(j+1)/2}(x) \int_0^{\infty} (is)^{(2j-1-n)/2} e^{-im^2 s} i ds. \tag{6.276}$$

The precise form of the coefficients b depends on the boundary conditions imposed on the field on ∂M. For Dirichlet boundary conditions imposed on a scalar field ϕ,

$$\phi(x) = 0, \quad x \in \partial M, \tag{6.277}$$

one obtains for the first few coefficients

$$\begin{aligned} b_{\frac{1}{2}} &= -\tfrac{1}{2}\pi^{\frac{1}{2}} \\ b_1 &= \tfrac{1}{3}\chi \\ b_{\frac{3}{2}} &= \tfrac{1}{192}\pi^{\frac{1}{2}}[3(3-32\xi)\chi^2 + 6(16\xi - 1)\chi_{\mu\nu}\chi^{\mu\nu} \\ &\quad - 16(1-6\xi)\hat{R} - 24(8\xi - 1)R_{\mu\nu}n^{\mu}n^{\nu}], \end{aligned} \tag{6.278}$$

where n_μ is the inward pointing unit vector on $\partial M, \chi_{\mu\nu}$ and $\chi = \chi_\mu{}^\mu$ are respectively the second fundamental form of ∂M and its trace, and $\hat{R}$ is the Ricci scalar of the induced metric $h_{\mu\nu}$ (see, for example, Hawking & Ellis 1973, §2.7). For Robin boundary conditions

$$[\psi(x) + n^{\mu}\nabla_{\mu}]\phi(x) = 0, \tag{6.279}$$

one obtains

$$\begin{aligned} b_{\frac{1}{2}} &= \tfrac{1}{2}\pi^{\frac{1}{2}} \\ b_1 &= \tfrac{1}{3}(\chi - 6\psi) \\ b_{\frac{3}{2}} &= \tfrac{1}{192}\pi^{\frac{1}{2}}[192\psi^2 + 96\psi\chi + 3(32\xi - 1)\chi^2 + 6(3-16\xi)\chi_{\mu\nu}\chi^{\mu\nu} \\ &\quad + 16(1-6\xi)\hat{R} - 24(1-8\xi)R_{\mu\nu}n^{\mu}n^{\nu}]. \end{aligned} \tag{6.280}$$

It is clear that the terms in (6.276) with $j \leq (n-1)/2$ are divergent. In the foregoing sections, divergent terms have been dealt with by first introducing a regularization scheme, and then removing the divergent terms by renormalization. The surface effective action (6.275) can be regularized using the methods introduced previously, and one must consider whether one can remove the divergent terms by renormalization of constants in the gravitational action.

The generalized gravitational action, which was renormalized in §6.2, makes no provision for surface divergences, so it appears at first sight that the presence of a boundary ∂M, on which boundary conditions on the field are imposed, gives rise to truly infinite vacuum stress there. It has been argued by Deutsch & Candelas (1979) that this surface divergence represents a real physical effect of the quantum field. A boundary to spacetime is in any case such a pathological feature that the occurrence of a divergence in $\langle T_{\mu\nu}\rangle$ there is not especially surprising. In the case of material surfaces, one must take into account that treating them as manifold boundaries is an idealization. For example, in the electromagnetic case, the finite conductivity of a real conductor would render the material transparent at very high frequencies, thus providing an ultraviolet cut-off to the mode integrals. (The cut-off parameter used in §4.1 as a mathematical device has, in this case, some foundation in physics.) Consequently, the apparent divergences at a material surface are really only very large, but finite, contributions to $\langle T_{\mu\nu}\rangle$. Deutsch & Candelas point out that any change in the conductivity of the material would lead to a large, and possibly measurable shift in $\langle T_{\mu\nu}\rangle$.

On the other hand, if one allows the addition to the generalized Einstein action of a surface action involving terms appearing in (6.278) and (6.280), then one can remove the divergences arising from the matter fields' surface effective action by renormalization. This approach has been advocated by Kennedy, Critchley & Dowker (1980).

The necessity of adding a surface action to the conventional Einstein action S_g given by (6.11) has been pointed out in another context by Gibbons & Hawking (1977*b*). They note that the action (6.11) is no longer appropriate if one requires that under all variations of the metric that vanish on ∂M, stationarity of S_g leads to the Einstein equation. The reason is that the variation of terms in S_g that arise from terms in R that are linear in the second derivatives of the metric can be converted by integration by parts into an integral over ∂M that involves the normal derivative of the metric at the surface. Unless the normal derivative is required to vanish at ∂M, one must augment (6.11) by a term that will cancel this normal derivative surface integral. Gibbons & Hawking find such a term to be

$$(-1/8\pi G)\int_{\partial M} \chi(\pm h)^{\frac{1}{2}}\, \mathrm{d}^{n-1}x + C, \tag{6.281}$$

where C depends only on h, not on g. If the boundary can be embedded in flat space, with second fundamental form $\chi^0_{\mu\nu}$, a natural choice of C is

$$(1/8\pi G)\int_{\partial M} \chi^0(\pm h)^{\frac{1}{2}}\mathrm{d}^{n-1}x, \tag{6.282}$$

so that the surface action vanishes when the spacetime is flat. The importance of the inclusion of (6.281) in considerations of quantum gravity has been stressed by Hawking (1979) (see also §8.5).

7

Applications of renormalization techniques

This short chapter presents some explicit examples of the theory of regularization and renormalization discussed in chapter 6. The number of spacetimes for which one may compute $\langle T_{\mu\nu} \rangle$ in terms of simple functions is extremely limited, and we think it probable that all such cases have been included either here, in chapter 6, or in our references.

Special importance is attached to the Robertson–Walker models, both because of their cosmological significance, and also because, being conformally flat, they provide a good illustration of conformal anomalies at work. However, precisely because of their simplicity, these models do not display the full non-local structure of the stress-tensor, and in §7.3 we turn briefly to the less elegant but more realistic example of an anisotropic, homogeneous cosmological model.

Although the primary subject of this book is the theory of quantum fields propagating in a prescribed background spacetime, the motivation for much of this work rests with its possible application to cosmological and astrophysical situations, where the gravitational dynamics must be taken into account. Many cosmologists, for example, believe that the back-reaction of quantum effects induced by the background gravitational field could have a profound effect on the dynamical evolution of the early universe, such as bringing about isotropization. We do not dwell in detail on this important extension of the theory, but note that the results presented here constitute the starting point for such investigations. A short discussion of the wider cosmological implications is given in §7.4.

7.1 Two-dimensional examples

The renormalized stress-tensor for the conformally trivial case of a massless scalar field with $\xi = 0$ propagating in two-dimensional spacetime has been derived in §§6.3 and 6.4.

We work with the line element

$$ds^2 = C(u, v)\, du\, dv \tag{7.1}$$

which has scalar curvature

$$R = \Box \ln C. \tag{7.2}$$

In flat spacetime, $R = 0$, so

$$\Box \ln C = 4C^{-1}\partial_u \partial_v \ln C = 0. \tag{7.3}$$

Thus

$$C = F(u)G(v) \tag{7.4}$$

where F and G are arbitrary differentiable functions.

One solution of (7.4) is $F = G = 1$, which recovers the familiar Minkowski space metric:

$$\mathrm{d}s^2 = \mathrm{d}u\,\mathrm{d}v. \tag{7.5}$$

Another can be obtained from (7.5) by a coordinate transformation $u \to \bar{u}$, $v \to \bar{v}$, such as

$$\left.\begin{aligned} u &= f(\bar{u}) \\ v &= \bar{v} \end{aligned}\right\}, \tag{7.6}$$

which changes (7.5) to

$$\mathrm{d}s^2 = f'(\bar{u})\,\mathrm{d}\bar{u}\,\mathrm{d}\bar{v}. \tag{7.7}$$

Here a prime is used to denote differentiation of a function with respect to its argument.

If we choose the function f as follows

$$\tfrac{1}{2}[\bar{t} - f(\bar{t})] = z\{\tfrac{1}{2}[\bar{t} + f(\bar{t})]\}, \tag{7.8}$$

where

$$\bar{t} = \tfrac{1}{2}(\bar{u} + \bar{v}) \tag{7.9}$$

is the time coordinate in the $\bar{u}$, $\bar{v}$ system, then (7.8) is the restriction to $\bar{x} = 0$ of

$$\tfrac{1}{2}[\bar{v} - f(\bar{u})] = z\{\tfrac{1}{2}[\bar{v} + f(\bar{u})]\}, \tag{7.10}$$

where

$$\bar{x} = \tfrac{1}{2}(\bar{v} - \bar{u}) \tag{7.11}$$

is the space coordinate in the $\bar{u}$, $\bar{v}$ system.

In the (u, v) system, (7.10) reduces to

$$x = z(t), \tag{7.12}$$

which can be taken as the trajectory of the moving mirror investigated in §4.4. In the $(\bar{u}, \bar{v})$ coordinate system the mirror remains at $\bar{x} = 0$. Because of (7.7), the moving mirror system is conformally related to a static mirror system, so we may employ the general relation (6.136) which connects $\langle T_{\mu\nu} \rangle$ evaluated in conformally related vacuum states. In §4.3 we showed that, in four dimensions with a conformally coupled massless field, $\langle T_{\mu\nu} \rangle$ in the half space bounded by a static plate vanishes. The same result is readily recovered for the the two-dimensional case. Hence we may set to zero the first term of the right of (6.136). The relevant conformal factor C is found from (7.7) to be $f'(\bar{u})$, so recalling that $R = 0$, (6.137) yields (Davies & Fulling 1977*b*) the expectation value in the 'conformal vacuum' state:

$$\langle T_{\bar{u}\bar{u}} \rangle_{\text{ren}} = -(1/12\pi)(f')^{\frac{1}{2}} \partial_{\bar{u}}^2 (f')^{-\frac{1}{2}}$$

$$\langle T_{\bar{u}\bar{v}} \rangle_{\text{ren}} = \langle T_{\bar{v}\bar{u}} \rangle_{\text{ren}} = \langle T_{\bar{v}\bar{v}} \rangle_{\text{ren}} = 0.$$

Transforming to the (u, v) coordinate system, this gives

$$\langle T_{uu} \rangle_{\text{ren}} = -(1/12\pi)(f')^{-\frac{3}{2}} \partial_{\bar{u}}^2 (f')^{-\frac{1}{2}} \tag{7.13}$$

$$\langle T_{uv} \rangle_{\text{ren}} = \langle T_{vu} \rangle_{\text{ren}} = \langle T_{vv} \rangle_{\text{ren}} = 0. \tag{7.14}$$

In (7.13) the argument of f is

$$\bar{u} = f^{-1}(u) = 2\tau_u - u \equiv p(u) \tag{7.15}$$

(see (4.44) and (4.50)). Writing (7.13) in terms of $p(u)$ one obtains

$$\langle T_{uu} \rangle_{\text{ren}} = (1/12\pi)(p')^{\frac{1}{2}} \partial_u^2 (p')^{-\frac{1}{2}}. \tag{7.16}$$

As $\langle T_{uu} \rangle_{\text{ren}}$ is a function of u only, it is constant along the retarded null rays $u = \text{constant}$ from $\mathscr{I}^+$ back to the surface of the mirror. Thus energy is created at the mirror itself and flows away undiminished to the right. Notice that although one cannot pin down where the individual quanta are created, the source of energy is unambiguous. There is no energy flux from right to left. This asymmetry results from the choice of retarded boundary conditions implied through (7.6). In a situation in which static in and out regions exist, such a choice is equivalent to taking $|0, \text{in}\rangle$ rather than $|0, \text{out}\rangle$ as the quantum state. To see this, suppose that $z(t) = \text{constant}$ for $t < t_0$, thereby defining an in region $(t < t_0)$. Then from (7.8) $f'(\bar{u}) = 1$ for $t < t_0$ and it is clear that the 'conformal vacuum' defined from (7.7) agrees with the in vacuum, which is the conventional vacuum state for a static system. A similar relation between the advanced boundary condition and the out vacuum can be achieved by reversing the roles of u and v in (7.6) (Fulling & Davies 1976).

Transforming (7.14) and (7.16) to (t, x) coordinates, and using (4.44) and (7.15) to write the result in terms of the mirror trajectory $z(\tau_u)$, one obtains

$$\langle T_{tt}\rangle_{\text{ren}} = \langle T_{xx}\rangle_{\text{ren}} = -\langle T_{xt}\rangle_{\text{ren}} = -\langle T_{tx}\rangle_{\text{ren}}$$

$$= -\frac{1}{12\pi}\frac{(1-\dot{z}^2)^{\frac{1}{2}}}{(1-\dot{z})^2}\frac{\mathrm{d}}{\mathrm{d}\tau_u}\left[\frac{\ddot{z}}{(1-\dot{z}^2)^{\frac{3}{2}}}\right]. \tag{7.17}$$

Here $z = z(\tau_u)$ and the dot denotes differentiation with respect to τ_u. The parameter τ_u (defined by (4.44)) is, in fact, the time coordinate of the mirror when its trajectory intersects the retarded null ray u. Note that for non-relativistic motion, (7.17) is simply $-(12\pi)^{-1}\dddot{z}$. The right-hand side of (7.17) may be written

$$-\frac{1}{12\pi}\frac{(1-V^2)^{\frac{1}{2}}}{(1-V)^2}\frac{\mathrm{d}\alpha}{\mathrm{d}\tau_u} \tag{7.18}$$

where V is the mirror velocity $\dot{z}$ and α is the proper acceleration. As $|V| < 1$, (7.18) changes sign according to whether α is increasing or decreasing in time. Thus, the mirror may radiate *negative* energy if its acceleration is increasing to the right, or decreasing to the left. The emission of negative energy is a purely quantum phenomenon. It opens up the possibility of unusual new physical processes not encountered in classical theory. We shall see in the next chapter that negative energy fluxes play a role in the quantum evaporation of black holes. It might be supposed that a moving mirror could be used to violate the second law of thermodynamics by radiating negative energy into a hot body, thereby cooling it. However, if (7.17) is integrated over time between periods of mirror stasis, it is always positive definite, i.e., the negative flux is restricted to finite intervals (Fulling & Davies 1976). Ford (1978) has shown that negative energy fluxes cannot be sustained for long enough to reduce the entropy of a hot body by more than would be expected on the basis of ordinary thermal fluctuations.

One trajectory of interest occurs if $\alpha = \text{constant}$, i.e., uniform proper acceleration, corresponding to a hyperbolic trajectory. From (7.18) it follows that $\langle T_{\mu\nu}\rangle_{\text{ren}} = 0$ which implies that no energy at all is radiated during intervals when the mirror acceleration is uniform. However in §4.4 we treated a special case of uniform acceleration and computed the non-trivial Bogolubov transformation (4.64) and (4.65). Clearly, the mirror emits *particles*, and from the discussion associated with (3.80) it is also clear that a particle detector would respond to the presence of quanta. Nevertheless no *energy* is radiated. This example beautifully illustrates the looseness of the relation between particles and energy–momentum. The presence of quanta need not imply the presence of energy.

There appears to be a paradox concerning how the particle detector can, in the absence of field energy, absorb quanta and make a transition to an excited state. The resolution is that in so doing, the detector emits negative energy into the field to compensate. The emission of negative energy by mirrors and detectors is not without precedent in quantum field theory. It is possible to construct many-particle states with negative or zero fluxes even in the absence of mirrors or detectors (Epstein, Gaser & Jaffe 1965; see also Appendix A of Davies & Fulling 1977*a*).

Another mirror trajectory of interest is (4.51) which gives rise to a thermal flux at late times. Applying (7.16) to (4.52) yields

$$\langle T_{tt} \rangle_{\text{ren}} = \frac{\kappa^2}{48\pi}, \quad t \to \infty \tag{7.19}$$

which is the energy of a thermal flux of radiation with temperature $\kappa/2\pi$. This result is in complete agreement with the computation of the thermal spectrum given in §4.4. Indeed, using the thermal spectrum (4.61)

$$\langle T_{tt} \rangle = \frac{1}{2\pi} \int_0^\infty \frac{\omega}{e^{2\pi\omega/\kappa} - 1} d\omega = \frac{\kappa^2}{48\pi}, \tag{7.20}$$

which shows that the rate of creation of particles in the mode ω is $(1/2\pi)(e^{2\pi\omega/\kappa} - 1)^{-1}$. (This result is deduced from (4.61) by constructing wave packets to convert the particle *number* per mode, $|\beta_{\omega\omega'}|^2$, to a number rate. See §8.1.)

An interesting conclusion from (7.20) is that, in the case of thermal emission, the energy emitted is given by the particle number per mode, weighted by the energy of one quantum ω, integrated over all modes. We have seen that in general there is no simple relation between the number of quanta present and the field energy. For thermal radiation, however, the naive relation of '$\hbar\omega$ per quantum' survives. This can be traced to the complete absence of correlations between the modes in the thermal case (see also chapter 8). It will not be true for all mirror trajectories.

It is also possible to treat a problem of two mirrors, one static and one moving, which involves a 'disturbed Casimir' contribution from the first term on the right-hand side of (6.136). The results (Fulling & Davies 1976) are

$$\left.\begin{aligned} \langle T_{tt}\rangle_{\text{ren}} = \langle T_{xx}\rangle_{\text{ren}} &= \Lambda(u) + \Lambda(v) \\ \langle T_{tx}\rangle_{\text{ren}} = \langle T_{xt}\rangle_{\text{ren}} &= \Lambda(v) - \Lambda(u) \end{aligned}\right\} \tag{7.21}$$

where

$$\Lambda = -(1/24\pi)[(R'''/R') - \tfrac{3}{2}(R''/R') + \pi^2(R')^2/2L^2]. \tag{7.22}$$

The prime denotes differentiation with respect to the argument and the function R is determined by the equation

$$R[t + z(t)] = R[t - z(t)] + 2L \tag{7.23}$$

in terms of the moving mirror trajectory $z(t)$. The static mirror is at $x = 0$, and before $t = 0$ the other mirror is static at $x = L$. The vacuum is defined in this 'in' region ($t < 0$). Note that (7.21) represent a right- and left-moving flux superimposed. This arises because radiation created by the moving mirror reflects from the static mirror. In the special case $z(t) = L$ for all t, (7.23) yields $R(u) = u$ and (7.21) gives $\langle T_{tt}\rangle = -\pi/(24L^2)$, which is the Casimir energy for plate separation L. This result is one-quarter of (4.8), the difference arising due to the use of vanishing, rather than periodic, boundary conditions.

Another interesting case occurs for $F(x) = G(x) = \exp(ax)$, $a =$ constant, in (7.4). Once again this is flat spacetime, but with metric

$$\mathrm{d}s^2 = \mathrm{e}^{2a\eta}(\mathrm{d}\eta^2 - \mathrm{d}x^2)$$

i.e., the Milne universe discussed in §5.3. Substituting $C = \mathrm{e}^{2a\eta}$ into the general formula (6.136) and, as the two-dimensional Milne universe is conformal to the whole of two-dimensional Minkowski space, setting the first term on the right-hand side equal to zero, gives

$$\langle T_t{}^t\rangle_{\text{ren}} = -\langle T_x{}^x\rangle_{\text{ren}} = -1/24\pi t^2, \tag{7.24}$$

in agreement with (5.46).

Conformally non-trivial two-dimensional exactly soluble examples are few in number. One case is de Sitter space, which can be treated in the same way as its four-dimensional counterpart considered in §6.4. Another is the Robertson–Walker spacetime with conformal factor

$$C(\eta) \propto \mathrm{e}^{\alpha\eta^2}, \quad \alpha \text{ constant}. \tag{7.25}$$

With the choice (7.25), $RC = 2\alpha$ and the normalized massless mode solutions of (5.4), with $\xi(n)=\xi(2)=0$, and arbitrary ξ are

$$(4\pi\omega)^{-\frac{1}{2}}e^{i(kx-\omega\eta)} \tag{7.26}$$

where

$$\omega^2 = k^2 + \beta^2 \tag{7.27}$$

and

$$\beta^2 \equiv 2\alpha\xi.$$

Thus, using a vacuum state based on positive frequency modes (7.26) we have

$$G^{(1)}(x'',x') = (1/2\pi)\,\mathrm{Re}\int_{-\infty}^{\infty} \omega^{-1} e^{ik\Delta x - i\omega\Delta\eta}\,dk \tag{7.28}$$

$$= (1/\pi)\,\mathrm{Re}\,K_0[i\beta(\Delta u\,\Delta v)^{\frac{1}{2}}] \tag{7.29}$$

where $\Delta u = (\eta''-\eta')-(x''-x')$, $\Delta v = (\eta''-\eta')+(x''-x')$. Then expanding (7.29) in powers of $\Delta u\,\Delta v$ yields

$$G^{(1)}(x'',x') = (1/\pi)\{-\gamma - \tfrac{1}{2}\ln|\tfrac{1}{4}\beta^2\Delta u\,\Delta v| + \tfrac{1}{4}\beta^2\Delta u\,\Delta v[\gamma + \tfrac{1}{2}\ln|\tfrac{1}{4}\beta^2\Delta u\,\Delta v| + 1] + \ldots\}. \tag{7.30}$$

We may now use the expansion (6.200), and the corresponding expansion with u replaced by v, to obtain in explicitly geometrical form,

$$G^{(1)}(x'',x') = -(1/\pi)\{(1-\varepsilon^2\Sigma\xi R)[\gamma + \tfrac{1}{2}\ln|\varepsilon^2\xi R|] + \varepsilon^2\Sigma(A_{\alpha\beta}t^\alpha t^\beta\Sigma^{-1} - \xi R)\} \tag{7.31}$$

where $A_{\alpha\beta}$ is a tensor with components

$$\left.\begin{aligned} A_{uu} &= A_{vv} = \tfrac{1}{48}(-\dot{D} + 2D^2) \\ A_{uv} &= A_{vu} = -\tfrac{1}{24}\dot{D}. \end{aligned}\right\} \tag{7.32}$$

According to our established renormalization procedure, it is now necessary to subtract from (7.31) terms up to adiabatic order two in the DeWitt–Schwinger expansion, i.e., the terms displayed in (6.203). The logarithmically divergent terms cancel, leaving

$$\begin{aligned} &-(1/\pi)\{\tfrac{1}{2}(1-\varepsilon^2\Sigma\xi R)\ln|\xi R m^{-2}| \\ &+ \varepsilon^2\Sigma(A_{\alpha\beta}t^\alpha t^\beta\Sigma^{-1} - \tfrac{3}{2}\xi R + \tfrac{1}{12}R) \\ &- \tfrac{1}{2}(\xi - \tfrac{1}{6})m^{-2}R + O(m^2)\}. \end{aligned} \tag{7.33}$$

Comparison of (7.33) with (6.205) leads to the following identifications, dropping terms of $O(m^2)$,

$$c = -(1/2\pi)[\ln|\xi R m^{-2}| - (\xi - \tfrac{1}{6})m^{-2}R]$$

$$e_{\alpha\beta} = -(1/\pi)A_{\alpha\beta}$$

$$f = (1/2\pi)(3\xi - \tfrac{1}{6})R, \quad q = r = 0,$$

so (6.206) yields the following renormalized stress-tensor in the massless limit

$$\langle 0|T_{\mu\nu}|0\rangle_{\text{ren}} = -(1/48\pi)Rg_{\mu\nu} + \theta_{\mu\nu} + (\xi/4\pi)[(R_{;\mu\nu}/R) - (R_{;\mu}R_{;\nu}/R^2) + \tfrac{3}{2}Rg_{\mu\nu}], \tag{7.34}$$

where $\theta_{\mu\nu}$ is the traceless tensor

$$\theta_{\mu\nu} = (1/4\pi)\{A_{\mu\nu} - \tfrac{1}{2}A_\alpha{}^\alpha g_{\mu\nu} - \tfrac{1}{4}[(R_{;\mu\nu}/R) - (R_{;\mu}R_{;\nu}/R^2)] - \tfrac{1}{8}Rg_{\mu\nu}\} \tag{7.35}$$

and we have dropped terms of adiabatic order greater than two coming from differentiation of terms in the DeWitt–Schwinger expansion as explained in chapter 6. (This eliminates a term proportional to m^{-2}.) In arriving at (7.34) we have also used the fact that $R_{\mu\nu} - \tfrac{1}{2}g_{\mu\nu}R = 0$ in two-dimensional spacetime and that $\Box \ln R = -\Box \ln C = -R$ in this example. Note that for $\xi = 0$, the trace of (7.34) is $-R/24\pi$ as expected, while $\theta_{\mu\nu}$ in (7.35) is, of course, identical to $\theta_{\mu\nu}$ given by (6.137). Thus the term proportional to ξ in (7.34) may be regarded as a correction to (6.136) due to non-conformal coupling.

7.2 Robertson–Walker models

Much attention has been devoted to computing stress-tensors for quantum fields propagating in Robertson–Walker background spacetimes. Interest is due to the high degree of symmetry present in these models, as well as their cosmological relevance.

As a first illustration, consider the four-dimensional static, hyperbolic ($K = -1$) Robertson–Walker spacetime, which has been discussed in §5.2. We treat a massless conformally coupled scalar field (Bunch 1978*a*). The $D^{(1)}$ Green function is given from (5.33) by deleting the $i\varepsilon$ and multiplying by 2. Expanding $\sinh \Delta\chi$ to order $(\Delta\chi)^6$ and using the expansions (6.201) one readily obtains

$$D^{(1)}(x'', x') = -\frac{1}{(8\pi^2\varepsilon^2\Sigma)} + \frac{1}{(12\pi^2\Sigma)}(t^1)^2 - \frac{7}{180\pi^2\Sigma^2}\varepsilon^2(t^1)^4 + O(\varepsilon^4). \tag{7.36}$$

With the help of (5.13) we find $R_{\eta\eta} = 0$, $R_{\chi\chi} = 2$ and so may rewrite (7.36) in manifestly geometrical form

$$D^{(1)}(x'', x') = -\frac{1}{8\pi^2\varepsilon^2\Sigma} + \frac{1}{24\pi^2}R_{\alpha\beta}\frac{t^\alpha t^\beta}{\Sigma} - \frac{7\varepsilon^2}{720\pi^2}R_{\alpha\beta}R_{\gamma\delta}\frac{t^\alpha t^\beta t^\gamma t^\delta}{\Sigma^2} + O(\varepsilon^4). \tag{7.37}$$

Comparison of (7.37) with (6.204) for the static hyperbolic model reveals that their massless terms are identical. Moreover, the coefficient of the m^{-2} (final) term in (6.204) vanishes for this spacetime, so there is no conformal anomaly. Hence, when $G^{(1)}$ is subtracted from $D^{(1)}$ to yield the renormalized Green function, one obtains precisely zero.

We therefore have the result

$$\langle 0|T_{\mu\nu}|0\rangle_{\text{ren}} = 0, \tag{7.38}$$

where the vacuum state here is the conformal vacuum, defined by positive frequency modes (5.28), and is associated with the conformal Killing vector related to the ∂_η Killing vector on Rindler spacetime (see §5.5), This result was arrived at indirectly, using the Rindler stress-tensor, in §6.3 (6.158). We could alternatively have used the above calculation together with (6.158) to determine the Rindler stress-tensor (6.157).

As a second illustration, consider the $K = 0$ and $+1$ models, but with C chosen to be a function of time. The $D^{(1)}$ function in the conformal vacuum is given by conformal scaling of (2.79) and (5.32) respectively (putting $\varepsilon = 0$ and doubling the latter). The two results may be written

$$D^{(1)}(x'', x') = \frac{KC^{-\frac{1}{2}}(\eta'')C^{-\frac{1}{2}}(\eta')}{4\pi^2[\cos(K^{\frac{1}{2}}\Delta\eta) - \cos(K^{\frac{1}{2}}\Delta\chi)]}, \tag{7.39}$$

in which, without loss of generality, we have chosen the separation between x'' and x' to lie in the (η, χ) plane.

Using the expansions (6.201), and the result

$$C^{-\frac{1}{2}}(\eta'')C^{-\frac{1}{2}}(\eta') = C^{-1}\{1 + \varepsilon^2[(-\tfrac{1}{2}\dot{D} + \tfrac{1}{4}D^2)(t_1^0)^2 + \tfrac{1}{4}D^2\Upsilon(t_1^1)^2]$$
$$+ \varepsilon^4[(-\tfrac{1}{24}\dddot{D} + \tfrac{7}{48}\ddot{D}D + \tfrac{5}{24}\dot{D}^2 - \tfrac{5}{16}\dot{D}D^2 + \tfrac{1}{16}D^4)(t_1^0)^4$$
$$+ (\tfrac{7}{48}\ddot{D}D + \tfrac{1}{12}\dot{D}^2 - \tfrac{31}{48}\dot{D}D^2 + \tfrac{1}{4}D^4)\Upsilon(t_1^0t_1^1)^2$$
$$+ (-\tfrac{1}{24}\dot{D}D^2 + \tfrac{1}{16}D^4)\Upsilon^2(t_1^1)^4]\} + O(\varepsilon^6), \qquad (7.40)$$

which is valid in any Robertson–Walker spacetime, one obtains

$$D^{(1)}(x'', x') = -(1/8\pi^2\varepsilon^2\Sigma)\{1 + \varepsilon^2[(-\tfrac{1}{3}\dot{D} + \tfrac{1}{12}D^2 + \tfrac{1}{3}K)(t_1^0)^2$$
$$+ (\tfrac{1}{12}D^2 + \tfrac{1}{3}K)\Upsilon(t_1^1)^2] + \varepsilon^4[(-\tfrac{1}{30}\dddot{D} + \tfrac{1}{10}\ddot{D}D$$
$$+ \tfrac{7}{60}\dot{D}^2 - \tfrac{2}{15}\dot{D}D^2 + \tfrac{1}{60}D^4 - \tfrac{1}{6}\dot{D}K + \tfrac{1}{12}D^2K + \tfrac{3}{45}K^2)(t_1^0)^4$$
$$+ (\tfrac{1}{10}\ddot{D}D + \tfrac{11}{180}\dot{D}^2 - \tfrac{23}{72}\dot{D}D^2 + \tfrac{31}{360}D^4 - \tfrac{5}{18}\dot{D}K + \tfrac{7}{18}D^2K$$
$$+ \tfrac{8}{45}K^2)\Upsilon(t_1^0t_1^1)^2 + (-\tfrac{1}{40}\dot{D}D^2 + \tfrac{1}{60}D^4 + \tfrac{1}{12}D^2K$$
$$+ \tfrac{3}{45}K^2)(t_1^1)^4\Upsilon^2]\}. \qquad (7.41)$$

If one now takes linear combinations of geometrical tensors with the appropriate adiabatic order, it is possible to express (7.41) entirely in terms of geometrical tensors:

$$D^{(1)}(x'', x') = -\frac{1}{8\pi^2\varepsilon^2\Sigma} + \frac{1}{24\pi^2}\left[R_{\alpha\beta}\frac{t^\alpha t^\beta}{\Sigma} - \tfrac{1}{6}R\right]$$
$$+ \frac{\varepsilon^2\Sigma}{1440\pi^2}\left[2R_\alpha{}^\lambda R_{\lambda\beta}\frac{t^\alpha t^\beta}{\Sigma} + 4RR_{\alpha\beta}\frac{t^\alpha t^\beta}{\Sigma} - \tfrac{1}{3}R^2\right.$$
$$\left. - R_{;\alpha\beta}\frac{t^\alpha t^\beta}{\Sigma} - 14R_{\alpha\beta}R_{\gamma\delta}\frac{t^\alpha t^\beta t^\gamma t^\delta}{\Sigma^2} + 6R_{\alpha\beta;\gamma\delta}\frac{t^\alpha t^\beta t^\gamma t^\delta}{\Sigma^2}\right]. \qquad (7.42)$$

Note that the two other available terms $R^{\alpha\beta}R_{\alpha\beta}$ and $\Box R$ do not occur here. The fact that (7.42) is purely geometrical is a consequence of the conformal triviality of this example.

To renormalize, we now subtract terms up to adiabatic order four in the DeWitt–Schwinger expansion (6.204). The divergences and the terms involving four t^α vectors cancel, leaving

$$-\frac{R}{144\pi^2} + \frac{\varepsilon^2\Sigma}{1440\pi^2}\left[(-2R_\alpha{}^\lambda R_{\lambda\beta} + \tfrac{16}{3}RR_{\alpha\beta} - R_{;\alpha\beta})\frac{t^\alpha t^\beta}{\Sigma}\right.$$
$$\left. - \tfrac{2}{3}R^2 + \Box R + R^{\alpha\beta}R_{\alpha\beta}\right] - \frac{m^{-2}}{1440\pi^2}(R^{\alpha\beta}R_{\alpha\beta} - \tfrac{1}{3}R^2 - \Box R). \qquad (7.43)$$

Comparison with (6.205) reveals that we have terms of the form c, $e_{\alpha\beta}$ and f, so (6.206) yields for the renormalized stress-tensor (after dropping terms of order greater than four coming from differentiation of $G^{(1)}_{\mathrm{DS}}$)

$$\begin{aligned}\langle 0|T_{\mu\nu}|0\rangle_{\mathrm{ren}} &= (1/2880\pi^2)[(-\tfrac{1}{3}R_{;\mu\nu} + R_{\mu}{}^{\rho}R_{\rho\nu} - RR_{\mu\nu}) \\ &\quad + g_{\mu\nu}(\tfrac{1}{3}\Box R - \tfrac{1}{2}R^{\rho\tau}R_{\rho\tau} + \tfrac{1}{3}R^2)] \\ &= -(1/2880\pi^2)[\tfrac{1}{6}\,{}^{(1)}H_{\mu\nu} - {}^{(3)}H_{\mu\nu}], \end{aligned} \tag{7.44}$$

which is in agreement with (6.141), with the first term on the right-hand side set to zero because the $K = 0$ and $+1$ spacetimes have a conformal vacuum based on the Minkowski vacuum (see §5.5 and fig. 20). The result (7.44) was originally obtained by the method used above for $K = 0$ by Davies, Fulling, Christensen & Bunch (1977) and for $K = 0$ and $+1$ by Bunch & Davies (1977*a*).

A similar calculation may be carried out for the $K = -1$ case in the same conformal vacuum as in (7.38). The purely geometrical $D^{(1)}$ in (7.42) is found to be augmented by a local, but non-geometrical term, which, when substituted in (6.206), yields a term $(\zeta/a(\eta))^4\langle T_{\mu\nu}\rangle_{\mathrm{Rindler}}$ that appears in (6.141) in this case (see (6.158) and accompanying discussion).

More ambitious calculations have been given by Bunch & Davies (1978*b*), involving a non-conformally coupled massless scalar field in Robertson–Walker spacetime with special choices of the scale factor that enable the Green function $D^{(1)}$ to be computed in terms of known functions. For example, with $K = 0$ and

$$a(t) = \alpha t^c \tag{7.45}$$

(α, c constant) one has

$$C(\eta) = \alpha^{2/(1-c)}(1-c)^{2c/(1-c)}\eta^{2c/(1-c)} \tag{7.46}$$

and the *minimally* coupled field equation $\Box\phi = 0$ possesses mode solutions (5.3) with $n = 4$ and (Ford & Parker 1977)

$$\chi_k(\eta) = C^{\frac{1}{4}}(\eta)(|b|\eta/a_0^2)^{1/2b}[c_1 H_\nu^{(1)}(k\eta) + c_2 H_\nu^{(2)}(k\eta)], \tag{7.47}$$

where $H^{(1)}, H^{(2)}$ are Hankel functions, $k = |\mathbf{k}|$ and

$$\left.\begin{aligned} b &= (1-c)/(1-3c) \\ \nu &= 1/(2|b|) \\ a_0 &= \alpha[\alpha^3(1-3c)]^{c/(1-3c)}. \end{aligned}\right\} \tag{7.48}$$

The coefficients c_1 and c_2 are complex numbers subject to the Wronskian condition (5.6), which here reduces to

$$|c_2|^2 - |c_1|^2 = \pi/4b. \tag{7.49}$$

A minimally coupled scalar field can be used to describe linearized gravitons in a Robertson–Walker model universe (Grishchuk 1974, 1975).

We choose the vacuum state defined by $c_1 = 0$, as this reduces to the standard Minkowski space vacuum in the limit $c \to 0$. The Green function $D^{(1)}(x'', x')$ is easily evaluated as a mode integral

$$\begin{aligned} D^{(1)}(x'', x') &= \int [u_{\mathbf{k}}(x'')u_{\mathbf{k}}^*(x') + u_{\mathbf{k}}^*(x'')u_{\mathbf{k}}(x')]\,\mathrm{d}^3k \\ &= (1/8\pi\eta''\eta')C^{-\frac{1}{2}}(\eta'')C^{-\frac{1}{2}}(\eta')(\tfrac{1}{4} - \nu^2)\sec \pi\nu \\ &\quad \times F(\tfrac{3}{2} + \nu, \tfrac{3}{2} - \nu; 2 : 1 + (\Delta\eta^2 - \Delta z^2)/4\eta''\eta') \end{aligned} \tag{7.50}$$

where F is a hypergeometric function and we have separated the points x'', x' in the (η, z) plane, obtaining $\Delta\eta = \eta'' - \eta'$, $\Delta z = z'' - z' = x^{3''} - x^{3'}$.

It is now necessary to expand (7.50) in powers of ε up to order ε^2 using (6.201), with r replaced by z throughout, and using (7.40). By taking linear combinations of all available geometrical tensors of correct adiabatic order, the result can be cast in the following form

$$\begin{aligned} D^{(1)}(x'', x') = &-\frac{C^{-\frac{1}{2}}(\eta'')C^{-\frac{1}{2}}(\eta')}{2\pi^2(\Delta\eta^2 - \Delta z^2)} + \left[\tfrac{1}{2}\ln\left|\frac{\varepsilon^2}{C\eta^2}\right| + \gamma + \tfrac{1}{2}\psi(\tfrac{3}{2} + \nu) + \tfrac{1}{2}\psi(\tfrac{3}{2} - \nu)\right] \\ &\times \left[-\frac{R}{24\pi^2} + \frac{\varepsilon^2\Sigma}{288\pi^2}\left(4RR_{\alpha\beta}\frac{t^\alpha t^\beta}{\Sigma} - 2R_{;\alpha\beta}\frac{t^\alpha t^\beta}{\Sigma} + 2\Box R - R^2\right)\right] \\ &+ \frac{R}{48\pi^2} - \frac{\varepsilon^2\Sigma}{144\pi^2}\left[RR_{\alpha\beta}\frac{t^\alpha t^\beta}{\Sigma} - \tfrac{19}{24}R^2 + \frac{3R}{C\eta^2}\right] + O(\varepsilon^4). \end{aligned} \tag{7.51}$$

The first term on the right is the expression for $D^{(1)}(x'', x')$ in the conformally coupled case ($\xi = 1/6$), so we see that the effect of putting $\xi = 0$ is to add a correction term. Although still entirely a local bi-scalar, $D^{(1)}$ does now contain *non-geometrical* terms, e.g., the η^{-2} term, which arises because of the departure from conformal triviality. Note that when $R = 0$, the correction term vanishes. This is the case where there is no distinction between conformal and minimal coupling.

Renormalization is effected by subtracting the DeWitt–Schwinger terms (6.204) with $\xi = 0$ from (7.51). Because we already know the result of subtracting (6.204) with $\xi = 1/6$ from the *conformal* $D^{(1)}(x'', x')$, i.e., the first term of (7.51), the present result is simply (7.44) plus the correction terms resulting from the remainder of (7.51), and from the difference between the conformal and minimal $G_{DS}^{(1)}$. The result is

$$\begin{aligned}\langle 0|T_{\mu\nu}|0\rangle_{ren} = &-(1/2880\pi^2)[\tfrac{1}{6}{}^{(1)}H_{\mu\nu} - {}^{(3)}H_{\mu\nu}]\\ &-(1/1152\pi^2){}^{(1)}H_{\mu\nu}[\ln(R/m^2) + \psi(\tfrac{3}{2}+\nu) + \psi(\tfrac{3}{2}-\nu) + \tfrac{4}{3}]\\ &+(1/13824\pi^2)[24\Box Rg_{\mu\nu} + 24RR_{\mu\nu} + 3R^2 g_{\mu\nu}]\\ &-Rg_{\mu\nu}/192\pi^2 C\eta^2. \end{aligned} \tag{7.52}$$

The logarithmic term, which always arises when conform triviality is broken, contains an arbitrary mass scale m. Rescaling m merely adds multiples of the conserved tensor ${}^{(1)}H_{\mu\nu}$ to (7.52). However, as pointed out in §6.2 all renormalized stress-tensors are ambiguous up to multiples of ${}^{(1)}H_{\mu\nu}$, because such a term appears on the *left*-hand side of the gravitational field equations. Thus, one could remove all ${}^{(1)}H_{\mu\nu}$ terms from (7.52) by renormalizing α in (6.220). There is then no problem in the massless limit ($m \to 0$); one merely has an additional (infrared this time) renormalization of α. Note that one could absorb the non-geometrical ψ-function terms and the factor $\frac{4}{3}$ into a rescaling of m.

The remaining terms of (7.52) are all local, some being 'accidently' geometrical, due to the high degree of symmetry in the model, and the final term being non-geometrical. There is no possibility of placing all the extra geometrical terms on the left-hand side of the gravitational field equations, as they are not conserved without the ln R and η^{-2} terms.

In a more general, or less symmetric example, one would expect not only non-geometrical contributions to $\langle T_{\mu\nu}\rangle_{ren}$, but *non-local* terms as well; for example, had a more complicated state been chosen, or if the spacetime were anisotropic. In §7.3 an example will be discussed where non-local terms appear explicitly.

7.3 Perturbation calculation of the stress-tensor

Exactly soluble models, whilst invaluable for pedagogic purposes, are of little value in practical calculations. To proceed beyond these models, one must turn either to numerical computation, or, as in §5.6, to approximation methods.

Indeed, the perturbation technique discussed in §5.6 is particularly useful in the calculation of $\langle T_{\mu\nu}\rangle_{ren}$. One simply iterates (5.105) to form approximate functions $\chi_{\mathbf{k}}$, which reduce in the in region to functions $\chi_{\mathbf{k}}^{in}$, these being positive frequency with respect to the conformal Killing vector ∂_η in the in region (see (5.103)). Then one can use the associated modes to construct $G^{(1)}(x'',x') = \langle \text{in}, 0|\{\phi(x''), \phi(x')\}|0, \text{in}\rangle$, where $|0, \text{in}\rangle$ is the conformal vacuum in the in region, which is renormalized and differentiated to form the vacuum expectation value of the stress-tensor in the usual way. Davies & Unruh (1979) have computed $\langle \text{in}, 0|T_{\mu\nu}|0, \text{in}\rangle_{\text{ren}}$ for $h_i = 0, m = 0$, up to second order in $(\xi - \frac{1}{6})$. They find

$$\begin{aligned}
&\langle \text{in}, 0|T_{\mu\nu}|0, \text{in}\rangle_{\text{ren}} \\
&\quad = (1/2880\pi^2)\{-\tfrac{1}{6}{}^{(1)}H_{\mu\nu} + {}^{(3)}H_{\mu\nu} + 10(\xi - \tfrac{1}{6})^{(1)}H_{\mu\nu} \\
&\qquad + 180(\xi - \tfrac{1}{6})^2[{}^{(1)}H_{\mu\nu}(1 + \ln C^{\frac{1}{2}}) + C_{\mu\nu} \\
&\qquad + \mathscr{H}_{\mu\nu}(C^{-1}\int_{-\infty}^{\eta} \tilde{V}'(\eta_1)\ln(\mu|\eta - \eta_1|)\mathrm{d}\eta_1 \\
&\qquad + \tfrac{1}{2}e_{\mu\nu}C^{-1}\tilde{V}(\eta)\int_{-\infty}^{\eta} \tilde{V}'(\eta_1)\ln(\mu|\eta - \eta_1|)\mathrm{d}\eta_1 \\
&\qquad - \tfrac{1}{2}e_{\mu\nu}C^{-1}\int_{-\infty}^{\eta}\mathrm{d}\eta_1\int_{-\infty}^{\eta}\mathrm{d}\eta_2\tilde{V}'(\eta_1)\tilde{V}'(\eta_2)\ln(\mu|\eta_1 - \eta_2|)]\}, \qquad (7.53)
\end{aligned}$$

where $\tilde{V}(\eta) = R(\eta)C(\eta)$, μ is an arbitrary mass scale, $e_{\mu\nu}$ is a tensor, diagonal in Cartesian coordinates, with constant components $e_{00} = 1$, $e_{ii} = \frac{1}{3}$, and primes denote differentiation of a function with respect to its argument. The tensor operator $\mathscr{H}_{\mu\nu}$ is defined by

$$2(\nabla_\mu\nabla_\nu - g_{\mu\nu}\Box + R_{\mu\nu} - \tfrac{1}{4}Rg_{\mu\nu})$$

so that $\mathscr{H}_{\mu\nu}[R(\eta)] = {}^{(1)}H_{\mu\nu}(\eta)$ (see (6.53)). Finally, $C_{\mu\nu}$ is a local, but non-geometrical tensor with components

$$\begin{aligned}
C_{\eta\eta} &= C^{-1}(-\tfrac{9}{2}\dot{D}D^2 - \tfrac{9}{4}D^4) \\
C_{xx} = C_{yy} = C_{zz} &= C^{-1}(6\ddot{D}D + \tfrac{9}{2}\dot{D}^2 + \tfrac{9}{2}\dot{D}D^2 - \tfrac{15}{8}D^4).
\end{aligned}$$

Expression (7.53), which is non-local only at second order in $(\xi - \frac{1}{6})$, reduces to (5.122) (with $h_i = m = 0$) in the out region $\eta \to \infty$, if one uses (5.102) and $\rho = \langle T_0{}^0\rangle = C^{-1}\langle T_{00}\rangle$.

One advantage of the perturbation technique is that, with each higher order of perturbation, an additional factor of ω^{-1} appears multiplying the correction to χ_k (see (5.105)). This means that divergences only occur in $\langle T_{\mu\nu}\rangle$ up to the second order of the perturbation. The higher orders all yield finite additional contributions, without the need for regularization, making them particularly amenable to numerical calculation.

An alternative approach to perturbation theory is to work with the effective action (Hartle 1977, Fischetti, Hartle & Hu 1979, Hartle & Hu 1979,

Hu 1980, Hartle 1981). Consider the slightly more general metric

$$ds^2 = C(\eta)[d\eta^2 - (e^{2\beta(\eta)})_{ij} dx^i dx^j] \tag{7.54}$$

where $\beta(\eta)$ is a symmetric, traceless, 3×3 matrix. Note that to first order $2\beta_{ii} = h_i$.

For this spacetime the Feynman Green function for a massless, conformally coupled field satisfies an equation of the form

$$(\Box_x + \tfrac{1}{6}R_I)D_F(x, x') = -[-g(x)]^{-\frac{1}{2}}\delta^4(x - x') - VD_F(x, x') \tag{7.55}$$

where $\Box_x$ is the wave operator and R_I is the Ricci scalar for the case of exact isotropy, and the operator V symbolizes the (small) correction to these quantities due to the anisotropy.

Equation (7.55) possesses the formal solution

$$D_F = D_F^{(0)} + D_F^{(0)} V D_F \tag{7.56}$$

where $D_F^{(0)}$ is the Feynman propagator in the conformal vacuum (i.e., (3.154) with $n = 4$); this being chosen as the natural propagator in the limit of exact isotropy. This equation for G_F may be approximated iteratively

$$D_F = D_F^{(0)} + D_F^{(0)} V D_F^{(0)} + D_F^{(0)} V D_F^{(0)} V D_F^{(0)} + \ldots \tag{7.57}$$

and used in (6.25) to expand $\ln(-D_F)$, giving a perturbation series for the effective action

$$W = \sum_{i=0}^{\infty} W^{(i)} \tag{7.58}$$

where

$$W^{(0)} = -\tfrac{1}{2}\mathrm{i}\,\mathrm{tr}[\ln(-D_F^{(0)})] \tag{7.59}$$

$$W^{(1)} = -\tfrac{1}{2}\mathrm{i}\,\mathrm{tr}(VD_F^{(0)}) \tag{7.60}$$

$$W^{(2)} = -\tfrac{1}{4}\mathrm{i}\,\mathrm{tr}(VD_F^{(0)}VD_F^{(0)}). \tag{7.61}$$

The zeroth order term $W^{(0)}$ was studied in §6.3, and gave rise to the stress-tensor (6.141). The dimensionally regularized $D_F^{(0)}(x, x)$ vanishes (see page 170), so that $W^{(1)} = 0$. However, $W^{(2)}$, being quadratic in $D_F^{(0)}$, has a nonzero dimensionally regularized value. Writing out the part of the dimensionally regularized $W^{(2)}$ which is of second order in β (there are also third and fourth order terms in (7.61)), one obtains (Hartle & Hu 1979)

$$W^{(2)} = \int d^n x \int d^n x' \beta^{ij}(\eta) K_{ijkl}(x - x') \beta^{kl}(\eta') \tag{7.62}$$

where

$$K_{ijkl}(x) = (2\pi)^{-n}\int d^n k\, e^{i\mathbf{k}\cdot\mathbf{x} - ik_0\eta_0}\hat{K}_{ijkl}(k) \tag{7.63}$$

and

$$\begin{aligned}\hat{K}_{ijkl}(k) = &-(1/1920\pi^2)[\delta_{ij}\delta_{kl} + \delta_{ik}\delta_{jl} + \delta_{il}\delta_{jk}](\mathbf{k}^2 - k_0^2)^2 \\ &\times [(1/(n-4) + \tfrac{1}{2}\ln(\mathbf{k}^2 - k_0^2) - \text{constant} + O(n-4)] \\ &+ (\text{terms which give vanishing contribution to } W^{(2)}).\end{aligned} \tag{7.64}$$

In arriving at (7.64), the explicit forms of $G_F^{(0)}$ (in n dimensions) and V have been used, and the result has been expanded about $n = 4$. The pole term can be evaluated as

$$-[1/1920\pi^2(n-4)]\int d^n x(-g)^{\frac{1}{2}} C^{\alpha\beta\gamma\delta} C_{\alpha\beta\gamma\delta}$$

in which $(-g)^{\frac{1}{2}}$ and $C_{\alpha\beta\gamma\delta}$ are four-dimensional quantities. This can be rewritten as (cf. (6.138))

$$-[1/16\pi^2(n-4)]\int d^n x(-g)^{\frac{1}{2}}\alpha F(x) + O(1), \tag{7.65}$$

with $(-g)^{\frac{1}{2}}$ and F (see (6.104)) constructed from n-dimensional quantities and α being given by (6.106). The pole term in (7.65) can be removed by renormalization as discussed in §§6.2 and 6.3, leaving a finite $W^{(2)}$ with contributions from the logarithm in (7.64) and the finite (O(1)) terms in (7.65). This can be calculated to be

$$\begin{aligned}W^{(2)}_{\text{ren}} = (V/2880\pi^2)\Bigg\{&\int_{-\infty}^{\infty} d\eta[-(\ddot{a}/a + \dot{a}^2/a^2)\dot{\beta}_{ij}\dot{\beta}^{ij} + 3[\tfrac{1}{2}i\pi + \ln a]\ddot{\beta}_{ij}\ddot{\beta}^{ij}] \\ &-3\int_{-\infty}^{\infty} d\eta\int_{-\infty}^{\infty} d\eta'\ddot{\beta}_{ij}(\eta)K(\eta-\eta')\ddot{\beta}^{ij}(\eta')\Bigg\}\end{aligned} \tag{7.66}$$

where V is the volume of space, $a^2 = C(\eta)$ and

$$K(\eta) = (1/\pi)\int_0^{\infty}\cos(\omega\eta)\ln(\omega/\mu)\,d\omega,$$

μ being the arbitrary mass scale introduced in renormalization.

Functional differentiation of the perturbation series for W can be used to give the vacuum expectation value of the stress-tensor. This will consist of the usual isotropic anomaly terms (6.141) from $W^{(0)}$, together with anisotropy correction terms to this anomalous piece arising from (7.65), and finally the non-anomalous terms from (7.66), which may be regarded as representing particle creation induced by the perturbation. The explicit results have been given by Hartle & Hu (1979).

As yet we have not discussed the boundary conditions that have been built into the formal (integral) equation (7.56) by the choice of $D_F^{(0)}$ as the Feynman propagator in the conformal vacuum. This boundary condition determines which vacuum expectation value of the stress-tensor will be arrived at by variation of (7.58)

To examine this equation let us suppose that the anisotropy vanishes in the distant past and future, so that there exist in and out conformally flat regions and associated in and out conformal vacua. To establish the boundary conditions built into (7.56) we derive the equation in terms of field operators $\phi(x)$ rather than Green functions.

Using Gauss' theorem one can write

$$
\begin{aligned}
-\int \mathrm{d}^n y[-g(y)]^{\frac{1}{2}}[K_y D_F^{(0)}(x,y)]T(\phi(y)\phi(x')) \\
= -\int \mathrm{d}^n y[-g(y)]^{\frac{1}{2}} D_F^{(0)}(x,y) K_y T(\phi(y)\phi(x')) \\
+ \int_{\substack{\Sigma \\ y^0 \to \infty}} \mathrm{d}\Sigma_y^\mu [-g(y)]^{\frac{1}{2}} D_F^{(0)}(x,y) \overleftrightarrow{\nabla}_\mu{}^y(\phi(y)\phi(x')) \\
- \int_{\substack{\Sigma \\ y^0 \to -\infty}} \mathrm{d}\Sigma_y^\mu [-g(y)]^{\frac{1}{2}} D_F^{(0)}(x,y) \overleftrightarrow{\nabla}_\mu{}^y(\phi x')\phi(y))
\end{aligned}
\tag{7.67}
$$

(for notation see (3.28)) where for generality we work in n dimensions, and

$$K_y \equiv \Box_y + \tfrac{1}{4}[(n-2)/(n-1)]R(y). \tag{7.68}$$

Thus $D_F^{(0)}$, defined by (3.154), satisfies

$$K_y D_F^{(0)}(x,y) = -[-g(y)]^{-\frac{1}{2}}\delta^n(x-y). \tag{7.69}$$

We are also assuming, for simplicity, that the fields vanish at spacelike infinity, so that spatial boundary terms may be ignored.

In the conformally flat in and out regions, the field ϕ will have mode decompositions (3.153), i.e.,

$$\phi(y) \underset{y^0 \to -\infty}{\longrightarrow} \sum_{\mathbf{k}} (a_{\mathbf{k}} u_{\mathbf{k}}^{(0)} + a_{\mathbf{k}}^{\dagger} u_{\mathbf{k}}^{(0)*}) \tag{7.70}$$

$$\phi(y) \underset{y^0 \to +\infty}{\longrightarrow} \sum_{\mathbf{k}} (b_{\mathbf{k}} u_{\mathbf{k}}^{(0)} + b_{\mathbf{k}}^{\dagger} u_{\mathbf{k}}^{(0)*}) \tag{7.71}$$

where the modes $u_{\mathbf{k}}^{(0)}$ are positive frequency with respect to conformal time η and are conformally related to the flat spacetime modes (3.151)

$$u_{\mathbf{k}}^{(0)} = C^{(2-n)/4} \bar{u}_{\mathbf{k}}. \tag{7.72}$$

Because of the anisotropy between the in and out regions, $a_{\mathbf{k}}$ and $b_{\mathbf{k}}$ will not be equal. If they are used to define in and out conformal vacua

$$a_{\mathbf{k}}|0, \text{in}\rangle = 0, \quad b_{\mathbf{k}}|0, \text{out}\rangle = 0, \tag{7.73}$$

then these will not be the same state.

Using the mode decomposition for the flat space Feynman propagator, (3.154) can be written as

$$\begin{aligned} \mathrm{i}D_{\mathrm{F}}^{(0)}(x, y) &= \theta(x^0 - y^0) \sum_{\mathbf{k}} u_{\mathbf{k}}^{(0)}(x) u_{\mathbf{k}}^{(0)*}(y) \\ &\quad + \theta(y^0 - x^0) \sum_{\mathbf{k}} u_{\mathbf{k}}^{(0)*}(x) u_{\mathbf{k}}^{(0)}(y) \\ &\left. \begin{aligned} &\underset{y^0 \to -\infty}{\longrightarrow} \sum_{\mathbf{k}} u_{\mathbf{k}}^{(0)}(x) u_{\mathbf{k}}^{(0)*}(y) \\ &\underset{y^0 \to +\infty}{\longrightarrow} \sum_{\mathbf{k}} u_{\mathbf{k}}^{(0)*}(x) u_{\mathbf{k}}^{(0)}(y). \end{aligned} \right\} \end{aligned} \tag{7.74}$$

Employing (7.70), (7.71) and (7.74) in the surface terms of (7.67), applying the orthonormality conditions (3.29), and substituting (7.69) into the left-hand side of (7.67), one obtains

$$\begin{aligned} T(\phi(x)\phi(x')) = &- \int \mathrm{d}^n y [-g(y)]^{\frac{1}{2}} D_{\mathrm{F}}^{(0)}(x, y) K_y T(\phi(y)\phi(x')) \\ &+ \sum_{\mathbf{k}} u_{\mathbf{k}}^{(0)*}(x) b_{\mathbf{k}}^{\dagger} \phi(x') + \sum_{\mathbf{k}} u_{\mathbf{k}}^{(0)}(x) \phi(x') a_{\mathbf{k}}. \end{aligned} \tag{7.75}$$

If one now takes the $\langle \text{out}, 0|\ldots|0, \text{in}\rangle$ vacuum expectation value of (7.75), the final two terms vanish by virtue of (7.73), and one is left with

$$D_{\mathrm{F}}(x, x') = - \int \mathrm{d}^n y [-g(y)]^{\frac{1}{2}} D_{\mathrm{F}}^{(0)}(x, y) K_y D_{\mathrm{F}}(y, x'), \tag{7.76}$$

where

$$D_F(x, x') = \langle \text{out}, 0|T(\phi(x)\,\phi(x'))|0, \text{in}\rangle. \tag{7.77}$$

This will satisfy the n-dimensional generalization of (7.55), namely

$$K_x D_F(x, x') = -[-g(x)]^{-\frac{1}{2}}\,\delta^n(x - x') - V D_F(x, x'). \tag{7.78}$$

Application of (7.78) to (7.76) finally yields

$$D_F(x, x') = D_F^{(0)}(x, x') + \int d^n y [-g(y)]^{\frac{1}{2}}\, D_F^{(0)}(x, y)\, V D_F(y, x'), \tag{7.79}$$

which is precisely the explicit form of (7.56). Had we taken the $\langle$in, 0|...|0, in$\rangle$ or $\langle$out, 0|...|0, out$\rangle$ expectation values of (7.75), then one of the two surface terms would have survived and the resulting equation would differ from (7.56).

The above argument can be generalized to more complicated spacetimes without changing the basic result, i.e., that an equation of the form (7.56) determines a propagator that is of the $\langle$out, 0|...|0, in$\rangle$ variety. The reason can be traced to the use of the Feynman propagator, which is formally achieved by the addition of a $(-i\varepsilon)$ term to the left-hand side of (7.55), (see the discussion on page 76).

Methods for solving the integral equations for either $\langle$out|...|in$\rangle$ or $\langle$in|...|in$\rangle$ propagators using momentum space techniques have been devised by Birrell (1979*c*) (see also Birrell 1979*a* for $\langle$out|...|in$\rangle$ only).

Finally, a completely different approach to perturbation theory has been given by Horowitz (1980), who has attempted to use the Wald axioms of §6.6 to restrict the structure of the first order (general) perturbation about Minkowski space of the stress-tensor for a massless, conformally coupled scalar field. If it is assumed that $\langle T_{\mu\nu}\rangle$ does not contain terms of adiabatic order four or greater (e.g., terms of the $\Box R g_{\mu\nu}$ type) then a unique, non-local result is obtained in terms of an integral over the past null cone through the spacetime point of interest. This perturbation result for $\langle T_{\mu\nu}\rangle$ had previously been derived in momentum space by Capper, Duff & Halpern (1974), and has been extended to cover arbitrary perturbations about conformally flat spacetimes by Horowitz & Wald (1980).

7.4 Cosmological considerations

As remarked in chapter 1, gravitational effects in quantum field theory are likely to be of observational significance only close to microscopic black holes, or in the early phases of the primeval universe. In the absence of observational data concerning black holes, we shall restrict attention to the consequences for physical cosmology of quantum field theory in curved spacetime.

Most cosmologists today assume that the universe had a singular origin about 15 billion years ago, and that the epoch immediately after about one second was characterized by isotropic expansion and thermodynamic equilibrium. As we have seen in previous sections, quantum field energy densities induced by the cosmological gravitational field are characterized by terms of the sort R^2, which become comparable with the gravitational terms in Einstein's equation only in the region of the Planck era (10^{-43}s). Reliable information about the primeval universe does not extend back before about one second after the hypothesized singular origin, so that the effects of interest in this book are unlikely to ever be directly observable. Nevertheless, there are indirect ways in which some of the model predictions can be tested.

Quantum gravitational effects in the primeval universe will result in the production of entropy through particle creation. In addition, back-reaction of the gravitationally induced stress-tensor will modify the cosmological dynamics. In particular, any initial anisotropy and inhomogeneity is likely to result in prolific particle creation (see §5.6) and vacuum stress (see §7.3). Back-reaction would then be expected to result in strong damping of the initial turbulence and irregularity. The fact that the presently observed universe is highly uniform on the large scale suggests that either initial anisotropy and inhomogeneity has been efficiently damped away, or that the universe began with a degree of uniformity that is *a priori* exceedingly improbable (Penrose 1979).

It has been pointed out by Barrow & Matzner (1977) that, on quite general grounds, the entropy per baryon produced by cosmological anisotropy dissipation is highly sensitive to the epoch of dissipation. Anisotropy perturbations can be regarded as behaving like a cosmological fluid with an energy density that varies with the cosmological scale factor as a^{-6} (Misner 1968). On the other hand, the energy density of radiation varies like a^{-4}. Consequently, the radiation entropy resulting in the damping of a given amount of anisotropy varies like a^{-2}. The earlier the dissipation epoch, the more the entropy one obtains.

As explained, the efficiency of quantum gravitational processes rises as the initial singularity is approached ($a \to 0$), which suggests that these effects will result in prolific entropy production at very early times. (For estimates see, for example, Mamaev, Mostepanenko & Starobinsky 1976, Frolov, Mamaev & Mostepanenko 1976.) As the second law of thermodynamics forbids this cosmic entropy from subsequently declining, we may use the currently observed entropy per baryon ratio in the universe to constrain the primeval anisotropy. (We are assuming that the universe contains a fixed nonzero baryon number. Certain recent gauge theories suggest that baryon number may not be conserved under the conditions found in the primeval universe. In this case the entropy per baryon ceases to be a meaningful parameter for judging the initial anisotropy.)

To obtain some idea of the numbers involved, we can use the results of the perturbation analysis of §5.6. Equation (5.122) gives the energy density of created massless particles in the out region where the anisotropy h_i and the non-conformal coupling V have fallen to zero. If h_i and V are appreciable only in some brief interval at conformal time η_0, we may neglect the slow logarithmic factor in the integrand and integrate (5.122) to obtain

$$\rho a^4 = \text{constant} \times [60(\xi - \tfrac{1}{6})^2 R^2(\eta_0) + C^{\alpha\beta\gamma\delta}(\eta_0) C_{\alpha\beta\gamma\delta}(\eta_0)],$$

where the constant is some appropriate numerical factor (not $\gg 1$).

From elementary considerations it is hard to see how a spacetime emerging from the Planck era could avoid attaining values of R^2 and $C^{\alpha\beta\gamma\delta}C_{\alpha\beta\gamma\delta}$ comparable with the Planck values, purely as a result of quantum metric fluctuations. Thus, even ignoring subsequent conformal symmetry breaking, one expects ρ to be comparable with the Planck energy density at the Planck time, redshifted to the present epoch (at least for gravitons; see Berger 1974, Grishchuk 1977). Expressed in thermodynamic language, the Planck temperature (10^{32} K) at the Planck time, redshifted to now, yields a thermal background of a few degrees K (Parker 1976), which is what is actually observed in the case of the photon background. We may therefore conclude that the observed cosmic entropy per baryon is not greatly in excess of the minimum value that is consistent with quantum metric fluctuations in the Planck era.

Further confirmation that large scale irregularities did not survive much beyond the Planck era comes from the calculations of a number of authors on damping due to back-reaction effects (Zel'dovich & Starobinsky 1971, Lukash & Starobinsky 1974, Lukash, Novikov & Starobinsky 1975, Lukash, Novikov, Starobinsky & Zel'dovich 1976, Hu & Parker 1978, Hartle & Hu 1980; see also Hu 1980 and Hartle 1981). Their work on model anisotropic spacetimes indicates that the damping occurs extremely rapidly over less than one Planck time.

Sometimes quantum field effects are invoked as an attempt to find a mechanism whereby the universe might avoid an initial singularity (Ruzmaikina & Ruzmaikin 1970, Nariai 1971, Parker & Fulling 1973,

Davies 1977*a*, Melnikov & Orlov 1979, Parnovsky 1979). Of course, such a universe would then be of infinite age, with all the associated thermodynamic problems that implies (see, for example Davies 1974). The Hawking–Penrose singularity theorems (for a review and references see Hawking & Ellis 1973) depend crucially upon so-called energy conditions. In Robertson–Walker spacetimes one of these conditions reduces to requirement that

$$\rho + 3p > 0.$$

However, as we have seen, quantum field effects permit p and ρ to become negative under some circumstances, so the possibility arises that the universe may 'bounce' at some highly dense epoch, rather than encounter a singularity. Indeed we considered such models in §§3.4, 3.5.

In the absence of another length scale in the theory, it seems probable that any 'bounce' would occur close to the Planck regime. Unfortunately, this is precisely where the one-loop approximation is no longer reliable. Attempts have been made to solve the gravitational field equation with a source term given by $\langle T_{\mu\nu} \rangle$ for a massless conformally invariant field given in §6.3 (Davies 1977*a*, Fischetti, Hartle & Hu 1979, Starobinsky 1980). The analysis is greatly complicated by the presence of terms of the form ${}^{(1)}H_{\mu\nu}$ and ${}^{(2)}H_{\mu\nu}$.

If the theory contains a fundamental length scale in addition to the Planck length, then it may be that a 'bounce' can occur well away from the Planck regime. At first sight it appears that a nonzero field mass (typically $\sim 10^{-24}$ gm corresponding to a Compton time of 10^{-23} s) might produce a profound modification of the gravitational dynamics over length and time scales many orders of magnitude in excess of the Planck values. Indeed, early work (Parker & Fulling 1973) with specially contrived quantum states shows that this is a possibility. However, in general, the effect of a mass is negligible compared with other mechanisms for breaking conformal invariance (Birrell & Davies 1980*a*). Basically, this is because terms in $\langle T_{\mu\nu} \rangle$ such as $m^2 R$ only exceed terms like R^2 when $m^2 > R \sim t^{-2}$, i.e. for t greater than the particle Compton time. But at this late epoch, quantum gravitational effects are negligible anyway. There is thus a sort of resonance effect near $t \sim m^{-1}$.

This resonance effect can be seen explicitly by using the perturbation method of §5.6 to treat the simple model of a massive, conformally coupled scalar field in a spatially flat Robertson–Walker spacetime with scale factor

$$a(\eta) = 1 - \tfrac{1}{2}\alpha^2/(\alpha^2 + \eta^2), \quad \alpha \text{ constant} \tag{7.80}$$

The spacetime contracts to a small value of the scale factor, 'bounces' and then expands again, undergoing a period of most rapid expansion (with rate $\sim \alpha^{-1}$) when $\eta \sim \alpha$. Using (5.106) with $C = a^2(h_i = 0, \xi = \frac{1}{6})$ in (5.111) to calculate the Bogolubov coefficient β and subsequently the energy density, one obtains (Birrell & Davies 1980*a*)

$$\rho = \frac{\alpha m^5}{2048}\left\{\tfrac{1}{4}\alpha^3 \frac{\partial^2}{\partial \alpha^2}[\alpha^{-1}K_1(4m\alpha)] + 7\frac{\partial}{\partial \alpha}[\alpha K_1(4m\alpha)] + 49K_1(4m\alpha)\right\}. \quad (7.81)$$

The graph of $\alpha^4 \rho$ against $m\alpha$ is shown in fig. 23 and exhibits a distinct resonance about $m\alpha \sim 1$. That is, the final energy density is greatest when the time of most rapid expansion (i.e., most particle production) occurs around the particle Compton time.

An alternative way in which a fundamental length can enter the theory is if one allows for non-gravitational interactions. For example, in the phenomenological theory of the decay of the neutral pi-meson, a coupling

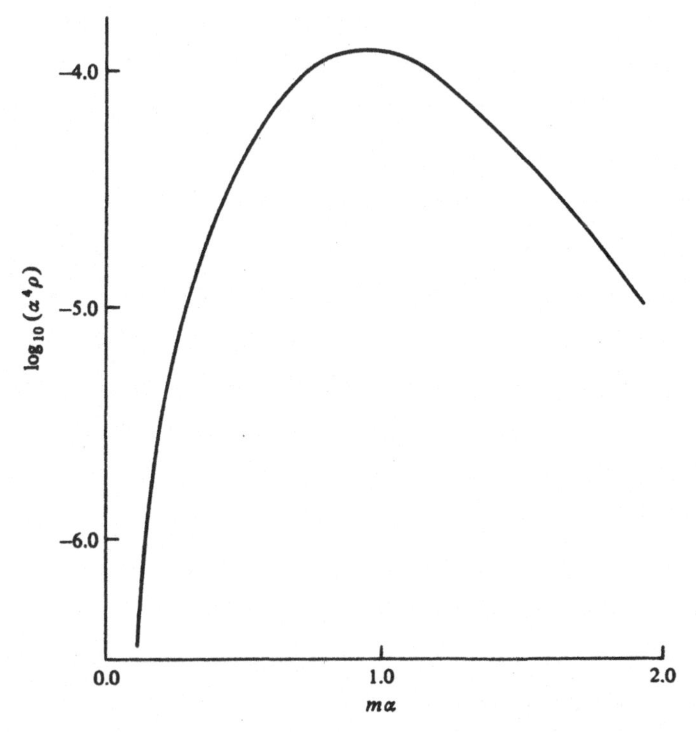

Fig. 23. The energy density of conformally coupled scalar particles created in a cosmological model with time dependence given by (7.80), as a function of particle mass.

constant with units of length is introduced. The back-reaction effects on the gravitational field of such interactions might become large at this characteristic length scale rather than the Planck length (Birrell, Davies & Ford 1980).

To a certain extent the magnitude of quantum effects in the early universe will depend on the actual quantum state, which we do not know. In the absence of any observational guide one is obliged to resort to mathematical criteria (see, for example, Chitre & Hartle 1977, who use analytic continuation techniques to select a particular quantum state).

At the present epoch particle creation and vacuum effects are utterly negligible (Parker 1968, 1969, 1971).

8
Quantum black holes

In January 1974, Hawking (1974) announced his celebrated result that black holes are not, after all, completely black, but emit radiation with a thermal spectrum due to quantum effects. This announcement proved to be a pivotal event in the development of the theory of quantum fields in curved spacetime, and greatly increased the attention given to this subject by other workers. In devoting an entire chapter to the topic of quantum black holes, we are reflecting the widespread interest in Hawking's remarkable discovery.

With the presentation of all the major aspects of free quantum field theory in curved spacetime complete, we here deploy all the various techniques described in the foregoing chapters. The basic result – that the gravitational disturbance produced by a collapsing star induces the creation of an outgoing thermal flux of radiation – is not hard to reproduce. The wavelength of radiation leaving the surface of a star undergoing gravitational collapse to form a black hole is well known to increase exponentially. It therefore seems plausible that the standard incoming complex exponential field modes should, after passing through the interior of the collapsing star and emerging on the remote side, also be exponentially redshifted. It is then a simple matter to demonstrate that the Bogolubov transformation between these exponentially redshifted modes and standard outgoing complex exponential modes is Planckian in structure. This implies that the 'in vacuum' state contains a thermal flux of outgoing particles.

Unfortunately, solutions of the wave equation in the background metric of a collapsing star cannot be written down in terms of simple functions, so we establish the exponential redshift in the outgoing modes using two simplified models. The first is a two-dimensional analogue of gravitational collapse, in which a metric is chosen to correspond to a spherically symmetric ball of matter imploding across its event horizon in an arbitrary way. The second is by ignoring the effects of backscattering in the four-dimensional picture.

The two-dimensional model has the added advantage of permitting a complete solution of the renormalized stress-tensor at all spacetime points, which greatly assists our investigations of the physics of the Hawking effect close to, and inside, the black hole. It also leads to a curious connection with the conformal anomaly. Some space is devoted to issues such as where the particles are created and how the black hole loses mass.

We also present a more elegant treatment of the quantum black hole which omits the collapse phase and replaces it with appropriate boundary conditions on the past horizon of a maximally extended manifold. This enables the intimate relation between the event horizons and the thermal properties of black holes to be more readily discerned, and also leads to a close connection between the black hole and the Rindler system discussed in chapter 4.

Although much of the interest in quantum black holes rests with their astrophysical and thermodynamic implications, we have restricted discussion of these topics to some brief remarks. Extensive references are, however, given in the text.

8.1 Particle creation by a collapsing spherical body

Consider a spherically symmetric ball of matter surrounded by empty space. In the exterior region the unique spherically symmetric solution of Einstein's equation is the Schwarzschild spacetime, described by the metric (3.18). We do not worry here about the interior metric, as it will turn out to be unimportant.

It is known (see, for example, Misner, Thorne & Wheeler 1973, chapter 31) that when sufficiently compact, the ball will implode catastrophically to form a Schwarzschild black hole. The exterior metric remains undisturbed by the collapse, but the modes of any quantum field propagating through the *interior* of the ball will be severely disrupted. Consequently, we expect particle production to take place.

If it is assumed that in the remote past the ball is sufficiently distended that the spacetime is approximately flat, then one may construct the standard Minkowski space quantum vacuum state. After the collapse, the spacetime will have the Schwarzschild form and, in this out region, the vacuum will no longer correspond to the Minkowski space vacuum constructed in the in region. To calculate the particle production, one must compute the Bogolubov transformation between the in and out vacuum states in the usual way.

Our treatment follows that of Hawking (1975) and Parker (1975, 1977);

we restrict our attention to a massless scalar field. As $R=0$ for Schwarzschild spacetime (and as the results do not depend on the detailed metric inside the ball), no distinction need be made between conformal and minimal coupling in the computation of Bogolubov coefficients.

Mode solutions of the wave equation $\Box\phi=0$ in the Schwarzschild spacetime have the form

$$r^{-1}R_{\omega l}(r)Y_{lm}(\theta,\phi)\mathrm{e}^{-i\omega t} \tag{8.1}$$

where Y_{lm} is a spherical harmonic, and the radial function R satisfies the equation

$$\frac{\mathrm{d}^2R_{\omega l}}{\mathrm{d}r^{*2}}+\{\omega^2-[l(l+1)r^{-2}+2Mr^{-3}][1-2Mr^{-1}]\}R_{\omega l}=0 \tag{8.2}$$

(r^* is defined on page 41). In the asymptotic region $r\to\infty$, (8.2) possesses the solutions $\mathrm{e}^{\pm i\omega r}$ and (8.1) reduces to

$$r^{-1}Y_{lm}\mathrm{e}^{-i\omega u} \tag{8.3}$$

and

$$r^{-1}Y_{lm}\mathrm{e}^{-i\omega v} \tag{8.4}$$

in terms of the null coordinates $u=t-r^*, v=t+r^*$.

Because of the 'potential' term in square brackets in (8.2), the standard incoming waves (8.4) will partially scatter back off the gravitational field to become a superposition of incoming and outgoing waves. Fortunately, it is not necessary to solve (8.2) in detail, so long as attention is restricted to observations made in the asymptotic ($r\to\infty$) region.

Following the general theory of chapter 3 we decompose ϕ into a complete set of positive frequency modes denoted $f_{\omega lm}$:

$$\phi=\sum_{l,m}\int\mathrm{d}\omega(a_{\omega lm}f_{\omega lm}+a^{\dagger}_{\omega lm}f^*{}_{\omega lm}) \tag{8.5}$$

where $f_{\omega lm}$ are normalized according to the condition (3.29), i.e.,

$$(f_{\omega_1l_1m_1},f_{\omega_2l_2m_2})=\delta(\omega_1-\omega_2)\delta_{l_1l_2}\delta_{m_1m_2}, \tag{8.6}$$

and are chosen to reduce to the incoming spherical modes (8.4) in the remote past. The quantum state is chosen to be the in vacuum, defined by

$$a_{\omega lm}|0\rangle=0,\quad\forall\omega,l,m, \tag{8.7}$$

which corresponds to the absence of incoming (advanced) radiation from $\mathscr{I}^-$.

To find the form of $f_{\omega l m}$ in the remote future, we first note that the incoming waves (8.4) will converge on the centre of the ball, where they will pass on through to become outgoing spherical waves. As the incoming waves approach the surface of the ball they will suffer a blueshift, but when they re-emerge after passing through the ball, there will be a redshift. If the ball is static, these two effects exactly compensate and the waves would reach $\mathscr{I}^+$ with the form (8.3). However, if the ball is collapsing, during the time that the waves spend in transit through the ball, the ball will shrink somewhat, thereby raising its surface gravity. The emerging waves will therefore suffer a redshift that is in excess of the blueshift acquired during the infall. For a ball undergoing complete gravitational collapse to a black hole, the shrinkage timescale is comparable to the light transit time across the ball, and the net redshift becomes appreciable. We shall find below that it increases exponentially with an e-folding time comparable to the transit time.

The situation is depicted schematically in fig. 24, which shows the ball collapsing to form a singularity. Some ingoing null rays are shown passing through the centre of the collapsing ball and out the other side. There exists a latest advanced ray marked γ, that just manages to penetrate the ball and reach $\mathscr{I}^+$ on the far side. This null ray forms the event horizon around the black hole. Later rays pass across the horizon and do not reach $\mathscr{I}^+$, but fall into the singularity instead. It is important to note that the direct interaction between the quantum field and the collapsing matter is being ignored. The presence of the matter in the model is used simply to produce an appropriate gravitational field.

To compute the form of the redshifted modes reaching $\mathscr{I}^+$, we shall suppress the angular variables and work with a two-dimensional model of a collapsing ball that will prove useful later. Outside the ball, the line element is taken to be

$$\mathrm{d}s^2 = C(r)\,\mathrm{d}u\,\mathrm{d}v \tag{8.8}$$

where now

$$\left.\begin{aligned} u &= t - r^* + R_0^* \\ v &= t + r^* - R_0^* \end{aligned}\right\} \tag{8.9}$$

$$r^* = \int C^{-1}\mathrm{d}r, \tag{8.10}$$

$R_0^* =$ constant. Inside the ball, we take the general line element

$$\mathrm{d}s^2 = A(U, V)\,\mathrm{d}U\,\mathrm{d}V, \tag{8.11}$$

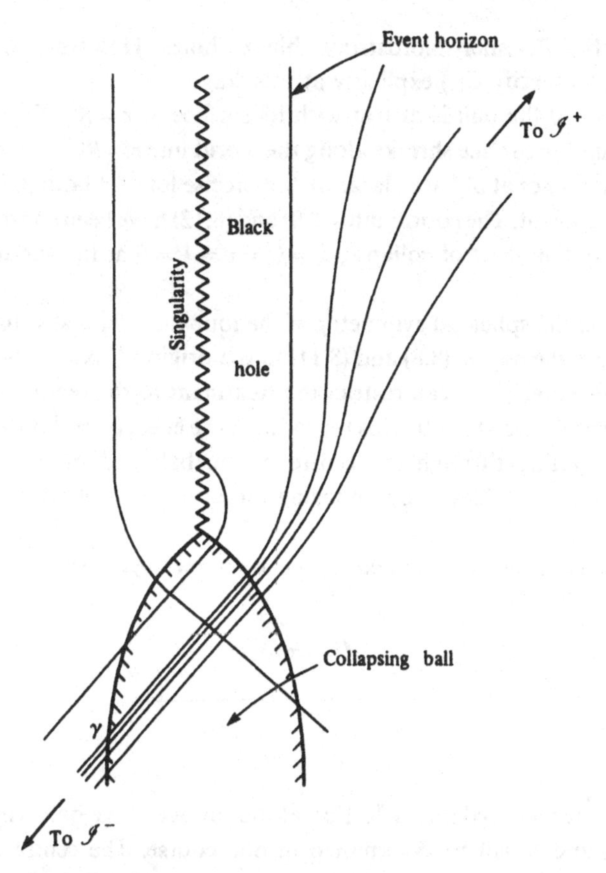

Fig. 24. As the ball collapses to a singularity, null rays converging on the ball's centre from $\mathscr{I}^-$ are distorted. One such ray, labelled, γ, forms the event horizon that marks the boundary between those rays that precede it and reach $\mathscr{I}^+$, and later rays which are trapped by the black hole and drawn into the singularity. Rays which are equispaced along $\mathscr{I}^+$ at late times crowd up along γ on $\mathscr{I}^-$.

where A is an arbitrary smooth, non-singular function and

$$\left.\begin{aligned} U &= \tau - r + R_0 \\ V &= \tau + r - R_0. \end{aligned}\right\} \tag{8.12}$$

The relation between R_0 and R_0^* is the same as that between r and r^* given by (8.10). The exterior metric (8.8) is chosen so that $C \to 1$ and $\partial C/\partial r \to 0$ as $r \to \infty$. There is an event horizon at some value of r for which $C = 0$ (we assume the spacetime to be non-singular outside the horizon). For example $C = 1 - 2Mr^{-1}$ has an event horizon at $r = 2M$, and models in two dimensions the Schwarzschild black hole. Similarly, $C = 1 - 2Mr^{-1} + e^2r^{-2}$

models the Reissner–Nordstrom black hole. However, it is not necessary to specify $C(r)$ explicitly at this stage.

Before $\tau = 0$ the ball is at rest with its surface at $r = R_0$. For $\tau > 0$ we assume that the surface shrinks along the world line $r = R(\tau)$. We shall find that for late times at $\mathscr{I}^+$ (i.e., large u), the precise form of both $A(U, V)$ and $R(\tau)$ are irrelevant. The coordinates (8.9) and (8.12) have been chosen so that at $\tau = t = 0$, the onset of collapse, $u = U = v = V = 0$ at the surface of the ball.

To model the spherical symmetry of the four-dimensional situation, we could reflect the metric (8.8) and (8.11) in the origin of spatial coordinates $r = 0$. Alternatively, we can restrict the treatment to the region $r \geq 0$, and reflect the null rays at $r = 0$. This reproduces the effect of radially incoming rays propagating through the centre of the ball and out again. Such reflection can be achieved by imposing the boundary condition $\phi = 0$ at $r = 0$.

Denote the transformation equations between the interior and exterior coordinates by

$$U = \alpha(u) \tag{8.13}$$

and

$$v = \beta(V) \tag{8.14}$$

where we have ignored any reflection at the surface of the ball. The precise form of α and β will be determined in due course. The centre of radial coordinates is the line

$$V = U - 2R_0. \tag{8.15}$$

We desire solutions of the two-dimensional wave equation

$$\Box\phi = 0$$

that vanish along (8.15) and reduce to standard exponentials on $\mathscr{I}^-$. Conformal symmetry facilitates the solution. Noting from (8.15) that at $r = 0$

$$v = \beta(V) = \beta(U - 2R_0) = \beta[\alpha(u) - 2R_0],$$

one is led to the mode solutions (cf. (4.43))

$$\mathrm{i}(4\pi\omega)^{-\frac{1}{2}}(\mathrm{e}^{-\mathrm{i}\omega v} - \mathrm{e}^{-\mathrm{i}\omega\beta[\alpha(u) - 2R_0]}). \tag{8.16}$$

Thus we see explicitly that the simple 'incoming' (left-moving) wave $\mathrm{e}^{-\mathrm{i}\omega v}$ is converted by the collapsing ball to the complicated 'outgoing' (right-moving) wave $\exp\{-i\omega\beta[\alpha(u)-2R_0]\}$.

On physical grounds, one expects the complicated phase factor $\beta[\alpha(u)-2R_0]$ to reduce to a steady escalating redshift as the surface of the ball approaches the event horizon. To determine the form of this redshift factor, we have to match the interior and exterior metrics across the collapsing surface $r=R(\tau)$. This yields

$$\alpha'(u)=\frac{\mathrm{d}U}{\mathrm{d}u}=(1-\dot{R})C\{[AC(1-\dot{R}^2)+\dot{R}^2]^{\frac{1}{2}}-\dot{R}\}^{-1} \tag{8.17}$$

$$\beta'(V)=\frac{\mathrm{d}v}{\mathrm{d}V}=C^{-1}(1+\dot{R})^{-1}\{[AC(1-\dot{R}^2)+\dot{R}^2]^{\frac{1}{2}}+\dot{R}\} \tag{8.18}$$

where $\dot{R}$ denotes $\mathrm{d}R/\mathrm{d}\tau$, and U, V and C are here evaluated at $r=R(\tau)$. Note $\dot{R}<0$ for a collapsing surface, so $(\dot{R}^2)^{\frac{1}{2}}=-\dot{R}$.

As the surface of the ball approaches the event horizon, $C=0$, (8.17) and (8.18) simplify to

$$\frac{\mathrm{d}U}{\mathrm{d}u}\sim\frac{(\dot{R}-1)}{2\dot{R}}C(R) \tag{8.19}$$

$$\frac{\mathrm{d}v}{\mathrm{d}V}\sim\frac{A(1-\dot{R})}{2\dot{R}}. \tag{8.20}$$

Close to $C=0$ we may expand $R(\tau)$:

$$R(\tau)=R_h+v(\tau_h-\tau)+O((\tau_h-\tau)^2) \tag{8.21}$$

where $R=R_h$ at the horizon, $\tau=\tau_h$ when $R(\tau)=R_h$, and $-v=\dot{R}(\tau_h)$. Then (8.19) integrates to give, to $O(\tau_h-\tau)$,

$$\kappa u=-\ln|U+R_h-R_0-\tau_h|+\text{constant} \tag{8.22}$$

where the quantity

$$\kappa=\frac{1}{2}\frac{\partial C}{\partial r}\bigg|_{r=R_h} \tag{8.23}$$

is defined to be the surface gravity of the black hole (Carter 1975).

From (8.22) we see that as $U\to\tau_h+R_0-R_h$, $u\to\infty$. Inverting this relation yields

$$U\propto \mathrm{e}^{-\kappa u}+\text{constant} \tag{8.24}$$

for late times on $\mathscr{I}^+$.

The situation is represented in the Penrose diagram (fig. 25). The null ray

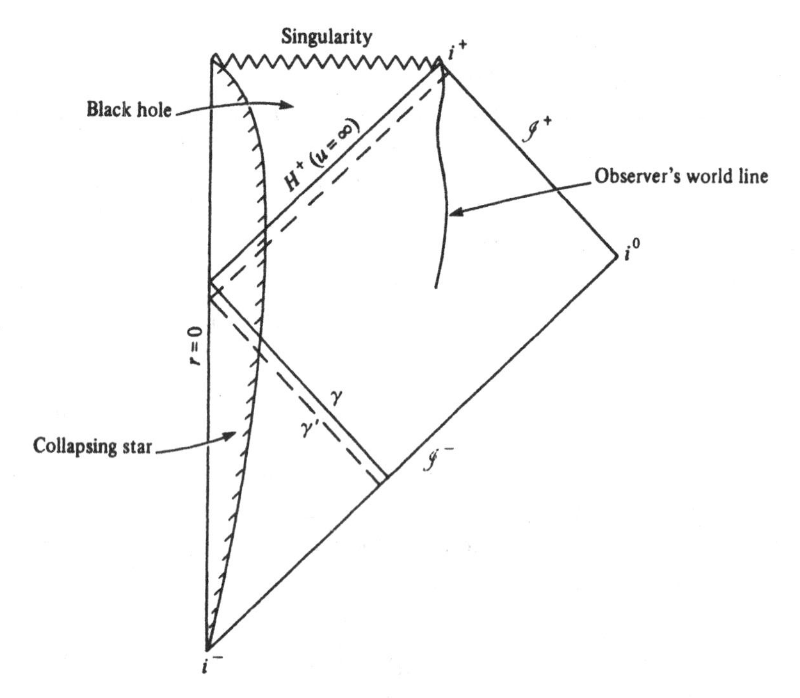

Fig. 25. Penrose diagram of a star that collapses to form a black hole. The exterior region is a fragment of fig. 6, including portions of regions I and III. The null ray γ passes through the centre of the star and emerges to form the event horizon ($u = \infty$), marked H^+, as in fig. 24. A ray γ' (broken line) immediately prior to γ on $\mathscr{I}^-$ reaches $\mathscr{I}^+$ at a finite value of u. Recalling the conformal compression of $\mathscr{I}$, one sees that the whole of the infinite future of the observer outside the hole corresponds to the narrow strip of null rays between γ and γ', representing a brief duration on $\mathscr{I}^-$. Hence the infinity of equispaced late time null rays $u =$ constant, when extended back through the star to $\mathscr{I}^-$, pile up next to γ inside this narrow strip.

$u = \infty$, that marks the horizon of the black hole, corresponds within the collapsing ball to the null ray $U = \tau_h + R_0 - R_h$. The associated 'incoming' ray from $\mathscr{I}^-$, marked γ, is the latest ray that can rebound off $r = 0$ and still reach $\mathscr{I}^+$. Later rays strike the singularity. The null rays $u =$ constant, for large values of u, when traced backwards in time and reflected out to $\mathscr{I}^-$, pile up densely along the latest advanced ray γ (this is reminiscent of fig. 12). Conversely, a very narrow range of values for v and V correspond to the whole late time asymptotic region of $\mathscr{I}^+$. Consequently, to compute the experiences of an asymptotic observer at late time u, we may treat A as approximately constant in (8.20), and integrate to give

$$v \sim \text{constant} - AV(1 + \nu)/2\nu. \tag{8.25}$$

Substituting (8.24) and (8.25) into (8.16), one obtains the late time asymptotic modes

$$\mathrm{i}(4\pi\omega)^{-\frac{1}{2}}(\mathrm{e}^{-\mathrm{i}\omega v} - \mathrm{e}^{\mathrm{i}\omega(c\mathrm{e}^{-\kappa u} + d)}) \tag{8.26}$$

where c and d are constants. Evidently the 'outgoing' null rays suffer an exponentially increasing redshift, with an e-folding time of κ^{-1}, which is precisely the same as the redshift of the surface luminosity of the collapsing ball (see, for example, Misner, Thorne & Wheeler 1973, §32.3).

The result (8.26) has the same form as that obtained by substituting (4.52) into (4.43), i.e., for a moving mirror receding on a trajectory with asymptotic form (4.51). The Bogolubov transformations in the two cases are thus essentially identical. The reason for this can be traced to the fact that the above analysis in the black hole case is based on geometric optics. The propagation of the field modes through the interior of a collapsing 'star' is similar geometrically to the reflection of the modes from a receding mirror. In the latter case the Doppler shift reproduces the same effect that the gravitational redshift here produces on the field modes. Moreover, the functional form ln cosh, which gives a mirror trajectory with asymptotic behaviour (4.51), is precisely the form of the trajectory in (r^*, t) coordinates of a particle falling across the black hole event horizon (Davies & Fulling 1977*a*).

It follows from the general theory of chapter 3 that because the field modes reduce to standard form at $\mathscr{I}^-$, an inertial particle detector at $\mathscr{I}^-$ will register no particles for the vacuum state based on these modes (i.e., the in vacuum). However, because of the complicated form (8.26), an inertial detector at $\mathscr{I}^+$, in the out region, will register particles for this state. Moreover, because only the u-dependent part of the field modes involves the complicated redshift factor, these particles will be outgoing (i.e., right-moving in the two-dimensional model) corresponding to a flux of particles leaving the vicinity of the black hole and streaming outwards.

The spectrum of this flux may be determined in the usual way by computing the Bogolubov transformation between the modes (8.26) and standard exponential modes in the out region. This was first done by Hawking (1974, 1975). We do not need to carry out this calculation here, as it is mathematically identical to the Bogolubov transformation computed in §4.4. in connection with the receding mirror problem (see remarks above). From (4.61) we find that the expected spectrum is Planckian, corresponding to a thermal spectrum from a black body of temperature

$$T = \kappa/(2\pi k_{\mathrm{B}}). \tag{8.27}$$

For the purpose of computing the Bogolubov transformation in §4.4, it was found to be convenient to take the surface of integration to lie in the in region. This meant inverting the function $p(u)$ to give the function $f(v)$ in (4.58). Physically this corresponds to selecting modes that are standard outgoing waves at $\mathscr{I}^+$, but which become complicated functions of v on $\mathscr{I}^-$. Functionally inverting (8.26) (to within a phase factor), one obtains for these latter modes

$$\mathrm{i}(4\pi\omega)^{-\frac{1}{2}}\{\mathrm{e}^{\mathrm{i}\omega\kappa^{-1}\ln[(v_0-v)/c]}-\mathrm{e}^{-\mathrm{i}\omega u}\}, \quad v<v_0, \tag{8.28}$$

where v_0 is a constant corresponding to the final null ray γ. The pileup of advanced rays along γ is manifested in the rapid variation of phase in (8.28) as $v \to v_0$.

The four-dimensional calculation, given in Hawking's paper (Hawking 1975), is essentially the same as for the two-dimensional model described here. The central result – a flux of particles from the vicinity of the hole with a thermal spectrum corresponding to the temperature given by (8.27) – is the same. There are, however, some technical complications in the four-dimensional case, and we shall briefly summarize it here. It is not possible to write the solutions $R_{\omega l}(r)$ to the radial equation (8.2) in terms of known functions, though the properties of the solutions have been extensively investigated (Press & Teukolsky 1972, Starobinsky & Churilov 1973, Teukolsky 1973, Teukolsky & Press 1974, Boulware 1975*a*,*b*, Page 1976*a*, Rowan & Stephenson 1976, Candelas 1980: see also Boulware 1975*b* for the spin 1/2 case).

Equation (8.2) has the form of a one-dimensional wave equation with a potential term (square brackets). Physically this term will give rise to reflected waves which may be envisaged as backscattering of the field modes from the spacetime curvature. In the case of the collapsing ball, some of the ingoing field disturbance will be converted to outgoing disturbance as a result of backscattering rather than from passage through the interior of the ball to the opposite side. As the interesting thermal effects arise only from the latter contribution, we shall temporarily remove the backscattering by artificially deleting the potential term in (8.2). The radial functions then reduce to ordinary exponentials, and the normalized field modes become simply

$$\frac{Y_{lm}(\theta,\phi)}{(8\pi^2\omega)^{\frac{1}{2}}r}\times\begin{cases}\mathrm{e}^{-\mathrm{i}\omega u} & (8.29)\\ \mathrm{e}^{-\mathrm{i}\omega v} & (8.30)\end{cases}$$

which reduce to the usual flat space form at large r, where

$$u \equiv t - r^* \to t - r, \quad v \equiv t + r^* \to t + r.$$

We are interested in that particular linear combination of modes (8.30) that corresponds to standard modes at $\mathscr{I}^+$. Tracing these modes backwards in time (Hawking 1975) through the collapsing ball and out along the advanced null ray to $\mathscr{I}^-$, the mode which has the form (8.29) on $\mathscr{I}^+$ looks like

$$\begin{array}{ll} \dfrac{Y_{lm}(\theta,\phi)}{(8\pi^2\omega)^{\frac{1}{2}}r}\exp\{4Mi\omega\ln[(v_0-v)/c]\}, & v<v_0 \\ 0 & v>v_0 \end{array} \tag{8.31}$$

(c constant) close to the latest advanced ray γ in fig. 24. The similarity between (8.31) and (8.28) is obvious.

The ordinary in vacuum is defined with respect to modes (8.30). The Bogolubov coefficients relating (8.30) and (8.31) are given by

$$\left.\begin{array}{l}\alpha_{\omega\omega'}\\ \beta_{\omega\omega'}\end{array}\right\}=(1/2\pi)\int_{-\infty}^{v_0}\mathrm{d}v(\omega'/\omega)^{\frac{1}{2}}\mathrm{e}^{\pm i\omega' v}\exp\{4Mi\omega\ln[(v_0-v)/c]\}, \tag{8.32}$$

which is readily evaluated in terms of Γ-functions (Hawking 1975). The answer is essentially the same as (4.60). As in that case, the factor $(\omega')^{-\frac{1}{2}}$ implies that

$$\int|\beta_{\omega\omega'}|^2\mathrm{d}\omega'$$

diverges logarithmically. There are an infinite number of particles in each mode at $\mathscr{I}^+$. The divergence here is connected with the normalization of the continuous modes (8.29)–(8.31). The collapsing ball produces a steady flux of radiation to $\mathscr{I}^+$, so the total flux for all time is infinite. Of greater interest is the number of particles emitted per unit time. This may be found by localizing the modes in some way; e.g., by confining the system to a box with periodic boundary conditions so that the modes become discrete (Page 1976*a*).

In this case, one may use the Wronskian condition (3.39) to write

$$\sum_{\omega}(|\alpha_{\omega\omega'}|^2-|\beta_{\omega\omega'}|^2)=1. \tag{8.33}$$

From the evaluation of (8.32) (see (4.60)) one notes that

$$|\alpha_{\omega\omega'}|^2=\mathrm{e}^{8\pi M\omega}|\beta_{\omega\omega'}|^2 \tag{8.34}$$

(the analyticity argument leading to the factor $\mathrm{e}^{8\pi M\omega}$ is similar to that discussed on page 116).

To compute the particle flux going to $\mathscr{I}^+$ at late times we note that the density of states inside a sphere of radius R centred on the collapsing ball is

$R\,\mathrm{d}\omega/2\pi$ (for fixed l, m). The particle number per mode is, from (8.33) and (8.34)

$$N_{\omega l m} = \sum_{\omega'} |\beta_{\omega\omega'}|^2 = 1/(\mathrm{e}^{8\pi M\omega} - 1) \tag{8.35}$$

so the number of particles per unit time in the frequency range ω to $\omega + \mathrm{d}\omega$, passing out through the surface of the sphere is

$$(\mathrm{d}\omega/2\pi)(\mathrm{e}^{8\pi M\omega} - 1)^{-1}, \tag{8.36}$$

where we have used the fact that a particle takes a time R to reach the surface of the sphere. This is a Planck (black body) spectrum with temperature given by (8.27), exactly as in the two-dimensional case. Alternatively, one can construct localized wave packets to use as a basis in place of (8.29)–(8.31) (Hawking 1975).

At this stage we must take into account the neglect of backscattering. Equation (8.33) can be interpreted as the conservation of probability. The effect of backscattering is to deplete the outgoing flux by a factor $1 - \Gamma_\omega$ representing reflection back down the black hole. Thus the right-hand side of (8.33) must be replaced by Γ_ω, which introduces a factor Γ_ω into (8.36). Because this backscattering is a function of ω, the spectrum is not precisely Planckian. Nevertheless, it may still be regarded as 'thermal' in the following sense. If the black hole is immersed in a heat bath at the temperature (8.27), then the same fraction $1 - \Gamma_\omega$ of incoming radiation will be backscattered by the hole as was removed from the outgoing flux by backscattering. Thus only a fraction Γ_ω of the incoming radiation in the mode ω is absorbed by the hole. It follows that the ratio of emission to absorption per mode is independent of Γ_ω and identical to that of a black body replacing the black hole. The hole therefore remains in thermal equilibrium with the surrounding heat bath in spite of the spectral distortion.

The total luminosity of the hole is given by integrating (8.36) over all modes

$$L = (1/2\pi) \sum_{l=0}^{\infty} (2l + 1) \int_0^\infty \mathrm{d}\omega\, \omega \Gamma_{\omega l}/(\mathrm{e}^{8\pi M\omega} - 1), \tag{8.37}$$

where we have introduced a Γ factor, as well as l-dependence.

If we were dealing with fermions rather than bosons, because of their anticommuting nature, the normalization condition (8.33) would require a + sign in front of $|\beta_{\omega\omega'}|^2$ (see, for example, Parker 1971, DeWitt (1975)). One would therefore obtain a factor $\mathrm{e}^{8\pi M\omega} + 1$ in the expression for the spectrum, which is the appropriate Planck factor for Fermi statistics.

The computation of L for a realistic black hole model involves consideration of real quantum fields. Although the discussion so far has been restricted to a massless scalar field, the essential argument applies to any field. In particular, quanta of the electromagnetic, neutrino and linearized graviton fields will all be radiated thermally with the temperature (8.27). The chief difference lies in their respective Γ factors, which are rather sensitive to the spin of the field. Detailed results require a study of the radial equations such as (8.2) and numerical computations.

Page (1976*a*) (see also Page 1976*b*, 1977) has estimated

$$L = (3.4 \times 10^{46})(M/1\,\mathrm{gm})^{-2}\,\mathrm{ergs}^{-1} \tag{8.38}$$

for a Schwarzchild black hole of mass $\gg 10^{17}$ gm, which consists of 81% neutrinos (four types), 17% photons and 2% gravitons. (The suppression of higher spins is connected with a greater angular momentum barrier.)

The temperature (8.27) may be written

$$T = (1.2 \times 10^{26}\ \mathrm{K})(1\ \mathrm{gm}/M), \tag{8.39}$$

which for a solar mass object gives 6×10^{-8} K. For such an object, only massless quantum emission is relevant. However, for $M \lesssim 10^{17}$ gm, $T \gtrsim 10^9$ K, and the creation of thermal electron–positron pairs becomes possible. At lower masses, other species of elementary particles will be emitted. The details of this high temperature regime will be complicated by the presence of interactions between the particles. Note that a hole of mass 10^{15} gm has a radius of about one fermi, i.e., within the range of the strong interaction. However, although these complications preclude the analysis of the rate of energy emission, one may still appeal to the fundamentally thermodynamic nature of quantum particle production.

So far we have restricted attention to the spherically symmetric case. If rotation is permitted, however, the black hole belongs to the axisymmetric Kerr family, with a metric parametrized by the angular speed of the event horizon, Ω.

The solution of the wave equation in the Kerr background is more complicated than for the Schwarzschild case (see, for example, DeWitt 1975 for a detailed discussion). The spherical harmonics Y_{lm} of (8.1) are replaced by axisymmetric spheroidal harmonics, while the potential term in the radial equation (8.2) acquires some complicated new terms. However, at large r, the effect is merely to replace ω by $\omega - m\Omega$, where m is the azimuthal quantum number of the spheroidal harmonics (playing the role of m in Y_{lm}). Thus, as far as the analysis of radiation at $\mathscr{I}^+$ is concerned, one merely replaces ω by $\omega - m\Omega$ in the Planck factor in (8.36) or (8.37):

$$1/\{\exp[2\pi\kappa^{-1}(\omega - m\Omega)] \pm 1\}. \tag{8.40}$$

The rotation therefore enters into the thermal spectrum in much the same way as a chemical potential. The effect of this alteration on the luminosity of the hole has also been computed by Page (1976*b*), who finds that the rotation of the hole greatly enhances the emission of higher-spin particles.

Because the emission probability depends on the azimuthal quantum number m, it will be asymmetric around the hole. The factor (8.40) is larger for positive m than negative, thus favouring the emission of quanta with angular momenta oriented towards that of the hole rather than away. This implies that the emission will steadily deplete the hole's angular momentum, causing its rotation rate to slow.

In the boson case ($-$ sign in (8.40)), when $\omega < m\Omega$, (8.40) is negative. Moreover, even in the limit $M \to \infty$ ($T \to 0$), when the Hawking thermal emission due to the collapsing body dies away, (8.40) remains finite, and equal to -1, for $\omega < m\Omega$ (it vanishes for $\omega > m\Omega$, and hence for all ω in the Schwarzschild case, $\Omega = 0$). The meaning of this negative flux is the following: when a low frequency, classical wave (with positive m) impinges on a rotating black hole, the effect of rotation is to amplify the wave, causing more energy to emerge than went in originally – a phenomenon known as super-radiance (Misner 1972, Press & Teukolsky 1972, Zel'dovich 1971, 1972). In quantum language the hole induces stimulated emission. Thus, the absorption probability of the hole for these modes is negative. Super-radiance is unconnected with the Hawking mechanism (which depends on mode propagation through the collapsing body) and even occurs around ordinary rotating stars (Ashtekar & Magnon 1975*b*). Quantum super-radiance was calculated before the Hawking effect by Starobinsky (1973) and Unruh (1974) (see also Ford 1975).

Also of interest is the effect of electric charge on the hole. First consider the case of an electrically neutral field propagating in the background of a Reissner–Nordstrom (non-rotating) hole with charge e. This has the line element

$$\mathrm{d}s^2 = [1 - (2M/r) + (e^2/r^2)]\,\mathrm{d}t^2 - [1 - (2M/r) + (e^2/r^2)]^{-1}\,\mathrm{d}r^2 \\ - r^2(\mathrm{d}\theta^2 + \sin^2\theta\,\mathrm{d}\phi^2), \tag{8.41}$$

which possesses an event horizon at

$$r = r_+ = M + (M^2 - e^2)^{1/2}. \tag{8.42}$$

Using (8.42) for R_h in (8.23) one obtains the surface gravity κ for the metric

(8.41). Equation (8.27) then yields the temperature of the charged black hole:

$$T = (1/8\pi k_B M)(1 - 16\pi^2 e^4/\mathscr{A}^2) \tag{8.43}$$

where $\mathscr{A} = 4\pi r_+^2$ is the area of the event horizon.

It is clear from (8.43) that the presence of a charge depresses the temperature of the hole. In the extreme case $e^2 = M^2$ (maximally charged hole), $T = 0$. Inspection of (8.41) and (8.42) shows that for $e^2 > M^2$ no event horizon exists: there is a naked singularity at $r = 0$. Thus, the cosmic censorship hypothesis (Penrose 1969) that naked singularities cannot form from gravitational collapse can be upheld by requiring $T > 0$, which may be interpreted as the third law of thermodynamics applied to black holes (Carter 1975). However, closer inspection (Davies 1978, Farrugia & Hájiček 1980) leaves open the question as to whether $T = 0$ could in principle be achieved.

If the black hole is sufficiently small, it will be hot enough to produce electron–positron pairs. The creation of charged particles in the field of a Reissner–Nordstrom black hole is complicated by the presence of the electric field of the hole itself. A detailed treatment has been given by Gibbons (1975).

It has long been known that the presence of a background electric field opens up the possibility of the spontaneous creation of charged particle pairs even in the absence of a background gravitational field. This may be seen even in Dirac's old semiclassical model of pair creation. The positive and negative energy states of a free Dirac field are separated by a gap $2m$. For particles of charge q, the presence of an external electric field of strength E will lead to a potential energy gradient qE. If a virtual particle–antiparticle pair is created, the two particles will accelerate apart under the action of the electric field. If their separation Δx becomes sufficiently large before their expected annihilation, then the energy gained from the field will be enough to promote the pair to real particles. This will happen if $qE\Delta x \simeq 2m$. This may equivalently be regarded as particles tunnelling through the energy gap $2m$, to appear as ordinary particles accompanied by antiparticles. This phenomenon became known as the Klein paradox (Klein 1929) and led to controversy concerning the stability of the vacuum.

When the gravitational field of a black hole is also present, the situation is more complicated, especially if it is rotating (Gibbons 1975, Deruelle & Ruffini 1976, Damour 1977). There will be a tendency for the hole to discharge itself rather quickly by preferentially emitting charged particles of the same sign as the charge of the hole, rather than the oppositely charged antiparticles. There will also be a 'charge super-radiance' phenomenon, similar to the effect of the hole's rotation, for particles with $\omega < m\Omega + q\Phi$, where Φ is the electric potential at the horizon (Hawking 1975). Thus, electric charge also contributes to the 'chemical potential' of the hole.

Because of the presence of the electric field, which encourages the production of particle pairs, electron–position emission can occur even for very massive black holes with very low temperatures. In this regime, the electric (Klein paradox) effect dominates. Only for $M \gtrsim 10^5 M_\odot$ is the charged particle creation suppressed. At the other extreme, for $M \lesssim 10^{15}$ gm, the gravitational (Hawking) process dominates over the electrical effects.

8.2 Physical aspects of black hole emission

At first sight, black hole radiance seems paradoxical, for nothing can apparently escape from within the event horizon. However, inspection of (8.36) shows that the average wavelength of the emitted quanta is $\sim M$, i.e., comparable with the size of the hole. As it is not possible to localize a quantum to within one wavelength, it is therefore meaningless to trace the origin of the particles to any particular region near the horizon. The particle concept, which is basically global, is only useful near $\mathscr{I}^+$. In the vicinity of the hole, the spacetime curvature is comparable with the radiation wavelength in the energy range of interest, and the concept of locally-defined particles breaks down.

Heuristically, one can envisage the emergent quanta as 'tunnelling' out through the event horizon (Hawking 1977*b*). Alternatively, the continuous, spontaneous creation of virtual particle–antiparticle pairs around the black hole can be used to explain the Hawking radiation. Virtual particle pairs created with wavelength λ separate temporarily to a distance $\simeq \lambda$. For $\lambda \simeq M$, the size of the hole, strong tidal forces operate to prevent re-annihilation. One particle escapes to infinity with positive energy to contribute to the Hawking flux, while its corresponding antiparticle enters the black hole trapped by the deep gravitational potential well on a timelike path of negative energy relative to infinity (see, for example, Gibbons 1979). Thus the hole radiates quanta with wavelength $\simeq M$.

In spite of the ill-defined nature of particles near the horizon, it is clear that the thermal emission will carry away energy to $\mathscr{I}^+$, and the question arises as to the source of this energy. It can only come from the gravitational field itself, which must lose mass as a consequence. To study this steady depletion of mass–energy, we can investigate the stress-tensor expectation value, $\langle T_{\mu\nu}(x) \rangle$, in the vicinity of the hole. Unlike the particle concept, the stress-tensor is a *local* object, and hence may be used to probe the physics close to, and within, the black hole itself.

In a two-dimensional model, $\langle T_{\mu\nu} \rangle$ may be evaluated explicitly for a conformally invariant quantum field (see §6.3). Consider a static ball described by the general static metric (8.8). The wave equation may be solved in terms of the modes $e^{-i\omega u}$ and $e^{-i\omega v}$ (with u and v given by (8.9)), which are of positive frequency with respect to the global Killing vector ∂_t. The vacuum state constructed with respect to these modes is stable; no particles are produced and the system is time-symmetric.

There will, however, be a nonzero vacuum 'polarization' stress due to spacetime curvature, given by substituting $C(r)$ into (6.136):

$$\langle 0|T_{uu}|0\rangle = \langle 0|T_{vv}|0\rangle = -F_u(C) = (1/192\pi)[2CC'' - C'^2] \quad (8.44)$$

$$\langle 0|T_{uv}|0\rangle = (1/96\pi)CC'', \quad (8.45)$$

where the functional F is defined by

$$F_x(y) = \frac{1}{12\pi} y^{\frac{1}{2}} \frac{\partial^2}{\partial x^2}(y^{-\frac{1}{2}}), \quad (8.46)$$

and a prime denotes differentiation with respect to r.

To take a specific case, the two-dimensional analogue of the Reissner–Nordstrom spacetime (8.41) is described by $C(r) = (1 - 2Mr^{-1} + e^2r^{-2})$, from which (8.44) gives

$$\langle 0|T_{uu}|0\rangle = \langle 0|T_{vv}|0\rangle = \frac{1}{24\pi}\left[-\frac{M}{r^3} + \frac{3M^2}{2\,r^4} + \frac{3e^2}{2r^4} - \frac{3Me^2}{r^5} + \frac{e^4}{r^6}\right] \quad (8.47)$$

which describes a static cloud of energy, singular at $r = 0$, and *negative* outside the event horizon $r = r_+$.

Suppose now that the 'star' undergoes gravitational collapse from radius R_0 in the fashion described on page 254. Although the metric (8.8) still correctly describes the geometry in the exterior of the 'star', the coordinates r^* and t (or u and v) are no longer appropriate for the (simple, exponential) solution of the wave equation, because 'outgoing' modes $e^{-i\omega u}$ degenerate to the complicated function (8.28) on $\mathscr{I}^-$. Instead, we wish to choose coordinates such that the 'incoming' modes, and hence the vacuum state (denoted by $|\hat{0}\rangle$), are of the standard Minkowski form on $\mathscr{I}^-$ (i.e., the usual 'in' vacuum state). As explained in the previous section, this means using, in place of u and v as defined above, the coordinates

$$\hat{u} = \beta[\alpha(u) - 2R_0] \tag{8.48}$$

$$\hat{v} = v, \tag{8.49}$$

where α and β are defined by (8.13) and (8.14). The metric (8.8) in these coordinates becomes

$$ds^2 = C(\hat{u}, \hat{v})d\hat{u}d\hat{v} \tag{8.50}$$

with

$$\hat{C}(\hat{u}, \hat{v}) = C(r)\frac{du}{d\hat{u}}\frac{dv}{d\hat{v}}. \tag{8.51}$$

Using (8.17) and (8.18) to evaluate $du/d\hat{u}$ and (8.49) to give $dv/d\hat{v}$, and then substituting into (6.136) one obtains (Davies 1976; see also Davies, Fulling & Unruh 1976, Hiscock 1977*a*, *b*, 1979, 1980)

$$\langle \hat{0}| T_{uu} |\hat{0}\rangle = (8.44) + (\alpha')^2 F_U(\beta') + F_u(\alpha'), \tag{8.52}$$

where $\alpha = \alpha(u)$, $\beta = \beta(U - 2R_0)$ and primes denote differentiation with respect to the argument of a function. The expressions for $\langle \hat{0}| T_{vv} |\hat{0}\rangle$ and $\langle \hat{0}| T_{uv} |\hat{0}\rangle$ are the same as (8.44) and (8.45) respectively.

Equation (8.52) applies to the region outside the collapsing body. In the interior region, described by the metric (8.11), we have

$$\langle \hat{0}| T_{UU} |\hat{0}\rangle = F_U(\beta') - F_U(A) \tag{8.53}$$

$$\langle \hat{0}| T_{VV} |\hat{0}\rangle = F_V(\beta') - F_V(A), \tag{8.54}$$

where $\beta = \beta(U - 2R_0)$ and $\beta = \beta(V)$, respectively.

The general formula (6.136) yields $\langle \hat{0}| T_{\mu\nu} |\hat{0}\rangle$ in the $\hat{u}, \hat{v}$ coordinates (8.48) and (8.49). In arriving at (8.52)–(8.54), we have converted to u, v, or U, V coordinates by using the ordinary tensor transformation relations. The results show that the effect of collapse is to augment the static vacuum energy (8.44) with an outgoing (retarded) flux of radiation that is constant along the retarded null rays u, or U (there is no backscattering here). In the interior of the body, both ingoing and outgoing radiation fluxes occur that depend in a complicated way upon the interior metric A and the collapse trajectory $R(\tau)$. This represents particles created inside the material of the collapsing 'star', and at its surface. For example, $F_V(\beta')$ in (8.54) describes radiation created at the surface that propagates towards $r = 0$, while $F_U(\beta')$ in (8.53) simply describes the same radiation propagating outward again after passage through the centre of the body. When this outgoing flux emerges (unchanged) from the body, it is described by the term proportional to $(\alpha')^2$ in (8.52). In addition to this 'rebounded' outgoing flux, there will be the contribution described by the final term of (8.52), which can also be traced back to the surface of the collapsing body, but which is *outgoing* from the outset.

If the function $\alpha'(u)$ is expressed in terms of τ as a function $\gamma(\tau)$, then we have from (8.17) the following limits as the surface of the body approaches the horizon ($R \to R_h$, $C \to 0$, $u \to \infty$)

$$\gamma \to 0 \text{ as } e^{-\kappa u}$$

$$\dot{\gamma} \to -(1-\dot{R})\kappa$$

$$\ddot{\gamma}\gamma \to 0$$

(here a dot denotes ∂_τ), from which it follows that the middle term on the right of (8.52) vanishes exponentially fast as measured by a distant observer using u, v coordinates. This term therefore represents a sort of prompt radiation emitted as a transient pulse, and which decays exponentially in an identical fashion to the surface luminosity of the collapsing 'star'.

The final term of (8.52) may be written

$$\frac{1}{24\pi}\left[\frac{-\gamma\ddot{\gamma}}{(1-\dot{R})^2} - \frac{\dot{\gamma}\gamma\ddot{R}}{(1-\dot{R})^3} + \frac{\frac{1}{2}\dot{\gamma}^2}{(1-\dot{R})^2}\right] \to \frac{\kappa^2}{48\pi}, \tag{8.55}$$

which is precisely the flux expected (in two spacetime dimensions) from a thermal radiator with temperature $T = \kappa/2\pi k_B$ (see (4.27) for $L \to \infty$), which is the Hawking temperature (8.27). This term therefore represents the Hawking radiation. Note that it is independent of the details of the collapse, depending only on the single parameter κ, the surface gravity of the final black hole. The collapse-dependent radiation is all contained in the exponentially decaying middle term of (8.52).

For late times, in the asymptotic region, $r \to \infty$, the vacuum polarization term (8.47) vanishes, and the stress-tensor reduces to

$$\langle \hat{0}|T_{uu}|\hat{0}\rangle = \kappa^2/48\pi, \quad \langle \hat{0}|T_{vv}|\hat{0}\rangle = \langle \hat{0}|T_{uv}|\hat{0}\rangle = 0. \tag{8.56}$$

The manifest time asymmetry present in (8.56) (retarded radiation only) reflects the irreversible nature of collapse across an event horizon.

Although the various terms of (8.52)–(8.54) have readily-identifiable origins, it is important not to attach too much physical significance to them individually. For instance, it would be wrong to conclude that the Hawking flux arises from the surface of the collapsing body. It must be remembered that an experimenter could not distinguish between the various terms operationally; only the total stress-tensor components can be measured. The decomposition given here is peculiar to the special u, v (or U, V) coordinate system. However, this system is singular at the event horizon.

To determine what an observer would actually measure, the world line of the apparatus must be specified. If the observer's instantaneous velocity vector is u^μ, then his measuring apparatus records the energy density $\langle T_{\mu\nu}\rangle u^\mu u^\nu$ and energy flux $\langle T_{\mu\nu}\rangle u^\mu n^\nu$, where $u^\mu n_\mu = 0$. For example, an observer constrained to fixed r would have 2-velocity $C^{-\frac{1}{2}}(1,0)$ (in(t, r^*) coordinates) and measure the energy density $C^{-1}\langle \hat{0}|T_{tt}|\hat{0}\rangle$, where T_{tt} is given using (4.17). This diverges at the horizon, where $C=0$, which is a reflection of the fact that to remain at fixed r, the observer's world line deviates ever further from a free-falling world line as the horizon is approached.

A more realistic procedure would be to consider a freely-falling observer. In the Schwarzschild case, the Kruskal coordinates (which are non-singular at the horizon – see §3.1) would then be more appropriate. Using the transformation equations (3.19) and the conformal factor

$$\bar{C} = 2Mr^{-1}e^{-r/2M}$$

from (3.20), one finds that an observer moving along a line of constant Kruskal position with two-velocity $\bar{C}^{-\frac{1}{2}}(1,0)$ (in coordinates $(\bar{t}, \bar{r})$; $\bar{t} = \frac{1}{2}(\bar{v}+\bar{u})$, $\bar{r} = \frac{1}{2}(\bar{v}-\bar{u})$), would measure the energy density

$$\bar{C}^{-1}\langle \hat{0}|T_{tt}|\hat{0}\rangle = \langle \hat{0}|T_{\bar{u}}{}^{\bar{v}} + T_{\bar{v}}{}^{\bar{u}} + 2T_{\bar{u}}{}^{\bar{u}}|\hat{0}\rangle, \tag{8.57}$$

where

$$\langle \hat{0}|T_{\bar{u}}{}^{\bar{v}}|\hat{0}\rangle = \frac{(24\pi)^{-1}(\bar{v})^2 e^{-r/2M}}{8Mr}\left(1 + \frac{4M}{r} + \frac{12M^2}{r^2}\right), \tag{8.58}$$

$$\langle \hat{0}|T_{\bar{v}}{}^{\bar{u}}|\hat{0}\rangle = \frac{(24\pi)^{-1}e^{r/2M}}{\bar{v}^2}\left(\frac{1}{r^2} - \frac{3}{2}\frac{M}{r^3}\right), \tag{8.59}$$

$$\langle \hat{0}|T_{\bar{u}}{}^{\bar{u}}|\hat{0}\rangle = -\frac{R}{48\pi} = -\frac{M}{12\pi r^3}. \tag{8.60}$$

It is clear that (8.57) is finite as $\bar{u} \to \infty$, because $\bar{v}$ remains finite on the future horizon. Hence we conclude that an observer who crosses the event horizon along a constant Kruskal position line measures a finite energy density.

These considerations resolve an apparent paradox concerning the Hawking effect. The proper time for a freely-falling observer to reach the event horizon from a finite distance is finite, yet the free-fall time as measured at infinity (in u, v coordinates) is infinite. Ignoring back-reaction, the black hole will emit an infinite amount of radiation during the time that the falling observer is seen, from a distance, to reach the event horizon. Hence it would appear that, in the falling frame, the observer should encounter an infinite amount of radiation in a finite time, and so be destroyed. On the other hand, the event horizon is a global construct, and has no local significance (Hawking 1973), so it is absurd to conclude that it acts as a physical barrier to the falling observer.

The paradox is resolved when a careful distinction is made between particle number and energy density. When the observer approaches the horizon, the notion of a well-defined particle number loses its meaning at the wavelengths of interest in the Hawking radiation (see page 264); the observer is 'inside' the particles. We need not, therefore, worry about the observer encountering an infinite quantity of particles. On the other hand, energy does have a local significance. In this case, however, although the Hawking flux does diverge as the horizon is approached, so does the static vacuum polarization, and the latter is *negative*. The falling observer cannot distinguish operationally between the energy flux due to the oncoming Hawking radiation and that due to the fact that he is sweeping through the cloud of vacuum polarization. The net effect is to cancel the divergence on the event horizon, and yield a finite result, such as that given above by (8.57)–(8.60). (The precise value depends on the infall trajectory.)

The present analysis also solves the mystery of how the black hole can lose mass without matter crossing from the interior of the black hole into the outside universe. Inspection of $\langle \hat{0}|T_{vv}|\hat{0}\rangle$ at the event horizon shows that it is given by

$$-\frac{1}{192\pi}\left(\frac{\partial C}{\partial r}\right)^2\bigg|_{r=R_h} = -\frac{\kappa^2}{48\pi},$$

which is always negative, and equal to minus the Hawking flux at infinity. This is necessarily true because covariant conservation was built into the construction of $\langle T_{\mu\nu}\rangle$. As $\langle T_{vv}\rangle$ represents a null flux crossing the event horizon, one can see that the steady loss of mass–energy by the Hawking flux is balanced by an equal negative energy flux crossing into the black hole from outside. The hole therefore loses mass, not by emitting quanta, but by absorbing negative energy. The idea of a flux of negative energy has already been encountered in connection with moving mirrors in §7.1.

In the above analysis, we have assumed that the quantum state is the conventional vacuum state in the in region. It might be wondered to what extent the presence of quanta initially will change the Hawking effect (Wald

1976). From (2.44) (which is also valid in curved spacetime), one finds the contribution to $\langle T_{\mu\nu} \rangle$ in the out region due to the initial presence of n quanta in the mode σ. In the case of bosons, as a result of stimulated emission, there will be additional energy present in the out region (Bekenstein & Meisels 1977). The form of the outgoing modes in the out region is given by the second term of (8.26), from which one readily sees that the final term of (2.44) decays exponentially in u with e-folding time $(2\kappa)^{-1}$, i.e., the collapse timescale. Thus, the effect of initial quanta fades out exponentially on the same timescale as any surface luminosity, and the black hole soon settles down to thermal equilibrium, having 'forgotten' the details of the initial state. We may therefore conclude that the Hawking effect is extremely general, and independent of any physically reasonable initial quantum state.

These general features of black hole radiance persist in the full four-dimensional treatment.

The fact that the black hole loses mass raises the question of the back-reaction on the structure of the hole itself. Assuming that one may continue to treat the background metric as classical, then the stream of negative energy across the event horizon will cause the area of the horizon to shrink. This is shown schematically in fig. 26 for a Schwarzschild hole. As the area decreases, so does the mass, implying that the temperature (8.39) and luminosity (8.38) rise. Schwarzschild black holes therefore have a negative specific heat – they radiate and get hotter, behaviour which is typical of self-gravitating systems (Lynden-Bell & Wood 1967).

Having an expression for $\langle T_{\mu\nu} \rangle$ at the event horizon enables the features of the back-reaction to be examined in more detail, by attempting to integrate the Einstein equations. This provides a description of the evolution of the event horizon. In particular, one expects that, in the adiabatic approximation that the black hole's shrinkage is ignored in computing $\langle T_{\mu\nu} \rangle$, the horizon should shrink at a steady rate appropriate to the loss of mass in the Hawking radiation. Unfortunately the horizon evolution equation is nonlinear, and has not been solved. If some higher order terms are neglected, however, the rate of shrinkage is indeed what would be naively expected (Candelas 1980). Nevertheless, it has yet to be determined whether the effect of the higher order terms is in fact negligible, and whether the shape of the shrinking horizon in fig. 26 is correct (Tipler 1980, Hájíček & Israel 1980).

As a hole gets hotter, so it will radiate subatomic particles of greater mass. Once above about 10^{10} K, electrons and positrons will emerge, and any residual charge on the hole will rapidly disappear. Moreover, super-

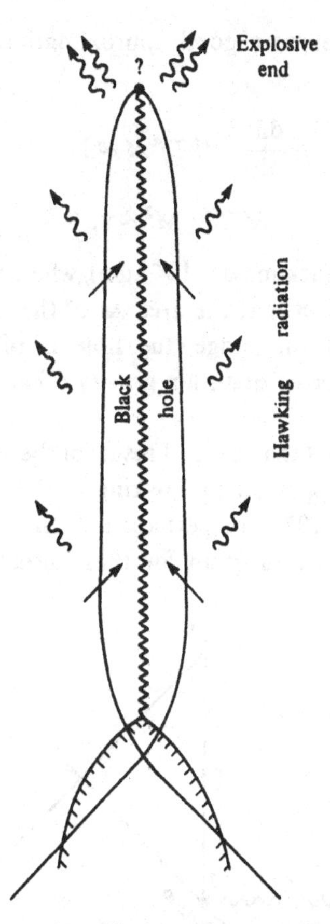

Fig. 26. Black hole evaporation. The lower portion of the diagram corresponds to fig. 24. Gradually the flux of negative energy crossing the horizon into the hole (straight arrows) causes the horizon area to shrink. The process escalates until the horizon collapses rapidly onto the singularity amid an explosive radiation of quanta.

radiance effects will tend to deplete the angular momentum, so the black hole has a tendency to slowly approach the Schwarzschild form.

The continuation of the Hawking process seems to imply that the hole will evaporate away ever faster. Its ultimate fate cannot be decided within the context of the present theory, for when

$$\frac{1}{M}\frac{\mathrm{d}M}{\mathrm{d}t} \simeq k_{\mathrm{B}}T \simeq M^{-1} \tag{8.61}$$

the hole is shrinking at a rate comparable with the frequency of the radiation. Thermal equilibrium no longer applies, nor is the notion of a fixed background spacetime a good approximation. Condition (8.61) is reached when

$$\frac{dM}{dt} \sim (kT)^4 \mathscr{A} \sim 1$$

or

$$M^{-4} \times M^2 \sim 1,$$

i.e., when $M \sim 1$ (the Planck mass $\sim 10^{-5}$ gm), where we have used Stefan's radiation law and the fact that the area $\mathscr{A}$ of the hole is $16\pi M^2$ in the Schwarzschild case. At this stage, the hole is of Planck dimensions (10^{-33} cm) and higher order quantum gravity effects will undoubtedly be important.

It has been conjectured that the end result of the Hawking evaporation process is explosive disappearance (Hawking 1977*a*) or a naked singularity (DeWitt 1975, Penrose 1979), or perhaps a Planck mass object. Fig. 27 shows a possible Penrose diagram for these situations, where the dot

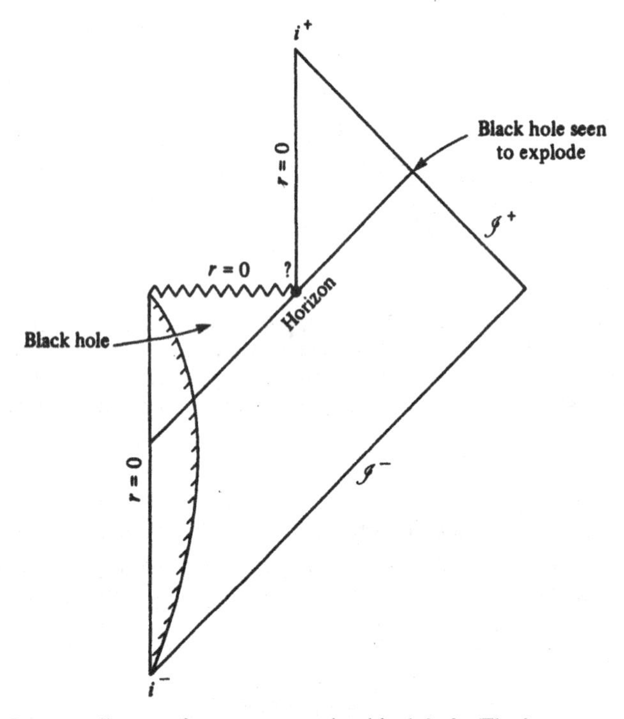

Fig. 27. Penrose diagram for an evaporating black hole. The lower portion of the diagram corresponds to fig. 25. However, when back-reaction is included, the hole evaporates, and the horizon intersects $\mathscr{I}^+$ at a finite time, after passing through a naked singularity (marked ?) Presumably the region above this is Minkowski space.

represents one of the above three alternatives. It should be noted, however, that Gerlach (1976) claims the horizon will not form at all. Whatever the outcome, the detailed behaviour of the hole in the final stages will depend on the nature of subatomic particles at high energies. For example, if there are a small number of truly elementary particles, then the emission rate does not escalate as fast as if the number of particle species rises rapidly at high masses as suggested by Hagedorn (1965). These could affect the observational consequences (Blandford 1977, Rees 1977). Thus, a study of black hole evaporation could provide a unique opportunity for us to probe the physics of ultra-high energy particles.

During the final tenth of a second of its life, a black hole will emit in excess of 10^{30} ergs, or the equivalent of 10^6 megatonne thermonuclear bombs, a significant fraction of which will be in the form of γ-radiation. Existing γ-ray telescopes have failed to reveal evidence for such bursts, although the flux necessary for their detection with present equipment is far greater than expected on other grounds (Blandford & Thorne 1979). More promising is a search for radio bursts caused by the explosive ejection of plasma from the hole into the interstellar magnetic field (Rees 1977). The fact that such bursts have not been detected with current equipment enables an upper limit to be placed on the black hole explosion rate of about $10^{-5}\,\mathrm{pc}^{-3}\,\mathrm{yr}^{-1}$ (Meikle, 1977).

Assuming that the higher order gravitational terms can be neglected, the lifetime of an evaporating hole can be computed from the luminosity (8.38). One obtains $10^{-26}\,(M/1\,\mathrm{gm})^3$ sec, so for a black hole of fermi size (10^{-13} cm), about 10^{15} gm, the lifetime is comparable to the age of the universe. As it is most unlikely that such mini-holes would form anywhere other than the primeval universe, this implies that black holes with an initial mass less than 10^{15} gm would have evaporated away by now.

In spite of the fact that a quantum black hole creates elementary particles and antiparticles in pairs, some of the subatomic conservation laws are violated. For example, a hole that forms from the collapse of a star swallows up mainly baryons, but emits mainly neutrinos and photons, as for the greater part of its life its temperature is too low for massive particle emission. Thus, the law of baryon number conservation is transcended. (It must be remembered that in spite of the evaporation of the hole there is still a singularity present at which baryons may leave spacetime.)

The fact that the quantum black hole enables some of the laws of particle physics to be transcended means that the presence of a hole allows certain subatomic processes to occur that would otherwise be forbidden. One may even suppose that reactions could take place via virtual black hole intermediate states. For example, Hawking (1981) has discussed

$$\mu^{\pm} \rightarrow \text{black hole} \rightarrow e^{\pm} + \gamma$$

(the reaction $\mu \rightarrow e + \gamma$ is otherwise forbidden by conservation of muon lepton number).

Undoubtedly the most persuasive evidence that the Hawking black hole radiance should be taken seriously is the strong connection that it provides between black holes and thermodynamics. Even before Hawking's paper it was appreciated that black holes conform to four laws closely analogous to the four laws of classical thermodynamics (Bardeen, Carter & Hawking 1973). The existence of a surface gravity parameter κ, constant over the event horizon, is reminiscent of the zeroth law of thermodynamics that requires the existence of a temperature parameter constant throughout a system in thermal equilibrium. The conservation of energy during black hole encounters and processes such as the Penrose mechanism (see page 262) and super-radiance, where energy is extracted from, or delivered to, a black hole, is equivalent to the first law of thermodynamics. Moreover, it has already been mentioned (page 263) that a kind of third law, forbidding approach to an extreme Kerr–Newman black hole, was known.

The second law of thermodynamics, requiring the irreversible increase of entropy, finds a natural analogue in Hawking's area theorem (Hawking 1972), which requires the event horizon area $\mathscr{A}$ to be non-decreasing

$$\mathrm{d}\mathscr{A} \geq 0 \tag{8.62}$$

in all black hole processes for which the weak energy condition (Hawking & Penrose 1970) is satisfied. This strongly suggests the identification of $\mathscr{A}$ with entropy $\mathscr{S}$.

The existence of black hole entropy is also implied by the well known connection between entropy and information (Shannon & Weaver 1949; see also the review of Wehrl 1978). When a star implodes to form a black hole, all information about the internal microstates of the star is hidden by the event horizon. Assuming roughly one bit of information per subatomic particle, the total information lost down the hole is about M/m, where M is the mass of the hole and m is the mass of a typical subatomic constituent of the sacrificed body. The associated entropy is

$$\mathscr{S} \sim Mk_{\mathrm{B}}/m.$$

It may appear that, in principle, there is no lower limit to m, and $\mathscr{S}$ should be unbounded. (This accords with the fact that a classical black hole, being black, has zero temperature.) However, as originally pointed out by Bekenstein (1972, 1973) there is a lower bound on m due to the fact that, in order to 'fit' into the hole, the Compton wavelength of the constituent particles should be $\lesssim M$ (the black hole radius). Hence, the maximum information loss is $\sim M^2/\hbar$ and so

$$\mathscr{S} \sim M^2 k_B/\hbar = M^2 k_B \propto k_B \mathscr{A}$$

in our units. Clearly $\mathscr{S}$ becomes infinite in the classical limit ($\hbar \to 0$).

Hawking's work places Bekenstein's conjecture on a firm foundation, and supplies the precise relation (see, for example, DeWitt 1975)

$$\mathscr{S} = \tfrac{1}{4} k_B \mathscr{A}. \tag{8.63}$$

The area law (8.62) is thus seen to be merely a special case of the second law of thermodynamics, $d\mathscr{S} \geq 0$.

The evaporation of the hole, during which the area shrinks, is in violation of (8.62). This is because the existence of negative energy vacuum stress violates the weak energy condition on which the theorem is based. There is, however, no violation of the second law of thermodynamics, because account must also be taken of the increase of entropy in the environment of the hole brought about by the emission of thermal radiation. The *total* entropy, therefore, still increases. Nor does it appear possible to violate the second law by deliberately shooting negative energy (such as from a moving mirror – see §7.1) down a black hole (Ford 1978).

It is possible to develop a complete theory of black hole thermodynamics (Davies 1977*c*, 1978, Hawking 1976*a*, Hut 1977) including features such as phase transitions, Carnot cycles, stability analysis, etc., and even extend the theory to non-equilibrium situations (Candelas & Sciama 1977, Sciama 1976; see also Zurek 1980). Some problems remain, however: What is the relation between information loss due to the imploded star, and the internal microstates of an eternal black hole, such as described by the Kruskal solution (see the next section) that is everywhere a *vacuum* solution of Einstein's equation (Bekenstein 1975)? Can the notion of black hole entropy be extended to arbitrary gravitational fields (Davies 1974, 1981, Penrose 1979)? Will the Hawking radiation always be precisely thermal, even in the presence of interactions (see §9.3) and recoil of the hole (Page 1980)? Does the Hawking process imply time-reversal symmetry violation in quantum gravity (Wald 1980)? Many of these questions are still under active investigation.

8.3 Eternal black holes

In the previous sections we have considered the behaviour of a quantum field in the background spacetime of a collapsing body. However, the features of the Hawking effect turned out to be independent of the details of collapse, which suggests that the effect is more a consequence of the causal and topological structure of spacetime than the specific geometry. This indeed turns out to be the case. One is prompted to dispense entirely with the collapsing body and examine quantum field theory on the maximally extended manifold which is everywhere a solution of the vacuum Einstein equation (see §3.1).

We begin by treating the two-dimensional model of a Schwarzschild black hole, by putting $C=(1-2Mr^{-1})$ in (8.8). The results can easily be extended to the Reissner–Nordstrom case (Davies 1978). The Penrose diagram for this manifold is shown in fig. 6. In terms of Kruskal coordinates $\bar{u}, \bar{v}$ defined by (3.19), the line element is

$$ds^2 = 2Mr^{-1}e^{-r/2M}d\bar{u}\,d\bar{v}, \tag{8.64}$$

which is regular everywhere except at the physical singularity $r=0$. Though this spacetime is symmetric under time reversal, the quantum state imposed on it need not be.

Two natural basis modes for the massless scalar wave equation exist, being either proportional to $e^{-i\omega u}, e^{-i\omega v}$, in terms of Schwarzschild null coordinates u and v, or $e^{-i\omega\bar{u}}$, $e^{-i\omega\bar{v}}$ in terms of Kruskal null coordinates (3.19). The former set oscillate infinitely rapidly on the event horizon, while the latter set are regular on the entire manifold. Each will have its associated vacuum, which we denote by $|0_S\rangle$ and $|0_K\rangle$ respectively. The scalar Green functions in each case will be given in terms of the Minkowski space Green function (4.23) with $L\to\infty$ by a relation of the form (3.154) with $n=2$. Discarding, as usual, an infinite constant (infrared divergence) one obtains

$$D_S^{(1)}(x'',x') = -(1/2\pi)\ln \Delta u\ \Delta v \tag{8.65}$$

$$D_K^{(1)}(x'',x') = -(1/2\pi)\ln \Delta\bar{u}\ \Delta\bar{v}. \tag{8.66}$$

If we transform $D_K^{(1)}$ into Schwarzschild coordinates we obtain

$$D_K^{(1)}(x'',x') = -(1/2\pi)\ln[\cosh\kappa(t''-t') - \cosh\kappa(r^{*''}-r^{*'})] + \text{function of } (r'',r') \tag{8.67}$$

which is manifestly invariant under the transformation $t''\to t''+2\pi in/\kappa$ (n integer), where here the surface gravity $\kappa=(4M)^{-1}$. That is, $D_K^{(1)}$ is periodic in imaginary Schwarzschild time, with period $2\pi/\kappa$. As explained in §2.7 (see (2.104)), this is a feature characteristic of thermal Green functions. In this case the temperature is $\kappa/2\pi k_B$, the Hawking temperature (8.27). A similar periodicity is found in four dimensions, even when the black hole is rotating

(Hartle & Hawking 1976). For example, for modes with azimuthal quantum number m

$$D_K^{(1)}(t''-t'+2\pi in/\kappa;r'',r')=e^{2\pi m\Omega/\kappa}D_K^{(1)}(t''-t';r'',r') \qquad (8.68)$$

confirming that $m\Omega$ behaves like a chemical potential (see remark following (2.104)). These concepts also generalize readily to the case of electrically charged holes and quanta (Gibbons & Perry 1976, 1978).

Far from the hole, the Green function (8.65) reduces to the usual $D^{(1)}$ for flat space quantum field theory; in this region $|0_S\rangle$ is the conventional vacuum state. Evidently, an observer in this region would regard $|0_K\rangle$ as a thermal state at temperature $\kappa/2\pi k_B$ and $D_K^{(1)}$ as a thermal Green function, i.e., the black hole is immersed in a bath of thermal radiation at the Hawking temperature. Indeed, careful analysis (Birrell & Davies 1978*a*) shows that far from the hole $D_K^{(1)}$ can be written as an infinite sum (2.111) of Green functions $D_S^{(1)}$, confirming that it is precisely a thermal Green function. Clearly (8.67) is invariant under time reversal, so the thermal bath represents a flux of thermal radiation passing into the hole from $\mathscr{I}^-$ at a rate equal to that of the Hawking flux reaching $\mathscr{I}^+$. This may be readily confirmed by using $D_K^{(1)}$ to compute $\langle 0_K|T_{\mu\nu}|0_K\rangle$ in the standard way, to find an energy density at $\mathscr{I}^\pm$ of $\kappa^2/24\pi$. The vacuum $|0_K\rangle$ therefore describes a steady-state thermal equilibrium between the black hole and its surroundings, such as would be obtained by confining the hole to the interior of a perfectly reflecting cavity. (If the cavity is too large, the equilibrium is unstable – see for example, Davies (1978)). The state $|0_K\rangle$ is known as the Hartle–Hawking, or Israel vacuum (Hartle & Hawking 1976, Israel 1976).

A comparison of (3.19) with (4.69) and (4.70), putting $a=\kappa=(4M)^{-1}$, reveals that the state $|0_S\rangle$ is closely analogous to the state $|0_R\rangle$ associated with an accelerated observer in the Rindler wedge of Minkowski space. Likewise $|0_K\rangle$ corresponds to $|0_M\rangle$. This is no surprise, as the horizon structures of the Rindler wedge and the Schwarzschild black hole are identical. Comparison of figs. 6 and 14 shows that the region R of Minkowski space possesses the same causal relationship to L as the region I (representing the universe external to the black hole) has to II (the 'mirror' universe). In the same way as in §4.5, we may here relate the modes defined on the extended manifold to those defined in regions I and II only (Unruh 1976). Because the mathematical relation between the coordinates u, v and $\bar{u}, \bar{v}$ is identical in both examples, we do not need to repeat the analysis here, but may write down straight away the Bogolubov transformations connecting both sets of modes: these will be formally identical to (4.95) and (4.96), where $b_k^{(2)}$ now represents an annihilation operator for excitation of the Schwarzschild modes in region I (annihilation of a particle in 'our' universe outside the hole) and $b_k^{(1)}$ represents an annihilation operator for quanta in region II. Similarly the d_k operators are associated with the Kruskal modes.

The Bogolubov transformation leads, as demonstrated in §4.5, to a thermal spectrum (4.97), with temperature $a/2\pi k_B = \kappa/2\pi k_B$, as expected. However, the transformation contains more information than this. Putting $\tanh \phi_\omega = e^{-\pi\omega/a}$, (4.95) and (4.96) give

$$b_k^{(1)} = e^{iJ} d_k^{(1)} e^{-iJ} \tag{8.69}$$

where $\omega = |k|$,

$$J = \sum_k i\phi_\omega (b_{-k}^{(1)\dagger} b_k^{(2)\dagger} - b_{-k}^{(1)} b_k^{(2)}) \tag{8.70}$$

and we have used the commutation relations.

From the definitions

$$b_k^{(1)}|0_S\rangle = b_k^{(2)}|0_S\rangle = 0 \tag{8.71}$$

$$d_k^{(1)}|0_K\rangle = d_k^{(2)}|0_K\rangle = 0 \tag{8.72}$$

and
$$e^{-iJ} b_k^{(1)}|0_S\rangle = d_k^{(1)} e^{-iJ}|0_S\rangle,$$

which follows from (8.69), one has

$$|0_K\rangle = e^{-iJ}|0_S\rangle. \tag{8.73}$$

This can be expanded and rearranged to give

$$\begin{aligned}|0_K\rangle &= \exp\{\sum_k [-\ln \cosh \phi_\omega + \tanh \phi_\omega b_k^{(1)\dagger} b_k^{(2)\dagger}]\}|0_S\rangle \\ &= \prod_k (\cosh \phi_\omega)^{-1} \sum_{n_k=0}^{\infty} e^{-n_k\pi\omega/\kappa} |n_k^{(1)}\rangle |n_k^{(2)}\rangle \end{aligned} \tag{8.74}$$

in terms of states with n_k quanta in region I and n_k quanta in region II (Unruh 1976, Israel 1976).

If an observer is restricted to the region I (outside the black hole) then he will not have access to the modes in region II and will not be able to measure $|n_k^{(1)}\rangle$. When this observer measures an observable A, with associated operator $\hat{A}$, in the quantum state $|0_K\rangle$, then the $|n_k^{(1)}\rangle$ part of (8.74) just factors out:

$$\begin{aligned}\langle 0_K|\hat{A}|0_K\rangle &= \sum_{n_k} \prod_k \langle n_k^{(2)}|\hat{A}|n_k^{(2)}\rangle \exp(-2n_k\pi\omega/\kappa) \\ &\quad \times [1 - \exp(-2\pi\omega/\kappa)] \\ &= \mathrm{tr}(\hat{A}\rho) \end{aligned} \tag{8.75}$$

where ρ can be written

$$\sum_n \prod_k \left\{ \frac{e^{-\beta E_n}}{\sum_{m=0}^{\infty} e^{-\beta E_m}} \right\} |n_k\rangle\langle n_k| \tag{8.76}$$

with $E_n = n\omega$, $\beta = 2\pi/\kappa$, and regarded as a density matrix corresponding to a *thermal* average (see §2.7) at temperature $\kappa/2\pi k_B$. In particular, if $\hat{A}$ is the particle number operator for mode σ

$$\begin{aligned}\langle 0_K | N_\sigma | 0_K \rangle &= \sum_{n=0}^{\infty} n_\sigma e^{-\beta E_n} \Big/ \sum_{m=0}^{\infty} e^{-\beta E_m} \\ &= 1/(e^{\beta\sigma} - 1),\end{aligned} \tag{8.77}$$

i.e., a Planck spectrum.

The fact that the pure state $|0_K\rangle$, defined on the whole manifold, appears to an observer confined to region I as a mixed state to be described by a density matrix (8.76), is no surprise. The presence of the event horizon hides the information about the modes in region II, and this loss of information is naturally associated with a nonzero entropy in region I (the entropy of a pure state is zero). Indeed, not only does (8.76) imply a Planck spectrum, it also implies completely thermal radiation, i.e., the complete absence of correlations between the emitted quanta (Parker 1975, Wald 1975, Hawking 1976*b*). For example, the probability that $|0_K\rangle$ contains ${}^1n_{k_1}$ quanta in mode k_1, ${}^2n_{k_2}$ quanta in mode k_2, etc. in region I, is

$$\begin{aligned}|\langle {}^1n_{k_1}, {}^2n_{k_2}, \ldots, | 0_K \rangle|^2 &= \prod_j \exp(-2\pi\, {}^jn_{k_j} |k_j|/k)[1 - \exp(-2\pi/k_j|/\kappa)] \\ &= \prod_j P({}^jn_{k_j}),\end{aligned} \tag{8.78}$$

where $P({}^jn_{k_j})$ is the probability that the mode k_j contains ${}^jn$ particles. The fact that all $P({}^jn_{k_j})$ are multiplied shows that these individual probabilities are independent of each other.

The details concerning the density matrix and thermal states are immediately generalized to four dimensions and also apply to the case of uniformly accelerated observers. Attempts have also been made (Gibbons & Hawking 1977*a*, Lapedes 1978*b*, Lohiya & Panchapakesan 1978, 1979, Denardo & Spallucci 1980*a*, *b*) to extend them to non-black hole horizons, such as de Sitter space, though in the latter case the physical significance of the thermal structure is not so clear.

In arriving at these results we have neglected the back-reaction of the particle production on the background spacetime. The recoil of emission from the black hole will necessarily introduce some correlations on the emitted quanta when only a small fraction of the total energy has been radiated (Page 1980). This, of course, would occur with any hot body, and is a consequence of the hole not being in thermodynamic equilibrium with its environment.

Although the above treatment has been based on the maximally extended Kruskal manifold, the thermal nature of the radiation and the absence of correlations is also a feature of the more realistic model involving collapse. In the case of a massless field, $\mathscr{I}^-$ is a Cauchy surface for the collapse spacetime, but $\mathscr{I}^+$ is not; one must take into account the propagation of the field across the future event horizon H^+. The union of $\mathscr{I}^+$ and H^+ does form a Cauchy surface (see fig. 25). Thus, to construct a complete set of modes in the out region, it is necessary to augment the standard exponential modes on $\mathscr{I}^+$ with modes on H^+. An explicit construction of such a complete set has been given by Wald (1975). However, the results of measurements made on $\mathscr{I}^+$ are independent of the detailed form of the horizon modes. Restricting measurement to $\mathscr{I}^+$ and forsaking information about the modes on the horizon once again introduces a density matrix of the form (8.76). One may envisage the horizon modes as representing quanta that cross into the black hole. Thus, one arrives at the model of the Hawking process mentioned on page 264: particle pairs are created close to the horizon, one particle of which travels to $\mathscr{I}^+$ (ignoring back scattering) and the other of which enters the hole. By relinquishing information about the latter quanta, the former must be described by a mixed state (i.e., using a density matrix). Thus, whether the Hawking effect is examined on the full Kruskal manifold or for the collapsing body, measurements at $\mathscr{I}^+$ will acquire a thermal nature.

We have so far discussed two 'vacuum' states, $|0_K\rangle$ and $|0_S\rangle$, the former corresponding to a bath of thermal radiation, the latter reducing to the conventional Minkowski vacuum state at large distance from the hole. As always in quantum theory, additional physical criteria are necessary to decide which quantum state corresponds to the physical situation of interest. If the universe contains an eternal black hole, only observation can reveal what quantum state is actually realized. However, the state $|0_K\rangle$ clearly reproduces the features outside the hole that would be present if a black hole that was formed from a collapsing body were subsequently confined in a box and allowed to come into thermal equilibrium.

It is possible to construct yet another 'vacuum' state on the maximally extended manifold, that will reproduce the effects of a collapsing body, i.e., yield a time-asymmetric thermal *flux* from the hole, rather than a thermal bath. This state was discovered by Unruh (1976).

Note that $|0_S\rangle$ is associated with modes that are positive frequency with respect to the Killing vector ∂_t (cf. 4.77)), while $|0_K\rangle$ is defined using modes which are positive frequency with respect to the vector $\partial_{\bar{t}}$, where $\bar{t}$ is the Kruskal time ($\bar{t} = \frac{1}{2}(\bar{u} + \bar{v})$) (cf. (4.82)). Inspection of fig. 6 shows that the null rays u = constant, that mark the surfaces of constant phase of the standard (Minkowski space) exponential modes ($e^{-i\omega u}$) at $\mathscr{I}^+$, when followed backwards in time would, if the collapsing body had been absent, cross the *past* horizon, H^-, of the extended Kruskal manifold. In the collapse picture (fig. 25) the null ray γ (the latest advanced ray to travel through the collapsing body, and emerge to form the future horizon H^+) plays a role analogous to H^- in fig. 6. Incoming rays immediately prior to γ reflect off the centre of coordinates and pass across γ on their way to $\mathscr{I}^+$, close to H^+. These rays represent waves that are positive frequency with respect to the affine parameter along γ (Hawking 1975; see also Parker 1977, Gibbons 1980). For rays that emerge very close to H^+, this affine parameter coincides to good approximation with the affine parameter $\bar{u}$ along the past horizon in the extended manifold, fig. 6. Hence, the geometrical effects of the collapse may be mocked up by choosing the in modes to be of the form $e^{-i\omega\bar{u}}$ along H^- and $e^{-i\omega v}$ along $\mathscr{I}^-$. The modes $e^{-i\omega\bar{u}}$ are of positive frequency with respect to the vector $\partial_{\bar{u}}$, which is a Killing vector on H^- (Unruh 1976). The collapsing body may then be dispensed with, and we may work with the full Kruskal manifold. The vacuum state associated with this choice of modes in the past is called the Unruh vacuum, and we denote it by $|0_U\rangle$. It corresponds to a thermal flux of particles leaving the region of the black hole. The time reversed state, with modes $e^{-i\omega\bar{v}}$ along H^+ and $e^{-i\omega u}$ along $\mathscr{I}^+$, would describe a steady flux of thermal radiation going into the hole.

Further insight into the properties of the vacua $|0_S\rangle$ (sometimes called the Boulware vacuum (Boulware 1975*a*, *b*)), $|0_K\rangle$ and $|0_U\rangle$ can be obtained by studying the experiences of a model particle detector at a fixed Schwarzschild radius $r = R$. For massless fields in two dimensions, the response function (3.55) can be evaluated explicitly using the Wightman functions

$$D_S^+(x, x') = -(1/4\pi)\ln[(\Delta u - i\varepsilon)(\Delta v - i\varepsilon)] \tag{8.79}$$

$$D_K^+(x, x') = -(1/4\pi)\ln[(\Delta\bar{u} - i\varepsilon)(\Delta\bar{v} - i\varepsilon)] \tag{8.80}$$

$$D_U^+(x, x') = -(1/4\pi)\ln[(\Delta\bar{u} - i\varepsilon)(\Delta v - i\varepsilon)]. \tag{8.81}$$

For a detector at $r = R$, the proper time is given by

$$d\tau = (1 - 2M/R)^{\frac{1}{2}} dt, \tag{8.82}$$

while $\Delta u = \Delta v = \Delta t$ and, from (3.19),

$$\Delta \bar{u} = -4M e^{R^*/4M}(e^{-t/4M} - e^{-t'/4M}), \tag{8.83}$$

$$\Delta \bar{v} = 4M e^{R^*/4M}(e^{t/4M} - e^{t'/4M}). \tag{8.84}$$

For the Boulware vacuum, substituting (8.79) into the response function (3.55), gives an expression which is essentially the same as for a detector at rest in two-dimensional Minkowski space in the usual vacuum. As in that case one obtains $\mathscr{F}(E) = 0$ for $E > 0$, i.e., the detector registers no particles in $|0_S\rangle$.

For the Unruh vacuum, substitution of (8.81) into (3.55), noting (8.82) and (8.84), gives a response function per unit proper time which is identical to (4.54) with $w = 0$ and $\kappa = [16M^2(1 - 2M/R)]^{-\frac{1}{2}}$. Thus,

$$\mathscr{F}_U(E)/\text{unit proper time} = 1/E(e^{E/kT} - 1) \tag{8.85}$$

where $kT = [64\pi^2 M^2(1 - 2M/R)]^{-\frac{1}{2}}$. As in the case of the accelerating mirror, which led to (4.54), the detector registers a flux of particles at apparent temperature $T_0 = 1/8\pi k_B M$ given by the Tolman relation (4.98). This is in agreement with (8.27) (noting $\kappa = (4M)^{-1}$ for a Schwarzschild black hole). As the detector approaches the horizon ($R \to 2M$), the temperature of the flux determined by the detector diverges. This is due to the fact that the detector must be non-inertial to maintain a fixed distance from the black hole. The magnitude of the acceleration relative to the local freely-falling frame is $M/[R^2(1 - 2M/R)^{\frac{1}{2}}]$. Such acceleration gives rise to the detection of additional particles, as in Minkowski space (see §3.3). As the horizon is approached, the acceleration diverges, as does the temperature (cf. (3.68)).

The calculation for the vacuum $|0_K\rangle$ is similar, and one obtains twice the result (8.85), corresponding to a bath of radiation at apparent temperature T.

The approximate response for a particle detector in four dimensions has been studied by Unruh (1976) and Candelas (1980). In particular, Candelas has evaluated the response function as $R \to 2M$ and $R \to \infty$ for each of the three vacua studied above, and obtains results consistent with those in two dimensions.

An extensive analysis of the properties of $|0_S\rangle, |0_K\rangle$ and $|0_U\rangle$ has been given by Fulling (1977).

8.4 Analysis of the stress-tensor

It is a straightforward matter to compute $\langle T_{\mu\nu}\rangle$ in the above 'vacuum' states using the theory of chapter 6. In the two-dimensional case the results may be obtained in closed form. However, it is instructive to carry out a general analysis of $\langle T_{\mu\nu}\rangle$ in the spirit of the Wald axioms (see §6.6). This was done by Christensen & Fulling (1977).

We begin by treating first the two-dimensional 'Schwarzschild' case. Putting $C = (1 - 2M/r)$ in (8.8) the covariant conservation equation (6.142) reduces to

$$\langle T^r_t\rangle = \text{constant} \tag{8.86}$$

$$\frac{\partial}{\partial r}\left[\left(1-\frac{2M}{r}\right)\langle T_r{}^r\rangle\right] = \frac{M}{r^2}\mathscr{T}, \tag{8.87}$$

where $\mathscr{T}$ is the trace $\langle T_\mu{}^\mu\rangle$ and we assume that all the components of $\langle T_{\mu\nu}\rangle$ are independent of time. Integrating (8.87) and using $\langle T_t{}^t\rangle = \mathscr{T} - \langle T_r{}^r\rangle$ yields

$$\langle T_\mu{}^\nu\rangle = \langle T^{(1)}{}_\mu{}^\nu\rangle + \langle T^{(2)}{}_\mu{}^\nu\rangle + \langle T^{(3)}{}_\mu{}^\nu\rangle, \tag{8.88}$$

where, in (t, r^*) coordinates,

$$\langle T^{(1)\nu}_{\mu}\rangle = \begin{bmatrix} \dfrac{-1}{(1-2M/r)}H(r) + \mathscr{T}(r) & 0 \\ 0 & \dfrac{1}{(1-2M/r)}H(r) \end{bmatrix} \tag{8.89}$$

$$\langle T^{(2)\nu}_{\mu}\rangle = \frac{K}{M^2}\frac{1}{(1-2M/r)}\begin{pmatrix} 1 & -1 \\ 1 & -1 \end{pmatrix} \tag{8.90}$$

$$\langle T^{(3)\nu}_{\mu}\rangle = \frac{Q}{M^2}\frac{1}{(1-2M/r)}\begin{pmatrix} -1 & 0 \\ 0 & 1 \end{pmatrix}, \tag{8.91}$$

and

$$H(r) = M\int_{2M}^{r} \mathscr{T}(\rho)\rho^{-2}\mathrm{d}\rho, \tag{8.92}$$

while K and Q are constants to be determined according to which quantum state is chosen.

Consider first the state $|0_S\rangle$. This has been investigated in detail by Blum (1973) and Boulware (1975*a*, 1976). At $\mathscr{I}^\pm$ it coincides with the conventional Minkowski space vacuum $|0_M\rangle$ so there can be no radiation at infinity. Hence we must choose $K=0$ and $Q=-M^2H(\infty)$, so that (8.88) vanishes at large r. Using (6.121) and (7.2) to give

$$\mathscr{T}(r) = -M/6\pi r^3, \tag{8.93}$$

one obtains from (8.92)

$$H(r) = -(1/384\pi M^2) + (M^2/24\pi r^4). \tag{8.94}$$

Thus, for the state $|0_S\rangle$, $Q = 1/384\pi$ and $\langle T_\mu{}^\nu\rangle$ reduces to the expression given by (8.47) (with $e=0$). Thus, $|0_S\rangle$ is the state appropriate to a vacuum around a static star.

It cannot represent the state of a black hole, however, as may be seen from the following analysis: Noting that (cf. (4.17)–(4.19))

$$T_{uu} = \tfrac{1}{4}(T_{tt} + T_{r^*r^*} - 2T_{tr^*}), \tag{8.95}$$

one finds from (8.88)–(8.91) that

$$\begin{aligned}\langle T_{uu}\rangle &= -\tfrac{1}{2}(H + Q/M^2) + \tfrac{1}{4}(1-2M/r)\mathscr{T}\\ &\rightarrow -Q/2M^2 \quad \text{as} \quad r\rightarrow 2M.\end{aligned} \tag{8.96}$$

Now the Schwarzschild coordinates are singular at the horizon. To investigate the behaviour of the stress-tensor there, we transform to Kruskal coordinates (3.19), which are regular at the horizon, and obtain

$$\langle T_{\bar{u}\bar{u}}\rangle = -(Q/32M^4)\mathrm{e}^{-r/M}\bar{v}^2(1-r/2M)^{-2}. \tag{8.97}$$

This diverges as $r\rightarrow 2M$ unless $Q=0$. The quantity $\langle T_{\bar{u}\bar{u}}\rangle$ is more indicative of the physical situation at the horizon, i.e., to what a freely-falling observer would measure (Fulling 1977). Hence the state $|0_S\rangle$ is unphysical if the metric has the Schwarzschild form as far as $r=2M$. In practice, if one attempted to set up such a quantum state around a black hole, the back-reaction near $r=2M$ would greatly modify the gravitational field.

Turning to another choice of constants consider the time symmetric case, where $\langle T_\mu{}^\nu\rangle$ is required to be regular on both the past and future horizons. An analysis of $\langle T_{vv}\rangle$ as $r\rightarrow 2M$ shows that one requires both $K=Q=0$. Hence $\langle T_\mu{}^\nu\rangle = \langle T^{(1)}{}_\mu{}^\nu\rangle$, and at $r\rightarrow\infty$ this reduces to

$$\tfrac{1}{12}\pi(k_B T)^2\begin{pmatrix}2 & 0\\ 0 & -2\end{pmatrix} \tag{8.98}$$

with $T = 1/8\pi k_B M = \kappa/2\pi k_B$. This is the stress-tensor for a *bath* of thermal radiation at temperature T, and therefore reproduces the properties of the Israel–Hartle–Hawking vacuum $|0_K\rangle$.

Finally, to describe the Hawking evaporation process one requires an outward flux of thermal radiation only, for which the stress-tensor at large r is

$$\tfrac{1}{12}\pi(k_B T)^2\begin{pmatrix} 1 & 1 \\ -1 & -1 \end{pmatrix}, \tag{8.99}$$

with $T = 1/8\pi k_B M$. Thus the energy density and flux are numerically equal, which demands that (with $Q = 0$)

$$K = \tfrac{1}{2}M^2[H(\infty) - \mathscr{T}(\infty)]$$

$$= \tfrac{1}{2}M^3 \int_{2M}^{\infty} \mathscr{T}(\rho)\rho^{-2}\mathrm{d}\rho, \tag{8.100}$$

noting that $\mathscr{T}(\infty) = 0$. This is a remarkable relation, giving the Hawking flux at $r = \infty$ (i.e., K/M^2) in terms of an integral over the trace of the stress-tensor. The fact that we know from the thermal nature of the Bogolubov transformation that $K \neq 0$ (there must be an energy flux if the spectrum is thermal) proves that there exists an anomalous trace, $\mathscr{T}$. In chapter 6 the existence of a trace anomaly was deduced from subtle arguments involving renormalization of $\langle T_{\mu\nu}\rangle$. Now, following Christensen & Fulling (1977), we have derived its existence by an entirely independent route, and one which makes no reference to renormalization.

One cannot, of course, obtain the form of $\mathscr{T}$ from (8.100), although, given that it must be geometrical, one immediately concludes $\mathscr{T} \propto R$, as R is the only geometrical scalar with dimensions $(\text{length})^{-2}$ in two dimensions. Hence, knowledge of K fixes the anomaly coefficient or vice versa. From either (8.99) or (8.93), we arrive at $K = -(768\pi)^{-1}$ and $\mathscr{T} = -R/24\pi$. We have demanded that $Q = 0$ so that the stress-tensor is regular on the future horizon H^+ and the situation describes the Unruh vacuum $|0_U\rangle$. The stress-tensor is not, however, regular on H^-.

Note that, in the case of a massless fermion field, the alteration of the Planck factor to $(e^{\omega/k_B T} + 1)$ introduces a factor $\frac{1}{2}$ into (8.98) and (8.99). When we take into account the two helicity states, the results are identical to the scalar case. Thus, we deduce that, in two dimensions, the conformal anomaly and the Hawking flux are the same for spins 0 and $\frac{1}{2}$, which confirms the results of chapter 6.

Turning to the four-dimensional case, the requirements of spherical symmetry, time-independence and covariant conservation do not suffice to uniquely relate the trace to the Hawking flux; there is an additional arbitrariness in the angular components $\langle T_\theta{}^\theta \rangle = \langle T_\phi{}^\phi \rangle$. Christensen & Fulling (1977) have analysed this case in detail and present arguments that place qualitative restrictions on $\langle T_\theta{}^\theta \rangle$. Their conjectures on the asymptotic form of the stress-tensor as $r \to \infty$ or $2M$ in each of the three 'vacuum' states have been confirmed by the explicit calculations of Candelas (1980).

We shall finish this section by briefly returning to the two-dimensional Reissner–Nordstrom black hole with $C = (1 - 2Mr^{-1} + e^2 r^{-2})$. This spacetime has two horizons, r_+, defined by (8.42), and another defined by

$$r_- = M - (M^2 - e^2)^{\frac{1}{2}}, \tag{8.101}$$

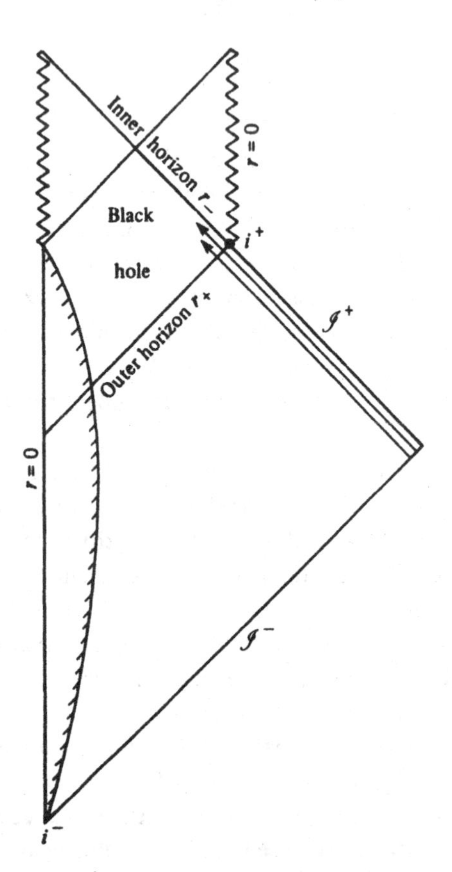

Fig. 28. Penrose diagram for a spherical star that collapses to form a charged, Reissner–Nordstrom, black hole. (Compare fig. 25. The manifold is drawn incomplete and can be extended vertically. This extension is irrelevant for the present discussion.) The singularity is timelike. Late time, infalling null rays from $\mathscr{I}^-$ crowd up, blueshifted, along r_-.

at which C also vanishes. This inner horizon is a Cauchy horizon, but not an event horizon for the spacetime (see, for example, Hawking & Ellis 1973, chapter 5). A conformal diagram is shown in fig. 28, from which one may see at a glance, recalling the conformal compression of $\mathscr{I}$, that r_- represents a surface of infinite blueshift, because null rays from $\mathscr{I}^-$ crowd along r_- inside the black hole.

It is a simple matter to evaluate $\langle T_\mu{}^\nu \rangle$ for the interior of the hole in the states $|0_U\rangle$ and $|0_K\rangle$ (Davies 1978). One finds (Hiscock 1977*a*, *b*, 1979) that $\langle T_\mu{}^\nu \rangle$ diverges on r_- even in a coordinate system that is regular there. General arguments (Birrell & Davies 1978*b*, Hiscock 1979) suggest that this result remains true in four dimensions. One concludes that the back-reaction effects of this quantum stress would disrupt the interior geometry of the hole and presumably give rise to a singularity along the Cauchy horizon, that would prevent analytic extension of the manifold to other asymptotically flat spacetime regions (see also Simpson & Penrose 1973, McNamara 1978).

Finally, it has been shown by Curir & Francaviglia (1978) that in the case of a rotating black hole, the inner horizon may be interpreted as a sort of negative temperature surface (in the Hawking sense), satisfying a type of area theorem similar to (8.62), and bearing some relation to the concept of spin temperature. (See also Calvani & Francaviglia 1978.)

8.5 Further developments

The detailed analysis to which the quantum black hole has been subjected, and the remarkable consistency of the results from widely different theoretical approaches, has encouraged a number of authors to use the quantum black hole as the starting point for a more searching analysis of quantum field theory in curved space, and even quantum gravity. Hawking and coworkers, in particular, have developed an extensive programme in which thermal aspects of black hole physics play a central role (for a review see Hawking 1979).

Much of this further work depends on the observation that the Green function which describes a black hole in the Israel–Hartle–Hawking vacuum state $|0_K\rangle$ (or its generalization to other spacetimes) is periodic in imaginary Schwarzschild time (see page 276). This suggests defining $\tau = it$ and considering, in place of the Schwarzschild spacetime, a related Riemannian space (i.e., negative definite metric, as opposed to pseudo-Riemannian, with indefinite metric) with a Kruskal-like line element

$$ds^2 = -(2M/r)e^{r/2M}(dX^2 + dY^2) - r^2(d\theta^2 + \sin^2\theta\, d\phi^2) \quad (8.102)$$

where

$$X = 4M[r/2M - 1]^{\frac{1}{2}}\, e^{r/4M} \cos \kappa\tau$$

$$Y = 4M[r/2M - 1]^{\frac{1}{2}}\, e^{r/4M} \sin \kappa\tau.$$

This space is Ricci flat and has topology $R^2 \times S^2$. The Killing vector ∂_τ generates rotations about the fixed point $X = Y = 0$. This rotational symmetry endows τ with the properties of an angular coordinate, thus building in the periodicity requirement; the points whose τ coordinates differ by $2\pi/\kappa$ are identified. The origin, $X = Y = 0$, corresponds to $r = 2M$, i.e., the event horizon. The region corresponding to the inside of the black hole, and in particular the singularity at $r = 0$, are absent from this Riemannian space. Hartle & Hawking (1976) use this as a starting point for defining a natural Green function which, when analytically continued back to the pseudo-Riemannian Schwarzschild spacetime, coincides with the Green function associated with $|0_K\rangle$. There is thus an elegant connection established between the geometrical symmetry of the Riemannian manifold, as implemented by rotations in the $X - Y$ plane, and the thermal character of the Hawking radiation. Generalization to Kerr–Newman black holes is straightforward (Gibbons & Perry 1976, 1978).

The replacement of t by $i\tau$ in the above is reminiscent of the way in which the Feynman propagator is constructed in ordinary Minkowski space quantum field theory (see page 24). In §2.8 it was explained how the convergence problems of the path-integral approach to quantization are alleviated by the formal device of replacing t by $-i\tau$ in the functional integral (2.115). The exponent then becomes $-\hat{S}$ in place of iS, where $\hat{S}$ is the action in Euclidean spacetime. This treatment can be extended to the Riemannian case.

If the action $\hat{S}$ of the Riemannian manifold is used to construct the generating functional $Z[J]$, as the functional integral over fields which are periodic in the variable τ with period β, the Green functions obtained using (2.117) will not be the usual vacuum Green functions, but rather thermal Green functions at temperature $T = 1/k_B\beta$.

The connection between the path-integral approach and thermodynamics is strengthened by noting that the amplitude for field configuration ϕ_1 at time t_1 to propagate to ϕ_2 at time t_2 is given by

$$\langle \phi_2, t_2 | \phi_1, t_1 \rangle = \int \mathscr{D}[\phi] e^{iS[\phi]}, \tag{8.103}$$

where the integral is over all fields satisfying the given initial and final conditions. In the Schrödinger picture this can be written as

$$\langle \phi_2 | \exp[-\mathrm{i}H(t_2 - t_1)] | \phi_1 \rangle,$$

where H is the Hamiltonian. Putting $t_2 - t_1 = -\mathrm{i}\beta$ and $\phi_2 = \phi_1$, and summing over a complete set of field configurations gives

$$Z = \sum_n \mathrm{e}^{-\beta E_n},$$

where E_n is the energy of the field configuration ϕ_n. From (2.95), Z is identified as the thermodynamic partition function at temperature $T = 1/k_B\beta$ and zero chemical potential. On the other hand, from (8.103), we have

$$Z = \int \mathscr{D}[\phi] \mathrm{e}^{-\hat{S}[\phi]}, \tag{8.104}$$

where the integral is over all fields which are periodic in τ with periodicity β. Thus, using this path integral on the Riemannian manifold as a representation of the partition function, the thermodynamics of the system may be evaluated. (For a rigorous discussion of the connection between thermodynamics and path integrals in Minkowski space, see, for example, Ginibre 1971).

Gibbons & Hawking (1977*a*) have applied this result to the gravitational field itself (see also Hawking 1978). If the quantization of the gravitational field is carried out using the background field method (see chapter 1) the metric is written as

$$g_{\mu\nu} = g_{c\mu\nu} + \bar{g}_{\mu\nu}$$

where g_c is a solution of the classical Einstein equation and $\bar{g}$ is a field giving the quantum fluctuations about this background. Expanding the action in a functional Taylor series about the classical background as

$$\hat{S}[g] = \hat{S}[g_c] + S_2[\bar{g}] + \text{ higher order terms,}$$

where S_2 is quadratic in $\bar{g}$, the partition function is given by

$$\ln Z = -\hat{S}[g_c] + \ln \int \mathscr{D}[\bar{g}] \exp(-S_2[\bar{g}]) + \text{higher order terms.} \tag{8.105}$$

The first term on the right-hand side is the contribution of the classical gravitational field, while the second term gives the 'one-loop' correction due to the quantization of gravitons, and can be treated in the manner discussed for matter fields in chapter 6. The higher order terms give the contribution of graviton Feynman diagrams with more than one loop, and their meaningful evaluation is barred by the non-renormalizable nature of quantum gravity.

Ignoring for now all but the first (classical) term on the right-hand side of (8.105), the action is given by (see §6.6, setting $G = 1$ in the units of this chapter)

$$\hat{S} = -(1/16\pi)\int(R - 2\Lambda)g^{\frac{1}{2}}\,\mathrm{d}^4x - (1/8\pi)\int(\chi - \chi^0)(h)^{\frac{1}{2}}\mathrm{d}^3x, \quad (8.106)$$

where g is now the determinant of a Riemannian metric, h is the determinant of the induced metric on the boundary of the manifold, and χ and χ^0 are defined on page 224. Taking the background spacetime to be Schwarzschild spacetime, with Riemannianized metric (8.102), then $R = \Lambda = 0$ and $\hat{S}[g_c]$ reduces to $4\pi M^2$ (Gibbons & Hawking 1977*b*; see also Gibbons 1977*b*). Thus the classical contribution to the partition function is given by

$$\ln Z = -4\pi M^2, \quad (8.107)$$

from which the average value at temperature $T = 1/k_B\beta = 1/8\pi k_B M$ of the energy, $\langle E\rangle_\beta$, and the entropy, $\mathscr{S}$, are computed by the standard procedure (see, for example, Isihara 1971, §3.2)

$$\langle E\rangle_\beta = -\frac{\partial}{\partial\beta}\ln Z = M$$

$$S = k_B\beta\langle E\rangle_\beta + k_B\ln Z = 4\pi k_B M^2 = \tfrac{1}{4}k_B\mathscr{A},$$

which is identical to (8.63).

This result is all the more remarkable because it has been derived by appeal to the action of the classical *gravitational* field, not the quantum matter fields as before. Indeed, the fact that $\mathscr{S}$ is a purely *geometrical* quantity (the event horizon area) rather than dependent on the types of matter fields present, as is usual for entropy, indicates that the concept of black hole entropy is more fundamental than quantum field theory in curved spacetime, and is really an *intrinsic* quality of the black hole, a feature confirmed by the present treatment. This is the first time that the notion of objective, instrinsic entropy has appeared in physics. In the traditional treatment, entropy has an element of subjectivity associated with coarse-graining and our inability to distinguish between members of various classes of microstates.

The development of Riemannianized quantization of the gravitational field has led to a connection between black holes and the subject of instantons (see, for example, Hawking 1977*c*, Charap & Duff 1977, Gibbons & Hawking 1979; see also Gibbons & Pope 1978 for the extension of these ideas to other spacetimes), and a better understanding of the relationship between spacetime topology and quantum field theory (Hawking 1979). Similar techniques to those discussed in this section have been applied to de Sitter space (Gibbons & Hawking 1977*b*) and Robinson–Bertotti spacetime (Lapedes 1978*a*) where thermal effects also arise. Further discussion of these topics is beyond the scope of this book, and we refer the reader to the literature.

Some attention has also been given to the creation of particles near spacetime singularities. The intense quantum effects that accompany escalating spacetime curvature might be expected to produce a strong back-reaction that could be relevant to the question of cosmic censorship. Ford & Parker (1978) have investigated whether quantum back-reaction will disrupt the formation of an idealized naked singularity brought about by the implosion of a highly charged ($e^2 > M^2$) body. Quantum field theory near 'white holes' has also been discussed by Zel'dovich, Novikov & Starobinsky (1974) and Wald & Ramaswamy (1980).

If a singularity is allowed to form, the question of particle creation in its vicinity is complicated by the breakdown of predictability (Hawking & Ellis 1973). In particular, one does not know what boundary conditions to place on the quantum fields at the singularity. This problem occurs also in the cosmological case, as discussed in §7.4.

Hawking (1976*b*) has attempted to relate the randomness of black hole thermal emission to the boundary conditions at singularities by arguing that the creation of a particle pair just outside the horizon, with one particle going to infinity and its antiparticle 'tunnelling' into the hole eventually to strike the singularity, can be viewed in the spirit of Feynman's picture that treats an antiparticle as a particle travelling 'backwards in time'. The emitted particle can thereby be envisaged as originating on the singularity, travelling backwards in time to just outside the horizon, then scattering off the gravitational field into the future direction, to become part of the Hawking flux. In this way Hawking attributes the thermal character of the radiation to the singularity itself. Extending this idea to all singularities led to a new 'principle of ignorance' according to which the boundary conditions at a singularity should correspond to completely random influences emerging therefrom. This idea could have implications for cosmology as well as black holes.

9
Interacting fields

Once the theory of free quantum fields in curved spacetime had been worked out, the most natural extension was to include the effects of non-gravitational self and mutual interactions. Although this topic is still being developed, the basic framework is well established, and in this final chapter we outline the formal steps necessary for the computation of particle creation effects and the renormalization of $\langle T_{\mu\nu} \rangle$.

Two questions immediately spring to mind once interactions are included. The first is to what extent interactions can stimulate or inhibit particle creation by gravity over and above the free field case. Of course, interactions can lead to non-gravitational creation too, but we are more interested in processes that would be forbidden in Minkowski space, such as the simultaneous creation of a photon with an electron–positron pair.

The second question concerns renormalization theory. Will a field theory (e.g. Q.E.D.) that is renormalizable in Minkowski space remain so when the spacetime has a non-trivial topology or curvature? This question is of vital importance, for if a field theory is likely to lose its predictive power as soon as a small gravitational perturbation occurs, then its physical utility is suspect. It turns out to be remarkably difficult to establish general renormalizability, and significant progress has so far been limited to the so-called $\lambda\phi^4$ theory.

A third issue of great interest concerns black hole radiance. Is the Hawking flux precisely thermal even in the presence of field interactions? If not, a violation of the second law of thermodynamics seems possible. We discuss a particular model calculation in which the thermal character does indeed survive.

This chapter is intended only to introduce the reader to the topic of field interactions in curved space. No attempt is made at a comprehensive coverage, and only a few concrete examples are given.

9.1 Calculation of S-matrix elements

We consider a general interacting field theory defined by the Lagrangian density

$$\mathscr{L} = \mathscr{L}_0 + \mathscr{L}_I, \tag{9.1}$$

where $\mathscr{L}_0$ is a free field Lagrangian density as, for example, discussed in §§3.2 and 3.8, and $\mathscr{L}_I$ is an interaction Lagrangian density, containing field products of higher than quadratic order. We assume, as before, that the spacetime under consideration is globally hyperbolic, and that a choice of time parameter x^0 has been made.

It is further assumed that the interaction in (9.1) is switched off adiabatically in the distant past and future and that the fields, which will generically be denoted as ϕ, reduce to free fields ϕ_{in} and ϕ_{out} respectively in these regions:

$$\lim_{x^0 \to -\infty} \phi(x) = \phi_{\text{in}}(x), \tag{9.2}$$

$$\lim_{x^0 \to +\infty} \phi(x) = \phi_{\text{out}}(x), \tag{9.3}$$

where the limit is defined in the sense of weak operator convergence. These asymptotic conditions, which are necessary if particle states are to be defined, are justified in Minkowski space quantum field theory by consideration of a typical scattering situation in which the particles are initially and finally well separated, outside the effective range of interaction. It is only at intermediate times, as the particles approach one another, that the interaction becomes important. Such an argument can frequently be extended to curved spacetime. However, it is not difficult to envisage spacetimes in which the asymptotic conditions (9.2) and (9.3) cannot be justified. For example, in a spacetime with closed spatial sections, the particles can never be infinitely far apart, and the asymptotic conditions must be regarded as only approximate. In addition, it may be that the spacetime model of interest contains singularities which bound the space in the past or future, so that asymptotic time regions do no exist. In this case one may try appealing to other physical or mathematical criteria, or attempt an alternative formulation of quantum field theory (see, for example Kay 1980)

Since ϕ_{in} and ϕ_{out} are free fields, they can be treated in exactly the manner described in §3.2. In particular, they can be expanded in terms of complete sets of modes as in (3.30). As discussed extensively in chapter 3, the definition of positive frequency modes in the distant past and future need not necessarily agree, so let us assume that $u_{\mathbf{k}}^{\text{in}}$ and $u_{\mathbf{k}}^{\text{out}}$ are positive frequency modes in the far past and future, respectively, chosen, for example, using the adiabatic definition of positive frequency. Then ϕ_{in} and ϕ_{out} can be expanded in terms of either set of modes:

$$\phi_{\text{in}}(x) = \sum_i (a_i^{\text{in}} u_i^{\text{in}}(x) + a_i^{\text{in}\dagger} u_i^{\text{in}*}(x)) \tag{9.4a}$$

$$= \sum_i (\bar{a}_i^{\text{in}} u_i^{\text{out}}(x) + \bar{a}_i^{\text{in}\dagger} u_i^{\text{out}*}(x)) \tag{9.4b}$$

$$\phi_{\text{out}}(x) = \sum_i (a_i^{\text{out}} u_i^{\text{in}}(x) + a_i^{\text{out}\dagger} u_i^{\text{in}*}(x)) \tag{9.5a}$$

$$= \sum_i (\bar{a}_i^{\text{out}} u_i^{\text{out}}(x) + \bar{a}_i^{\text{out}\dagger} u_i^{\text{out}*}(x)). \tag{9.5b}$$

Since, in general, u_i^{in} will not be equal to u_i^{out}, but rather will be related by a Bogolubov transformation of the form (3.34), so too will a_i^{in} be related to $\bar{a}_i^{\text{in}}$, and a_i^{out} to $\bar{a}_i^{\text{out}}$ by Bogolubov transformations of the form (3.37). Thus, there will in general exist four inequivalent vacua defined by (cf. (2.19), (3.33))

$$a_i^{\text{in}}|0, \text{in}\rangle = 0, \quad \forall i, \tag{9.6}$$

$$\bar{a}_i^{\text{in}}|\bar{0}, \text{in}\rangle = 0, \quad \forall i, \tag{9.7}$$

$$a_i^{\text{out}}|0, \text{out}\rangle = 0, \quad \forall i, \tag{9.8}$$

$$\bar{a}_i^{\text{out}}|\bar{0}, \text{out}\rangle = 0, \quad \forall i, \tag{9.9}$$

each with associated Fock spaces. The Fock space based on $|0,\text{in}\rangle$ will be related to that based on $|\bar{0}, \text{in}\rangle$ by S-matrix elements such as those in (3.45), which can be calculated purely in terms of the Bogolubov coefficients. The $|0, \text{out}\rangle$ and $|\bar{0}, \text{out}\rangle$ based Fock spaces will also be related in such a manner. However the relationship between the Fock spaces based on $|0, \text{in}\rangle$ and $|0, \text{out}\rangle$ will depend on the interaction. If it is removed, these Fock spaces will be equivalent.

Since the vacuum $|0, \text{in}\rangle$ is defined with respect to modes that are positive frequency in the distant past where the full interacting field has the form ϕ_{in}, it defines the 'physical' Fock space in the far past. Similarly, the vacuum $|\bar{0}, \text{out}\rangle$ being defined with respect to modes which are positive frequency in the far future, where ϕ reduces to ϕ_{out}, defines the 'physical' Fock space in this region. It is the S-matrix elements relating the Fock space based on $|0, \text{in}\rangle$ to that based on $|\bar{0}, \text{out}\rangle$ that will describe the physics of the interactions of 'particles' in curved spacetime.

Notice that one is still faced with the problems discussed in chapter 3 of the meaning of particle states in curved spacetime, and it will only be in special cases, such as when the adiabatic definition of positive frequency, or conformal symmetry can be exploited, that the choice of u_i^{in} and u_i^{out} will be unambiguous.

Assuming that a definition of particle states has been chosen, then we are interested in computing the S-matrix elements (scattering amplitudes) between in and out states (see §3.2). An in state such as $|{}^1n_{i_1}, {}^2n_{i_2}, \ldots, \text{in}\rangle$ is likely to develop into an out state $|{}^1\bar{m}_{i_1}, {}^2\bar{m}_{i_2}, \ldots, \text{out}\rangle$ as a result of spacetime curvature and the interaction. It proves helpful to separate these two effects in the formalism that we shall give. This can be achieved by expanding the state $|{}^1\bar{m}_{i_1}, {}^2\bar{m}_{i_2}, \ldots, \text{out}\rangle$ in terms of a complete set of unbarred out states, after the fashion of (3.44). Then the relevant S-matrix element is

$$
\begin{aligned}
&\langle \text{out}, \ldots, {}^2\bar{m}_{i_2}, {}^1\bar{m}_{i_1} | {}^1n_{i_1}, {}^2n_{i_2}, \ldots, \text{in}\rangle \\
&= \sum_{k=0}^{\infty} \frac{1}{k!} \sum_{j_1 \ldots j_k} \langle \text{out}, \ldots, {}^2\bar{m}_{i_2}, {}^1\bar{m}_{i_1} | 1_{j_1}, 1_{j_2}, \ldots, 1_{j_k}, \text{out}\rangle \\
&\quad \times \langle \text{out}, 1_{j_k}, \ldots, 1_{j_2}, 1_{j_1} | {}^1n_{i_1}, {}^2n_{i_2}, \ldots, \text{in}\rangle. \qquad (9.10)
\end{aligned}
$$

Wald (1979a, b) discusses the existence of the S-matrix in curved spacetime. The probability for a transition between these particular in and out states is given by

$$
|\langle \text{out}, \ldots, {}^2\bar{m}_{i_2}, {}^1\bar{m}_{i_1} | {}^1n_{i_1}, {}^2n_{i_2}, \ldots, \text{in}\rangle|^2. \qquad (9.11)
$$

The amplitude $\langle \text{out}, \ldots, {}^2\bar{m}_{i_2}, {}^1\bar{m}_{i_1} | 1_{j_1}, 1_{j_2}, \ldots, 1_{j_k}, \text{out}\rangle$ is given entirely in terms of Bogolubov coefficients (see (3.44)), and is independent of details of the interaction. On the other hand, the amplitude $\langle \text{out}, 1_{j_k}, \ldots, 1_{j_2}, 1_{j_1} | {}^1n_{i_1}, {}^2n_{i_2}, \ldots, \text{in}\rangle$ depends on the interaction, but is an S-matrix element formed from elements of two Fock spaces defined using the same definition (u_i) of positive frequency modes. It can thus be calculated using methods familiar in Minkowski space quantum field theory. We shall now discuss two such methods.

One technique is the so-called LSZ (Lehmann, Symanzik & Zimmermann 1955) method of reducing S-matrix elements to expressions given in terms of Green functions. The derivation of these reduction formulae proceeds essentially as in Minkowski space (see, for example Bjorken & Drell 1965, §16.7). In particular, for a self-interacting scalar field with $\mathscr{L}_0$ given by (3.24), one obtains (cf. Bjorken & Drell, equation (16.81))

$$
\begin{aligned}
&\langle \text{out}, 1_{p_1}, \ldots, 1_{p_l} | 1_{q_1}, \ldots, 1_{q_m}, \text{in}\rangle / \langle \text{out}, 0 | 0, \text{in}\rangle \\
&= \mathrm{i}^{m+l} \prod_{i=1}^{m} \int \mathrm{d}^n x_i [-g(x_i)]^{\frac{1}{2}} \prod_{j=1}^{l} \int \mathrm{d}^n y_j [-g(y_j)]^{\frac{1}{2}} u_{q_i}(x_i) u^*_{p_j}(y_j) \\
&\quad \times [\Box_{x_i} + m^2 + \xi R(x_i)][\Box_{y_j} + m^2 + \xi R(y_j)] \tau(y_1 \ldots y_l, x_1 \ldots x_m), \qquad (9.12)
\end{aligned}
$$

where it has been assumed that $p_i \neq q_j, \forall i,j$, and we have defined the Green function

$$\tau(x_1, x_2 \ldots x_m) = \frac{\langle \text{out}, 0| T(\phi(x_1)\phi(x_2)\ldots\phi(x_m))|0, \text{in}\rangle}{\langle \text{out}, 0|0, \text{in}\rangle}. \tag{9.13}$$

Details of the derivation of equations such as (9.12) are given in Birrell (1979*c*) and Birrell & Taylor (1980), where reduction formulae for the full amplitude (9.11) in terms of $\langle \text{out}, \bar{0}|\ldots|0, \text{in}\rangle$ Green functions and Bogolubov coefficients have been given.

There are two points in the derivation of (9.12) which warrant special mention. The first is that, unlike the case of Minkowski space where $|0, \text{out}\rangle = |0, \text{in}\rangle$ (up to a phase factor) the vacuum $|0, \text{in}\rangle$ in curved spacetime will not in general be stable: $\langle \text{out}, 0|0, \text{in}\rangle \neq 1$. This possible instability is not due to different definitions of positive frequency ($|0, \text{out}\rangle$ and $|0, \text{in}\rangle$ are based on the same definition), but rather to the lack of Poincaré invariance in curved spacetime, which can give rise to particle production additional to that discussed in chapter 3. This topic will form the subject of §9.3.

The second comment to be made on the derivation of (9.12) concerns the fact that in arriving at this equation Gauss' theorem has been used, with surface terms at spacelike infinity being discarded. If the spacetime has spacelike boundaries, it will be necessary to retain these surface terms and apply physically motivated boundary conditions to them.

The Green functions (9.13) must now be computed. Exact expressions are almost impossible to obtain in a practical calculation, and some approximation scheme, such as perturbation theory, must be used. This may be implemented by first constructing the so-called evolution matrix U (Dyson 1949; see, for example, Bjorken & Drell 1965, chapter 17), which leads to a perturbation series for S which is essentially the same as the interaction picture approach to be described shortly.

An alternative technique is to use the path-integral formulation (§2.8), and work with the Green function generating functional which in the interacting case is given by

$$Z[J] = \int \mathscr{D}[\phi] \exp\left\{ i \int \mathscr{L}_I[\phi] d^n x \right\} \exp\left\{ i \int [\mathscr{L}_0[\phi] + J\phi] d^n x \right\}. \tag{9.14}$$

If the interaction Lagrangian density is a polynomial in the fields, then (9.14) can be rewritten as

$$Z[J] = \exp\left\{ i \int \mathscr{L}_I \left[\frac{1}{i} \frac{\delta}{\delta J} \right] d^n x \right\} \int \mathscr{D}[\phi] \exp\left\{ i \int [\mathscr{L}_0[\phi] + J\phi] d^n x \right\}. \tag{9.15}$$

The functional integral in (9.15) is now simply the free generating functional, which we shall denote by $Z_0[J]$. Thus

$$Z[J]=\exp\left\{\mathrm{i}\int\mathscr{L}_{\mathrm{I}}\left[\frac{1}{\mathrm{i}}\frac{\delta}{\delta J}\right]\mathrm{d}^n x\right\}Z_0[J]. \tag{9.16}$$

For scalar fields, Z_0 is given by (2.126), which also holds in curved spacetime. Hence, in this case,

$$Z[J]\propto[\det(-G_{\mathrm{F}})]^{\frac{1}{2}}\exp\left\{\mathrm{i}\int\mathscr{L}_{\mathrm{I}}\left[\frac{1}{\mathrm{i}}\frac{\delta}{\delta J(x)}\right]\mathrm{d}^n x\right\}$$
$$\times\exp\left\{-\tfrac{1}{2}\mathrm{i}\int J(y)G_{\mathrm{F}}(y,z)J(z)\mathrm{d}^n y\,\mathrm{d}^n z\right\} \tag{9.17}$$

where the proportionality constant is independent of the metric and J, and does not affect the Green functions. By expanding the exponentials in (9.17) and performing the functional differentiations one obtains a perturbation expansion for Z entirely in terms of the free field Feynman propagator. This expansion can then be used in (2.117) to generate expansions for the Green functions.

It is important to note that the boundary conditions that are encoded in the interacting Green functions generated by (9.17) will depend on the choice of Feynman propagator G_{F}. Since we wish to generate the Green functions (9.13), which are defined with respect to only one definition of positive frequency modes u_i, the associated G_{F} must be calculated from these modes:

$$\mathrm{i}G_{\mathrm{F}}(x,y)=\theta(x^0-y^0)\sum_i u_i(x)u_i^*(y)+\theta(y^0-x^0)\sum_i u_i^*(x)u_i(y) \tag{9.18}$$

(cf. (7.74)). With this choice we have

$$\tau_{\mathrm{c}}(x_1,x_2,\ldots,x_m)=\mathrm{i}^{-m}\left[\frac{\delta^m\ln Z[J]}{\delta J(x_1)\delta J(x_2)\ldots\delta J(x_m)}\right]_{J=0}, \tag{9.19}$$

where the subscript 'c' on τ_{c} means that only connected Feynman diagrams will be generated by the use of the perturbation expansion (9.17) in (9.19), as may be explicitly verified for specific interaction Lagrangians. We shall consider a particular example in the next section.

One method of studying interacting quantum fields in curved spacetime which has been used for some time (Utiyama 1962, Freedman & Pi 1975), is the expansion of the generating functional (or Green functions) in a functional Taylor series (Volterra series) in the metric about Minkowski space. Each term in this series can then be examined using ordinary Minkowski space perturbation theory in the interaction coupling constant.

Rather than expand $Z[J]$ itself, it is more convenient, in practice, to expand $\ln Z$, which appears in (9.19). Writing (cf. (6.25))

$$W[J;\, g_{\mu\nu}] = -\mathrm{i}\ln Z[J;g_{\mu\nu}], \tag{9.20}$$

where

$$g_{\mu\nu}(x) = \eta_{\mu\nu} + h_{\mu\nu}(x) \tag{9.21}$$

we can expand W as a Volterra series (see, for example, Rzewuski 1969, §I, 2):

$$W[J;g_{\mu\nu}] = \sum_{m=0}^{\infty} \frac{1}{m!}\left(\prod_{i=1}^{m} \mathrm{d}^n x_i h^{\mu_i\nu_i}(x_i)\, W_{\mu_1\nu_1,\mu_2\nu_2,\ldots,\mu_m\nu_m}[J;x_1,x_2,\ldots,x_m]\right), \tag{9.22}$$

with

$$W_{\mu_1\nu_1,\mu_2\nu_2,\ldots,\mu_m\nu_m}[J;x_1,x_2,\ldots,x_m] = \left[\frac{\delta^m W[J;g_{\mu\nu}]}{\delta g^{\mu_1\nu_1}(x_1)\delta g^{\mu_2\nu_2}(x_2)\ldots\delta g^{\mu_m\nu_m}(x_m)}\right]_{g^{\mu_i\nu_i}=\eta^{\mu_i\nu_i}}. \tag{9.23}$$

Setting $J=0$ in (9.22) gives an expansion for the effective action for the interacting theory, and we see from (6.13) that the first order term in this expansion involves

$$W_{\mu_1\nu_1}[0] = \langle 0|T_{\mu_1\nu_1}|0\rangle, \tag{9.24}$$

the vacuum expectation value of the Minkowski space stress-tensor for the interacting theory. This quantity has been studied by several authors (Callan, Coleman & Jackiw 1970, Freedman, Muzinich & Weinberg 1974, Freedman & Weinberg 1974, Collins 1976). The higher order terms are more complicated, involving not only terms such as $\langle 0|T_{\mu_1\nu_1}T_{\mu_2\nu_2}\ldots T_{\mu_m\nu_m}|0\rangle$, but also the so-called 'seagull' contributions arising from the functional differentiation of $T_{\mu\nu}[g_{\mu\nu}]$ (Freedman & Pi 1975).

The disadvantages of treating interacting theories in curved spacetime by this method are three-fold: (i) Calculationally it is only useful for small deviations from Minkowski space. (ii) It gives no information about the effects of spacetime topology (see, for example, Drummond & Hathrell 1980). (iii) It is not possible, without great difficulty, to prescribe the vacuum state which is used in the calculation. We shall not, therefore, pursue this approach here, but, when using the path-integral formulation, shall consider only perturbation series in the interaction coupling strength, as given by (9.17).

As already mentioned, there is a second technique, which can easily be adapted from Minkowski space, for the calculation of the amplitudes $\langle \text{out}, {}^k j_k, \ldots, {}^2 j_2, {}^1 j_1 | {}^1 n_{i_1}, {}^2 n_{i_2}, \ldots, \text{in} \rangle$ appearing in (9.11). This is the interaction picture approach in which the fields satisfy the free equations, and the dynamical information is carried by the states of the system. (See, for example, Roman 1969, chapter 4, for a covariant treatment in Minkowski space. Formulation of the interaction picture method in curved spacetime has been given by Birrell & Ford 1979, and Bunch, Panangaden & Parker 1980.)

In the interaction picture, the states $|\psi\rangle$ satisfy the Schrödinger equation

$$\mathscr{H}_{\rm I}(x)|\psi[\Sigma]\rangle = \mathrm{i}\frac{\delta|\psi[\Sigma]\rangle}{\delta\Sigma(x)}, \tag{9.25}$$

where $\mathscr{H}_{\rm I}$ is the interaction Hamiltonian density, which, for non-derivative coupling, is given by

$$\mathscr{H}_{\rm I} = -\mathscr{L}_{\rm I}, \tag{9.26}$$

and $\Sigma(x)$ is a spacelike Cauchy hypersurface through x. (For constant time slicing, (9.25) becomes

$$\mathscr{H}_{\rm I}(x)|\psi(x^0)\rangle = \mathrm{i}\frac{\partial|\psi(x^0)\rangle}{\partial x^0}.) \tag{9.27}$$

The solution to equation (9.25) is determined in terms of a unitary operator U, defined by

$$|\psi[\Sigma]\rangle = U[\Sigma, \Sigma_0]|\psi[\Sigma_0]\rangle, \tag{9.28}$$

which, using (9.25), is seen to satisfy the so-called Tomonaga–Schwinger equation (Tomonaga 1946, Schwinger 1948, 1949*a*, *b*)

$$\mathscr{H}_{\rm I}(x)U[\Sigma, \Sigma_0] = \mathrm{i}\frac{\delta U[\Sigma, \Sigma_0]}{\delta\Sigma(x)}, \tag{9.29a}$$

with initial condition

$$U[\Sigma_0, \Sigma_0] = 1. \tag{9.29b}$$

Writing (9.29) as an integral equation

$$U[\Sigma, \Sigma_0] = 1 - \mathrm{i}\int_{\Sigma_0}^{\Sigma} \mathscr{H}_{\rm I}(x')U[\Sigma', \Sigma_0]\mathrm{d}^n x', \tag{9.30}$$

the following closed form solution is obtained by iteration:

$$U[\Sigma, \Sigma_0] = P\exp[-\mathrm{i}\int_{\Sigma_0}^{\Sigma} \mathscr{H}_{\mathrm{I}}(x')\mathrm{d}^n x'], \tag{9.31}$$

where the ordering symbol P is the same as the time-ordering symbol T, except that no sign changes are made under transpositions in the case of fermion fields.

A theorem due to Haag (1955) implies that U cannot be a well-defined operator on the Hilbert space of states (see, for example, Roman 1969, §8.4), so we consider (9.31) as purely a formal expression for the purpose of computing the S-matrix operator defined by

$$S = U[\Sigma^{\mathrm{out}}, \Sigma^{\mathrm{in}}]. \tag{9.32}$$

The surface Σ^{out} lies in the out region in the infinitely far future, while Σ^{in} is a surface in the in region in the infinitely remote past. Then, from (9.28), one has

$$|\psi[\Sigma^{\mathrm{out}}]\rangle = S|\psi[\Sigma^{\mathrm{in}}]\rangle, \tag{9.33}$$

and, from (9.31), S possesses the perturbation expansion

$$S = \sum_{m=0}^{\infty} S^{(m)} \tag{9.34}$$

with

$$S^{(0)} = 1 \tag{9.35}$$

$$S^{(m)} = \frac{(-\mathrm{i})^m}{m!}\int P(\mathscr{H}_{\mathrm{I}}(x_1)\mathscr{H}_{\mathrm{I}}(x_2)\ldots\mathscr{H}_{\mathrm{I}}(x_m))\,\mathrm{d}^n x_1\,\mathrm{d}^n x_2\ldots\mathrm{d}^n x_m, \tag{9.36}$$

Let us now choose $|\psi[\Sigma^{\mathrm{in}}]\rangle$ to be a member of the Fock space based on the Heisenberg vacuum $|0, \mathrm{in}\rangle$ (see(9.6)). This immediately implies that on the surface Σ^{in}, the interaction picture field ϕ, and the Heisenberg picture field ϕ must be equal. But on Σ^{in}, the Heisenberg field reduces to the free field ϕ_{in} (see(9.2)). So therefore does the interaction picture field. Since the interaction picture field obeys the free field equation, it follows that it will be equal to ϕ_{in} for all time.

Consider, in particular, the choice of state

$$|\psi[\Sigma^{\mathrm{in}}]\rangle = |1_{j_1}, 1_{j_2}, \ldots, 1_{j_k}, \mathrm{in}\rangle. \tag{9.37}$$

Then (9.33) becomes

$$|1_{j_1}, 1_{j_2}, \ldots, 1_{j_k}, \mathrm{out}\rangle = S|1_{j_1}, 1_{j_2} \ldots . 1_{j_k}, \mathrm{in}\rangle. \tag{9.38}$$

Note that, because S represents only the effects of interaction, $|\psi[\Sigma^{\text{out}}]\rangle$ as given by the left-hand side of (9.38), is defined in terms of the unbarred out states. We therefore possess a second means of computing that part of the amplitude (9.10) independent of the Bogolubov transformation. From (9.38):

$$\langle \text{out}, 1_{j_k}, \ldots, 1_{j_2}, 1_{j_1} | {}^1n_{i_1}, {}^2n_{i_2}, \ldots, \text{in} \rangle$$
$$= \langle \text{in}, 1_{j_k}, \ldots 1_{j_2}, 1_{j_1} | S^\dagger | {}^1n_{i_1}, {}^2n_{i_2}, \ldots, \text{in} \rangle. \tag{9.39}$$

The right-hand side of (9.39) can be calculated to any order of perturbation theory, by using (9.34), with $\mathscr{H}_\text{I}$ in (9.36) constructed from ϕ_{in} as given by (9.4).

Whether one uses the interaction picture approach or reduction formulae and path-integral quantization, as in Minkowski space, one encounters infinities which must be removed by renormalization. The application of the computational methods defined above, and the regularization and renormalization of the resulting infinite quantities are best described by a specific example, which we shall now give.

9.2 Self-interacting scalar field in curved spacetime

We consider a self-interacting scalar field theory with Lagrangian density (9.1); $\mathscr{L}_0$ being given by (3.24), and $\mathscr{L}_\text{I}$ by

$$\mathscr{L}_\text{I} = -\frac{1}{4!}(-g)^{\frac{1}{2}}\lambda\phi^4. \tag{9.40}$$

This is the so-called $\lambda\phi^4$ theory.

Of particular interest are amplitudes representing scattering from an initial vacuum to a final many-particle state, since these processes cannot occur in Minkowski space. The examination of such amplitudes is sufficient to demonstrate the renormalization techniques required in the calculation of other amplitudes. Setting $|{}^1n_{i_1}, {}^2n_{i_2} \ldots, \text{in}\rangle = |0, \text{in}\rangle$ in (9.10), one obtains

$$\langle \text{out}, \ldots, {}^2\bar{m}_{i_2}, {}^1\bar{m}_{i_1} | 0, \text{in} \rangle$$
$$= \sum_{k=0}^{\infty} \frac{1}{k!} \sum_{j_1 \ldots j_k} \langle \text{out}, \ldots, {}^2\bar{m}_{i_2}, {}^1\bar{m}_{i_1} | 1_{j_1}, 1_{j_2}, \ldots, 1_{j_k}, \text{out} \rangle$$
$$\times \langle \text{out}, 1_{j_k}, \ldots, 1_{j_2}, 1_{j_1} | 0, \text{in} \rangle. \tag{9.41}$$

The first amplitude on the right-hand side is easily constructed from Bogolubov coefficients and, in particular, is finite. We thus consider the first few contributions from the second amplitude.

To illustrate in detail the two calculational methods described in the previous section, we shall first describe the perturbation theory calculation of $\langle \text{out}, 1_{p_1}, 1_{p_2}|0, \text{in}\rangle$. Starting with the LSZ method, we note that the reduction formula (9.12) gives for this amplitude

$$\frac{\langle \text{out}, 1_{p_1}, 1_{p_2}|0, \text{in}\rangle}{\langle \text{out}, 0|0, \text{in}\rangle}$$

$$= \int \mathrm{d}^n y_1\, \mathrm{d}^n y_2 [-g(y_1)]^{\frac{1}{2}} [-g(y_2)]^{\frac{1}{2}} u^*_{p_1}(y_1) u^*_{p_2}(y_2)$$

$$\times \mathrm{i}[\Box_{y_1} + m^2 + \xi R(y_1)]\mathrm{i}[\Box_{y_2} + m^2 + \xi R(y_2)]\tau(y_1, y_2). \quad (9.42)$$

We thus need to calculate the Green function $\tau(y_1, y_2)$, which is also known as the complete propagator (for the interacting theory).

Substituting (9.40) into (9.17), and expanding the exponentials, it is not difficult to compute, to a given order in λ, the terms which will contribute only two factors of J to the expression of $\ln Z[J]$. It is these terms which will give non-vanishing contributions to

$$\tau_c(y_1, y_2) = -\left[\frac{\delta^2 \ln Z[J]}{\delta J(y_1) J(y_2)}\right]_{J=0} \quad (9.43)$$

(see (9.19)). Disconnected contributions to τ can be calculated by simply combining together connected components, and we shall not discuss them further except to say that the disconnected components will contribute to the total amplitude.

To first order in λ, use of (9.17) and (9.43) gives

$$\tau_c(y_1, y_2) = \mathrm{i}G_F(y_1, y_2)$$

$$-\tfrac{1}{2}\lambda \int G_F(y_1, x) G_F(x, x) G_F(x, y_2)[-g(x)]^{\frac{1}{2}} \mathrm{d}^n x \quad (9.44)$$

which reduces to $\mathrm{i}\, G_F(y_1, y_2)$ when $\lambda = 0$, as one would expect (cf. (9.13) and (2.69)). It is convenient to introduce Feynman diagrams to represent expressions such as (9.44). Representing λ by a vertex in the diagram, and $\mathrm{i}G_F(x, y)$ by a line, the diagrammatic form of (9.44) is given in fig. 29, where integration over the point at a vertex is understood.

The quantity involving the Green function that actually appears in the reduction formula (9.42) is

$$\gamma(y_1, y_2) \equiv \mathrm{i}[\Box_{y_1} + m^2 + \xi R(y_1)]\mathrm{i}[\Box_{y_2} + m^2 + \xi R(y_2)]\tau(y_1, y_2). \quad (9.45)$$

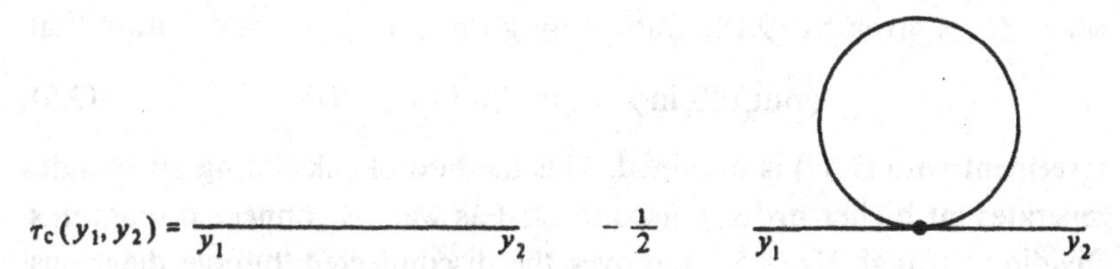

Fig. 29. Feynman diagram of the zeroth and first order contributions to the complete, connected propagator $\tau_c(y_1, y_2)$ for $\lambda\phi^4$ theory.

Substituting (9.44) into (9.45) and using (3.49), one obtains

$$\gamma_c(y_1, y_2) = \mathrm{i}K_{y_1 y_2} + \tfrac{1}{2}\lambda G_F(y_1, y_1)\delta(y_1 - y_2)/[-g(y_1)]^{\frac{1}{2}}, \quad (9.46)$$

where K is defined by (6.21) or (6.23). The function γ is known as the *amputated Green function*, because its diagrammatic representation is obtained from that of τ_c in fig. 29 by amputation of the external 'legs'. Finally, substituting (9.46) into (9.42), one finds

$$\frac{\langle \text{out}, 1_{p_1}, 1_{p_2}|0, \text{in}\rangle}{\langle \text{out}, 0|0, \text{in}\rangle} = \tfrac{1}{2}\lambda \int u^*_{p_1}(y)u^*_{p_2}(y)G_F(y, y)[-g(y)]^{\frac{1}{2}}\mathrm{d}^n y + O(\lambda^2). \quad (9.47)$$

The first term in (9.46) does not contribute to (9.47) because the modes satisfy the free field equation.

Before discussing this result, higher order corrections and other amplitudes, we show how it can be reproduced using the interaction picture approach. One starts from (9.39), which in this case gives

$$\langle \text{out}, 1_{p_1}, 1_{p_2}|0, \text{in}\rangle = \langle \text{in}, 1_{p_1}, 1_{p_2}|S^\dagger|0, \text{in}\rangle, \quad (9.48)$$

and, to obtain a result to first order in λ, one substitutes the first two terms of (9.34) for S. The first term does not contribute, and one is left with (using (9.26) and (9.40))

$$\langle \text{out}, 1_{p_1}, 1_{p_2}|0, \text{in}\rangle = -\frac{\mathrm{i}\lambda}{4!}\int \langle \text{in}, 1_{p_1}, 1_{p_2}|T(\phi^4(x))|0, \text{in}\rangle[-g(x)]^{\frac{1}{2}}\mathrm{d}^n x + O(\lambda^2). \quad (9.49)$$

The matrix element in the integrand of (9.49) can be evaluated by substituting (9.4*a*) for ϕ (recall $\phi = \phi_{\text{in}}$ in the interaction picture), or, alternatively, using Wick's theorem (Wick 1950; see, for example, Roman 1969, chapter 4). Either way one obtains

$$\langle \text{in}, 1_{p_1}, 1_{p_2}|T(\phi^4(x))|0, \text{in}\rangle = 12u^*_{p_1}(x)u^*_{p_2}(x)\mathrm{i}G_F(x, x)\langle \text{in}, 0|0, \text{in}\rangle, \quad (9.50)$$

where G_F is given by (9.18). Substituting this into (9.49) and noting that

$$\langle \text{out}, 0|0, \text{in}\rangle = \langle \text{in}, 0|0, \text{in}\rangle + O(\lambda), \tag{9.51}$$

agreement with (9.47) is obtained. This method of calculating amplitudes generates (at higher orders) disconnected as well as connected diagrams. Dividing through by (9.51) removes the disconnected 'bubble diagrams' with no external legs (see, for example, Bjorken & Drell 1965, §17.6; or Schweber 1961, §14b).

Inspection of (9.47) immediately reveals a problem; $G_F(y, y)$ is infinite as $n \to 4$. Indeed, we have kept n arbitrary in (9.47) in anticipation of the need to regularize the expression; a procedure most easily carried out using dimensional regularization. The nature of the pole terms can be determined using the DeWitt–Schwinger expansion for G_F. From (3.138), or (6.29), one obtains

$$G_F(y, y) \approx -\frac{i}{(4\pi)^{n/2}} \sum_{j=0}^{\infty} (m^2)^{\frac{1}{2}n-j-1} a_j(y) \Gamma(j - \tfrac{1}{2}n + 1)$$

$$= -\frac{2i}{(4\pi)^2} \frac{[m^2 + (\xi - \frac{1}{6})R]}{(n-4)} + G_F^{\text{finite}}(y, y) \tag{9.52}$$

where $G_F^{\text{finite}}(y, y)$ is finite as $n \to 4$. The pole terms must be removed by renormalization of constants in the Lagrangian, as in Minkowski space theory. However, there is an evident difference, i.e., the appearance of a pole term proportional to the Ricci scalar R. We shall see that this extra pole term can be removed by renormalization of the constant ξ, which is absent in Minkowski space theory.

A method of systematically performing renormalization in dimensional regularization has been devised by 't Hooft (1973) and studied by Collins & Macfarlane (1974), and Collins (1974), who carries out the explicit second order renormalization of $\lambda\phi^4$ in Minkowski space. (See also the textbook treatment of Nash 1978.) The total Lagrangian (9.1) is

$$\mathscr{L} = \tfrac{1}{2}(-g)^{\frac{1}{2}}[g^{\mu\nu}\phi_{,\mu}\phi_{,\nu} - (m_R^2 + \xi_R R(x))\phi_R^2]$$

$$-(-g)^{\frac{1}{2}}\left[\frac{1}{4!}\lambda\phi^4 + \tfrac{1}{2}(\delta m^2 + \delta\xi R)\phi^2\right], \tag{9.53}$$

where m_R and ξ_R are 'renormalized constants' related to the 'bare constants' m and ξ by

$$m_R^2 = m^2 - \delta m^2 \tag{9.54}$$

$$\xi_R = \xi - \delta\xi. \tag{9.55}$$

For convenience we have not used a subscript B for the bare quantities. We now treat the first term in (9.53) involving renormalized quantities as a free Lagrangian density $\mathscr{L}_0$, and the second term as a new interaction Lagrangian density. Our aim is to cancel pole terms appearing in the S-matrix by appropriate choices of δm^2 and $\delta\xi$. In anticipation of this we write

$$\delta m^2 = m_R^2 \sum_{\nu=1}^{\infty} \sum_{j=\nu}^{\infty} b_{\nu j}\lambda_R^j (n-4)^{-\nu} \equiv m_R^2 \sum_{\nu=1}^{\infty} b_\nu(\lambda_R)(n-4)^{-\nu} \tag{9.56}$$

$$\delta\xi = \sum_{\nu=1}^{\infty} \sum_{j=\nu}^{\infty} d_{\nu j}\lambda_R^j (n-4)^{-\nu} \equiv \sum_{\nu=1}^{\infty} d_\nu(\lambda_R)(n-4)^{-\nu} \tag{9.57}$$

where we also allow for the possibility of coupling constant renormalization by writing

$$\lambda = \lambda_R + \delta\lambda = \mu^{4-n}\left[\lambda_R + \sum_{\nu=1}^{\infty} \sum_{j=\nu}^{\infty} a_{\nu j}\lambda_R^j (n-4)^{-\nu}\right]$$

$$= \mu^{4-n}\left[\lambda_R + \sum_{\nu=1}^{\infty} a_\nu(\lambda_R)(n-4)^{-\nu}\right]. \tag{9.58}$$

As in §6.2, an arbitrary mass scale μ has been introduced into (9.58) so as to maintain the dimensionless nature of the total action.

One must also take into account a possible renormalization of the fields ('wavefunction' renormalization) in which

$$\phi \to \phi_R = Z^{-\frac{1}{2}}\phi, \tag{9.59}$$

where

$$Z = 1 + \sum_{\nu=1}^{\infty} \sum_{j=\nu}^{\infty} c_{\nu j}\lambda_R^j (n-4)^{-\nu} \equiv \sum_{\nu=1}^{\infty} c_\nu(\lambda_R)(n-4)^{-\nu}. \tag{9.60}$$

The theory is renormalizable if all pole terms in S-matrix elements can be removed by appropriate choices of $a_{\nu j}, b_{\nu j}, c_{\nu j}$ and $d_{\nu j}$.

If the second term in (9.53) is used as the interaction Lagrangian i.e.,

$$\mathscr{L}_1 = -(-g)^{\frac{1}{2}}\left[\frac{1}{4!}\lambda\phi^4 + \tfrac{1}{2}(\delta m^2 + \delta\xi R)\phi^2\right], \tag{9.61}$$

then the amputated Green function γ_c is easily computed to be

$$\gamma_c(y_1, y_2) = \mathrm{i}K_{y_1y_1} + \lambda_R\Big\{\tfrac{1}{2}\mu^{4-n}G_F(y_1, y_1)$$
$$- \mathrm{i}\sum_{\nu=1}^{\infty}[m_R^2 b_{\nu 1} + d_{\nu 1}R(y_1)](n-4)^{-\nu}\Big\}$$
$$\times[-g(y_1)]^{-\frac{1}{2}}\delta^n(y_1 - y_2)] + O(\lambda_R^2), \tag{9.62}$$

in place of (9.46). In obtaining this expression we have used (9.56)–(9.58), and have only retained terms up to order λ_R. All quantities in (9.62) are expressed in terms of the renormalized constants m_R^2 and ξ_R, rather than the bare parameters as before. Since $\tau_c(y_1, y_2)$ is the vacuum expectation value of two fields ϕ (see (9.13)), we expect to have to renormalize the fields by multiplying τ_c by Z^{-2}. However, in forming the amputated Green function, two factors of G_F, and thus a factor Z^{-4}, are removed. It follows that (9.62) should be multiplied by Z^2. The only contribution at order λ_R to S-matrix elements that this multiplication will make comes from the first term on the right in (9.62), resulting in

$$\lambda_R \sum_{\nu=1}^{\infty} c_{\nu 1}(n-4)^{-\nu}\mathrm{i}K_{y_1y_2}$$
$$= \mathrm{i}\lambda_R \sum_{\nu=1}^{\infty} c_{\nu 1}(n-4)^{-\nu}[\Box_{y_1} + m_R^2 + \xi_R R(y_1)][-g(y_1)]^{-\frac{1}{2}}\delta^n(y_1 - y_2). \tag{9.63}$$

Substituting (9.52) (with the replacements $m \to m_R$, $\xi \to \xi_R$) into (9.62), one observes that all of the pole terms in $\gamma_c(y_1, y_2)$ can be removed by the choice

$$\left.\begin{aligned} &c_{\nu 1} = 0,\ \forall\nu; \quad d_{\nu 1} = b_{\nu 1} = 0,\ \forall\nu \neq 1; \\ &b_{11} = -1/16\pi^2; \quad d_{11} = -(\xi_R - \tfrac{1}{6})/16\pi^2. \end{aligned}\right\} \tag{9.64}$$

With this choice, the use of (9.62) in the reduction formulae now yields a finite amplitude as $n \to 4$:

$$\left[\frac{\langle \mathrm{out}, 1_{p_1}, 1_{p_2}|0, \mathrm{in}\rangle}{\langle \mathrm{out}, 0|0, \mathrm{in}\rangle}\right]_{\mathrm{ren}}$$
$$= \tfrac{1}{2}\lambda_R \int u_{p_1}^*(y)u_{p_2}^*(y)\{G_F^{\mathrm{finite}}(y, y)$$
$$+ (\mathrm{i}/8\pi^2)[m_R^2 + (\xi_R - \tfrac{1}{6})R(y)]\ln\mu\}[-g(y)]^{\frac{1}{2}}\mathrm{d}^4y. \tag{9.65}$$

The appearance of the arbitrary mass scale μ in this amplitude expresses the renormalization ambiguity inherent in any renormalization scheme. Rescaling of μ readjusts the relationship between the bare and renormalized constants, and reflects the fact that measurements must be made to fix the values of λ_R, m_R and ξ_R. (See the discussion on page 163, where this point was discussed in connection with stress-tensor renormalization. We shall return to this issue below.)

Evaluation of other many-particle amplitudes proceeds similarly (Birrell 1980, Birrell & Ford 1980, Bunch, Panangaden & Parker 1980, Bunch & Panangaden 1980, Bunch & Parker 1979, Bunch 1980*b*). Of these only the four-particle amplitude is divergent, and its renormalization along the above lines fixes the coefficients $a_{\nu j}$ in (9.58). Therefore, the renormalizations of the two- and four-particle amplitudes have fixed all the coefficients in (9.56)–(9.58) and (9.60), and hence in the scalar field Lagrangian. However, as we shall see, it turns out that there exists a divergence in the vacuum-to-vacuum amplitude also.

To see how this further divergence can be removed by renormalization, recall that the vacuum-to-vacuum amplitude is related to the effective action $W[g_{\mu\nu}] = W[0; g_{\mu\nu}]$(see (9.20)) by

$$W[g_{\mu\nu}] = -\mathrm{i}\ln(\langle \mathrm{out}, 0|0, \mathrm{in}\rangle) = -\mathrm{i}\ln Z[0]. \tag{9.66}$$

Hence one expects that this additional divergence can be removed by renormalization of constants in the generalized Einstein action, just as in the case of the free field effective action discussed in §6.2. That this is indeed the case has been verified to second order in λ_R by Bunch & Panangaden (1980), while similar renormalizations have been found necessary in massless quantum electrodynamics in de Sitter space (Shore 1980*a*, *b*; see also Drummond & Hathrell 1980, Brown & Collins 1980).

We treat here the renormalization of the vacuum-to-vacuum amplitude to first order in λ_R. Working in the interaction picture, we have

$$\begin{aligned}\langle \mathrm{out}, 0|0, \mathrm{in}\rangle &= \langle \mathrm{in}, 0|S|0, \mathrm{in}\rangle \\ &= \langle \mathrm{in}, 0|0, \mathrm{in}\rangle + \langle \mathrm{in}, 0|S^{(1)}|0, \mathrm{in}\rangle + O(\lambda_R^2),\end{aligned} \tag{9.67}$$

where we have used (9.34). Substituting this into (9.66) and expanding the logarithm gives

$$W[g_{\mu\nu}] = -\mathrm{i}\ln(\langle \mathrm{in}, 0|0, \mathrm{in}\rangle) - \mathrm{i}\frac{\langle \mathrm{in}, 0|S^{(1)}|0, \mathrm{in}\rangle}{\langle \mathrm{in}, 0|0, \mathrm{in}\rangle} + O(\lambda_R^2). \tag{9.68}$$

Notice that we have not assumed that $\langle \mathrm{in}, 0|0, \mathrm{in}\rangle$ is normalized to unity. This is merely a convenience; if it is taken to have its un-normalized value

$$\langle \mathrm{in}, 0|0, \mathrm{in}\rangle = [\det(-G_F)] , \tag{9.69}$$

given by (9.17) (with no interaction), then

$$W_0 \equiv -\mathrm{i}\ln(\langle \mathrm{in}, 0|0, \mathrm{in}\rangle) \tag{9.70}$$

and when used in (6.13) W_0 will yield $\langle \text{in}, 0|T_{\mu\nu}|0, \text{in}\rangle / \langle \text{in}, 0|0, \text{in}\rangle$.

However, if one wishes to use W for a purpose other than the evaluation of the stress-tensor expectation value, one can normalize $\langle \text{in}, 0|0, \text{in}\rangle$ to have value unity, in which case the first term in (9.68) vanishes.

Using (9.26), (9.36) and the interaction Lagrangian density (9.61) one obtains

$$\frac{\langle \text{in}, 0|S^{(1)}|0, \text{in}\rangle}{\langle \text{in}, 0|0, \text{in}\rangle}$$

$$= \mathrm{i}\int \mathrm{d}^n x [-g(x)]^{\frac{1}{2}} \left\{ \frac{\lambda}{8} G_{\mathrm{F}}^2(x,x) - \frac{\mathrm{i}}{2}(\delta m^2 + \delta\xi R) G_{\mathrm{F}}(x,x) \right\}$$

$$= \mathrm{i}\int \mathrm{d}^n x [-g(x)]^{\frac{1}{2}} \left\{ \frac{\lambda_{\mathrm{R}}}{8} G_{\mathrm{F}}^2(x,x) \right.$$

$$\left. + \frac{\mathrm{i}\lambda_{\mathrm{R}}}{32\pi^2} \frac{[m_{\mathrm{R}}^2 + (\xi_{\mathrm{R}} - \frac{1}{6})R]}{(n-4)} G_{\mathrm{F}}(x,x) \right\} + O(\lambda_{\mathrm{R}}^2), \tag{9.71}$$

where in the second term we have used (9.56)–(9.58) with the coefficients (9.64). Now substituting (9.52) for G_{F} gives

$$\frac{\langle \text{in}, 0|S^{(1)}|0, \text{in}\rangle}{\langle \text{in}, 0|0, \text{in}\rangle} = \mathrm{i}\int \mathrm{d}^n x [-g(x)]^{\frac{1}{2}} \left\{ \frac{\lambda_{\mathrm{R}}}{8} [G_{\mathrm{F}}^{\text{finite}}(x,x)]^2 \right.$$

$$\left. + \frac{\lambda_{\mathrm{R}}}{512\pi^4} \frac{[m_{\mathrm{R}}^2 + (\xi_{\mathrm{R}} - \frac{1}{6})R]^2}{(n-4)^2} \right\}. \tag{9.72}$$

The pole terms proportional to $(n-4)^{-1}$ have cancelled from this expression because of the presence of the m and ξ renormalizations. There is still, however, a double pole, which has not been removed in this way, and which will give a double pole term in the effective action (9.68). This remaining divergent term can be removed by renormalization of constants in the generalized Einstein Lagrangian. In this case, renormalization of Newton's constant, the cosmological constant and the constant multiplying R^2 will be required.

The proof that the renormalization scheme outlined above will remove all the poles that arise in $\lambda\phi^4$ theory, to all orders of perturbation, is already non-trivial in Minkowski space (see, for example, Breitenlohner & Maison 1977 and Manoukian 1979, who shows that the subtractions prescribed in methods such as Breitenlohner & Maison's, are equivalent to renormalization of constants in the action). In curved spacetime, the construction of such proofs is much more difficult and is beset by technical as well as more fundamental problems (Birrell & Taylor 1980). In particular, arguments involving Poincaré invariance, which can be used to good effect in demonstrating renormalizability in Minkowski space (see, for example, Roman 1969, §§4.2 & 5.1), can no longer be applied in curved spacetime. One immediate consequence of this already encountered above is the appearance of pole terms proportional to the Ricci scalar, and the consequent necessity of renormalizing ξ. However, R is not the only new scalar that could appear. For example a pole term proportional of $G_{\mathrm{F}}^{\mathrm{finite}}$ arises from the term G_{F}^2 in (9.71). This scalar quantity will in general be an extremely complicated non-local function of the spacetime geometry. (Of course, the asymptotic expansion of $G_{\mathrm{F}}^{\mathrm{finite}}(y, y)$ in the first line of (9.52) is local.)

In terms of Feynman diagrams, the G_{F}^2 term in (9.71) is represented by fig. 30. The pole proportional to $G_{\mathrm{F}}^{\mathrm{finite}}$ arises because the pole associated with one of the loops multiplies the finite part of the other loop. This is known as an 'overlapping' divergence, and its cancellation is vital for the renormalizability of the effective action W in terms of purely local quantities. In the case of (9.71), the overlapping divergence was cancelled by a term arising from m and ξ renormalization. Cancellation of overlapping divergences is also necessary in the two-particle S-matrix elements if they are to be made finite by (local) m and ξ renormalization. Bunch (1980) has proved that such cancellations do occur, and that $\lambda\phi^4$ theory remains renormalizable in curved spacetime.

The fact that the presence of an interaction will, in general, require additional λ-dependent renormalizations of constants in the generalized Einstein action, immediately suggests that there will be a λ-dependent contribution to the conformal anomaly. To study this it proves convenient to use renormalization group methods as adapted to dimensional

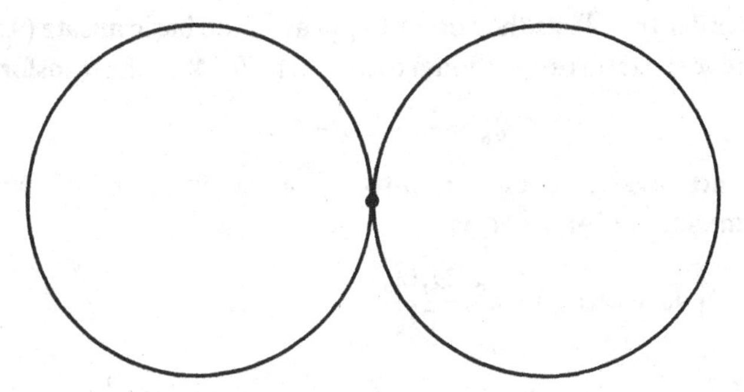

Fig. 30. Feynman diagram for the first order contribution to the vacuum-to-vacuum amplitude in $\lambda\phi^4$ theory.

regularization by 't Hooft (1973) (see also Collins & Macfarlane 1974).

The essential idea has already been encountered in §6.2. The theory can only predict precise S-matrix elements for a fixed value of the arbitrary mass scale μ. This introduces an inherent ambiguity into the values of the renormalized constants, which cannot be determined in practice by experiment. If μ is varied, it is necessary for the values of the renormalized constants to vary also in such a way that the S-matrix elements retain their physical, observed values.

Following 't Hooft, such changes in, for example, λ_R and m_R (the treatment of ξ_R and any other constant is similar to that of m_R) can be deduced by expressing the (fixed) bare constants λ and m given by (9.54), (9.55) and (9.58) in terms of a new mass scale μ', related to μ by the infinitesimal transformation

$$\mu' = \mu(1+\varepsilon). \tag{9.73}$$

One obtains from (9.58) (to first order in ε)

$$\begin{aligned}\lambda &= (\mu')^{4-n}[1+\varepsilon(n-4)]\{\lambda_R + \sum_{\nu=1}^{\infty} a_\nu(\lambda_R)(n-4)^{-\nu}\}\\ &= (\mu')^{4-n}\{\varepsilon(n-4)\lambda_R + \lambda_R + \varepsilon a_1(\lambda_R)\\ &\quad + \sum_{\nu=1}^{\infty}[a_\nu(\lambda_R)+\varepsilon a_{\nu+1}(\lambda_R)](n-4)^{-\nu}\}.\end{aligned} \tag{9.74}$$

Because this expansion contains a term of order $(n-4)$, it is not of the same form as the original expansion (9.58). One is always free to add such terms to the expansions of the bare constants in terms of the renormalized ones, because they vanish at the physical dimension $n=4$; they reflect the usual renormalization ambiguity. However, if we had started the renormalization process with the mass scale μ' instead of μ, we should not have had terms of order $(n-4)$, as they do not appear in the basic ansatz (9.58). We thus remove the term proportional to $(n-4)$ in (9.74) by the transformation

$$\lambda_R = \tilde{\lambda}_R - \varepsilon(n-4)\tilde{\lambda}_R, \tag{9.75}$$

which, once again, is only non-trivial away from $n=4$. With this transformation, (9.74) becomes

$$\begin{aligned}\lambda = (\mu')^{4-n}\Bigg[&\tilde{\lambda}_R + \varepsilon a_1(\tilde{\lambda}_R) - \varepsilon\tilde{\lambda}_R\frac{\partial a_1(\tilde{\lambda}_R)}{\partial\tilde{\lambda}_R}\\ &+ \sum_{\nu=1}^{\infty}(n-4)^{-\nu}\left(a_\nu(\tilde{\lambda}_R)+\varepsilon a_{\nu+1}(\tilde{\lambda}_R)-\varepsilon\tilde{\lambda}_R\frac{\partial a_{\nu+1}(\tilde{\lambda}_R)}{\partial\tilde{\lambda}_R}\right)\Bigg] + O(\varepsilon^2).\end{aligned} \tag{9.76}$$

Also, making the transformation (9.75) in (9.56) gives a new expansion for m:

$$m^2 = m_R^2 - \varepsilon m_R^2 \tilde{\lambda}_R \frac{\partial b_1(\tilde{\lambda}_R)}{\partial \tilde{\lambda}_R} + m_R^2 \sum_{\nu=1}^{\infty} (n-4)^{-\nu} \left(b_\nu(\tilde{\lambda}_R) - \varepsilon \tilde{\lambda}_R \frac{\partial b_{\nu+1}(\tilde{\lambda}_R)}{\partial \tilde{\lambda}_R} \right) + O(\varepsilon^2). \quad (9.77)$$

To recast (9.76) and (9.77) into the standard forms (9.56) and (9.58) (with μ replaced by μ') which we would have obtained had we used the mass scale μ' *ab initio*, define

$$\lambda_R' = \lambda_R + \varepsilon \left(a_1(\tilde{\lambda}_R) - \tilde{\lambda}_R \frac{\partial a_1(\tilde{\lambda}_R)}{\partial \tilde{\lambda}_R} \right) \quad (9.78)$$

and

$$(m_R')^2 = m_R^2 - \varepsilon m_R^2 \tilde{\lambda}_R \frac{\partial b_1(\tilde{\lambda}_R)}{\partial \tilde{\lambda}_R} \quad (9.79)$$

respectively. Since λ and m are considered fixed, if the renormalized S-matrix elements are to have the same value when the mass scale μ' is used as when μ is used, then λ_R and m_R must undergo the transformations (9.78) and (9.79), which reduce at $n = 4$ to

$$\lambda_R' = \lambda_R + \varepsilon \left(a_1(\lambda_R) - \lambda_R \frac{\partial a_1(\lambda_R)}{\partial \lambda_R} \right), \quad (9.80)$$

$$(m_R')^2 = m_R^2 - \varepsilon m_R^2 \lambda_R \frac{\partial b_1(\lambda_R)}{\partial \lambda_R}. \quad (9.81)$$

From these infinitesimal transformations, differential equations for the change of λ_R and m_R^2 with respect to μ are obtained:

$$\mu \frac{\partial \lambda_R}{\partial \mu} = a_1(\lambda_R) - \lambda_R \frac{\partial a_1(\lambda_R)}{\partial \lambda_R} \equiv \beta(\lambda_R), \quad (9.82)$$

$$\mu \frac{\partial m_R^2}{\partial \mu} = - m_R^2 \lambda_R \frac{\partial b_1(\lambda_R)}{\partial \lambda_R} \equiv - m_R^2 \gamma_m(\lambda_R). \quad (9.83)$$

The functions β and γ_m, are sometimes called the Callan–Symanzik β- and γ-functions, respectively, because of the role that they play in the renormalization group equations originally derived by Callan (1970) and Symanzik (1970). We shall now see that they play a similar role in connection with conformal anomalies (Drummond & Shore 1979, Birrell & Davies 1980*b*, Brown & Collins 1980).

The free scalar Lagrangian density (3.24) is invariant in four dimensions under the conformal transformations (3.1) and (3.7), provided $m = 0$ and $\xi = \frac{1}{6}$. The interaction Lagrangian density (9.40) is also invariant under these transformations in four dimensions. As discussed in §6.3, such invariance formally guarantees the vanishing of the trace of the vacuum expectation value of the stress-tensor. We have already seen that renormalization of constants in generalized Einstein action breaks this conformal invariance, and gives rise, even in the free field case, to an anomalous trace. In fact the free field trace anomaly can be separated from any interacting contributions, because the effective action (9.66) for the interacting field can, from (9.17), be written as

$$W = W_0 + W_1, \tag{9.84}$$

where W_0 is the free field effective action (6.25), and W_1 is the interaction-dependent part:

$$W_1 = \mathrm{i}\ln\left[\left\{\exp \mathrm{i}\int \mathrm{d}^n x \mathscr{L}_1\left(\frac{1}{\mathrm{i}}\frac{\delta}{\delta J(x)}\right)\right\}\right.$$
$$\left.\times \exp\left[-\tfrac{1}{2}\mathrm{i}\int J(y) G_{\mathrm{F}}(y, z) J(z) \mathrm{d}^n y \mathrm{d}^n z\right]\right]_{J=0} \tag{9.85}$$

Thus, using (cf. (6.117))

$$\langle T_\mu{}^\mu(x)\rangle = -\frac{1}{[-g(x)]^{\frac{1}{2}}}\frac{\delta W[\Omega^2 g_{\mu\nu}]}{\delta\Omega(x)}\bigg|_{\Omega=1}, \tag{9.86}$$

one obtains,

$$\langle T_\mu{}^\mu\rangle = \langle T_\mu{}^\mu\rangle_0 + \langle T_\mu{}^\mu\rangle_1 \tag{9.87}$$

where $\langle T_\mu{}^\mu\rangle_0$ is the free field anomaly already discussed, and $\langle T_\mu{}^\mu\rangle_1$ is an interaction-dependent part obtained by substituting W_1 for W in (9.86).

In the case of interacting fields, it is not generally possible to maintain the conformal invariance of the action, let alone the effective action. The reason for this is that it is necessary to renormalize m and ξ, so that they cannot maintain the values $m = 0$ and $\xi = \frac{1}{6}$ needed for conformal invariance. Rather, these constants become indeterminate and it is only the renormalized constants m_{R} and ξ_{R} whose values can be determined by experiment. In a given regularization scheme, the bare and renormalized constants can be related by (9.54) and (9.55), where δm^2 and $\delta\xi$ have well-defined values. For example, in 't Hooft's scheme, provided one can show that the ansatz (9.56) successfully implements mass renormalization, then $m_R = 0$ implies $m = 0$. However, in general, δm^2 and $\delta\xi$ are totally arbitrary, and, in particular, if a different regularization scheme were used, $m_R = 0$ would not necessarily imply $\delta m = 0$. This is the case if point-splitting is used (Birrell & Ford 1979).

Since $m_R = 0$ implies $\delta m^2 = 0$ in the 't Hooft scheme, one might ask whether there is also a choice of ξ_R for which $\delta\xi = 0$. We see from (9.64) that if $\xi_R = \frac{1}{6}$, then $\delta\xi = 0$ to first order in λ_R. However, lengthy calculations (Birrell 1980, Bunch & Panangaden 1980) show that this relation breaks down at second order. If one is prepared to abandon the strict 't Hooft renormalization ansatz by allowing ξ_R to be a function of n, then Collins (1976) and Brown & Collins (1980) have shown that the choice $\xi_R = \xi(n)$ (see (3.27)) gives $\delta\xi = 0$ up to third order in λ_R, but not for higher orders. This choice of $\xi_R = \xi(n)$ is equivalent to taking $\xi_R = \frac{1}{6}$ in (9.57) and adding terms of order $n - 4$, thereby exploiting a form of the renormalization ambiguity discussed previously.

Since $\delta\xi \neq 0$ at fourth and higher order and hence the action is not conformally invariant, one does not expect the trace of the stress-tensor vacuum expectation value to be even formally traceless. However, the resulting trace can be divided into an anomalous and a non-anomalous part. The anomalous part comes from three sources: (i) The free field trace anomaly, which we have shown can be treated separately. (ii) Coupling-constant (λ_R) dependent renormalization of constants in the generalized Einstein action. There will also be a non-anomalous part arising from this source, but the anomalous part comes about in a similar fashion to the free field anomaly. (iii) The third source of the anomalous part of the trace is perhaps the most interesting, and arises from the fact that (9.40) is only conformally invariant in four dimensions. In n dimensions, under the transformations (3.1) and (3.7), (9.40) transforms to

$$\begin{aligned}\mathscr{L}_I[\bar{g}_{\mu\nu}] &= \Omega^{4-n}\mathscr{L}_I[g_{\mu\nu}] \\ &= -[-g(x)]^{\frac{1}{2}}(\mu\Omega)^{4-n}\frac{1}{4!}\Big[\lambda_R + \sum_{\nu=1}^{\infty} a_\nu(\lambda_R)(n-4)^{-\nu}\Big]\phi^4,\end{aligned} \tag{9.88}$$

where we have used (9.58). Because $\Omega(x)$ only enters (9.88) in the combination $\mu\Omega$ we may replace the variation of Ω by the variation of μ in (9.86), treating μ temporarily as a function of x:

$$\langle T_\mu^{\ \mu}(x)\rangle_I = -\frac{1}{[-g(x)]^{\frac{1}{2}}}\mu\frac{\delta W_I[\Omega^2 g_{\mu\nu}]}{\delta\mu(x)}\bigg|_{\substack{\mu(x)=\mu\\ \Omega=1}} \tag{9.89}$$

We can calculate (9.89) by using our knowledge of how λ_R and ξ_R must vary in order to exactly compensate for changes in μ. Suppose that $\lambda_R(x)$ and $\xi_R(x)$ have been chosen to compensate for changes in $\mu(x)$. Then

$$\left\{\left[\mu\frac{\delta}{\delta\mu(x)}+\mu\frac{\partial\lambda_R}{\partial\mu}\frac{\delta}{\delta\lambda_R(x)}+\mu\frac{\partial\xi_R}{\partial\mu}\frac{\delta}{\delta\xi_R(x)}\right]W_I[\Omega^2 g_{\mu\nu}]\right\}_{\Omega=1}=0, \quad (9.90)$$

where $\Omega=1$ also signifies $\mu(x)=\mu$, $\lambda_R(x)=\lambda_R$, $\xi_R(x)=\xi_R$. Using (9.82), and additionally defining in analogy to (9.83)

$$\gamma_\xi(\lambda_R)\equiv-\mu\frac{\partial\xi_R}{\delta\mu}=\lambda_R\frac{\partial d_1(\lambda_R)}{\partial\lambda_R}, \quad (9.91)$$

(9.90) becomes

$$\left\{\left[\mu\frac{\delta}{\delta\mu(x)}+\beta(\lambda_R)\frac{\delta}{\delta\lambda_R(x)}-\gamma_\xi(\lambda_R)\frac{\delta}{\delta\xi_R(x)}\right]W_I[\Omega^2 g_{\mu\nu}]\right\}_{\Omega=1}=0, \quad (9.92)$$

which is similar in form to the usual renormalization group equation (see, for example, Collins & Macfarlane 1974, equation (11)). Employing (9.92) in (9.89), one obtains for this particular contribution to $\langle T_\mu{}^\mu\rangle_I$

$$\left\{\frac{1}{[-g(x)]^{\frac{1}{2}}}\left[\beta(\lambda_R)\frac{\delta}{\delta\lambda_R(x)}-\gamma_\xi(\lambda_R)\frac{\delta}{\delta\xi_R(x)}\right]W_I[\lambda_R(x),\xi_R(x)]\right\}_{\substack{\lambda_R(x)=\lambda_R\\ \xi_R(x)=\xi_R}} \quad (9.93)$$

In practice, one would not calculate the trace in a piecewise manner as above, but rather would endeavour to calculate the entire quantity. Equation (9.93) does, however, illustrate a very important difference between the free and interacting 'anomalous' traces: While the free field anomaly is state-independent and local, the interacting trace depends on the entire interaction part of the effective action, which will in general be both state-dependent and non-local. In particular, the use of $\langle T_\mu{}^\mu\rangle_I$ in (6.166) will generally give rise to an increase in energy density with time, representing particle production, unlike the case of the free field anomaly discussed on page 189. Particle production due to interactions has been discussed using this approach by Birrell & Davies (1980*b*). In the next section we consider the effects of interactions on particle production from an alternative point of view.

9.3 Particle production due to interaction

When interactions are present, particle production can occur for two reasons. Firstly, as in the free field case, the definition of positive and negative frequency differs between the in and out regions. Secondly, lack of Poincaré invariance permits energy and momentum to appear from the vacuum. Using the results of the previous two sections, it is not difficult to calculate the number of particles created by these effects.

If we work in the Heisenberg picture, then the state of the system is fixed for all time. Let us choose it to be the state which is the physical vacuum at early times, i.e., $|0, \text{in}\rangle$. In this picture, the operators carry the dynamics, and, in particular, the number operators at early and late times will be different. At early times, the operator representing the number of particles in the mode i, associated with the positive frequency mode function u_i^{in} (see (9.4a)) is

$$N_i^{\text{in}} = a_i^{\dagger\text{in}} a_i^{\text{in}}, \tag{9.94}$$

and $\langle \text{in}, 0 | N_i^{\text{in}} | 0, \text{in} \rangle = 0$. While at late times, the field has the mode decomposition (9.5), and the modes u_i^{out} are positive frequency, so the physical number operator is

$$\bar{N}_i^{\text{out}} = \bar{a}_i^{\text{out}\dagger} \bar{a}_i^{\text{out}}. \tag{9.95}$$

The number of particles which have been created from the $|0, \text{in}\rangle$ vacuum is thus

$$\langle \text{in}, 0 | \bar{N}_i^{\text{out}} | 0, \text{in} \rangle. \tag{9.96}$$

We can evaluate this expectation value in several ways. Firstly, we could write

$$|0, \text{in}\rangle = \sum_{l=0}^{\infty} \frac{1}{l!} \sum_{j_1, \dots, j_l} |\bar{1}_{j_1}, \dots, \bar{1}_{j_l}, \text{out}\rangle \langle \text{out}, \bar{1}_{j_l}, \dots, \bar{1}_{j_1} | 0, \text{in}\rangle, \tag{9.97}$$

which, when inserted in (9.96), gives products of expectation values

$$\langle \text{out}, \bar{1}_{j_l}, \dots, \bar{1}_{j_1} | \bar{N}_i^{\text{out}} | \bar{1}_{j'_1}, \dots, \bar{1}_{j'_l}, \text{out}\rangle \tag{9.98}$$

with amplitudes of the form (9.41). The expectation value (9.98) is trivially evaluated, since the barred out-Fock space is defined with respect to the $\bar{a}_i^{\text{out}}$ operators appearing in (9.95), and all of the interest lies in the amplitude (9.41). In (9.41), the two sources (i) and (ii) of particle production are distinctly separated; all the effects of mixing of positive and negative frequencies being confined to the first amplitude on the right-hand side, while it is only because of the lack of Poincaré invariance that the second amplitude is nonzero in more than just the trivial vacuum-to-vacuum case.

Since the amplitudes formed from the Bogolubov coefficients are fairly complicated expressions, in practice it is more convenient to follow a different procedure, expanding

$$|0, \text{in}\rangle = \sum_{l=0}^{\infty} \frac{1}{l!} \sum_{j_1, \ldots, j_l} |1_{j_1}, \ldots, 1_{j_l}, \text{out}\rangle\langle \text{out}, 1_{j_l}, \ldots, 1_{j_1}|0, \text{in}\rangle, \quad (9.99)$$

whence

$$\begin{aligned}\langle \text{in}, 0|\bar{N}_i^{\text{out}}|0, \text{in}\rangle = \sum_{l=0}^{\infty} \frac{1}{l!} \sum_{m=0}^{\infty} \frac{1}{m!} \sum_{j_1, \ldots, j_l} \sum_{k_1, \ldots, k_m} &\langle \text{in}, 0|1_{k_1}, \ldots, 1_{k_m}, \text{out}\rangle \\ &\times \langle \text{out}, 1_{j_l}, \ldots, 1_{j_1}|0, \text{in}\rangle \\ &\times \langle \text{out}, 1_{k_m}, \ldots, 1_{k_1}|\bar{N}_i^{\text{out}}|1_{j_1}, \ldots, 1_{j_l}, \text{out}\rangle. \quad (9.100)\end{aligned}$$

Now, the first two amplitudes on the right-hand side are, in general, nonzero because of the lack of Poincaré invariance, while it is the final amplitude which carries information about particle creation due to mixing of positive and negative frequencies. This final amplitude is most easily evaluated by using the Bogolubov transformation (3.38) to write (9.95) in terms of the operators a_i^{out}, with respect to which the unbarred out-Fock space is defined. For simplicity, we consider only spatially flat Robertson–Walker spacetimes, in which the Bogolubov coefficients have the form (3.76), (3.77). Then (3.38) gives

$$\bar{a}_i^{\text{out}} = \alpha_i^* a_i^{\text{out}} - \beta_i^* a_{-i}^{\text{out}\dagger}, \quad (9.101)$$

and thus

$$\begin{aligned}\bar{N}_i^{\text{out}} = |\alpha_i|^2 N_i^{\text{out}} + |\beta_i|^2 a_{-i}^{\text{out}} a_{-i}^{\text{out}\dagger} \\ - \alpha_i \beta_i^* a_i^{\text{out}\dagger} a_{-i}^{\text{out}\dagger} - \alpha_i^* \beta_i a_{-i}^{\text{out}} a_i^{\text{out}}, \quad (9.102)\end{aligned}$$

with

$$N_i^{\text{out}} = a_i^{\text{out}\dagger} a_i^{\text{out}}. \quad (9.103)$$

Substituting (9.102) in (9.100), and including only those terms which give a nonzero contribution to the sum, one obtains

$$\begin{aligned}&\langle \text{in}, 0|\bar{N}_i^{\text{out}}|0, \text{in}\rangle \\ &= |\beta_i|^2 + \sum_{m=0}^{\infty} \frac{1}{m!} \sum_{k_1, \ldots, k_m} \{|\alpha_i \langle \text{out}, 1_{k_m}, \ldots, 1_{k_1}|0, \text{in}\rangle|^2 \\ &\quad \times \langle \text{out}, 1_{k_m}, \ldots, 1_{k_1}|N_i^{\text{out}}|1_{k_1}, \ldots, 1_{k_m}, \text{out}\rangle \\ &\quad - 2\text{Re}[\alpha_i \beta_i^* \langle \text{in}, 0|1_i, 1_{-i}, 1_{k_1}, \ldots, 1_{k_m}, \text{out}\rangle\langle \text{out}, 1_{k_m}, \ldots, 1_{k_1}|0, \text{in}\rangle \\ &\quad \times \langle \text{out}, 1_{k_m}, \ldots, 1_{k_1}, 1_{-i}, 1_i|a_i^{\text{out}\dagger} a_{-i}^{\text{out}\dagger}|1_{k_1}, \ldots, 1_{k_m}, \text{out}\rangle]\}. \quad (9.104)\end{aligned}$$

In arriving at the first term on the right-hand side we have assumed that $\langle \text{in}, 0|0, \text{in}\rangle$ is normalized to unity (see page 307). This first term is precisely the number of particles produced in the absence of interaction (cf. (3.42)).

The remaining terms arise from the interaction, and must be calculated in perturbation theory. We shall give expressions only to first order in λ_R. The term involving $|\langle \text{out}, 1_{k_m}, \dots, 1_{k_1}|0, \text{in}\rangle|^2$ in (9.104) does not contribute until order λ_R^2 so we shall ignore it. On the other hand, the term involving the β Bogolubov coefficient gives an order λ_R contribution coming from the $m = 0$ term in the sum:

$$-2\text{Re}[\alpha_i\beta_i^*\langle \text{in}, 0|1_i, 1_{-i}, \text{out}\rangle\langle \text{out}, 0|0, \text{in}\rangle\langle \text{out}, 1_{-i}, 1_i|a_i^{\text{out}\dagger}a_{-i}^{\text{out}\dagger}|0, \text{out}\rangle]$$

$$= -2\text{Re}[\alpha_i\beta_i^*\langle \text{in}, 0|1_i, 1_{-i}, \text{out}\rangle\langle \text{out}, 0|0, \text{in}\rangle]. \qquad (9.105)$$

We have seen in the previous section that the first amplitude in (9.105) is of order λ_R, while $\langle \text{out}, 0|0, \text{in}\rangle = 1 + O(\lambda_R)$ (with the normalization assumed above). Hence, to first order in λ_R, use of (9.65) in (9.105) gives the renormalized expectation value

$$\langle \text{in}, 0|\bar{N}_i^{\text{out}}|0, \text{in}\rangle_{\text{ren}}$$

$$= |\beta_i|^2 - \text{Re}\{\alpha_i\beta_i^*\lambda_R \int u_i(y)u_{-i}(y)[G_F^{\text{finite}*}(y, y)$$

$$-(\text{i}/8\pi^2)(m_R^2 + (\xi_R - \tfrac{1}{6})R(y))\ln\mu][-g(y)]^{\frac{1}{2}}\text{d}^4y\}. \qquad (9.106)$$

Note that the first order contribution comes about because of the combined effect of the Bogolubov transformation and the interaction. In particular, if $\beta_i = 0$, then there is no particle production to first order.

The occurrence of particle production to first order suggests that, even for fairly weak coupling (i.e., small λ_R), the contribution of the interaction to the total number of particles will be of importance compared with the free field production resulting from $|\beta_i|^2$. The existence of such first order terms has been noted by Lotze (1978) and Birrell & Ford (1979), and by Bunch, Panangaden & Parker (1980), who calculate the expectation value of the stress-tensor in the out region. Numerical examples of particle production due to interactions have been given by Birrell & Ford (1979) and Birrell, Davies & Ford (1980).

9.4 Other effects of interactions

The formalism developed in the preceding section can be used in the case of an eternal black hole with little difficulty (Birrell & Taylor 1980), but it does not immediately indicate whether the Hawking radiation will still be thermal in the presence of interactions. In view of the importance of the thermal nature of the Hawking radiation to the second law of thermodynamics (see §8.2), it is desirable to verify that interactions do not destroy this property.

An argument that interacting Hawking radiation is indeed thermal has been given by Gibbons & Perry (1976, 1978) and developed by Hawking (1981). Recall from §8.3 that, in the case of the free field, the scalar Green function G evaluated in the Kruskal vacuum is the same as a thermal Green function associated with the Schwarzschild coordinates. In particular, it is periodic in imaginary Schwarzschild time. Gibbons & Perry note that, in perturbation theory, the corresponding interacting Green functions are constructed from the free field Green functions, and so the Kruskal-associated one has this same time periodicity automatically built into it. Hence, as the time periodicity argument on page 26 made no reference to whether H and ϕ are free or interacting, this latter Green function would appear to share the thermal properties of the free field case.

If one wishes to forsake perturbation theory then formidable technical problems arise. Even in Minkowski space, the properties of exact (non-perturbative) quantum field theories can only be discussed in general terms (e.g., using spectral representations for the propagators). In curved spacetime, without even Poincaré invariance as a guide, progress is still harder.

Fortunately, some very special model field theories exist that can be extended to curved spacetime without use of perturbative or approximation techniques. One of these is the so-called massless Thirring model (Thirring 1958*a*, *b*), which has been extensively studied in the Minkowski space case (see, for example, Klaiber 1968). This model possesses exact solutions that can readily be extended to curved spacetime (Scarf 1962, Birrell & Davies 1978*a*).

The Thirring model is a theory of a massless, self-interacting, spin-$\frac{1}{2}$ field in two-dimensional spacetime. The theory is transcribed from Minkowski space to curved spacetime by applying the method used for free spinor fields in §3.8. One obtains for the Lagrangian density the sum of the free Lagrangian density (3.176) with m set to zero, and an interaction term

$$\mathscr{L}_I = (\det V)\lambda J^\mu J_\mu, \tag{9.107}$$

where

$$J^\mu = \bar{\psi}\gamma^\mu\psi \tag{9.108}$$

(γ^μ being the curved spacetime Dirac gamma matrices; see page 85).

The theory is easily solved in two-dimensional curved spacetime because it is conformally invariant, with conformal weight $\frac{1}{2}$. Thus, the operator solution ψ of the field equation obtained by variation of the action is given in the line element (7.1) by

$$\psi = C^{-\frac{1}{4}}\tilde{\psi}, \tag{9.109}$$

where $\tilde{\psi}$ is the solution in Minkowski space (see Klaiber 1968).

Using (9.109), it is a simple matter to write down the Green functions in the spacetime of an eternal black hole in terms of the Minkowski spacetime Green functions given by Klaiber. In particular, using Kruskal coordinates to evaluate the Green functions in the vacuum $|0_K\rangle$ (as in the free field case, see page 276), one finds (Birrell & Davies 1978*a*) that, far from the black hole, these Green functions agree, as functions of the Schwarzschild coordinates, with the Thirring model thermal Green functions constructed by Dubin (1976). This confirms in the case of the Thirring model that interactions do not destroy the thermal nature of the Hawking radiation.

The Thirring model is exceptional for an interacting field theory in curved spacetime because it can be solved exactly. In general, it is extremely difficult to obtain solutions even using perturbation theory, because of the complicated nature of Feynman propagators in a general curved spacetime. However, the effects of topology on interacting fields can be readily investigated in the case where the geometry remains flat. For example, in §4.1, free quantum field theory in two-dimensional, flat spacetime with topology $R^1 \times S^1$ was studied. It is not difficult to extend that study to the perturbation solution of interacting theories, by using momentum space in much the same way as in Minkowski space theory.

To demonstrate this we consider flat spacetime with topology $R^3 \times S^1$. The Feynman propagator in this spacetime has the same form as in Minkowski space, namely $(p^2 - m^2 + i\varepsilon)^{-1}$, but the momentum component p^3 (corresponding to the dimension with S^1 topology) takes on discrete values

$$p^3 = \begin{cases} 2\pi j/L, & \text{untwisted fields} \\ 2\pi(j+\frac{1}{2})/L, & \text{twisted fields} \end{cases} \qquad j = 0, \pm 1, \pm 2 \ldots \tag{9.110}$$

where L is the periodicity length (as in §4.1). Thus, the evaluation of Feynman diagrams can be carried out in momentum space using techniques (such as Feynman parametrization) which are familiar from Minkowski space theory.

Consider, for example, the evaluation of the first order contribution to the amputated Green function (9.62) using dimensional regularization. To calculate $G_F(x, x)$ in n dimensions, we assume that the additional $n - 4$

dimensions have topology R^1, i.e., we compute the Feynman propagator in the flat spacetime with topology $R^3 \times S^1 \times R^{n-4}$. There is no fundamental reason why we must add dimensions with topology R^1; it is equally as natural to add dimensions with S^1 topology. This is, of course, always arbitrary in dimensional regularization. However, the difference in the topology of the extra dimensions will only make a difference of order $(n-4)$ in $G_F(x, x)$, which leaves (9.62) unchanged. At higher orders of perturbation theory, one can show that the difference will only result in additions to terms which must in any case be renormalized, i.e., to terms which must be measured by experiment anyway (the usual renormalization ambiguity–see page 310).

With the above choice of n-dimensional topology, one has

$$G_F(x,x) = \frac{1}{L}\sum_{j=-\infty}^{\infty}\int\frac{d^{n-1}p}{(2\pi)^{n-1}}(p^2 - m_R^2 + i\varepsilon)^{-1}. \tag{9.111}$$

The integral can be evaluated using the useful formula ('t Hooft & Veltman 1972)

$$\int d^n p(m^2 - i\varepsilon - 2pk - p^2)^{-\alpha} = \frac{i\pi^{n/2}}{(m^2+k^2)^{\alpha-n/2}}\frac{\Gamma(\alpha-\frac{1}{2}n)}{\Gamma(\alpha)}. \tag{9.112}$$

The sum can then be simplified by various techniques (see, for example, Ford 1980*a*, Appendix B). In the limit in which $m_R = 0$, after performing the integral, (9.111) becomes

$$\frac{i\pi^{(n-5)/2}}{2L^{n-2}}\Gamma\left(\frac{3-n}{2}\right)\times\begin{cases}\sum_{j=-\infty}^{\infty}|j|^{n-3}, & \text{untwisted fields}\\ \sum_{j=-\infty}^{\infty}|j+\frac{1}{2}|^{n-3}, & \text{twisted fields.}\end{cases}$$

In this case, the sum can be evaluated in terms of Riemann's ζ-function, giving

$$G_F(x,x) = -\frac{i\pi^{(n-5)/2}}{2L^{n-2}}\Gamma\left(\frac{3-n}{2}\right)\zeta(3-n)\times\begin{cases}1, & \text{untwisted fields}\\ (2^{3-n}-1), & \text{twisted fields,}\end{cases} \tag{9.113}$$

which have finite limits as $n \to 4$:

$$G_F(x,x)\underset{n\to 4}{\longrightarrow}\begin{cases}-i/(12L^2), & \text{untwisted fields}\\ i/(24L^2), & \text{twisted fields.}\end{cases} \tag{9.114}$$

The finite nature of the result is characteristic of the massless theory, and is anticipated by the fact that the pole term involving b_{v1} in (9.62) vanishes in this case (R is, of course, also zero in this case).

Substituting (9.114) into (9.62) (and setting $n = 4$), one observes that the effect is exactly the same as that of including a finite addition to δm^2 (recall (9.56)). In the case of untwisted fields this addition is $\lambda_R/(24L^2)$, while, for twisted fields the addition is $-\lambda_R/(48L^2)$. Thus, the field, which is massless at the zero-loop level, has developed an effective mass m_L at the one-loop level, due to the effect of the topology:

$$\left.\begin{aligned} m_L^2 &= \lambda_R/(24L^2), && \text{untwisted} \\ m_L^2 &= -\lambda_R/(48L^2), && \text{twisted.} \end{aligned}\right\} \tag{9.115}$$

This mass generation phenomenon in $\lambda\phi^4$ theory has been noted and studied by Ford & Yoshimura (1979), Birrell & Ford (1980), Denardo & Spallucci (1980*a*) and Toms (1980*a*, *b*, *c*), while similar phenomena in other theories have been investigated by Denardo & Spallucci (1980*b*), Ford (1980*a*), Omero & Percacci (1980) and Toms (1980*d*). It is closely related to the first order correction to the Casimir energy (Ford 1979, Kay 1979).

For $\lambda_R > 0$, the mass generated for untwisted fields is real, while that generated for twisted fields is imaginary (tachyonic). The presence of a tachyonic mass often means that the theory is unstable (Ford 1980*b*) and that spontaneous symmetry breaking will occur (see, for example, Goldstone 1961).

In the case of untwisted fields in flat spacetime with $R^3 \times S^1$ topology, the effects of the topology are very similar to those of nonzero temperature. In particular nonzero temperature $\lambda\phi^4$ theory leads to temperature-dependent mass generation and to restoration, at sufficiently high temperature, of broken symmetries (Kirzhnits & Linde 1972, 1976, Dolan & Jackiw 1974, Weinberg 1974, Kislinger & Morley 1976).

In the case of twisted fields, there is no simple relation between topological and nonzero temperature effects, and the study of spontaneous symmetry breaking is considerably complicated by the fact that the classical solutions to the field equation with a tachyonic mass cannot be constant (as in the untwisted case) because of the need to maintain antiperiodic boundary conditions (Avis & Isham 1978).

The combined effect of topology, nonzero temperature, and curvature on theories involving spontaneously broken symmetries is likely to have been of considerable importance in the early stages of the evolution of the universe. Some possible consequences are discussed in, for example, the papers of Dreitlein (1974), Kobsarev, Okun & Zel'dovich (1974), Domokos, Janson & Kovesi-Domokos (1975), Bludman & Ruderman (1977), Canuto & Lee (1977), Frolov, Grib & Mostepanenko (1977, 1978) and Gibbons (1978), within which references to earlier works can be found (see also page 128).

References*

Abers, E.S. and Lee, B.W. (1973) *Phys. Rep.*, **9C**, 1.
Abramowitz, M. and Stegun, I.A. (eds.) (1965) *Handbook of Mathematical Functions* (New York: Dover).
Abrikosov, A.A., Gorkov, L.P. and Dzyaloskinskii, I.Ye. (1963) *Quantum Field Theory Methods in Statistical Physics* (Englewood Cliffs: Prentice-Hall).
Adler, S.L. (1969) *Phys. Rev.*, **177**, 2426.
Adler, S.L., Lieberman, J. and Ng, Y.J. (1977) *Ann. Phys.* (NY), **106**, 279.
Adler, S.L., Lieberman, J. and Ng, Y.J. (1978) *Ann. Phys.* (NY), **113**, 294.
Ashmore, J.F. (1972) *Lett. Nuovo Cimento*, **4**, 289.
Ashtekar, A. and Magnon, A. (1975*a*) *Proc. R. Soc. London*, **A346**, 375.
Ashtekar, A. and Magnon, A. (1975*b*) *C.R. Acad. Sci. Ser. A*, **281**, 875.
Audretsch, J. and Schäfer, G. (1978*a*) *Phys. Lett.*, **66A**, 459.
Audretsch, J. and Schäfer, G. (1978*b*) *J. Phys. A: Gen. Phys.*, **11**, 1583.
Avis, S.J. and Isham, C.J. (1978) *Proc. R. Soc. London*, **A363**, 581.
Avis, S.J. and Isham, C.J. (1979*a*) *Nucl. Phys.*, **B156**, 441.
Avis, S.J. and Isham, C.J. (1979*b*) 'Quantum field theory and fibre bundles in a general spacetime' in *Recent Developments in Gravitation–Cargese 1978*, eds. M. Levy and S. Deser (New York: Plenum).
Avis, S.J. and Isham, C.J. (1980) *Commun. Math. Phys.*, **72**, 103.
Avis, S.J., Isham, C.J. and Storey, D. (1978) *Phys. Rev. D*, **18**, 3565.
Balazs N.L. (1958) *Astrophys. J.*, **128**, 398.
Balian, R. and Duplantier, B. (1977) *Ann. Phys.* (NY), **104**, 300.
Balian, R. and Duplantier, B. (1978) *Ann. Phys.* (NY), **112**, 165.
Banach, R. and Dowker, J.S. (1979) *J. Phys. A: Gen. Phys.*, **12**, 2527, 2545.
Bander, M. and Itzykson, C. (1966) *Rev. Mod. Phys.*, **38**, 346.
Bardeen, J.W., Carter, B. and Hawking, S.W. (1973) *Commun. Math. Phys.*, **31**, 161.
Bargmann, V. (1932) *Sitz. Preuss. Akad. Wiss.*, 346.
Barrow, J.D. and Matzner, R.A. (1977) *Mon. Not. R. Astron. Soc.*, **181**, 719.
Barshay, S. and Troost, W. (1978) *Phys. Lett.*, **73B**, 437.
Barut, A.O., Muzinich, I.J. and Williams, D.N. (1963) *Phys. Rev.*, **130**, 442.
Bekenstein, J.D. (1972) *Lett. Nuovo Cimento*, **4**, 7371.
Bekenstein, J.D. (1973) *Phys. Rev. D*, **7**, 2333.
Bekenstein, J.D. (1975) *Phys. Rev. D*, **12**, 3077.
Bekenstein, J.D. and Meisels, A. (1977) *Phys. Rev. D*, **15**, 2775.
Belifante, F.J. (1940) *Physica* (Utrecht), **7**, 305.
Berger, B. (1974) *Ann. Phys.* (NY), **83**, 458.
Berger, B. (1975) *Phys. Rev. D*, **12**, 368.

*In what follows we adhere to the American Physical Society approved style, as detailed in the APS *Style Manual* (3rd edition, 1978), reproduced in the *Bulletin of the American Physical Society* (1979), volume 24.

Bernard, C. and Duncan, A. (1977) *Ann. Phys.* (NY), **107**, 201.
Bianchi, L. (1918) *Lezioni Sulla Teoria dei Gruppi Continui Finiti Transformazioni* (Pisa: Spoerri).
Birrell, N.D. (1978) *Proc. R. Soc. London*, **A361**, 513.
Birrell, N.D. (1979*a*) *Proc. R. Soc. London*, **A367**, 123.
Birrell, N.D. (1979*b*) *J. Phys. A: Gen. Phys.*, **12**, 337.
Birrell, N.D. (1979*c*) Ph.D. Thesis, King's College, London (unpublished).
Birrell, N.D. (1980) *J. Phys. A: Gen. Phys.*, **13**, 569.
Birrell, N.D. and Davies, P.C.W. (1978*a*) *Phys. Rev. D*, **18**, 4408.
Birrell, N.D. and Davies, P.C.W. (1978*b*) *Nature* (London), **272**, 35.
Birrell, N.D. and Davies, P.C.W. (1980*a*) *J. Phys. A: Gen. Phys.*, **13**, 2109.
Birrell, N.D. and Davies, P.C.W. (1980*b*) *Phys. Rev. D*, **22**, 322.
Birrell, N.D., Davies, P.C.W. and Ford, L.H. (1980) *J. Phys. A: Gen. Phys.*, **13**, 961.
Birrell, N.D. and Ford, L.H. (1979) *Ann. Phys.* (NY), **122**, 1.
Birrell, N.D. and Ford, L.H. (1980) *Phys. Rev. D*, **22**, 330.
Birrell, N.D. and Taylor, J.G. (1980) *J. Math. Phys.* (NY), **21**, 1740.
Bjorken, J.D. and Drell, S.D. (1965) *Relativistic Quantum Mechanics* (Chs. 1–10) and *Relativistic Quantum Fields* (Chs. 11–19) (New York: McGraw-Hill).
Blandford, R.D. (1977) *Mon. Not. R. Astron. Soc.*, **181**, 489.
Blandford, R.D. and Thorne, K.S. (1979) in *General Relativity: An Einstein Centenary Survey*, eds. S.W. Hawking and W. Israel (Cambridge: Cambridge University Press).
Bleuler, K. (1950) *Helv. Phys. Acta*, **23**, 567.
Bludman, S.A. and Ruderman, M.A. (1977) *Phys. Rev. Lett.*, **38**, 255.
Blum, B.S. (1973) Ph.D. Thesis, Brandeis University (unpublished).
Bogolubov, N.N. (1958) *Zh. Eksp. Teor. Fiz.*, **34**, 58 (*Sov. Phys. JETP*, **7**, 51 (1958)).
Bogolubov, N.N. and Shirkov, D.V. (1959) *Introduction to the Theory of Quantized Fields* (New York: Interscience).
Bollini, C.G. and Giambiagi, J.J. (1972) *Phys. Lett.*, **40B**, 566.
Bondi, H. and Gold, T. (1948) *Mon. Not. R. Astron. Soc.*, **108**, 252.
Boulware, D.G. (1975*a*) *Phys. Rev. D*, **11**, 1404.
Boulware, D.G. (1975*b*) *Phys. Rev. D*, **12**, 350.
Boulware, D.G. (1976) *Phys. Rev. D*, **13**, 2169.
Breitenlohner, P. and Maison, D. (1977) *Commun. Math. Phys.*, **52**, 11.
Brown, L.S. and Cassidy, J.P. (1977*a*) *Phys. Rev. D*, **16**, 1712.
Brown, L.S. and Cassidy, J.P. (1977*b*) *Phys. Rev. D*, **15**, 2810.
Brown, L.S. and Collins, J.C. (1980) *Ann. Phys.* (NY), **130**, 215.
Brown, L.S. and Maclay, G.J. (1969) *Phys. Rev.*, **184**, 1272.
Brown, M.R. and Dutton, C.R. (1978) *Phys. Rev. D*, **18**, 4422.
Bucholz, H. (1969) *The Confluent Hypergeometric Function* (Berlin: Springer-Verlag).
Bunch, T.S. (1977) Ph.D. Thesis, King's College, London (unpublished).
Bunch, T.S. (1978*a*) *Phys. Rev. D*, **18**, 1844.
Bunch, T.S. (1978*b*) *J. Phys. A: Gen. Phys.*, **11**, 603.
Bunch, T.S. (1979) *J. Phys. A: Gen. Phys.*, **12**, 517.
Bunch, T.S. (1980) *J. Phys. A: Gen. Phys.*, **13**, 1297.
Bunch, T.S. (1981*a*) *Gen. Relativ. Gravit.*
Bunch, T.S. (1981*b*) *Ann. Phys.* (NY), **131**, 118.
Bunch, T.S., Christensen, S.M. and Fulling, S.A. (1978) *Phys. Rev. D*, **18**, 4435.

Bunch, T.S. and Davies, P.C.W. (1977*a*) *Proc. R. Soc. London*, **A357**, 381.
Bunch, T.S. and Davies, P.C.W. (1977*b*) *Proc. R. Soc. London*, **A356**, 569.
Bunch, T.S. and Davies, P.C.W. (1978*a*) *Proc. R. Soc. London*, **A360**, 117.
Bunch, T.S. and Davies, P.C.W. (1978*b*) *J. Phys. A: Gen. Phys.*, **11**, 1315.
Bunch, T.S. and Panangaden, P. (1980) *J. Phys. A: Gen. Phys.*, **13**, 919.
Bunch, T.S., Panangaden, P. and Parker, L. (1980) *J. Phys. A: Gen. Phys.*, **13**, 901.
Bunch, T.S. and Parker, L. (1979) *Phys. Rev. D*, **20**, 2499.
Caianello, E.R. (1973) *Combinatorics and Renormalization in Quantum Field Theory* (New York: W.A. Benjamin).
Callan, C.G. (1970) *Phys. Rev. D*, **2**, 1541.
Callan, C.G., Coleman, S. and Jackiw, R. (1970) *Ann. Phys.* (NY), **59**, 42.
Calvani, M. and Francaviglia, M. (1978) *Acta Phys. Pol.*, **B9**, 11.
Candelas, P. (1980) *Phys. Rev. D*, **21**, 2185.
Candelas, P. and Deutsch, D. (1977) *Proc. R. Soc. London*, **A354**, 79.
Candelas, P. and Deutsch, D. (1978) *Proc. R. Soc. London*, **A362**, 251.
Candelas, P. and Dowker, J.S. (1979) *Phys. Rev. D*, **19**, 2902.
Candelas, P. and Raine, D.J. (1975) *Phys. Rev. D*, **12**, 965.
Candelas, P. and Raine, D.J. (1976) *J. Math. Phys.* (NY) **17**, 2101.
Candelas, P. and Raine, D.J. (1977*a*) *Phys. Rev. D*, **15**, 1494.
Candelas, P. and Raine, D.J. (1977*b*) *Proc. R. Soc. London*, **A354**, 79.
Candelas, P. and Sciama, D.W. (1977) *Phys. Rev. Lett.*, **38**, 1372.
Canuto, V. and Lee, J.F. (1977) *Phys. Lett.*, **B72**, 281.
Capper, D.M. and Duff, M.J. (1974) *Nuovo Cimento*, **A23**, 173.
Capper, D.M. and Duff, M.J. (1975) *Phys. Lett.*, **53A**, 361.
Capper, D.M., Duff, M.J. and Halpern, L. (1974) *Phys. Rev. D*, **10**, 461.
Carter, B. (1975) in *Black Holes*, eds. C. DeWitt and B.S. DeWitt (New York: Gordon & Breach).
Casimir, H.B.G. (1948) *Proc. Kon. Ned. Akad. Wet.*, **51**, 793.
Castagnino, M., Verbeure, A. and Weder, R.A. (1974) *Phys. Lett.*, **8A**, 99.
Castagnino, M., Verbeure, A. and Weder, R.A. (1975) *Nuovo Cimento*, **26B**, 396.
Chakraborty, B. (1973) *J. Math. Phys.* (NY), **14**, 188.
Chandrasekhar, S. (1958) in *The Plasma in a Magnetic Field*, ed. R.K.M. Landshoff (Stanford, California: Stanford University Press).
Charap, J.M. and Duff, M.J. (1977) *Phys. Lett.*, **69B**, 445; **71B**, 219.
Chern, S.S. (1955) *Hamburg Abh.*, **20**, 177.
Chern, S.S. (1962) *J. Soc. Indust. Appl. Math.*, **10**, 751.
Chernikov, N.A. and Tagirov, E.A. (1968) *Ann. Inst. Henri Poincaré*, **9A**, 109.
Chitre, D.M. and Hartle, J.B. (1977) *Phys. Rev. D*, **16**, 251.
Christensen, S.M. (1975) *Bull. Am. Phys. Soc.*, **20**, 99.
Christensen, S.M. (1976) *Phys. Rev. D*, **14**, 2490.
Christensen, S.M. (1978) *Phys. Rev. D*, **17**, 946.
Christensen, S.M. and Duff, M.J. (1978*a*) *Phys. Lett.*, **76B**, 571.
Christensen, S.M. and Duff, M.J. (1978*b*) *Nucl. Phys.*, **B146**, 11.
Christensen, S.M. and Duff, M.J. (1979) *Nucl. Phys.*, **B154**, 301.
Christensen, S.M. and Duff, M.J. (1980) *Nucl. Phys.*, **B170** [FSI], 480.
Christensen, S.M. and Fulling, S.A. (1977) *Phys. Rev. D*, **15**, 2088.
Cline, D.B. and Mills, F.E. (eds.) (1978) *Unification of Elementary Forces and Gauge Theories* (London: Harwood Academic Publ.).
Collins, J.C. (1974) *Phys. Rev. D*, **10**, 1213.
Collins, J.C. (1976) *Phys. Rev. D*, **14**, 1965.
Collins, J.C. and Macfarlane, A.J. (1974) *Phys. Rev. D*, **10**, 1201.

Critchley, R. (1976) Ph.D. Thesis, University of Manchester (unpublished).
Curir, A. and Francaviglia, M. (1978) *Acta Phys. Pol.*, **B9**, 3.
Damour, T. (1977) in *Recent Developments in the Fundamentals of General Relativity*, ed. R. Ruffini (Amsterdam: North Holland).
Davies, P.C.W. (1974) *The Physics of Time Asymmetry* (London: Surrey University Press; Berkeley & Los Angeles: University of California Press).
Davies, P.C.W. (1975) *J. Phys. A: Gen. Phys.*, **8**, 609.
Davies, P.C.W. (1976) *Proc. R. Soc. London*, **A351**, 129.
Davies, P.C.W. (1977*a*) *Phys. Lett.*, **68B**, 402.
Davies, P.C.W. (1977*b*) *Proc. R. Soc. London*, **A354**, 529.
Davies, P.C.W. (1977*c*) *Proc. R. Soc. London*, **A353**, 499.
Davies, P.C.W. (1978) *Rep. Prog. Phys.*, **41**, 1313.
Davies, P.C.W. (1981) in *Quantum Gravity II: A Second Oxford Symposium*, eds. C.J. Isham, R. Penrose and D.W. Sciama (Oxford: Clarendon).
Davies, P.C.W. and Fulling, S.A. (1977*a*) *Proc. R. Soc. London*, **A356**, 237.
Davies, P.C.W. and Fulling, S.A. (1977*b*) *Proc. R. Soc. London*, **A354**, 59.
Davies, P.C.W., Fulling, S.A., Christensen, S.M. and Bunch, T.S. (1977) *Ann. Phys.* (NY), **109**, 108.
Davies, P.C.W., Fulling, S.A. and Unruh, W.G. (1976) *Phys. Rev. D.*, **13**, 2720.
Davies, P.C.W. and Unruh, W.G. (1977) *Proc. R. Soc. London*, **A356**, 569.
Davies, P.C.W. and Unruh, W.G. (1979) *Phys. Rev. D.*, **20**, 388.
Delbourgo, R. and Prasad, V.B. (1974) *Nuovo Cimento*, **21A**, 32.
Denardo, G. and Spallucci, E. (1979) *Nuovo Cimento*, **53B**, 334.
Denardo, G. and Spallucci, E. (1980*a*) *Nuovo Cimento*, **55B**, 97.
Denardo, G. and Spallucci, E. (1980*b*) *Nucl. Phys.*, **B169**, 514.
Denardo, G. and Spallucci, E. (1980*c*) *Nuovo Cimento*, **59A**, 1.
Deruelle, N. and Ruffini, R. (1976) *Phys. Lett.*, **57B**, 248.
Deser, S., Duff, M.J. and Isham, C.J. (1976) *Nucl. Phys.*, **B111**, 45.
Deser, S. and Zumino, B. (1976) *Phys. Lett.*, **62B**, 335.
de Sitter, W. (1917*a*) *Proc. Kon. Ned. Akad. Wet.*, **19**, 1217.
de Sitter, W. (1917*b*) *Proc. Kon. Ned. Akad. Wet.*, **20**, 229.
Deutsch, D. & Candelas, P. (1979) *Phys. Rev. D*, **20**, 3063.
DeWitt, B.S. (1953) *Phys. Rev.*, **90**, 357.
DeWitt, B.S. (1965) 'The dynamical theory of groups and fields' in *Relativity, groups and Topology*, eds. B.S. DeWitt and C. DeWitt (New York: Gordon & Breach).
DeWitt, B.S. (1967*a*) *Phys. Rev.*, **160**, 1113.
DeWitt, B.S. (1967*b*) *Phys. Rev.*, **162**, 1195, 1239.
DeWitt, B.S. (1975) *Phys. Rep.*, **19C**, 297.
DeWitt, B.S. (1979) in *General Relativity*, eds. S.W. Hawking and W. Israel (Cambridge: Cambridge University Press).
DeWitt, B.S. and Brehme, R.W. (1960) *Ann. Phys.* (NY), **9**, 220.
DeWitt, B.S., Hart, C.F. and Isham, C.J. (1979) in *Themes in Contemporary Physics*, ed. S. Deser (Amsterdam: North Holland).
Dirac, P.A.M. (1936) *Proc. R. Soc. London*, **A155**, 447.
di Sessa, A. (1974) *J. Math. Phys.* (NY), **15**, 1892.
Dolan, L. and Jackiw, R. (1974) *Phys. Rev. D*, **9**, 3320.
Dolginov, A.Z. and Toptygin, I.N. (1959) *Zh. Eksp. Teor. Fiz.*, **37**, 1411 (*Sov. Phys. JETP*, **10**, 1022 (1960)).
Domokos, G., Janson, M.M. and Kovesi-Domokos, S. (1975) *Nature* (London), **257**, 203.
Dowker, J.S. (1967*a*) *Proc. R. Soc. London*, **A297**, 351.
Dowker, J.S. (1967*b*) *Supp. Nuovo Cimento*, **5**, 734.

Dowker, J.S. (1971) *Ann. Phys.* (NY), **62**, 361.
Dowker, J.S. (1972) *Ann. Phys.* (NY), **71**, 577.
Dowker, J.S. (1977) *J. Phys. A: Gen. Phys.*, **10**, 115.
Dowker, J.S. and Banach, R. (1978) *J. Phys. A: Gen. Phys.*, **11**, 2255.
Dowker, J.S. and Critchley, R. (1976*a*) *Phys. Rev. D*, **13**, 224.
Dowker, J.S. and Critchley, R. (1976*b*) *Phys. Rev. D*, **13**, 3224.
Dowker, J.S. and Critchley, R. (1976*c*) *J. Phys. A: Gen. Phys.*, **9**, 535.
Dowker, J.S. and Critchley, R. (1977*a*) *Phys. Rev. D*, **15**, 1484.
Dowker, J.S. and Critchley, R. (1977*b*) *Phys. Rev. D*, **16**, 3390.
Dowker, J.S. and Dowker, Y.P. (1966*a*) *Proc. R. Soc. London*, **A294**, 175.
Dowker, J.S. and Dowker, Y.P. (1966*b*) *Proc. R. Soc. London*, **87**, 65.
Dowker, J.S. and Kennedy, G. (1978) *J. Phys. A: Gen. Phys.*, **11**, 895.
Dreitlein, J. (1974) *Phys. Rev. Lett.*, **33**, 1243.
Drummond, I.T. and Hathrell, S.J. (1980) *Phys. Rev D*, **21**, 958.
Drummond, I.T. and Shore, G.M. (1979) *Phys. Rev. D*, **19**, 1134.
Duan', I.S. (1956) *Zh. Eksp. Teor. Fiz.*, **31**, 1098 (*Sov. Phys. JETP*, **7**, 437 (1956)).
Dubin, D.A. (1976) *Ann. Phys.* (NY), **102**, 71.
Duff, M.J. (1975) in *Quantum Gravity: An Oxford Symposium*, eds. C.J. Isham, R. Penrose, and D.W. Sciama (Oxford: Clarendon).
Duff, M.J. (1977). *Nucl. Phys.*, **B125**, 334.
Duff, M.J. (1981) in *Quantum Gravity II: A Second Oxford Symposium*, eds. C.J. Isham, R. Penrose & D.W. Sciama (Oxford: Clarendon).
Dyson, F.J. (1949) *Phys. Rev.*, **75**, 486, 1736.
Einstein, A. (1917) *Sitz. Preuss. Akad. Wiss.*, 142. For an English translation, see *The Principle of Relativity* (New York: Dover 1952).
Epstein, H., Gaser, V. and Jaffe, A. (1965) *Nuovo Cimento*, **36**, 1016.
Fadeev, L.D. and Popov, V.N. (1967) *Phys. Lett.*, **25B**, 29.
Farrugia, C.J. and Hájícek, P. (1980) *Commun. Math. Phys.*, **68**, 291.
Fetter, A.L. and Walecka, J.D. (1971) *Quantum Theory of Many-Particle Systems* (New York: McGraw-Hill).
Feynman, R.P. (1963) *Acta Phys. Pol.*, **24**, 697.
Feynman, R.P. and Hibbs, A.R. (1965) *Quantum Mechanics and Path Integrals* (New York: McGraw-Hill).
Fierz, M. (1939) *Helv. Phys. Acta*, **12**, 3.
Fierz, M. and Pauli, W. (1939) *Proc. R. Soc. London*, **A173**, 211.
Fischetti, M.V., Hartle, J.B. and Hu, B.L. (1979) *Phys. Rev. D.*, **20**, 1757.
Fock, V.A. (1929) *Z. Phys.*, **57**, 261.
Fock, V.A. (1937) *Phys. Z. Sowjetunion*, **12**, 404.
Fock, V.A. and Ivanenko, D. (1929) *Z. Phys.*, **54**, 798.
Ford, L.H. (1975) *Phys. Rev. D.*, **12**, 2963.
Ford, L.H. (1976) *Phys. Rev. D*, **14**, 3304.
Ford, L.H. (1978) *Proc. R. Soc. London*, **A364**, 227.
Ford, L.H. (1979) *Proc. R. Soc. London*, **A368**, 305.
Ford, L.H. (1980*a*) *Phys. Rev. D*, **21**, 933.
Ford, L.H. (1980*b*) *Phys. Rev. D*, **22**, 3003.
Ford, L.H. and Parker, L. (1977) *Phys. Rev. D*, **16**, 245.
Ford, L.H. and Parker, L. (1978) *Phys. Rev. D*, **17**, 1485.
Ford, L.H. and Yoshimura T. (1979) *Phys. Lett.*, **70A**, 89.
Fradkin, E.S. and Vilkovisky, G.S. (1978) *Phys. Lett.*, **73B**, 209.
Frampton, P.H. (1977) *Lectures on Gauge Field Theories. Part Two: Quantization* University of California at Los Angeles publication UCLA/77/TEP/21.

Freedman, D.Z., Muzinich, I.J. and Weinberg, E.J. (1974) *Ann. Phys.* (NY), **87**, 95.
Freedman, D.Z. and Pi, S-Y. (1975) *Ann. Phys.* (NY), **91**, 442.
Freedman, D.Z., van Nieuwenhuizen, P. and Ferrara, S. (1976) *Phys. Rev. D*, **13**, 3214.
Freedman, D.Z. and Weinberg, E.J. (1974) *Ann. Phys.* (NY) **87**, 354.
Friedlander, F.G. (1975) *The Wave Equation in Curved Space-Time* (Cambridge: Cambridge University Press).
Frolov, V.M., Grib, A.A. and Mostepanenko, V.M. (1977) *Teor. & Math. Fiz.* (USSR), **33**, 42 (*Theor. & Math. Phys.*, **33**, 869 (1977)).
Frolov, V.M., Grib, A.A. and Mostepanenko, V.M. (1978) *Phys. Lett.*, **65A**, 282.
Frolov, V.M., Mamaev, S.G. and Mostepanenko, V.M. (1976) *Phys. Lett.*, **55A**, 389.
Fulling, S.A. (1972) Ph.D. Thesis, Princeton University (unpublished).
Fulling, S.A. (1973) *Phys. Rev. D*, **7**, 2850.
Fulling, S.A. (1977) *J. Phys. A: Gen. Phys.*, **10**, 917.
Fulling, S.A. (1979) *Gen. Relativ. Gravit.*, **10**, 807.
Fulling, S.A. and Davies, P.C.W. (1976) *Proc. R. Soc. London*, **A348**, 393.
Fulling, S.A., Parker, L. and Hu, B.L. (1974) *Phys. Rev. D*, **10**, 3905; erratum *ibid* **11**, 1714.
Fulling, S.A., Sweeny, M. and Wald, R.M. (1978) *Commun. Math. Phys.* **63**, 257.
Garabedian, P.R. (1964) *Partial Differential Equations* (New York: Wiley).
Georgi, H. and Glashow, S.L. (1974) *Phys. Rev. Lett.*, **32**, 438.
Gerlach, U. (1976) *Phys. Rev. D*, **14**, 1479.
Gibbons, G.W. (1975) *Commun. Math. Phys.*, **44**, 245.
Gibbons, G.W. (1977*a*) *Phys. Lett.*, **60A**, 385.
Gibbons, G.W. (1977*b*) *Phys. Lett.*, **61A**, 3.
Gibbons, G.W. (1978) *J. Phys. A:Gen. Phys.*, **11**, 1341.
Gibbons, G.W. (1979) in *General Relativity: An Einstein Centenary Survey*, eds. S.W. Hawking and W. Israel (Cambridge: Cambridge University Press).
Gibbons, G.W. (1980) *Ann. Phys.* (NY), **125**, 98.
Gibbons, G.W. and Hawking, S.W. (1977*a*) *Phys. Rev. D*, **15**, 2738.
Gibbons, G.W. and Hawking, S.W. (1977*b*) *Phys. Rev. D*, **15**, 2752.
Gibbons, G.W. and Hawking, S.W. (1979) *Commun. Math. Phys.*, **66**, 291.
Gibbons, G.W. and Perry, M.J. (1976) *Phys. Rev. Lett.*, **36**, 985.
Gibbons, G.W. and Perry, M.J. (1978) *Proc. R. Soc. London*, **A358**, 467.
Gibbons, G.W. and Pope, C.N. (1978) *Commun. Math. Phys.*, **61**, 239.
Gilkey, P.B. (1975) *Adv. Math.*, **15**, 334.
Ginibre, J. (1971) in *Statistical Mechanics and Quantum Field Theory*, eds. C. DeWitt and R. Stora (New York: Gordon and Breach).
Ginzburg, V.L., Kirzhnits, D.A. and Lyubushin, A.A. (1971) *Zh. Eksp. Teor. Fiz.*, **60**, 451 (*Sov. Phys. JETP*, **33**, 242 (1971)).
Gitman, D.M. (1977) *J. Phys. A: Gen. Phys.*, **10**, 2007.
Goldstone, J. (1961) *Nuovo Cimento*, **19**, 154.
Gradshteyn, I.S. and Ryzhik, I.M. (1965) *Tables of Integrals, Series and Products* (New York: Academic Press).
Greiner, P. (1971) *Archs. Ration. Mech. Analysis*, **41**, 163.
Grensing, G. (1977) *J. Phys. A: Gen. Phys.*, **10**, 1687.
Grib, A.A. and Mamaev, S.G. (1969) *Yad. Fiz.*, **10**, 1276 (*Sov. J. Nucl. Phys.*, **10**, 722 (1970)).
Grib, A.A. and Mamaev, S.G. (1971) *Yad. Fiz.*, **14**, 800 (*Sov. J. Nucl. Phys.*, **14**, 450 (1972)).

Grib, A.A., Mamaev, S.G. and Mostepanenko, V.M. (1976) *Gen. Relativ. Gravit*,. **7**, 535.
Grib, A.A., Mamaev, S.G. and Mostepanenko, V.M. (1980*a*) *J. Phys. A: Gen: Phys.*, **13**, 2057.
Grib, A.A., Mamaev, S.G. and Mostepanenko, V.M. (1980*b*) *Quantum Effects in Strong External Fields* (Moscow: Atomizdat). In Russian.
Grishchuk, L.P. (1974) *Zh. Eksp. Teor. Fiz.*, **67**, 825 (*Sov. Phys. JETP*, **40**, 409 (1975)).
Grishchuk, L.P. (1975) *Lett. Nuovo Cimento*, **12**, 60.
Grishchuk, L.P. (1977) 'Graviton creation in the early universe' in *Eighth Texas Symposium on Relativistic Astrophysics*, ed. M.D. Papagiannis (New York: New York Academy of Sciences).
Gupta, S.N. (1950) *Proc. Phys. Soc.*, (London) **A63**, 681.
Gürsey, F. (1964) 'Introduction to group theory' in *Relativity, Groups and Topology*, eds. B.S. DeWitt and C. DeWitt (New York: Gordon and Breach).
Gutzwiller, M. (1956) *Helv. Phys. Acta*, **29**, 313.
Haag, R. (1955) *Kgl. Danske Videnskab Selskab, Mat.-Fys. Medd.*, **29**, No. 12.
Hadamard, J. (1923) *Lectures on Cauchy's Problem in Linear Partial Differential Equations* (New Haven: Yale University Press).
Hagedorn, R. (1965) *Supp. Nuovo Cimento*, **3**, 147.
Hájiček, P. (1976) *Nuovo Cimento*, **33B**, 597.
Hájiček, P. (1977) *Phys. Rev. D*, **15**, 2757.
Hájiček, P. and Israel, W. (1980) *Phys. Lett.*, **80A**, 9.
Halpern, L. (1967) *Ark. Fys.*, **34**, 539.
Hartle, J.B. (1977) *Phys. Rev. Lett.*, **39**, 1373.
Hartle, J.B. (1981) in *Quantum Gravity II: A Second Oxford Symposium*, eds. C.J. Isham, R. Penrose and D.W. Sciama (Oxford: Clarendon).
Hartle, J.B. and Hawking, S.W. (1976) *Phys. Rev. D*, **13**, 2188.
Hartle, J.B. and Hu, B.L. (1979) *Phys. Rev. D*, **20**, 1772.
Hartle, J.B. and Hu, B.L. (1980) *Phys. Rev. D*, **21**, 2756.
Hatalkar, M.M. (1954) *Phys. Rev.*, **94**, 1472.
Hawking, S.W. (1970) *Commun. Math. Phys.*, **18**, 301.
Hawking, S.W. (1972) *Commun. Math. Phys.*, **25**, 152.
Hawking, S.W. (1973) 'The event horizon' in *Black Holes*, eds. C. DeWitt and B.S. DeWitt (New York: Gordon and Breach).
Hawking, S.W. (1974) *Nature* (London), **248**, 30.
Hawking, S.W. (1975) *Commun. Math. Phys.*, **43**, 199.
Hawking, S.W. (1976*a*) *Phys. Rev. D*, **13**, 191.
Hawking, S.W. (1976*b*) *Phys. Rev. D*, **14**, 2460.
Hawking, S.W. (1977*a*) *Commun. Math. Phys.*, **55**, 133.
Hawking, S.W. (1977*b*) *Sci. Am.*, **236**, 34.
Hawking, S.W. (1977*c*) *Phys. Lett.*, **60A**, 81.
Hawking, S.W. (1978) *Phys. Rev. D*, **18**, 1747.
Hawking, S.W. (1979) 'The path-integral approach to quantum gravity' in *General Relativity: An Einstein Centenary Survey*, eds. S.W. Hawking and W. Israel (Cambridge: Cambridge University Press).
Hawking, S.W. (1981) 'Acausal propagation in quantum gravity' in *Quantum Gravity II: A Second Oxford Symposium*, eds. C.J. Isham, R. Penrose and D. W. Sciama (Oxford: Clarendon).
Hawking, S.W. and Ellis, G.F.R. (1973) *The Large Scale Structure of Space-Time* (Cambridge: Cambridge University Press).
Hawking, S.W. and Penrose, R. (1970) *Proc. R. Soc. London*, **A314**, 529.
Hiscock, W.A. (1977*a*) *Phys. Rev. D*, **15**, 3054.

Hiscock, W.A. (1977*b*) *Phys. Rev. D*, **16**, 2673.
Hiscock, W.A. (1979) Ph.D. Thesis, University of Maryland (unpublished).
Hiscock, W.A. (1980) *Phys. Rev. D*, **21**, 2063.
Horibe, M. (1979) *Prog. Theor. Phys.*, **61**, 661.
Horowitz, G.T. (1980) *Phys. Rev. D*, **21**, 1445.
Horowitz, G.T. and Wald, R.M. (1978) *Phys. Rev. D*, **17**, 414.
Horowitz, G.T. and Wald, R.M. (1980) *Phys. Rev. D*, **21**, 1462.
Hosoya, A. (1979) *Prog. Theor. Phys.*, **61**, 280.
Hoyle, F. (1948) *Mon. Not. R. Astron. Soc.*, **108**, 372.
Hu, B.L. (1972) Ph.D. Thesis, Princeton University (unpublished).
Hu, B.L. (1973) *Phys. Rev. D*, **8**, 1048.
Hu, B.L. (1974) *Phys. Rev. D*, **9**, 3263.
Hu, B.L. (1978) *Phys. Rev. D*, **18**, 4460.
Hu, B.L. (1979) *Phys. Lett.*, **71A**, 169.
Hu, B.L. (1980) 'Quantum field theories and relativistic cosmology' in *Recent Developments in General Relativity*, ed. R. Ruffini, (Amsterdam: North Holland).
Hu, B.L., Fulling, S.A. and Parker, L. (1973) *Phys. Rev. D*, **8**, 2377.
Hu, B.L. and Parker, L. (1978) *Phys. Rev. D*, **17**, 933.
Hut, P. (1977) *Mon. Not. R. Astron. Soc.*, **180**, 379.
Imamura, T. (1960) *Phys. Rev.*, **118**, 1430.
Isham, C.J. (1975) 'An introduction to quantum gravity' in *Quantum Gravity: An Oxford Symposium*, eds. C.J. Isham, R. Penrose and D.W. Sciama (Oxford: Clarendon).
Isham, C.J. (1978*a*) 'Quantum field theory in a curved spacetime – a general mathematical framework' in *Proceedings of the Bonn Conference on Differential Geometrical Methods in Mathematical Physics*, eds. K. Bleuler, H.R. Petry and A. Reetz (New York: Springer Verlag).
Isham, C.J. (1978*b*) *Proc. R. Soc. London*, **A362**, 383.
Isham, C.J. (1978*c*) *Proc. R. Soc. London*, **A364**, 591.
Isham, C.J. (1979) *Proc. R. Soc. London*, **A368**, 33.
Isham, C.J. (1981) 'Quantum gravity – an overview' in *Quantum Gravity II: A Second Oxford Symposium*, eds. C.J. Isham, R. Penrose and D.W. Sciama (Oxford: Clarendon).
Isihara, A. (1971) *Statistical Physics* (New York: Academic Press).
Israel, W. (1976) *Phys. Lett.*, **57A**, 107.
Itzykson, C. and Zuber, J.-C. (1980) *Quantum Field Theory* (New York: McGraw-Hill).
Jauch, J.M. and Rohrlich, F. (1955) *The Theory of Photons and Electrons* (Reading, Mass.: Addison-Wesley).
Kadanoff, L.P. and Baym, G. (1962) *Quantum Statistical Mechanics* (Menlo Park, California: Benjamin).
Kay, B.S. (1979) *Phys. Rev. D*, **20**, 3052.
Kay, B.S. (1980) *Commun. Math. Phys.*, **71**, 29.
Kennedy, G. (1978) *J. Phys. A: Gen. Phys.*, **11**, L173.
Kennedy, G., Critchley, R. and Dowker, J.S. (1980) *Ann. Phys.* (NY), **125**, 346.
Kennedy, G. and Unwin, S.D. (1980) *J. Phys. A: Gen. Phys.*, **13**, L253.
Kibble, T.W.B. (1978) *Commun. Math. Phys.*, **64**, 73.
Kibble, T.W.B. and Randjbar-Daemi, S. (1980) *J. Phys. A: Gen. Phys.*, **13**, 141.
Kirzhnits, D.A. and Linde, A.D. (1972) *Phys. Lett.*, **42B**, 471.

Kirzhnits, D.A. and Linde, A.D. (1976) *Ann. Phys.*, (NY), **101**, 195.
Kislinger, M.B. and Morley, P.D. (1976) *Phys. Rev. D*, **13**, 2771.
Klaiber, B. (1968) in *Lectures in Theoretical Physics*, Vol. XA, eds. A.O. Barut and W.E. Britten (New York: Gordon and Breach).
Klein, O. (1929) *Z. Phys.*, **53**, 157.
Kobsarev, I.Yu., Okun, L.B. and Zel'dovich, Ya.B. (1974) *Phys. Lett.*, **50B**, 340.
Kreyszig, E. (1968) *Introduction to Differential Geometry and Riemannian Geometry* (Toronto: University of Toronto Press).
Kruskal, M.D. (1960) *Phys. Rev.*, **119**, 1743.
Landau, L.D. and Lifshitz, E.M. (1958) *Statistical Physics* (London: Pergamon).
Lapedes, A.S. (1978*a*) *Phys. Rev. D*, **17**, 2556.
Lapedes, A.S. (1978*b*) *J. Math. Phys.* (NY), **19**, 2289.
Lehmann, H., Symanzik, K. and Zimmermann, W. (1955) *Nuovo Cimento*, **1**, 1425.
Lifshitz, E.M. (1955) *Zh. Eksp. Teor. Fiz.*, **29**, 94 (*Sov. Phys. JETP*, **2**, 73 (1956)).
Lifshitz, E.M. and Khalatnikov, I.M. (1963) *Adv. Phys.*, **12**, 185.
Liouville, J. (1837) *Journal de Math.*, **2**, 24.
Lohiya, D. and Panchapakesan, N. (1978) *J. Phys. A*: *Gen. Phys.*, **11**, 1963.
Lohiya, D. and Panchapakesan, N. (1979) *J. Phys. A*: *Gen. Phys.*, **12**, 523.
Lotze, K.-H. (1978) *Acta Phys. Pol.*, **B9**, 665, 677.
Lukash, V.N., Novikov, I.D. and Starobinsky, A.A. (1975) *Zh, Eksp. Teor. Fiz.*, **69**, 1484 (*Sov. Phys. JETP*, **42**, 751 (1976)).
Lukash, V.N., Novikov, I.D., Starobinsky, A.A. and Zel'dovich, Ya.B. (1976) *Nuovo Cimento*, **35B**, 293.
Lukash, V.N. and Starobinsky, A.A. (1974) *Zh. Eksp. Teor. Fiz.*, **66**, 515 (*Sov. Phys. JETP*, **39**, 742 (1974)).
Lynden-Bell, D. and Wood, R. (1967) *Mon. Not. R. Astron. Soc.*, **136**, 107.
MacCallum, M.A.H. (1979) 'Anisotropic and inhomogenous relativistic cosmologies' in *General Relativity: An Einstein Centenary Survey*, eds. S.W. Hawking and W. Israel (Cambridge: Cambridge University Press).
Mamaev, S.G., Mostepanenko, V.M. and Starobinsky, A.A. (1976) *Zh. Eksp. Teor. Fiz.*, **70**, 1577 (*Sov. Phys. JETP*, **43**, 823 (1976)).
Manoukian, E.B. (1979) *Nuovo Cimento*, **53A**, 345.
Martellini, M., Sodano, P. and Vitiello, G. (1978) *Nuovo Cimento*, **48A**, 341.
Mashhoon, B. (1973) *Phys. Rev. D*, **8**, 4297.
Mattuck, R.D. (1967) *A Guide to Feynman Diagrams in the Many-Body Problem* (London: McGraw-Hill).
McKean, H.P. and Singer, I.M. (1967) *J. Diff. Geom.*, **1**, 43.
McNamara, J.M. (1978) *Proc. R. Soc. London*, **A358**, 499.
Meikle, W.P.S. (1977) *Nature* (London), **269**, 41.
Melnikov, V.N. and Orlov, S.V. (1979) *Phys. Lett.*, **70A**, 263.
Milne, E.A. (1932) *Nature* (London), **130**, 9.
Minakshisundaram, S. and Pleijel, A. (1949) *Can. J. Math.*, **1**, 242.
Misner, C.W. (1968) *Astrophys. J.*, **151**, 431.
Misner, C.W. (1972) *Phys. Rev. Lett.*, **28**, 994.
Misner, C.W., Thorne, K.S. and Wheeler, J.A. (1973) *Gravitation* (San Francisco: Freeman).
Mohan, G. (1968) in *Lectures in Theoretical Physics*, Vol. XB, eds. A.O. Barut and W.E. Britten (New York: Gordon and Breach).
Moore, G.T. (1970) *J. Math. Phys.* (NY), **9**, 2679.
Nambu, Y. (1950) *Prog. Theor. Phys.*, **5**, 82.

Nambu, Y. (1966) *Phys. Lett.*, **26B**, 626.
Nariai, H. (1971) *Prog. Theor. Phys.*, **46**, 433 and 776.
Nariai, H. (1976) *Nuovo Cimento*, **35B**, 259.
Nariai, H. (1977*a*) *Prog. Theor. Phys.*, **57**, 67.
Nariai, H. (1977*b*) *Prog. Theor. Phys.*, **58**, 560.
Nariai, H. (1977*c*) *Prog. Theor. Phys.*, **58**, 842.
Nash, C. (1978) *Relativistic Quantum Fields* (London: Academic Press).
Omero, C. and Percacci, R. (1980) *Nucl. Phys.*, **B165**, 351.
Page, D.N. (1976*a*) *Phys. Rev. D*, **13**, 198.
Page, D.N. (1976*b*) *Phys. Rev. D*, **14**, 3260.
Page, D.N. (1977) *Phys. Rev., D*, **16**, 2402.
Page, D.N. (1980) *Phys. Rev. Lett.*, **44**, 301.
Parker, L. (1966) 'The Creation of Particles in an Expanding Universe', Ph.D. Thesis, Harvard University (available from: University Microfilms Library Service, Xerox Corp., Ann Arbor, Michigan, USA).
Parker, L. (1968) *Phys. Rev. Lett.*, **21**, 562.
Parker, L. (1969) *Phys. Rev.*, **183**, 1057.
Parker, L. (1971) *Phys. Rev. D*, **3**, 346.
Parker, L. (1972) *Phys. Rev. D*, **5**, 2905.
Parker, L. (1973) *Phys. Rev. D*, **7**, 976.
Parker, L. (1975) *Phys. Rev. D*, **12**, 1519.
Parker, L. (1976) *Nature* (London), **261**, 20.
Parker, L. (1977) 'The production of elementary particles by strong gravitational fields' in *Asymptotic Structure of Space-Time*, eds. F.P. Esposito and L. Witten (New York: Plenum).
Parker, L. (1979) 'Aspects of quantum field theory in curved spacetime: effective action and energy–momentum tensor' in *Recent Developments in Gravitation*, eds. S. Deser and M. Levy (New York: Plenum).
Parker, L. and Fulling, S.A. (1973) *Phys. Rev. D*, **7**, 2357.
Parker, L. and Fulling, S.A. (1974) *Phys. Rev. D*, **9**, 341.
Parker, L., Fulling and Hu, B.L. (1974) *Phys. Rev. D*, **10**, 3905.
Parnovsky, S.L. (1979) *Phys. Lett.*, **73A**, 153.
Pauli, W. and Villars, F. (1949) *Rev. Mod. Phys.*, **21**, 434.
Penrose, R. (1964)'Conformal treatment of infinity' in *Relativity, Groups and Topology*, eds. B.S. DeWitt and C. DeWitt, New York: Gordon and Breach).
Penrose, R. (1965) *Proc. R. Soc. London*, **A284**, 159.
Penrose, R. (1969) *Rev. Nuovo Cimento*, **1**, 252.
Penrose, R. (1979) 'Singularities and time-asymmetry' in *General Relativity: An Einstein Centenary Survey*, eds. S.W. Hawking and W. Israel (Cambridge: Cambridge University Press).
Petrov, A.Z. (1969) *Einstein Spaces* (Oxford: Pergamon).
Planck, M.K.E.L. (1899) *Sitz. Deut. Akad. Wiss. Berlin, Kl. Math. -Phys Tech.*, 440.
Press, W.H. and Teukolsky, S.A. (1972) *Nature* (London), **238**, 211.
Raine, D.J. and Winlove, C.P. (1975) *Phys. Rev. D*, **12**, 946.
Randjbar-Daemi, S., Kay, B.S. and Kibble, T.W.B. (1980) *Phys. Lett.*, **91B**, 417.
Rees, M.J. (1977) *Nature* (London), **266**, 333.
Rindler, W. (1966) *Am. J. Phys.*, **34**, 1174. Many features of this work were anticipated in a paper by Max Born (1909) *Ann. Phys.* (Leipzig), **30**, 1.
Rindler, W. (1969) *Essential Relativity* (New York: Van Nostrand).
Roman, P. (1969) *Introduction to Quantum Field Theory* (New York: Wiley).

Rowan, D.J. and Stephenson, G. (1976) *J. Phys. A: Gen. Phys.*, **9**, 1631.
Rumpf, H. (1976*a*) *Phys. Lett.*, **61B**, 272.
Rumpf, H. (1976*b*) *Nuovo Cimento*, **35B**, 321.
Ruzmaikina, T.V. and Ruzmaikin, A.A. (1970) *Zh. Eksp. Teor. Fiz.*, **57**, 680 (*Sov. Phys. JETP*, **30**, 372 (1970)).
Rzewuski, J. (1969) *Field Theory*, Vol. II (London: Iliffe Books).
Sakharov, A.D. (1967) *Dok. Akad. Nauk. SSR*, **177**, 70 (*Sov. Phys. Dok.*, **12**, 1040 (1968)).
Salam, A. (1968) 'Weak and electromagnetic interactions' in *Elementary Particle Theory*, ed. N. Svartholm, (Stockholm: Almquist, Forlag AB), p. 367.
Sanchez, N. (1979) *Phys. Lett.*, **87B**, 212.
Scarf, F.L. (1962) 'A soluble quantum field theory in curved space' in *Les Théories Relativistes de la Gravitation* (Paris: Centre National de Recherche Scientifique).
Schäfer, G. (1978) *J. Phys. A: Gen. Phys.*, **11**, L179.
Schäfer, G. and Dehnen, H. (1977) *Astron. Astrophys.*, **54**, 823.
Schiff, L.I. (1949) *Quantum Mechanics* (New York: McGraw-Hill).
Schomblond, C. and Spindel, P. (1976) *Ann. Inst. Henri Poincaré*, **25A**, 67.
Schrödinger, E. (1932) *Sitz. Preuss. Akad. Wiss.*, 105.
Schrödinger, E. (1939) *Physica* (Utrecht), **6**, 899.
Schrödinger, E. (1956) *Expanding Universes* (Cambridge: Cambridge University Press).
Schweber, S. (1961) *An Introduction to Relativistic Quantum Field Theory* (New York: Harper & Row).
Schwinger, J. (1948) *Phys. Rev.*, **74**, 1439.
Schwinger, J. (1949*a*) *Phys. Rev.*, **75**, 651.
Schwinger, J. (1949*b*) *Phys. Rev.*, **76**, 790.
Schwinger, J. (1951*a*) *Phys. Rev.*, **82**, 664.
Schwinger, J. (1951*b*) *Proc. Nat. Acad. Sci.* (USA), **37**, 452.
Sciama, D.W. (1976) *Vistas in Astron.*, **19**, 385.
Segal, I.E. (1967) in *Applications of Mathematics to Problems in Theoretical Physics*, ed. F. Luciat (New York: Gordon and Breach).
Sexl. R.U. and Urbantke, H.K. (1967) *Acta Phys. Austriaca*, **26**, 339.
Sexl, R.U. and Urbantke, H.K. (1969) *Phys. Rev.*, **179**, 1247.
Shannon, C. and Weaver, W. (1949) *The Mathematical Theory of Communication*, (Urbana: University of Illinois Press).
Shore, G.M. (1980*a*) *Phys. Rev. D*, **21**, 2226.
Shore, G.M. (1980*b*) *Ann. Phys.* (NY), **128**, 376.
Simpson, M. and Penrose, R. (1973) *Int. J. Theor. Phys.*, **7**, 183.
Sommerfield, C.M. (1974) *Ann. Phys.* (NY), **84**, 285.
Sparnaay, M.J. (1958) *Physica* (Utrecht), **24**, 751.
Starobinsky, A.A. (1973) *Zh. Eksp. Teor. Fiz.*, **64**, 48 (*Sov. Phys. JETP*, **37**, 28(1973))
Starobinsky, A.A. (1980) *Phys. Lett.*, **91B**, 99.
Starobinsky, A.A. and Churilov, S.M. (1973) *Zh. Eksp. Teor. Fiz.*, **65**, 3 (*Sov. Phys. JETP*, **38**, 1 (1974)).
Stelle, K.S. (1977) *Phys. Rev. D*, **16**, 953.
Stelle, K.S. (1978) *Gen. Relativ. Gravit.*, **9**, 353.
Stewartson, K. and Waechter, R.T. (1971) *Proc. Camb. Phil. Soc.*, **69**, 353.
Symanzik, K. (1970) *Commun. Math. Phys.*, **18**, 48.
Synge, J.L. (1960) *Relativity: The General Theory* (Amsterdam: North Holland).
Tabor, D. and Winterton, R.H.S. (1969) *Proc. R. Soc. London*, **A312**, 435.

Tagirov, E.A. (1973) *Ann. Phys.* (NY), **76**, 561.
Takahashi, Y. and Umezawa, H. (1957) *Nuovo Cimento*, **6**, 1324.
Taylor, J.G. (1960) *Nuovo Cimento*, **17**, 695.
Taylor, J.G. (1963) *Supp. Nuovo Cimento*, **1**, 857.
Taylor, J.C. (1976) *Gauge Theories of Weak Interactions* (Cambridge: Cambridge University Press).
Teukolsky, S.A. (1973) *Astrophys. J.*, **185**, 635.
Teukolsky, S.A. and Press, W.H. (1974) *Astrophys. J.*, **193**, 443.
Thirring, W.E. (1958*a*) *Ann. Phys.* (NY), **3**, 91.
Thirring, W.E. (1958*b*) *Nuovo Cimento*, **9**, 1007.
't Hooft, G. (1973) *Nucl. Phys.*, **B61**, 455.
't Hooft, G. and Veltman, M. (1972) *Nucl. Phys*, **B44**, 189.
Tipler, F. (1980) *Phys. Rev. Lett.*, **45**, 949.
Tolman, R.C. (1934) *Relativity, Thermodynamics, and Cosmology* (Oxford: Clarendon).
Tomonaga, S. (1946) *Prog. Theor. Phys.*, **1**, 27.
Toms, D.J. (1980*a*) *Phys. Rev. D*, **21**, 928.
Toms, D.J. (1980*b*) *Phys. Rev. D*, **21**, 2805.
Toms, D.J. (1980*c*) *Ann. Phys.* (NY), **129**, 334.
Toms, D.J. (1980*d*) *Phys. Lett.*, **77A**, 303.
Troost, W. and van Dam H. (1977) *Phys. Lett.*, **71B**, 149.
Unruh, W.G. (1974) *Phys. Rev. D*, **10**, 3194.
Unruh, W.G. (1976) *Phys. Rev. D*, **14**, 870.
Urbantke, H.K. (1969) *Nuovo Cimento*, **63B**, 203.
Utiyama, R. (1962) *Phys. Rev.*, **125**, 1727.
Utiyama, R. and DeWitt, B.S. (1962) *J. Math. Phys.* (NY), **3**, 608.
Valatin, J.G. (1954*a*) *Proc. R. Soc. London*, **A222**, 93.
Valatin, J.G. (1954*b*) *Proc. R. Soc. London*, **A222**, 228.
Valatin, J.G. (1954*c*) *Proc. R. Soc. London*, **A225**, 535.
van Nieuwenhuizen, P. and Freedman, D.Z. (1979) (eds.) *Supergravity* (Amsterdam: North Holland).
Van Vleck (1928) *Proc. Nat. Acad. Sci.* (USA), **14**, 178.
Vilenkin, A. (1978) *Nuovo Cimento*, **44A**, 441.
Volovich, I.V., Zagrebnov, V.A. and Frolov, V.P. (1977) *Teor. & Math. Fiz.* (USSR), **33**, 3 (*Theor. & Math. Phys.*, **33**, 843 (1977)).
Waechter, R.T. (1972) *Proc. Camb. Phil. Soc.*, **72**, 439.
Wald, R.M. (1975) *Commun. Math. Phys.*, **45**, 9.
Wald, R.M. (1976) *Phys. Rev. D*, **13**, 3176.
Wald, R.M. (1977) *Commun. Math. Phys.*, **54**, 1.
Wald, R.M. (1978*a*) *Phys. Rev. D*, **17**, 1477.
Wald, R.M. (1978*b*) *Ann. Phys.* (NY), **110**, 472.
Wald, R.M. (1979*a*) *Ann. Phys.* (NY), **118**, 490.
Wald, R.M. (1979*b*) *Commun. Math. Phys.*, **70**, 221.
Wald, R.M. (1980) *Phys. Rev. D*, **21**, 2742.
Wald, R.M. and Ramaswamy, S. (1980) *Phys. Rev. D*, **21**, 2736.
Wehrl, A. (1978) *Rev. Mod. Phys.*, **50**, 221.
Weinberg, S. (1964*a*) *Phys. Rev.*, **133**, B1318.
Weinberg, S. (1964*b*) *Phys. Rev.*, **134**, B882.
Weinberg, S. (1964*c*) 'The quantum theory of massless particles' in *Lectures on Particles and Field Theory*. Vol. II, eds. K.A. Johnson and others (Englewood Cliffs, N.J.: Prentice-Hall).
Weinberg, S. (1967) *Phys. Rev. Lett.*, **19**, 1264.

Weinberg, S. (1972) *Gravitation and Cosmology: Principles and Applications of the General Theory of Relativity* (New York: Wiley).
Weinberg, S. (1974) *Phys. Rev. D*, **9**, 3357.
Wichmann, E.H. (1962) *Selected Topics in the Theory of Particles and Fields* (Copenhagen: Nordita).
Wick, G.C. (1950) *Phys. Rev.*, **80**, 268.
Zel'dovich, Ya,B. (1970) *Pis'ma Zh. Eksp. Teor. Fiz.*, **12**, 443 (*JETP Lett.*, **12**, 307(1970)).
Zel'dovich, Ya.B. (1971) *Pis'ma Zh. Eksp. Teor. Fiz.*, **14**, 270 (*JETP Lett.*, **14**, 180(1971)).
Zel'dovich, Ya.B. (1972) *Zh. Eksp. Teor. Fiz.*, **62**, 2076 (*Sov. Phys. JETP*, **35**, 1085 (1972)).
Zel'dovich, Ya.B., Novikov, I.D. and Starobinsky, A.A. (1974) *Zh. Eksp. Teor. Fiz.*, **66**, 1897 (*Sov. Phys. JETP*, **39**, 933 (1974)).
Zel'dovich, Ya.B. and Pitaevsky, L.P. (1971) *Commun. Math. Phys.*, **23**, 185.
Zel'dovich, Ya.B. and Starobinsky, A.A. (1971) *Zh. Eksp. Teor. Fiz.*, **61**, 2161 (*Sov. Phys. JETP*, **34**, 1159 (1972)).
Zel'dovich, Ya.B. and Starobinsky, A.A. (1977) *Pis'ma Zh. Eksp. Teor. Fiz.*, **26**, 373 (*JETP Lett.*, **26**, 252 (1977)).
Zurek, W.H. (1980) *Phys. Lett.*, **77A**, 399.

CONFORMAL ANOMALIES

Adler, S.L., Lieberman, J. and Ng, Y.J. (1977) *Ann. Phys.* (NY), **106**, 279.
Bernard, C. and Duncan, A. (1977) *Ann. Phys.* (NY), **107**, 201.
Brown, L.S. (1977) *Phys. Rev. D*, **15**, 1469.
Brown, L.S. and Cassidy, J.P. (1977*a*) *Phys.Rev. D*, **15**, 2810.
Brown, L.S. and Cassidy, J.P. (1977*b*) *Phys. Rev. D*, **16**, 1712.
Bunch, T.S. (1978) *J. Phys.*, **A11**, 603.
Bunch, T.S. (1979) *J. Phys.*, **A12**, 517.
Capper, D.M. and Duff, M.J. (1974) *Nuovo Cimento*, **A23**, 173.
Capper, D.M. and Duff, M.J. (1975) *Phys. Lett.*, **53A**, 361.
Christensen, S.M. (1976) *Phys. Rev. D*, **14**, 2490.
Christensen, S.M. (1978) *Phys. Rev. D*, **17**, 946.
Christensen, S.M. and Duff, M.J. (1978*a*) *Phys. Lett.*, **76B**, 571.
Christensen, S.M. and Duff, M.J. (1978*b*) *Nucl. Phys.*, **B146**, 11.
Christensen, S.M. and Fulling, S.A. (1977) *Phys. Rev. D*, **15**, 2088.
Critchley, R. (1978*a*) *Phys. Rev. D*, **18** 1849.
Critchley, R. (1978*b*) *Phys. Lett.*, **78B**, 410.
Davies, P.C.W. (1977) 'Stress tensor calculations and conformal anomalies' in *Eighth Texas Symposium on Relativistic Astrophysics*, ed. M.D. Papagiannis (New York: New York Academy of Sciences).
Davies, P.C.W. and Fulling, S.A. (1977) *Proc. R. Soc. London*, **A354**, 59.
Davies, P.C.W., Fulling, S.A. and Unruh, W.G. (1976) *Phys. Rev. D*, **13**, 2720.
Deser, S., Duff, M.J. and Isham, C.J. (1976) *Nucl. Phys.*, **B111**, 45.
Dowker, J.S. and Critchley, R. (1976) *Phys. Rev. D*, **13**, 3224.
Dowker, J.S. and Critchley, R. (1977) *Phys. Rev. D*, **16**, 3390.
Duncan, A. (1977) *Phys. Lett.*, **66B**, 170.
Duff, M.J. (1978) *Nucl. Phys.*, **B125**, 334.
Eguchi, T. and Freund, P.G.O. (1976) *Phys. Rev. Lett.*, **37**, 125.
Fradkin, E.S. and Vilkovisky, G.A. (1978) *Phys. Lett.*, **73B**, 209.

Hawking, S.W. (1977) *Commun. Math. Phys.*, **55**, 133.
Hu, B.L. (1978) *Phys. Rev. D.*, **18**, 4460.
Hu, B.L. (1979) *Phys. Lett.*, **71A**, 169.
Perry, M.J. (1978) *Nucl. Phys.*, **B143**, 114.
Tsao, H.S. (1977) *Phys. Lett.*, **68B**, 79.
Wald, R.M. (1977) *Commun. Math. Phys.*, **54**, 1.
Wald, R.M. (1978a) *Ann. Phys.* (NY), **110**, 472.
Wald, R.M. (1978b) *Phys. Rev. D*, **17**, 1477.
Vilenkin, A. (1978) *Nuovo Cimento*, **44A**, 441.

DE SITTER SPACE

Adler, S.L. (1972) *Phys. Rev. D*, **6**, 3445.
Adler, S.L. (1973) *Phys. Rev. D*, **8**, 2400.
Birrell, N.D. (1978) *Proc. R. Soc. London*, **A361**, 513.
Birrell, N.D. (1979) *J. Phys. A: Gen. Phys.*, **12**, 337.
Borner, G. and Durr, H.P. (1969) *Nuovo Cimento*, **64**, 669.
Bunch, T.S. and Davies, P.C.W. (1978) *Proc. R. Soc. London*, **360**, 117.
Candelas, P. and Raine D.J. (1975) *Phys. Rev. D*, **12**, 965.
Chernikov, N.A. and Tagirov. E.A. (1968) *Ann. Inst. Henri Poincaré*, **9A**, 109.
Dirac, P.A.M. (1935) *Ann. Math.*, **36**, 657.
Dowker, J.S. and Critchley, R. (1976*a*) *Phys. Rev. D*, **13**, 224.
Dowker, J.S. and Critchley, R. (1976*b*) *Phys. Rev. D*, **13**, 3224.
Drummond, I.T. (1975) *Nucl. Phys.*, **B94**, 115.
Drummond, I.T. and Shore, G.M. (1979) *Phys. Rev. D*, **19**, 1134.
Figari, R., Hoegh-Krohn, R. and Nappi, C. (1975) *Commun. Math. Phys.*, **44**, 265.
Fronsdal, C. (1965) *Rev. Mod. Phys.*, **37**, 221.
Fronsdal, C. (1974) *Phys. Rev. D*, **10**, 589.
Fronsdal, C. and Haugen, R.B. (1975) *Phys. Rev. D*, **12**, 3810.
Fulling, S.A. (1972) Ph.D. Thesis, Princeton University (unpublished).
Geheniau, J. and Schomblond, C. (1968) *Acad. R. de Belgique, Bull. Cl. des Sciences*, **54**, 1147.
Gibbons, G.W. and Hawking, S.W. (1977) *Phys. Rev. D*, **15**, 2738.
Grensing, G. (1977) *J. Phys. A: Gen. Phys.*, **10**, 1687.
Lohiya, D. (1978) *J. Phys. A: Gen Phys.*, **11**, 1335.
Lohiya, D. and Panchapakesan, N. (1978) *J. Phys. A: Gen. Phys.*, **11**, 1963.
Lohiya, D. and Panchapakesan, N. (1979) *J. Phys. A: Gen. Phys.*, **12**, 533.
Nachtmann, O. (1967) *Commun. Math. Phys.*, **6**, 1.
Rumpf, H. (1979) *Gen. Relativ. Gravit.*, **10**, 647.
Schomblond, C. and Spindel, P. (1976) *Ann. Inst. Henri Poincaré*, **25A**, 67.
Shore, G.M. (1979) *Ann. Phys.* (NY), **117**, 89.
Shore, G.M. (1980) *Phys. Rev. D*, **21**, 2226.
Tagirov, E.A. (1973) *Ann. Phys.* (NY), **76** 561.

Index

Advanced Textbooks in Control and Signal Processing

Series editors

Michael J. Grimble, Glasgow, UK
Michael A. Johnson, Kidlington, UK

More information about this series at http://www.springer.com/series/4045

Basil Kouvaritakis · Mark Cannon

Model Predictive Control

Classical, Robust and Stochastic

Basil Kouvaritakis
University of Oxford
Oxford
UK

Mark Cannon
University of Oxford
Oxford
UK

ISSN 1439-2232
Advanced Textbooks in Control and Signal Processing
ISBN 978-3-319-24851-6 ISBN 978-3-319-24853-0 (eBook)
DOI 10.1007/978-3-319-24853-0

Library of Congress Control Number: 2015951764

Springer Cham Heidelberg New York Dordrecht London

Printed on acid-free paper

Springer International Publishing AG Switzerland is part of Springer Science+Business Media
(www.springer.com)

To Niko Kouvaritakis, who introduced us to the sustainable development problem and initiated our interest in stochastic predictive control.

Series Editors' Foreword

The *Advanced Textbooks in Control and Signal Processing* series is designed as a vehicle for the systematic textbook presentation of both fundamental and innovative topics in the control and signal processing disciplines. It is hoped that prospective authors will welcome the opportunity to publish a more rounded and structured presentation of some of the newer emerging control and signal processing technologies in this textbook series. However, it is useful to note that there will always be a place in the series for contemporary presentations of foundational material in these important engineering areas.

In 1995, our monograph series *Advances in Industrial Control* published *Model Predictive Control in the Process Industries* by Eduardo F. Camacho and Carlos Bordons (ISBN 978-3-540-19924-3, 1995). The subject of model predictive control in all its different varieties is a popular control technique and the original monograph benefited from that popularity and consequently moved to the *Advanced Textbooks in Control and Signal Processing* series. In 2004, it was republished in a thoroughly updated second edition now simply entitled *Model Predictive Control* (ISBN 978-1-85233-694-3, 2004). A decade on, the new edition is a successful and well-received textbook within the textbook series.

As demonstrated by the continuing demand for Prof. Camacho and Prof. Bordon's textbook, the technique of model predictive control or "MPC" has been startlingly successful in both the academic and industrial control communities. If the reader considers the various concepts and principles that are combined in MPC, the reasons for this success are not so difficult to identify.

"M $\sim$ Model" From an early beginning with transfer-function models using the Laplace transform through the 1960s' revolution of state-space system descriptions leading on to the science of system identification, the use of a system or process model in control design is now very well accepted.

"P $\sim$ Predictive" The art of looking forward from a current situation and planning ahead to achieve an objective is simply a natural human activity. Thus, once a system model is available it can be used to predict ahead from a currently measured position to anticipate the future and avoid constraints and other restrictions.

"C ~ Control" This is the computation of the control action to be taken. The enabling idea here is automated computation achieved using optimization. A balance between output error and control effort used is captured in a cost function that is usually quadratic for mathematical tractability.

There may be further reasons for its success connected with nonlinearities, the future process output values to be attained, and any control signal restrictions; these combine to require constrained optimization. In applying the forward-looking control signal, the one-step-at-a-time receding-horizon principle is implemented.

Academic researchers have investigated so many theoretical aspects of MPC that it is a staple ingredient of innumerable journal and conference papers, and monographs. However, looking at the industrial world, the two control techniques that appear to find extensive real application seem to be PID control for simple applications and MPC for the more complicated situations. In common with the ubiquitous PID controller, MPC has intuitive depth that makes it easily understood and used by industrial control engineers. For these reasons alone, the study of PID control and MPC cannot be omitted from today's modern control course.

This is not to imply that all the theoretical or computational problems in MPC have been solved or are even straightforward. But it is the adding in of more process properties that leads to a need for a careful analysis of the MPC technique. This is the approach of this *Advanced Textbooks in Control and Signal Processing* entry entitled *Predictive Control: Classical, Robust and Stochastic* by Basil Kouvaritakis and Mark Cannon. The authors' work on predictive control at Oxford has been carried out over a long period and they have been very influential in stimulating interest in new algorithms for both linear and nonlinear systems.

Divided into three parts, the text considers linear system models subject to process-output and control constraints. The three parts are as follows: "Classical" refers to deterministic formulations, "Robust" incorporates uncertainty to the system description and "Stochastic" considers system uncertainty that has probabilistic properties. In the presence of constraints, the authors seek out conditions for closed-loop system stability, control feasibility, convergence and algorithmic computable control solutions. The series editors welcome this significant contribution to the MPC textbook literature that is also a valuable entry to the *Advanced Textbooks in Control and Signal Processing* series.

August 2015

Michael J. Grimble
Michael A. Johnson
Industrial Control Centre
Glasgow, Scotland, UK

Preface

One of the motivations behind this book was to collect together the many results of the Oxford University predictive control group. For this reason we have, rather unashamedly, included a number of ideas that were developed at Oxford and in this sense some of the discussions in the book are included as background material that some readers may wish to skip on an initial reading. Elsewhere, however, the preference for our own methodology is quite deliberate on account of the distinctive nature of some of the Oxford results. Thus, for example, in Stochastic MPC our attention is focussed on algorithms with guaranteed control theoretic properties, including that of recurrent feasibility. On account of this, contrary to common practice, we often eschew the normal distribution, which despite its mathematical convenience neither lends itself to the proof of stability and feasibility, nor does it allow accurate representations of model and measurement uncertainties, as these rarely assume arbitrarily large values. On the other hand, we have clearly attempted to incorporate all the major developments in the field, some of which are rather recent and as yet may not be widely known. We apologise to colleagues whose work did not get a mention in our account of the development of MPC; mostly this is due to fact that we had to be selective of our material so as to give a fuller description over a narrower range of concepts and techniques.

Over the past few decades the state of the art in MPC has come closer to the optimum trade-off between computation and performance. But it is still nowhere near close enough for many control problems. In this respect the field is wide open for researchers to come up with fresh ideas that will help bridge the gap between ideal performance and what is achievable in practice.

October 2014

Basil Kouvaritakis
Mark Cannon

Contents

Chapter 1
Introduction

The benefits of feedback control have been known to mankind for more than 2,000 years and examples of its use can be found in ancient Greece, notably the float regulator of the water clock invented by Ktesibios in about 270 BC [1]. The formal development of the field as a mathematical tool for the analysis of the behaviour of dynamical systems is much more recent, beginning around 150 years ago when Maxwell published his work on governors [2]. Since then the field has seen spectacular developments, promoted by the work of mathematicians, engineers and physicists. Laplace, Lyapunov, Kolmogorov, Wiener, Nyquist, Bode, Bellman are just a few of the major contributors to the edifice of what is known today as control theory.

A development of particular interest, both from a theoretical point of view but also one that has enjoyed considerable success in terms of practical applications, is that of optimal control. Largely based on the work of Pontryagin [3] and Bellman [4], optimal control theory is an extension of the calculus of variations [5, 6] addressing the problem of optimizing a cost index that measures system performance through the choice of system parameters that are designated as control inputs. The appeal of this work from a control engineering perspective is obvious because it provides a systematic approach to the design of strategies that achieve optimal performance. Crucially, the optimal control solution has the particularly simple form of linear state feedback for the case of linear systems and quadratic cost functions, and the feedback gains can be computed by solving an equation known as the steady-state Riccati equation [7]. This applies to both continuous time systems described by sets of differential equations and to discrete time systems formulated in terms of difference equation models. The latter description is, of course, of special importance to modern applications, which are almost entirely implemented using digital microprocessors.

The benefits of optimal control are, however, difficult to achieve in the case of systems with nonlinear models and systems that are subject to constraints on input variables or model states. For both these cases, in general, it is not possible to derive analytic expressions for the optimal control solution. Given the continuing

B. Kouvaritakis and M. Cannon, *Model Predictive Control*,
Advanced Textbooks in Control and Signal Processing,
DOI 10.1007/978-3-319-24853-0_1

improvements in the processing power of inexpensive microprocessors, one might hope that optimal solutions could be computed numerically. However, the associated optimization problem is difficult to solve for all but the simplest cases, and hence it is impracticable for the majority of realistic control problems. In the pursuit of optimality one is therefore forced to consider approximate solutions, and this is perhaps the single most important reason behind the phenomenal success of model predictive control (MPC). MPC is arguably the most widely accepted modern control strategy because it offers, through its receding horizon implementation, an eminently sensible compromise between optimality and speed of computation.

The philosophy of MPC can be described simply as follows. Predict future behaviour using a system model, given measurements or estimates of the current state of the system and a hypothetical future input trajectory or feedback control policy. In this framework future inputs are characterized by a finite number of degrees of freedom, which are used to optimize a predicted cost. Only the first control input of the optimal control sequence is implemented, and, to introduce feedback into this strategy, the process is repeated at the next time instant using newly available information on the system state. This repetition is instrumental in reducing the gap between the predicted and the actual system response (in closed-loop operation). It also provides a certain degree of inherent robustness to the uncertainty that can arise from imperfect knowledge or unknown variations in the model parameters (referred to as multiplicative uncertainty), as well as to model uncertainty in the form of disturbances appearing additively in the system dynamics (referred to as additive uncertainty).

Many early MPC strategies took account of predicted system behaviour over a finite horizon only and therefore lacked guarantees of nominal stability (i.e. closed-loop stability in the absence of any uncertainty). This difficulty was initially overcome by imposing additional conditions, known as equality terminal constraints, on the predicted model states. Such conditions were chosen in order to ensure that the desired steady state was reached at the end of a finite prediction horizon. The effect of these constraints was to render a finite horizon equivalent to an infinite horizon, thereby ensuring various stability and convergence properties.

Imposing the requirement that predicted behaviour reaches steady state over a finite future horizon is in general an overly-stringent requirement, and furthermore it presents computational challenges in the case of systems described by nonlinear models. Instead it was proposed that a stabilizing feedback law could be used to define the predicted control inputs at future times beyond the initial, finite prediction horizon. This feedback law is known as a terminal control law, and is often taken to be the optimal control law for the actual system dynamics in the absence of constraints (if that is available), or otherwise can be chosen as the optimal control law for the unconstrained, linearized dynamics about the desired steady state. To ensure that this control law meets the system constraints, thereby ensuring the future feasibility of the receding horizon strategy, additional constraints known as terminal constraints are imposed. These typically require that the system state at the end of the initial finite prediction horizon should belong to a subset of state space with the property that once entered, the state of the constrained system will never leave it.

We refer to the MPC algorithms that are derived from the collection of ideas discussed above as Classical MPC. These cause the controlled system, in closed-loop operation, to be stable, to meet constraints and to converge (asymptotically) to the desired steady state. However, it is often of paramount importance that a controller should have an acceptable degree of robustness to model uncertainty. This constitutes a much more challenging control problem. We refer to the case in which the uncertainty has known bounds but no further information is assumed as Robust MPC (or RMPC), and to the case in which model uncertainty is assumed to be random with known probability distribution, and where some or all of the constraints are probabilistic in nature, as Stochastic MPC (or SMPC).

Thus let the state model of a system be $x^+ = Ax + Bu + Dw$, where x and x^+ denote, respectively, the current model state and the successor state (i.e. the state at the next time instant), u is the vector of control inputs and w represents an unknown vector of external disturbance inputs. For such a model it may be the case that the numerical values of the elements of the matrices A, B, D are not known (or indeed knowable) precisely; this corresponds to the case of multiplicative uncertainty. The model parameters may however be known to lie in particular intervals, whether they are constant or vary with time, in which case the uncertainty is bounded by a known set of values. Typically these uncertainty sets will be polytopic sets defined by known vertices or by a number of linear inequalities. To give an example of this, consider the payloads of a robotic arm that may differ from time to time, depending on the task performed. This naturally leads to uncertainty which can be modelled (albeit conservatively) as multiplicative bounded uncertainty in a linear model. Similarly, the additive disturbance representing, for example, torques arising from static friction in the robot arm system discussed above, though unknown, will lie in a bounded set of additive uncertainty.

Such a problem would almost certainly be subject to constraints implied by maxima on possible applied motor torques, or consideration of safety and/or singularities which impose limits on the angular positions of the various links of the robotic arm. A general linear representation of such constraints, whether they apply only to the control input, or to the state, or are mixed input and state constraints, is $Fx + Bu \leq 1$. The concern then for RMPC would be to guarantee stability, constraint satisfaction and convergence of the state vector to a given steady-state condition or set of states, for all possible realizations of uncertainty.

In a number of applications, however, uncertainty is subject to some statistical regularity and can be modelled as random but with known probability distribution. Thus consider the problem of controlling the pitch of the blades of a wind turbine with the aim of maximizing electrical power generation while at the same time limiting the fatigue damage to the turbine tower due to fore-and-aft movement caused by fluctuations in the aerodynamic forces experienced by the blades. Although the wind speed is subject to random variations, it can be modelled in terms of given probability distributions. Furthermore, fatigue damage occurs when tower movement exceeds given limits frequently. Therefore the implied constraint is not that extreme tower movement is not allowed but rather that it happens with a probability which is below a given threshold. This situation cannot be described by a hard constraint such as

$Fx + Gu \leq 1$, but can more conveniently be modelled by probabilistic constraints of the form $\Pr\{Fx + Gu \leq 1\} \leq p$ where p represents a given probability. It is the object of Stochastic MPC to ensure that such constraints, together with any additional hard constraints that may be present, are met in closed-loop operation, and to simultaneously stabilize the system, for example by causing the state to converge to a given steady-state set.

Classical, Robust and Stochastic MPC are the main topics of this book; the three distinct parts of the book discuss each of these in turn. Our tendency has been in each part of the book to start with background material that helps define the basic concepts and then progressively present more sophisticated algorithms. By and large these are capable of affording advantages over the earlier presented results in terms of ease of computation, or breadth of applicability in terms of the size of their allowable set of initial conditions or degree of optimality of performance. This in a way is a reflection of the development of the field as a whole, which continuously aspires for optimality tempered with ease of implementation.

This book only explicitly addresses systems with linear dynamics, but the reader should be aware that most of the results presented have obvious extensions to the nonlinear case, provided certain assumptions about convexity are satisfied. There are clear implications of course in terms of ease of online computation, but the hope is that with the increasing speed and storage capabilities of computing hardware for control, this aspect will become less significant. Below, we give a brief description of the salient features to be presented in each of the three parts of the book.

1.1 Classical MPC

The first part comprises a single chapter that describes developments concerning the nominal case only, but simultaneously lays the foundations for the remainder of the book by introducing key concepts (e.g. set invariance, recurrent feasibility, Lyapunov stability, etc.). It begins with the problem definition and an unconventional derivation of the optimal unconstrained control law; this is done deliberately so as to avoid repeating the classical presentation of ideas of calculus of variations and Pontryagin's maximum principle. It then moves on to describe the dual mode prediction setting. Some readers may object to the use of the term "dual mode" here since it is sometimes reserved to describe a control strategy that switches between two different control laws (e.g. when the system state transitions into a particular region of state space around the desired steady state). Our use of the term "dual mode" refers instead to the split of the predicted control sequence into the control inputs at the first N predicted time-steps (which are not predetermined) and those that apply to the remainder of the prediction horizon (which are fixed by a terminal feedback law). The terms mode 1 and mode 2 constitute in our opinion a useful and intuitive shorthand for the two modes of predicted operation.

Discussion then turns to set invariance, which in the first instance is introduced in connection with a terminal control law. Subsequently this is developed into the concept of controlled invariance, which is a convenient concept for describing the feasibility properties of the system constraints combined with terminal constraints. Instrumental in this and in the attendant proof of stability is the idea of what we call, for lack of a better descriptive terminology, the "tail". This constitutes a method of extending a future trajectory computed at any given time instant to a subsequent time instant, and provides a convenient tool for establishing both recurrent feasibility and a monotonic non-increasing property of the cost. The former relates to the property that feasibility at current time implies feasibility at the next and subsequent instants, whereas the latter can be used to establish closed-loop stability.

A formulation of interest concerns an alternative representation of the prediction dynamics that makes use of a lifted autonomous state space model. Here the state is augmented to include the degrees of freedom in the control sequence of mode 1; this provides a framework that expedites many of the arguments presented in the subsequent parts of the book, and it also leads to a particularly efficient implementation of nominal MPC. Moreover, it enables the design of prediction dynamics that can be optimized, for example with the aim of maximizing the size of the set of allowable initial conditions.

The chapter also considers aspects of computation and presents early MPC algorithms, one of which enables the introduction of a Youla parameter into the MPC problem.

1.2 Robust MPC

The presence of bounded uncertainty leads to a generally more challenging control problem which is addressed by RMPC and is discussed in the second part of the book. We begin by considering the case of additive disturbances, examining first a class of MPC strategies that perform prediction optimization over open-loop input trajectories. These are distinct from strategies employing optimization over control policies, which in essence are closed-loop prediction strategies in that they take account of future realizations of uncertainty which, though not known to the controller at current time, will be available when the control law is computed at a future time.

We begin our account by describing a state decomposition into nominal and uncertain components. This, in conjunction with an augmented predicted state model, enables the treatment of robust invariance and recursive feasibility and also suggests convenient ways to define and compute maximal and minimal robust invariant sets. Then, using the dual mode prediction paradigm and a nominal predicted cost, it is possible to develop an RMPC strategy with guaranteed stability and convergence properties.

Next we consider a game theoretic approach, a control strategy that is revisited later when the case of mixed additive and multiplicative uncertainty is examined. This approach uses a min-max optimization in which the cost is defined so as to set up a dynamic game between the uncertainty, over which RMPC performs a maximization, and the controller, which selects the control input by minimizing the maximum cost. Over and above the usual control theoretic properties of recursive feasibility and convergence to some steady-state set, this approach also provides a quantification of the disturbance rejection properties of RMPC.

The exposition then moves on to the tube RMPC methodology that appears to have dominated the relevant literature over the past 15 years or so. According to this, constraints are enforced by inclusion conditions that ensure the uncertain future state and input trajectories lie in sequences of sets, known as tubes, which are contained entirely within sets in which constraints of the control problem are satisfied. The sets defining the tubes were originally taken to be low-complexity polytopic sets (i.e. affine transformations of hypercubes), but were later replaced by more general sets which are either fixed or scalable. Such tubes can be used to guarantee recursive feasibility through the use of suitably defined constraints. The topic of open-loop strategies for the additive uncertainty case is brought to a close through a review of some early Tube RMPC strategies that deployed tubes with low-complexity polytopic cross sections as well as a review of an RMPC strategy that achieved recurrent feasibility with respect to the entire class of additive uncertainty through an artificial tightening of constraints.

Open-loop strategies are computationally convenient but can be conservative since they ignore information about future uncertainty that, though not available at current time, will be available to the controller. For a given prediction horizon and terminal control law, the best possible performance and the largest set of admissible initial conditions is obtained by optimizing a multistage min-max control problem over all feedback policies. This problem, and its solution through the use of dynamic programming (DP), is considered next in the book. The drawback of this approach and its implementation within an MPC framework is that computation grows exponentially with the prediction horizon and system dimension.

The optimal solution can be shown to be an affine function of the system state which is dictated by the set of active constraints. Thus, in theory, one could potentially compute this function offline at the regions of the state space defined by different sets of active constraints. The number of such regions however typically grows exponentially with the size of the system and therefore this approach is not practicable for anything other than low order systems and short prediction horizons. To avoid this problem it is possible to use an approach based on an online interpolation between the current state and a state at which the optimal control law is known. The active constraint set is updated during this interpolation and an equality constrained optimization problem is solved at each active set change. Although this active set dynamic programming approach can lead to significant reductions in online computation, it still requires the offline computation of controllability sets which may be computationally demanding. In such cases it may be preferable to perform the optimization over a restricted class of control policies, since this may provide a good approximation of the optimal solution at a fraction of the computational cost.

One such policy employs a feedback parameterization with an affine dependence on past disturbance inputs, resulting in a convex optimization problem in a number of variables that grows quadratically with the prediction horizon. Through the use of a separable prediction scheme with a triangular structure it is possible to provide disturbance feedback that is piecewise affine (rather than simply affine) in the disturbance. The resulting RMPC algorithm, Parameterized Tube MPC (PTMPC), provides a greater degree of optimality for a similar computational load, with the number of optimization variables again depending quadratically on the prediction horizon. Computation can be reduced by using a scheme with a striped triangular structure (for which the dependence of the number of optimization variables on horizon length is linear) rather than a triangular prediction scheme. With this approach it is possible to outperform PTMPC since the effects of the striped prediction structure are allowed to extend into mode 2, thus effectively replacing the fixed terminal law by one that is optimized online.

We next consider RMPC in the presence of multiplicative uncertainty. Early work on this topic used a parameterization of predicted control inputs in terms of a linear state feedback law in which the feedback gain is taken to be an optimization variable computed online at each time instant. The approach uses quadratic constraints, expressed as linear matrix inequalities, to ensure that the predicted state is contained in ellipsoids within which constraints are satisfied. These ellipsoidal sets also guarantee a monotonic non-increasing property for the predicted cost for a polytopic class of uncertainty. The prediction structure of this approach was subsequently enriched by introducing additional optimization variables in the form of a perturbation sequence applied to the predicted linear feedback law. However, this increases the required online computation considerably, making it overly demanding for high order systems or systems with many uncertain parameters. A very significant reduction in online computation can be achieved by using a lifted autonomous model for the prediction dynamics. This latter approach also enables the dynamics defining the predicted state and input trajectories to be optimized through an offline optimization, thus maximizing the volume of an ellipsoidal region of attraction.

As in the case of additive uncertainty, multiplicative uncertainty can be handled conveniently through the introduction of tubes defining the evolution of the predicted state and control trajectories. This is considered next, first with tubes consisting of low-complexity polytopic sets, and then with general polytopic tube cross sections through appropriate use of Farkas' lemma. A combination of these ideas with the lifted optimized dynamics is discussed for the derivation of a min-max RMPC algorithm for the case of mixed additive and multiplicative uncertainty.

1.3 Stochastic MPC

The final part of this book is dedicated to SMPC which deals with the case when uncertainty is random, with known probability distribution, rather than simply being known to lie in a given set. This opens up the possibility of incorporating statistical information in the definition of optimal performance, and of allowing some or all of the constraints to be of a probabilistic nature. Early versions of SMPC were concerned with the unconstrained case and considered the minimization of the one-step-ahead variance or the expected value of a quadratic cost over a given prediction horizon. However, use of only expected values removes much of the stochastic nature of the problem, so these early algorithms, although historically important, cannot really be classified as SMPC.

We begin our treatment of SMPC by defining the stochastic system models together with the probabilistic constraints and discussing the basic assumption of mean-square stability. Use is made of the dual mode prediction paradigm and the unconstrained optimal control law is developed for a particular form of predicted cost. This is then followed by the treatment of a mean-variance predicted cost SMPC. The discussion is concluded by a review of some earlier stochastic predictive control work, namely Minimum Variance Control and MPC through the use of moving average models in conjunction with chance constraints. Also included is a description of earlier work based on a fully stochastic formulation in which both the constraints and the cost are defined using the probabilistic information about the uncertainty. This approach was developed in the context of a sustainable development problem which is also discussed. Such a formulation appears to be eminently appropriate for a problem with such a strong stochastic element: it involves a prediction horizon, which by definition has to be the inter-generational gap, and over which it is unrealistic to model the uncertainties of world economy in a deterministic manner.

The following chapter presents useful tools for the construction of SMPC algorithms such as recursive feasibility, supermartingale convergence analysis, probabilistic invariance and Markov chain models. Using these ingredients, SMPC algorithms are proposed using an expectation cost and a mean variance cost as well as algorithms based on probabilistically invariant ellipsoids and algorithms based on tubes with polytopic cross sections constructed on the basis of Markov chain models.

Our discussion of SMPC concludes by considering algorithms that use tubes with polytopic and elliptical cross sections constructed on the basis of information on the probability distributions of additive and multiplicative uncertainty. One feature of these algorithms is that they achieve recursive feasibility by treating, at any prediction time, the uncertainty up to the previous prediction time robustly. This is because, at any given time, all earlier realizations of uncertainty will have already occurred and will have taken any allowable value in the uncertainty class. We emphasize that our preference is for uncertainty with finite support, despite the mathematical convenience of distributions such as the Gaussian distribution. In general uncertainty distributions with infinite support do not accord well with realistic applications, where model uncertainty never assumes arbitrarily large values. Moreover, assumptions of unbounded model uncertainty preclude the possibility of establishing control theoretic properties such as stability and feasibility. Consideration is also given to SMPC which addresses constraints on the average number of constraint violations. The case of multiplicative uncertainty poses interesting problems in respect of the online calculation of probability distributions of predicted states. We discuss how this difficulty can be overcome through the use of techniques based on numerical integration and random sampling.

1.4 Concluding Remarks and Comments on the Intended Readership

In summary, predictive control has experienced a phenomenal amount of development, and this has been matched by wide acceptability in practice. Classical MPC is now a mature research area in which further major developments are, in our opinion, unlikely. The same however is certainly not true of RMPC, where the race is still on for the development of approaches that provide an improved balance between optimality and practicability of implementation. It is to be expected that fresh ideas about the structure of predictions will emerge that will narrow the gap between what can realistically be implemented and the ideal optimal solution. SMPC has itself seen some significant advances but is still more in a state of flux, even in respect of what its aims ought to be. This area will undoubtedly see in the future several major stages of development. It is our hope that this book will seed some of the ideas that will make this possible.

The levels of difficulty and technical detail of the book differ from chapter to chapter. The intention is that Chap. 2 should be accessible to all undergraduates specializing in control, whereas Chaps. 2, 3 and 5 should be of interest to graduate students of control. With that in mind we have provided exercises (with solutions) to these chapters in the hope that the students will be able to test their understanding by solving the given problems. This can be done either as a paper-and-pencil exercise, or with the aid of mathematical software such as MATLAB. It is anticipated that Chaps. 4 and 6–8 would be read mostly by research students, researchers and academics. The technical level here is more involved and testing one's understanding could only be attempted using rather sophisticated suites of (MATLAB) programs and for this reason we have refrained from providing exercises for these chapters.

References

1. Vitruvius, *The Ten Books on Architecture* (Harvard University Press, Cambridge, 1914)
2. J.C. Maxwell, On governors. Proc. R. Soc. Lond. **16**, 270–283 (1868)
3. L.S. Pontryagin, V.G. Boltyanskii, R.V. Gamkrelidze, E.F. Mishchenko, *The Mathematical Theory of Optimal Processes* (Interscience, Wiley, 1962)
4. L.S. Pontryagin, V.G. Boltyanskii, R.V. Gamkrelidze, E.F. Mishchenko, *Dynamic Programming* (Princeton University Press, Princeton, 1957)
5. H.H. Goldstine, *A History of the Calculus of Variations From the 17th Through the 19th Century* (Springer, Berlin, 1980)
6. A.E. Bryson, Optimal control—1950 to 1985. IEEE Control Syst. Mag. **16**(3), 26–33 (1996)
7. R.E. Kalman, Contributions to the theory of optimal control. Bol. Soc. Mat. Mexicana **5**, 102–119 (1960)

Part I
Classical MPC

Chapter 2
MPC with No Model Uncertainty

2.1 Problem Description

This section provides a review of some of the key concepts and techniques in classical MPC. Here the term "classical MPC" refers to a class of control problems involving linear time invariant (LTI) systems whose dynamics are described by a discrete time model that is not subject to any uncertainty, either in the form of unknown additive disturbances or imprecise knowledge of the system parameters. In the first instance the assumption will be made that the system dynamics can be described in terms of the LTI state-space model

$$x_{k+1} = Ax_k + Bu_k \tag{2.1a}$$

$$y_k = Cx_k \tag{2.1b}$$

where $x_k \in \mathbb{R}^{n_x}$, $u_k \in \mathbb{R}^{n_u}$, $y_k \in \mathbb{R}^{n_y}$ are, respectively, the system state, the control input and the system output, and k is the discrete time index. If the system to be controlled is described by a model with continuous time dynamics (such as an ordinary differential equation), then the implicit assumption is made here that the controller can be implemented as a sampled data system and that (2.1a) defines the discrete time dynamics relating the samples of the system state to those of its control inputs.

Assumption 2.1 Unless otherwise stated, the state x_k of the system (2.1a) is assumed to be measured and made available to the controller at each sampling instant $k = 0, 1, \ldots$

The controlled system is also assumed to be subject to linear constraints. In general these may involve both states and inputs and are expressed as a set of linear inequalities

$$Fx + Gu \leq \mathbf{1} \tag{2.2}$$

B. Kouvaritakis and M. Cannon, *Model Predictive Control*,
Advanced Textbooks in Control and Signal Processing,
DOI 10.1007/978-3-319-24853-0_2

where $F \in \mathbb{R}^{n_C \times n_x}$, $G \in \mathbb{R}^{n_C \times n_u}$ and the inequality applies elementwise. We denote by $\mathbf{1}$ a vector with elements equal to unity, the dimension of which is context dependent, i.e. $\mathbf{1} = [1 \cdots 1]^T \in \mathbb{R}^{n_C}$ in (2.2). Setting F or G to zero results in constraints on inputs or states alone. A feasible pair (x_k, u_k) or feasible sequence $\{(x_0, u_0), (x_1, u_1), \ldots\}$ for (2.2) is any pair or sequence satisfying (2.2). The constraints in (2.2) are symmetric if $(-x_k, -u_k)$ is feasible whenever (x_k, u_k) is feasible, and non-symmetric otherwise. Although the form of (2.2) does not encompass constraints involving states or inputs at more than one sampling instant (such as, for example rate constraints or more general dynamic constraints), these can be handled through a suitable and obvious extension of the results to be presented.

The classical regulation problem is concerned with the design of a controller that drives the system state to some desired reference point using an acceptable amount of control effort. For the case that the state is to be steered to the origin, the controller performance is quantified conveniently for this type of problem by a quadratic cost index of the form

$$J\left(x_0, \{u_0, u_1, u_2 \ldots\}\right) \doteq \sum_{k=0}^{\infty} \left(\|x_k\|_Q^2 + \|u_k\|_R^2\right). \tag{2.3}$$

Here $\|v\|_S^2$ denotes the quadratic form $v^T S v$ for any $v \in \mathbb{R}^{n_v}$ and $S = S^T \in \mathbb{R}^{n_v \times n_v}$, and Q, R are weighting matrices that specify the emphasis placed on particular states and inputs in the cost. We assume that R is a symmetric positive-definite matrix (i.e. the eigenvalues of R are real and strictly positive, denoted $R \succ 0$) and that Q is symmetric and positive semidefinite (all eigenvalues of Q are real and non-negative, denoted $Q \succeq 0$). This allows, for example, the choice $Q = C^T Q_y C$ for some positive-definite matrix Q_y, which corresponds to the case that the output vector, y, rather than the state, x, is to be steered to the origin. At time k, the optimal value of the cost (2.3) with respect to minimization over admissible control sequences $\{u_k, u_{k+1}, u_{k+2}, \ldots\}$ is denoted

$$J^*(x_k) \doteq \min_{u_k, u_{k+1}, u_{k+2}, \ldots} J\left(x_k, \{u_k, u_{k+1}, u_{k+2} \ldots\}\right).$$

This problem formulation leads to an optimal control problem whereby the controller is required to minimize at time k the performance cost (2.3) subject to the constraints (2.2). To ensure that the optimal value of the cost is well defined, we assume that the state of the model (2.1) is stabilizable and observable.

Assumption 2.2 In the system model (2.1) and cost (2.3), the pair (A, B) is stabilizable, the pair (A, Q) is observable, and R is positive-definite.

Given the linear nature of the controlled system, the problem of setpoint tracking (in which the output y is to be steered to a given constant setpoint) can be converted into the regulation problem considered here by redefining the state of (2.1a) in terms of the deviation from a desired steady-state value. The more general case of tracking a time-varying setpoint (e.g. a ramp or sinusoidal signal) can also be tackled within the framework outlined here provided the setpoint can itself be generated by applying a constant reference signal to a system with known LTI dynamics.

2.2 The Unconstrained Optimum

The problem of minimizing the quadratic cost of (2.3) in the unconstrained case (i.e. when $F = 0$ and $G = 0$ in (2.2)) is addressed by Linear Quadratic (LQ) optimal control, which forms an extension of the calculus of variations. The solution is usually obtained either using Pontryagin's Maximum Principle [1] or Dynamic Programming and the recursive Bellman equation [2]. Rather than replicating these solution methods, here we first characterize the optimal linear state feedback law that minimizes the cost of (2.3), and later show (in Sect. 2.7) through a lifting formulation that this control law is indeed optimal over all input sequences.

We first obtain an expression for the cost under linear feedback, $u = Kx$, for an arbitrary stabilizing gain matrix $K \in \mathbb{R}^{n_u \times n_x}$, using the closed-loop system dynamics

$$x_{k+1} = (A + BK)x_k$$

to write $x_k = (A + BK)^k x_0$ and $u_k = K(A + BK)^k x_0$, for all k. Therefore $J(x_0) = J(x_0, \{Kx_0, Kx_1, \ldots\})$ is a quadratic function of x_0,

$$J(x_0) = x_0^T W x_0, \tag{2.4a}$$

$$W = \sum_{k=0}^{\infty} (A + BK)^{k^T} (Q + K^T RK)(A + BK)^k. \tag{2.4b}$$

If $A + BK$ is strictly stable (i.e. each eigenvalue of $A + BK$ is strictly less than unity in absolute value), then it can easily be shown that the elements of the matrix W defined in (2.4b) are necessarily finite. Furthermore, if R is positive-definite and (A, Q) is observable, then $J(x_0)$ is a positive-definite function of x_0 (since then $J(x_0) \geq 0$, for all x_0, and $J(x_0) = 0$ only if $x_0 = 0$), which implies that W is a positive-definite matrix.

The unique matrix W satisfying (2.4) can be obtained by solving a set of linear equations rather than by evaluating the infinite sum in (2.4b). This is demonstrated by the following result, which also shows that $(A + BK)$ is necessarily stable if W in (2.4) exists.

Lemma 2.1 (Lyapunov matrix equation) *Under Assumption 2.2, the matrix W in (2.4) is the unique positive definite solution of the Lyapunov matrix equation*

$$W = (A + BK)^T W(A + BK) + Q + K^T RK \tag{2.5}$$

if and only if $A + BK$ *is strictly stable.*

Proof Let W_n denote the sum of the first n terms in (2.4b), so that

$$W_n \doteq \sum_{k=0}^{n-1} (A + BK)^{k^T} (Q + K^T RK)(A + BK)^k.$$

Then $W_1 = Q + K^T RK$ and $W_{n+1} = (A + BK)^T W_n (A + BK) + Q + K^T RK$ for all $n > 0$. Assuming that $A + BK$ is strictly stable and taking the limit as $n \to \infty$, we obtain (2.5) with $W = \lim_{n\to\infty} W_n$. The uniqueness of W satisfying (2.5) is implied by the uniqueness of W_{n+1} in this recursion for each $n > 0$, and $W \succ 0$ follows from the positive-definiteness of $J(x_0)$.

If we relax the assumption that $A + BK$ is strictly stable, then the existence of $W \succ 0$ satisfying (2.5) implies that there exists a Lyapunov function demonstrating that the system $x_{k+1} = (A + BK)x_k$ is asymptotically stable, since (A, Q) is observable and $R \succ 0$ by Assumption 2.2. Hence $A + BK$ must be strictly stable if (2.5) has a solution $W \succ 0$. □

The optimal unconstrained linear feedback control law is defined by the stabilizing feedback gain K that minimizes the cost in (2.3) for all initial conditions $x_0 \in \mathbb{R}^{n_x}$. The conditions for an optimal solution to this problem can be obtained by considering the effect of perturbing the value of K on the solution, W, of the Lyapunov equation (2.5). Let $W + \delta W$ denote the sum in (2.4b) when K is replaced by $K + \delta K$. Then $W + \delta W$ and $K + \delta K$ satisfy the Lyapunov equation

$$\begin{aligned} W + \delta W = {} & \big[A + B(K + \delta K)\big]^T (W + \delta W)\big[A + B(K + \delta K)\big] \\ & + Q + (K + \delta K)^T R(K + \delta K) \end{aligned}$$

which, together with (2.5), implies that δW satisfies

$$\begin{aligned} \delta W = {} & \delta K^T \big[B^T W(A + BK) + RK\big] + \big[(A + BK)^T WB + K^T R\big]\delta K \\ & + (A + BK)^T \delta W(A + BK) + \delta K^T (B^T WB + R)\delta K \\ & + \delta K^T B^T \delta W(A + BK) + (A + BK)^T \delta W\ B\ \delta K + \delta K^T B^T \delta W\ B\ \delta K. \end{aligned} \tag{2.6}$$

For given $\delta K_1 \in \mathbb{R}^{n_u \times n_x}$, consider a perturbation of the form

$$\delta K = \epsilon\, \delta K_1,$$

and consider the effect on δW of varying the scaling parameter $\epsilon \in \mathbb{R}$. Clearly K is optimal if and only if $x_0^T (W + \delta W)x_0 \geq x_0^T W x_0$, for all $x_0 \in \mathbb{R}^{n_x}$, for all

$\delta K_1 \in \mathbb{R}^{n_u \times n_x}$ and for all sufficiently small ϵ. It follows that K is optimal if and only if the solution of (2.6) has the form

$$\delta W = \epsilon^2\, \delta W_2 + \epsilon^3\, \delta W_3 + \cdots$$

for all $\epsilon \in \mathbb{R}$, where δW_2 is a positive semidefinite matrix. Considering terms in (2.6) of order ϵ and order ϵ^2, we thus obtain the following necessary and sufficient conditions for optimality:

$$B^T W(A + BK) + RK = 0, \tag{2.7a}$$

$$\delta W_2 \succeq 0, \tag{2.7b}$$

$$\delta W_2 = (A + BK)^T \delta W_2 (A + BK) + \delta K_1^T (B^T W B + R)\delta K_1. \tag{2.7c}$$

Solving (2.7a) for K gives $K = -(B^T W B + R)^{-1} B^T W A$ as the optimal feedback gain, whereas Lemma 2.1 and (2.7c) imply that

$$\delta W_2 = \sum_{k=0}^{\infty} (A + BK)^{k^T} \delta K_1^T (B^T W B + R)\delta K_1 (A + BK)^k$$

and therefore (2.7b) is necessarily satisfied since $A + BK$ is strictly stable and $B^T W B + R$ is positive-definite.

These arguments are summarized by the following result.

Theorem 2.1 (Discrete time algebraic Riccati equation) *The feedback gain matrix K for which the control law*

$$u = Kx$$

minimizes the cost of (2.3) for any initial condition x_0 under the dynamics of (2.1a) is given by

$$K = -(B^T W B + R)^{-1} B^T W A, \tag{2.8}$$

where $W \succ 0$ is the unique solution of

$$W = A^T W A + Q - A^T W B (B^T W B + R)^{-1} B^T W A. \tag{2.9}$$

Under Assumption 2.2, $A + BK$ is strictly stable whenever there exists $W \succ 0$ satisfying (2.9).

Proof The optimality of (2.8) is a consequence of the necessity and sufficiency of the optimality conditions in (2.7a), (2.7b) and (2.7c). Equation (2.9) (which is known as the discrete time algebraic Riccati equation) is obtained by substituting K in (2.8) into (2.5). From Lemma 2.1, we can conclude that, under Assumption 2.2, the solution of (2.9) for W is unique and positive-definite if and only if $A + BK$ is strictly stable. □

2.3 The Dual-Mode Prediction Paradigm

The control law that minimizes the cost (2.3) is not in general a linear feedback law when constraints (2.2) are present. Moreover, it may not be computationally tractable to determine the optimal controller as an explicit state feedback law. Predictive control strategies overcome this difficulty by minimizing, subject to constraints, a predicted cost that is computed for a particular initial state, namely the current plant state. This constrained minimization of the predicted cost is solved online at each time step in order to derive a feedback control law. The predicted cost corresponding to (2.3) can be expressed

$$J(x_k, \{u_{0|k}, u_{1|k}, \ldots\}) = \sum_{i=0}^{\infty} \left(\|x_{i|k}\|_Q^2 + \|u_{i|k}\|_R^2 \right) \tag{2.10}$$

where $x_{i|k}$ and $u_{i|k}$ denote the predicted values of the model state and input, respectively, at time $k + i$ based on the information that is available at time k, and where $x_{0|k} = x_k$ is assumed.

The prediction horizon employed in (2.10) is infinite. Hence if every element of the infinite sequence of predicted inputs $\{u_{0|k}, u_{1|k}, \ldots\}$ were considered to be a free variable, then the constrained minimization of this cost would be an infinite-dimensional optimization problem, which is in principle intractable. However predictive control strategies provide effective approximations to the optimal control law that can be computed efficiently and in real time. This is possible because of a parameterization of predictions known as the dual-mode prediction paradigm, which enables the MPC optimization to be specified as a finite-dimensional problem.

The dual-mode prediction paradigm divides the prediction horizon into two intervals. Mode 1 refers to the predicted control inputs over the first N prediction time steps for some finite horizon N (chosen by the designer), while mode 2 denotes the control law over the subsequent infinite interval. The mode 2 predicted inputs are specified by a fixed feedback law, which is usually taken to be the optimum for the problem of minimizing the cost in the absence of constraints [3–6]. Therefore the predicted cost (2.10) can be written as

$$J(x_k, \{u_{0|k}, u_{1|k}, \ldots\}) = \sum_{i=0}^{N-1} \left(\|x_{i|k}\|_Q^2 + \|u_{i|k}\|_R^2 \right) + \|x_{N|k}\|_W^2 \tag{2.11}$$

where, by Theorem 2.1, W is the solution of the Riccati equation (2.9). The term $\|x_{N|k}\|_W^2$ is referred to as a terminal penalty term and accounts for the cost-to-go after N prediction time steps under the mode 2 feedback law.

To simplify notation we express the predicted cost as an explicit function of the initial state of the prediction model and the degrees of freedom in predictions. Hence for the dual-mode prediction paradigm in which the control inputs over the prediction horizon of mode 1 are optimization variables, we write (2.11) as

$$J(x_k, \mathbf{u}_k) = \sum_{i=0}^{N-1} \left(\|x_{i|k}\|_Q^2 + \|u_{i|k}\|_R^2 \right) + \|x_{N|k}\|_W^2. \tag{2.12}$$

where $\mathbf{u}_k = \{u_{0|k}, u_{1|k}, \ldots, u_{N-1|k}\}$.

The receding horizon implementation of MPC stipulates that at each time instant k the optimal mode 1 control sequence $\mathbf{u}_k^* = \{u_{0|k}^*, \ldots, u_{N-1|k}^*\}$ is computed, and only the first element of this sequence is implemented, namely $u_k = u_{0|k}^*$. Thus at each time step the most up-to-date measurement information (embodied in the state x_k) is employed. This creates a feedback mechanism that provides some compensation for any uncertainty present in the model of (2.1a). It also reduces the gap between the optimal value of the predicted cost $J(x_k, \mathbf{u}_k)$ in (2.12) and the optimal cost for the infinite-dimensional problem of minimizing (2.10) over the infinite sequence of future inputs $\{u_{0|k}, u_{1|k}, \ldots\}$.

The rationale behind the dual-mode prediction paradigm is as follows. Let $\{u_{0|k}^0, u_{1|k}^0, \ldots\}$ denote the optimal control sequence for the problem of minimizing the cost (2.10) over the infinite sequence $\{u_{0|k}, u_{1|k}, \ldots\}$ subject to the constraints $Fx_{i|k} + Gu_{i|k} \leq \mathbf{1}$, for all $i \geq 0$, for an initial condition $x_{0|k} = x_k$ such that this problem is feasible. If the weights Q and R satisfy Assumption 2.2, then this notional optimal control sequence drives the predicted state of the model (2.1a) asymptotically to the origin, i.e. $x_{i|k} \to 0$ as $i \to \infty$. Since $(x, u) = (0, 0)$ is strictly feasible for the constraints $Fx + Gu \leq \mathbf{1}$, there exists a neighbourhood, $\mathcal{S}$, of $x = 0$ with the property that these constraints are satisfied at all times along trajectories of the model (2.1a) under the unconstrained optimal feedback law, $u = Kx$, starting from any initial condition in $\mathcal{S}$. Hence there necessarily exists a horizon N_∞ (which depends on x_k) such that $x_{i|k} \in \mathcal{S}$, for all $i \geq N_\infty$. Since the optimal trajectory for $i \geq N_\infty$ is necessarily optimal for the problem with initial condition $x_{N_\infty|k}$ (by Bellman's Principle of Optimality [7]), the constrained optimal sequence must therefore coincide with the unconstrained optimal feedback law, i.e. $u_{i|k}^0 = Kx_{i|k}$, for all $i \geq N_\infty$. It follows that if the mode 1 horizon is chosen to be sufficiently long, namely if $N \geq N_\infty$, then the mode 1 control sequence, $\mathbf{u}_k^*$, that minimizes the cost of (2.12) subject to the constraints $Fx_{i|k} + Gu_{i|k} \leq \mathbf{1}$ for $i = 0, 1, \ldots, N-1$ must be equal to the first N elements of the infinite sequence that minimizes the cost (2.10), namely $u_{i|k}^* = u_{i|k}^0$ for $i = 0, \ldots, N-1$.

For completeness we next give a statement of this result; for a detailed proof and further discussion we refer the interested reader to [4, 5].

Theorem 2.2 *There exists a finite horizon N_∞, which depends on x_k, with the property that, whenever $N \geq N_\infty$: (i). the sequence $\mathbf{u}_k^*$ that achieves the minimum of $J(x_k, \mathbf{u}_k)$ in (2.12) subject to $Fx_{i|k} + Gu_{i|k} \leq \mathbf{1}$ for $i = 0, 1, \ldots, N-1$ is equal to the first N terms of the infinite sequence $\{u_{0|k}^0, u_{1|k}^0, \ldots\}$ that minimizes $J(x_k, \{u_{0|k}, u_{1|k}, \ldots\})$ in (2.10) subject to $Fx_{i|k} + Gu_{i|k} \leq \mathbf{1}$, for all $i \geq 0$; and (ii). $J(x_k, \mathbf{u}_k^*) = J(x_k, \{u_{0|k}^0, u_{1|k}^0, \ldots\})$.*

It is generally convenient to consider the LQ optimal feedback law $u = Kx$ as underlying both mode 1 and mode 2, and to introduce perturbations $c_{i|k} \in \mathbb{R}^{n_u}$, $i = 0, 1, \ldots, N-1$ over the horizon of mode 1 in order to meet constraints. Then the predicted sequence of control inputs is given by

$$u_{i|k} = Kx_{i|k} + c_{i|k}, \qquad i = 0, 1, \ldots, N-1 \tag{2.13a}$$

$$u_{i|k} = Kx_{i|k}, \qquad i = N, N+1, \ldots \tag{2.13b}$$

with $x_{0|k} = x_k$. This prediction scheme is sometimes referred to as the closed-loop paradigm because the term Kx provides feedback in the horizons of both modes 1 and 2.

We argue in Sect. 3.1 (in the context of robustness to model uncertainty) that (2.13) should be classified as an open-loop prediction scheme because K is fixed rather than computed on the basis of measured information (namely x_k). Nevertheless, the feedback term Kx forms a pre-stabilizing feedback loop around the dynamics of (2.1a), which assume the form

$$x_{i+1|k} = \Phi x_{i|k} + Bc_{i|k}, \qquad i = 0, 1, \ldots, N-1 \tag{2.14a}$$

$$x_{i+1|k} = \Phi x_{i|k}, \qquad i = N, N+1, \ldots \tag{2.14b}$$

where $\Phi = A + BK$, with $x_{0|k} = x_k$. The strict stability property of Φ prevents numerical ill-conditioning that could arise in the prediction equations and the associated MPC optimization problem in the case of open-loop unstable models [8].

For the closed-loop paradigm formulation in (2.13), the predicted state trajectory can be generated by simulating (2.14a) forwards over the mode 1 prediction horizon, giving

$$\mathbf{x}_k = M_x x_k + M_c \mathbf{c}_k, \tag{2.14c}$$

where

$$\mathbf{x}_k \doteq \begin{bmatrix} x_{1|k} \\ \vdots \\ x_{N|k} \end{bmatrix}, \quad \mathbf{c}_k \doteq \begin{bmatrix} c_{0|k} \\ \vdots \\ c_{N-1|k} \end{bmatrix}$$

$$M_x = \begin{bmatrix} \Phi \\ \Phi^2 \\ \vdots \\ \Phi^N \end{bmatrix}, \quad M_c = \begin{bmatrix} B & 0 & \cdots & 0 \\ \Phi B & B & \cdots & 0 \\ \vdots & \vdots & \ddots & \vdots \\ \Phi^{N-1} B & \Phi^{N-2} B & \cdots & B \end{bmatrix}.$$

On the basis of these prediction equations and the fact that the predicted cost over mode 2 is given by $\|x_{N|k}\|_W^2$ (where W is the solution of the Lyapunov equation (2.5)), the predicted cost of (2.11) can be written as a quadratic function of the degrees of freedom, namely the vector of predicted perturbations $\mathbf{c}_k$. The details of this computation are straightforward and will not be given here. Instead we derive an equivalent but more convenient form for the predicted cost in Sect. 2.7. For simplicity (but with a slight abuse of notation) in the following development, we denote the cost of (2.11) evaluated along the predicted trajectories of (2.13a) and (2.14a) as $J(x_k, \mathbf{c}_k)$, thus making explicit the dependence of the cost on the optimization variables $\mathbf{c}_k$.

2.4 Invariant Sets

The determination of the minimum prediction horizon N which ensures that the predicted state and input trajectories in mode 2 meet constraints (2.2) is not a trivial matter. Instead lower bounds for this horizon were proposed in [4, 5]. However such bounds could be conservative, leading to the use of unnecessarily long prediction horizons. This in turn could make the online optimization of the predicted cost computationally intractable as a result of large numbers of free variables and large numbers of constraints in the minimization of predicted cost. In such cases it becomes necessary to use a shorter horizon N while retaining the guarantee that predictions over mode 2 satisfy constraints on states and inputs. This can be done by imposing a terminal constraint which requires that the state at the end of the mode 1 horizon should lie in a set which is positively invariant under the dynamics defined by (2.13b) and (2.14b) and under the constraints (2.2).

Definition 2.1 (*Positively invariant set*) A set $\mathcal{X} \subseteq \mathbb{R}^{n_x}$ is positively invariant under the dynamics defined by (2.13b) and (2.14b) and the constraints (2.2) if and only if $(F + GK)x \leq \mathbf{1}$ and $\Phi x \in \mathcal{X}$, for all $x \in \mathcal{X}$.

The use of invariant sets within the dual prediction mode paradigm is illustrated in Fig. 2.1 for a second-order system. The predicted state at the end of mode 1 is constrained to lie in an invariant set $\mathcal{X}_T$ via the constraint $x_{N|k} \in \mathcal{X}_T$. Thereafter, in

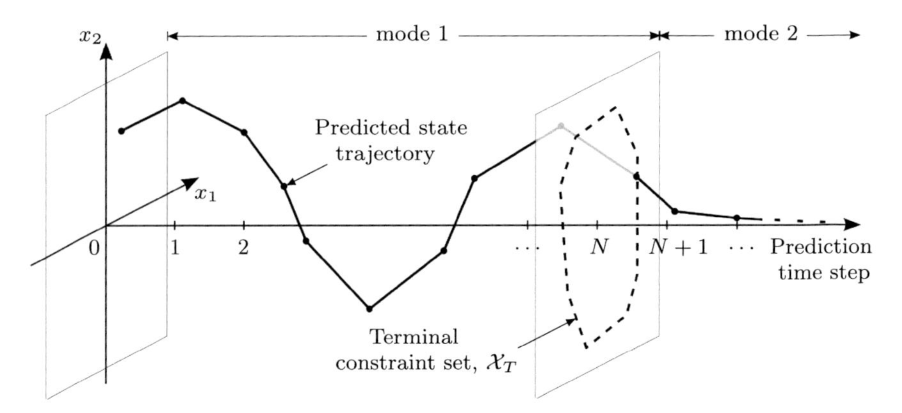

Fig. 2.1 The dual-mode prediction paradigm with terminal constraint. The control inputs in mode 1 are chosen so as to satisfy the system constraints as well as the constraint that the N step ahead predicted state should be inside the invariant set $\mathcal{X}_T$. Over the infinite mode 2 prediction horizon the predicted state trajectory is dictated by the prescribed feedback control law $u = Kx$

mode 2, the evolution of the state trajectory is that prescribed by the state feedback control law $u_k = Kx_k$.

In order to increase the applicability of the MPC algorithm, and in particular to increase the size of the set of initial conditions $x_{0|k}$ for which the terminal condition $x_{N|k} \in \mathcal{X}_T$ can be met, it is important to choose the maximal positively invariant set as the terminal constraint set. This set is defined as follows.

Definition 2.2 (*Maximal positively invariant set*) The maximal positively invariant (MPI) set under the dynamics of (2.13b) and (2.14b) and the constraints (2.2) is the union of all sets that are positively invariant under these dynamics and constraints.

It was shown in [9] that, for the case of linear dynamics and linear constraints considered here, the MPI set is defined by a finite number of linear inequalities. This result is summarized next.

Theorem 2.3 ([9]) *The MPI set for the dynamics defined by (2.13b) and (2.14b) and the constraints (2.2) can be expressed*

$$\mathcal{X}^{\text{MPI}} \doteq \{x : (F + GK)\Phi^i x \le \mathbf{1},\ i = 0, \dots, \nu\} \tag{2.15}$$

where ν is the smallest positive integer such that $(F + GK)\Phi^{\nu+1}x \le \mathbf{1}$, for all x satisfying $(F + GK)\Phi^i x \le \mathbf{1}$, $i = 0, \dots, \nu$. If Φ is strictly stable and $(\Phi, F + GK)$ is observable, then ν is necessarily finite.

Proof Let $\mathcal{X}^{(n)} = \{x : (F + GK)\Phi^i x \le \mathbf{1},\ i = 0, \dots, n\}$ for $n \ge 0$, then it can be shown that (2.15) holds for some finite ν using Definition 2.2 to show that the MPI set $\mathcal{X}^{\text{MPI}}$ is equal to $\mathcal{X}^{(\nu)}$ for finite ν.

In particular, if $x_{0|k} \notin \mathcal{X}^{(n)}$ for given n, then the constraint (2.2) must be violated under the dynamics of (2.13b) and (2.14b). By Definition 2.2 therefore, any $x \notin \mathcal{X}^{(n)}$ cannot lie in $\mathcal{X}^{\text{MPI}}$, so $\mathcal{X}^{(n)}$ must contain $\mathcal{X}^{\text{MPI}}$, for all $n \geq 0$.

Furthermore, if $(F + GK)\Phi^{\nu+1}x \leq \mathbf{1}$, for all $x \in \mathcal{X}^{(\nu)}$, then $\Phi x \in \mathcal{X}^{(\nu)}$ must hold whenever $x \in \mathcal{X}^{(\nu)}$ (since $x \in \mathcal{X}^{(\nu)}$ and $(F + GK)\Phi^{\nu+1}x \leq \mathbf{1}$ imply $(F + GK)\Phi^{i}(\Phi x) \leq \mathbf{1}$ for $i = 0, \ldots \nu$). But from the definition of $\mathcal{X}^{(\nu)}$ we have $(F + GK)x \leq \mathbf{1}$ for all $x \in \mathcal{X}^{(\nu)}$, and therefore $\mathcal{X}^{(\nu)}$ is positively invariant under (2.13b), (2.14b) and (2.2). From Definition 2.2 it can be concluded that $\mathcal{X}^{(\nu)}$ is a subset of, and therefore equal to $\mathcal{X}^{\text{MPI}}$.

Finally, for $\nu \geq n_x$, the set $\mathcal{X}^{(\nu)}$ is necessarily bounded if $(\Phi, F + GK)$ is observable, and, since Φ is strictly stable, the set $\{x : (F + GK)\Phi^{(\nu+1)}x \leq \mathbf{1}\}$ must contain $\mathcal{X}^{(\nu)}$ for finite ν; therefore $\mathcal{X}^{\text{MPI}}$ must be defined by (2.15) for some finite ν. □

The value of ν satisfying the conditions of Theorem 2.3 can be computed by solving at most νn_C linear programs (LPs), namely

$$\underset{x}{\text{maximize}}\ (F + GK)_j\Phi^{n+1}x \ \text{ subject to } \ (F + GK)\Phi^{i}x \leq \mathbf{1},\ i = 0, \ldots, n$$

for $j = 1, \ldots, n_C, n = 1, \ldots, \nu$, where $(F+GK)_j$ denotes the jth row of $F+GK$. The value of ν clearly does not depend on the system state, and this procedure can therefore be performed offline. In general $\nu \geq n_x$, and (2.15) defines the MPI set as a polytope. Therefore if $\mathcal{X}_T$ is equal to the MPI set, the terminal constraint $x_{N|k} \in \mathcal{X}_T$ can be invoked via linear inequalities on the degrees of freedom in mode 1 predictions. It will be convenient to represent the terminal set $\mathcal{X}_T$ in matrix form

$$\mathcal{X}_T = \{x : V_T x \leq \mathbf{1}\},$$

so that with $\mathcal{X}_T$ chosen as the MPI set (2.15), V_T is given by

$$V_T = \begin{bmatrix} F + GK \\ (F + GK)\Phi \\ \vdots \\ (F + GK)\Phi^{\nu} \end{bmatrix}.$$

Example 2.1 Figure 2.2 gives an illustration of the MPI set for a second-order system with state-space matrices

$$A = \begin{bmatrix} 1.1 & 2 \\ 0 & 0.95 \end{bmatrix}, \quad B = \begin{bmatrix} 0 \\ 0.0787 \end{bmatrix}, \quad C = \begin{bmatrix} -1 & 1 \end{bmatrix} \tag{2.16a}$$

and constraints $-\mathbf{1} \leq x/8 \leq \mathbf{1}$, $-1 \leq u \leq 1$, which correspond to the following constraint matrices in (2.2),

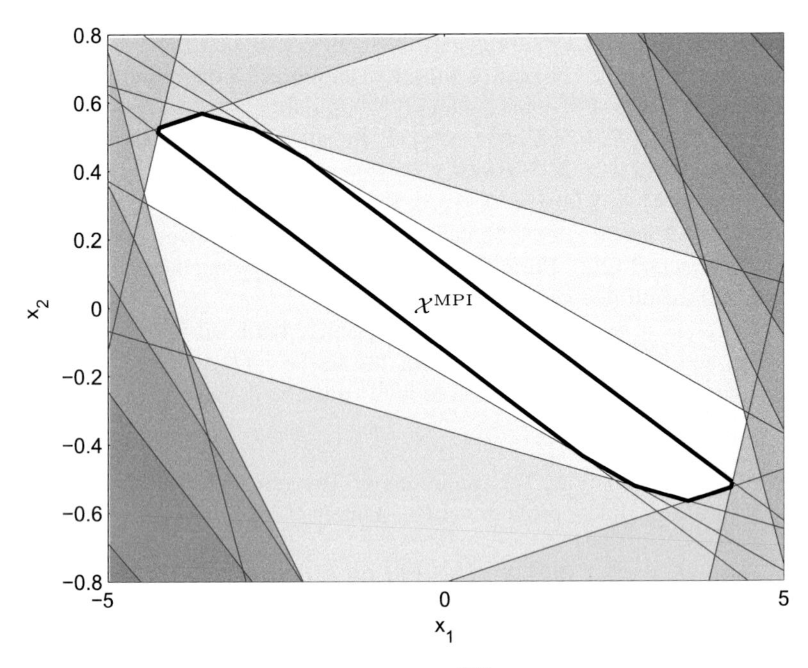

Fig. 2.2 The maximal positively invariant (MPI) set, $\mathcal{X}^{\text{MPI}}$, for the system of (2.16a), (2.16b). Each of the inequalities defining $\mathcal{X}^{\text{MPI}}$ is represented by a straight line on the diagram

$$F = \begin{bmatrix} 0 & 1/8 \\ 1/8 & 0 \\ 0 & -1/8 \\ -1/8 & 0 \\ 0 & 0 \\ 0 & 0 \end{bmatrix}, \quad G = \begin{bmatrix} 0 \\ 0 \\ 0 \\ 0 \\ 1 \\ -1 \end{bmatrix}. \tag{2.16b}$$

The mode 2 feedback law is taken to be the optimal unconstrained linear feedback law $u = Kx$, with cost weights $Q = C^T C$ and $R = 1$, for which $K = -\begin{bmatrix}1.19 & 7.88\end{bmatrix}$. The MPI set is given by (2.15) with $\nu = 5$. After removing redundant constraints, this set is defined by 10 inequalities corresponding to the 10 straight lines that intersect the boundary of the MPI set, marked $\mathcal{X}^{\text{MPI}}$ in Fig. 2.2. ♢

2.5 Controlled Invariant Sets and Recursive Feasibility

Collecting the ideas discussed in the previous sections we can state the following MPC algorithm:

Algorithm 2.1 (*MPC*) At each time instant $k = 0, 1, \ldots$:

(i) Perform the optimization

$$\underset{\mathbf{c}_k}{\text{minimize}} \quad J(x_k, \mathbf{c}_k) \tag{2.17a}$$

$$\text{subject to} \quad (F + GK)x_{i|k} + Gc_{i|k} \leq \mathbf{1}, i = 0, \dots, N-1 \tag{2.17b}$$

$$V_T x_{N|k} \leq \mathbf{1} \tag{2.17c}$$

where $J(x_k, \mathbf{c}_k)$ is the cost of (2.11) evaluated for the predicted trajectories of (2.13a) and (2.14a).

(ii) Apply the control law $u_k = Kx_k + c^*_{0|k}$, where $\mathbf{c}^*_k = (c^*_{0|k}, \dots, c^*_{N-1|k})$ is the optimal value of $\mathbf{c}_k$ for problem (2.17). ◁

The terminal condition (2.17c) is sometimes referred to as a stability constraint because it provides a means of guaranteeing the closed-loop stability of the MPC law. It does this by ensuring that the mode 2 predicted trajectories (2.13b) and (2.14b) satisfy the constraint $(F + GK)x_{i|k} \leq \mathbf{1}$, thus ensuring that the predicted cost over mode 2 is indeed given by $\|x_{N|k}\|^2_W$, and also by guaranteeing that Algorithm 2.1 is feasible at all time instants if it is feasible at initial time. The latter property of recursive feasibility is a fundamental requirement for closed-loop stability since it guarantees that the optimization problem (2.17) is solvable and hence that the control law of Algorithm 2.1 is defined at every time instant if (2.17) is initially feasible.

Recall that the feasibility of predicted trajectories in mode 2 is ensured by constraining the terminal state to lie in a set which is positively invariant. The feasibility of Algorithm 2.1 can be similarly ensured by requiring that the state x_k lies in an invariant set. However, since there are degrees of freedom in the predicted trajectories of (2.13a) and (2.14a), the relevant form of invariance is controlled positive invariance.

Definition 2.3 (*Controlled positively invariant set*) A set $\mathcal{X} \subseteq \mathbb{R}^{n_x}$ is controlled positively invariant (CPI) for the dynamics of (2.1a) and constraints (2.2) if, for all $x \in \mathcal{X}$, there exists $u \in \mathbb{R}^{n_u}$ such that $Fx + Gu \leq \mathbf{1}$ and $Ax + Bu \in \mathcal{X}$. Furthermore $\mathcal{X}$ is the maximal controlled positively invariant (MCPI) set if it is CPI and contains all other CPI sets.

To show that Algorithm 2.1 is recursively feasible, we demonstrate next that its feasible set is a CPI set. Algorithm 2.1 is feasible whenever x_k belongs to the feasible set $\mathcal{F}_N$ defined by

$$\mathcal{F}_N \doteq \big\{x_k : \exists\, \mathbf{c}_k \text{ such that } (F + GK)x_{i|k} + Gc_{i|k} \leq \mathbf{1},\ i = 0, \dots, N-1 \text{ and } V_T x_{N|k} \leq \mathbf{1}\big\}. \tag{2.18}$$

Clearly this is the same as the set of states of (2.1a) that can be driven to the terminal set $\mathcal{X}_T = \{x : V_T x \le \mathbf{1}\}$ in N steps subject to the constraints (2.2), and it therefore has the following equivalent definition:

$$\mathcal{F}_N = \big\{x_0 : \exists\,\{u_0, \ldots, u_{N-1}\} \text{ such that } Fx_i + Gu_i \le \mathbf{1},\ i = 0, \ldots, N-1, \\ \text{and } x_N \in \mathcal{X}_T\big\}. \tag{2.19}$$

Theorem 2.4 *If $\mathcal{X}_T$ in (2.19) is positively invariant for (2.13b), (2.14b) and (2.2), then $\mathcal{F}_N \subseteq \mathcal{F}_{N+1}$, for all $N > 0$, and $\mathcal{F}_N$ is a CPI set for the dynamics of (2.1a) and constraints (2.2).*

Proof If $x_0 \in \mathcal{F}_N$, then by definition there exists a sequence $\{u_0, \ldots, u_{N-1}\}$ such that $Fx_i + Gu_i \le \mathbf{1}$, $i = 0, \ldots, N-1$ and $x_N \in \mathcal{X}_T$. Also, since $\mathcal{X}_T$ is positively invariant, the choice $u_N = Kx_N$ would ensure $Fx_N + Gu_N \le \mathbf{1}$ and $x_{N+1} \in \mathcal{X}_T$, and this in turn implies $x_0 \in \mathcal{F}_{N+1}$ whenever $x_0 \in \mathcal{F}_N$. Furthermore if $x_0 \in \mathcal{F}_N$, then by definition u_0 exists such that $Fx_0 + Gu_0 \le \mathbf{1}$ and $x_1 \in \mathcal{F}_{N-1}$, and since $\mathcal{F}_{N-1} \subset \mathcal{F}_N$, it follows that $\mathcal{F}_N$ is CPI. □

Although the proof of Theorem 2.4 considers the sequence of control inputs $\{u_0, \ldots, u_{N-1}\}$, the same arguments apply to the optimization variables $\mathbf{c}_k$ in (2.17), since for each feasible u_k, $k = 0, \ldots, N-1$, there exists a feasible c_k such that $u_k = Kx_k + c_k$. Therefore, the fact that $\mathcal{F}_N$ is a CPI set for (2.1a) and (2.2) also implies that $\mathcal{F}_N$ is CPI for the dynamics (2.14a) and constraints (2.17b). Hence for any $x_k \in \mathcal{F}_N$ there must exist c_k such that $(F + GK)x_k + Gc_k \le \mathbf{1}$ and $x_{k+1} = \Phi x_k + Bc_k \in \mathcal{F}_N$. Furthermore, the proof of Theorem 2.4 shows that if $c_k = c^*_{0|k}$ (where $\mathbf{c}^*_k = (c^*_{0|k}, \ldots, c^*_{N-1|k})$ is the optimal value of $\mathbf{c}_k$ in step (ii) of Algorithm 2.1), then the sequence

$$\mathbf{c}_{k+1} = (c^*_{1|k}, \ldots, c^*_{N-1|k}, 0) \tag{2.20}$$

is necessarily feasible for the optimization (2.17) at time $k+1$, and therefore Algorithm 2.1 is recursively feasible.

The candidate feasible sequence in (2.20) can be thought of as the extension to time $k+1$ of the optimal sequence at time k. It is in fact the sequence that generates, via (2.13a), the input sequence

$$\{u_{1|k}, \ldots, u_{N-1|k}, Kx_{N|k}\}$$

at time $k+1$. For this reason, it is sometimes referred to as the *tail of the solution of the MPC optimization problem at time k*, or simply the *tail*. As well as demonstrating recursive feasibility, the tail is often used to construct a suboptimal solution at time $k+1$ based on the optimal solution at time k. This enables a comparison of the optimal costs at successive time steps, which is instrumental in the analysis of the closed-loop stability properties of MPC laws.

Theorem 2.4 shows that the feasible sets corresponding to increasing values of N are nested, so that the feasible set $\mathcal{F}_N$ necessarily grows as N is increased. In practice the length of the mode 1 horizon is likely to be limited by the growth in computation that is required to solve Algorithm 2.1 (this is discussed in Sect. 2.8). However, given that $\mathcal{F}_N$ increases as N grows, the question arises as to whether there exists a finite value of N such that $\mathcal{F}_N$ is equal to the maximal feasible set defined by

$$\mathcal{F}_\infty \doteq \bigcup_{N=1}^{\infty} \mathcal{F}_N.$$

Here $\mathcal{F}_\infty$ is defined as the set of initial conditions that can be steered to $\mathcal{X}_T$ over an infinite horizon subject to constraints. However, $\mathcal{F}_\infty$ is independent of the choice of $\mathcal{X}_T$; this is a consequence of the fact that, for any bounded positively invariant set $\mathcal{X}_T$, the system (2.1a) can be steered from any initial state in $\mathcal{X}_T$ to the origin subject to the constraints (2.2) in finite time, as demonstrated by the following result.

Theorem 2.5 *Let* $\mathcal{F}_N^0 \doteq \big\{x_0 : \exists\{u_0, \ldots, u_{N-1}\}$ *such that* $Fx_i + Gu_i \leq \mathbf{1},\ i = 0, \ldots, N-1,$ *and* $x_N = 0\big\}$. *If* $\mathcal{X}_T$ *in (2.19) is positively invariant for (2.13b), (2.14b) and (2.2), where* Φ *is strictly stable and* $(\Phi, F + GK)$ *is observable, then* $\mathcal{F}_\infty = \bigcup_{N=1}^{\infty} \mathcal{F}_N = \bigcup_{N=1}^{\infty} \mathcal{F}_N^0$.

Proof First, note that any positively invariant set $\mathcal{X}_T$ must contain the origin because Φ is strictly stable. Second, strict stability of Φ and boundedness of $\mathcal{X}_T$ (which follows from observability of $(\Phi, F + GK)$) also implies that, for any $\epsilon > 0$, the set $\mathcal{B}_\epsilon \doteq \{x : \|x\| \leq \epsilon\}$ is reachable from any point in $\mathcal{X}_T$ in a finite number of steps (namely for all $x_0 \in \mathcal{X}_T$ there exists a sequence $\{u_0, \ldots, u_{n-1}\}$ such that $Fx_i + Gu_i \leq \mathbf{1}$ for $i = 0, \ldots, n-1$ and $x_n \in \mathcal{B}_\epsilon$) since $\|\Phi^n x\| \leq \epsilon$, for all $x \in \mathcal{X}_T$ for some finite n. Third, since (A, B) is controllable and $(0, 0)$ lies in the interior of the constraint set $\{(x, u) : Fx + Gu \leq \mathbf{1}\}$, there must exist $\epsilon > 0$ such that the origin is reachable in n_x steps from any point in $\mathcal{B}_\epsilon$, i.e. $\mathcal{B}_\epsilon \subseteq \mathcal{F}_{n_x}^0$. Combining these observations we obtain $\{0\} \subseteq \mathcal{X}_T \subseteq \mathcal{F}_{n+n_x}^0$ and hence $\mathcal{F}_N^0 \subseteq \mathcal{F}_N \subseteq \mathcal{F}_{n+n_x+N}^0$ for some finite n and all $N \geq 0$. From this we conclude that $\bigcup_{N=1}^{\infty} \mathcal{F}_N = \bigcup_{N=1}^{\infty} \mathcal{F}_N^0$. □

A consequence of Theorem 2.5 is that replacing the terminal set $\mathcal{X}_T$ by any bounded positively invariant set (or in fact any CPI set) in (2.18) results in the same set $\mathcal{F}_\infty$. Therefore $\mathcal{F}_\infty$ is identical to the maximal CPI set or *infinite time reachability set* [10, 11], which by definition is the largest possible feasible set for any stabilizing control law for the dynamics (2.1a) and constraints (2.2). In general $\mathcal{F}_N$ does not necessarily tend to a finite limit[1] as $N \to \infty$, but the following result shows that under certain conditions $\mathcal{F}_\infty$ is equal to $\mathcal{F}_N$ for finite N.

---

[1] If for example the system (2.1a) is open-loop stable and $F = 0$, then clearly the MCPI set is the entire state space and $\mathcal{F}_N$ grows without bound as N increases. In general the MCPI set is finite if and only if the system (A, B, F, G), mapping input u_k to output $Fx_k + Gu_k$ has no transmission zeros inside the unit circle (see, e.g. [11, 12]).

Theorem 2.6 *If $\mathcal{F}_{N+1} = \mathcal{F}_N$ for finite $N > 0$, then $\mathcal{F}_\infty = \mathcal{F}_N$.*

Proof An alternative definition of $\mathcal{F}_{N+1}$ (which is nonetheless equivalent to (2.18)) is that $\mathcal{F}_{N+1}$ is the set of states x for which there exists a control input u such that $Fx + Gu \le \mathbf{1}$ and $Ax + Bu \in \mathcal{F}_N$. If $\mathcal{F}_{N+1} = \mathcal{F}_N$, then it immediately follows from this definition that $\mathcal{F}_{N+2} = \mathcal{F}_{N+1}$. Applying this argument repeatedly we get $\mathcal{F}_{N+i} = \mathcal{F}_N$, for all $i = 1, 2, \ldots$ and hence $\mathcal{F}_\infty = \mathcal{F}_N$. □

Example 2.2 Figure 2.3 shows the feasible sets $\mathcal{F}_N$ of Algorithm 2.1 for the system model and constraints of Example 2.1, for a range of values of mode 1 horizon N. Here the terminal set $\mathcal{X}_T$ is the maximal positively invariant set $\mathcal{X}^{\text{MPI}}$ of Fig. 2.2; this is shown in Fig. 2.3 as the feasible set for $N = 0$. As expected the feasible sets $\mathcal{F}_N$ for increasing N are nested. For this example, the maximal CPI set is given by $\mathcal{F}_\infty = \mathcal{F}_N$ for $N = 26$ and the minimal description of $\mathcal{F}_\infty$ involves 100 inequalities. ◊

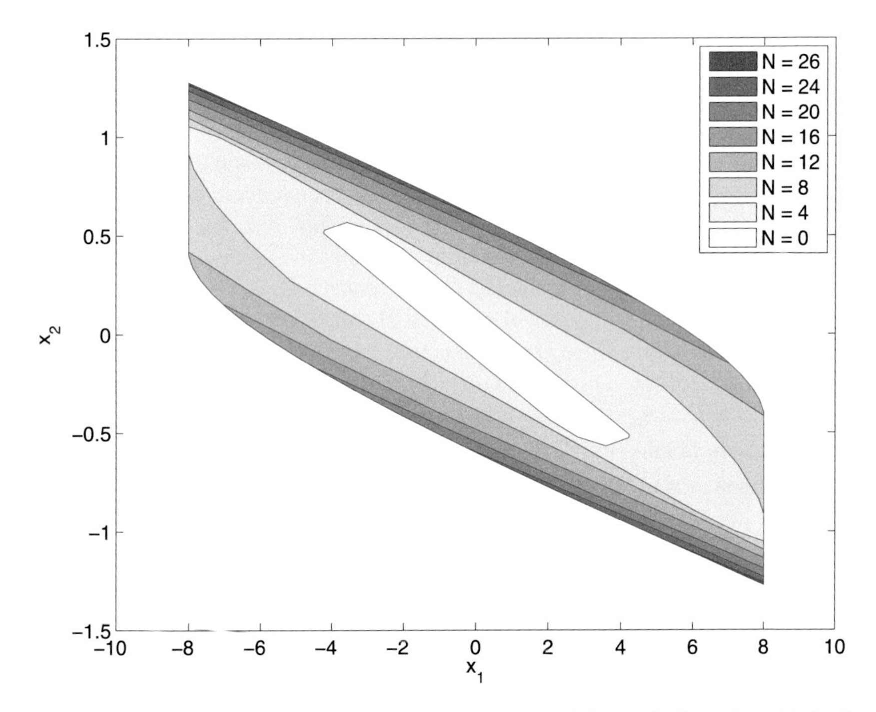

Fig. 2.3 The feasible sets $\mathcal{F}_N$, $N = 4, 8, 12, 16, 20, 24, 26$ and the terminal set $\mathcal{F}_0 = \mathcal{X}_T$ for the example of (2.16a), (2.16b). The maximal controlled invariant set is $\mathcal{F}_\infty = \mathcal{F}_{26}$

2.6 Stability and Convergence

This section introduces the main tools for analysing closed-loop stability under the MPC law of Algorithm 2.1 for the ideal case of no model uncertainty or unmodeled disturbances. The control law is nonlinear because of the inequality constraints in the optimization (2.17), and the natural framework for the stability analysis is therefore Lyapunov stability theory. Using the feasible but suboptimal tail sequence that was introduced in Sect. 2.5, we show that the optimal value of the cost function in (2.17) is non-increasing along trajectories of the closed-loop system. This provides guarantees of asymptotic convergence of the state and Lyapunov stability under Assumption 2.2. Where possible, we keep the discussion in this section non-technical and refer to the literature on stability theory for technical details.

The feasibility of the tail of the optimal sequence $\mathbf{c}_k^*$ implies that the sequence $\mathbf{c}_{k+1}$ defined in (2.20) is feasible but not necessarily an optimal solution of (2.17) at time $k+1$. Using (2.20) it is easy to show that the corresponding cost $J(x_{k+1}, \mathbf{c}_{k+1})$ is equal to $J^*(x_k) - \|x_k\|_Q^2 - \|u_k\|_R^2$. After optimization at time $k+1$, we therefore have

$$J^*(x_{k+1}) \leq J^*(x_k) - \|x_k\|_Q^2 - \|u_k\|_R^2. \tag{2.21}$$

Summing both sides of this inequality over all $k \geq 0$ gives the closed-loop performance bound

$$\sum_{k=0}^{\infty} \left(\|x_k\|_Q^2 + \|u_k\|_R^2\right) \leq J^*(x_0) - \lim_{k\to\infty} J^*(x_k). \tag{2.22}$$

The quantity appearing on the LHS of this inequality is the cost evaluated along the closed-loop trajectories of (2.1) under Algorithm 2.1. Since $J^*(x_k)$ is non-negative for all k, the bound (2.22) implies that the closed-loop cost can be no greater than the initial optimal cost value, $J^*(x_0)$.

Given that the optimal cost is necessarily finite if (2.17) is feasible, and since each term in the sum on the LHS of (2.22) is non-negative, the closed-loop performance bound in (2.22) implies the following convergence result

$$\lim_{k\to\infty} \left(\|x_k\|_Q^2 + \|u_k\|_R^2\right) = 0 \tag{2.23}$$

along the trajectories of the closed-loop system. We now give the basic results concerning closed-loop stability.

Theorem 2.7 *If (2.17) feasible at $k = 0$, then the state and input trajectories of (2.1a) under Algorithm 2.1 satisfy* $\lim_{k\to\infty}(x_k, u_k) = (0, 0)$.

Proof This follows from (2.23) and Assumption 2.2 since $R \succ 0$ implies $u_k \to 0$ as $k \to \infty$; hence from the observability of (Q, A) and $\|x_k\|_Q \to 0$ we conclude that $x_k \to 0$ as $k \to \infty$. □

Theorem 2.8 *Under the control law of Algorithm 2.1, the origin* $x = 0$ *of the system (2.1a) is asymptotically stable and its region of attraction is equal to the feasible set* $\mathcal{F}_N$. *If* $Q \succ 0$, *then* $x = 0$ *is exponentially stable.*

Proof The conditions on Q and R in Assumption 2.2 ensure that the optimal cost $J^*(x_k)$ is a positive-definite function of x_k since $J^*(x_k) = 0$ if and only if $x_k = 0$, and $J^*(x_k) > 0$ whenever $x_k \neq 0$. Therefore (2.21) implies that $J^*(x_k)$ is a Lyapunov function which demonstrates that $x = 0$ is a stable equilibrium (in the sense of Lyapunov) of the closed-loop system [13]. Combined with the convergence result of Theorem 2.7, this shows that $x = 0$ is an asymptotically stable equilibrium point, and since Theorem 2.7 applies to all feasible initial conditions, the region of attraction is $\mathcal{F}_N$.

To show that the rate of convergence is exponential if $Q \succ 0$ we first note that the optimal value of (2.17) is a continuous piecewise quadratic function of x_k [14]. Therefore, $J^*(x_k)$ can be bounded above and below for all $x_k \in \mathcal{F}_N$ by

$$\alpha \|x_k\|^2 \leq J^*(x_k) \leq \beta \|x_k\|^2 \tag{2.24}$$

where α and β are necessarily positive scalars since $J^*(x_k)$ is positive-definite. If the smallest eigenvalue of Q is $\underline{\lambda}(Q)$, then from (2.24) and (2.21) we get

$$\|x_k\|^2 \leq \frac{1}{\alpha}\left|1 - \frac{\underline{\lambda}(Q)}{\beta}\right|^k J^*(x_0)$$

for all $k = 0, 1, \ldots$, and hence $x = 0$ is exponentially stable. □

Example 2.3 For the same system dynamics, constraints and cost as in Example 2.1 the predicted and closed-loop state trajectories under the MPC law of Algorithm 2.1 with $N = 6$ and initial state $x(0) = (-7.5, 0.5)$ are shown in Fig. 2.4. Figure 2.5 gives the corresponding predicted and closed-loop input trajectories. The jump in the predicted input trajectory at $N = 6$ is due to the switch to the mode 2 feedback law at that time step.

Table 2.1 gives the variation with mode 1 horizon N of predicted cost J_0^* and closed-loop cost $J_{\text{cl}}(x_0) \doteq \sum_{k=0}^{\infty}(\|x_k\|_Q^2 + \|u_k\|_R^2)$ for $x(0) = (-7.5, 0.5)$. The infinite-dimensional optimal performance is obtained with $N = N_\infty$, where $N_\infty = 11$ for this initial condition, so there is no further decrease in predicted cost for values of $N > 11$. However, because of the receding horizon implementation, the closed-loop response of the MPC law for $N = 6$ is indistinguishable from the ideal optimal response for this initial condition. ◊

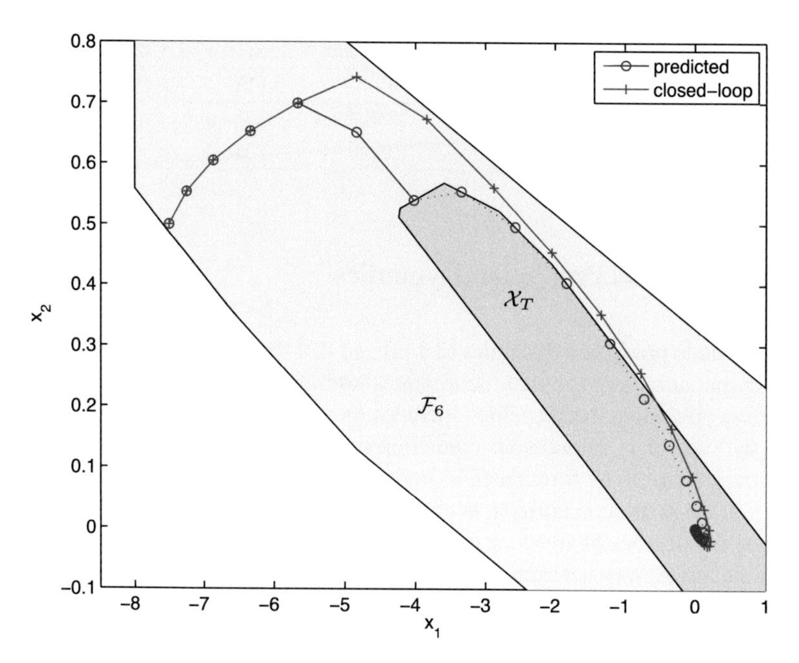

Fig. 2.4 Predicted and closed-loop state trajectories for Algorithm 2.1 with $N = 6$

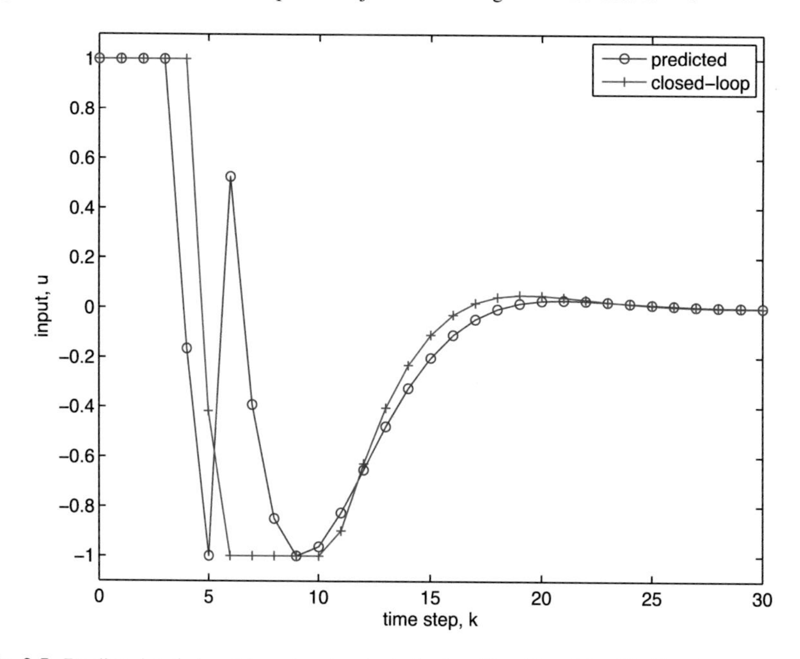

Fig. 2.5 Predicted and closed-loop input trajectories for Algorithm 2.1 with $N = 6$

Table 2.1 Variation of predicted and closed-loop cost with N for $x_0 = (-7.5, 0.5)$ in Example 2.3

N	6	7	8	11	>11
$J^*(x_0)$	364.2	357.0	356.3	356.0	356.0
$J_{\text{cl}}(x_0)$	356.0	356.0	356.0	356.0	356.0

2.7 Autonomous Prediction Dynamics

The dual-mode prediction dynamics (2.14a) and (2.14b) can be expressed in a more compact autonomous form that incorporates both prediction modes [15, 16]. This alternative prediction model, which includes the degrees of freedom in predictions within the state of an autonomous prediction system, enables the constraints on predicted trajectories to be formulated as constraints on the prediction system state at the start of the prediction horizon. With this approach the feasible sets for the model state and the degrees of freedom in predictions are determined simultaneously by computing an invariant set (rather than a controlled invariant set) for the autonomous system state. This can result in significant reductions in computation for the case that the system model is uncertain since, as discussed in Chap. 5, it greatly simplifies handling the the effects of uncertainty over the prediction horizon. In this section we show that an autonomous formulation is also convenient in the case of nominal MPC.

An autonomous prediction system that generates the predictions of (2.13a), (2.13b) and (2.14a), (2.14b) can be expressed as

$$z_{i+1|k} = \Psi z_{i|k}, \quad i = 0, 1, \ldots \tag{2.25}$$

where the initial state $z_{0|k} \in \mathbb{R}^{n_x + Nn_u}$ consists of the state x_k of the model (2.1a) appended by the vector $\mathbf{c}_k$ of degrees of freedom,

$$z_{0|k} = \begin{bmatrix} x_k \\ c_{0|k} \\ \vdots \\ c_{N-1|k} \end{bmatrix}.$$

The state transition matrix in (2.25) is given by

$$\Psi = \begin{bmatrix} \Phi & BE \\ 0 & M \end{bmatrix} \tag{2.26a}$$

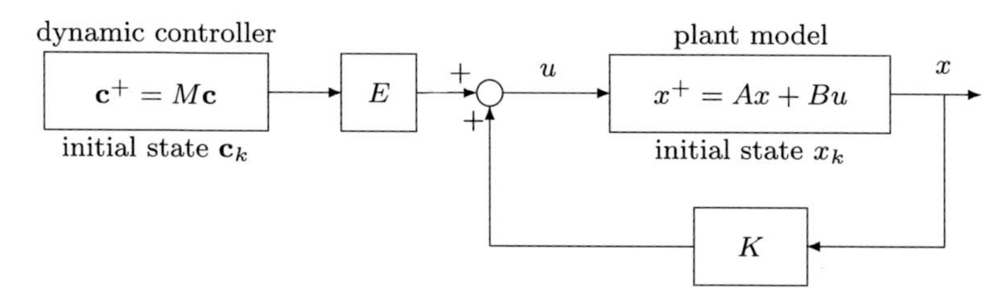

Fig. 2.6 Block diagram representation of the autonomous prediction systems (2.25) and (2.26). The free variables in the state and input predictions at time k are contained in the initial controller state $\mathbf{c}_k$; the signals marked x and u are the i steps ahead predicted state and control input, and x^+, $\mathbf{c}^+$ denote their successor states

where $\Phi = A + BK$ and

$$E = \begin{bmatrix} I_{n_u} & 0 \cdots 0 \end{bmatrix}, \quad M = \begin{bmatrix} 0 & I_{n_u} & 0 & \cdots & 0 \\ 0 & 0 & I_{n_u} & \cdots & 0 \\ \vdots & \vdots & \vdots & \ddots & \vdots \\ 0 & 0 & 0 & \cdots & I_{n_u} \\ 0 & 0 & 0 & \cdots & 0 \end{bmatrix}. \tag{2.26b}$$

The state and input predictions of (2.13a), (2.13b) and (2.14a), (2.14b) are then given by

$$u_{i|k} = \begin{bmatrix} K & E \end{bmatrix} z_{i|k} \tag{2.27a}$$

$$x_{i|k} = \begin{bmatrix} I_{n_x} & 0 \end{bmatrix} z_{i|k} \tag{2.27b}$$

for $i = 0, 1, \ldots$. The prediction systems (2.25) and (2.26) can be interpreted as a dynamic feedback law applied to (2.1a), with the controller state at the beginning of the prediction horizon containing the degrees of freedom, $\mathbf{c}_k$, in predictions (Fig. 2.6).

2.7.1 Polytopic and Ellipsoidal Constraint Sets

The constraints (2.2) applied to the predictions of (2.27a), (2.27b) are equivalent to the following constraints on the initial prediction system state $z_k = z_{0|k}$:

$$\begin{bmatrix} F + GK & GE \end{bmatrix} \Psi^i z_k \le \mathbf{1}, \quad i = 0, 1, \ldots \tag{2.28}$$

Clearly this implies an infinite number of constraints that apply across an infinite prediction horizon. However, analogously to the definition of terminal invariant sets in Sect. 2.4, a feasible set for z_k satisfying (2.28) can be constructed by determining a positively invariant set for the dynamics $z_{k+1} = \Psi z_k$ and constraints

$\begin{bmatrix}F + GK & GE\end{bmatrix} z_k \leq \mathbf{1}$. Theorem 2.3 shows that the maximal positively invariant set for these dynamics and constraints is given by

$$\mathcal{Z} \doteq \{z : \begin{bmatrix}F + GK & GE\end{bmatrix} \Psi^i z \leq \mathbf{1}, \ \ i = 0, 1, \ldots, \nu_z\} \tag{2.29}$$

where ν_z is a positive integer such that $\begin{bmatrix}F + GK & GE\end{bmatrix} \Psi^{\nu_z+1} z \leq \mathbf{1}$, for all z satisfying $\begin{bmatrix}F + GK & GE\end{bmatrix} \Psi^i z \leq \mathbf{1}, i = 0, 1, \ldots, \nu_z$. Since $\mathcal{Z}$ is the MPI set, every state z_k for which (2.28) is satisfied must lie in $\mathcal{Z}$. Given that a mode 1 prediction horizon of N steps is implicit in the augmented prediction dynamics (2.25), the projection of $\mathcal{Z}$ onto the x-subspace is therefore equal to the feasible set $\mathcal{F}_N$ defined in (2.18), i.e.

$$\mathcal{F}_N = \left\{x : \exists\, \mathbf{c} \text{ such that } \begin{bmatrix}F + GK & GE\end{bmatrix} \Psi^i \begin{bmatrix} x \\ \mathbf{c} \end{bmatrix} \leq \mathbf{1}, \ i = 0, 1, \ldots, \nu_z\right\}.$$

The value of ν_z defining the MPI set in (2.29) grows as the mode 1 prediction horizon N is increased. Furthermore, it can be seen from (2.26) that every eigenvalue of Ψ is equal either to 0 or to an eigenvalue of Φ, so if one or more of the eigenvalues of Φ lies close to the unit circle in the complex plane, then ν_z in (2.29) could be large even for short horizons N. The equivalence of (2.27a), (2.27b) with (2.13a), (2.13b) and (2.14a), (2.14b) implies that the online MPC optimization in (2.17) is equivalent to

$$\underset{\mathbf{c}_k}{\text{minimize}} \ \ J(x_k, \mathbf{c}_k) \ \ \text{subject to} \ \begin{bmatrix} x_k \\ \mathbf{c}_k \end{bmatrix} \in \mathcal{Z}. \tag{2.30}$$

which is a quadratic programming problem with $\nu_z n_C$ constraints.

A large value of ν_z could therefore make the implementation of Algorithm 2.1 computationally demanding. If this is the case, and in particular for applications with very high sampling rates, it may be advantageous to replace the polyhedral invariant set $\mathcal{Z}$ with an ellipsoidal invariant set, $\mathcal{E}_z$:

$$\underset{\mathbf{c}_k}{\text{minimize}} \ \ J(x_k, \mathbf{c}_k) \ \ \text{subject to} \ \begin{bmatrix} x_k \\ \mathbf{c}_k \end{bmatrix} \in \mathcal{E}_z. \tag{2.31}$$

This represents a simplification of the online optimization to a problem that involves just a single constraint, thus allowing for significant computational savings. Furthermore, using an ellipsoidal set that is positively invariant for the autonomous prediction dynamics (2.25) and constraints $\begin{bmatrix}F + GK & GE\end{bmatrix} z \leq \mathbf{1}$, the resulting MPC law retains the recursive feasibility and stability properties of Algorithm 2.1. Approximating the MPI set $\mathcal{Z}$ (which is by definition maximal) using a smaller ellipsoidal set necessarily introduces suboptimality into the resulting MPC law; but as discussed in Sect. 2.8, the degree of suboptimality is in many cases negligible.

The invariant ellipsoidal set $\mathcal{E}_z$ can be computed offline by solving an appropriate convex optimization problem. The design of these sets is particularly convenient computationally because the conditions for invariance with respect to the linear autonomous dynamics (2.25) and linear constraints $\begin{bmatrix}F + GK & GE\end{bmatrix} z_k \leq \mathbf{1}$

may be written in terms of linear matrix inequalities (LMIs), which are necessarily convex and can be handled using semidefinite programming (SDP) [17]. Linear matrix inequalities and the offline optimization of $\mathcal{E}_z$ are considered in more detail in Sect. 2.7.3; here we simply summarize the conditions for invariance of $\mathcal{E}_z$ in the following theorem:

Theorem 2.9 *The ellipsoidal set defined by* $\mathcal{E}_z \doteq \{z : z^T P_z z \le 1\}$ *for* $P_z \succ 0$ *is positively invariant for the dynamics* $z_{k+1} = \Psi z_k$ *and constraints* $\begin{bmatrix} F + GK & GE \end{bmatrix} z_k \le \mathbf{1}$ *if and only if* P_z *satisfies*

$$P_z - \Psi^T P_z \Psi \succeq 0 \tag{2.32}$$

and

$$\begin{bmatrix} H & \begin{bmatrix} F + GK & GE \end{bmatrix} \\ \begin{bmatrix} (F + GK)^T \\ (GE)^T \end{bmatrix} & P_z \end{bmatrix} \succeq 0, \quad e_i^T H e_i \le 1, \quad i = 1, 2, \ldots, n_C \tag{2.33}$$

for some symmetric matrix H*, where* e_i *is the* i*th column of the identity matrix.*

Proof The inequality in (2.32) implies $z^T \Psi^T P \Psi z \le z^T P z \le 1$, which is a sufficient condition for invariance of the ellipsoidal set $\mathcal{E}_z$ under $z_{k+1} = \Psi z_k$. Conversely, (2.32) is also necessary for invariance since if $P_z - \Psi^T P_z \Psi \not\succeq 0$, then there would exist z satisfying $z^T \Psi^T P_z \Psi z > z^T P_z z$ and $z^T P_z z = 1$, which would imply that $\Psi z \notin \mathcal{E}_z$ for some $z \in \mathcal{E}_z$.

We next show that (2.33) provides necessary and sufficient conditions for satisfaction of the constraints $\begin{bmatrix} F + GK & GE \end{bmatrix} z \le \mathbf{1}$, for all $z \in \mathcal{E}_z$. To simplify notation, let $\tilde{F} \doteq \begin{bmatrix} F + GK & GE \end{bmatrix}$ and let $\tilde{F}_i$ denote the ith row of $\tilde{F}$. Since

$$\max_z \{\tilde{F}_i z \text{ subject to } z^T P_z z \le 1\} = \big(\tilde{F}_i P_z^{-1} \tilde{F}_i^T\big)^{1/2}$$

it follows that $\tilde{F} x \le \mathbf{1}$, for all $x \in \mathcal{E}_z$ if and only if $\tilde{F}_i P_z^{-1} \tilde{F}_i^T \le 1$ for each row $i = 1, \ldots, n_C$. These conditions can be expressed equivalently in terms of a condition on a positive-definite diagonal matrix:

$$\begin{bmatrix} H_{1,1} - \tilde{F}_1 P_z^{-1} \tilde{F}_1^T & & \\ & \ddots & \\ & & H_{n_C,n_C} - \tilde{F}_{n_C} P_z^{-1} \tilde{F}_{n_C}^T \end{bmatrix} \succeq 0$$

for some scalars $H_{i,i} \le 1$, $i = 1, \ldots, n_C$, and this in turn is equivalent to

$$H - \tilde{F} P^{-1} \tilde{F}^T \succeq 0$$

for some symmetric matrix H with $e_i^T H e_i \le 1$, for all i. Using Schur complements (as discussed in Sect. 2.7.3), this condition is equivalent to

$$\begin{bmatrix} H & \tilde{F} \\ \tilde{F}^T & P_z \end{bmatrix} \succeq 0, \quad e_i^T H e_i \le 1, \quad i = 1, \ldots, n_C$$

which implies the necessity and sufficiency of (2.33). □

2.7.2 The Predicted Cost and MPC Algorithm

Given the autonomous form of the prediction dynamics of (2.25) it is possible to use a Lyapunov equation similar to (2.5) to evaluate the predicted cost $J(x_k, \mathbf{c}_k)$ of (2.11) along the predicted trajectories of (2.27a), (2.27b). The stage cost (namely the part of the cost incurred at each prediction time step) has the general form

$$\begin{aligned} \|x\|_Q^2 + \|u\|_R^2 &= \|x\|_Q^2 + \|Kx + c\|_R^2 = x^T(Q + K^T RK)x + \mathbf{c}^T E^T RE\mathbf{c} \\ &= \|z\|_{\hat{Q}}^2, \quad \hat{Q} = \begin{bmatrix} Q + K^T RK & K^T RE \\ E^T RK & E^T RE \end{bmatrix}. \end{aligned}$$

Hence $J(x_k, \mathbf{c}_k)$ can be written as

$$J(x_k, \mathbf{c}_k) = \sum_{i=0}^{\infty} \left(\|x_{i|k}\|_Q^2 + \|u_{i|k}\|_R^2 \right) = \sum_{i=0}^{\infty} \|z_{i|k}\|_{\hat{Q}}^2 = \|z_{0|k}\|_W^2$$

where, by Lemma 2.1, W is the (positive-definite) solution of the Lyapunov equation

$$W = \Psi^T W \Psi + \hat{Q}. \tag{2.34}$$

The special structure of Ψ and $\hat{Q}$ in this Lyapunov equation implies that its solution also has a specific structure, as we describe next.

Theorem 2.10 *If K is the optimal unconstrained linear feedback gain for the dynamics of (2.1a), then the cost (2.11) for the predicted trajectories of (2.27a), (2.27b) can be written as*

$$J(x_k, \mathbf{c}_k) = x_k^T W_x x_k + \mathbf{c}_k^T W_c \mathbf{c}_k$$

$$W_c = \begin{bmatrix} B^T W_x B + R & 0 & \cdots & 0 \\ 0 & B^T W_x B + R & \cdots & 0 \\ \vdots & \vdots & \ddots & \\ 0 & 0 & & B^T W_x B + R \end{bmatrix} \tag{2.35}$$

where W_x is the solution of the Riccati equation (2.9).

Proof Let $W = \begin{bmatrix} W_x & W_{xc} \\ W_{cx} & W_c \end{bmatrix}$, then substituting for W, Ψ and $\hat{Q}$ in (2.34) gives

$$W_x = \Phi^T W_x \Phi + Q + K^T R K \tag{2.36a}$$

$$W_{cx} = M^T W_{cx} \Phi + E^T (B^T W_x \Phi + RK) \tag{2.36b}$$

$$W_c = (BE)^T W_x (BE) + (BE)^T W_{xc} M + M^T W_{cx} BE + M^T W_c M + E^T RE \tag{2.36c}$$

The predicted cost for $\mathbf{c}_k = 0$ is $\|x_k\|^2_{W_x}$, and since K is the unconstrained optimal linear feedback gain, it follows from (2.36a) and Theorem 2.1 that W_x is the solution of the Riccati equation (2.9). Furthermore, from Theorem 2.1 we have $K = -(B^T W_x B + R)^{-1} B^T W_x A$, so that $B^T W_x \Phi + RK = 0$ and hence (2.36b) gives $W_{cx} - M^T W_{cx} \Phi = 0$, which implies that $W_{cx} = 0$. Therefore,

$$W = \begin{bmatrix} W_x & 0 \\ 0 & W_c \end{bmatrix}, \tag{2.37}$$

and from (2.36c) we have $W_c - M^T W_c M = E^T (B^T W_x B + R) E$. Hence from the structure of M and E in (2.26b), W_c is given by (2.35). □

Corollary 2.1 *The unconstrained LQ optimal control law is given by the feedback law $u = Kx$, where $K = -(B^T W_x B + R)^{-1} B^T W_x A$ and W_x is the solution of the Riccati equation (2.9).*

Proof Theorem 2.1 has already established that the unconstrained optimal linear feedback gain is as given in the corollary. The question remains as to whether it is possible to obtain a smaller cost by perturbing this feedback law. Equation (2.35) implies that this cannot be the case because the minimum cost is obtained for $\mathbf{c}_k = 0$. This argument applies for arbitrary N and hence for perturbation sequences of any length. □

Using the autonomous prediction system formulation of this section, Algorithm 2.1 can be restated as follows:

Algorithm 2.2 At each time instant $k = 0, 1, \ldots$:

(i) Perform the optimization

$$\underset{\mathbf{c}_k}{\text{minimize}} \; \|\mathbf{c}_k\|^2_{W_c} \; \text{subject to} \; \begin{bmatrix} x_k \\ \mathbf{c}_k \end{bmatrix} \in \mathcal{S} \tag{2.38}$$

where $\mathcal{S} = \mathcal{Z}$ defined in (2.29) ($\nu_z n_C$ linear constraints), or $\mathcal{S} = \mathcal{E}_z$ defined by the solution of (2.32) and (2.33) (a single quadratic constraint).

(ii) Apply the control law $u_k = Kx_k + c^*_{0|k}$, where $\mathbf{c}^*_k = (c^*_{0|k}, \ldots, c^*_{N-1|k})$ is the optimal value of $\mathbf{c}_k$ for problem (2.38). ◁

Theorem 2.11 *Under the MPC law of Algorithm 2.2, the origin $x = 0$ of system (2.1a) is an asymptotically stable equilibrium with a region of attraction equal to the set of states that are feasible for the constraints in (2.38).*

Proof The constraint set in (2.38) is by assumption positively invariant. Therefore, the tail $\mathbf{c}_{k+1} = M\mathbf{c}^*_k$ provides a feasible but suboptimal solution for (2.38) at time $k+1$. Stability and asymptotic convergence of x_k to the origin is then shown by applying the arguments of the proofs of Theorems 2.7 and 2.8 to the optimal value of the cost $J(x_k, \mathbf{c}^*_k)$ at the solution of (2.38). □

2.7.3 *Offline Computation of Ellipsoidal Invariant Sets*

In order to determine the invariant ellipsoidal set $\mathcal{E}_z$ for the autonomous prediction dynamics (2.25), the matrices P_z and H must be considered as variables in the conditions of Theorem 2.9. These conditions then constitute Linear Matrix Inequalities (LMIs) in the elements of P_z and H. Linear matrix inequalities are used extensively throughout this book; for an introduction to the properties of LMIs and LMI-based techniques that are commonly used in systems analysis and control design problems, we refer the reader to [17].

In its most general form a linear matrix inequality is a condition on the positive definiteness of a linear combination of matrices, where the coefficients of this combination are considered as variables. Thus a (strict) LMI in the variable $x \doteq (x_1, \ldots, x_n) \in \mathbb{R}^n$ can be expressed

$$M(x) \doteq M_0 + M_1 x_1 + \ldots + M_n x_n \succ 0 \tag{2.39}$$

where $M_0, \ldots, M_n$ are given matrices.[2] The convenience of LMIs lies in the convexity of (2.39) (see also Questions 1–3 on page 233). This property makes it possible to include conditions, such as those defining an invariant ellipsoidal set in Theorem 2.9, in convex optimization problems that can be solved efficiently using semidefinite programming.

[2]A non-strict LMI is similarly defined by $M(x) \succeq 0$. Any non-strict LMI can be expressed equivalently as a combination of a linear equality constraint and a strict LMI (see, e.g. [17]). However, none of the non-strict LMIs encountered in this chapter or in Chap. 5 carry implicit equality constraints, and hence non-strict LMIs may be assumed to be either strictly feasible or infeasible. We therefore make use of both strict and non-strict LMIs with the understanding that $M(x) \succeq 0$ can be replaced with $M(x) \succ 0$ for the purposes of numerical implementation.

A suitable criterion for selecting P_z is to maximize the region of attraction of Algorithm 2.2, namely the feasible set for the constraint $z_k^T P_z z_k \le 1$. This region is equal to the projection of $\mathcal{E}_z = \{z : z^T P_z z \le 1\}$ onto the x-subspace:

$$\{x : \exists \mathbf{c} \text{ such that } x^T P_{xx} x + 2\mathbf{c}^T P_{cx} x + \mathbf{c}^T P_{cc} \mathbf{c} \le 1\}$$

where the matrices P_{xx}, P_{xc}, P_{cx}, P_{cc} are blocks of P_z partitioned according to

$$P_z = \begin{bmatrix} P_{xx} & P_{xc} \\ P_{cx} & P_{cc} \end{bmatrix}. \tag{2.40}$$

By considering the minimum value of $z^T P_z z$ over all $\mathbf{c}$ for given x, it is easy to show that the projection of $\mathcal{E}_z$ onto the x-subspace is given by

$$\mathcal{E}_x \doteq \big\{x : x^T (P_{xx} - P_{xc} P_{cc}^{-1} P_{xc}) x \le 1\big\}.$$

Inverting the partitioned matrix P_z we obtain

$$P_z^{-1} = S \doteq \begin{bmatrix} S_{xx} & S_{xc} \\ S_{cx} & S_{cc} \end{bmatrix},$$

where

$$S_{xx} = \big(P_{xx} - P_{xc} P_{cc}^{-1} P_{xc}\big)^{-1},$$

and hence the volume of the projected ellipsoidal set $\mathcal{E}_x$ is proportional to $1/\det(S_{xx}^{-1}) = \det(S_{xx})$. The volume of the region of attraction of Algorithm 2.2 is therefore maximized by the optimization

$$\underset{S, P_z, H}{\text{maximize}} \ \det(S_{xx}) \text{ subject to (2.32), (2.33)} \tag{2.41}$$

Maximizing the objective in (2.41) is equivalent to maximizing $\log\det(S_{xx})$, which is a concave function of the elements of S (see, e.g. [18]). But this is not yet a semidefinite programming problem since (2.32) and (2.33) are LMIs in P_z rather than S. These constraints can however be expressed as Linear Matrix Inequalities in S using Schur complements.

In particular, the positive definiteness of a partitioned matrix

$$\begin{bmatrix} U & V^T \\ V & W \end{bmatrix} \succ 0$$

where U, V, W are real matrices of conformal dimensions, is equivalent to positive definiteness of the Schur complements

$$U \succ 0 \text{ and } W - VU^{-1}V^T \succ 0,$$

or

$$W \succ 0 \text{ and } U - V^T W^{-1} V \succ 0$$

(the proof of this result is discussed in Question 1 in Chap. 5 on page 233). Therefore, after pre- and post-multiplying (2.32) by S, using Schur complements we obtain the following condition:

$$\begin{bmatrix} S & \Psi S \\ S\Psi^T & S \end{bmatrix} \succeq 0, \tag{2.42}$$

which is an LMI in S. Similarly, pre- and post-multiplying the matrix inequality in (2.33) by $\left[\begin{smallmatrix} I & 0 \\ 0 & S \end{smallmatrix}\right]$ yields the condition

$$\begin{bmatrix} H & \begin{bmatrix} F + GK & GE \end{bmatrix} S \\ S \begin{bmatrix} (F+GK)^T \\ (GE)^T \end{bmatrix} & S \end{bmatrix} \succeq 0 \tag{2.43}$$

which is an LMI in S and H. Therefore $\mathcal{E}_z$ can be computed by solving the SDP problem

$$\underset{S,H}{\text{maximize}} \ \log\det(S_{xx}) \ \text{ subject to (2.42), (2.43)} \\ \text{and } e_i^T H e_i \le 1, \ i = 1, \dots, n_C. \tag{2.44}$$

Example 2.4 For the system model, constraints and cost of Example 2.1, Fig. 2.7 shows the ellipsoidal regions of attraction $\mathcal{E}_x$ of Algorithm 2.2 for values of N in the range 5–40 and compares these with the polytopic feasible set $\mathcal{F}_N$ for $N = 10$. As expected, the ellipsoidal feasible sets are smaller than the polytopic feasible sets of Fig. 2.3, but the difference in area is small; the area of $\mathcal{E}_x$ for $N = 40$ is 13.4 while that of $\mathcal{F}_{10}$ is 13.6, a difference of only 1 %. On the other hand 36 linear constraints are needed to define the polytopic set $\mathcal{Z}$ for $N = 10$ whereas $\mathcal{E}_z$ is a single (quadratic) constraint.

Figure 2.8 shows closed-loop state and input responses for Algorithm 2.2, comparing the responses obtained with the ellipsoidal constraint $z_k \in \mathcal{E}_z$ against the responses obtained with the linear constraint set $z_k \in \mathcal{Z}$ for $N = 10$. The difference in the closed-loop costs of the two controllers for the initial condition $x_0 = (-7.5, 0.5)$ is 17 %. ◊

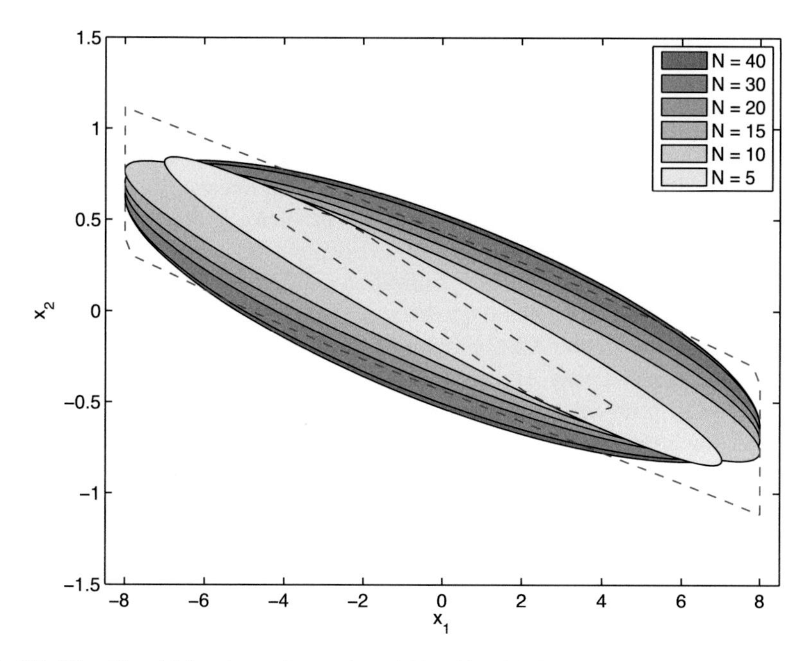

Fig. 2.7 The ellipsoidal regions of attraction of Algorithm 2.2 for $N = 5, 10, 15, 20, 30, 40$. The polytopic sets $\mathcal{F}_{10}$ and $\mathcal{X}_T$ are shown (*dashed lines*) for comparison

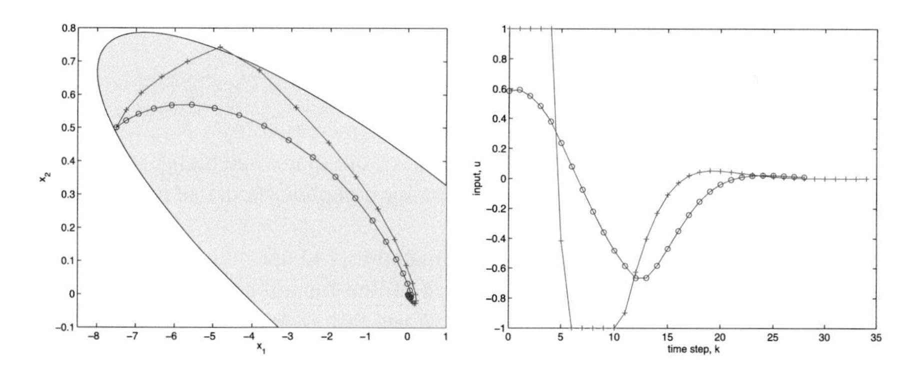

Fig. 2.8 Closed-loop responses of Algorithm 2.2 for the example of (2.16a), (2.16b) for the quadratic constraint $z_k \in \mathcal{E}_z$ with $N = 20$ (*blue* o) and the linear constraints $z_k \in \mathcal{Z}$ with $N = 10$ (*red* +). *Left* state trajectories and the feasible set $\mathcal{E}_x$ for $N = 20$. *Right* control inputs

2.8 Computational Issues

The optimization problem to be solved online in Algorithm 2.1 has a convex quadratic objective function and linear constraints, and is therefore a convex Quadratic Program (QP). Likewise if Algorithm 2.2 is formulated in terms of linear constraints, then this also requires the online solution of a convex QP problem. A variety of general QP solvers (based on active set methods [19] or interior point methods [20]), can therefore be used to perform the online MPC optimization required by these algorithms.

However algorithms for general quadratic programming problems do not exploit the special structure of the MPC problem considered here, and as a result their computational demand may exceed allowable limits. In particular they may not be applicable to problems with high sample rates, high-dimensional models, or long prediction horizons. For example the computational load of both interior point and active set methods grows approximately cubically with the mode 1 prediction horizon N.

The rate of growth with N of the required computation can be reduced however if the predicted model states are considered to be optimization variables. Thus redefining the vector of degrees of freedom as $d_k \in \mathbb{R}^{Nn_x+Nn_u}$:

$$d_k = (c_{0|k}, x_{1|k}, c_{1|k}, x_{2|k}, \ldots, c_{N-1|k}, x_{N|k})$$

and introducing the predicted dynamics of (2.14) as equality constraints results in an online optimization of the form

$$\underset{d_k}{\text{minimize}}\ d_k^T H_d d_k \ \text{ subject to } \ D_d d_k = h_h,\ C_c d_k \leq h_c.$$

Although the number of optimization variables has increased from Nn_u to $Nn_u + Nn_x$, the key benefit is that the matrices H_d, D_d, C_c are sparse and highly structured. This structure can be exploited to reduce the online computation so that it grows only linearly with N (e.g. see [19, 20]).

An alternative to reducing the online computation is to use multiparametric programming to solve the optimization problem offline for initial conditions that lie in different regions of the state space. Thus, given that x_k is a known constant, the minimization of the cost of (2.35) is equivalent to the minimization of

$$J(d) = d^T H_0 d \tag{2.45}$$

where for simplicity, the vector of degrees of freedom $\mathbf{c}$ has been substituted by d and the cost is renamed as simply J. The minimization of J is subject to the linear constraints implied by the dynamics (2.14) and system constraints (2.2), together with the terminal constraints of (2.35); the totality of these constraints can be written as

$$C_0 d \leq h_0 + V_0 x \tag{2.46}$$

Then adjoining the constraints (2.46) with the cost of (2.45) through the use of a vector of Lagrange multipliers λ, we obtain the first-order Karush–Kuhn–Tucker (KKT) conditions [19]

$$H_0 d + C_0^T \lambda = 0 \tag{2.47a}$$

$$\lambda^T (C_0 d - h_0 - V_0 x) = 0 \tag{2.47b}$$

$$C_0 d \leq h_0 + V_0 x \tag{2.47c}$$

$$\lambda \geq 0 \tag{2.47d}$$

Now suppose that at the given x only a subset of (2.46) is active, so that gathering all these active constraints and the corresponding Lagrange multipliers we can write

$$\tilde{C}_0 d - \tilde{h}_0 - \tilde{V}_0 x = 0 \tag{2.48a}$$

$$\tilde{\lambda} \geq 0 \tag{2.48b}$$

In addition, the Lagrange multipliers corresponding to inactive constraints will be zero so that from (2.47) it follows that

$$d = -H_0^{-1} \tilde{C}_0^T \tilde{\lambda}. \tag{2.49}$$

The solution for $\tilde{\lambda}$ can be derived by substituting (2.49) into (2.48a) as

$$\tilde{\lambda} = -(\tilde{C}_0 H_0^{-1} \tilde{C}^T)^{-1} (\tilde{h}_0 + \tilde{V}_0 x). \tag{2.50}$$

and substituting this into (2.49) produces the optimal solution as

$$d = H_0^{-1} \tilde{C}_0^T - (\tilde{C}_0 H_0^{-1} \tilde{C}_0^T)^{-1} (\tilde{h}_0 + \tilde{V}_0 x). \tag{2.51}$$

Thus for given active constraints, the optimal solution is a known affine function of the state. Clearly the optimal solution must satisfy the constraints (2.46) as well as the Lagrange multipliers of (2.50) must satisfy (2.48a):

$$C_o [H_o^{-1} \tilde{C}_o^T - (\tilde{C}_o H_o^{-1} \tilde{C}_o^T)^{-1} (\tilde{h}_o + \tilde{V}_o x)] \leq h_o + V_o x$$

and

$$-(\tilde{C}_o H_o^{-1} \tilde{C}_o^T)^{-1} (\tilde{h}_o + \tilde{V}_o x) > 0.$$

These two conditions give a characterization of the polyhedral region in which x must lie in order that (2.48a) is the active constraint set.

A procedure based on these considerations is given in [14] for partitioning the controllable set of Algorithms 2.1 and 2.2 into the union of a number of non-overlapping polyhedral regions. Then the MPC optimization can be implemented online by identifying the particular polyhedral region in which the current state lies. In this approach the associated optimal solution (2.51) is then recovered from a lookup table, and the first element of this is used to compute and implement the current optimal control input.

A disadvantage of this multiparametric approach is that the number of regions grows exponentially with the dimension of the state and the length of the mode 1 prediction horizon N, and this can make the approach impractical for anything other than small-scale problems with small values of N. Indeed in most other cases, the computational and storage demands of the multiparametric approach exceed those required by the QP solvers that exploit the MPC structure described above. Methods have been proposed (e.g. [21]) for improving the efficiency with which the polyhedral state-space partition is computed by merging regions that have the same control law, however the complexity of the polyhedral partition remains prohibitive in this approach.

Example 2.5 For the second-order system defined in (2.16a), (2.16b), with the cost and terminal constraints of Example 2.3 the MPC optimization problem (2.17) can be solved using multiparametric programming. For a mode 1 horizon of $N = 10$ this results in a partition of the state space into 243 polytopic regions (Fig. 2.9), each of which corresponds to a different active constraint set at the solution of the MPC optimization problem (2.17). $\diamondsuit$

A further alternative [15, 16] which results in significant reduction in the online computation replaces the polytopic constraints $z_k \in \mathcal{Z}$ defined (2.29) by the ellipsoidal constraint $z_k \in \mathcal{E}_z$ defined in (2.44) and thus addresses the optimization

$$\underset{\mathbf{c}_k}{\text{minimize}} \ \|z_k\|_W^2 \quad \text{subject to } z_k^T P_z z_k \leq 1, \quad z_k = \begin{bmatrix} x_k \\ \mathbf{c}_k \end{bmatrix} \tag{2.52}$$

As discussed in Sect. 2.7, this results in a certain degree of conservativeness because the ellipsoidal constraint $z_k \in \mathcal{E}_z$ gives an inner approximation to the polytopic constraint $z_k \in \mathcal{Z}$ of (2.29). The problem defined in (2.52) can be formulated as a second-order cone program (SOCP) in $Nn_u + 1$ variables.[3] If a generic solution method is employed, then this problem could turn out to be more computationally demanding than the QP that arises when the constraints are linear. However, the simple form of the cost and constraint in (2.38) allow for a particularly efficient solution, which is to be discussed next.

---

[3]Second-order cone programs are convex optimization problems that can be solved using interior point methods. See [22] for details and further applications of SOCP.

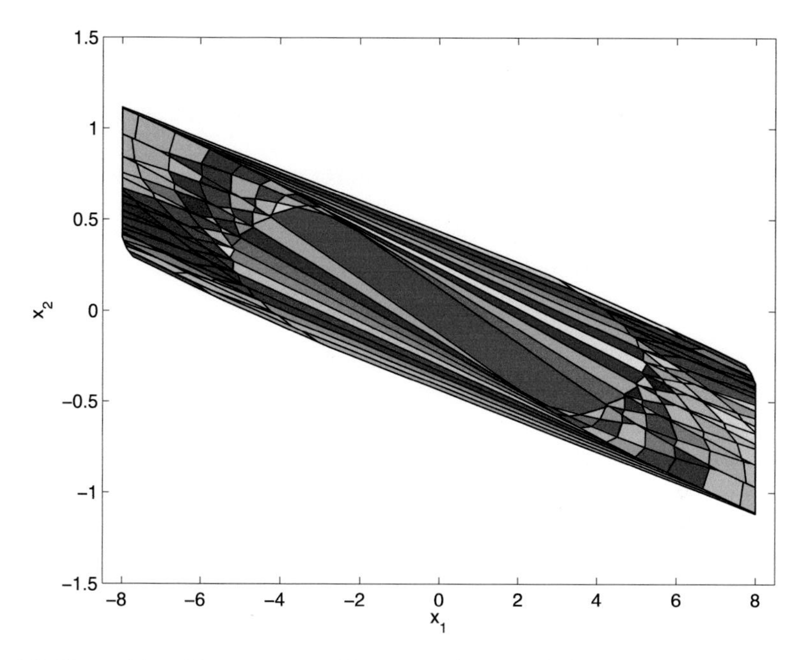

Fig. 2.9 The partition of the state space of the system of Example 2.5 into regions in which different constraint sets are active at the solution of the online MPC optimization problem

To exploit the structure of the cost and constraint in (2.52), we use the partitions of (2.37) and (2.40) to write $z_k^T W z_k = x_k^T W_x x + \mathbf{c}_k^T W_c \mathbf{c}_k$ and $z_k^T P_z z_k = x_k^T P_{xx} x_k + 2\mathbf{c}_k^T P_{cx} x_k + \mathbf{c}_k^T P_{cc} \mathbf{c}_k$, where use has been made of the fact that $P_{xc} = P_{cx}^T$. The minimizing value of $\mathbf{c}_k$ in (2.52) can only occur at a point at which the two ellipsoidal boundaries, $\partial\mathcal{E}_J \doteq \{z_k : z_k^T W z_k = \alpha\}$ and $\partial\mathcal{E}_z \doteq \{z_k : z_K^T P_z z_k = 1\}$, are tangential to one another for some constant $\alpha > 0$, namely when the gradients (with respect to $\mathbf{c}$) are parallel, i.e.

$$W_c \mathbf{c}_k = \mu(P_{cx} x_k + P_{cc} \mathbf{c}_k), \quad \mu \leq 0 \tag{2.53}$$

for some scalar μ, or equivalently

$$\mathbf{c}_k = \mu M_\mu P_{cx} x_k, \quad M_\mu = (W_c - \mu P_{cc})^{-1}. \tag{2.54}$$

At the solution therefore, the inequality constraint in (2.52) will hold with equality so that μ can be obtained as the solution of $x_k^T P_{xx} x_k + 2\mathbf{c}_k^T P_{cx} + \mathbf{c}_k^T P_{cc} \mathbf{c}_k = 1$, which after some algebraic manipulation gives μ as a root of

$$\phi(\mu) = x_k^T P_{xc} \left(M_\mu W_c P_{cc}^{-1} W_c M_\mu - P_{cc}^{-1} \right) P_{cx} x_k + x_k^T P_{xx} x_k - 1 = 0. \tag{2.55}$$

Equation (2.55) is equivalent to a polynomial equation in μ which can be shown (using straightforward algebra) to have $2N$ roots, all corresponding to points of tangency of $\partial\mathcal{E}_J$ and $\partial\mathcal{E}_z$. However (2.52) has a unique minimum, and it follows that only one of these roots can be negative, as is required by (2.53).

By repeatedly differentiating $\phi(\mu)$ with respect to μ it is easy to show that the derivatives of this polynomial satisfy

$$\frac{d^r\phi}{d\mu^r} > 0 \ \ \forall\mu \leq 0.$$

This implies that the Newton–Raphson method, when initialised at $\mu = 0$, is guaranteed to converge to the unique negative root of (2.55), and that the rate of its convergence is quadratic.

Thus the optimal solution to (2.52) is obtained extremely efficiently by substituting the negative root of (2.55) into (2.54); in fact the computation required is equivalent to solving a univariate polynomial with monotonic derivatives. The price that must be paid for this gain in computational efficiency is a degree of suboptimality that results from the use of the ellipsoidal constraint $z_k \in \mathcal{E}_z$, which provides only an inner approximation to the actual polytopic constraint of (2.29). However, simulation results [16] show that in most cases the degree of suboptimality is not significant. Furthermore predicted performance can be improved by a subsequent univariate search over $\alpha \in [0, 1]$ with $z_k = (x_k, \alpha\mathbf{c}_k^*)$ where $\mathbf{c}_k^*$ is the solution of (2.52). To retain the guarantee of closed-loop stability this is performed subject to the constraints that the vector Ψz_k defining the tail of the predicted sequence at time k should lie in the ellipsoid $\mathcal{E}_z$ and subject to the constraint $Fx_k + Gu_k \leq 1$. This modification requires negligible additional computation.

2.9 Optimized Prediction Dynamics

The MPC algorithms described thus far parameterize the predicted inputs in terms of a projection onto the standard basis vectors e_i, so, for example

$$\mathbf{c}_k = \sum_{i=0}^{N-1} c_{i|k} e_{i+1}$$

in the case that if $n_u = 1$. As a consequence the degrees of freedom have a direct effect on the predictions only over the N-step mode 1 prediction horizon, which therefore has to be taken to be sufficiently long to ensure that constraints are met during the transients of the prediction system response. Combined with the additional requirement that the terminal constraint is met at the end of the mode 1 horizon for as large a set of initial conditions as possible, this places demands on N that can make the computational load of MPC prohibitive for applications with high sampling rates.

To overcome this problem an extra mode can be introduced into the predicted control trajectories, as is done for example in triple mode MPC [23]. This additional mode introduces degrees of freedom into predictions after the end of the mode 1 horizon but allows efficient handling of the constraints at these prediction instants, thus allowing the mode 1 horizon to be shortened without adversely affecting optimality and the size of the feasible set. Alternatively in the context of dual-mode predictions it is possible to consider parameterizing predicted control trajectories as an expansion over a finite set of basis functions. Exponential basis functions, which allow the use of arguments based on the tail for analysing stability and convergence (e.g. [24]), are most commonly employed in MPC, a special case being expansion over Laguerre functions (e.g. [25]).

A framework that encompasses projection onto a general set of exponential basis functions was developed in [26]. In this approach, the matrices E and M appearing in the transition matrix Ψ of the augmented prediction dynamics (2.25) are not chosen as prescribed by (2.26b), but instead are replaced by variables, denoted A_c and C_c that are optimized offline as we discuss later in this section. With this modification the prediction dynamics are given by

$$z_{i+1|k} = \Psi z_{i|k}, \quad i = 0, 1, \ldots \tag{2.56a}$$

where

$$z_{0|k} = \begin{bmatrix} x_k \\ \mathbf{c}_k \end{bmatrix}, \quad \Psi = \begin{bmatrix} \Phi & BC_c \\ 0 & A_c \end{bmatrix} \tag{2.56b}$$

and the predicted state and control trajectories are generated by

$$u_{i|k} = \begin{bmatrix} K & C_c \end{bmatrix} z_{i|k} \tag{2.56c}$$

$$x_{i|k} = \begin{bmatrix} I & 0 \end{bmatrix} z_{i|k}. \tag{2.56d}$$

As in Sect. 2.7, the predicted control law of (2.56c) has the form of a dynamic feedback controller, the initial state of which is given by $\mathbf{c}_k$. However in Sect. 2.7 the matrix M of (2.26) is nilpotent, so that $M^N \mathbf{c}_k = 0$ and hence $u_{i|k} = K x_{i|k}$, for all $i = N, N+1, \ldots$. For the general case considered in (2.56), A_c is not necessarily nilpotent, which implies that the direct effect of the elements of $\mathbf{c}_k$ can extend beyond the initial N steps of the prediction horizon in this setting.

Following a development analogous to that of Sect. 2.7, the predicted cost (2.11) can be expressed as $J(x_k, \mathbf{c}_k) = \|z_{0|k}\|_W^2$ where W satisfies the Lyapunov matrix equation

$$W = \Psi^T W \Psi + \hat{Q}, \quad \hat{Q} = \begin{bmatrix} Q + K^T R K & K^T R C_c \\ C_c^T R K & C_c^T R C_c \end{bmatrix}. \tag{2.57}$$

By examining the partitioned blocks of this equation, it can be shown (using the same approach as the proof of Theorem 2.10) that its solution is block diagonal

$$W = \begin{bmatrix} W_x & 0 \\ 0 & W_c \end{bmatrix}$$

whenever K is the unconstrained optimal feedback gain. Here W_x is the solution of the Riccati equation (2.9) and W_c is the solution of the Lyapunov equation $W_c = A_c^T W A_c + C_c^T (B^T W_x B + R) C_c$. By Lemma 2.1, the solution is unique and satisfies $W_c \succ 0$ whenever A_c is strictly stable.

The constraints (2.2) applied to the predictions of (2.56) require that $z_{0|k}$ lies in the polytopic set

$$\mathcal{Z} = \{z : \begin{bmatrix} F + GK & GC_c \end{bmatrix} \Psi^i z \le \mathbf{1}, \;\; i = 0, 1, \ldots, \nu_z\}, \tag{2.58}$$

where $\begin{bmatrix} F + GK & GC_c \end{bmatrix} \Psi^{\nu_z+1} z \le \mathbf{1}$, for all z satisfying $\begin{bmatrix} F + GK & GC_c \end{bmatrix} \Psi^i z \le \mathbf{1}$, $i = 0, 1, \ldots, \nu_z$. By Theorem 2.3 this is the MPI set for the dynamics of (2.56) and constraints (2.2), and its projection onto the x-subspace is therefore equal to the feasible set for x_k for the prediction system (2.56) and constraints $\begin{bmatrix} F + GK & GC_c \end{bmatrix} z \le \mathbf{1}$. The MPC law of Algorithm 2.2 with the cost matrix W defined in (2.57) and constraint set $\mathcal{Z}$ defined in (2.58) has the stability and convergence properties stated in Theorem 2.11.

Alternatively, and similarly to the discussion in Sect. 2.7, it is possible to replace the linear constraints $z_{0|k} \in \mathcal{Z}$ by a single quadratic constraint $z_{0|k} \in \mathcal{E}_z$ in order to reduce the online computational load of Algorithm 2.2. As in Sect. 2.7, we require that $\mathcal{E}_z = \{z : z^T P_z z \le 1\}$ is positively invariant for the dynamics $z_{k+1} = \Psi z_k$ and constraints $\begin{bmatrix} F + GK & GC_c \end{bmatrix} z_k \le \mathbf{1}$, which by Theorem 2.9 requires that there exists a symmetric matrix H such that P_z, A_c and C_c satisfy

$$P_z - \Psi^T P_z \Psi \succeq 0 \tag{2.59a}$$

$$\begin{bmatrix} H & \begin{bmatrix} F + GK & GC_c \end{bmatrix} \\ \begin{bmatrix} (F + GK)^T \\ (GC_c)^T \end{bmatrix} & P_z \end{bmatrix} \succeq 0, \quad e_i^T H e_i \le 1, \;\; i = 1, \ldots n_C. \tag{2.59b}$$

Under these conditions the stability and convergence properties specified by Theorem 2.11 again apply.

Using $\mathcal{E}_z$ as the constraint set in the online optimization in place of $\mathcal{Z}$ reduces the region of attraction of the MPC law. However, to compensate for this effect it is possible to design the prediction system parameters A_c and C_c so as to maximize the projection of $\mathcal{E}_z$ onto the x-subspace. Analogously to (2.44), this is achieved by maximizing the determinant of $[I_{n_x} \;\; 0] P_z^{-1} [I_{n_x} \;\; 0]^T$ subject to (2.59a), (2.59b). Unlike the case considered in Sect. 2.7, this is performed with A_c and C_c as optimization variables. Viewed as inequalities in these variables, (2.59a), (2.59b) represent non-convex constraints. The problem can however be convexified provided the dimension of $\mathbf{c}_k$ is at least as large as that of n_x [26] using a technique introduced by [27] in the context of $\mathcal{H}_\infty$ control, as we discuss next.

Introducing variables $U, V \in \mathbb{R}^{n_x \times \nu_c}$ (where ν_c is the length of $\mathbf{c}_k$), $\Xi \in \mathbb{R}^{n_x \times n_x}$, $\Gamma \in \mathbb{R}^{n_u \times n_x}$ and symmetric $X, Y \in \mathbb{R}^{n_x \times n_x}$, we re-parameterize the problem by defining

$$P_z = \begin{bmatrix} X^{-1} & X^{-1}U \\ U^T X^{-1} & \bullet \end{bmatrix} \quad P_z^{-1} = \begin{bmatrix} Y & V \\ V^T & \bullet \end{bmatrix}, \quad \Xi = U A_c V^T, \quad \Gamma = C_c V^T \tag{2.60}$$

(where $\bullet$ indicates blocks of P_z and P_z^{-1} that are determined uniquely by X, Y, U, V). Since $P_z P_z^{-1} = I$, we also require that

$$UV^T = X - Y. \tag{2.61}$$

The constraints (2.59a), (2.59b) can then be expressed as LMIs in Ξ, Γ, X and Y. Specifically, using Schur complements, (2.59a) is equivalent to

$$\begin{bmatrix} P_z & P_z \Psi \\ \Psi^T P_z & P_z \end{bmatrix} \succeq 0,$$

and multiplying the LHS of this inequality by $\operatorname{diag}\{\Pi^T, \Pi^T\}$ on the left and $\operatorname{diag}\{\Pi, \Pi\}$ on the right, where $\Pi = \begin{bmatrix} Y & X \\ V^T & 0 \end{bmatrix}$, yields the equivalent condition

$$\begin{bmatrix} \begin{bmatrix} Y & X \\ X & X \end{bmatrix} & \begin{bmatrix} \Phi Y + B\Gamma & \Phi X \\ \Xi + \Phi Y + B\Gamma & \Phi X \end{bmatrix} \\ \star & \begin{bmatrix} Y & X \\ X & X \end{bmatrix} \end{bmatrix} \succeq 0 \tag{2.62a}$$

(where the block marked $\star$ is omitted as the matrix is symmetric). Similarly, pre- and post-multiplying the matrix inequality in (2.59b) by $\operatorname{diag}\{I, \Pi^T\}$ and $\operatorname{diag}\{I, \Pi\}$, respectively, yields

$$\begin{bmatrix} H & \begin{bmatrix} (F + GK)Y + G\Gamma & (F + GK)X \end{bmatrix} \\ \star & \begin{bmatrix} Y & X \\ X & X \end{bmatrix} \end{bmatrix} \succeq 0, \quad e_i^T H e_i \le 1, \quad i = 1, \ldots, n_C. \tag{2.62b}$$

Therefore matrices P_z, A_c and C_c can exist satisfying (2.59a), (2.59b) only if the conditions (2.62a), (2.62b) are feasible. Moreover, (2.62a), (2.62b) are both necessary and sufficient for feasibility of (2.59a), (2.59b) if $\nu_c \geq n_x$ since (2.61) then imposes no additional constraints on X and Y (in the sense that U and V then exist satisfying (2.61), for all $X, Y \in \mathbb{R}^{n_x \times n_x}$). The volume of the projection of $\mathcal{E}_z$ onto the x-subspace is proportional to $\det(Y)$, which is maximized by solving the convex optimization:

$$\underset{\Xi,\Gamma,X,Y}{\text{maximize}} \ \log\det(Y) \ \text{ subject to (2.62a), (2.62b).} \qquad (2.63)$$

Finally, we note that the conditions (2.62a), (2.62b) do not depend on the value of ν_c, and since there is no advantage to be gained using a larger value, we set $\nu_c = n_x$. From the solution of (2.63), A_c and C_c are given uniquely by

$$A_c = U^{-1}\Xi V^{-T}, \quad C_c = \Gamma V^{-T}.$$

while P_z can be recovered from (2.60).

A remarkable property of the optimized prediction dynamics is that the maximal projection of $\mathcal{E}_z$ onto the x-subspace is as large as the maximal positively invariant ellipsoidal set under any linear state feedback control law [26]. The importance of this is that it overcomes the trade-off that exists in the conventional MPC formulations of Sects. 2.7 and 2.5 between performance and the size of the feasible set. Thus, in the interests of enlarging the terminal invariant set (and hence the overall region of attraction), it may be tempting to de-tune the terminal control law. But this has an adverse effect on predicted performance, and potentially also reduces closed-loop performance. Such loss of performance is however avoided if the optimized prediction dynamics are used since K can be chosen to be the unconstrained LQ optimal gain, without any detriment to the size of the region of attraction.

Example 2.6 The maximal ellipsoidal region of attraction of Algorithm 2.2 for the same system model, constraints and cost as Example 2.1 is shown in Fig. 2.10. Since this is obtained by optimizing the prediction dynamics using (2.63), the number of degrees of freedom in the resulting prediction system (i.e. the length of $\mathbf{c}_k$ in (2.56)) is the same as n_x, which here is 2. The area of this maximal ellipsoid is 13.5, whereas the area of the ellipsoidal region of attraction obtained from (2.44) for the non-optimized prediction system (2.25) and the same number of degrees of freedom in predictions (i.e. $N = 2$) is just 2.3.

Figure 2.10 also shows the polytopic feasible set for x_k in Algorithm 2.2 when the optimized prediction dynamics are used to define the polytopic constraint set $\mathcal{Z}$ in (2.58). Despite having only 2 degrees of freedom, the optimized prediction dynamics result in a polytopic feasible set covering 97 % of the area of the maximal feasible set $\mathcal{F}_\infty$, which for this example is equal to the polytopic feasible set for the non-optimized dynamics with $N = 26$ degrees of freedom (also shown in Fig. 2.10). For the initial condition $x_0 = (-7.5, 0.5)$, the closed-loop cost of Algorithm 2.2 with the optimized prediction dynamics containing 2 degrees of freedom and polytopic constraint set $\mathcal{Z}$ is 357.7, which from Table 2.1 is only 0.5 % suboptimal relative to the ideal optimal cost with $N = 11$. ◊

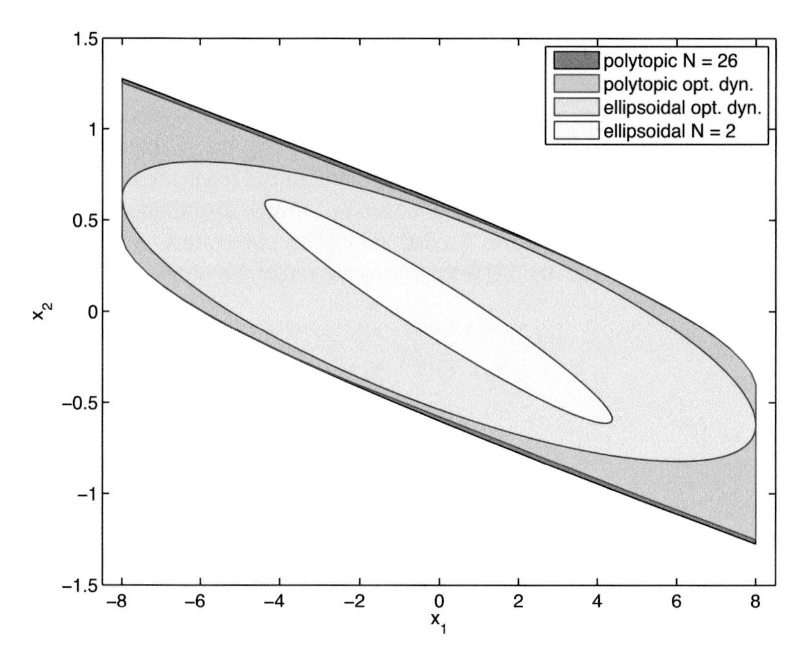

Fig. 2.10 Ellipsoidal region of attraction for optimized dynamics (with 2 degrees of freedom) and ellipsoidal region of attraction for $N = 2$. Also shown are the maximal polytopic region of attraction ($\mathcal{F}_{26}$) and the polytopic region of attraction for the optimized dynamics

2.10 Early MPC Algorithms

Perhaps the earliest reference to MPC strategies is [28], although the ideas of rolling horizons and decision making based on forecasts had been used earlier in different contexts (e.g. production scheduling). There have since been thousands of MPC papers published in the open literature, including a plethora of reports on applications of MPC to industrial problems. Early contributions (e.g. [29, 30]) were based on finite horizon predictive costs and as such did not carry guarantees of closed-loop stability.

The most cited of the early papers on predictive control is the seminal work [31, 32] on Generalized Predictive Control (GPC). This uses an input–output model to express the vector of output predictions as an affine function of the vector of predicted inputs

$$\mathbf{y}_k = \begin{bmatrix} y_{1|k} \\ \vdots \\ y_{N|k} \end{bmatrix} = C_G \Delta\mathbf{u}_k + \mathbf{y}_k^f, \quad \Delta\mathbf{u}_k = \begin{bmatrix} \Delta u_{0|k} \\ \vdots \\ \Delta u_{N_u-1|k} \end{bmatrix}$$

Here N_u denotes an input prediction horizon which is chosen to be less than or equal to the prediction horizon N. The matrix C_G is the block striped (Toeplitz) lower triangular matrix comprising the coefficients of the system step response, $C_G \Delta \mathbf{u}_k$ denotes the predicted forced response at time k, and $\mathbf{y}_k^f$ denotes the free response at time k due to non-zero initial conditions. The notation Δu is used to denote the control increments (i.e. $\Delta u_{i|k} = u_{i|k} - u_{i-1|k}$). Posing the problem in terms of control increments implies the automatic inclusion in the feedback loop of integral action which rejects (in the steady state) constant additive disturbances.

The GPC algorithm minimizes a cost, subject to constraints, which penalizes predicted output errors (deviations from a constant reference vector r) and predicted control increments

$$J_k = (\mathbf{r} - \mathbf{y}_k)^T \hat{Q} (\mathbf{r} - \mathbf{y}_k) + \Delta \mathbf{u}_k^T \hat{R} \Delta \mathbf{u}_k \tag{2.64}$$

where $\mathbf{r} = [r^T \; \cdots \; r^T]^T$, $\hat{Q} = \text{diag}\{Q, \ldots, Q\}$ and $\hat{R} = \text{diag}\{R, \ldots, R\}$. By setting the derivative of this cost with respect to $\Delta \mathbf{u}_k$ equal to zero, the unconstrained optimum vector of predicted control increments can be derived as

$$\Delta \mathbf{u}_k = \left(C_G^T \hat{Q} C_G + \hat{R} \right)^{-1} C_G^T \hat{Q} (\mathbf{r} - \mathbf{y}_k^f) \tag{2.65}$$

The optimal current control move $\Delta u_{0|k}$ is then computed from the first element of this vector, and the control input $u_k = \Delta u_{0|k} + u_{k-1}$ is applied to the plant.

GPC has proven effective in a wide range of applications and is the basis of a number of commercially successful MPC algorithms. There are several reasons for the success of the approach, principal among these are: the simplicity and generality of the plant model, and the lack of sensitivity of the controller to variable or unknown plant dead time and unknown model order; the fact that the approach lends itself to self-tuning and adaptive control, output feedback control and stochastic control problems; and the ability of GPC to approximate various well-known control laws through appropriate definition of the cost (2.64), for example LQ optimal control, minimum variance and dead-beat control laws. For further discussion of these aspects of GPC and its industrial applications we refer the reader to [31–34].

Although widely used in industry, the original formulation of GPC did not guarantee closed-loop stability except in limiting cases of the input and output horizons (for example, in the limit as both the prediction and control horizons tend to infinity, or when the control horizon is $N_u = 1$, the prediction horizon is $N = \infty$ and the open-loop system is stable). However, the missing stability guarantee can be established by imposing a suitable terminal constraint on predictions.

Terminal equality constraints that force the predicted tracking errors to be zero at all prediction times beyond the N-step prediction horizon were proposed for receding horizon controllers in the context of continuous time, time-varying unconstrained systems in [35], time invariant discrete time unconstrained systems [36], and nonlinear constrained systems [37]. This constraint effectively turns the cost of (2.64) into an infinite horizon cost which can be shown to be monotonically non-increasing using an argument based on the prediction tail. As a result it can be shown that tracking errors are steered asymptotically to zero. The terminal equality constraint need only to be applied over n_x prediction steps after the end of an initial N-step horizon. Under the assumption that $N > n_x$, the general solution of the equality constraints will contain, implicitly, $(N - n_x)n_u$ degrees of freedom and these can be used to minimize the resulting predicted cost (i.e. the cost of (2.64) after the expression for the general solution of the equality constraints has been substituted into (2.64)). A closely related algorithm to GPC that addresses the case of constrained systems is Stable GPC (SGPC) [38], which establishes closed-loop stability by ensuring that optimal predicted cost is a Lyapunov function for the closed-loop system. Related approaches [36, 39] use terminal equality constraints explicitly, however SGPC implements the equality constraints implicitly while preserving an explicit representation of the degrees of freedom in predictions.

The decision variables in the SGPC predicted control trajectories appear as perturbations of a stabilizing feedback law, and in terms of a left factorization of transfer function matrices, the predicted control sequence is given by

$$u_k = \tilde{Y}^{-1}(z^{-1})\left(c_k - z^{-1}\tilde{X}(z^{-1})y_{k+1}\right). \tag{2.66}$$

Here z is the z-transform variable (z^{-1} can be thought of as the backward shift operator, namely $z^{-1}f_k = f_{k-1}$), and $\tilde{X}(z^{-1})$, $\tilde{Y}(z^{-1})$ are polynomial solutions (expressed in powers of z^{-1}) of the matrix Bezout identity

$$\tilde{Y}(z^{-1})A(z^{-1}) + z^{-1}\tilde{X}(z^{-1})B(z^{-1}) = I. \tag{2.67}$$

For simplicity, we use u_k instead of Δu_k and consider the regulation rather than the setpoint tracking problem (i.e. we take $r = 0$). Here $B(z^{-1})$, $A(z^{-1})$ are the polynomial matrices (in powers of z^{-1}) defining right coprime factors of the system transfer function matrix, $G(z^{-1})$, where

$$y_{k+1} = G(z^{-1})u_k = B(z^{-1})A^{-1}(z^{-1})u_k \tag{2.68}$$

The determination of the coprime factors can be achieved through the computation of the Smith–McMillan form of the transfer function matrix, $G(z^{-1}) = L(z^{-1})S(z^{-1})R(z^{-1})$ where $S(z^{-1}) = \mathcal{E}(z^{-1})\Psi^{-1}(z^{-1})$ with both $\mathcal{E}(z^{-1})$ and $\Psi(z^{-1})$ being diagonal polynomial matrix functions of z^{-1}. The right coprime factors can then be chosen as $B(z^{-1}) = L(z^{-1})\mathcal{E}(z^{-1})$, $A(z^{-1}) = R^{-1}(z^{-1})\Psi(z^{-1})$. Alternatively, $B(z^{-1})$, $A(z^{-1})$ can be computed through an iterative procedure, which we describe now.

Assuming that $G(z^{-1})$ is given as

$$G(z^{-1}) = \frac{1}{d\left(z^{-1}\right)}N(z^{-1})$$

we need to find the solution, $A(z^{-1})$, $B(z^{-1})$, of the Bezout identity

$$N(z^{-1})A(z^{-1}) = B(z^{-1})d(z^{-1}) \tag{2.69}$$

for which (2.67) admits a solution for $\tilde{X}(z^{-1})$, $\tilde{Y}(z^{-1})$. This solution can be shown to be unique under the assumption that the coefficient of z^0 in $A(z^{-1})$ is the identity, and that $A(z^{-1})$ and $B(z^{-1})$ are of minimal degree. Equation (2.69) defines a set of under-determined linear conditions on the coefficients of $B(z^{-1})$, $A(z^{-1})$. Thus the coefficients of $B(z^{-1})$, $A(z^{-1})$ can be expressed as an affine function of a matrix, say R, where R defines the degrees of freedom which are to be given up so that (2.67) admits a solution. The determination of R constitutes a nonlinear problem which, nevertheless, can be solved to any desired degree of accuracy by solving (2.67) iteratively. The iteration consists of using the least squares solution for R of (2.67) to update the choice for the coefficients of $A(z^{-1})$, $B(z^{-1})$; these updated values are then used in (2.67) to update the solution for $\tilde{Y}(z^{-1})$, $\tilde{X}(z^{-1})$, and so on. Each cycle of this iteration reduces the norm of the error in the solution of (2.67) and the iterative process can be terminated when the norm of the error is below a practically desirable threshold.

Substituting (2.68) into (2.66), pre-multiplying by $\tilde{Y}(z^{-1})$ and using the Bezout identity (2.67) provides the prediction model:

$$\begin{aligned} y_{k+1} &= B(z^{-1})c_k + y^f_{k+1} \\ u_k &= A(z^{-1})c_k + u^f_k. \end{aligned} \tag{2.70}$$

Here y^f_k and u^f_k denote the components of the predicted output and input trajectories corresponding to the free response of the model due to non-zero initial conditions. Consider now the dual coprime factorizations $B(z^{-1})A^{-1}(z^{-1}) = \tilde{A}^{-1}(z^{-1})\tilde{B}(z^{-1})$, $X(z^{-1})Y^{-1}(z^{-1}) = \tilde{Y}^{-1}(z^{-1})\tilde{X}(z^{-1})$ satisfying the Bezout identity

$$\begin{bmatrix} z^{-1}\tilde{X}(z^{-1}) & \tilde{Y}(z^{-1}) \\ \tilde{A}(z^{-1}) & -\tilde{B}(z^{-1}) \end{bmatrix} \begin{bmatrix} B(z^{-1}) & Y(z^{-1}) \\ A(z^{-1}) & -z^{-1}X(z^{-1}) \end{bmatrix} = \begin{bmatrix} I & 0 \\ 0 & I \end{bmatrix} \tag{2.71}$$

Detailed calculation, based on simulating forward in time the relationships $\tilde{Y}(z^{-1})u_k = c_k - z^{-1}\tilde{X}(z^{-1})y_{k+1}$ and $\tilde{A}(z^{-1})y_{k+1} = \tilde{B}(z^{-1})u_k$, leads to the following affine relationship from the vector of predicted controller perturbations, $\mathbf{c}_k = (c_{0|k}, \dots, c_{N-1|k})$ (with $c_{i|k} = 0$, for all $i \geq \nu$), to the vectors of predicted outputs, $\mathbf{y}_k = (y_{1|k}, \dots, y_{N|k})$, and inputs, $\mathbf{u}_k = (u_{0|k}, \dots, u_{N-1|k})$:

$$\begin{bmatrix} C_{z^{-1}\tilde{X}} & C_{\tilde{Y}} \\ C_{\tilde{A}} & -C_{\tilde{B}} \end{bmatrix} \begin{bmatrix} \mathbf{y}_k \\ \mathbf{u}_k \end{bmatrix} = \begin{bmatrix} \mathbf{c}_k \\ 0 \end{bmatrix} - \begin{bmatrix} H_{z^{-1}\tilde{X}} & C_{\tilde{Y}} \\ H_{\tilde{A}} & -H_{\tilde{B}} \end{bmatrix} \begin{bmatrix} \mathbf{y}^p_k \\ \mathbf{u}^p_k \end{bmatrix} \tag{2.72}$$

where $N = \nu + n_A$, $\mathbf{y}^p_k = (y_{k-n_X-1}, \dots, y_k)$ and $\mathbf{u}^p_k = (u_{k-n_Y}, \dots, u_{k-1})$ denote vectors of past input and output values and n_A, n_X, n_Y are the degrees of the polynomials $A(z^{-1}), X(z^{-1}), Y(z^{-1})$. The C and H matrices are block Toeplitz convolution matrices, which are defined for any given matrix polynomial $F(z^{-1}) = F_0 + F_1 z^{-1} + \cdots + F_m z^{-m}$ by

$$C_F \doteq \begin{bmatrix} F_0 & 0 & \cdots & 0 & 0 & \cdots & 0 \\ F_1 & F_0 & \cdots & 0 & 0 & \cdots & 0 \\ \vdots & \vdots & \ddots & \vdots & \vdots & \ddots & \vdots \\ F_m & F_{m-1} & \cdots & F_0 & 0 & \cdots & 0 \\ 0 & F_m & \cdots & F_1 & F_0 & \cdots & 0 \\ \vdots & \vdots & \ddots & \vdots & \vdots & \ddots & \vdots \\ 0 & 0 & \cdots & F_m & F_{m-1} & \cdots & F_0 \end{bmatrix}, \quad H_F \doteq \begin{bmatrix} F_m & F_{m-1} & \cdots & F_1 \\ 0 & F_m & \cdots & F_2 \\ \vdots & \vdots & \ddots & \vdots \\ 0 & 0 & \cdots & F_m \\ 0 & 0 & \cdots & 0 \\ \vdots & \vdots & \ddots & \vdots \\ 0 & 0 & \cdots & 0 \end{bmatrix}$$

where the row-blocks of C_F and H_F consist, respectively, of N and m blocks.

The solution of (2.72) for the vectors, $\mathbf{y}_k$ and $\mathbf{u}_k$, of output and input predictions is affine in the vector of the degrees of freedom $\mathbf{c}_k$, and hence the predicted cost is quadratic in $\mathbf{c}_k$. In particular the Bezout identity (2.71) implies an explicit expression for the inverse of the matrix on the LHS of (2.72), which in turn implies the solution

$$\begin{bmatrix} \mathbf{y}_k \\ \mathbf{u}_k \end{bmatrix} = \begin{bmatrix} C_B \\ C_A \end{bmatrix} \mathbf{c}_k - \begin{bmatrix} C_B & C_Y \\ C_A & -C_{z^{-1}X} \end{bmatrix} \begin{bmatrix} H_{z^{-1}\tilde{X}} & H_{\tilde{Y}} \\ H_{\tilde{A}} & -H_{\tilde{B}} \end{bmatrix} \begin{bmatrix} \mathbf{y}_k^p \\ \mathbf{u}_k^p \end{bmatrix}.$$

The second term on the RHS of this expression corresponds to the free responses of the output and input predictions, and, on account of the structure of the convolution matrices in (2.72) and the Bezout identity (2.71), these free responses are zero at the end of the prediction horizon consisting of $N = \nu + N_A$ steps. From this observation and the finite impulse response of the filters $B(z^{-1})$ and $A(z^{-1})$ in (2.70), it follows that SGPC imposes an implicit terminal equality constraint, namely that both the predicted input and output vectors reach the steady value of zero at the end of the horizon of $N = \nu + N_A$ prediction time steps, and this gives the algorithm a guarantee of closed-loop stability.

Equality terminal constraints can be overly stringent but it is possible to modify SGPC so that the predicted control law of (2.66) imitates what is obtained using the predicted control law, $u_{i|k} = Kx_{i|k} + c_{i|k}$, of the closed-loop paradigm. This can be achieved through the use of the Bezout identity

$$\tilde{Y}(z^{-1})A(z^{-1}) + z^{-1}\tilde{X}(z^{-1})B(z^{-1}) = A_{cl}(z^{-1}) \tag{2.73}$$

where $A_{cl}(z^{-1})$ is such that $B(z^{-1})$ and $A_{cl}(z^{-1})$ define right coprime factors of the closed-loop transfer function matrix (under the control law $u = Kx + c$). The fact that the same $B(z^{-1})$ can be used for both the open and closed-loop transfer function matrices can be argued as follows. Let $\hat{B}(z^{-1})$, $\hat{A}(z^{-1})$ be the right coprime factors of $(zI - A)^{-1}B$ such that $B\hat{A}(z^{-1}) = (zI - A)\hat{B}(z^{-1})$. The consistency condition for this equation is $N(zI - A)\hat{B}(z^{-1}) = 0$ where N is the full-rank left annihilator of B (satisfying the condition $NB = 0$). This is however is also the consistency condition for the equation

$$BA_{cl}(z^{-1}) = (zI - A - BK)^{-1}\,\hat{B}(z^{-1}),$$

which implies that $\hat{B}(z^{-1})$ can also be used in the right coprime factorization of $(zI - A - BK)^{-1}B$. Thus the same $B(z^{-1}) = C\hat{B}(z^{-1})$ can be used for both the open and closed-loop transfer function matrices given that these transfer function matrices are obtained by the pre-multiplication by C of $(zI - A)^{-1}B$ and $(zI - A - BK)^{-1}B$, respectively. The property that a common $B(z^{-1})$ can be used in the factorization of the open and closed-loop transfer function matrices can also be used to prove that the control law of (2.66) guarantees the internal stability of the closed-loop system [40] (when $\tilde{Y}(z^{-1})$, $\tilde{X}(z^{-1})$ satisfy either of (2.67) or (2.73)).

SGPC introduced a Youla parameter into the MPC problem and this provides an alternative way to that described in Sect. 2.9 to endow the prediction structure with control dynamics. This can be achieved by replacing the polynomial matrices $\tilde{Y}(z^{-1})$, $\tilde{X}(z^{-1})$, respectively by

$$\begin{aligned}\tilde{M}(z^{-1}) &= \tilde{Y}(z^{-1}) - z^{-1}Q(z^{-1})B(z^{-1})\\ \tilde{N}(z^{-1}) &= \tilde{X}(z^{-1}) + A(z^{-1})Q(z^{-1})\end{aligned}$$

where $Q(z^{-1})$ represents a free parameter (which can be chosen to be any polynomial matrix, or stable transfer function matrix). If $\tilde{Y}(z^{-1})$ and $\tilde{X}(z^{-1})$ satisfy the Bezout identity (either (2.67) or (2.73)), then so will $\tilde{M}(z^{-1})$ and $\tilde{N}(z^{-1})$, which therefore can be used in the control law of (2.66) in place of $\tilde{Y}(z^{-1})$ and $\tilde{X}(z^{-1})$. The advantage of this is that the degrees of freedom in $Q(z^{-1})$ can be used to enhance the robustness of the closed-loop system to model parameter uncertainty or to enlarge the region of attraction of the algorithm [38].

At first sight it may appear that the relationships above will not hold in the presence of constraints. However this is not so, because the perturbations c_k have been introduced in order to ensure that constraints are respected and therefore the predicted trajectories are generated by the system operating within its linear range. These prediction equations can be used to express the vector of predicted outputs and inputs as functions of the vector of predicted degrees of freedom, $\mathbf{c}_k = (c_{0k}, \ldots, c_{N-1|k}, c_\infty, c_\infty, \ldots)$ where c_∞ denotes the constant value of c which ensures that the steady-state predicted output is equal to the desired setpoint vector r and the vector $\mathbf{c}_k$ contains Nn_u degrees of freedom. Clearly for a regulation problem with $r = 0$, c_∞ would be chosen to be zero. SGPC then proceeds to minimize the cost of (2.65) over the degrees of freedom $(c_{0k}, \ldots, c_{N-1|k})$ subject to constraints and implements the control move indicated by (2.66).

The algorithms discussed in this section are based on output feedback and are appropriate in cases where the assumption that the states are measurable and available for the purposes of feedback does not hold true. In instances like this one can, instead, revert to a state-space system representation constructed using current and past inputs and outputs as states (e.g. [41]) or a state-space description of the combination of the system dynamics together with the dynamics of a state observer (e.g. [42], which established invariance using low-complexity polytopes, namely polytopes with $2n_x$ vertices).

2.11 Exercises

1 A first-order system with the discrete time model

$$x_{k+1} = 1.5x_k + u_k$$

is to be controlled using a predictive controller that minimizes the predicted performance index

$$J(x_k, u_{0|k}, u_{1|k}) = \sum_{i=0}^{1} \left(x_{i|k}^2 + 10u_{i|k}^2\right) + qx_{2|k}^2$$

where q is a positive constant.

(a) Show that the unconstrained predictive control law is $u_k = -0.35x_k$ if $q = 1$.
(b) The unconstrained optimal control law with respect to the infinite horizon cost $\sum_{k=0}^{\infty}(x_k^2 + 10u_k^2)$ is $u_k = -0.88x_k$. Determine the value of q so that the unconstrained predictive control law coincides with this LQ optimal control law.
(c) The predicted cost is to be minimized subject to input constraints

$$-0.5 \le u_{i|k} \le 1.$$

If the predicted inputs are defined as $u_{i|k} = -0.88x_{i|k}$, for all $i \ge 2$, show that the MPC optimization problem is guaranteed to be recursively feasible if $u_{i|k}$ satisfies these constraints for $i = 0, 1$ and 2.

2 (a) A discrete time system is defined by

$$x_{k+1} = \begin{bmatrix} 0 & 1 \\ 0 & \alpha \end{bmatrix} x_k, \quad y_k = \begin{bmatrix} 1 & 0 \end{bmatrix} x_k$$

where α is a constant. Show that $-1 \le y_k \le 1$, for all $k \ge 0$ if and only if $|\alpha| < 1$ and

$$\begin{bmatrix} -1 \\ -1 \end{bmatrix} \le x_0 \le \begin{bmatrix} 1 \\ 1 \end{bmatrix}.$$

(b) A model predictive control strategy is to be designed for the system

$$x_{k+1} = \begin{bmatrix} \beta & 1 \\ 0 & \alpha \end{bmatrix} x_k + \begin{bmatrix} 1 \\ 0 \end{bmatrix} u_k, \quad y_k = \begin{bmatrix} 1 & 0 \end{bmatrix} x_k, \quad -1 \le u_k \le 1$$

where α and β are constants, with $|\alpha| < 1$. Assuming that, for $i \ge N$, the i steps ahead predicted input is defined as

$$u_{i|k} = \begin{bmatrix} -\beta & 0 \end{bmatrix} x_{i|k},$$

show that:

(i) $\displaystyle\sum_{i=0}^{\infty} (y_{i|k}^2 + u_{i|k}^2) = \sum_{i=0}^{N-1} (y_{i|k}^2 + u_{i|k}^2) + (\beta^2 + 1) x_{N|k}^T \begin{bmatrix} 1 & 0 \\ 0 & \frac{1}{1-\alpha^2} \end{bmatrix} x_{N|k}.$

(ii) $-1 \le u_{i|k} \le 1$ for all $i \ge N$ if

$$\begin{bmatrix} -1 \\ -1 \end{bmatrix} \le |\beta| \, x_{N|k} \le \begin{bmatrix} 1 \\ 1 \end{bmatrix}.$$

(c) Comment on the suggestion that an MPC law based on minimizing the cost in (b)(i) subject to $-1 \le u_{i|k} \le 1$ for $i = 0, \ldots, N-1$ and the terminal constraint $x_{N|k} = 0$ would be stable. Why would it be preferable to use the terminal inequality constraints of (b)(ii) instead of this terminal equality constraint.

3 A system has the model

$$x_{k+1} = \begin{bmatrix} 0 & 1 \\ -1 & 0 \end{bmatrix} x_k + \frac{1}{2} \begin{bmatrix} -1 \\ 1 \end{bmatrix} u_k, \quad y_k = \frac{1}{\sqrt{2}} \begin{bmatrix} 1 & 1 \end{bmatrix} x_k.$$

(a) Show that, if $u_k = \frac{1}{\sqrt{2}} y_k$, then

$$\sum_{k=0}^{\infty} \tfrac{1}{2} \left(y_k^2 + u_k^2 \right) = \|x_0\|^2.$$

(b) A predictive control law is defined at each time step k by $u_k = u_{0|k}^*$, where $(u_{0|k}^*, \ldots, u_{N-1|k}^*)$ is the minimizing argument of

$$\min_{u_{0|k}, \ldots, u_{N-1|k}} \sum_{i=0}^{N-1} \tfrac{1}{2} \left(y_{i|k}^2 + u_{i|k}^2 \right) + \|x_{N|k}\|^2.$$

Show that the closed-loop system is stable.

(c) The system is now subject to the constraint $-1 \le y_k \le 1$, for all k. Will the closed-loop system necessarily be stable if the optimization in part (b) includes the constraints $-1 \le y_{i|k} \le 1$, for $i = 1, 2, \ldots, N+1$?

4 A discrete time system is described by the model $x_{k+1} = Ax_k + Bu_k$ with

$$A = \begin{bmatrix} 0.3 & -0.9 \\ -0.4 & -2.1 \end{bmatrix}, \quad B = \begin{bmatrix} 0.5 \\ 1 \end{bmatrix}$$

where $u_k = Kx_k$ for $K = \begin{bmatrix} 0.244 & 1.751 \end{bmatrix}$, and for all $k = 0, 1 \ldots$ the state x_k is subject to the constraints

$$\left|\begin{bmatrix} 1 & -1 \end{bmatrix} x_k\right| \le 1.$$

(a) Describe a procedure based on linear programming for determining the largest invariant set compatible with constraints $|\begin{bmatrix} 1 & -1 \end{bmatrix} x| \le 1$.

(b) Demonstrate by solving a linear program that the maximal invariant set is defined by

$$\{x : Fx \le \mathbf{1} \text{ and } F\Phi x \le \mathbf{1}\},$$

where $F = \begin{bmatrix} 1 & -1 \\ -1 & 1 \end{bmatrix}$ and $\Phi = \begin{bmatrix} 0.42 & -0.025 \\ -0.16 & -0.35 \end{bmatrix}$.

5 Consider the system of Question 4 with the cost $\sum_{k=0}^{\infty}\left(\|x_k\|_Q^2 + \|u_k\|_R^2\right)$, with $Q = I$ and $R = 1$.

(a) For $K = \begin{bmatrix} 0.244 & 1.751 \end{bmatrix}$, solve the Lyapunov matrix equation (2.5) to find W and hence verify using Theorem 2.1 that K is the optimal unconstrained feedback gain.

(b) Use the maximal invariant set given in Question 4(b) to prove that $x_{i|k} = \begin{bmatrix} I & 0 \end{bmatrix} \Psi^i z_k$ satisfies the constraints $|\begin{bmatrix} 1 & -1 \end{bmatrix} x_{i|k}| \le 1$, for all $i \ge 0$ if $\begin{bmatrix} F & 0 \end{bmatrix} \Psi^i z_k \le \mathbf{1}$ for $i = 0, 1, \ldots, N+1$, where

$$F = \begin{bmatrix} 1 & -1 \\ -1 & 1 \end{bmatrix}, \quad \Psi = \begin{bmatrix} A + BK & BE \\ 0 & M \end{bmatrix}, \quad z_k = \begin{bmatrix} x_k \\ \mathbf{c}_k \end{bmatrix}$$

$$E = \begin{bmatrix} 1 & 0 & \cdots & 0 \end{bmatrix} \in \mathbb{R}^{1\times N}, \quad M = \begin{bmatrix} 0 & 1 & 0 & \cdots & 0 & 0 \\ 0 & 0 & 1 & \cdots & 0 & 0 \\ \vdots & \vdots & \vdots & & \vdots & \vdots \\ 0 & 0 & 0 & \cdots & 0 & 1 \\ 0 & 0 & 0 & \cdots & 0 & 0 \end{bmatrix} \in \mathbb{R}^{N\times N}.$$

(c) Show that the predicted cost is given by

$$J(x_k, \mathbf{c}_k) = \|x_k\|_W^2 + \rho\|\mathbf{c}_k\|^2, \quad W = \begin{bmatrix} 1.33 & 0.58 \\ 0.58 & 4.64 \end{bmatrix}, \quad \rho = 6.56.$$

(d) For the initial condition $x_0 = (3.8, 3.8)$, the optimal predicted cost,

$$J_N^*(x_0) \doteq \min_{\mathbf{c}\in\mathbb{R}^N} J(x_0, \mathbf{c}) \text{ subject to } \begin{bmatrix} F & 0 \end{bmatrix} \Psi^i \begin{bmatrix} x_0 \\ \mathbf{c} \end{bmatrix} \leq \mathbf{1},\ i = 1, \ldots, N+1$$

varies with N as follows:

N	8	9	10	11
$J_N^*(x_0)$	∞	826.6	826.6	826.6

(the problem is infeasible for $N \leq 8$). Suggest why $J_N^*(x_0)$ is likely to be equal to 826.6, for all $N > 9$ and state the likely value of the infinite horizon cost for the closed loop state and control sequence starting from x_0 under $u_k = Kx_k + c_{0|k}^*$ if $N = 9$.

6 For the system and constraints of Question 4 with $K = \begin{bmatrix} 0.244 & 1.751 \end{bmatrix}$:

(a) Taking $N = 2$, solve the optimization (2.41) to determine, for the prediction dynamics $z_{k+1} = \Psi z_k$, the ellipsoidal invariant set $\{z : z^T P_z z \leq 1\}$ that has the maximum area projection onto the x-subspace. Hence show that the greatest scalar α such that $x_0 = (\alpha, \alpha)$ satisfies $z_0^T P_z z_0 \leq 1$ for $z_0 = (x_0, \mathbf{c}_0)$, for some $\mathbf{c}_0 \in \mathbb{R}^2$, is $\alpha = 1.79$.

(b) Show that, for $N = 2$, the greatest α such that $x_0 = (\alpha, \alpha)$ is feasible for the constraints $\begin{bmatrix} F & 0 \end{bmatrix} \Psi^i z_0 \leq \mathbf{1}$, $i = 0, \ldots, N+1$, for $z_0 = (x_0, \mathbf{c}_0)$, for some $\mathbf{c}_0 \in \mathbb{R}^2$, is $\alpha = 2.41$. Explain why this value is necessarily greater than the value of α in (a).

(c) Determine the optimized prediction dynamics by solving (2.63) and verify that

$$C_c = \begin{bmatrix} -1.22 & -0.45 \end{bmatrix}, \quad A_c = \begin{bmatrix} 0.96 & 0.32 \\ -0.015 & -0.063 \end{bmatrix},$$

and also that the maximum scaling α such that $x_0 = (\alpha, \alpha)$ is feasible for $z_0^T P_z z_0 \leq 1$ for $z_0 = (x_0, \mathbf{c}_0)$, for some $\mathbf{c}_0 \in \mathbb{R}^2$, is $\alpha = 2.32$.

(d) Using the optimized prediction dynamics computed in part (c), define

$$\hat{\Psi} = \begin{bmatrix} A + BK & BC_c \\ 0 & A_c \end{bmatrix}$$

and show that $x_{i|k} = \begin{bmatrix} I & 0 \end{bmatrix} \hat{\Psi}^i z_k$ satisfies constraints $|\begin{bmatrix} 1 & -1 \end{bmatrix} x_{i|k}| \leq 1$, for all $i \geq 0$ if $\begin{bmatrix} F & 0 \end{bmatrix} \hat{\Psi}^i z_k \leq \mathbf{1}$ for $i = 0, \ldots, 5$. Hence show that the maximum scaling α such that $x_0 = (\alpha, \alpha)$ satisfies these constraints for some $\mathbf{c}_0 \in \mathbb{R}^2$ is $\alpha = 3.82$.

(e) Show that the optimal value of the predicted cost for the prediction dynamics and constraints determined in (d) with $x_0 = (3.8, 3.8)$ is $J^*(x_0) = 1686$.

Explain why this value is greater than the predicted cost in Question 5(d) for $N = 9$. What is the advantage of the MPC law based on the optimized prediction dynamics?

7 With $K = \begin{bmatrix}0.067 & 2\end{bmatrix}$, the model of Question 4 gives

$$A + BK = \begin{bmatrix} 0.33 & 0.1 \\ -0.33 & -0.1 \end{bmatrix}.$$

(a) Explain the significance of this for the size of the feasible initial condition set of an MPC law which is subject to the state constraints $|\begin{bmatrix}1 & 1\end{bmatrix}x| \le 1$ rather than the constraints of Question 4?

(b) Explain why the feasible set of the MPC algorithm in Question 5(d) (which is subject to the constraints $|\begin{bmatrix}1 & -1\end{bmatrix}x| \le 1$) is finite for all N.

8 GPC can be cast in terms of state-space models, through which the predicted output sequence $\mathbf{y}_k = (y_{1|k}, \ldots, y_{N|k})$ can be expressed as an affine function of the predicted input sequence $\mathbf{u}_k = (u_{0|k}, \ldots, u_{N_u-1|k})$ as $\mathbf{y}_k = C_x x_k + C_u \mathbf{u}_k$. Using this expression show that the unconstrained optimum for the minimization of the regulation cost $J_k = \mathbf{y}_k^T \hat{Q} \mathbf{y}_k + \mathbf{u}_k^T \hat{R} \mathbf{u}_k$, with $\hat{Q} = \operatorname{diag}\{Q, \ldots, Q\}$ and $\hat{R} = \operatorname{diag}\{R, \ldots, R\}$, is given by

$$\mathbf{u}_k^* = -\left(\hat{R} + C_u^T \hat{Q} C_u\right)^{-1} C_u^T \hat{Q} C_x x_k.$$

Hence show that for

$$A = \begin{bmatrix} 0.83 & -0.46 \\ -0.05 & 0.86 \end{bmatrix}, \quad B = \begin{bmatrix} 0.26 \\ 0.55 \end{bmatrix}, \quad C = \begin{bmatrix}0.67 & 0.71\end{bmatrix},$$

and in the absence of constraints, GPC results in an unstable closed loop system for all prediction horizons $N \le 9$ and input horizons $N_u \le N$. Confirm that the open-loop system is stable but that its zero is non-minimum phase. Construct an argument which explains the instability observed above.

9 (a) Compute the transfer function of the system of Question 8 and show that the polynomials

$$\tilde{X}(z^{-1}) = 21.0529z^{-1} - 32.2308, \quad \tilde{Y}(z^{-1}) = 19.8907z^{-1} + 1$$

are solutions of the Bezout identity (2.67).

(b) It is proposed to use SGPC to regulate the system of part (a) about the origin (i.e. the reference setpoint is taken to be $r = 0$) using two degrees of freedom, $\mathbf{c}_k = (c_{0|k}, c_{1|k})$, in the predicted state and input sequences, the implicit assumption being that $c_{i|k} = 0$, for all $i \ge 2$. Form the 4×4 convolution matrices $C_{z^{-1}\tilde{X}}$, $C_{\tilde{Y}}$, $C_{\tilde{A}}$, $C_{\tilde{B}}$ and confirm that

$$\begin{bmatrix} C_{z^{-1}\tilde{X}} & C_{\tilde{Y}} \\ C_{\tilde{A}} & -C_{\tilde{B}} \end{bmatrix}^{-1} = \begin{bmatrix} C_A & C_Y \\ C_B & -C_{z^{-1}X} \end{bmatrix}.$$

Hence show that the prediction equation giving the vectors of predicted outputs $\mathbf{y}_k = (y_{1|k}, \ldots, y_{4|k})$ and inputs $\mathbf{u}_k = (u_{0|k}, \ldots, u_{3|k})$ is

$$\begin{bmatrix} \mathbf{y}_k \\ \mathbf{u}_k \end{bmatrix} = \begin{bmatrix} C_B \\ C_A \end{bmatrix} \begin{bmatrix} \mathbf{c}_k \\ 0_{2\times 1} \end{bmatrix} - \begin{bmatrix} 12.6 & -19.9 & 11.9 \\ & 0_{3\times 3} & \\ 21.1 & -32.2 & 19.9 \\ -13.31 & 21.1 & -12.6 \\ & 0_{2\times 3} & \end{bmatrix} \begin{bmatrix} \mathbf{y}_k^p \\ \mathbf{u}_k^p \end{bmatrix}.$$

(c) Show that the predicted sequences in (b) implicitly satisfy a terminal constraint. Hence explain why the closed-loop system under SGPC is necessarily stable.

10 For the data of Question 9 plot the frequency response of the modulus of $K(z^{-1})/\big(1 + G(z^{-1})K(z^{-1})\big)$ where

$$K(z^{-1}) = \frac{\tilde{X}(z^{-1}) + A(z^{-1})Q(z^{-1})}{\tilde{Y}(z^{-1}) - z^{-1}B(z^{-1})Q(z^{-1})}$$

for the following two cases:

(a) $Q(z^{-1}) = 0$
(b) $Q(z^{-1}) = -11.7z^{-1} + 43$

Hence suggest what might be the benefit of introducing a Youla parameter into SGPC in terms of robustness to additive uncertainty in the system transfer function.

References

1. L.S. Pontryagin, V.G. Boltyanskii, R.V. Gamkrelidze, E.F. Mishchenko, *The Mathematical Theory of Optimal Processes* (Interscience, Wiley, 1962)
2. R. Bellman, *Dynamic Programming* (Princeton University Press, Princeton, 1957)
3. M. Sznaier, M.J. Damborg, Suboptimal control of linear systems with state and control inequality constraints, in *Proceedings of the 26th IEEE Conference on Decision and Control*, Los Angeles, USA, vol. 26 (1987), pp. 761–762
4. P.O.M. Scokaert, J.B. Rawlings, Constrained linear quadratic regulation. IEEE Trans. Autom. Control **43**(8), 1163–1169 (1998)
5. D. Chmielewski, V. Manousiouthakis, On constrained infinite-time linear quadratic optimal control. Syst. Control Lett. **29**(3), 121–129 (1996)
6. P.O.M. Scokaert, D.Q. Mayne, J.B. Rawlings, Suboptimal model predictive control (feasibility implies stability). IEEE Trans. Autom. Control **44**(3), 648–654 (1999)
7. R.E. Bellman, On the theory of dynamic programming. Proc. Natl. Acad. Sci. **38**(8), 716–719 (1952)

8. J.A. Rossiter, B. Kouvaritakis, M.J. Rice, A numerically robust state-space approach to stable-predictive control strategies. Automatica **34**(1), 65–73 (1998)
9. E.G. Gilbert, K.T. Tan, Linear systems with state and control constraints: the theory and application of maximal output admissible sets. IEEE Trans. Autom. Control **36**(9), 1008–1020 (1991)
10. D.P. Bertsekas, Infinite time reachability of state-space regions by using feedback control. IEEE Trans. Autom. Control **17**(5), 604–613 (1972)
11. F. Blanchini, S. Miani, *Set-theoretic Methods in Control* (Birkhäuser, Switzerland, 2008)
12. C.E.T. Dórea, J.C. Hennet, (A, B)-Invariant polyhedral sets of linear discrete-time systems. J. Optim. Theory Appl. **103**(3), 521–542 (1999)
13. M. Vidyasagar, *Nonlinear Systems Analysis*, 2nd edn. (Prentice Hall, New Jersey, 1993)
14. A. Bemporad, M. Morari, V. Dua, E.N. Pistikopoulos, The explicit linear quadratic regulator for constrained systems. Automatica **38**(1), 3–20 (2002)
15. B. Kouvaritakis, J.A. Rossiter, J. Schuurmans, Efficient robust predictive control. IEEE Trans. Autom. Control **45**(8), 1545–1549 (2000)
16. B. Kouvaritakis, M. Cannon, J.A. Rossiter, Who needs QP for linear MPC anyway? Automatica **38**(5), 879–884 (2002)
17. S.P. Boyd, L El Ghaoui, E. Feron, V. Balakrishnan, *Linear Matrix Inequalities in System and Control Theory* (Society for Industrial and Applied Mathematics, Philadelphia, 1994)
18. S. Boyd, L. Vandenberghe, *Convex Optimization* (Cambridge University Press, Cambridge, 2004)
19. R. Fletcher, *Practical Methods of Optimization*, 2nd edn. (Wiley, New York, 1987)
20. C.V. Rao, S.J. Wright, J.B. Rawlings, Application of interior-point methods to model predictive control. J. Optim. Theory Appl. **99**(3), 723–757 (1998)
21. P. Tøndel, T.A. Johansen, A. Bemporad, An algorithm for multi-parametric quadratic programming and explicit MPC solutions. Automatica **39**(3), 489–497 (2003)
22. M. Lobo, L. Vandenberghe, S. Boyd, H. Lebret, Applications of second-order cone programming. Linear Algebra Appl. **284**(1–3), 193–228 (1998)
23. M. Cannon, B. Kouvaritakis, J.A. Rossiter, Efficient active set optimization in triple mode MPC. IEEE Trans. Autom. Control **46**(8), 1307–1312 (2001)
24. M. Cannon, B. Kouvaritakis, Efficient constrained model predictive control with asymptotic optimality. SIAM J. Control Optim. **41**(1), 60–82 (2002)
25. L. Wang, Discrete model predictive controller design using Laguerre functions. J. Process Control **14**(2), 131–142 (2004)
26. M. Cannon, B. Kouvaritakis, Optimizing prediction dynamics for robust MPC. IEEE Trans. Autom. Control **50**(11), 1892–1897 (2005)
27. C. Scherer, P. Gahinet, M. Chilali, Multiobjective output-feedback control via LMI optimization. IEEE Trans. Autom. Control **42**(7), 896–911 (1997)
28. A.I. Propoi, Use of linear programming methods for synthesizing sampled-data automatic systems. Autom. Remote Control **24**(7), 837–844 (1963)
29. C. Cutler, B. Ramaker, Dynamic matrix control—a computer control algorithm, in *Proceedings of the Joint Automatic Control Conference, San Francisco*, (1980)
30. J. Richalet, S. Abu el Ata-Doss, C. Arber, H.B. Kuntze, A. Jacubash, W. Schill, Predictive functional control: application to fast and accurate robots, in *Proceedings of the 10th IFAC World Congress, Munich*, (1987), pp. 251–258
31. D.W. Clarke, C. Mohtadi, P.S. Tuffs, Generalized predictive control–part I. The basic algorithm. Automatica **23**(2), 137–148 (1987)

32. D.W. Clarke, C. Mohtadi, P.S. Tuffs, Generalized predictive control–part II. Extensions and interpretations. Automatica **23**(2), 149–160 (1987)
33. R.R. Bitmead, M. Gevers, V. Wertz, *Adaptive Optimal Control: The Thinking Man's GPC* (Prentice Hall, New Jersey, 1990)
34. S.J. Qin, T.A. Badgwell, A survey of industrial model predictive control technology. Control Eng. Pract. **11**(7), 733–764 (2003)
35. W.H. Kwon, A.E. Pearson, A modified quadratic cost problem and feedback stabilization of a linear system. IEEE Trans. Autom. Control **22**(5), 838–842 (1977)
36. D.W. Clarke, R. Scattolini, Constrained receding-horizon predictive control. Control Theory Appl. IEEE Proc. D **138**(4), 347–354 (1991)
37. S.S. Keerthi, E.G. Gilbert, Optimal infinite-horizon feedback laws for a general class of constrained discrete-time systems: stability and moving-horizon approximations. J. Optim. Theory Appl. **57**(2), 265–293 (1988)
38. B. Kouvaritakis, J.A. Rossiter, A.O.T. Chang, Stable generalised predictive control: an algorithm with guaranteed stability. Control Theory Appl. IEEE Proc. D **139**(4), 349–362 (1992)
39. E. Mosca, J. Zhang, Stable redesign of predictive control. Automatica **28**(6), 1229–1233 (1992)
40. J.M. Maciejowski, *Multivariable Feedback Design* (Addison-Wesley, Boston, 1989)
41. E. Mosca, *Optimal, Predictive, and Adaptive Control*. Prentice Hall Information and System Sciences Series (Prentice Hall, New Jersey, 1995)
42. Y.I. Lee, B. Kouvaritakis, Receding horizon output feedback control for linear systems with input saturation. IEE Proc. Control Theory Appl. **148**(2), 109–115 (2001)

Part II
Robust MPC

Chapter 3
Open-Loop Optimization Strategies for Additive Uncertainty

The essential components of the classical predictive control algorithms considered in Chap. 2 also underpin the design of algorithms for robust MPC. Guarantees of closed-loop properties such as stability and convergence rely on appropriately defined terminal control laws, terminal sets and cost functions. Likewise, to ensure that constraints can be met in the future, the initial plant state must belong to a suitable controllable set. However the design of these constituents and the analysis of their effects on the performance of MPC algorithms becomes more complex in the case where the system dynamics are subject to uncertainty. The main difficulty is that properties such as invariance, controlled invariance (including recursive feasibility) and monotonicity of the predicted cost must be guaranteed for all possible uncertainty realizations. In many cases this leads to computation which grows rapidly with the problem size and the prediction horizon.

Uncertainty is a ubiquitous feature of control applications. It can arise as a result of the presence of additive disturbances in the system model and can also be multiplicative in nature, for example as a result of an imprecise knowledge of the model parameters. In either case it is essential that certain properties (including closed-loop stability and performance) are preserved despite the presence of uncertainty, and this is the main preoccupation of robust MPC (RMPC). In this chapter and in Chap. 4, consideration will be given to the additive case, whereas the multiplicative case will be examined in Chap. 5. These topics will be re-examined in later chapters in the context of stochastic MPC.

Within the range of approaches that have been proposed for robust MPC, there is a fundamental difference between strategies in which optimization is performed over open-loop prediction strategies and those that optimize the parameters of predicted feedback laws. This chapter discusses open-loop optimization algorithms; these are often conceptually simpler and generally have lower computational complexity than their counterparts employing closed-loop strategies. However the techniques introduced here for determining feasibility of robust constraints and closed-loop stability analysis carry over to the closed-loop strategies considered in Chap. 4. The chapter begins with a discussion of robust constraint handling, then describes cost functions

B. Kouvaritakis and M. Cannon, *Model Predictive Control*,
Advanced Textbooks in Control and Signal Processing,
DOI 10.1007/978-3-319-24853-0_3

and stability analyses, before continuing to describe alternative approaches based on the online optimization of tubes that bound predicted state and input trajectories. The chapter concludes with a discussion of early robust MPC algorithms.

3.1 The Control Problem

We consider the system model that is obtained when a disturbance term representing additive model uncertainty is introduced into the linear dynamics of (2.1):

$$x_{k+1} = Ax_k + Bu_k + Dw_k. \tag{3.1}$$

Here D is a matrix of known parameters and $w_k \in \mathbb{R}^{n_w}$ is a vector of disturbance inputs, unknown at time k.

As in Chap. 2, the system matrices A, B are assumed known, $u_k \in \mathbb{R}^{n_u}$ is the control input at time k and the state $x_k \in \mathbb{R}^{n_x}$ is known at time k. The disturbance w_k is assumed to belong to a known set $\mathcal{W}$, namely, at each time instant k, we require that

$$w_k \in \mathcal{W}.$$

The disturbance set $\mathcal{W}$ is assumed to be full dimensional (i.e. not restricted to a subspace of $\mathbb{R}^{n_w}$) and D is assumed to be full rank, with $\text{rank}(D) = n_w$. We also assume that $\mathcal{W}$ contains the origin in its interior. Clearly, if the origin did not lie in the interior of $\mathcal{W}$, then the model uncertainty in (3.1) could be represented as the sum of a known, constant disturbance and an unknown, time-varying disturbance belonging to a known disturbance set that does contain the origin in its interior. A constant disturbance would result in a constant offset in the evolution of the state of (3.1), which could be accounted for by a translation of the origin of state space; with this modification it can always be assumed that the origin lies in the interior of $\mathcal{W}$.

The disturbance set $\mathcal{W}$ is further assumed to be a convex polytopic set. Any compact convex polytope may be represented by its vertices, for example

$$\mathcal{W} = \text{Co}\{w^{(j)},\ j = 1, \ldots, m\}, \tag{3.2}$$

where $w^{(j)}$, $j = 1, \ldots, m$ are the vertices (extreme points) of $\mathcal{W}$ and $\text{Co}\{\cdot\}$ denotes the convex hull. An alternative description in terms of linear inequalities is given by

$$\mathcal{W} = \{w : Vw \leq \mathbf{1}\}, \tag{3.3}$$

for some matrix $V \in \mathbb{R}^{n_V \times n_w}$ that specifies the hyperplanes bounding $\mathcal{W}$. Here V is assumed to be minimal in the sense that n_V is the smallest number hyperplanes that define the boundary of $\mathcal{W}$. The linear inequality representation (3.3) is usually more parsimonious than the vertex representation (3.2). This is because the number, m, of vertices of $\mathcal{W}$ must be at least as large as the minimal number, n_V, of rows of V because $\mathcal{W}$ is by assumption full dimensional, and m is typically much greater than n_V. However both representations are employed in different RMPC formulations, and depending on the context we will make use of either (3.2) or (3.3).

In this chapter, we introduce some notation specific to sets.

- The *Minkowski sum* of a pair of sets $\mathcal{X}, \mathcal{Y} \subseteq \mathbb{R}^n$ is denoted $\mathcal{X} \oplus \mathcal{Y}$, and is defined as the set (see Fig. 3.1)

$$\mathcal{X} \oplus \mathcal{Y} \doteq \{z \in \mathbb{R}^n : z = x + y, \text{ for any } x \in \mathcal{X} \text{ and } y \in \mathcal{Y}\}.$$

- The *Pontryagin difference* of two sets $\mathcal{X}, \mathcal{Y} \subset \mathbb{R}^n$, denoted $\mathcal{X} \ominus \mathcal{Y}$, is the set

$$\mathcal{X} \ominus \mathcal{Y} = \{z \in \mathbb{R}^n : z + y \in \mathcal{X}, \text{ for all } y \in \mathcal{Y}\},$$

  so that $\mathcal{Z} = \mathcal{X} \ominus \mathcal{Y}$ if and only if $\mathcal{X} = \mathcal{Y} \oplus \mathcal{Z}$.
- The image of $\mathcal{X} \subset \mathbb{R}^n$ under a matrix $H \in \mathbb{R}^{m \times n}$ is defined as the set

$$H\mathcal{X} \doteq \{z \in \mathbb{R}^m : z = Hx, \text{ for any } x \in \mathcal{X}\}.$$

  Thus, for example $H\mathcal{X} = \mathrm{Co}\{Hx^{(j)}, \ j = 1, \ldots, q\}$ if $\mathcal{X}$ is a compact convex polytope described in terms of its vertices as $\mathcal{X} = \mathrm{Co}\{x^{(j)}, \ j = 1, \ldots, q\}$.
- For a closed set $\mathcal{X} \subset \mathbb{R}^n$ and $F \in \mathbb{R}^{m \times n}$, $h \in \mathbb{R}^m$, we use $F\mathcal{X} \leq h$ to denote the conditions

$$\max_{x \in \mathcal{X}} Fx \leq h.$$

  Here and throughout this chapter the maximization of a vector-valued function is to be performed elementwise, so that $\max_{x \in \mathcal{X}} Fx$ denotes the vector in $\mathbb{R}^m$ with ith element equal to $\max_{x \in \mathcal{X}} F_i x$, where F_i is the ith row of the matrix F.

As in Chap. 2 the system is assumed to be subject to mixed constraints on the states and control inputs

$$Fx_k + Gu_k \leq \mathbf{1} \tag{3.4}$$

and performance is again judged in terms of a quadratic cost of the form of (2.3). However, in the uncertain case considered here, the quantity

$$\sum_{i=0}^{\infty} \left(\|x_{k+i}\|_Q^2 + \|u_{k+i}\|_R^2\right) \tag{3.5}$$

evaluated along trajectories of (3.1) is uncertain at time k since it depends on the disturbance sequence $\{w_k, w_{k+1}, \ldots\}$, which is unknown at time k.

To account for this, we consider two alternative definitions of the cost when evaluating predicted performance: the nominal cost corresponding to the case of no model uncertainty (i.e. $w_k = 0$ for all k); and the worst-case cost over all admissible

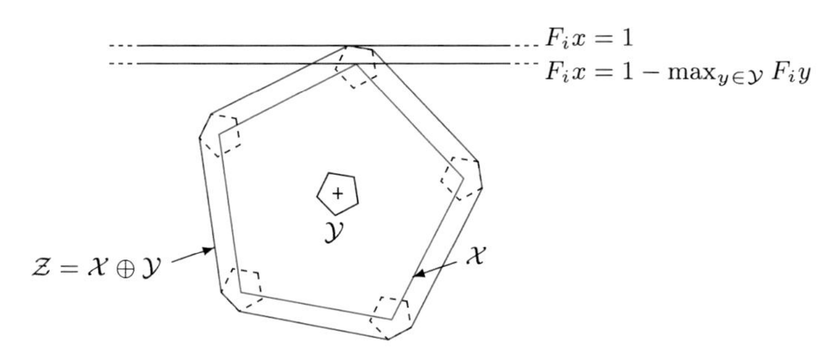

Fig. 3.1 A graphical representation of the Minkowski sum $\mathcal{X} \oplus \mathcal{Y}$ of a pair of closed convex polytopic sets $\mathcal{X}$, $\mathcal{Y}$, and of the linear constraint $F_i x \leq \mathbf{1}$ applied to $\mathcal{X} \oplus \mathcal{Y}$. If $\mathcal{Z} = \mathcal{X} \oplus \mathcal{Y}$, then $F_i \mathcal{Z} \leq \mathbf{1}$ is equivalent to $F_i \mathcal{X} \leq \mathbf{1} - h_i$ for $h_i = \max_{y \in \mathcal{Y}} F_i \mathcal{Y}$

disturbance sequences ($w_k \in \mathcal{W}$ for all k). In the latter case, we introduce into the stage cost of (3.5) an additional quadratic term which is negative definite in w_k to ensure a finite worst-case cost; this results in the $\mathcal{H}_\infty$ cost of classical robust control.

Robust MPC algorithms employing open-loop and closed-loop optimization strategies possess fundamentally different properties. To make this distinction clear we introduce the following definition.

Definition 3.1 (*Open-loop and closed-loop strategies*) In an open-loop optimization strategy the sequence of control inputs, $\{u_{0|k}, u_{1|k}, \ldots\}$, predicted at time k is independent of the realization of the disturbance sequence $\{w_{0|k}, w_{1|k}, \ldots\}$. In a closed-loop optimization strategy the predicted control sequence $\{u_{0|k}, u_{1|k}, \ldots\}$ is a function of the realization of the disturbance sequence $\{w_{0|k}, w_{1|k}, \ldots\}$.

By a slight abuse of terminology, we refer to the closed-loop paradigm introduced in Sect. 2.3 as an open-loop optimization strategy when it is used in the context of RMPC. In (2.13) the control inputs predicted at time k are specified by the closed-loop paradigm for $i = 0, 1, \ldots$ as

$$u_{i|k} = K x_{i|k} + c_{i|k} \tag{3.6}$$

where K is a fixed stabilizing feedback gain, and $c_{i|k}$ for $i = 0, \ldots, N-1$ are optimization variables with $c_{i|k} = 0$ for $i \geq N$. Clearly $u_{i|k}$ depends on the state $x_{i|k}$ and hence on the realization of $w_{0|k}, \ldots, w_{i-1|k}$. Strictly speaking therefore (3.6) is a closed-loop strategy. However the optimization variables $c_{0|k}, \ldots, c_{N-1|k}$ do not depend on the realization of future uncertainty and hence they appear in the state predictions generated by (3.1) as an open-loop control sequence applied to pre-stabilized dynamics:

$$x_{i+1|k} = \Phi x_{i|k} + B c_{i|k} + D w_{i|k} \tag{3.7}$$

for $i = 0, 1, \ldots$, where $\Phi = A + BK$. Therefore the closed-loop paradigm is equivalent to an open-loop strategy applied to this system. Throughout this chapter, as in

Chap. 2, we assume that K is the optimal feedback gain in the absence of inequality constraints on the system states and control inputs.

A closed-loop optimization strategy optimizes predicted performance over a class of feedback policies which is parameterized by the degrees of freedom in the optimization problem. The approach benefits from information on future disturbances which, though unknown to an observer at current time, will be available to the controller when the future control input is applied to the plant. We note here that robust MPC based on closed-loop optimization has the distinct advantage that it can provide the optimal achievable performance in respect of both the region of attraction and the worst-case performance objective. Closed-loop optimization strategies are considered in detail in Chap. 4 and their advantages over open-loop strategies are discussed in Sect. 4.1.

3.2 State Decomposition and Constraint Handling

The future values of the state of the system (3.1) are uncertain because of the unknown future disturbances acting on the system. However, given knowledge of a set containing all realizations of the disturbance input, sequences of sets can be determined that necessarily contain the future state and control input, and this is the basis of methods for guaranteeing robust satisfaction of constraints. Because of the linearity of the model (3.1), the component of the predicted state that is generated by the disturbance input evolves independently of the optimization variables when an open-loop optimization strategy is employed. Since the constraints (3.4) are also linear, the worst-case disturbances with respect to these constraints do not depend on the optimization variables and can therefore be determined offline. This leads to a computationally convenient method of handling constraints for open-loop optimization strategies. In fact the resulting constraints on predicted states and inputs are of the same form as those of the nominal (uncertainty-free) MPC problem, and are simply tightened to account for the uncertainty in predictions.

This section discusses the application of the constraints (3.4) to the predictions of the model (3.1) under the open-loop strategy (3.6). We first decompose the predicted state into nominal and uncertain components, denoted s and e, respectively. Thus, let $x_{i|k} = s_{i|k} + e_{i|k}$, where the nominal and uncertain components evolve for $i = 0, 1, \ldots$ according to

$$s_{i+1|k} = \Phi s_{i|k} + Bc_{i|k}, \tag{3.8a}$$

$$e_{i+1|k} = \Phi e_{i|k} + Dw_{i|k}, \tag{3.8b}$$

with initial conditions $e_{0|k} = 0$ and $s_{0|k} = x_k$.

As in Sect. 2.7, it is convenient to augment the predicted model state with the degrees of freedom in predictions by defining an augmented state variable $z \in \mathbb{R}^{n_z}$, $n_z = n_x + Nn_u$,

$$z = \begin{bmatrix} s \\ \mathbf{c} \end{bmatrix},$$

where $\mathbf{c}$ is the vector of optimization variables, with $\mathbf{c}_k = (c_{0|k}, \dots, c_{N-1|k})$ at time k. Then the nominal predicted state is given by $s_{i|k} = \begin{bmatrix} I_{n_x} & 0 \end{bmatrix} z_{i|k}$, where $z_{i+1|k}$ evolves for $i = 0, 1, \dots$ according to the autonomous dynamics

$$z_{i+1|k} = \Psi z_{i|k}. \tag{3.9}$$

Here, the matrix Ψ and the initial condition $z_{0|k}$ are defined by

$$\Psi = \begin{bmatrix} \Phi & BE \\ 0 & M \end{bmatrix} \quad \text{and} \quad z_{0|k} = \begin{bmatrix} x_k \\ \mathbf{c}_k \end{bmatrix}$$

with E and M given by (2.26b). In terms of the augmented state $z_{i|k}$ we obtain the predicted future values of the state and control input generated by (3.6) and (3.7) as

$$x_{i|k} = \begin{bmatrix} I & 0 \end{bmatrix} z_{i|k} + e_{i|k} \tag{3.10a}$$

$$u_{i|k} = \begin{bmatrix} K & E \end{bmatrix} z_{i|k} + K e_{i|k}. \tag{3.10b}$$

The predicted state and input sequences in (3.10) satisfy the constraints (3.4) if and only if the following condition is satisfied for all $i = 0, 1, \dots,$

$$F x_{i|k} + G u_{i|k} \leq \mathbf{1} \quad \forall \{w_{0|k}, \dots, w_{i-1|k}\} \in \mathcal{W} \times \cdots \times \mathcal{W}.$$

Therefore the constraints (3.4) are equivalent to the following conditions

$$\bar{F} \Psi^i z_{0|k} \leq \mathbf{1} - h_i, \quad i = 0, 1, \dots \tag{3.11}$$

where

$$\bar{F} = \begin{bmatrix} F + GK & GE \end{bmatrix}$$

and the vectors h_i are defined for all $i \geq 0$ by

$$h_0 \doteq 0 \tag{3.12a}$$

$$h_i \doteq \max_{\{w_{0|k}, \dots, w_{i-1|k}\} \in \mathcal{W} \times \cdots \times \mathcal{W}} (F + GK) e_{i|k}, \quad i = 1, 2, \dots \tag{3.12b}$$

From (3.8b) we obtain $e_{i|k} = w_{i-1|k} + \cdots + \Phi^{i-2} w_{1|k} + \Phi^{i-1} w_{0|k}$, and hence h_i in (3.12b) can be expressed

$$h_i = \sum_{j=0}^{i-1} \max_{w_j \in \mathcal{W}} (F + GK) \Phi^j D w_j, \quad i = 1, 2, \dots \tag{3.13}$$

or equivalently by the recursion

$$h_i = h_{i-1} + \max_{w \in \mathcal{W}} (F + GK)\Phi^{i-1} D w, \quad i = 1, 2, \ldots$$

Since the maximization in this expression applies elementwise, h_i is determined by the solution of $i n_u$ linear programs, each of which determines the maximizing value of w for an element of an individual term in (3.13).

Comparing (3.11) with (2.28) it can be seen that the robust constraints for additive model uncertainty are almost identical to the constraints for the case of no uncertainty that was considered in Chap. 2. The difference between these two cases is the vector h_i appearing in (3.11). The definition of h_i in (3.12) implies that this term simply tightens the constraint set by the minimum that is required to accommodate the worst-case value of $e_{i|k}$, namely the worst-case future uncertainty with respect to the constraints.

3.2.1 Robustly Invariant Sets and Recursive Feasibility

The conditions in (3.11) are given in terms of an infinite number of constraints, and from (3.13) these depend on the solution of an infinite number of linear programs. Clearly, this infinite set of conditions is not implementable, and it is necessary to consider whether (3.11) can be equivalently stated in terms of a finite number of constraints. A further question relates to recursive feasibility of (3.11). Specifically, whether there necessarily exists $\mathbf{c}_{k+1}$ so that, if the conditions of (3.11) are satisfied by $z_{0|k} = (x_k, \mathbf{c}_k)$, then they will also hold when $z_{0|k}$ is replaced by $z_{0|k+1} = (x_{k+1}, \mathbf{c}_{k+1})$.

Both of these issues can be addressed using the concept of robust positive invariance. We define this in the context of the uncertain dynamics and constraints given by

$$z_{i+1} = \Psi z_i + \bar{D} w_i, \quad w_i \in \mathcal{W} \tag{3.14a}$$

$$\bar{F} z_i \leq \mathbf{1}, \quad i = 0, 1, \ldots \tag{3.14b}$$

Note that the constraints of (3.11) are satisfied if and only if (3.14b) holds for all $w_i \in \mathcal{W}$, $i = 0, 1, \ldots$, if $\bar{D}$ is defined as

$$\bar{D} = \begin{bmatrix} D \\ 0 \end{bmatrix}.$$

Definition 3.2 (*Robustly positively invariant set*) A set $\mathcal{Z} \subset \mathbb{R}^{n_z}$ is robustly positively invariant (RPI) under the dynamics (3.14a) and constraints (3.14b) if and only if $\bar{F} z \leq \mathbf{1}$ and $\Psi z + \bar{D} w \in \mathcal{Z}$ for all $w \in \mathcal{W}$, for all $z \in \mathcal{Z}$.

It is often desirable to determine the largest possible RPI set for a given system. This is defined analogously to the case of systems with no model uncertainty considered in Sect. 2.4.

Definition 3.3 (*Maximal robustly positively invariant set*) The maximal robustly positively invariant (MRPI) set under (3.14a) and (3.14b) is the union of all RPI sets under these dynamics and constraints.

Unlike the case of no model uncertainty, for which the maximal invariant set is necessarily non-empty if Φ is strictly stable, the MRPI set for (3.14a) and (3.14b) will be empty whenever the disturbance set $\mathcal{W}$ is sufficiently large. Since the constraints (3.14b) are equivalent to (3.11), it is clear that the MRPI set can be non-empty only if the constraint tightening parameters h_i defined in (3.12) satisfy $h_i < \mathbf{1}$ for all i. From the expression for h_i in (3.13), where Φ is by assumption a stable matrix (i.e. all of its eigenvalues lie inside the unit circle), it follows that h_i has a limit as $i \to \infty$ and hence, we require that

$$\lim_{i\to\infty} h_i < \mathbf{1}. \tag{3.15}$$

The conditions under which this inequality holds will be examined in Sect. 3.2.2. Assuming that (3.15) is satisfied, the following theorem (which is a simplified version of a result from [1]) shows that the MRPI set is defined in terms of a finite number of linear inequalities.

Theorem 3.1 *If (3.15) holds, then the MRPI set $\mathcal{Z}^{\text{MRPI}}$ for the dynamics defined by (3.14a) and the constraints (3.14b) can be expressed*

$$\mathcal{Z}^{\text{MRPI}} = \{z : \bar{F}\Psi^i z \le \mathbf{1} - h_i,\ i = 0, \ldots, \nu\} \tag{3.16}$$

where ν is the smallest positive integer such that $\bar{F}\Psi^{\nu+1} z \le \mathbf{1} - h_{\nu+1}$ for all z satisfying $\bar{F}\Psi^i z \le \mathbf{1} - h_i$, $i = 0, \ldots, \nu$. Furthermore ν is necessarily finite if Ψ is strictly stable and $(\Psi, \bar{F})$ is observable.

Proof For any nonnegative integer n, let $\mathcal{Z}^{(n)}$ denote the set

$$\mathcal{Z}^{(n)} \doteq \{z : \bar{F}\Psi^i z \le \mathbf{1} - h_i,\ i = 0, \ldots, n\}.$$

If $\bar{F}\Psi^{\nu+1} z \le \mathbf{1} - h_{\nu+1}$ for all $z \in \mathcal{Z}^{(\nu)}$, then, since $h_{i+1} \ge h_i + \bar{F}\Psi^i \bar{D} w$ for all $w \in \mathcal{W}$, the following conditions hold for all $z \in \mathcal{Z}^{(\nu)}$,

(a) $\bar{F}\Psi^i(\Psi z + \bar{D} w) \le \mathbf{1} - h_i$ for all $w \in \mathcal{W}$, for $i = 0, \ldots, \nu$,
(b) $\bar{F} z \le \mathbf{1}$.

These conditions imply that $\mathcal{Z}^{(\nu)}$ is RPI under the dynamics (3.14a) and constraints (3.14b), and hence $\mathcal{Z}^{(\nu)} \subseteq \mathcal{Z}^{\text{MRPI}}$. Furthermore $\mathcal{Z}^{\text{MRPI}}$ must be a subset of $\mathcal{Z}^{(n)}$ for all $n \ge 0$, since if $z \notin \mathcal{Z}^{(n)}$, then z cannot belong to any set that is RPI under (3.14a) and (3.14b). Hence $\mathcal{Z}^{\text{MRPI}} = \mathcal{Z}^{(\nu)}$ if $\bar{F}\Psi^{\nu+1} z \le \mathbf{1} - h_{\nu+1}$ for all $z \in \mathcal{Z}^{(\nu)}$.

It can moreover be concluded that $\mathcal{Z}^{\text{MRPI}} = \mathcal{Z}^{(\nu)}$ for some finite ν since $\mathcal{Z}^{\text{MRPI}}$ is necessarily bounded given that $(\Psi, \bar{F})$ is observable, and because $\mathcal{Z}^{(n+1)} = \mathcal{Z}^{(n)} \cap \{z : \bar{F}\Psi^{(n+1)}z \leq \mathbf{1} - h_{n+1}\}$, where $\{z : \bar{F}\Psi^{(\nu+1)}z \leq \mathbf{1} - h_{\nu+1}\}$ must contain any bounded set for some finite n since Ψ is strictly stable and since (3.15) implies that the elements of h_n are strictly less than unity for all $n \geq 0$. □

Since the conditions (3.11) hold if and only if (3.14b) holds for all $w_i \in \mathcal{W}$, $i = 0, 1, \ldots$, Theorem 3.1 implies that the constraints of (3.11) are equivalent to the condition that $z_{0|k} \in \mathcal{Z}^{\text{MRPI}}$, which is determined by a finite set of linear constraints. In addition, the MRPI set for the lifted dynamics provides the largest possible feasible set for the open-loop strategy (3.6) applied to (3.7). To see this, consider for example the projection of $\mathcal{Z}^{\text{MRPI}}$ in (3.16) onto the x-subspace,

$$\mathcal{F}_N \doteq \left\{x : \exists\, \mathbf{c} \text{ such that } \bar{F}\Psi^i \begin{bmatrix} x \\ \mathbf{c} \end{bmatrix} \leq \mathbf{1} - h_i,\ i = 0, \ldots, \nu\right\}. \tag{3.17}$$

Analogously to the case of no model uncertainty considered in Sect. 2.7.1, $\mathcal{F}_N$ has an interpretation as the set of all feasible initial conditions for the predictions generated by (3.6) and (3.7) subject to constraints (3.4):

$$\begin{aligned} \mathcal{F}_N = \Big\{x_0 : \exists\, \{c_0, \ldots, c_{N-1}\} \\ &\text{such that } (F + GK)x_i + Gc_i \leq \mathbf{1},\ i = 0, \ldots, N-1, \\ &\text{and } x_N \in \mathcal{X}_T,\ \forall \{w_0, \ldots, w_{N-1}\} \in \mathcal{W} \times \cdots \times \mathcal{W}\Big\}. \end{aligned} \tag{3.18}$$

(Here the use of an open-loop strategy means that, for given x_0, the sequence $\mathbf{c} = \{c_0, \ldots, c_{N-1}\}$ must ensure that the constraints are satisfied for all possible disturbance sequences $\{w_0, \ldots, w_{N-1}\} \in \mathcal{W} \times \cdots \times \mathcal{W}$.) This interpretation of $\mathcal{F}_N$ implies that the predicted state N steps ahead must lie in a terminal set, $\mathcal{X}_T$, which is RPI for (3.1) under $u_k = Kx_k$. Furthermore, from (3.17) and the definition of Ψ, the terminal set $\mathcal{X}_T$ must be the intersection of $\mathcal{Z}^{\text{MRPI}}$ with the subspace on which $\mathbf{c} = 0$, so that

$$\begin{aligned} \mathcal{X}_T &= \left\{x : \begin{bmatrix}(F + GK) & GE\end{bmatrix}\Psi^i \begin{bmatrix} x \\ 0 \end{bmatrix} \leq \mathbf{1} - h_i,\ i = 0, 1, \ldots\right\} \\ &= \big\{x_0 : (F + GK)x_i \leq \mathbf{1} - h_i,\ x_{i+1} = \Phi x_i,\ i = 0, 1, \ldots\big\}. \end{aligned}$$

It follows that $\mathcal{X}_T$ is the maximal RPI set for (3.1) under $u_k = Kx_k$ subject to (3.4). Therefore $\mathcal{F}_N$ contains all initial conditions for which the constraints (3.4) can be satisfied over an infinite horizon with the open-loop strategy (3.6) and with $c_{i|k} = 0$ for all $i \geq N$.

Having established that the conditions in (3.11) can be expressed in terms of a finite number of constraints, and that these constraints allow the largest possible set of initial conditions $x_{0|k}$ under the open-loop strategy (3.6), we next show that these conditions are recursively feasible. This property, which is essential for ensuring the stability of a robust MPC strategy incorporating (3.11), can be established by defining a candidate vector $\mathbf{c}_{k+1}$ of optimization variables at time $k+1$ in terms of a vector $\mathbf{c}_k$ which satisfies, by assumption, the constraints (3.11) at time k. As in Sects. 2.5 and 2.7.2, we define, this candidate as $\mathbf{c}_{k+1} = M\mathbf{c}_k$, so that

$$c_{i|k+1} = \begin{cases} c_{i+1|k}, & i = 0, \ldots, N-2 \\ 0 & i \geq N-1 \end{cases}$$

Since $u_k = Kx_k + c_{0|k}$ implies $x_{k+1} = \Phi x_k + Bc_{0|k} + Dw_k$, we then obtain

$$s_{i|k+1} = s_{i|k} + \Phi^{i-1} Dw_k, \quad i = 0, 1, \ldots$$

and hence

$$z_{0|k+1} = \Psi z_{0|k} + \bar{D}w_k.$$

But $\mathbf{c}_k$ satisfies (3.11) at time k if and only if $z_{0|k} \in \mathcal{Z}^{\mathrm{MRPI}}$, so the requirement that $\mathbf{c}_{k+1} = M\mathbf{c}_k$ should satisfy constraints at time $k+1$ is equivalent to requiring $\Psi z + \bar{D}w \in \mathcal{Z}^{\mathrm{MRPI}}$ for all $w \in \mathcal{W}$ and all $z \in \mathcal{Z}^{\mathrm{MRPI}}$. This is ensured by the robust positive invariance of $\mathcal{Z}^{\mathrm{MRPI}}$ under (3.14a) and (3.14b). It follows that there necessarily exists $\mathbf{c}_{k+1}$ such that $z_{0|k+1} \in \mathcal{Z}^{\mathrm{MRPI}}$ whenever $z_{0|k} \in \mathcal{Z}^{\mathrm{MRPI}}$, and hence this constraint set is recursively feasible.

3.2.2 *Interpretation in Terms of Tubes*

Tubes provides an intuitive geometric interpretation of robust constraint handling and are convenient for analysing the asymptotic behaviour of uncertain systems. In particular, a tube formulation allows the condition (3.15) on the infinite sequence of constraint tightening parameters h_i to be checked by solving a finite number of linear programs. Because of the model uncertainty in (3.7), the predicted states are described by a tube comprising a sequence of sets, each of which contains the state at a given future time instant for all realizations of future uncertainty. The use of tubes in control is not new (e.g. see [2, 3]), and they have been used in the context of MPC for a couple of decades (e.g. [4, 5]); their use in MPC has led to specialized techniques such as Tube MPC (TMPC) (e.g. [6]) which are discussed in more detail in Sect. 3.5.

Denoting the tube containing the predicted states as the sequence of sets $\{\mathcal{X}_{0|k}, \mathcal{X}_{1|k}, \ldots\}$, where $x_{i|k} = s_{i|k} + e_{i|k} \in \mathcal{X}_{i|k}$, $i = 0, 1, \ldots$, and using the decomposition (3.8) yields

$$\mathcal{X}_{i|k} = \{s_{i|k}\} \oplus \mathcal{E}_{i|k}$$

where $e_{i|k} \in \mathcal{E}_{i|k}$ for all i. Here the Minkowski sum $\{s_{i|k}\} \oplus \mathcal{E}_{i|k}$ simply translates each element $e_{i|k}$ of the set $\mathcal{E}_{i|k}$ to $s_{i|k} + e_{i|k}$. The sets that form the tube $\{\mathcal{E}_{0|k}, \mathcal{E}_{1|k}, \ldots\}$ evolve, by (3.8b), according to

$$\mathcal{E}_{i+1|k} = \Phi \mathcal{E}_{i|k} \oplus D\mathcal{W} \tag{3.19}$$

for all $i \geq 0$, with initial condition $\mathcal{E}_{0|k} = \{0\}$. Thus $\mathcal{E}_{i|k}$ can be expressed

$$\mathcal{E}_{i|k} = D\mathcal{W} \oplus \Phi D\mathcal{W} \oplus \cdots \oplus \Phi^{i-1} D\mathcal{W} = \bigoplus_{j=0}^{i-1} \Phi^j D\mathcal{W}. \tag{3.20}$$

The state tube $\{\mathcal{X}_{0|k}, \mathcal{X}_{1|k}, \ldots\}$ implies a tube for the predicted control input, $\{\mathcal{U}_{0|k}, \mathcal{U}_{1|k}, \ldots\}$, where $u_{i|k} \in \mathcal{U}_{i|k}$ for all i. In accordance with (3.6), $\mathcal{U}_{i|k}$ is given for $i = 0, 1, \ldots$ by

$$\mathcal{U}_{i|k} = \{K s_{i|k} + c_{i|k}\} \oplus K \mathcal{E}_{i|k}.$$

In this setting, the constraints (3.4) are therefore equivalent to

$$\bar{F}\left(\{\Psi^i z_{0|k}\} \oplus \begin{bmatrix} I \\ 0 \end{bmatrix} \mathcal{E}_{i|k}\right) \leq \mathbf{1}, \quad i = 0, 1, \ldots \tag{3.21}$$

Comparing (3.21) with (3.11) it can be seen that the amount by which the constraints on the nominal predictions $z_{i|k}$ must be tightened in order that the constraints (3.4) are satisfied for all uncertainty realizations is

$$h_i = \max_{e_{i|k} \in \mathcal{E}_{i|k}} (F + GK) e_{i|k},$$

which is in agreement with (3.13). This is illustrated in Fig. 3.2.

A consequence of (3.21) is that the requirement, which must be met in order for the MRPI set to be non-empty, that $h_i \leq (1 - \epsilon)\mathbf{1}$ for all $i \geq 0$ and some $\epsilon > 0$ is equivalent to

$$(F + GK)\mathcal{E}_{i|k} \leq (1 - \epsilon)\mathbf{1}, \quad i = 0, 1, \ldots \tag{3.22}$$

for some $\epsilon > 0$. However (3.20) implies that $\mathcal{E}_{i+1|k} = \mathcal{E}_{i|k} \oplus \Phi^i D\mathcal{W}$ and hence $\mathcal{E}_{i|k}$ is necessarily a subset of $\mathcal{E}_{i+1|k}$. Therefore the conditions in (3.22) are satisfied if and only if they hold asymptotically as $i \to \infty$. This motivates the consideration of the *minimal robust invariant set*, which defines the asymptotic behaviour of $\mathcal{E}_{i|k}$ as $i \to \infty$.

Definition 3.4 (*Minimal robustly positively invariant set*) The minimal robustly invariant (mRPI) set under (3.8b) is the RPI set contained in every closed RPI set of (3.8b).

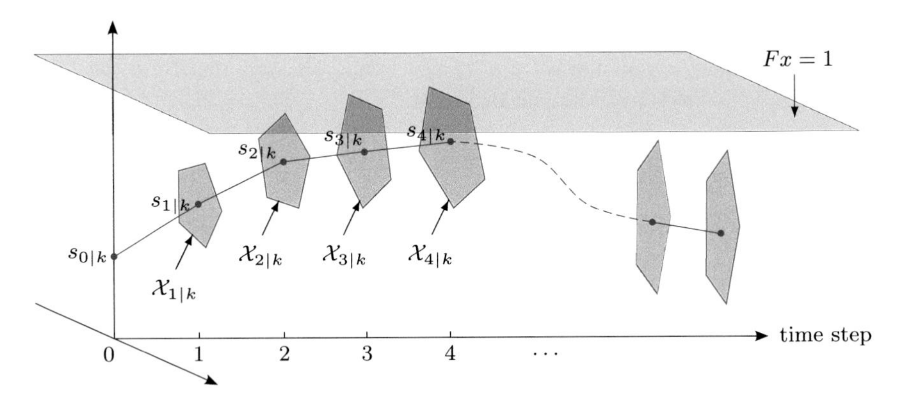

Fig. 3.2 An illustration of the state tube and constraints for the case that $G = 0$ and x is 2-dimensional

Because of the linearity of the dynamics in (3.8b), each set $\mathcal{E}_{i|k}$, $i = 0, 1, \ldots$ generated by (3.19) with initial condition $\mathcal{E}_{0|k} = \{0\}$ must be contained in an RPI set of (3.8b). Given also that $\mathcal{E}_{i|k} \subset \mathcal{E}_{i+1|k}$ and that, in the limit as $i \to \infty$, $\mathcal{E}_{i|k}$ in (3.20) is clearly RPI, the mRPI set for (3.8b) is given by

$$\mathcal{X}^{\text{mRPI}} \doteq \bigoplus_{j=0}^{\infty} \Phi^j D\mathcal{W}. \tag{3.23}$$

Unfortunately, unlike the maximal RPI set, the minimal RPI cannot generally be expressed either in terms of a finite number of linear inequalities or as the convex hull of a finite number of vertices. This is a consequence of the fact that, unless $\Phi^i = 0$ for some finite i, $\mathcal{E}_{i|k}$ is a proper subset of $\mathcal{E}_{i+1|k}$ for all $i \geq 0$. As a result it is not in general possible either to compute $\mathcal{X}^{\text{mRPI}}$ or to obtain an exact asymptotic value for h_i as $i \to \infty$. Instead it is necessary to characterize the asymptotic behaviour of $\mathcal{E}_{i|k}$ by finding an outer bound, $\hat{\mathcal{X}}^{\text{mRPI}}$, satisfying $\hat{\mathcal{X}}^{\text{mRPI}} \supseteq \mathcal{X}^{\text{mRPI}}$. Given a bounding set $\hat{\mathcal{X}}^{\text{mRPI}}$, a corresponding upper bound, $\hat{h}_\infty$, can be computed that satisfies $\hat{h}_\infty \geq h_i$ for all i, thus providing a sufficient condition for (3.15).

The bound $\hat{h}_\infty$ can be computed using several different approaches, however the method presented here is based on the mRPI set approximation of [7]. This approximation is derived from the observation that, if there exist a positive integer r and scalar $\rho \in [0, 1)$ satisfying

$$\Phi^r D\mathcal{W} \subseteq \rho D\mathcal{W} \tag{3.24}$$

then

$$\bigoplus_{j=0}^{\infty} \Phi^j D\mathcal{W} \subseteq \bigoplus_{j=0}^{r-1} \Phi^j D\mathcal{W} \oplus \rho \bigoplus_{j=0}^{r-1} \Phi^j D\mathcal{W} \oplus \rho^2 \bigoplus_{j=0}^{r-1} \Phi^j D\mathcal{W} \cdots$$

But $\bigoplus_{j=0}^{r-1} \Phi^j D\mathcal{W}$ is necessarily convex (since $\mathcal{W}$ is by assumption convex and hence $\Phi^j D\mathcal{W}$ is also convex), and for any convex set $\mathcal{X}$ and scalar $\alpha > 0$ we have $\mathcal{X} \oplus \alpha\mathcal{X} = (1+\alpha)\mathcal{X}$. It therefore follows that

$$\bigoplus_{j=0}^{\infty} \Phi^j D\mathcal{W} \subseteq (1+\rho+\rho^2+\cdots)\bigoplus_{j=0}^{r-1} \Phi^j D\mathcal{W}$$
$$= \frac{1}{1-\rho}\bigoplus_{j=0}^{r-1} \Phi^j D\mathcal{W}$$

Defining

$$\hat{\mathcal{X}}^{\text{mRPI}} \doteq \frac{1}{1-\rho}\bigoplus_{j=0}^{r-1} \Phi^j D\mathcal{W} \tag{3.25}$$

it can be concluded that $\mathcal{X}^{\text{mRPI}} \subseteq \hat{\mathcal{X}}^{\text{mRPI}}$.

The mRPI set approximation given by (3.25) has the desirable properties that it approaches the actual mRPI set arbitrarily closely if ρ is chosen to be sufficiently small. In addition, for any $\rho > 0$ there necessarily exists a finite r satisfying (3.24) since Φ is strictly stable by assumption. Most importantly, $\hat{\mathcal{X}}^{\text{mRPI}}$ is defined in terms of a finite number of inequalities (or vertices), and this allows the corresponding bound $\hat{h}_\infty$ to be determined as

$$\hat{h}_\infty \doteq \frac{1}{1-\rho}\sum_{j=0}^{r-1} \max_{w_j \in \mathcal{W}} (F+GK)\Phi^j w_j = \frac{1}{1-\rho} h_r$$

which gives a sufficient condition for (3.15) as

$$\hat{h}_\infty = \frac{1}{1-\rho} h_r < \mathbf{1}, \tag{3.26}$$

for any r and ρ satisfying (3.24).

In order to check whether the condition (3.24) is satisfied by a given disturbance set $\mathcal{W} = \{w : Vw \le \mathbf{1}\}$, matrix Φ and scalars r and ρ, note that $\theta \in \rho D\mathcal{W}$ if and only if $VD^\dagger\theta \le \rho\mathbf{1}$, where $D^\dagger$ is the Moore-Penrose pseudoinverse of D (i.e. $D^\dagger \doteq (D^T D)^{-1} D^T$). Therefore (3.24) is equivalent to

$$\max_{w \in \mathcal{W}} VD^\dagger \Phi^r Dw \le \rho\mathbf{1}. \tag{3.27}$$

This can be checked by solving n_V linear programs.

Example 3.1 A simple supply chain model contains a supplier, a production facility and a distributor (Fig. 3.3). At the beginning of the kth discrete-time interval, a

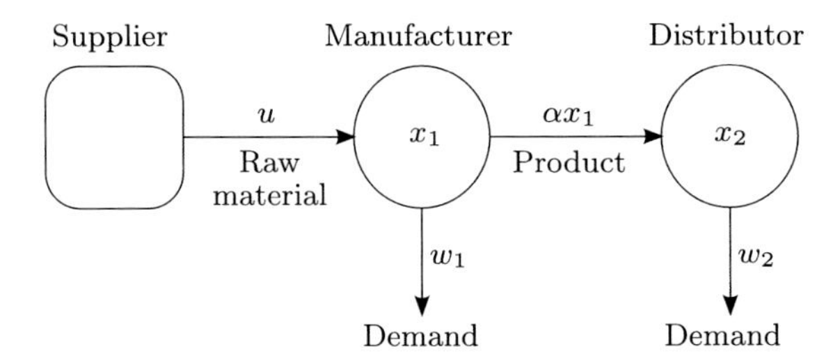

Fig. 3.3 A simple supply chain model

quantity u_k of raw material is delivered by the supplier to the manufacturer. Of this material, an amount $w_{1,k}$ is transferred to other manufacturers, and the remainder is added to the amount $x_{1,k}$ held in storage by the manufacturer. A fraction, $\alpha x_{1,k}$ of this is converted into product and transferred to the distributor, who stores an amount $x_{2,k}$ and supplies $w_{2,k}$ to customers. The value $\alpha = 0.5$ is assumed to be known, while the demand quantities $w_{1,k}$ and $w_{2,k}$ are unknown at time k but have known bounds.

The system can be represented by a model of the form (3.1) with state $x_k = (x_{1,k}, x_{2,k})$, control input u_k and disturbance input $w_k = (w_{1,k}, w_{2,k})$. As a result of limits on the supply rate, the storage capacities and the demand, we obtain the following input, state and disturbance constraints:

$$0 \le u_k \le 0.5, \quad (0, 0) \le x_k \le (1, 1), \quad (0.1, 0.1) \le w_k \le (0.2, 0.2).$$

The viability with respect to these constraints of the control strategy (3.6) can be determined using the robust invariant sets discussed in Sect. 3.2.1. To this end, we first determine a (non-zero) setpoint about which to regulate x_k. Given that only bounds on w_k are available, and in the absence of any statistical information on w, it is reasonable to define the setpoint in terms of the equilibrium values of states and inputs (denoted x^0 and u^0) that correspond to a constant disturbance (w^0) at the centroid of the disturbance set. Therefore, defining

$$w^0 = (0.15, 0.15), \quad u^0 = 0.3, \quad x^0 = (0.3, 0.5),$$

the system model can be expressed in terms of the transformed variables $x^\delta = x - x^0$, $u^\delta = u - u^0$ and $w^\delta = w - w^0$ as $x^\delta_{k+1} = Ax^\delta_k + Bu^\delta_k + Dw^\delta_k$ with

$$A = \begin{bmatrix} 0.5 & 0 \\ 0.5 & 1 \end{bmatrix}, \quad B = \begin{bmatrix} 1 \\ 0 \end{bmatrix}, \quad D = \begin{bmatrix} -1 & 0 \\ 0 & -1 \end{bmatrix},$$

and with constraints

$$\begin{gathered} -0.3 \le u^\delta_k \le 0.2, \quad (-0.3, -0.5) \le x^\delta_k \le (0.7, 0.5) \\ (-0.05, -0.05) \le w^\delta_k \le (0.05, 0.05). \end{gathered} \tag{3.28}$$

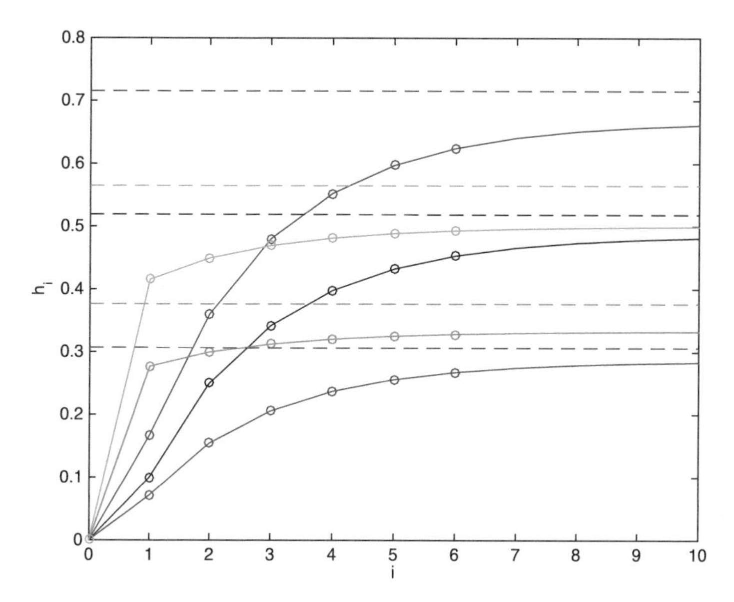

Fig. 3.4 The elements of the constraint tightening parameters h_i, for $i = 0, \ldots, 6$ (*circles*) and the elements of $\hat{h}_\infty = (1-\rho)^{-1}h_r$ (*dashed lines*) with $r = 6$ and $\rho = 0.127$ in Example 3.1. The *solid lines* show the evolution of h_i for $i > 6$ to give an indication of the asymptotic value, h_∞

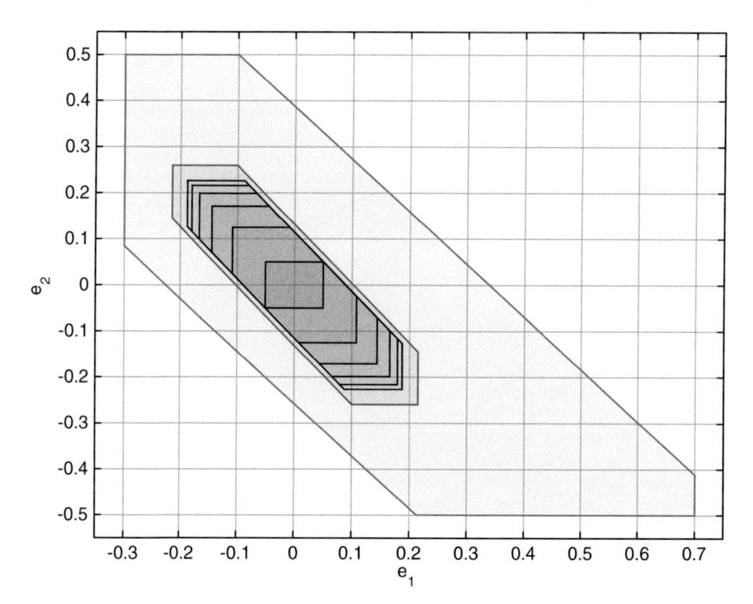

Fig. 3.5 The sets $\mathcal{E}_i$ in Example 3.1 for $i = 1, 2, 3, 4, 5, 6$ (*black lines*); the minimal robust invariant set approximation $\hat{\mathcal{X}}^{\mathrm{mRPI}} = (1-\rho)^{-1}\mathcal{E}_r$ for $r = 6$, $\rho = 0.127$ (*red line*); and the constraint set $\{e : (F + GK)e \leq \mathbf{1}\}$ (*green line*)

Using the open-loop strategy (3.6), with K chosen as the unconstrained LQ-optimal feedback gain for the nominal system and the cost (3.5) with $Q = I$ and $R = 0.01$, we obtain $K = [-0.89\ -0.78]$, which fixes the eigenvalues of $\Phi = A + BK$ at 0.61 and 0.005.

To determine whether the set of feasible initial conditions is non-empty, we first check whether (3.15) is satisfied by checking the sufficient condition (3.26). We therefore need values of r and $\rho < 1$ such that $\Phi^r D\mathcal{W} \subseteq \rho D\mathcal{W}$; these can be found by computing the minimum value of ρ that satisfies (3.27) for a given value of r, and then increasing r until ρ is judged to be sufficiently small. Here it is expected that a value of ρ around 0.1 will be small enough for $\hat{h}_\infty$ to provide an accurate estimate of h_∞. Taking $r = 6$, for which the minimum value of ρ satisfying (3.27) is $\rho = 0.127$, we obtain $\hat{h}_\infty = (0.72, 0.31, 0.52, 0.52, 0.38, 0.56)$ (Fig. 3.4)—note that h_i is a 6-dimensional vector because there are $n_C = 6$ individual constraints in (3.28). Hence $\hat{h}_\infty < \mathbf{1}$, which implies that the minimal RPI set approximation $\hat{\mathcal{X}}^{\text{mRPI}}$ is a proper subset of $\{e : (F + GK)e \leq \mathbf{1}\}$ (as shown in Fig. 3.5). Theorem 3.1 therefore indicates that the maximal RPI set defined in (3.16) is non-empty for all $N \geq 0$.

The MRPI sets $\mathcal{Z}^{\text{MRPI}}$ for this system under the open-loop strategy (3.6) can be computed for given N using Theorem 3.1. The set $\mathcal{F}_N$ of all feasible initial conditions, which by (3.18) is equal to the projection of the corresponding MRPI set onto the x-subspace, is shown in Fig. 3.6 for a range of values of N. For this example there is no increase in the x-subspace projection of the MRPI set for $N > 4$, since $\mathcal{F}_4$ is equal to the maximal stabilizable set $\mathcal{F}_\infty$. ◊

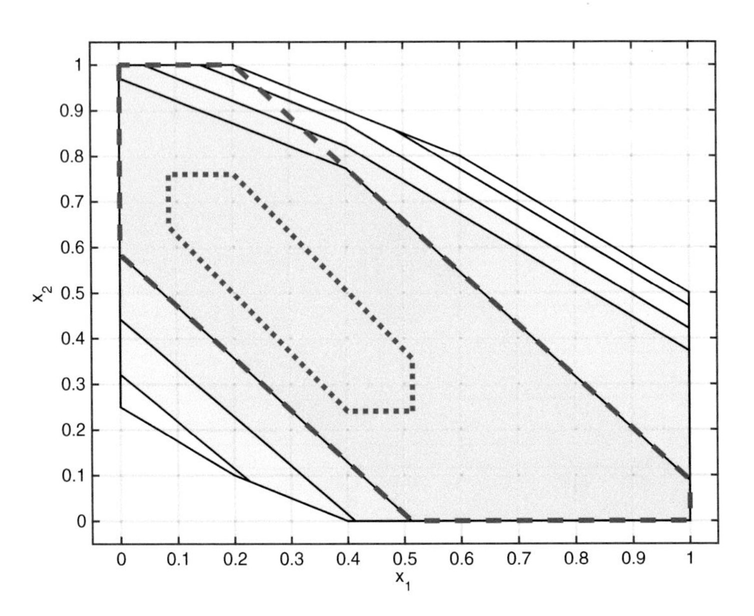

Fig. 3.6 The feasible initial condition sets $\mathcal{F}_N$, $N = 0, 1, 3, 4$, for Example 3.1; also shown are the sets $\{x : (F + GK)(x - x^0) \leq \mathbf{1}\}$ (*dashed line*) and $\hat{\mathcal{X}}^{\text{mRPI}} \oplus \{x^0\}$ (*dotted line*)

3.3 Nominal Predicted Cost: Stability and Convergence

When faced with the problem of defining a performance objective for robust MPC given only the knowledge of a nominal value and a bounding set for the model uncertainty, a nominal cost is a simple and obvious choice that can provide desirable closed-loop stability properties. This section defines a robust MPC strategy that combines a nominal cost with the robust constraints formulated in Sect. 3.2. We analyse closed-loop stability and convergence using a technique based on l_2 stability theory.

We use the term *nominal cost* when referring to a predicted performance index evaluated along the predicted trajectories that are obtained when the model uncertainty is equal to its nominal value. As in Sect. 3.2, the nominal value of the additive uncertainty w in the model (3.1) is taken to be $w = 0$. Then, assuming the open-loop strategy (3.6) and a quadratic cost index of the form (2.10), the nominal predicted cost $J(s_{0|k}, \{c_{0|k}, \ldots, c_{N-1|k}\}) = J(s_{0|k}, \mathbf{c}_k)$ is defined as

$$J(s_{0|k}, \mathbf{c}_k) \doteq \sum_{i=0}^{\infty} \big(\|s_{i|k}\|_Q^2 + \|v_{i|k}\|_R^2 \big). \tag{3.29}$$

where $\{s_{i|k}, i = 0, 1, \ldots\}$ is the nominal predicted state trajectory governed by (3.8a) with $s_{0|k} = x_k$, and $v_{i|k} = K s_{i|k} + c_{i|k}$. Throughout this section K is assumed to be the optimal unconstrained feedback gain for the nominal cost (3.29). However the methods discussed here are also applicable to the case that K is non-optimal but $A + BK$ is stable (see Question 4 on p. 114).

Expressing $s_{i|k}$ in terms of the augmented model state employed in Sect. 3.2, we obtain $s_{i|k} = [I \ 0] z_{i|k}$, where $z_{i|k}$ is generated by the autonomous dynamics (3.9) in an identical manner to the autonomous prediction system considered in Sect. 2.7. Therefore, using Theorem 2.10, the cost (3.29) is given by

$$J(s_{0|k}, \mathbf{c}_k) = \sum_{i=0}^{\infty} \|z_{i|k}\|_{\hat{Q}}^2 = \|z_{0|k}\|_W^2 \tag{3.30a}$$

$$\hat{Q} = \begin{bmatrix} Q + K^T R K & K^T R E \\ E^T R K & E^T R E \end{bmatrix}, \tag{3.30b}$$

and, by Lemma 2.1, the matrix W in the expression for $J(s_{0|k}, \mathbf{c}_k)$ can be determined by solving the Lyapunov equation (2.34). Furthermore, given that K is the unconstrained optimal feedback gain, Theorem 2.10 implies that W is block diagonal and hence

$$J(s_{0|k}, \mathbf{c}_k) = \|s_{0|k}\|_{W_x}^2 + \|\mathbf{c}_k\|_{W_c}^2,$$

where W_x is the solution of the Riccati equation (2.9), and where W_c is block diagonal: $W_c = \mathrm{diag}\{B^T W_x B + R, \ldots, B^T W_x B + R\}$. Combining the nominal predicted cost with the constraints constructed in Sect. 3.2 we obtain the following robust MPC strategy, which requires the online solution of a quadratic program with Nn_u variables and $n_C(\nu+1)$ constraints.

Algorithm 3.1 At each time instant $k = 0, 1, \ldots$:

(i) Perform the optimization

$$\underset{\mathbf{c}_k}{\text{minimize}} \ \|\mathbf{c}_k\|^2_{W_c} \ \text{subject to} \ \bar{F}\Psi^i \begin{bmatrix} x_k \\ \mathbf{c}_k \end{bmatrix} \leq \mathbf{1} - h_i, \ i = 0, \ldots, \nu \tag{3.31}$$

where ν satisfies the conditions of Theorem 3.1.

(ii) Apply the control law $u_k = Kx_k + c^*_{0|k}$, where $\mathbf{c}^*_k = (c^*_{0|k}, \ldots, c^*_{N-1|k})$ is the optimal value of $\mathbf{c}_k$ for problem (3.31). ◁

The assumption in step (i) that ν satisfies the conditions of Theorem 3.1 implies that the constraint set $\{z = (x_k, \mathbf{c}_k) : \bar{F}\Psi^i z \leq \mathbf{1} - h_i, \ i = 0, \ldots, \nu\}$ is robustly positive invariant, and, as discussed in Sect. 3.2.1, this ensures that the optimization (3.31) is recursively feasible. Therefore $c^*_{0|k}$ exists for all k and the state of the closed-loop system under Algorithm 3.1 is governed for $k = 0, 1, \ldots$ by

$$x_{k+1} = \Phi x_k + Bc^*_{0|k} + Dw_k. \tag{3.32}$$

However the MPC optimization (3.31) is equivalent to the minimization of the nominal cost (3.29) and, unlike the case in which there is no model uncertainty, there is no guarantee that the optimal value J^*_k will be monotonically non-increasing when the system is subject to unknown disturbances.

We therefore use an alternative method of analysing closed-loop stability. First we demonstrate that the sequence $\{\|c^*_{0|k}\|, \|c^*_{1|k}\|, \ldots\}$ is square-summable, and we then use l_2 stability theory to show that the closed-loop system imposes a finite l_2 gain between the disturbance sequence $\{w_0, w_1, \ldots\}$ and the sequence $\{x_0, x_1, \ldots\}$ of closed-loop plant states. Finally, we use this result to conclude that x_k converges asymptotically to the minimal RPI set $\mathcal{X}^{\text{mRPI}}$.

The discussion of recursive feasibility in Sect. 3.2.1 demonstrates that $\mathbf{c}_{k+1} = M\mathbf{c}^*_k$ is feasible but suboptimal for (3.31). Therefore $\mathbf{c}^*_{k+1}$ necessarily satisfies $\|\mathbf{c}^*_{k+1}\|_{W_c} \leq \|M\mathbf{c}^*_k\|_{W_c}$, which, from the definitions of M and W_c implies that

$$\|\mathbf{c}^*_{k+1}\|^2_{W_c} \leq \|\mathbf{c}^*_k\|^2_{W_c} - \|c^*_{0|k}\|^2_{R+B^TW_xB}. \tag{3.33}$$

From this bound we obtain the following result.

Lemma 3.1 *Let $\underline{\lambda}(R)$ denote the smallest eigenvalue of R in the cost (3.29), then*

$$\sum_{k=0}^{\infty} \|c^*_{0|k}\|^2 \leq \frac{1}{\underline{\lambda}(R)} \|\mathbf{c}^*_0\|^2_{W_c}. \tag{3.34}$$

Proof Summing both the sides of (3.33) over $k = 0, 1, \ldots$ and using the bound $\|c\|^2_{R+B^T W_x B} \geq \underline{\lambda}(R)\|c\|^2$, where $\underline{\lambda}(R) > 0$ due to $R \succ 0$, yields (3.34). □

We next give a version of a result from l_2 stability theory which states that the l_2 gains from the inputs $c^*_{0|k}$ and w_k to the state of the closed-loop system (3.32) are necessarily finite since Φ is assumed to be strictly stable.

Lemma 3.2 *All trajectories of the closed-loop system (3.32) satisfy the bound*

$$\sum_{k=0}^{\infty} \|x_k\|^2 \leq \|x_0\|^2_P + \gamma_1^2 \sum_{k=0}^{\infty} \|c^*_{0|k}\|^2 + \gamma_2^2 \sum_{k=0}^{\infty} \|w_k\|^2 \tag{3.35}$$

for some matrix $P \succ 0$ and some scalars γ_1, γ_2, provided Φ is strictly stable.

Proof There necessarily exists $P \succ 0$ satisfying $P - \Phi^T P \Phi \succ I_{n_x}$ since Φ is strictly stable (see, e.g. [8], Sect. 5.9). Using Schur complements it follows that there exists $P \succ 0$ and positive scalars γ_1, γ_2 satisfying

$$\begin{bmatrix} P & \Phi^T P \\ P\Phi & P \end{bmatrix} - \begin{bmatrix} 0 & 0 \\ PB & PD \end{bmatrix} \begin{bmatrix} \gamma_1^{-2} I_{n_u} & 0 \\ 0 & \gamma_2^{-2} I_{n_w} \end{bmatrix} \begin{bmatrix} 0 & B^T P \\ 0 & D^T P \end{bmatrix} \succ \begin{bmatrix} I_{n_x} & 0 \\ 0 & 0 \end{bmatrix}.$$

Using Schur complements again, we therefore have

$$\begin{bmatrix} P & 0 & 0 \\ 0 & \gamma_1^2 I_{n_u} & 0 \\ 0 & 0 & \gamma_2^2 I_{n_w} \end{bmatrix} - \begin{bmatrix} \Phi^T \\ B^T \\ D^T \end{bmatrix} P \begin{bmatrix} \Phi & B & D \end{bmatrix} \succeq \begin{bmatrix} I_{n_x} & 0 & 0 \\ 0 & 0 & 0 \\ 0 & 0 & 0 \end{bmatrix}.$$

Pre- and post-multiplying both sides of this inequality by $(x_k, c^*_{0|k}, w_k)$ and using (3.32) we obtain

$$\|x_k\|^2_P + \gamma_1^2 \|c^*_{0|k}\|^2 + \gamma_2^2 \|w_k\|^2 - \|x_{k+1}\|^2_P \geq \|x_k\|^2,$$

and summing both sides of this inequality over $k = 0, 1, \ldots$ gives (3.35). □

A consequence of inequalities (3.34) and (3.35) is that the closed-loop system under Algorithm 3.1 inherits the bound on the l_2 gain from w to x that is imposed by the linear feedback law $u = Kx$ in the absence of constraints. These inequalities can also be used to analyse the asymptotic convergence of x_k. Thus, let $x_k = s_k + e_k$, where s_k and e_k are the components of the state of the closed-loop system (3.32) that satisfy

$$s_{k+1} = \Phi s_k + B c^*_{0|k} \tag{3.36a}$$

$$e_{k+1} = \Phi e_k + D w_k \tag{3.36b}$$

with $s_0 = x_0$ and $e_0 = 0$. Then from (3.35) and Lemma 3.2 applied to (3.36a) we obtain

$$\sum_{k=0}^{\infty} \|s_k\|^2 \leq \|x_0\|_P^2 + \frac{\gamma_1^2}{\underline{\lambda}(R)} \|\mathbf{c}_0^*\|_{W_c}^2 \tag{3.37}$$

and it follows that $s_k \to 0$ as $k \to \infty$. Moreover, since $x_k = s_k + e_k$ where e_k lies for all k in the mRPI set $\mathcal{X}^{\text{mRPI}}$ defined in (3.23), the asymptotic convergence of s_k can be used to demonstrate convergence of x_k to $\mathcal{X}^{\text{mRPI}}$. This final point is explained in more detail in the proof of Theorem 3.2, which summarizes the results of this section.

Theorem 3.2 *For the system (3.1) with the control law of Algorithm 3.1:*

(a) the feasible set $\mathcal{F}_N$ defined in (3.18) is robustly positively invariant;
(b) for any $x_0 \in \mathcal{F}_N$, the closed-loop evolution of the state of (3.1) satisfies

$$\sum_{k=0}^{\infty} \|x_k\|^2 \leq \|x_0\|_P^2 + \frac{\gamma_1^2}{\underline{\lambda}(R)} \|\mathbf{c}_0^*\|_{W_c}^2 + \gamma_2^2 \sum_{k=0}^{\infty} \|w_k\|^2 \tag{3.38}$$

for some matrix $P \succ 0$ and scalars γ_1, γ_2;
(c) x_k converges asymptotically to the minimal RPI set $\mathcal{X}^{\text{mRPI}}$ of (3.23).

Proof The RPI property of the feasible set $\mathcal{F}_N$ follows from the definition of the constraint set in (3.31) as a RPI set, whereas the bound (3.38) is a direct consequence of bounds (3.34) and (3.35). Furthermore, by (3.37) $\|s_k\|$ is square-summable, so for any given $\epsilon > 0$ there must exist finite n such that $s_k \in \mathcal{B}_\epsilon = \{s : \|s\| \leq \epsilon\}$ for all $k \geq n$, and therefore

$$x_k \in \mathcal{E}_k \oplus \mathcal{B}_\epsilon, \ \forall k \geq n,$$

where $\mathcal{E}_k$ is the bounding set for e_k defined by $\mathcal{E}_{k+1} = \Phi\mathcal{E}_k \oplus D\mathcal{W}$ with $\mathcal{E}_0 = \{0\}$. Since $\mathcal{E}_k \subseteq \mathcal{X}^{\text{mRPI}}$ for all k, we have

$$x_k \in \mathcal{X}^{\text{mRPI}} \oplus \mathcal{B}_\epsilon, \ \forall k \geq n,$$

and it can be concluded that x_k converges to $\mathcal{X}^{\text{mRPI}}$ as $k \to \infty$. □

3.4 A Game Theoretic Approach

Robustly stabilizing controllers that guarantee limits on the response to additive disturbances can be designed using the linear quadratic game theory of optimal control [9, 10]. By choosing the control input so as to minimize a predicted cost that assumes the worst-case future model uncertainty, this approach is able to impose a specified bound on the l_2 gain from the disturbance input to a given system output. This strategy has its roots in game theory [11], which interprets the control input and the disturbance input as opposing players, each of which seeks to influence the behaviour of the system, one by minimizing and the other maximizing the performance index. For this reason it is also known as a min-max approach [12].

The idea has been exploited in unconstrained MPC (e.g. [13, 14]) but the concern here is with the linear discrete-time constrained case [15, 16]. Control laws that aim to optimize the worst-case performance with respect to model uncertainty can be overly cautious. On the other hand, game theoretic approaches to MPC based on the minimization of worst case predicted performance can avoid the possibility of poor sensitivity to disturbances which could be exhibited by MPC laws based on the nominal cost considered in Sect. 3.3. At the same time, the approach retains the feasibility and asymptotic convergence properties of robust MPC.

According to the game theoretic approach, the cost of (2.3) is replaced by

$$\check{J}(x_0, \{u_0, u_1, \ldots\}) \doteq \max_{\{w_0, w_1, \ldots\}} \sum_{i=0}^{\infty} \left(\|x_i\|_Q^2 + \|u_i\|_R^2 - \gamma^2 \|w_i\|^2\right). \tag{3.39}$$

The scalar parameter γ appearing in this cost limits the l_2 gain between the disturbance input w and the output $y = (Q^{1/2}x, R^{1/2}u)$ to γ. If there are no constraints on inputs and states, then the maximizing feedback law for w in (3.39) and the feedback law for u that minimizes $\check{J}$ are given by [17]

$$u = Kx, \quad w = Lx \tag{3.40}$$

where

$$\begin{bmatrix} K \\ L \end{bmatrix} = -\left(\begin{bmatrix} B^T \\ D^T \end{bmatrix} \check{W}_x \begin{bmatrix} B & D \end{bmatrix} + \begin{bmatrix} R & 0 \\ 0 & -\gamma^2 I \end{bmatrix} \right)^{-1} \begin{bmatrix} B^T \\ D^T \end{bmatrix} \check{W}_x A \tag{3.41}$$

and $\check{W}_x$ is the unique positive definite solution of the Riccati equation

$$\begin{aligned} \check{W}_x &= A^T \check{W}_x A + Q \\ &\quad - A^T \check{W}_x \begin{bmatrix} B & D \end{bmatrix} \left(\begin{bmatrix} B^T \\ D^T \end{bmatrix} \check{W}_x \begin{bmatrix} B & D \end{bmatrix} + \begin{bmatrix} R & 0 \\ 0 & -\gamma^2 I \end{bmatrix} \right)^{-1} \begin{bmatrix} B^T \\ D^T \end{bmatrix} \check{W}_x A. \end{aligned} \tag{3.42}$$

This result can be derived in a similar manner to the Riccati equation and optimal feedback gain for the unconstrained linear quadratic control problem with no model uncertainty in Theorem 2.1.

Under some mild assumptions on the system model (3.1) and the cost weights in (3.39) (see, e.g. [17] for details), the solution to the Riccati equation (3.42) exists whenever γ is sufficiently large that $\gamma^2 I - D^T \check{W}_x D$ is positive-definite, and moreover the resulting closed-loop system matrix $\Phi = A + BK$ is strictly stable. Clearly, it is important to have knowledge of the corresponding lower limit on γ^2, and it may therefore be more convenient to compute $\check{W}_x$ simultaneously with the minimum value of γ^2 using semidefinite programming. For example, minimizing γ^2 subject to the LMI

$$\begin{bmatrix} \begin{bmatrix} S & 0 \\ \star & \gamma^2 I \end{bmatrix} & \begin{bmatrix} (AS+BY)^T \\ D^T \end{bmatrix} & \begin{bmatrix} SQ^{1/2} & Y^T R^{1/2} \\ 0 & 0 \end{bmatrix} \\ \star & S & 0 \\ \star & \star & I \end{bmatrix} \succeq 0 \tag{3.43}$$

in variables S, Y and γ^2, yields the corresponding solutions for $\check{W}_x = S^{-1}$, $K = Y\check{W}_x$ and $L = (\gamma^2 I - D^T \check{W}_x D)^{-1} D^T \check{W}_x (A + BK)$. Throughout this section we assume that $\gamma^2 I - D^T \check{W}_x D \succ 0$ holds.

In order to formulate a predictive control law based on the cost (3.39), we adopt the open-loop strategy (3.6), with K defined via (3.41) as the optimal unconstrained state feedback gain for (3.39). By determining the maximizing disturbance sequence for any given x_k and optimization variables $\mathbf{c}_k$, the predicted cost, which we denote as $\check{J}(x_k, \mathbf{c}_k)$, along trajectories of (3.7) can be obtained as an explicit function of x_k and $\mathbf{c}_k$. This is consistent with the definition of an open-loop optimization strategy because it enables the entire sequence $\mathbf{c}_k = \{c_{0|k}, \ldots, c_{N-1|k}\}$ that achieves the minimum worst-case cost to be determined as a function of x_k.

The following lemma expresses $\check{J}(x_k, \mathbf{c}_k)$ as a quadratic function of x_k and $\mathbf{c}_k$ by considering the worst-case unconstrained disturbances in (3.39). Clearly, the resulting worst-case cost may be conservative since it ignores the information that the disturbance w_k lies in $\mathcal{W}$. An alternative cost definition that accounts for the constraints on the disturbance w by introducing extra optimization variables is discussed at the end of this section.

Lemma 3.3 *The worst-case cost (3.39) for the open-loop strategy (3.6) is given by*

$$\check{J}(x_k, \mathbf{c}_k) = \|x_k\|^2_{\check{W}_x} + \|\mathbf{c}_k\|^2_{\check{W}_c} \tag{3.44}$$

where $\check{W}_x$ is the solution of the Riccati equation (3.42) and $\check{W}_c$ is block diagonal:

$$\check{W}_c = \begin{bmatrix} B^T \check{W}'_x B + R & & 0 \\ & \ddots & \\ 0 & & B^T \check{W}'_x B + R \end{bmatrix} \tag{3.45a}$$

$$\check{W}'_x = \check{W}_x + \check{W}_x D(\gamma^2 I - D^T \check{W}_x D)^{-1} D^T \check{W}_x. \tag{3.45b}$$

Proof Let $z_{0|k} = (x_k, \mathbf{c}_k)$ and consider evaluating the cost (3.39) along the predicted trajectories generated by the dynamics $z_{i+1|k} = \Psi z_{i|k} + \bar{D} w_{i|k}, i = 0, 1, \ldots$ Clearly the cost (3.39) must be quadratic in $(x_k, \mathbf{c}_k)$. Furthermore, the minimizing control law and maximizing disturbance are given by the linear feedback laws (3.40) in the absence of constraints and so $\check{J}(0,0) = 0$ must be the minimum value of $\check{J}(x, \mathbf{c})$ over all x and $\mathbf{c}$. Therefore the cost must have the form $\check{J}(x_k, \mathbf{c}_k) = \|z_{0|k}\|^2_{\check{W}}$ for some matrix $\check{W}$, and hence

$$
\begin{aligned}
\|z_{0|k}\|^2_{\check{W}} &= \max_{\{w_{0|k}, w_{1|k}, \ldots\}} \sum_{i=0}^{\infty} \bigl(\|z_{i|k}\|^2_{\hat{Q}} - \gamma^2 \|w_{i|k}\|^2\bigr) \\
&= \max_{w_{0|k}} \Bigl\{ \|z_{0|k}\|^2_{\hat{Q}} - \gamma^2 \|w_{0|k}\|^2 + \max_{\{w_{1|k}, w_{2|k}, \ldots\}} \sum_{i=1}^{\infty} \bigl(\|z_{i|k}\|^2_{\hat{Q}} - \gamma^2 \|w_{i|k}\|^2\bigr) \Bigr\} \\
&= \|z_{0|k}\|^2_{\hat{Q}} + \max_{w_{0|k}} \bigl\{ \|\Psi z_{0|k} + \bar{D} w_{0|k}\|^2_{\check{W}} - \gamma^2 \|w_{0|k}\|^2 \bigr\} \qquad (3.46)
\end{aligned}
$$

with $\hat{Q}$ defined as in (3.30b). The maximizing disturbance is therefore $w_{0|k} = (\gamma^2 I - \bar{D}^T \check{W} \bar{D})^{-1} \bar{D}^T \check{W} \Psi z_{0|k}$, so that $\|z_{0|k}\|^2_{\check{W}} = z_{0|k}^T (\hat{Q} + \Psi^T \check{W}' \Psi) z_{0|k}$ where

$$
\check{W}' = \check{W} + \check{W} \bar{D} (\gamma^2 I - \bar{D}^T \check{W} \bar{D})^{-1} \bar{D}^T \check{W}.
$$

Invoking (3.46) for all $z_{0|k}$ then gives

$$
\check{W} = \Psi^T \check{W}' \Psi + \hat{Q},
$$

and the block-diagonal form of $\check{W}$ together with the expressions for its diagonal blocks in (3.45a, 3.45b) then follow from the definition of Ψ in terms of the unconstrained optimal feedback gain K. □

The cost of Lemma 3.3 can be used to form the objective of a min-max RMPC algorithm based on an open-loop optimization strategy. As in Algorithm 3.1, we use the set $\mathcal{Z}^{\mathrm{MRPI}}$ constructed in Sect. 3.2.1 to invoke constraints robustly and to guarantee recursive feasibility. Given the linearity of these constraints and the quadratic nature of the cost (3.44), the online optimization is again a quadratic programming problem with Nn_u variables and $n_C(\nu + 1)$ constraints.

Algorithm 3.2 At each time instant $k = 0, 1, \ldots$:

(i) Perform the optimization

$$
\underset{\mathbf{c}_k}{\text{minimize}} \ \|\mathbf{c}_k\|^2_{\check{W}_c} \ \text{ subject to } \bar{F} \Psi^i \begin{bmatrix} x_k \\ \mathbf{c}_k \end{bmatrix} \le \mathbf{1} - h_i, \ i = 0, \ldots, \nu \qquad (3.47)
$$

where ν satisfies the conditions of Theorem 3.1.

(ii) Apply the control law $u_k = K x_k + c^*_{0|k}$, where $\mathbf{c}^*_k = (c^*_{0|k}, \ldots, c^*_{N-1|k})$ is the optimal value of $\mathbf{c}_k$ for problem (3.47). ◁

The control theoretic properties of the nominal robust MPC law of Algorithm 3.1 apply to this min-max robust MPC strategy since the same method is used to construct the constraint set in each case, and because their respective cost matrices W_c and $\check{W}_c$ have the same structure. These properties can be summarized as follows.

Corollary 3.1 *(a) Recursive feasibility of the optimization (3.47) is ensured by robust invariance of $\mathcal{Z}^{\mathrm{MRPI}}$ under the dynamics $z_{0|k+1} = \Psi z_{0|k} + \bar{D} w_k$, $w_k \in \mathcal{W}$.*

(b) The bound (3.34) (with W_c replaced by $\check{W}_c$) holds along closed-loop trajectories of (3.1) under Algorithm 3.2 as a result of the block-diagonal structure of $\check{W}_c$.

(c) Lemma 3.2 applies to the closed-loop trajectories under Algorithm 3.2. since $\Phi = A + BK$ is by assumption strictly stable.

(d) From (a)–(c) it follows that the conclusions of Theorem 3.2 apply to Algorithm 3.2; thus the state of (3.1) converges asymptotically to the minimal RPI set $\mathcal{X}^{\mathrm{mRPI}}$ (3.23) associated with the control law that is defined by the solution of the Riccati equation (3.42).

In addition to these properties, the closed-loop system has a disturbance l_2 gain that is bounded from above by γ, as we show next.

Theorem 3.3 *For $x_0 \in \mathcal{F}_N$ and any nonnegative integer n, the control law of Algorithm 3.2 guarantees that the closed-loop trajectories of (3.1) satisfy*

$$\sum_{k=0}^{n} \big(\|x_k\|_Q^2 + \|u_k\|_R^2\big) \le \|x_0\|_{\check{W}_x}^2 + \|\mathbf{c}_0^*\|_{\check{W}_c}^2 + \gamma^2 \sum_{k=0}^{n} \|w_k\|^2. \tag{3.48}$$

Proof The effect of the actual disturbance at time k on the optimal value of the cost can be no greater than the worst case value predicted at time k:

$$\begin{aligned}
\check{J}(x_k, \mathbf{c}_k^*) &= \max_{\{w_{i|k}, w_{i+1|k}, \ldots\}} \sum_{i=0}^{\infty} \big(\|x_{i|k}\|_Q^2 + \|u_{i|k}\|_R^2 - \gamma^2 \|w_{i|k}\|^2\big) \\
&\ge \|x_k\|_Q^2 + \|u_k\|_R^2 - \gamma^2 \|w_k\|^2 \\
&\quad + \max_{\{w_{i+1|k}, w_{i+2|k}, \ldots\}} \sum_{i=0}^{\infty} \Big(\|x_{i|k+1}\|_Q^2 + \|u_{i|k+1}\|_R^2 - \gamma^2 \|w_{i+1|k}\|^2\Big)
\end{aligned}$$

where $u_{i|k+1} = K x_{i|k+1} + c_{i+1|k}^*$ for all $i = 0, 1, \ldots$ Hence

$$\check{J}(x_k, \mathbf{c}_k^*) \ge \|x_k\|_Q^2 + \|u_k\|_R^2 - \gamma^2 \|w_k\|^2 + \check{J}(x_{k+1}, M\mathbf{c}_k^*),$$

and since $(x_{k+1}, M\mathbf{c}_k^*)$ is feasible for (3.47), the minimization (3.47) at $k+1$ ensures that $\check{J}(x_{k+1}, M\mathbf{c}_k^*) \ge \check{J}(x_{k+1}, \mathbf{c}_{k+1}^*)$. For all k we therefore obtain

$$\check{J}_k^* \geq \|x_k\|_Q^2 + \|u_k\|_R^2 - \gamma^2\|w_k\|^2 + \check{J}_{k+1}^*$$

where $\check{J}_k^* \doteq \check{J}(x_k, \mathbf{c}_k^*)$. Summing this inequality over $k = 0, 1, \ldots, n$ yields the bound (3.48) since $\check{W}_x$ and $\check{W}_c$ are positive-definite matrices. □

In essence (3.48) defines an achievable upper bound on the induced $\mathcal{H}_\infty$ norm of the response of the output $y = (Q^{1/2}x, R^{1/2}u)$ to additive disturbances. This aspect will be revisited in a more general setting in Chap. 5, where consideration is given to the case in which both additive and multiplicative uncertainty are present in the model.

Example 3.2 Applying the min-max approach of Algorithm 3.2 to the supply chain model of Example 3.1 we find that, for cost weights $Q = I$ and $R = 0.01$, the minimum achievable disturbance l_2 gain is $\gamma^2 = 8.15$. Setting $\gamma^2 = 10$ in (3.39) results in the optimal unconstrained feedback gain $K = [-1.27\ -1.55]$. This places the eigenvalues of Φ at 0.22 and 0.005, which indicates that the auxiliary control law $u = Kx$ is more aggressive than its counterpart for the nominal cost in Example 3.1. As a result, for $r = 6$, the minimum value of ρ satisfying (3.27) is considerably smaller at $\rho = 1.3 \times 10^{-3}$, and the minimal RPI set and its outer approximation are also smaller (Fig. 3.7). The areas of $\mathcal{X}^{\text{mRPI}}$ and $\hat{X}^{\text{mRPI}}$ are 0.0594 and 0.0595, respectively; for comparison the areas for Example 3.1 are 0.073 and 0.096.

The more aggressive feedback gain K for the min-max approach is more likely to conflict with the input constraints of this example, and this is reflected in the maximum element of $\hat{h}_\infty$ being closer to unity (here $\hat{h}_\infty = (0.67, 0.29, 0.29, 0.29, 0.29, 0.61, 0.91)$). As a consequence, the feasible set given by the projection, $\mathcal{F}_N$, of the maximal RPI set $\mathcal{Z}^{\text{MRPI}}$ onto the x-subspace is smaller than the feasible set for the same horizon N in Example 3.1. This can be seen by comparing Figs. 3.6 and 3.7. Figure 3.7 also indicates that a horizon of $N = 18$ is required in order to achieve the maximal feasible set $\mathcal{F}_\infty$, whereas the maximal feasible set (which for this example is identical for the nominal and min-max approaches) is obtained with $N = 4$ in Example 3.1. ◊

To account for the disturbance constraints (3.3) in the definition of the MPC performance index, we replace the worst-case cost (3.39) with

$$\check{J}(x_0, \{u_0, u_1, \ldots\}) = \max_{\substack{w_i \in \mathcal{W} \\ i=0,\ldots,N-1}} \sum_{i=0}^{N-1} \left(\|x_i\|_Q^2 + \|u_i\|_R^2 - \gamma^2\|w_i\|^2\right) + \|x_N\|_{\check{W}_x}^2. \tag{3.49}$$

Given the definition of $\check{W}_x$ in (3.42), the cost (3.49) is equivalent to an infinite horizon worst-case cost in which the maximization is performed subject to $w_i \in \mathcal{W}$ for $i = 0, \ldots, N-1$ and without constraints on w_i for $i \geq N$. With the definitions

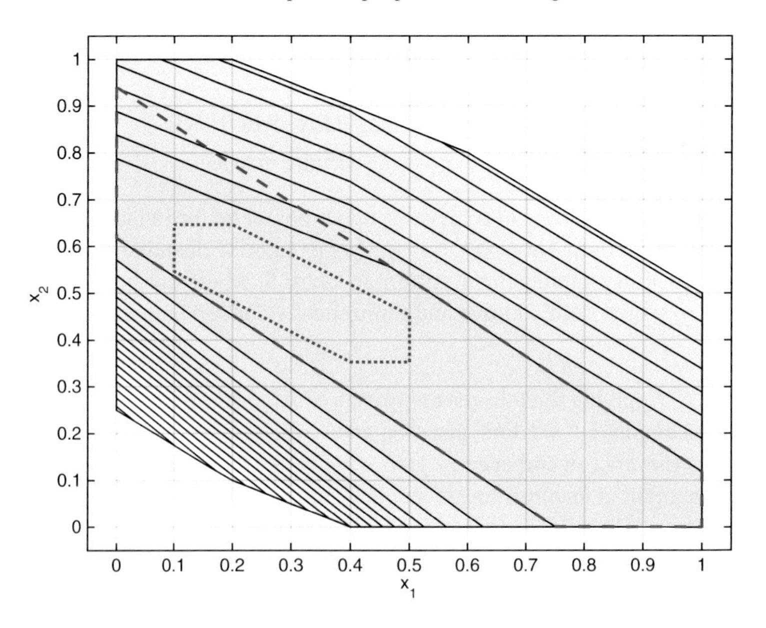

Fig. 3.7 The feasible initial condition sets $\mathcal{F}_N$, $N = 0, 1, \ldots, 18$, for Example 3.2; also shown are the sets $\{x : (F + GK)(x - x^0) \leq \mathbf{1}\}$ (*dashed line*) and $\hat{\mathcal{X}}^{\text{mRPI}} \oplus \{x^0\}$ (*dotted line*)

$$C_{xx} = \begin{bmatrix} I \\ \Phi \\ \vdots \\ \Phi^{N-1} \\ \Phi^N \end{bmatrix}, \ C_{xc} = \begin{bmatrix} 0 & \cdots & 0 & 0 \\ B & \cdots & 0 & 0 \\ \vdots & \ddots & \vdots & \vdots \\ \Phi^{N-2}B & \cdots & B & 0 \\ \Phi^{N-1}B & \cdots & \Phi B & B \end{bmatrix}, \ C_{xw} = \begin{bmatrix} 0 & \cdots & 0 & 0 \\ D & \cdots & 0 & 0 \\ \vdots & \ddots & \vdots & \vdots \\ \Phi^{N-2}D & \cdots & D & 0 \\ \Phi^{N-1}D & \cdots & \Phi D & D \end{bmatrix},$$

$$C_{ux} = \begin{bmatrix} K & & 0 \\ & \ddots & \vdots \\ & K & 0 \end{bmatrix}, \ \bar{Q} = \begin{bmatrix} Q & & & \\ & \ddots & & \\ & & Q & \\ & & & \check{W}_x \end{bmatrix}, \ \bar{R} = \begin{bmatrix} R & & \\ & \ddots & \\ & & R \end{bmatrix}, \ \bar{V} = \begin{bmatrix} V & & \\ & \ddots & \\ & & V \end{bmatrix},$$

the sequences of predicted states $\mathbf{x}_k = (x_{0|k}, \ldots, x_{N|k})$ and inputs $\mathbf{u}_k = (u_{0|k}, \ldots, u_{N-1|k})$ for the open-loop strategy (3.6) can be written explicitly in terms of x_k, $\mathbf{c}_k$ and the disturbance sequence $\mathbf{w}_k = (w_{0|k}, \ldots, w_{N-1|k})$ as

$$\begin{aligned} \mathbf{x}_k &= C_{xx}x_k + C_{xc}\mathbf{c}_k + C_{xw}\mathbf{w}_k, \\ \mathbf{u}_k &= C_{ux}\mathbf{x}_k + \mathbf{c}_k, \end{aligned}$$

and the cost (3.49) for the open-loop strategy (3.6) at time k can be expressed

$$\check{J}(x_k, \mathbf{c}_k) = \max_{\mathbf{w}_k \in \{\mathbf{w}:\, \bar{V}\mathbf{w} \le \mathbf{1}\}} \Big\{ \|C_{xx}x_k + C_{xc}\mathbf{c}_k + C_{xw}\mathbf{w}_k\|_{\bar{Q}}^2 \\ + \|C_{ux}(C_{xx}x_k + C_{xc}\mathbf{c}_k + C_{xw}\mathbf{w}_k) + \mathbf{c}\|_{\bar{R}}^2 - \gamma^2\|\mathbf{w}_k\|^2 \Big\}. \tag{3.50}$$

This maximization problem is a quadratic program, and it is convex if the matrix

$$\Delta \doteq \gamma^2 I - C_{xw}^T(\bar{Q} + C_{ux}^T\bar{R}C_{ux})C_{xw} \tag{3.51}$$

is positive-definite. Assuming that $\Delta \succ 0$, a more convenient but equivalent minimization can be derived from (3.50) using convex programming duality. This is based on the fact that, for $G \succ 0$, the optimal value of the QP:

$$\max_{x \in \{x:\, Ax \le b\}} g^T x - \tfrac{1}{2}x^T G x \tag{3.52}$$

is equal to the optimal value of the dual problem defined by the QP:

$$\min_{\lambda \in \{\lambda:\, \lambda \ge 0\}} b^T\lambda + \tfrac{1}{2}(g - A^T\lambda)^T G^{-1}(g - A^T\lambda), \tag{3.53}$$

(see e.g. [18] for a proof of this result).

Lemma 3.4 *For $\Delta \succ 0$, the worst-case cost (3.49) for the open-loop strategy (3.6) is equal to*

$$\check{J}(x_k, \mathbf{c}_k) = \min_{\mu \in \{\mu:\, \mu \ge 0\}} \begin{bmatrix} x_k \\ \mathbf{c}_k \end{bmatrix}^T \begin{bmatrix} \check{W}_x & 0 \\ 0 & \check{W}_c \end{bmatrix} \begin{bmatrix} x_k \\ \mathbf{c}_k \end{bmatrix} - 2\mu^T \check{W}_{\mu z} \begin{bmatrix} x_k \\ \mathbf{c}_k \end{bmatrix} \\ + 2\mu^T\mathbf{1} + \mu^T \check{W}_{\mu\mu}\mu \tag{3.54}$$

where $\check{W}_x$ satisfies the Riccati equation (3.42), $\check{W}_c$ is defined in (3.45), and

$$\check{W}_{\mu z} = \bar{V}\Delta^{-1}C_{xw}^T\Big((\bar{Q} + C_{ux}^T\bar{R}C_{ux})\big[C_{xx}\ C_{xc}\big] + C_{ux}^T\bar{R}\big[0\ I\big]\Big)$$
$$\check{W}_{\mu\mu} = \bar{V}\Delta^{-1}\bar{V}^T.$$

Proof This follows from (3.50) and the equivalence of (3.52) and (3.53). □

To use the worst-case cost (3.54) as the objective of the MPC optimization, we replace the optimization (3.47) in step (i) of Algorithm 3.2 by the following problem in $N(n_u + n_V)$ variables and $n_C(\nu + 1) + Nn_V$ constraints.

$$\begin{aligned} \underset{\mathbf{c}_k,\, \mu}{\text{minimize}} \quad & \begin{bmatrix} x_k \\ \mathbf{c}_k \end{bmatrix}^T \begin{bmatrix} \check{W}_x & 0 \\ 0 & \check{W}_c \end{bmatrix} \begin{bmatrix} x_k \\ \mathbf{c}_k \end{bmatrix} - 2\mu^T \check{W}_{\mu z} \begin{bmatrix} x_k \\ \mathbf{c}_k \end{bmatrix} + 2\mu^T\mathbf{1} + \mu^T \check{W}_{\mu\mu}\mu \\ \text{subject to} \quad & \bar{F}\Psi^i \begin{bmatrix} x_k \\ \mathbf{c}_k \end{bmatrix} \le \mathbf{1} - h_i,\ i = 0, \ldots, \nu \\ & \mu \ge 0 \end{aligned} \tag{3.55}$$

With this modification the online MPC optimization remains a convex quadratic program; note however that it involves a larger number of variables and constraints than the online optimization (3.47).

The presence of disturbance constraints implies that (3.54) gives a tighter bound on the worst-case performance of the MPC algorithm in closed loop operation than the cost of (3.44); hence the optimization (3.55) is likely to result in improved worst-case performance of the MPC law. However, although the guarantee of recursive feasibility is not affected, the stability and convergence results in (b)–(d) of Corollary 3.1 no longer apply when (3.47) is replaced by (3.55) in Algorithm 3.2. This is to be expected of course, since the presence of the disturbance constraints (3.3) in the definition of the worst-case cost (3.49) implies that $u = Kx$ is not necessarily optimal for this cost, even when the constraints on x and u in (3.4) are inactive. Therefore, in the general case of persistent disturbances, the MPC law will not necessarily converge asymptotically to this linear feedback law.

On the other hand a bound on the disturbance l_2 gain similar to (3.48) does hold for the closed-loop system when (3.55) replaces the optimization in step (i) of Algorithm 3.2. This can be shown by an extension of the argument that was used in the proof of Theorem 3.3.

Theorem 3.4 *If $x_0 \in \mathcal{F}_N$ and $\Delta \succ 0$, Algorithm 3.2 with (3.55) in place of (3.47) satisfies, for all $n \geq 0$, the bound:*

$$\sum_{k=0}^{n}\big(\|x_k\|_Q^2 + \|u_k\|_R^2\big) \leq \check{J}(x_0, \mathbf{c}_0^*) + \gamma^2 \sum_{k=0}^{n} \|w_k\|^2. \tag{3.56}$$

Proof From Lemma 3.4 the optimal value of the objective in (3.55) is equal to

$$\check{J}(x_k, \mathbf{c}_k^*) = \max_{\substack{w_{i|k}\in\mathcal{W}\\ i=0,\ldots,N-1}} \sum_{i=0}^{N-1}\big(\|x_{i|k}\|_Q^2 + \|u_{i|k}\|_R^2 - \gamma^2\|w_{i|k}\|^2\big) + \|x_{N|k}\|_{\check{W}_x}^2,$$

but $\|x_{N|k}\|_{\check{W}_x}^2 = \max_{w_{N|k}}(\|x_{N|k}\|_Q^2 + \|u_{N|k}\|_R^2 - \gamma^2\|w_{N|k}\|^2 + \|x_{N+1|k}\|_{\check{W}_x}^2)$ so that

$$\begin{aligned}\check{J}(x_k, \mathbf{c}_k^*) \geq{}& \max_{\substack{w_{i|k}\in\mathcal{W}\\ i=0,\ldots,N}} \sum_{i=0}^{N}\big(\|x_{i|k}\|_Q^2 + \|u_{i|k}\|_R^2 - \gamma^2\|w_{i|k}\|^2\big) + \|x_{N+1|k}\|_{\check{W}_x}^2\\ \geq{}& \|x_k\|_Q^2 + \|u_k\|_R^2 - \gamma^2\|w_k\|^2\\ &+ \max_{\substack{w_{i|k}\in\mathcal{W}\\ i=1,\ldots,N}} \sum_{i=0}^{N-1}\big(\|x_{i|k+1}\|_Q^2 + \|u_{i|k+1}\|_R^2 - \gamma^2\|w_{i|k+1}\|^2\big) + \|x_{N|k+1}\|_{\check{W}_x}^2\end{aligned}$$

where $u_{i|k+1} = Kx_{i|k+1} + \mathbf{c}^*_{i+1|k}$ for all $i = 0, 1, \ldots$ It follows that

$$\begin{aligned}\check{J}(x_k, \mathbf{c}^*_k) &\geq \|x_k\|^2_Q + \|u_k\|^2_R - \gamma^2\|w_k\|^2 + \check{J}(x_{k+1}, M\mathbf{c}^*_k) \\ &\geq \|x_k\|^2_Q + \|u_k\|^2_R - \gamma^2\|w_k\|^2 + \check{J}(x_{k+1}, \mathbf{c}^*_{k+1}),\end{aligned}$$

and hence

$$\sum_{k=0}^{n}\left(\|x_k\|^2_Q + \|u_k\|^2_R\right) \leq \check{J}(x_0, \mathbf{c}^*_0) - \check{J}(x_{n+1}, \mathbf{c}^*_{n+1}) + \gamma^2\sum_{k=0}^{n}\|w_k\|^2.$$

which implies the bound (3.56) since $\check{J}(x, \mathbf{c}) \geq 0$ for all $(x, \mathbf{c})$. The conclusion here that $\check{J}(x, \mathbf{c})$ is nonnegative follows from the convexity of $\check{J}(x, \mathbf{c})$ and from the fact that the optimal unconstrained feedback laws given by (3.40) are feasible for sufficiently small x; thus $\check{J}(0, 0) = 0$ is the global minimum of $\check{J}(x, \mathbf{c})$. Note also that $\check{J}(x, \mathbf{c})$ is necessarily convex whenever $\Delta \succ 0$ because the expression maximized on the RHS of (3.50) is convex in $(x, \mathbf{c})$ for any given $\mathbf{w}$, and since the pointwise maximum of convex functions is convex [19]. □

A numerical example comparing the minmax MPC strategies defined by the alternative online optimizations of (3.47) and (3.55) is provided in Question 7 on p. 116.

3.5 Rigid and Homothetic Tubes

The focus of this chapter has so far been on robust MPC laws with constraints derived from the decomposition (3.8) with initial conditions $s_{0|k} = x_k$ and $e_{0|k} = 0$ for the nominal and uncertain components of the predicted state. In this section, we consider alternative definitions of the nominal state and uncertainty tube that can provide alternative, potentially stronger, stability guarantees. The stability properties of the nominal cost and game theoretic approaches of Sects. 3.3 and 3.4 were stated in terms of a finite l_2 gain from the disturbance input to the state and control input in closed-loop operation, as well as a guarantee of asymptotic convergence of the system state to the mRPI set for the unconstrained optimal feedback law. Instead, by relaxing the requirement that $s_{0|k} = x_k$ and $e_{0|k} = 0$, the tube MPC approaches of this section, which are based on [6, 20], ensure exponential stability of an outer approximation of the mRPI set.

The guarantee of exponential stability of a given limit set for the closed-loop system state comes at a price. This is because the initial condition $\mathcal{E}_{0|k} = \{0\}$ in the uncertainty tube dynamics (3.19) results in an uncertainty tube $\{\mathcal{E}_{0|k}, \mathcal{E}_{1|k}, \ldots\}$ which is minimal in the sense that $\mathcal{E}_{i|k}$ is the smallest set that contains the uncertain component of the predicted state given the disturbance bounds. Consequently, if $e_{0|k} \neq 0$, so that the initial set $\mathcal{E}_{0|k}$ contains more points than just the origin, then the amount by which the constraints on the nominal predicted trajectories must be tightened in order to ensure that constraints are satisfied for all realizations of uncertainty will be overestimated. This leads to smaller sets of feasible initial conditions than are obtained using the methods of Sects. 3.3 and 3.4.

However the flexibility afforded by allowing non-zero initial conditions for the uncertainty tube enables the algorithms of this section to reduce the potential conservativeness of their constraint handling. This possibility is developed here by combining different linear feedback laws in the definition of the predicted input trajectories. For a robust MPC algorithm employing the open-loop strategy (3.6), the size of the feasible set of states depends on the sizes of the MPI set for the nominal model and the mRPI set in the presence of disturbances, both of which depend on the feedback gain K. For a small mRPI set we require good disturbance rejection, whereas a large MPI set for the nominal dynamics requires good tracking performance in the presence of constraints. These can be conflicting requirements, thus motivating the consideration of closed-loop strategies for which the uncertainty tube $\{\mathcal{E}_{0|k}, \mathcal{E}_{1|k}, \ldots\}$ may depend explicitly on the future state and constraints. Within the framework of open-loop strategies however, this issue can be addressed by incorporating different linear feedback gains in the nominal and uncertain components of the dynamics (3.8).

To this end, let $x_{i|k} = s_{i|k} + e_{i|k}$, and

$$u_{i|k} = Ks_{i|k} + K_e e_{i|k} + c_{i|k}, \tag{3.57a}$$

$$s_{i+1|k} = \Phi s_{i|k} + Bc_{i|k}, \tag{3.57b}$$

$$e_{i+1|k} = \Phi_e e_{i|k} + Dw_{i|k}, \quad w_{i|k} \in \mathcal{W} \tag{3.57c}$$

where $\Phi = A + BK$ and $\Phi_e = A + BK_e$. The freedom to choose different gains K_e and K can allow for improved disturbance rejection without adversely affecting the MPI set for the nominal state. However, in this framework, the initial conditions $e_{0|k} = 0$ and $s_{0|k} = x_k$ cannot be assumed because there is no single value of $c_{0|k+1}$ that makes $u_{1|k} = Ks_{1|k} + K_e e_{1|k} + c_{1|k}$ equal to $u_{0|k+1} = Kx_{k+1} + c_{0|k+1}$ for all $w_k \in \mathcal{W}$. Therefore the method described in Sect. 3.2.1 for ensuring recursive feasibility of an open-loop strategy would fail with this parameterization. However, if $\mathcal{E}_{0|k}$ is permitted to contain more points than just the origin, then it is possible to construct a feasible but suboptimal trajectory at time $k + 1$ by choosing $s_{i|k+1} = s_{i+1|k}$, $e_{i|k+1} = e_{i+1|k}$ and $c_{i|k+1} = c_{i+1|k}$ for $i = 0, 1, \ldots$ This is the case for tube MPC algorithms considered in this section that allow non-singleton initial uncertainty sets. Note that in this context, and throughout the current section, the term "nominal state" does not have the usual meaning (namely the state of the disturbance-free model) because $s_{0|k}$ is not necessarily chosen to coincide with the current state $x_{0|k} = x_k$.

3.5.1 Rigid Tube MPC

The convenience of the decomposition (3.57) is that, if $s_{0|k}$ is chosen so that $e_{0|k} = x_{0|k} - s_{0|k}$ belongs to a set $\mathcal{S}$ that is RPI for (3.57c), namely if

$$\Phi_e \mathcal{S} \oplus D\mathcal{W} \subseteq \mathcal{S}, \tag{3.58}$$

then $e_{i|k}$ must also lie in $\mathcal{S}$ for all $i = 1, 2, \ldots$ It is assumed that $\mathcal{S}$ is compact, convex and polytopic, and is described by $n_{\mathcal{S}}$ linear inequalities:

$$\mathcal{S} = \{e : V_{\mathcal{S}} e \leq 1\}.$$

Under these circumstances the predicted trajectory $\{e_{0|k}, e_{1|k}, \ldots\}$ lies in a tube of fixed cross section $\mathcal{S}$. Such tubes were used in [6]; we refer to them as *rigid* tubes to distinguish them from *homothetic* tubes [20] which allow for variable scaling of the tube cross sections.

This strategy simplifies the problem of ensuring robust satisfaction of the constraints (3.4). Applying these constraints to the predicted trajectories of (3.57) requires that

$$\bar{F}\Psi^i z_{0|k} \leq \mathbf{1} - h_{\mathcal{S}}, \quad i = 0, 1, \ldots \tag{3.59}$$

where $z_{0|k} = (s_{0|k}, \mathbf{c}_k)$ is the initial state of the autonomous dynamics (3.9) and the vector $h_{\mathcal{S}}$ can be computed offline by solving a set of linear programs:

$$h_{\mathcal{S}} = \max_{e \in \mathcal{S}} (F + GK_e)e.$$

The RHS of each constraint in (3.59) is independent of the time index i. By Theorem 2.3 therefore, the infinite sequence of inequalities in (3.59) can be reduced to an equivalent constraint set described by a finite number of inequalities: $z_{0|k} \in \mathcal{Z}^{\text{MPI}}$. Under the necessary assumption that $h_{\mathcal{S}} < \mathbf{1}$, $\mathcal{Z}^{\text{MPI}}$ is defined by

$$\mathcal{Z}^{\text{MPI}} \doteq \{z : \bar{F}\Psi^i z \leq \mathbf{1} - h_{\mathcal{S}}, \ i = 0, \ldots, \nu\},$$

where ν is the (necessarily finite) integer that satisfies the conditions of Theorem 2.3 with the RHS of (2.15) replaced by $\mathbf{1} - h_{\mathcal{S}}$.

Using rigid tubes to represent the evolution of the uncertain component of the predicted state of (3.1) causes the uncertainty in predicted trajectories to be overestimated, especially in the early part of the prediction horizon. This results in smaller feasible sets of initial conditions than the exact approach described in Sect. 3.2.1. Clearly $\mathcal{S}$ can be no smaller than the minimal RPI set for (3.57c), and to reduce the degree of conservatism it is therefore advantageous to define $\mathcal{S}$ as a close approximation of the minimal RPI. For example

$$\mathcal{S} \doteq \frac{1}{1-\rho} \bigoplus_{j=0}^{r-1} \Phi_e^j D\mathcal{W} \tag{3.60}$$

where r and $\rho \in [0, 1)$ satisfy $\Phi_e^r D\mathcal{W} \subseteq \rho D\mathcal{W}$. The following lemma shows that this choice of $\mathcal{S}$ meets the condition (3.58) for robust positive invariance.

Lemma 3.5 *The set $\mathcal{S}$ in (3.60) satisfies (3.58) and hence is RPI for (3.57c).*

Proof Under the assumption that $\Phi_e^r D\mathcal{W} \subseteq \rho D\mathcal{W}$, we obtain

$$
\begin{aligned}
\Phi_e \mathcal{S} \oplus D\mathcal{W} &= \frac{1}{1-\rho} \bigoplus_{j=1}^{r-1} \Phi_e^j D\mathcal{W} \oplus \frac{1}{1-\rho} \Phi_e^r D\mathcal{W} \oplus D\mathcal{W} \\
&\subseteq \frac{1}{1-\rho} \bigoplus_{j=1}^{r-1} \Phi_e^j D\mathcal{W} \oplus \frac{\rho}{1-\rho} D\mathcal{W} \oplus D\mathcal{W} \\
&= \frac{1}{1-\rho} \bigoplus_{j=0}^{r-1} \Phi_e^j D\mathcal{W} = \mathcal{S}.
\end{aligned}
$$

□

The final ingredient needed for the definition of a robust MPC algorithm is the choice of the cost function. This is taken to be the nominal cost of (3.29), with $s_{i|k}$ defined as the nominal predicted trajectory generated by (3.57b) and with $v_{i|k} = K s_{i|k} + c_{i|k}$ for all i. Therefore, we set $J(s_{0|k}, \mathbf{c}_k) \doteq \|z_{0|k}\|_W^2$ where W is the solution of the Lyapunov equation (2.34), and by Theorem 2.10 we obtain

$$J(s_{0|k}, \mathbf{c}_k) = \|s_{0|k}\|_{W_x}^2 + \|\mathbf{c}_k\|_{W_c}^2.$$

Unlike the nominal cost of Sect. 3.3, the initial condition, $s_{0|k}$, from which this cost is computed is not necessarily equal to the plant state x_k, but is instead treated as an optimization variable subject to the constraint that $e_{0|k} = x_k - s_{0|k} \in \mathcal{S}$. This choice of cost is justified by the argument that the primary objective of the MPC law is to steer the plant state into or close to the mRPI set associated with the linear feedback law $u = K_e x$ (which by assumption has been designed to provide good disturbance rejection), and that x_k necessarily converges to the mRPI approximation $\mathcal{S}$ if $s_{0|k}$ converges to zero. Since the aim is to enforce convergence of $s_{0|k}$, the cost weights Q and R in (3.29) are taken to be positive-definite matrices [6, 20].

The resulting online optimization problem is a quadratic program in $Nn_u + n_x$ variables and $n_C(\nu + 1) + n_{\mathcal{S}}$ constraints.

Algorithm 3.3 At each time instant $k = 0, 1, \ldots$:

(i) Perform the optimization

$$
\begin{aligned}
&\underset{s_{0|k}, \mathbf{c}_k}{\text{minimize}} \quad \|s_{0|k}\|_{W_x}^2 + \|\mathbf{c}_k\|_{W_c}^2 \\
&\text{subject to} \quad \bar{F}\Psi^i \begin{bmatrix} s_{0|k} \\ \mathbf{c}_k \end{bmatrix} \leq \mathbf{1} - h_{\mathcal{S}}, \ i = 0, \ldots, \nu \\
&\qquad\qquad\quad x_k - s_{0|k} \in \mathcal{S}
\end{aligned}
\tag{3.61}
$$

where ν satisfies the conditions of Theorem 2.3 with the RHS of (2.15) replaced by $\mathbf{1} - h_{\mathcal{S}}$.

(ii) Apply the control law $u_k = K s^*_{0|k} + K_e(x_k - s^*_{0|k}) + c^*_{0|k}$, where $(s^*_{0|k}, \mathbf{c}^*_k)$ is the optimizer of problem (3.61), and $\mathbf{c}^*_k = (c^*_{0|k}, \ldots, c^*_{N-1|k})$. ◁

To determine the feasible set for the state x_k in (3.61), let $\mathcal{F}^s_N$ denote the feasible set for $s_{0|k}$ in the optimization (3.61):

$$\mathcal{F}^s_N \doteq \left\{ s : \exists \mathbf{c} \text{ such that } \bar{F}\Psi^i \begin{bmatrix} s \\ \mathbf{c} \end{bmatrix} \leq \mathbf{1} - h_{\mathcal{S}},\ i = 0, \ldots, \nu \right\}.$$

Then, given that any feasible state x_k must satisfy $x_k = s_{0|k} + e_{0|k}$ for some $e_{0|k} \in \mathcal{S}$, the set of feasible initial conditions for Algorithm 3.3 can be expressed as

$$\mathcal{F}_N = \mathcal{F}^s_N \oplus \mathcal{S}.$$

Theorem 3.5 *For the system (3.1) with the control law of Algorithm 3.3, the feasible set $\mathcal{F}_N$ is RPI, and $\mathcal{S}$ is exponentially stable with region of attraction equal to $\mathcal{F}_N$ if $Q \succ 0$ and $R \succ 0$ in (3.29).*

Proof To demonstrate that $\mathcal{F}_N$ is RPI, suppose that $x_k \in \mathcal{F}_N$ so that $(s^*_{0|k}, \mathbf{c}^*_k) \in \mathcal{Z}^{\text{MPI}}$ and $x_k - s^*_{0|k} \in \mathcal{S}$. Let $s_{0|k+1} = \Phi s^*_{0|k} + BE\mathbf{c}^*_k$ and $\mathbf{c}_{k+1} = M\mathbf{c}^*_k$, then since $\mathcal{Z}^{\text{MPI}}$ is an invariant set for the autonomous dynamics (3.9) and $\mathcal{S}$ is RPI for (3.57) it follows that $(s_{0|k+1}, \mathbf{c}_{k+1}) \in \mathcal{Z}^{\text{MPI}}$ and $x_{k+1} - s_{0|k+1} = \Phi_e(x_k - s_{0|k}) + Dw_k \in \mathcal{S}$ for any disturbance $w_k \in \mathcal{W}$; therefore $(s_{0|k+1}, \mathbf{c}_{k+1})$ is feasible for (3.61), implying $x_{k+1} \in \mathcal{F}_N$.

The exponential stability of $\mathcal{S}$ follows from the definition (3.29) of the cost in (3.61), which implies the bound

$$J(s^*_{0|k+1}, \mathbf{c}^*_{k+1}) \leq J(s^*_{0|k}, \mathbf{c}^*_k) - (\|s_{0|k}\|^2_Q + \|K s_{0|k} + \mathbf{c}^*_k\|^2_R).$$

Therefore, by the same argument as was used the proof of Theorem 2.8, the closed-loop application of Algorithm 3.3 gives, for any $x_0 \in \mathcal{F}_N$ and all $k > 0$,

$$\|s^*_{0|k}\|^2 \leq \frac{b}{a}\left|1 - \frac{\underline{\lambda}(Q)}{b}\right|^k \|s^*_{0|0}\|^2,$$

for some constants $a, b > 0$ satisfying $a\|s_{0|k}\|^2 \leq J(s^*_{0|k}, \mathbf{c}^*_k) \leq b\|s_{0|k}\|^2$ for any $x_k \in \mathcal{F}_N$, and where $\underline{\lambda}(Q)$ is the smallest eigenvalue of Q. The constraint $s^*_{0|k} = x_k - e$ for $e \in \mathcal{S}$ implies that $\min_{e\in\mathcal{S}} \|x_k - e\| \leq \|s^*_{0|k}\|$; hence the distance of x_k from $\mathcal{S}$ is upper-bounded by an exponentially decaying function of time. □

Example 3.3 Considering again the supply chain model of Example 3.1, K and K_e are initially chosen to be equal, with $K = K_e = [-0.89\ -0.78]$, which is the unconstrained LQ-optimal feedback gain for the nominal cost with $Q = I$, $R = 0.01$.

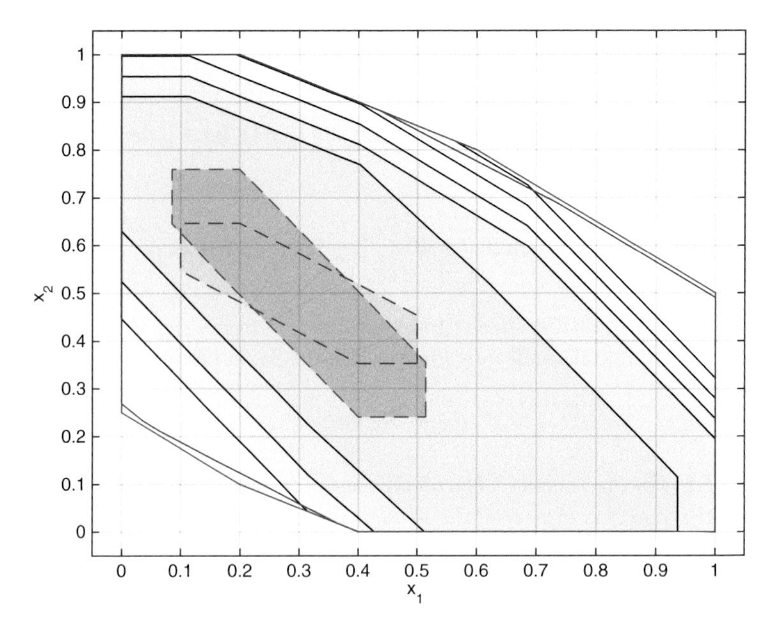

Fig. 3.8 The feasible initial condition sets $\mathcal{F}_N$, $N = 0, 1, 2, 3, 4$ for Algorithm 3.3 with $K_e = K$ chosen as the unconstrained LQ-optimal feedback gain (*black lines*). Also shown are the maximal feasible set for Algorithm 3.1 (*solid red line*) and the maximal feasible set for Algorithm 3.3 with K_e equal to the unconstrained min-max optimal feedback gain of Example 3.2 (*solid blue line*), and the mRPI sets for the unconstrained LQ-optimal and unconstrained min-max optimal feedback gains (*dotted red line* and *dotted blue line*, respectively)

The set $\mathcal{S}$ is taken to be the mRPI set approximation (3.60) for $r = 6$, $\rho = 0.127$, so that $h_\mathcal{S}$ is equal to the value of $\hat{h}_\infty$ in Example 3.1. The feasible initial condition sets, $\mathcal{F}_N$, $N = 0, 1, 2, 3, 4$, for Algorithm 3.3 are shown in Fig. 3.8; for these choices of K, K_e and $\mathcal{S}$, the largest possible feasible set for Algorithm 3.3 is obtained with $N = 4$. For comparison, the figure shows the maximal feasible set of states for Algorithm 3.1 (which is also obtained for $N = 4$). This contains and extends outside the feasible set for rigid tube MPC as a result of the conservative definition of the uncertainty tube in Algorithm 3.3. For this example, and for all values of N, Algorithm 3.1 has $6(N + 2)$ constraints whereas Algorithm 3.3 has $6(N + 2) + 8$ constraints.

Replacing $\mathcal{S}$ with a smaller mRPI set approximation reduces the degree of conservativeness of the constraints in Algorithm 3.3. In fact Fig. 3.8 shows that almost all of the discrepancy between the maximal feasible sets of Algorithms 3.1 and 3.3 disappears if $K_e = [-1.27\ -1.55]$ (which is the unconstrained min-max optimal feedback gain for $\gamma^2 = 10$) and, as before, $K = [-0.89\ -0.78]$. However, the maximal set is then obtained for a larger value of N; here $N = 20$ is needed to achieve the maximal feasible set. ♢

3.5.2 Homothetic Tube MPC

The rigid tube MPC described in Sect. 3.5.1 assumes that uncertainty in the error state e is uniform through the prediction horizon. This however is conservative given that e could initially be small (indeed $e_{0|k} = 0$ if $s_{0|k} = x_k$), and the set containing the uncertain component of the predicted future state only approaches the mRPI set asymptotically. Rather than tightening constraints on the nominal predicted trajectories by considering the worst case $e_{i|k} \in \mathcal{S}$ as is done in rigid tube MPC, it is therefore more reasonable to assume that

$$e_{i|k} \in \alpha_{i|k}\mathcal{S}^0 \tag{3.62}$$

for positive scalars $\alpha_{i|k}$, $i = 0, 1, \ldots$ that are taken to be variables in the online optimization, and where the set

$$\mathcal{S}^0 = \{e : V_{\mathcal{S}}^0 e \leq \mathbf{1}\} \tag{3.63}$$

is determined offline. This replaces the rigid tube $\{\mathcal{S}, \mathcal{S}, \ldots\}$ that is used in rigid tube MPC to bound the uncertainty tube $\{\mathcal{E}_{0|k}, \mathcal{E}_{1|k}, \ldots\}$ with an uncertainty tube given by $\{\alpha_{0|k}\mathcal{S}^0, \alpha_{1|k}\mathcal{S}^0, \ldots\}$. The sets $\alpha_{i|k}\mathcal{S}^0$ in this expression are homothetic to $\mathcal{S}^0$, and hence, the approach is known as homothetic tube MPC [20].

The presence of the scalar variables $\alpha_{i|k}$ implies that $\mathcal{S}^0$ (unlike $\mathcal{S}$) need not be RPI for the error dynamics (3.57c). Instead it is assumed that $\mathcal{S}^0$ is compact and satisfies the invariance condition $\Phi_e\mathcal{S}^0 \subseteq \mathcal{S}^0$. A convenient way to invoke the inclusion condition (3.62) is through a recursion relating $\alpha_{i+1|k}$ to $\alpha_{i|k}$ so as to ensure that $e_{i+1|k} \in \alpha_{i+1|k}\mathcal{S}^0$ whenever $e_{i|k} \in \alpha_{i|k}\mathcal{S}^0$. The required condition[1] can be expressed as

$$\Phi_e\alpha_{i|k}\mathcal{S}^0 \oplus D\mathcal{W} \subseteq \alpha_{i+1|k}\mathcal{S}^0, \tag{3.64}$$

or equivalently, given the representation (3.63), as

$$\alpha_{i|k} \max_{e\in\mathcal{S}^0} V_{\mathcal{S}}^0\Phi_e e + \max_{w\in\mathcal{W}} V_{\mathcal{S}}^0 D w \leq \alpha_{i+1|k}\mathbf{1}. \tag{3.65}$$

This condition is equivalent to

$$\alpha_{i|k}\bar{e} + \bar{w} \leq \alpha_{i+1|k}\mathbf{1} \tag{3.66}$$

[1] This is a simplified version of the more general inclusion condition that is considered in [20]: $\{\Phi s_{i|k} + Bc_{i|k}\} \oplus \Phi_e\alpha_{i|k}\mathcal{S}^0 \oplus D\mathcal{W} \subseteq \{s_{i+1|k}\} \oplus \alpha_{i+1|k}\mathcal{S}^0$.

where the vectors $\bar{e}$ and $\bar{w}$ can be computed offline by solving a pair of linear programs:

$$\bar{e} \doteq \max_{e} V_{\mathcal{S}}^{0} \Phi_e e \text{ subject to } V_{\mathcal{S}}^{0} e \leq \mathbf{1}$$

$$\bar{w} \doteq \max_{w} V_{\mathcal{S}}^{0} D w \text{ subject to } V w \leq \mathbf{1}.$$

Given that the condition (3.66) ensures that $e_{i|k} \in \alpha_{i|k}\mathcal{S}^0$ throughout the prediction horizon, the constraints (3.4) can be invoked as

$$\bar{F}\Psi^i z_{0|k} + \alpha_{i|k} h_{\mathcal{S}}^0 \leq \mathbf{1}, \quad i = 0, 1, \ldots \tag{3.67}$$

where $z_{0|k} = (s_{0|k}, \mathbf{c}_k)$ is the initial state of the autonomous dynamics (3.9) and where the vector $h_{\mathcal{S}}^0$ can be computed offline by solving a linear program:

$$h_{\mathcal{S}}^0 \doteq \max_{e} (F + GK_e) e \text{ subject to } V_{\mathcal{S}}^0 e \leq \mathbf{1}.$$

Hence, robust satisfaction of the constraints (3.4) by the predicted trajectories of (3.57) is ensured by conditions (3.66) and (3.67), which are linear in the variables $s_{0|k}$, $\mathbf{c}_k = (c_{0|k}, \ldots, c_{N-1|k})$ and $\alpha_{i|k}$, $i = 0, 1, \ldots$

To restrict the sequence $\{\alpha_{0|k},\ i = 0, 1, \ldots\}$ to a finite number of degrees of freedom, we invoke (3.66) for $i \geq N$ by the sufficient condition

$$\alpha_{i+1|k} = \lambda \alpha_{i|k} + \mu, \ i = N, N+1, \ldots \tag{3.68}$$

where $\lambda \doteq \|\bar{e}\|_\infty$ and $\mu \doteq \|\bar{w}\|_\infty$. For simplicity we assume here that the condition (3.68) is to be imposed after a horizon equal to N, but in general this could be replaced by any finite horizon. Under the necessary assumption that $\lambda < 1$ the dynamics of (3.68) are stable and converge to the limit

$$\bar{\alpha} = \frac{1}{1-\lambda}\mu$$

Thus, for $i \geq N$, $\alpha_{i|k}$ is given in terms of $\alpha_{N|k}$ by

$$\alpha_{i|k} = \lambda^{i-N}(\alpha_{N|k} - \bar{\alpha}) + \bar{\alpha},$$

and (3.66), (3.67) therefore constitute an infinite set of linear constraints in a finite number of variables: $s_{0|k}$, $\mathbf{c}_k = (c_{0|k}, \ldots, c_{N-1|k})$ and $\{\alpha_{0|k}, \ldots, \alpha_{N-1|k}\}$. These constraints are equivalent to a finite number of linear conditions by the following result, the proof of which is similar to the proof of Theorem 2.3.

Corollary 3.2 *Let ν be the smallest integer greater than or equal to N such that*

$$\bar{F}\Psi^{\nu+1} z + \big(\lambda^{\nu+1-N}(\alpha_N - \bar{\alpha}) + \bar{\alpha}\big) h_{\mathcal{S}}^0 \leq \mathbf{1}$$

for all z and $\{\alpha_0, \ldots, \alpha_N\}$ satisfying

$$\bar{F}\Psi^i z \leq \begin{cases} \mathbf{1} - \alpha_i h_{\mathcal{S}}^0, & i = 0, \ldots, N-1 \\ \mathbf{1} - \left(\lambda^{i-N}(\alpha_N - \bar{\alpha}) + \bar{\alpha}\right) h_{\mathcal{S}}^0, & i = N, \ldots, \nu \end{cases}$$

$$\alpha_i \bar{e} + \bar{w} \leq \alpha_{i+1}\mathbf{1}, \quad i = 0, \ldots, N-1$$

then (3.67) holds for all $i = 0, 1, \ldots$ Furthermore ν is necessarily finite if Ψ is strictly stable and $(\Psi, \bar{F})$ is observable.

The definition of an online predicted cost is needed before an algorithm can be stated. This is taken to be the same as the cost (3.29) employed by rigid tube MPC, but with the addition of terms that penalize the deviation of $\alpha_{i|k}$ from the asymptotic value $\bar{\alpha}$,

$$J(s_{0|k}, \mathbf{c}_k, \boldsymbol{\alpha}_k) = \|s_{0|k}\|_{W_x}^2 + \|\mathbf{c}_k\|_{W_c}^2 + \sum_{i=0}^{N-1} q_\alpha (\alpha_{i|k} - \bar{\alpha})^2 + p_\alpha (\alpha_{N|k} - \bar{\alpha})^2, \tag{3.69}$$

where $\boldsymbol{\alpha}_k = (\alpha_{0|k}, \ldots, \alpha_{N|k})$. In order to ensure the monotonic non-increasing property of the optimized cost, we assume that the weights $p_\alpha, q_\alpha > 0$ satisfy the condition

$$p_\alpha \geq (1 - \lambda^2)^{-1} q_\alpha. \tag{3.70}$$

This results in an online optimization consisting of a quadratic program in $Nn_u + n_x + N + 1$ variables and $n_C(\nu + 1) + N + n_{\mathcal{S}^0}$ constraints, where $n_{\mathcal{S}^0}$ is the number of rows of $V_{\mathcal{S}}^0$.

Algorithm 3.4 At each time instant $k = 0, 1, \ldots$:

(i) Perform the optimization

$$\begin{aligned} \underset{s_{0|k}, \mathbf{c}_k, \boldsymbol{\alpha}_k}{\text{minimize}} \quad & \|s_{0|k}\|_{W_x}^2 + \|\mathbf{c}_k\|_{W_c}^2 + \sum_{i=0}^{N-1} q_\alpha (\alpha_{i|k} - \bar{\alpha})^2 + p_\alpha (\alpha_{N|k} - \bar{\alpha})^2 \\ \text{subject to} \quad & \bar{F}\Psi^i \begin{bmatrix} s_{0|k} \\ \mathbf{c}_k \end{bmatrix} \leq \mathbf{1} - \alpha_{i|k} h_{\mathcal{S}}^0, \qquad i = 0, \ldots, \nu \\ & \alpha_{i|k}\bar{e} + \bar{w} \leq \alpha_{i+1|k}\mathbf{1}, \qquad i = 0, \ldots, N-1 \\ & \alpha_{i|k} = \lambda^{i-N}(\alpha_{N|k} - \bar{\alpha}) + \bar{\alpha} \quad i = N, \ldots, \nu \\ & x_k - s_{0|k} \in \alpha_{0|k}\mathcal{S}^0 \end{aligned} \tag{3.71}$$

where ν satisfies the conditions of Corollary 3.2.

(ii) Apply the control law $u_k = Ks_{0|k}^* + K_e(x_k - s_{0|k}^*) + c_{0|k}^*$, where $(s_{0|k}^*, \mathbf{c}_k^*, \boldsymbol{\alpha}_k^*)$ is the optimizer of (3.71), and $\mathbf{c}_k^* = (c_{0|k}^*, \ldots, c_{N-1|k}^*)$. ◁

Theorem 3.6 *For the system (3.1) and control law of Algorithm 3.4, the set $\mathcal{F}_N$ of feasible states x_k for (3.71) is RPI, and $\bar{\alpha}\mathcal{S}^0$ is exponentially stable with region of attraction equal to $\mathcal{F}_N$ if $Q, R \succ 0$ and $q_\alpha > 0$ in (3.69).*

Proof The recursive feasibility of the optimization (3.71) is demonstrated by the argument that was used to show recursive feasibility in the proof of Theorem 3.5. The exponential stability of $\bar{\alpha}\mathcal{S}^0$ can be shown using the feasible but suboptimal values for the optimization variables in (3.71) at time $k+1$ that are given by

$$s_{0|k+1} = \Phi s^*_{0|k} + BE\mathbf{c}^*_{0|k}, \quad \mathbf{c}_{k+1} = M\mathbf{c}^*_k,$$
$$\boldsymbol{\alpha}_{k+1} = \big(\alpha^*_{1|k}, \ldots, \alpha^*_{N|k}, \lambda(\alpha^*_{N|k} - \bar{\alpha}) + \bar{\alpha}\big).$$

These allow the optimal value of the cost in (3.71) at time $k+1$, denoted $J^*_{k+1} \doteq J(s^*_{0|k+1}, \mathbf{c}^*_{k+1}, \boldsymbol{\alpha}^*_{k+1})$, to be bounded as follows,

$$\begin{aligned}
J^*_{k+1} &\le \|\Phi s^*_{0|k} + BE\mathbf{c}^*_{0|k}\|^2_{W_x} + \|M\mathbf{c}^*_k\|^2_{W_c} + \sum_{i=1}^{N} q_\alpha(\alpha^*_{i|k} - \bar{\alpha})^2 + p_\alpha\lambda^2(\alpha^*_{N|k} - \bar{\alpha})^2 \\
&\le \|\Phi s^*_{0|k} + BE\mathbf{c}^*_{0|k}\|^2_{W_x} + \|M\mathbf{c}^*_k\|^2_{W_c} + \sum_{i=1}^{N-1} q_\alpha(\alpha^*_{i|k} - \bar{\alpha})^2 + p_\alpha(\alpha^*_{N|k} - \bar{\alpha})^2 \\
&\le J^*_k - (\|s^*_{0|k}\|^2_Q + \|Ks^*_{0|k} + \mathbf{c}^*_k\|^2_R) - q_\alpha(\alpha^*_{0|k} - \bar{\alpha})^2,
\end{aligned}$$

where (3.70) has been used. By the argument of the proof of Theorem 2.8 therefore, for any initial condition x_0 in the feasible set $\mathcal{F}_N$ for (3.71), we obtain, for all $k > 0$,

$$\|s^*_{0|k}\|^2 + |\alpha^*_{0|k} - \bar{\alpha}|^2 \le \frac{b}{a}\left|1 - \frac{\min\{\underline{\lambda}(Q), q_\alpha\}}{b}\right|^k \big(\|s^*_{0|0}\|^2 + |\alpha^*_{0|0} - \bar{\alpha}|^2\big), \quad (3.72)$$

where $a, b > 0$ are constants such that, for all $x_k \in \mathcal{F}_N$,

$$a\big(\|s_{0|k}\|^2 + |\alpha_{0|k} - \bar{\alpha}|^2\big) \le J(s^*_{0|k}, \mathbf{c}^*_k) \le b\big(\|s_{0|k}\|^2 + |\alpha_{0|k} - \bar{\alpha}|^2\big).$$

Since $x_k - e^*_{0|k} = s^*_{0|k}$ and $e^*_{0|k} \in \alpha^*_{0|k}\mathcal{S}^0$, the minimum Euclidean distance from x_k to any point in $\bar{\alpha}\mathcal{S}^0$ is bounded by

$$\begin{aligned}
\min_{e\in\bar{\alpha}\mathcal{S}^0} \|x_k - e\| \le \left\|x_k - \frac{\bar{\alpha}}{\alpha^*_{0|k}} e^*_{0|k}\right\| &\le \|s^*_{0|k}\| + \frac{|\alpha^*_{0|k} - \bar{\alpha}|}{\alpha^*_{0|k}} \max_{e\in\alpha^*_{0|k}\mathcal{S}^0} \|e\| \\
&= \|s^*_{0|k}\| + \beta|\alpha^*_{0|k} - \bar{\alpha}| \qquad (3.73)
\end{aligned}$$

where $\beta = \max_{e\in\bar{\alpha}\mathcal{S}^0} \|e\|/\bar{\alpha}$ is a constant, and it follows from (3.72) and (3.73) that the distance of x_k from $\bar{\alpha}\mathcal{S}^0$ is upper-bounded by an exponential decay. □

The cost and constraints of the HTMPC optimization (3.71) can be simplified (as discussed in Question 9 on p. 117) if the set $\mathcal{S}^0$ is robustly invariant for the error dynamics (3.57c). We note also that it is possible to relax the constraints of this approach using the equi-normalization technique described in [21]. This is achieved through exact scaling of the set $\mathcal{S}^0$, allowing for an expansion of the region attraction of Algorithm 3.4. Further improvements in the size of the feasible initial condition set can be achieved by formulating the degrees of freedom $\alpha_{i|k}$ as vectors rather than the scalars that, in Algorithm 3.4, scale the set $\mathcal{S}^0$ equally in all directions. This is possible through an appropriate use of Farkas' Lemma, and is discussed in detail in Chap. 5.

3.6 Early Robust MPC for Additive Uncertainty

To conclude this chapter we describe two of the main precursors of the robust MPC techniques described in Sects. 3.2, 3.3 and 3.5. The first of these is concerned with a robust extension of SGPC for systems with additive disturbances [22]. This approach imposes tightened constraints on nominal predicted trajectories to ensure robustness, and it also provides conditions for recursive feasibility analogous to those of Sect. 3.2.1, but in the context of input-output discrete time models and equality terminal constraints. We then discuss the tube MPC algorithms of [4, 23]. These use low-complexity polytopes to bound predicted trajectories, treating the parameters defining these sets as variables in the online MPC optimization. Similarly to the homothetic tubes considered in Sect. 3.5.2, the condition that these tubes should contain the predicted trajectories of the uncertain plant model is invoked through a recursive sequence of constraints.

3.6.1 Constraint Tightening

This section describes a formulation of robust MPC for the case of additive disturbances which is based on the SGPC algorithm described in Sect. 2.10. As in Sect. 3.3, a nominal cost is assumed. This is computed under the assumption that the nominal disturbance input is zero, and is therefore equal to the predicted cost when there are no future disturbances. In the constant setpoint problem considered in Sect. 2.10, SGPC with integral action steers the state asymptotically to a reference state (which for simplicity is taken to be zero in this section) whenever the future disturbance reaches a steady state. However, this presupposes recursive feasibility which is achieved in [22] through constraint tightening. The tightened constraints are derived in two stages, the first of which achieves an a posteriori feasibility that ensures the feasibility of predictions, and the second invokes feasibility a priori so that feasibility is retained recursively. The terms a posteriori and a priori are used in the sense that the former involves conditions based on past information whereas the latter anticipates the future in an attempt to ensure recursive feasibility.

To simplify presentation we consider here the case of a single-input single-output system. The convenience of the SGPC approach is that, for the disturbance-free case, it develops prediction dynamics which involve transfer functions described by finite impulse response (FIR) filters. Hence, a terminal (equality) stability constraint can be imposed on predicted trajectories implicitly, without the need to invoke any terminal constraints. This convenience can be preserved for the case when an additive disturbance is introduced into the system model (2.68):

$$\begin{aligned} y_k &= \frac{z^{-1}b(z^{-1})}{a(z^{-1})}u_k + \frac{1}{a(z^{-1})}\zeta_k \\ &= \frac{z^{-1}b(z^{-1})}{\alpha(z^{-1})}\Delta u_k + \frac{1}{\alpha(z^{-1})}\xi_k \end{aligned} \tag{3.74}$$

where $\alpha(z^{-1}) = \Delta(z^{-1})a(z^{-1})$, $\Delta(z^{-1}) = 1 - z^{-1}$ and $\zeta_k = \frac{1}{\Delta(z^{-1})}\xi_k$ with ξ_k denoting a zero mean white noise process, and where the polynomial matrices $A(z^{-1})$, $B(z^{-1})$ in (2.68) are replaced by polynomials $a(z^{-1})$, $b(z^{-1})$ for the single-input single-output case considered here. As explained in Chap. 2, consideration is given to the control increments, Δu_k (rather than the values of the control input, u_k), as a means of introducing integral action into the feedback loop.

Similarly to the decomposition of predicted trajectories in Sect. 3.2, we decompose the z-transforms, $y(z^{-1})$, $u(z^{-1})$, of the predicted output and control input sequences according to

$$y(z^{-1}) = y^{(1)}(z^{-1}) + y^{(2)}(z^{-1}), \quad u(z^{-1}) = u^{(1)}(z^{-1}) + u^{(2)}(z^{-1}).$$

Here, $y^{(1)}, u^{(1)}$ denote nominal predicted output and input sequences that correspond to the disturbance-free case, while $y^{(2)}, u^{(2)}$ model the effects of the additive disturbance in (3.74). Following the development of Sect. 2.10, the control law $\Delta u^{(1)}(z^{-1}) = \big(c(z^{-1}) - z^{-1}N(z^{-1})y^{(1)}(z^{-1})\big)/M(z^{-1})$ for some polynomials $N(z^{-1})$ and $M(z^{-1})$, results in predictions

$$y^{(1)}(z^{-1}) = z^{-1}b(z^{-1})c(z^{-1}) + y_f(z) \tag{3.75a}$$

$$\Delta u^{(1)}(z^{-1}) = \alpha(z^{-1})c(z^{-1}) + \Delta u_f(z) \tag{3.75b}$$

provided that $N(z^{-1})$ and $M(z^{-1})$ satisfy the Bezout identity

$$\alpha(z^{-1})M(z^{-1}) + z^{-1}b(z^{-1})N(z^{-1}) = 1. \tag{3.76}$$

Note that $y_f(z)$ and $\Delta u_f(z)$ in (3.75) are polynomials in positive powers of z that relate to past values of outputs and control increments, thereby taking account of non-zero initial conditions.

In the absence of disturbances, a terminal equality constraint requiring the nominal predicted outputs and inputs to be identically zero after a finite initial prediction horizon is imposed implicitly by setting $\Delta u(z^{-1}) = \Delta u^{(1)}(z^{-1})$. The case of non-zero disturbances can also be handled through use of the Bezout identity (3.76). From (3.74), $\Delta u^{(2)}(z^{-1})$ and $y^{(2)}(z^{-1})$ are related by

$$\alpha(z^{-1})y^{(2)}(z^{-1}) - z^{-1}b(z^{-1})\Delta u^{(2)}(z^{-1}) = \xi(z^{-1}),$$

and (3.76) therefore implies that the transfer functions from $\xi(z^{-1})$ to $y^{(2)}(z^{-1})$ and $\Delta u^{(2)}(z^{-1})$ have the form of FIR filters if the predicted control increments are defined by $\Delta u(z^{-1}) = \big(c(z^{-1}) - z^{-1}N(z^{-1})y(z^{-1})\big)/M(z^{-1})$ since from (3.75b) we then obtain

$$y^{(2)}(z^{-1}) = M(z^{-1})\xi(z^{-1}) \tag{3.77a}$$

$$\Delta u^{(2)}(z^{-1}) = -z^{-1}N(z^{-1})\xi(z^{-1}). \tag{3.77b}$$

The fixed order polynomials $M(z^{-1}), z^{-1}N(z^{-1})$ appearing in these expressions enable the worst-case values of the predicted values for $y^{(2)}$, $\Delta u^{(2)}$ to be computed conveniently and thus allow constraints to be applied robustly, for all allowable disturbance sequences $\xi(z^{-1})$.

To illustrate this point, consider the case where the system is subject to rate constraints only:

$$|\Delta u_{i|k}| \leq R, \quad i = 0, 1, \ldots$$

Then the implied constraints on the nominal control sequence must be tightened to give

$$|\Delta u^{(1)}_{i|k}| \leq R - R^{\#}_i, \quad i = 0, 1, \ldots \tag{3.78}$$

where $R^{\#}_i$ denotes the prediction i steps ahead of the worst-case absolute value of $\Delta u^{(2)}$ in (3.77b). To determine this, some assumption as to the size of uncertainty has to be made, and (analogously to the disturbance set (3.3)) a bound can be imposed through $\zeta_{i|k}$, which would in practice be limited as

$$|\zeta_{i|k}| \leq d, \quad i = 0, 1, \ldots.$$

Given this limit it is straightforward to compute $R^{\#}_i$, by writing (3.77b) in terms of $\zeta(z^{-1})$ as

$$\Delta u^{(2)}(z^{-1}) = -z^{-1}N(z^{-1})[\Delta(z^{-1})\zeta(z^{-1}) - \zeta_k],$$

and then extracting the worst-case value of the coefficient of z^{-i} in $\Delta u^{(2)}(z^{-1})$ over the allowable range of values of coefficients of $\zeta(z^{-1})$. Given that $M(z^{-1})$ and $\Delta u^{(1)}_{i|k}(z^{-1})$ are both finite-degree polynomials in z^{-1}, it is clear that only a finite number of terms: $R^{\#}_i$, $i = 0, 1, \ldots, \nu$, for finite ν, need to be evaluated in order to invoke the constraints (3.78) over an infinite prediction horizon.

Condition (3.78), imposed on the degrees of freedom in (3.75b), namely the coefficients of $c(z^{-1})$, defines the a posteriori conditions that ensure the feasibility of the control input increments, given the current measurements. However, this is not enough to guarantee recursive feasibility, as can be seen by considering the tail at time $k+1$ of a trajectory that was feasible at time k. This tail is generated by replacing the $c(z^{-1})$ polynomial in (3.75) with $z\big(c(z^{-1}) - c_{0|k}\big)$. In the absence of disturbances, such a tail would necessarily be feasible, but through the initial conditions term, Δu_f, of (3.75b), the effect on the sequence of predictions for $\Delta u^{(1)}$ at time $k+1$ due to non-zero ζ will be the addition of a term which can be shown to be

$$f_{\Delta u}(\zeta_k, \zeta_{k+1}) = -N(z^{-1})(\zeta_{k+1} - \zeta_k),$$

This term must be accommodated by the tightened constraints when they are applied at the next time instant, $k+1$. Thus application of (3.78) at time $k+1$ must ensure that for each prediction step i, $R^{\#}_i - R^{\#}_{i+1}$ is at least as large as the modulus of the corresponding element of $f_{\Delta u}(\zeta_k, \zeta_{k+1})$. Detailed calculation shows that this is indeed the case if the tightening parameters are

$$R^{\#}_1 = 0, \;\; R^{\#}_{i+1} = R^{\#}_i + 2d|N_{i-1}|, \quad i = 1, \ldots, \mu - 1 \tag{3.79}$$

where μ denotes the sum of the degrees of $\alpha(z^{-1})$ and $c(z^{-1})$ and N_i is the coefficient of z^{-i} in $N(z^{-1})$. The constraint (3.78) with (3.79) defines the a priori feasibility conditions which also satisfy the a posteriori conditions and are in fact the least conservative constraint tightening bounds that provide the guarantees of recursive feasibility.

The development presented in this section has obvious extensions to absolute input (rather than rate) constraints as well as mixed input/output constraints. Indeed, as shown in Sects. 3.3 and 3.5, a similar constraint tightening approach can also be extended to state space models and to algorithms other than SGPC. What makes the approach described here convenient is the fact that, by using the Bezout identity (3.76), the uncertainty tube converges to the mRPI set in a finite number of steps. This ensures that the sequence of constraint tightening parameters $R^{\#}_1, R^{\#}_2, \ldots$ converges to a limit in a finite number of steps, without the need for the more general, but computationally more demanding, theoretical framework of Sects. 3.2.1 and 3.2.2.

3.6.2 Early Tube MPC

Uncertainty tubes that are parameterized explicitly in terms of optimization variables were proposed in the robust MPC algorithms of [4, 24]. Although similar in this respect to the homothetic tubes of [20], the tubes of [4, 24] allow more variation between the shapes of the sets defining the tube cross section than simple translation and scaling of a given set. To avoid the need for large numbers of optimization variables, attention is restricted to tube cross sections defined as low-complexity polytopes. The approach is explained in this section within the context of additive model uncertainty; the application of low-complexity polytopes to the case of multiplicative model uncertainty (which was proposed in [4, 23]) is discussed in Chap. 5.

A low-complexity polytope is a linearly transformed hypercube. Let $\mathcal{S}_{i|k}$ denote the low-complexity polytope defined by

$$\mathcal{S}_{i|k} \doteq \{x : \underline{\alpha}_{i|k} \leq V_{\mathcal{S}} x \leq \overline{\alpha}_{i|k}\} \tag{3.80}$$

where $V_{\mathcal{S}} \in \mathbb{R}^{n_x \times n_x}$ is non-singular and $\underline{\alpha}_{i|k}, \overline{\alpha}_{i|k} \in \mathbb{R}^{n_x}$. Then the condition that the state $x \in \mathbb{R}^{n_x}$ of (3.1) belongs to $\mathcal{S}_{i|k}$ can be expressed as $2n_x$ linear inequalities. It is convenient to define a transformed state vector as

$$\xi = V_{\mathcal{S}} x.$$

In terms of this transformed state, the open-loop strategy of (3.6) and the corresponding prediction dynamics (3.7) can be re-written as

$$u_{i|k} = \tilde{K} \xi_{i|k} + c_{i|k}, \tag{3.81a}$$

$$\xi_{i|k} = \tilde{\Phi} \xi_{i|k} + \tilde{B} c_{i|k} + \tilde{D} w_{i|k}, \tag{3.81b}$$

where $\tilde{\Phi} = V_{\mathcal{S}} \Phi V_{\mathcal{S}}^{-1}$, $\tilde{B} = V_{\mathcal{S}} B$, $\tilde{D} = V_{\mathcal{S}} D$ and $\tilde{K} = K V_{\mathcal{S}}^{-1}$. As before, we assume that $c_{i|k} = 0$ for all $i \geq N$, where N is the mode 1 prediction horizon, and K is assumed to be the unconstrained LQ-optimal feedback gain.

Low-complexity polytopic tubes provide a compact and efficient means of bounding the predicted state as a function of the degrees of freedom $(c_{0|k}, \ldots, c_{N-1|k})$ in (3.81b), the initial state $\xi_{0|k} = V_{\mathcal{S}} x_k$, and the disturbance set $\mathcal{W}$. This can be achieved by propagating the tube cross sections recursively using the following result.

Lemma 3.6 *For $A \in \mathbb{R}^{n_A \times n_y}$, let $A^+ \doteq \max\{A, 0\}$ and $A^- \doteq \max\{-A, 0\}$, then $\underline{y} \leq y \leq \bar{y}$ implies*

$$A^+ \underline{y} - A^- \bar{y} \leq Ay \leq A^+ \bar{y} - A^- \underline{y}, \tag{3.82}$$

and, for each of the $2n_A$ elementwise bounds in (3.82), there exists y satisfying $\underline{y} \leq y \leq \bar{y}$ such that the bound holds with equality.

Consider the conditions on $\underline{\alpha}_{i|k}, \overline{\alpha}_{i|k}, c_{i|k}$ and $\mathcal{W}$ in order that $x_{i|k} \in \mathcal{S}_{i|k}$ implies that $x_{i+1|k} \in \mathcal{S}_{i+1|k}$. If we define

$$\underline{w}_D \doteq \min_{w \in \mathcal{W}} \tilde{D} w, \quad \bar{w}_D \doteq \max_{w \in \mathcal{W}} \tilde{D} w,$$

then, given that $x_{i|k} \in \mathcal{S}_{i|k}$, the elements of $\xi_{i+1|k}$ are bounded according to $\underline{\alpha}_{i+1|k} \leq \xi_{i+1|k} \leq \overline{\alpha}_{i+1|k}$ where

$$\underline{\alpha}_{i+1|k} \leq \tilde{\Phi}^{+}\underline{\alpha}_{i|k} - \tilde{\Phi}^{-}\overline{\alpha}_{i|k} + \tilde{B}c_{i|k} + \underline{w}_D \tag{3.83a}$$

$$\tilde{\Phi}^{+}\overline{\alpha}_{i|k} - \tilde{\Phi}^{-}\underline{\alpha}_{i|k} + \tilde{B}c_{i|k} + \bar{w}_D \leq \overline{\alpha}_{i|k}. \tag{3.83b}$$

Therefore the tube $\{\mathcal{S}_{0|k}, \ldots, \mathcal{S}_{N|k}\}$ contains the predicted state trajectories of (3.1) if and only if these conditions hold for $i = 0, \ldots, N-1$, starting from $\underline{\alpha}_{0|k} = \overline{\alpha}_{0|k} = V_{\mathcal{S}} x_k$.

Under the control law of (3.81a), the constraints of (3.4), expressed in terms of $\tilde{F} = F V_{\mathcal{S}}^{-1}$ and $\tilde{K}$, can be written as

$$(\tilde{F} + G\tilde{K})\xi_{i|k} + Gc_{i|k} \leq 1. \tag{3.84}$$

These constraints are satisfied for all $\xi_{i|k} \in \mathcal{S}_{i|k}$ if and only if

$$(\tilde{F} + G\tilde{K})^{+}\overline{\alpha}_{i|k} - (\tilde{F} + G\tilde{K})^{-}\underline{\alpha}_{i|k} + Gc_{i|k} \leq 1 \tag{3.85}$$

This takes care of constraints for $i = 0, \ldots, N-1$, whereas the constraints for $i \geq N$ are accounted for by imposing the condition that the terminal tube cross section should lie in a terminal set, $\mathcal{S}_T$, which is RPI. To allow for a guarantee of recursive feasibility, [4, 24] proposed a low-complexity polytopic terminal set:

$$\{x : |V_{\mathcal{S}} x| \leq \alpha_T\} \tag{3.86}$$

where the absolute value and inequality sign in (3.86) apply on an element by element basis. This terminal set is invariant under the dynamics of (3.81b) and the constraints of (3.84) for $c_{i|k} = 0$ if and only if

$$|\tilde{\Phi}|\alpha_T + \max\{\bar{w}_D, -\underline{w}_D\} \leq \alpha_T, \tag{3.87}$$

and

$$|\tilde{F} + G\tilde{K}|\alpha_T \leq 1 \tag{3.88}$$

The conditions (3.87) and (3.88) follow from the fact that, for $|\xi| \leq \alpha_T$, the achievable maximum (elementwise) values of $\tilde{\Phi}\xi$ and $(\tilde{F} + G\tilde{K})\xi$ are $|\tilde{\Phi}|\alpha_T$ and $|\tilde{F} + G\tilde{K}|\alpha_T$ respectively.

Combining the robust constraint handling of this section with the nominal cost of Sect. 3.3 gives the following online optimization problem which is a quadratic program in $N(n_u + 2n_x)$ variables and $2(N+1)n_x + Nn_C$ constraints.

$$\begin{aligned}
\underset{\mathbf{c}_k, \underline{\alpha}_{i|k}, \overline{\alpha}_{i|k}, i=0,\dots,N}{\text{minimize}} \quad & \|\mathbf{c}_k\|^2_{W_c} \\
\text{subject to} \quad & (3.83),\ (3.85) \text{ for } i = 0, \dots, N-1, \\
& \underline{\alpha}_{N|k} \geq -\alpha_T, \ \ \overline{\alpha}_{N|k} \leq \alpha_T \\
& \underline{\alpha}_{0|k} = \overline{\alpha}_{0|k} = V_{\mathcal{S}} x_k
\end{aligned} \tag{3.89}$$

The optimal value of $\mathbf{c}_k$ for this problem is clearly also optimal for the problem of minimizing, subject to the constraints of (3.89), the nominal cost $J(x_k, \mathbf{c}_k) = \|x_k\|^2_{W_x} + \|\mathbf{c}_k\|^2_{W_c}$ defined in (3.29). Hence the stability analysis of Sect. 3.3 applies to the MPC law $u_k = Kx_k + c^*_{0|k}$, where $\mathbf{c}^*_k = (c^*_{0|k}, \dots, c^*_{N-1|k})$ is the optimal value of $\mathbf{c}_k$ for problem (3.89). In particular, recursive feasibility is implied by the feasibility of the following values for the variables in (3.89) at time $k+1$:

$$\begin{aligned}
\mathbf{c}_{k+1} &= M\mathbf{c}^*_k \\
\underline{\alpha}_{i|k+1} = \underline{\alpha}^*_{i+1|k}, \quad \overline{\alpha}_{i|k+1} &= \overline{\alpha}^*_{i+1|k}, \quad i = 0, \dots, N-1 \\
\underline{\alpha}_{N|k+1} = \xi_T, \quad \overline{\alpha}_{N|k+1} &= -\xi_T.
\end{aligned}$$

where $\mathbf{c}^*_k$ and $\underline{\alpha}^*_{i|k}, \overline{\alpha}^*_{i|k}, i = 0, \dots, N$ are optimal for (3.89) at time k. The constraints of (3.89) must therefore be feasible at times $k = 1, 2, \dots$ if the initial condition x_0 lies in the feasible set for (3.89) at $k = 0$. In addition, the quadratic bound (3.38) holds for the closed-loop system, and asymptotic convergence of x_k to the minimal RPI set $\mathcal{X}^{\text{mRPI}}$ defined in (3.23) follow from Lemmas 3.1, 3.2 and the bound (3.37).

We close this section by noting that the offline computation that is required so that the MPC optimization (3.89) can be performed online concerns the selection of $V_{\mathcal{S}}$ and α_T defining the terminal set in (3.86). Since this terminal set must be invariant, a convenient choice for $V_{\mathcal{S}}$ is provided by the eigenvector matrix of Φ with the columns of $V_{\mathcal{S}}$ that correspond to complex conjugate eigenvalues of Φ replaced by the real and imaginary parts of the corresponding eigenvectors. With this choice of $V_{\mathcal{S}}$, a necessary and sufficient condition for existence of α_T satisfying (3.87) is that the eigenvalues of Φ should lie inside the box in the complex plane with corners at ± 1 and $\pm j$ (Fig. 3.9). Having defined $V_{\mathcal{S}}$, the elements of α can be determined so as to maximize the volume of the terminal set subject to (3.88) through the solution of a convex optimization problem.

Despite their computational convenience, the relatively inflexible geometry of low-complexity polytopes makes them rather restrictive. For this reason Chap. 5 replaces low-complexity polytopes with general polytopic sets through an appropriate use of Farkas' Lemma (see [25] or [26]).

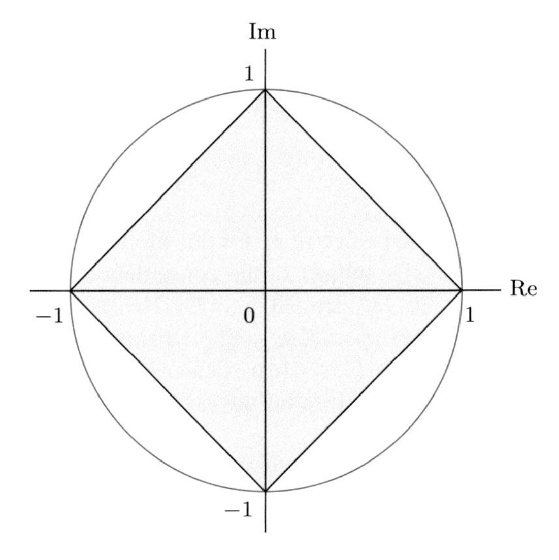

Fig. 3.9 The box in the complex plane with vertices at ± 1 and $\pm j$ (*shaded region*)

3.7 Exercises

1 A production planning problem requires the quantity u_k of product made in week k, for $k = 0, 1, \ldots$ to be optimized. The quantity w_k of product that is sold in week k is unknown in advance but lies in the interval $0 \le w_k \le W$ and has a nominal value of $\hat{w}$. The quantity x_{k+1} remaining unsold at the start of week $k+1$ is governed by

$$x_{k+1} = x_k + u_k - w_k, \quad k = 0, 1, \ldots$$

Limits on storage and manufacturing capacities imply that x and u can only take values in the intervals

$$0 \le x_k \le X, \quad 0 \le u_k \le U.$$

The desired level of x in storage is x^0, and the planned values $u_{0|k}, u_{1|k}, \ldots$ are to be optimized at the beginning of week k given a measurement of x_k.

(a) What are the advantages of using a receding horizon control strategy that is recomputed at $k = 0, 1, \ldots$ in this application instead of an open-loop control sequence computed at $k = 0$?
Let the planned production at time k be $u_{i|k} = \hat{w} - (x_{i|k} - x^0) + c_{i|k}$ for $i = 0, 1, \ldots$, where $(c_{0|k}, \ldots, c_{N-1|k}) = \mathbf{c}_k$ is a vector of optimization variables at time k, $c_{i|k} = 0$ for $i \ge N$, and $x_{i|k}$ is a prediction of x_{k+i}.
(b) Let $s_{i|k}$ be the nominal value of $x_{i|k} - x^0$ (namely the value that would be obtained if $w_{i|k} = \hat{w}$ for all $i \ge 0$, with $s_{0|k} = x_k - x^0$). Show that the nominal cost $J(x_k, \mathbf{c}_k) \doteq \sum_{i=0}^{\infty} s_{i|k}^2$ is given by

$$J(x_k, \mathbf{c}_k) = (x_k - x^0)^2 + \|\mathbf{c}_k\|^2.$$

(c) If $s_{i|k} + e_{i|k} = x_{i|k} - x^0$, verify that $e_{i|k} \in [\hat{w} - W, \hat{w}]$ for all $i \geq 1$. Hence show that $u_{i|k} \in [0, U]$ and $x_{i|k} \in [0, X]$ for all $w_{i|k} \in [0, W]$ and all $i \geq 0$ if and only if the following conditions hold,

$$\begin{aligned} c_{i|k} + \hat{w} + x^0 &\in [W, X] && i = 0, \ldots, N-1 \\ c_{i|k} - c_{i-1|k} &\in [0, U - W] && i = 1, \ldots, N-1 \\ c_{0|k} + \hat{w} + x^0 - x_k &\in [0, U - W] \\ c_{N-1|k} &\in [-(U - W), 0] \end{aligned}$$

and state the conditions on X, U, W that are required in order that the feasible set for x_0 is non-empty.

(d) Let $\mathbf{c}_k^*$ be the minimizer of the cost in (b) subject to the constraints in (c) at time k, and define the MPC law as $u_k = \hat{w} - (x_k - x^0) + c_{0|k}^*$, where $c_{0|k}^*$ is the first element of $\mathbf{c}_k^*$. Show that the constraints of (c) are recursively feasible and that $c_{0|k}^*$ converges to zero as $k \to \infty$.

(e) Explain why $x_0 \leq \hat{w} + x^0$ is needed for feasibility of the constraints of (c) at $k = 0$. How might this condition be relaxed?

2 If Φ is a nilpotent matrix, with $\Phi^n = 0$ for some integer $n > 0$, and Ψ is defined by

$$\Psi = \begin{bmatrix} \Phi & \Gamma \\ 0 & M \end{bmatrix}, \quad M = \begin{bmatrix} 0 & I_{n_u} & 0 & \cdots & 0 \\ 0 & 0 & I_{n_u} & \cdots & 0 \\ \vdots & \vdots & \vdots & & \vdots \\ 0 & 0 & 0 & \cdots & I_{n_u} \\ 0 & 0 & 0 & \cdots & 0 \end{bmatrix} \in \mathbb{R}^{Nn_u \times Nn_u}$$

for a given matrix Γ, prove that Ψ is nilpotent with $\Psi^m = 0$ for $m = n + N$.

Use this property to write down expressions for:

(a) the minimal RPI set for the dynamics $e_{k+1} = \Phi e_k + D w_k$,
(b) the set of inequalities defining the maximal RPI set for the system $z_{k+1} = \Psi z_k + \bar{D} w_k$ and constraints $\bar{F} z_k \leq \mathbf{1}$,
(c) the conditions under which the set of feasible initial states for the system $z_{k+1} = \Psi z_k + \bar{D} w_k$ and constraints $\bar{F} z_k \leq \mathbf{1}$ is non-empty,

where w_k lies in a compact polytopic set $\mathcal{W}$ for all k and $\bar{D} = \begin{bmatrix} D \\ 0 \end{bmatrix}$.

3 A system has dynamics $x_{k+1} = Ax_k + Bu_k + w_k$ and state constraints $Fx \le \mathbf{1}$, with

$$A = \begin{bmatrix} -1 & 0.2 \\ -0.25 & 0.65 \end{bmatrix}, \quad B = \begin{bmatrix} 1 \\ -0.5 \end{bmatrix}, \quad F = \begin{bmatrix} I \\ -I \end{bmatrix}$$

where the disturbance input w_k is unknown at time k and satisfies, for all $k \ge 0$

$$w_k \in \sigma \mathcal{W}_0, \quad \mathcal{W}_0 \doteq \left\{ w : \begin{bmatrix} -1 \\ -1 \end{bmatrix} \le \begin{bmatrix} 1 & 1 \\ 1 & -1 \end{bmatrix} w \le \begin{bmatrix} 1 \\ 1 \end{bmatrix} \right\}$$

for some constant scalar parameter $\sigma > 0$.

Verify that $\Phi = A + BK$ is nilpotent if $K = \begin{bmatrix} 0.5 & 0.3 \end{bmatrix}$. Hence determine the maximal RPI set, $\mathcal{Z}^{\text{MRPI}}(\sigma)$, for the system $z_{k+1} = \Psi z_k + \bar{D} w_k$, $w_k \in \sigma \mathcal{W}_0$, where Ψ is defined as in Question 2, with $N = 2$ and $\Gamma = B \begin{bmatrix} 1 & 0 \end{bmatrix}$. Show that $\mathcal{Z}^{\text{MRPI}}(\sigma)$ is non-empty if and only if $\sigma \le \frac{2}{3}$.

4 For the system considered in Question 3 an MPC law is defined at each time instant $k = 0, 1, \ldots$ as $u_k = Kx_k + c^*_{0|k}$, where $\mathbf{c}^*_k = (c^*_{0|k}, c^*_{1|k})$ is the solution of the QP:

$$\mathbf{c}^*_k = \arg\min_{\mathbf{c}_k} \ \|\mathbf{c}_k\|^2 \ \text{ subject to } \begin{bmatrix} x_k \\ \mathbf{c}_k \end{bmatrix} \in \mathcal{Z}^{\text{MRPI}}(\sigma)$$

and where K and $\mathcal{Z}^{\text{MRPI}}(\sigma)$ are as defined in Question 3. The controller is designed to operate with any σ in the interval $[0, \frac{2}{3})$, and σ can be assumed to be known and constant when the controller is in operation.

(a) Show that the closed-loop system is stable if the MPC optimization is feasible at time $k = 0$. What limit set will the closed-loop state converge to?

(b) Comment on the suggestion that better performance would be obtained with respect to the cost $\sum_{k=0}^{\infty}(\|x_k\|^2_Q + u_k^2)$, for a given matrix $Q \succeq 0$, if the MPC optimization at time k was defined

$$\mathbf{c}^*_k = \arg\min_{\mathbf{c}_k} \left\| \begin{bmatrix} x_k \\ \mathbf{c}_k \end{bmatrix} \right\|^2_W \ \text{ subject to } \begin{bmatrix} x_k \\ \mathbf{c}_k \end{bmatrix} \in \mathcal{Z}^{\text{MRPI}}(0.5),$$

where W is the solution of the Lyapunov equation

$$W - \Psi^T W \Psi = \hat{Q}, \quad \hat{Q} = \begin{bmatrix} Q + K^T K & K^T & 0 \\ K & 0 & 0 \\ 0 & 0 & 0 \end{bmatrix}.$$

5 (a) A matrix Φ and a compact convex polytopic set $\mathcal{W}$ satisfy the inclusion condition

$$\Phi^r \mathcal{W} \subseteq \rho \mathcal{W}$$

for some integer r and $\rho \in [0, 1)$. Show that $h_\infty \leq \hat{h}_\infty$, where

$$h_\infty \doteq \sum_{j=0}^{\infty} \max_{w_j \in \mathcal{W}} F\Phi^j w_j, \quad \hat{h}_\infty \doteq \frac{1}{1-\rho} \sum_{j=0}^{r-1} \max_{w_j \in \mathcal{W}} F\Phi^j w_j$$

and prove that the fractional error, $(\hat{h}_\infty - h_\infty)/h_\infty$, in this bound is no greater than $\rho/(1-\rho)$.

(b) For (A, B) and $\mathcal{W}_0$ as defined in Question 3 and $K = \begin{bmatrix}0.479 & 0.108\end{bmatrix}$, use the bounds in (a) to determine h_∞ to an accuracy of 1 % when $\Phi = A + BK$ and $\mathcal{W} \doteq 0.1\mathcal{W}_0$.

(c) Suggest an over-bounding approximation of the minimal RPI set for the system $e_{k+1} = (A + BK)e_k + w_k$, $w_k \in \mathcal{W}$ which is based on the inclusion condition in (a). What can be said about the accuracy of this approximation?

6 A robust MPC law is to be designed for the system $x_{k+1} = Ax_k + Bu_k + w_k$ with the state constraints $Fx \leq \mathbf{1}$ and disturbance bounds $w_k \in 0.5\mathcal{W}$, where (A, B), F and $\mathcal{W}_0$ are as defined in Question 3. The predicted control sequence is parameterized as $u_{i|k} = Kx_{i|k} + c_{i|k}$ with $K = \begin{bmatrix}0.479 & 0.108\end{bmatrix}$, where $c_{i|k}$ for $i < N$ are optimization variables and $c_{i|k} = 0$ for all $i \geq N$.

(a) For $N = 1$, construct matrices Ψ, $\bar{D}$ and $\bar{F}$ such that $Fx_{i|k} = \bar{F}z_{i|k}$ for all $i \geq 0$, where $z_{i+1|k} = \Psi z_{i|k} + \bar{D}w_{k+i}$ and hence determine the constraint set for $z_k = (x_k, \mathbf{c}_k)$ that gives the largest possible feasible set for x_k for this prediction system.

(b) Determine the matrix W_z defining the nominal cost

$$\sum_{i=0}^{\infty} (\|s_{i|k}\|^2 + v_{i|k}^2) = z_k^T W_z z_k$$

for the nominal predictions $s_{i+1|k} = (A + BK)s_{i|k} + Bc_{i|k}$ and $v_{i|k} = Ks_{i|k} + c_{i|k}$, with $s_{0|k} = x_k$. Verify that for $x_0 = (0, 1)$ the minimal value of this cost subject to the constraints computed in (a) is 1.707.

(c) Starting from the initial condition $x_0 = (0, 1)$, simulate the closed-loop system under the MPC law $u_k = Kx_k + c_{0|k}^*$, assuming that the disturbance sequence (which is unknown to the controller) is given by

$$\{w_0, w_1, w_2, w_3, \ldots\} = \left\{ \begin{bmatrix}0.5\\0\end{bmatrix}, \begin{bmatrix}-0.5\\0\end{bmatrix}, \begin{bmatrix}0\\0.5\end{bmatrix}, \begin{bmatrix}0\\-0.5\end{bmatrix} \right\}$$

and verify that the closed-loop state and control sequences satisfy

$$\sum_{k=0}^{3}\left(\|x_k\|^2 + u_k^2\right) = 4.49.$$

7 The worst case predicted cost:

$$\check{J}(x_k, \mathbf{c}_k) = \max_{\{w_k, w_{k+1}, \ldots\}} \sum_{i=0}^{\infty}\left(\|x_{i|k}\|^2 + u_{i|k}^2 - \gamma^2\|w_{k+i}\|^2\right)$$

is to be used to define a robust MPC law for the system with the model, disturbance bounds and state constraints of Question 6, with $N = 1$ and $\gamma^2 = 3.3$. For this value of γ^2 the unconstrained optimal feedback gain is $K = \begin{bmatrix}0.540 & 0.249\end{bmatrix}$ and the corresponding Riccati equation has the solution

$$\check{W}_x = \begin{bmatrix} 2.336 & -0.904 \\ -0.904 & 2.103 \end{bmatrix}.$$

(a) Determine W_x and W_c so that $\check{J}(x_k, \mathbf{c}_k) = \|x_k\|^2_{W_x} + \|\mathbf{c}_k\|^2_{W_c}$.

(b) Find the smallest integer ν such that $\bar{F}\Psi^{\nu+1}z \le \mathbf{1} - h_{\nu+1}$ for all z satisfying $\bar{F}\Psi^i z \le \mathbf{1} - h_i$, for $i = 0, \ldots, \nu$. Hence, solve the MPC optimization (3.47) at $k = 0$ for $x_0 = (0, 1)$, and verify that the optimal solution is $\mathbf{c}_0^* = 0.051$ and that $\check{J}^*(x_0) = 2.294$.

(c) Consider the alternative worst-case cost defined by

$$\check{J}(x_k, \mathbf{c}_k) = \max_{\substack{w_{k+i} \in \mathcal{W} \\ i=0,\ldots,N-1}} \sum_{i=0}^{N-1}\left(\|x_{i|k}\|^2 + u_{i|k}^2 - \gamma^2\|w_{k+i}\|^2\right) + \|x_{N|k}\|^2_{\check{W}_x}.$$

and determine the matrices $W_{\mu z}$, $W_{\mu\mu}$ in the online MPC optimization of (3.55). Hence verify that the optimum $\mathbf{c}_0^*$ is unchanged but $\check{J}^*(x_0) = 2.222$.

(d) Why is the predicted cost smaller in (c) than (b)? What are the advantages of (c) relative to (b), and what are the possible disadvantages?

8 In this problem the rigid tube MPC strategy (Sect. 3.5.1) is applied to the system with model $x_{k+1} = Ax_k + Bu_k + w_k$, state constraints $Fx_k \le \mathbf{1}$ and disturbance bounds $w_k \in \mathcal{W}$, $\mathcal{W} \doteq 0.4\mathcal{W}_0$, with A, B, F and $\mathcal{W}_0$ as given in Question 3, and feedback gain $K = K_e = \begin{bmatrix}0.479 & 0.108\end{bmatrix}$, which is optimal for the nominal cost $\sum_{k=0}^{\infty}(\|x_k\|^2 + u_k^2)$ in the absence of constraints.

(a) For $r = 2$ and $\Phi = A + BK$ find the smallest scalar ρ such that $\Phi^r\mathcal{W} \subseteq \rho\mathcal{W}$ and compute $h_{\mathcal{S}} \doteq \max_{e\in\mathcal{S}} Fe$ where

$$\mathcal{S} = \frac{1}{1-\rho}(\mathcal{W} \oplus \Phi \mathcal{W}).$$

(b) For predictions $u_{i|k} = Kx_{i|k} + c_{i|k}$, with $c_{i|k} = 0$ for $i \geq N$ and $N = 1$, verify that the maximal invariant set for the dynamics $z_{k+1} = \Psi z_k$ and constraints $\bar{F} z_k \leq \mathbf{1} - h_{\mathcal{S}}$, $k = 0, 1, \ldots$ is

$$\mathcal{Z}^{\text{MPI}} = \{z \,:\, \bar{F}\Psi^i z \leq \mathbf{1} - h_{\mathcal{S}},\ i = 0, 1, 2\}$$

where $\bar{F} = \begin{bmatrix} F & 0 \end{bmatrix}$ and $\Psi = \begin{bmatrix} \Phi & BE \\ 0 & M \end{bmatrix}$.

(c) The online MPC optimization is performed subject to the constraints $(s_{0|k}, \mathbf{c}_k) \in \mathcal{Z}^{\text{MPI}}$ and $x_k - s_{0|k} \in \mathcal{S}$. Explain the function of the optimization variable $s_{0|k}$ and the reason for including the constraint $x_k - s_{0|k} \in \mathcal{S}$. How can this constraint be expressed in terms of linear conditions on the optimization variables?

(d) Determine the matrices W_x, W_c that define the nominal predicted cost

$$\sum_{i=0}^{\infty} \left(\|s_{i|k}\|^2 + v_{i|k}^2 \right) = \|s_{0|k}\|_{W_x}^2 + \|\mathbf{c}_k\|_{W_c}^2$$

where $s_{i+1|k} = As_{i|k} + Bv_{i|k}$, $v_{i|k} = Ks_{i|k} + c_{i|k}$. Solve the MPC optimization for $x_0 = (0, 1)$ and verify that the optimal value of the objective function is $\|s_{0|0}^*\|_{W_x}^2 + \|\mathbf{c}_0^*\|_{W_c}^2 = 0.122$.

(e) What is the advantage of using a different feedback gain K_e in the definition of $\mathcal{S}$ and implementing the controller as $u_k = Ks_{0|k}^* + K_e(x_k - s_{0|k}^*) + c_{0|k}^*$?

9 The homothetic tube MPC strategy of Sect. 3.5.2 does not require the set $\mathcal{S}^0$ to be robustly invariant for the dynamics, $e_{k+1} = \Phi e_k + w_k$, $w_k \in \mathcal{W}$, of the uncertain component of the predicted state. This question concerns a simplification of the online HTMPC optimization that becomes possible when $\mathcal{S}^0$ is replaced by a set, $\mathcal{S} = \{s : V_{\mathcal{S}} s \leq \mathbf{1}\}$, which is robustly invariant for these dynamics.

(a) Show that, if $\Phi\mathcal{S} \oplus \mathcal{W} \subseteq \mathcal{S}$, then $\bar{e} + \bar{w} \leq \mathbf{1}$ where

$$\bar{e} = \max_{e \in \mathcal{S}} V_{\mathcal{S}} \Phi e \qquad \bar{w} = \max_{w \in \mathcal{W}} V_{\mathcal{S}} w.$$

(b) Show that, if $\nu \geq N - 1$ is an integer such that $\bar{F}\Psi^{\nu+1} z \leq \mathbf{1} - h_{\mathcal{S}}$ (where $h_{\mathcal{S}} = \max_{e \in \mathcal{S}} Fe$, and Ψ, $\bar{F}$ are as defined in Question 8(b)) for all z and $(\alpha_0, \ldots, \alpha_{N-1})$ satisfying

$$\begin{aligned} &\bar{F}\Psi^i z \leq \mathbf{1} - \alpha_i h_{\mathcal{S}}, && i = 0, \ldots, \nu \\ &\alpha_i \bar{e} + \bar{w} \leq \alpha_{i+1} \mathbf{1}, && i = 0, \ldots, N-1 \\ &\alpha_i = 1, && i \geq N, \end{aligned}$$

then, for the system $x_{k+1} = Ax_k + Bu_k + w_k$, $w_k \in \mathcal{W}$ and control law $u_k = Kx_k + c_{0|k}$, there exists $(s_{0|k}, \mathbf{c}_k, \boldsymbol{\alpha}_k)$ satisfying the following constraints for all $k > 0$ if they are feasible at $k = 0$:

$$\begin{aligned} &\bar{F}\Psi^i z_k \leq \mathbf{1} - \alpha_{i|k} h_{\mathcal{S}}, && i = 0, \ldots, \nu \\ &\alpha_{i|k}\bar{e} + \bar{w} \leq \alpha_{i+1|k}\mathbf{1}, && i = 0, \ldots, N-1 \\ &\alpha_{i|k} = 1, && i \geq N \\ &x_k - s_{0|k} \in \alpha_{0|k}\mathcal{S} && \end{aligned}$$

with $z_k = (s_{0|k}, \mathbf{c}_k)$, $\mathbf{c}_k = (c_{0|k}, \ldots, c_{N-1|k})$, $\boldsymbol{\alpha}_k = (\alpha_{0|k}, \ldots, \alpha_{N-1|k})$.

(c) Let $\mathbf{c}_k^* = (c_{0|k}^*, \ldots, c_{N-1|k}^*)$ be optimal for the problem of minimizing the predicted cost at time k:

$$J(s_{0|k}, \mathbf{c}_k, \boldsymbol{\alpha}_k) = \|s_{0|k}\|_{W_x}^2 + \|\mathbf{c}_k\|_{W_c}^2 + \sum_{i=0}^{N-1} q_\alpha(\alpha_{i|k} - 1)^2,$$

over $(s_{0|k}, \mathbf{c}_k, \boldsymbol{\alpha}_k)$ subject to the recursively feasible constraints of part (b), where W_x satisfies the Riccati equation (2.9) for $Q, R \succ 0$, $W_c = \text{diag}\{B^T W_x B + R, \ldots, B^T W_x B + R\}$ and q_α is any nonnegative scalar.

Show that, if this minimization is feasible at $k = 0$, then for any disturbance sequence with $w_k \in \mathcal{W}$ for all $k \geq 0$, the closed-loop system $x_{k+1} = Ax_k + Bu_k + w_k$ under the MPC law $u_k = Kx_k + c_{0|k}^*$ satisfies the constraints $Fx_k \leq \mathbf{1}$ for all $k \geq 0$, and its state converges asymptotically to the set $\mathcal{S}$.

10 This question considers the design and implementation of the homothetic tube MPC strategy of Question 9 for the system model, disturbance bounds and constraints of Question 8.

(a) Using the set $\mathcal{S}$ determined in Question 8, namely

$$\mathcal{S} = \frac{1}{1-\rho}(\mathcal{W} \oplus \Phi\mathcal{W})$$

where ρ is the smallest scalar such that $\Phi^2\mathcal{W} \subseteq \rho\mathcal{W}$, determine $\bar{e}$ and $\bar{w}$ defined in Question 9(a). Verify that, for $N = 1$, the smallest integer ν satisfying the conditions of Question 9(b) is $\nu = 2$.

(b) Taking $q_\alpha = 1$ solve the MPC optimization defined in Question 9(c) for $x_0 = (0, 1)$ and $N = 1$. Why is the optimal solution for $\mathbf{c}_0$ in this problem equal to the optimum for rigid tube MPC computed for the same x_0 in Question 8(d)?

(c) Compare HTMPC in terms of its closed-loop performance and the size of its feasible initial condition set with: (i) the robust MPC algorithm of Question 6 and (ii) the rigid tube MPC algorithm of Question 8.

References

1. I. Kolmanovsky, E.G. Gilbert, Theory and computation of disturbance invariant sets for discrete-time linear systems. Math. Probl. Eng. **4**(4), 317–367 (1998)
2. D.P. Bertsekas, I.B. Rhodes, On the minimax reachability of target sets and target tubes. Automatica **7**, 233–247 (1971)
3. D.P. Bertsekas, *Dynamic Programming and Optimal Control* (Academic Press, New York, 1976)
4. Y.I. Lee, B. Kouvaritakis, Constrained receding horizon predictive control for systems with disturbances. Int. J. Control **72**(11), 1027–1032 (1999)
5. J. Schuurmans, J.A. Rossiter, Robust predictive control using tight sets of predicted states. Control Theory Appl. IEE Proc. **147**(1), 13–18 (2000)
6. D.Q. Mayne, M.M. Seron, S.V. Raković, Robust model predictive control of constrained linear systems with bounded disturbances. Automatica **41**(2), 219–224 (2005)
7. S.V. Rakovic, E.C. Kerrigan, K.I. Kouramas, D.Q. Mayne, Invariant approximations of the minimal robust positively invariant set. IEEE Trans. Autom. Control **50**(3), 406–410 (2005)
8. M. Vidyasagar, *Nonlinear Systems Analysis*, 2nd edn. (Prentice Hall, Upper Saddle River, 1993)
9. I. Yaesh, U. Shaked, Minimum $\mathcal{H}_\infty$-norm regulation of linear discrete-time systems and its relation to linear quadratic discrete games. IEEE Trans. Autom. Control **35**(9), 1061–1064 (1990)
10. T. Başar, A dynamic games approach to controller design: disturbance rejection in discrete-time. IEEE Trans. Autom. Control **36**(8), 936–952 (1991)
11. J. von Neumann, O. Morgenstern, *Theory of Games and Economic Behavior* (Princeton University Press, Princeton, 1944)
12. P.O.M. Scokaert, D.Q. Mayne, Min-max feedback model predictive control for constrained linear systems. IEEE Trans. Autom. Control **43**(8), 1136–1142 (1998)
13. G. Tadmor, Receding horizon revisited: an easy way to robustly stabilize an LTV system. Syst. Control Lett. **18**(4), 285–294 (1992)
14. S. Lall, K. Glover, A game theoretic approach to moving horizon control, in *Advances in Model-Based Predictive Control*, ed. by D.W. Clarke (Oxford University Press, Oxford, 1994), pp. 131–144
15. Y.I. Lee, B. Kouvaritakis, Receding horizon $\mathcal{H}_\infty$ predictive control for systems with input saturation. Control Theory Appl. IEE Proc. **147**(2), 153–158 (2000)
16. L. Magni, H. Nijmeijer, A.J. van der Schaft, A receding-horizon approach to the nonlinear $\mathcal{H}_\infty$ control problem. Automatica **37**(3), 429–435 (2001)
17. A.A. Stoorvogel, A.J.T.M. Weeren, The discrete-time Riccati equation related to the H_∞ control problem. IEEE Trans. Autom. Control **39**(3), 686–691 (1994)
18. R. Fletcher, *Practical Methods of Optimization*, 2nd edn. (Wiley, New York, 1987)
19. S. Boyd, L. Vandenberghe, *Convex Optimization* (Cambridge University Press, Cambridge, 2004)
20. S.V. Raković, B. Kouvaritakis, R. Findeisen, M. Cannon, Homothetic tube model predictive control. Automatica **48**(8), 1631–1638 (2012)
21. S.V. Rakovic, B. Kouvaritakis, M. Cannon, Equi-normalization and exact scaling dynamics in homothetic tube MPC. Syst. Control Lett. **62**(2), 209–217 (2013)
22. J.R. Gossner, B. Kouvaritakis, J.A. Rossiter, Stable generalized predictive control with constraints and bounded disturbances. Automatica **33**(4), 551–568 (1997)
23. Y.I. Lee, B. Kouvaritakis, Linear matrix inequalities and polyhedral invariant sets in constrained robust predictive control. Int. J. Robust Nonlinear Control **10**(13), 1079–1090 (2000)
24. Y.I. Lee, B. Kouvaritakis, A linear programming approach to constrained robust predictive control. IEEE Trans. Autom. Control **45**(9), 1765–1770 (2000)
25. F. Blanchini, S. Miani, *Set-Theoretic Methods in Control* (Birkhäuser, Basel, 2008)
26. G. Bitsoris, On the positive invariance of polyhedral sets for discrete-time systems. Syst. Control Lett. **11**(3), 243–248 (1988)

Chapter 4
Closed-Loop Optimization Strategies for Additive Uncertainty

The performance and constraint handling capabilities of a robust predictive control law are limited by the amount of information on future model uncertainty that is made available to the controller. However, the manner in which the controller uses this information is equally important. Although the realization of future model uncertainty is by definition unknown when a predicted future control trajectory is optimized, this information may be available to the controller at the future instant of time when the control law is implemented. For example, the prediction, $u_{i|k}$, at time k of the control input i steps into the future should ideally depend on the predicted model state i steps ahead, but this is unknown at time k because of the unknown model uncertainty at times $k, \ldots, k+i-1$, even though the state x_{k+i} is, by assumption, available to the controller at time $k+i$. In order to fully exploit the potential benefits of this information, a closed-loop optimization strategy is needed that allows the degrees of freedom over which the predicted trajectories are optimized to depend on the realization of future model uncertainty.

This chapter considers again the control problem defined in terms of a linear system model in Sect. 3.1, which is restated here for convenience:

$$x_{k+1} = Ax_k + Bu_k + Dw_k. \tag{4.1}$$

The matrices A, B, D are known, the state x_k is known when the control input u_k is chosen, and w_k is an unknown additive disturbance input at time k. As in Chap. 3, w is assumed to lie in a compact convex polytopic set $\mathcal{W}$ containing the origin, described either by its vertices:

$$\mathcal{W} = \mathrm{Co}\{w^{(j)},\ j = 1, \ldots, m\}, \tag{4.2}$$

or by the intersection of half-spaces:

$$\mathcal{W} = \{w : Vw \leq \mathbf{1}\}, \tag{4.3}$$

B. Kouvaritakis and M. Cannon, *Model Predictive Control*,
Advanced Textbooks in Control and Signal Processing,
DOI 10.1007/978-3-319-24853-0_4

and the state and control input of (4.1) are subject to linear constraints:

$$Fx_k + Gu_k \leq \mathbf{1}, \quad k = 0, 1, \ldots \tag{4.4}$$

for given matrices $F \in \mathbb{R}^{n_c \times n_x}$ and $G \in \mathbb{R}^{n_c \times n_u}$.

The control objective is to steer the system state into a prescribed target set under any possible realization of model uncertainty, while minimizing an appropriately defined performance cost. Consideration is given to general class of control laws, and hence we remove the restriction to open-loop strategies that was imposed in Chap. 3.

A predictive control law that employs a closed-loop optimization strategy optimizes predicted performance over parameters that define predicted future feedback policies. In its most general setting, the problem can be solved using dynamic programming, and the first part of this chapter gives a brief overview of this technique and its application to robust MPC for systems with additive disturbances. In practice, the computation required by dynamic programming is often prohibitive. To render computation tractable, it may therefore be necessary to restrict the class of control policies considered in the MPC optimization, and the later sections of this chapter consider approaches that optimize over restricted classes of feedback policies in order to reduce computation. We discuss policies with affine dependence on disturbances, and then consider more general piecewise affine parameterizations. As in Chap. 3, the emphasis is on the tradeoff that can be achieved between computational tractability and performance while ensuring satisfaction of constraints for all uncertainty realizations.

4.1 General Feedback Strategies

Robust optimal control laws for constrained systems subject to unknown disturbances were proposed in [1–4]. The essence of these approaches is to construct a robust controller as a feedback solution to the problem of minimizing a worst-case (min-max) performance objective over all admissible realizations of model uncertainty. The description of model uncertainty in terms of sets of allowable parameter values leads naturally to set theoretic methods for robust constraint handling (e.g. [3, 4]). This is the starting point for this chapter's discussion of closed-loop optimization strategies for the system (4.1) with disturbance sets and constraints of the form of (4.3) and (4.4).

Multistage min-max control problems can be solved using Dynamic Programming (DP) techniques [5, 6]. In principle, the approach is able to determine the optimal feedback strategy in a very general setting, i.e. given only the performance objective, the system dynamics, the constraints on states and control and disturbance inputs, and knowledge of the information that is available to the controller at each time step. Consider for example the problem of robustly steering the state of (4.1) from an initial condition x_0, which belongs to an allowable set of initial conditions that is to be made as large as possible, to a pre-specified target set, $\mathcal{X}_T$, over N time steps. This problem can be formulated (see e.g. [4]) as an optimal control problem involving a terminal cost $\mathcal{I}(x_N)$ defined by

$$\mathcal{I}(x_N) \doteq \begin{cases} 0 & \text{if } x_N \in \mathcal{X}_T \\ 1 & \text{otherwise} \end{cases}$$

From the model (4.1) and the assumptions on the information that is available to the controller, it follows that the maximizing disturbance at time i depends on the control input u_i, whereas the minimizing control input must be computed without knowledge of the realization of the disturbance w_i at time i. Therefore the optimal sequence of feedback laws $u_i^*(x)$, $i = 0, \ldots, N-1$ can be obtained by solving the sequential min-max problem:

$$\mathcal{I}_N^*(x_0) \doteq \min_{\substack{u_0 \\ Fx_0+Gu_0 \le \mathbf{1}}} \max_{\substack{w_0 \\ w_0 \in \mathcal{W}}} \cdots \min_{\substack{u_{N-1} \\ Fx_{N-1}+Gu_{N-1} \le \mathbf{1}}} \max_{\substack{w_{N-1} \\ w_{N-1} \in \mathcal{W}}} \mathcal{I}(x_N). \tag{4.5}$$

Dynamic programming solves this problem by recursively determining the optimal costs and control laws for successive stages. Thus let $m = N - i$, where i is the time index and m denotes the number of stages to go until the target set is reached. Then set $\mathcal{I}_0^*(x) \doteq \mathcal{I}(x)$ and, for $m = 1, \ldots, N$, solve:

$$w_i^*(x, u) = \arg \max_{\substack{w \\ w \in \mathcal{W}}} \mathcal{I}_{m-1}^*(Ax + Bu + Dw) \tag{4.6a}$$

$$u_i^*(x) = \arg \min_{\substack{u \\ Fx+Gu \le \mathbf{1}}} \mathcal{I}_{m-1}^*\big(Ax + Bu + Dw_i^*(x, u)\big) \tag{4.6b}$$

with

$$\mathcal{I}_m^*(x) \doteq \mathcal{I}_{m-1}^*\Big(Ax + Bu_i^*(x) + Dw_i^*\big(x, u_i^*(x)\big)\Big) \tag{4.6c}$$

The sequence of feedback laws $u_i^*(x_i)$, $i = 0, \ldots, N-1$ satisfying (4.6b) necessarily steers the state of (4.1) into $\mathcal{X}_T$ over N steps, whenever this is possible for the given constraints, disturbance set and initial condition. The set of initial conditions from which $\mathcal{X}_T$ can be reached in N steps is referred to as the N-step controllable set to $\mathcal{X}_T$.

Definition 4.1 (*Controllable set*) The N-step controllable set to $\mathcal{X}_T$ is the set of all states x_0 of (4.1) for which there exists a sequence of feedback laws $u_i^*(x)$, $i = 0, \ldots, N-1$ such that, for any admissible disturbance sequence $\{w_0, \ldots, w_{N-1}\} \in \mathcal{W} \times \cdots \times \mathcal{W}$, we obtain $Fx_i + Gu_i \le \mathbf{1}$, $i = 0, \ldots, N-1$ and $x_N \in \mathcal{X}_T$ along trajectories of (4.1) under $u_i = u_i^*(x_i)$.

In (4.6), the problem of computing the N-step controllable set is split into N subproblems, each of which requires the computation of a 1-step controllable set. The conceptual description of the procedure given by (4.6) has the following geometric interpretation [4]. Let $\mathcal{X}^{(m)}$ denote the m-step controllable set to $\mathcal{X}_T$. Then clearly $\mathcal{X}^{(1)}$ is the set of states x such that there exists u satisfying $Fx + Gu \le \mathbf{1}$ and $Ax + Bu + Dw \in \mathcal{X}_T$ for all $w \in \mathcal{W}$, i.e.

$$\mathcal{X}^{(1)} = \{x : \exists u \text{ such that } Fx + Gu \le \mathbf{1}, \text{ and } Ax + Bu \in \mathcal{X}_T \ominus D\mathcal{W}\}.$$

By (4.6), the m-step controllable set $\mathcal{X}^{(m)}$ is defined in terms of $\mathcal{X}^{(m-1)}$ for $m = 1, 2, \ldots, N$ by

$$\mathcal{X}^{(m)} = \{x : \exists u \text{ such that } Fx + Gu \le \mathbf{1} \text{ and } Ax + Bu \in \mathcal{X}^{(m-1)} \ominus D\mathcal{W}\}$$

with $\mathcal{X}^{(0)} = \mathcal{X}_T$. Given the polytopic uncertainty description (4.2) and the linearity of the dynamics (4.1) and constraints (4.4), $\mathcal{X}^{(m)}$ is therefore convex and polytopic whenever it is non-empty. Hence, given the representation $\mathcal{X}^{(m-1)} = \{x : H^{(m-1)}x \le \mathbf{1}\}$, $\mathcal{X}^{(m)}$ can be determined for $m = 1, 2, \ldots, N$ using the following procedure.

Algorithm 4.1 (*Controllable sets*) Set $\mathcal{X}^{(0)} = \mathcal{X}_T$. For $m = 1, 2, \ldots, N$:

(i) Compute $\hat{\mathcal{X}}^{(m-1)} \doteq \mathcal{X}^{(m-1)} \ominus D\mathcal{W}$. Since $\mathcal{X}^{(m-1)} = \{x : H^{(m-1)}x \le \mathbf{1}\}$, this is given by $\hat{\mathcal{X}}^{(m-1)} = \{x : H^{(m-1)}x \le \mathbf{1} - h_{m-1}\}$ where each element of h_{m-1} is the solution of a linear program, namely

$$h_{m-1} = \max_{w \in \mathcal{W}} H^{(m-1)}w.$$

(ii) Define $\mathcal{Y}^{(m-1)} \subseteq \mathbb{R}^{n_x + n_u}$ as the set

$$\mathcal{Y}^{(m-1)} \doteq \left\{(x, u) : \begin{bmatrix} H^{(m-1)}A & H^{(m-1)}B \\ F & G \end{bmatrix} \begin{bmatrix} x \\ u \end{bmatrix} \le \mathbf{1} - \begin{bmatrix} h_{m-1} \\ 0 \end{bmatrix}\right\}.$$

(iii) Compute $\mathcal{X}^{(m)}$ as the projection of $\mathcal{Y}^{(m-1)}$ onto the x-subspace:

$$\mathcal{X}^{(m)} = \{x : \exists u \text{ such that } (x, u) \in \mathcal{Y}^{(m-1)}\}.$$

and determine $H^{(m)}$ such that $\{x : H^{(m)}x \le \mathbf{1}\} = \mathcal{X}^{(m)}$. ◁

In applications of dynamic programming to receding horizon control problems, the target set $\mathcal{X}_T$ is usually chosen as a robustly controlled positively invariant (RCPI) set, i.e. $\mathcal{X}_T$ is robustly positively invariant for (4.1) and (4.4) under some feedback law. In this case, it is easy to show that the controllable sets $\mathcal{X}^{(m)}$, $m = 1, \ldots, N$ are themselves RCPI and nested:

$$\mathcal{X}^{(N)} \supseteq \mathcal{X}^{(N-1)} \supseteq \cdots \supseteq \mathcal{X}^{(1)} \supseteq \mathcal{X}_T.$$

This nested property necessarily holds because, under the assumption that $\mathcal{X}_T$ is RCPI, $\mathcal{X}_T$ must lie in m-step controllable set to $\mathcal{X}_T$ for $m = 1, 2, \ldots$ It follows that $\mathcal{X}_T \subseteq \mathcal{X}^{(1)}$ and hence $\mathcal{X}^{(1)}$ is itself RCPI. By the same argument therefore, $\mathcal{X}^{(m)} \subseteq \mathcal{X}^{(m+1)}$, and hence $\mathcal{X}^{(m)}$ is RCPI for $m = 1, 2, \ldots$.

To illustrate the procedure of Algorithm 4.1, we next consider its application to a first-order system. For this simple example, the controllable set can be determined by straightforward algebra, providing some additional insight into each step of Algorithm 4.1. We also compare the controllable set for a general closed-loop strategy with the sets of initial conditions from which a given terminal set can be reached under either a fixed feedback law or an open-loop strategy.

Example 4.1 A first-order system with control and disturbance inputs is described by the dynamics

$$x_{k+1} = x_k + u_k - w_k.$$

The state x, control input u and disturbance input w are scalars that are constrained to lie in intervals:

$$x \in [0, X], \quad u \in [0, U], \quad w \in [0, W].$$

The control objective is to steer the state into a target set $\mathcal{X}_T = [\underline{x}^{(0)}, \bar{x}^{(0)}]$.

In this example, $\mathcal{X}^{(m)}$, the m-step controllable set to $\mathcal{X}_T$, is equal to an interval on the real line, i.e. for each $m = 1, \ldots, N$ we have

$$\mathcal{X}^{(m)} = [\underline{x}^{(m)}, \bar{x}^{(m)}].$$

It is therefore straightforward to determine the recursion relating the $(m-1)$-step controllable set to the m-step set without using Algorithm 4.1. Consider first the conditions defining the upper limit of $\mathcal{X}^{(m)}$ in terms of $\bar{x}^{(m-1)}$. By linearity, $\bar{x}^{(m-1)}$ must be the 1-step ahead state from $\bar{x}^{(m)}$ for some control input u and disturbance w. Since $u \geq 0$ and $w \geq 0$, the maximum over u of the minimum over w of $\bar{x}^{(m)} = \bar{x}^{(m-1)} - u + w$ is obtained with $u = w = 0$. Similarly, for the lower limit of $\mathcal{X}^{(m)}$, the worst case value of w (in the sense of maximizing $\underline{x}^{(m)} = \underline{x}^{(m-1)} - u + w$) is W, while the minimizing value of u is U. Taking into account the constraint that $0 \leq \underline{x}^{(m)} \leq \bar{x}^{(m)} \leq X$, we therefore obtain

$$\underline{x}^{(m)} = \max\{0,\ \underline{x}^{(m-1)} + W - U\}, \quad \bar{x}^{(m)} = \min\{X,\ \bar{x}^{(m-1)}\}. \qquad (4.7)$$

We can verify this result using Algorithm 4.1: $\hat{\mathcal{X}}^{(m-1)} = \mathcal{X}^{(m-1)} \ominus [-W, 0]$ in step (i) yields

$$\hat{\mathcal{X}}^{(m-1)} = [\underline{x}^{(m-1)} + W,\ \bar{x}^{(m-1)}],$$

so the conditions defining $\mathcal{Y}^{(m-1)}$ in step (ii) are

$$0 \leq x \leq X, \quad \underline{x}^{(m-1)} + W \leq x + u \leq \bar{x}^{(m-1)}, \quad 0 \leq u \leq U,$$

and, performing step (iii) by eliminating u using

$$
\begin{aligned}
x + u &\geq \underline{x}^{(m-1)} + W && \forall u \in [0, U] \text{ if and only if } x \geq \underline{x}^{(m-1)} + W - U \\
x + u &\leq \bar{x}^{(m-1)} && \forall u \in [0, U] \text{ if and only if } x \leq \bar{x}^{(m-1)}
\end{aligned}
$$

we therefore obtain $\mathcal{X}^{(m)} = [\underline{x}^{(m)}, \bar{x}^{(m)}]$ as defined in (4.7).

To demonstrate the improvement in the size of controllable set that is achievable using a general feedback law rather than a particular linear feedback law or open-loop control, consider a specific example with the constraints

$$x \in [0, 5], \quad u \in [0, 1], \quad w \in [0, \tfrac{1}{2}],$$

and target set

$$\mathcal{X}_T = [\underline{x}^{(0)}, \bar{x}^{(0)}] = [3, 4].$$

In this case the control input can exert a greater influence on the state than the disturbance input since $U > W$, and it is therefore to be expected that the m-step controllable set converges for finite m to the maximal controllable set for any horizon. Figure 4.1 shows that $\mathcal{X}^{(m)}$ for $m \geq 6$ is equal to the maximal set $\mathcal{X}^{(\infty)} = [0, 4]$. (Note that the upper limit of $\mathcal{X}^{(m)}$ cannot exceed the upper limit $\bar{x}^{(0)} = 4$ of the target interval because u is constrained to be non-negative.)

With an open-loop control law such as

$$u = \hat{w}, \quad \hat{w} = W/2 = \tfrac{1}{4},$$

we get $\mathcal{X}^{(m)} = [\underline{x}^{(m-1)} + W - \hat{w}, \bar{x}^{(m-1)} - \hat{w}] = [\underline{x}^{(m-1)} + \frac{1}{4}, \bar{x}^{(m-1)} - \frac{1}{4}]$. Consequently the set of states from which $\mathcal{X}_T$ can be reached in m steps necessarily shrinks as m increases and is in fact empty for $m > 2$ (Fig. 4.1).

Consider next the maximal controllable sets that can be obtained with a linear feedback law. A solution of (4.6b) for the optimal feedback is given by

$$u = \min\{U, \ \max\{0, \ K(x - \bar{x}^{(0)})\}\}$$

with $K = -1$ for all m, and it can moreover be shown that any control law capable of generating the m-step controllable set for $m > 2$ must depend nonlinearly on x. However it is possible to approximate the m-step controllable set using a linear feedback law. Thus, for example

$$u = K(x - \bar{x}^{(0)})$$

ensures $u = 0$ for $x = \bar{x}^{(0)} = 4$, which maximizes the upper limit of the set of initial conditions that can be driven in m steps to $\mathcal{X}_T$ under linear feedback. The gain must satisfy $K < -0.5$ in order that the lower limit of the m-step set to $\mathcal{X}_T$ extends below

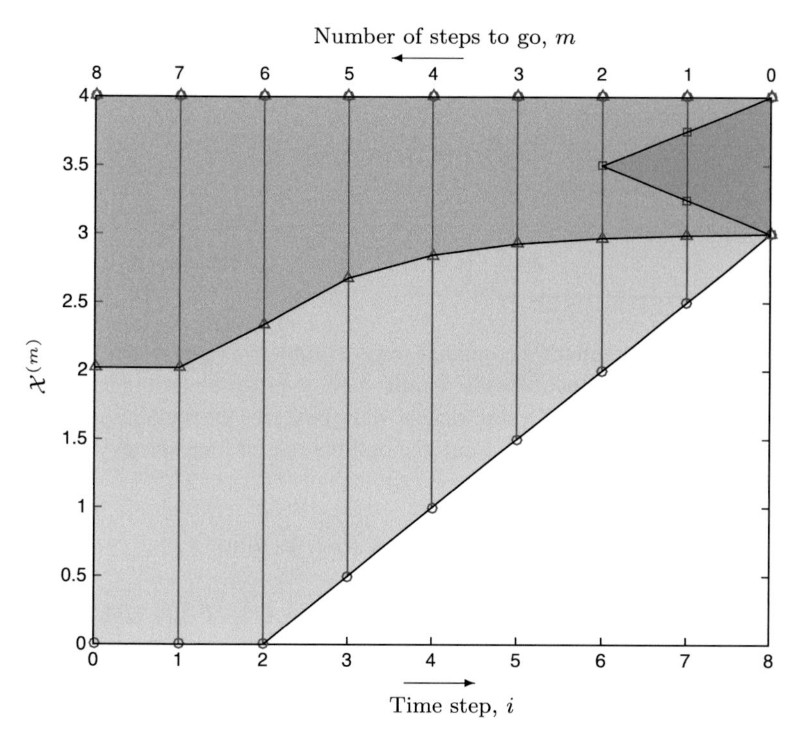

Fig. 4.1 The m-step controllable sets for Example 4.1 with $m \le 8$ for general feedback laws (*marked with* ○). Also shown are the sets of states that can be steered to $\mathcal{X}_T = [3, 4]$ in m steps under the linear feedback law $u = -0.51(x - 4)$ (*marked* △), and open-loop control $u = \hat{w}$ (*marked* □)

$\underline{x}^{(0)} = 3$, and for this range of gains, the asymptotic size of the m-step set for large m is limited by the constraint $u \le 1$. Hence the asymptotic m-step set under this linear feedback law necessarily increases as $|K|$ decreases, while the number of steps needed for convergence to this set increases as $|K|$ decreases. For a horizon $N = 8$, the 8-step set is maximized with $K = -0.51$ (to 2 decimal places); this is shown in Fig. 4.1. ◊

In addition to its uses in computing robust controllable sets, dynamic programming can also be used to solve problems involving the minimization of performance indices consisting of sums of stage costs over a horizon. For example, the optimal value of the quadratic min-max cost with horizon N considered in Sect. 3.4:

$$\check{J}_N^*(x_0) \doteq \min_{\substack{u_0 \\ Fx_0+Gu_0 \le \mathbf{1}}} \max_{\substack{w_0 \\ w_0 \in \mathcal{W}}} \cdots \min_{\substack{u_{N-1} \\ Fx_{N-1}+Gu_{N-1} \le \mathbf{1} \\ x_N \in \mathcal{X}_T}} \max_{\substack{w_{N-1} \\ w_{N-1} \in \mathcal{W}}} \sum_{i=0}^{N-1} (\|x_i\|_Q^2 + \|u_i\|_R^2 - \gamma^2 \|w_i\|^2) + \|x_N\|_{\check{W}_x}^2 \tag{4.8}$$

can be rewritten, using the fact that the optimal values of u_i and w_i depend, respectively, on x_i and (x_i, u_i), as

$$\check{J}_N^*(x_0) = \min_{\substack{u_0 \\ Fx_0+Gu_0\leq \mathbf{1}}} \max_{\substack{w_0 \\ w_0\in\mathcal{W}}} \Bigg\{ \|x_0\|_Q^2 + \|u_0\|_R^2 - \gamma^2\|w_0\|^2 + \cdots + \min_{\substack{u_{N-1} \\ Fx_{N-1}+Gu_{N-1}\leq \mathbf{1} \\ x_N\in\mathcal{X}_T}} \max_{\substack{w_{N-1} \\ w_{N-1}\in\mathcal{W}}} \Big\{ \|x_{N-1}\|_Q^2 + \|u_{N-1}\|_R^2 - \gamma^2\|w_{N-1}\|^2 + \|x_N\|_{\check{W}_x}^2 \Big\}\Bigg\}.$$

This exposes the structure of the optimal control problem as a sequence of subproblems. Each subproblem involves the optimization, over the control and disturbance inputs at a given time step, of a single stage of the cost plus the remaining cost-to-go. Therefore, analogously to (4.6a–4.6c), the optimal closed-loop strategy at time i is given by the solution of

$$\big(u_i^*(x), w_i^*(x,u)\big) = \arg \min_{\substack{u \\ Fx+Gu\leq \mathbf{1}}} \max_{\substack{w \\ w\in\mathcal{W}}} \check{J}_{N-i}(x,u,w) \quad \text{subject to } Ax+Bu \in \mathcal{X}^{(N-i-1)} \ominus D\mathcal{W}, \tag{4.9a}$$

where $\mathcal{X}^{(m)}$ is the m-step controllable set to a given target set $\mathcal{X}_T$, and where $m = N - i$ is the number of time steps until the end of the N-step horizon. The cost with m steps to go, $\check{J}_m$, is defined for $m = 1, 2, \ldots, N$ by the dynamic programming recursion

$$\check{J}_m(x,u,w) = \|x\|_Q^2 + \|u\|_R^2 - \gamma^2\|w\|^2 + \check{J}_{m-1}^*(Ax+Bu+Dw) \tag{4.9b}$$

with

$$\check{J}_{N-i}^*(x) = \check{J}_{N-i}\Big(x, u_i^*(x), w_i^*\big(x, u_i^*(x)\big)\Big) \tag{4.9c}$$

and the terminal conditions:

$$\check{J}_0^*(x) = \|x\|_{\check{W}_x}^2, \tag{4.9d}$$

$$\mathcal{X}^{(0)} = \mathcal{X}_T. \tag{4.9e}$$

The corresponding receding horizon control law for a prediction horizon of N time steps is given by $u = u_N^*(x)$.

The decomposition of (4.8) into a sequence of single-stage min-max problems in (4.9a–4.9e) enables the optimal feedback law to be determined without imposing a suboptimal controller parameterization on the problem. However, (4.9) shows an obvious difficulty with this approach, namely that at each stage $m = 1, 2, \ldots, N$ the function $\check{J}_{m-1}^*(x)$ giving the optimal cost for $m - 1$ stages must be known in order to be able to determine the optimal cost and control law for the m-stage problem. Conventional implementations of dynamic programming for this problem therefore require the optimal control problem to be solved globally, for all admissible states. This makes the method computationally intensive and, crucially, it results in poor scalability of the approach because the computational demand grows exponentially with the dimensions of the state and input variables.

Predictive controllers aim to avoid such computational difficulties by optimizing the predicted trajectories that emanate from a particular model state rather than globally, and this is the focus of the remainder of this chapter. We next describe two general feedback strategies before discussing parameterized feedback strategies that optimize over restricted classes of feedback law. The remainder of this chapter uses the following example problem to illustrate and compare robust control laws.

Example 4.2 A triple integrator with control and disturbance inputs is given by (4.1) with

$$A = \begin{bmatrix} 1 & 1 & 0 \\ 0 & 1 & 1 \\ 0 & 0 & 1 \end{bmatrix}, \quad B = \begin{bmatrix} 0 \\ 0 \\ 1 \end{bmatrix}, \quad D = \begin{bmatrix} 1 & 0 & 0 \\ 0 & 1 & 0 \\ 0 & 0 & 1 \end{bmatrix}.$$

The disturbance input w is constrained to lie in the set $\mathcal{W}$ defined by

$$\mathcal{W} = \left\{ w : \begin{bmatrix} -0.25 \\ -0.25 \\ -0.25 \end{bmatrix} \leq w \leq \begin{bmatrix} 0.25 \\ 0.25 \\ 0.25 \end{bmatrix} \right\}$$

and the state x and control input u are subject to constraints

$$-500 \leq \begin{bmatrix} 1 & 0 & 0 \end{bmatrix} x \leq 5, \quad -4 \leq u \leq 4$$

so that the constraints take the form of (4.4) with

$$F = \begin{bmatrix} 0.2 & 0 & 0 \\ -0.002 & 0 & 0 \\ 0 & 0 & 0 \\ 0 & 0 & 0 \end{bmatrix}, \quad G = \begin{bmatrix} 0 \\ 0 \\ 0.25 \\ -0.25 \end{bmatrix}.$$

The cost is defined by (4.8) with the weights

$$Q = \begin{bmatrix} 1 & 0 & 0 \\ 0 & 0 & 0 \\ 0 & 0 & 0 \end{bmatrix}, \quad R = 0.1, \quad \gamma^2 = 50,$$

and the terminal weighting matrix $\check{W}_x$ is defined as the solution of the Riccati equation (3.42). The terminal set $\mathcal{X}_T$ is defined as the maximal RPI set for (4.1) and (4.4) under $u = Kx$, where $K = [-0.77\ -2.40\ -2.59]$ is the optimal feedback gain for the infinite horizon cost (3.39).

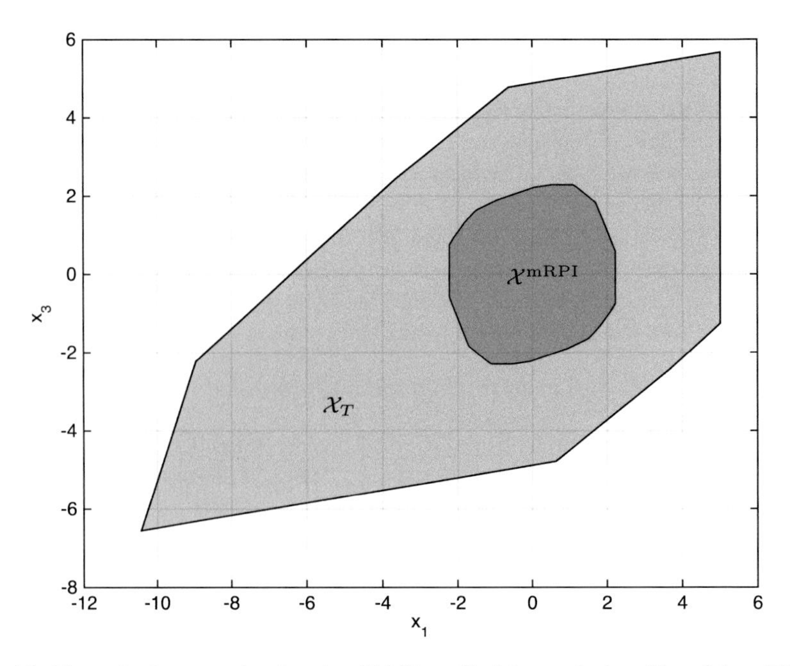

Fig. 4.2 The projections onto the plane $\{x : [0\ 1\ 0]x = 0\}$ of the terminal set $\mathcal{X}_T$ and the mRPI set $\mathcal{X}^{\text{mRPI}}$ under $u = Kx$

Figure 4.2 shows the projection of the terminal set $\mathcal{X}_T$ onto the plane on which $[0\ 1\ 0]x = 0$. For comparison, the projection of the minimal RPI set under $u = Kx$ is also shown in this figure. The projections of the N-step controllable sets to $\mathcal{X}_T$ onto the plane $[0\ 1\ 0]x = 0$, for $4 \leq N \leq 10$, are shown in Fig. 4.3. ◊

4.1.1 Active Set Dynamic Programming for Min-Max Receding Horizon Control

At each stage of the min-max problem defined in (4.9a–4.9e), the optimal control and worst case disturbance inputs are piecewise affine functions of the model state at that stage. This is a consequence of the quadratic nature of the cost and the linearity of the constraints, and it implies that the optimal control law for the full N-stage problem (4.8) is a piecewise affine function of the initial state x_0. Each constituent affine feedback law of this function depends on which of the inequality constraints are active (namely which constraints hold with equality) at the solution, and it follows that the regions of state space within which the optimal control law is given by a particular affine state feedback law are convex and polytopic.

A possible solution method consists of computing offline all of these polytopic regions and their associated affine feedback laws, in a similar manner to the multi-

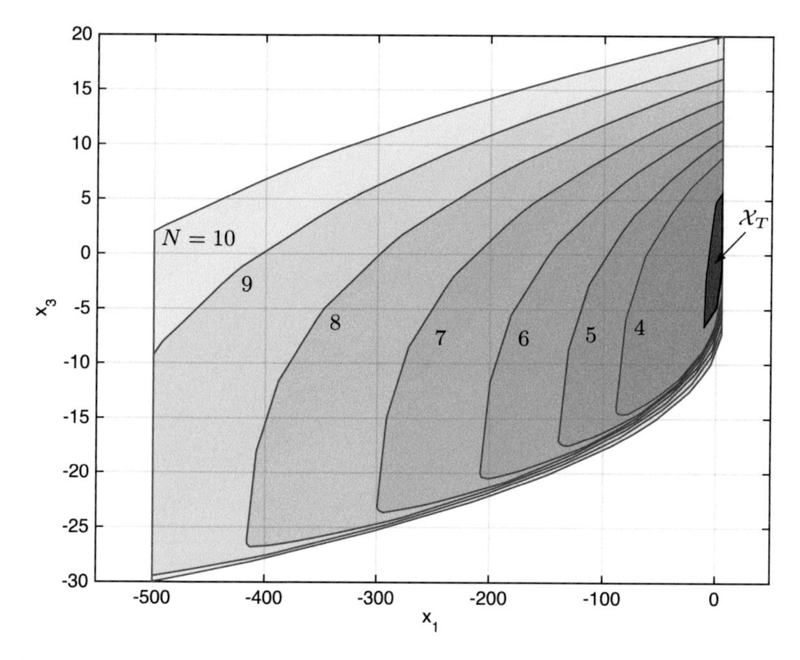

Fig. 4.3 The projections onto the plane $\{x : [0\ 1\ 0]x = 0\}$ of the N-step controllable sets to $\mathcal{X}_T$ for $N = 4, 5, \ldots, 10$ and the terminal set $\mathcal{X}_T$

parametric approach considered in Sect. 2.8. The optimal feedback law could then be determined online by identifying which region of state space contains the current state. This approach is described in the context of linear systems with unknown disturbances and piecewise linear cost indices in [7]. However, as in the case of no model uncertainty, the approach suffers from poor scalability of its computational requirements with the dimension of the state and the horizon length. An illustration of this is given in Example 4.3, where, for a horizon of $N = 4$ the number of affine functions and polytopic regions defining the optimal control law for (4.9a–4.9e) is around 100 (Fig. 4.4). For $N = 10$, the number of regions increases to around 10,000 (Fig. 4.5), implying a large increase in the online computation that is required to determine which region contains the current state, as well as large increases in the offline computation and storage requirements of the controller.

To avoid computing the optimal feedback law at all points in state space, it is possible instead to use knowledge of the solution of (4.9a–4.9e) at a particular point in state space to determine the optimal control law at the current plant state. This is the motivation behind the homotopy-based methods for constrained receding horizon control proposed for uncertainty-free control problems in [8, 9] and developed subsequently for min-max robust control problems with bounded additive disturbances in [10, 11].

Algorithms based on homotopy track the changes in the optimal control and worst case disturbance inputs as the system state varies in state space. For the problem

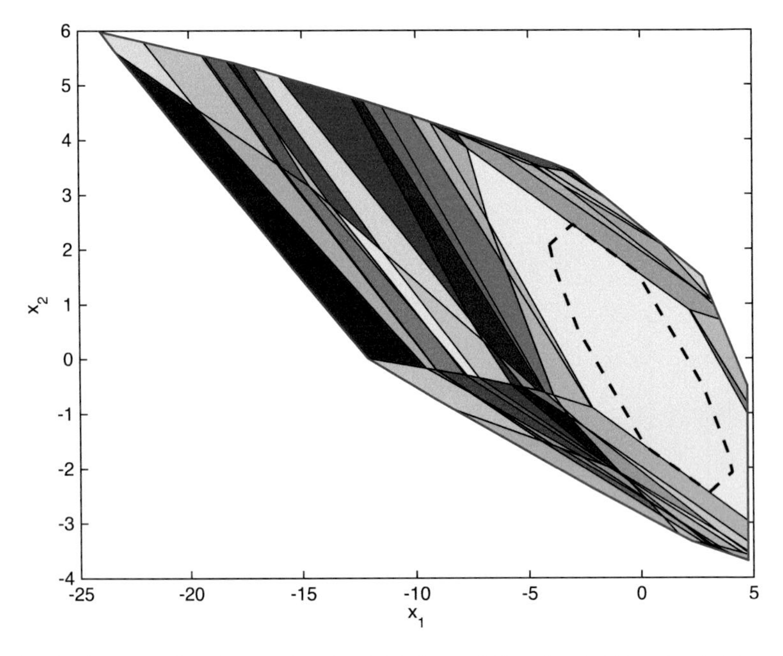

Fig. 4.4 Active set regions in the plane $\{x : [0\ 1\ 0]x = 0\}$ for $N = 4$ (74 regions). The *dashed lines* show the intersection of the terminal set $\mathcal{X}_T$ with this plane

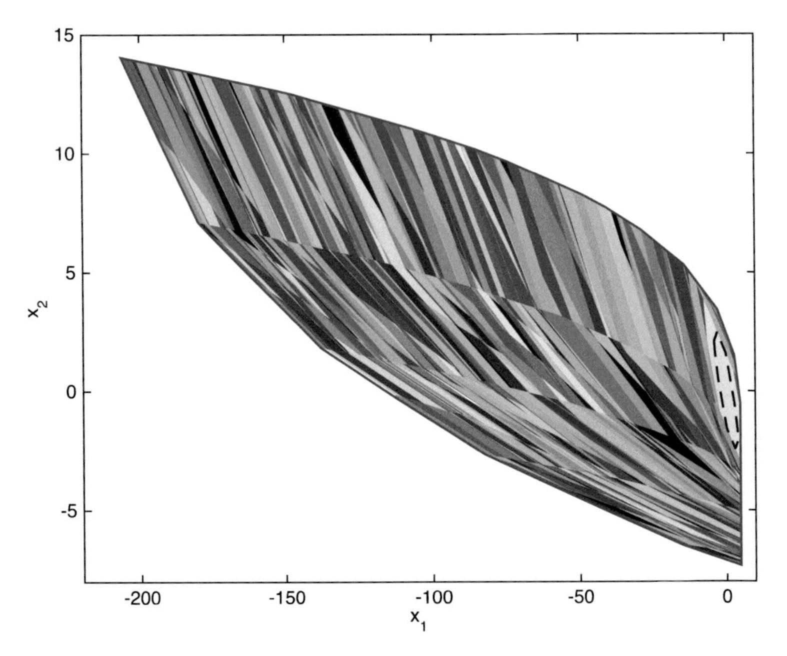

Fig. 4.5 Active set regions in the plane $\{x : [0\ 1\ 0]x = 0\}$ for $N = 10$ (6352 regions). The *dashed lines* show the intersection of the terminal set $\mathcal{X}_T$ with this plane

(4.9a–4.9e), this is done by determining, for a given set of active constraints, the optimal feedback laws at each stage of the problem and hence the optimal state, control and worst case disturbance trajectories as functions of the initial model state. As the initial model state x_0 moves along a search path in state space towards the location of the current plant state, a change in the active set is detected by determining the point at which any inactive constraint becomes active, or any active constraint becomes inactive. This information allows the active set to be updated, and a new piecewise affine control law to be determined, thus enabling the initial model state to move into a new region of state space while continuing to track the optimal solution.

With the search path for the initial model state x_0 defined as a straight line in state space, the detection of an active set change becomes a univariate feasibility problem with linear constraints. Furthermore the computational requirement of updating the optimal solution after an active set change scales linearly with the horizon length and polynomially with state and input dimensions. Clearly this is not the whole story because the overall computation also depends on the number of active set changes, and, as is usual for an active set solver for a constrained optimization problem, upper bounds on this grow exponentially with the problem size. In practice, however, the number of active set changes is likely to be small, and, when used in a receding horizon control setting, the algorithm can be initialized using knowledge of the active set corresponding to the optimal solution at a previous sampling instant.

A summary of the method is as follows. Suppose the optimal cost for the sub-problem (4.9a) with $N-i-1$ stages to go is given by

$$\check{J}^*_{N-i-1}(x) = x^T P_i x + 2x_i^T q_i + r_i \tag{4.10}$$

for some matrix P_i, vector q_i and scalar r_i, and let the active constraints in problem (4.9a) be given by

$$E_i(Ax + Bu) = \mathbf{1} \tag{4.11a}$$

$$F_i x + G_i u = \mathbf{1} \tag{4.11b}$$

$$V_i w = \mathbf{1} \tag{4.11c}$$

(where $E_i(Ax + Bu) = \mathbf{1}$ represents the active constraints in the condition $Ax + Bu \in \mathcal{X}^{N-i-1} \ominus D\mathcal{W}$). The first-order optimality conditions defining the maximizing function $w_i^*(x, u)$ are given by

$$\begin{bmatrix} \gamma^2 I - D^T P_i D & V_i^T \\ V_i & 0 \end{bmatrix} \begin{bmatrix} w_i \\ \eta_i \end{bmatrix} = \begin{bmatrix} D^T P_i \\ 0 \end{bmatrix} (Ax + Bu) + \begin{bmatrix} D^T q_i \\ \mathbf{1} \end{bmatrix},$$

where η_i is a vector of Lagrange multipliers for the constraints (4.11c). Under the assumption that γ is sufficiently large that the maximization subproblem in (4.9a) is concave, namely that

$$V_{i,\perp}^T \big(\gamma^2 I - D^T P_i D\big) V_{i,\perp} \succ 0,$$

where the columns of $V_{i,\perp}$ span the kernel of V_i, the maximizer $w_i^*(x,u)$ is unique and is given by

$$\begin{bmatrix} w_i^*(x,u) \\ \eta_i^*(x,u) \end{bmatrix} = M_i(Ax + Bu) + m_i$$

for some matrix M_i and vector m_i. Similarly, for the minimization subproblem in (4.9a), let

$$\begin{bmatrix} \hat{P}_i & \hat{q}_i \end{bmatrix} = \begin{bmatrix} P_i & q_i \end{bmatrix} + \begin{bmatrix} P_i D & 0 \end{bmatrix} \begin{bmatrix} M_i & m_i \end{bmatrix}.$$

Then the first-order optimality conditions defining the minimimizing function $u_i^*(x)$ are given by

$$\begin{bmatrix} R + B^T \hat{P}_i B & B^T E_i^T & G_i^T \\ E_i B & 0 & 0 \\ G_i & 0 & 0 \end{bmatrix} \begin{bmatrix} u_i \\ \nu_i \\ \mu_i \end{bmatrix} = - \begin{bmatrix} B^T \hat{P}_i A \\ E_i A \\ F_i \end{bmatrix} x_k + \begin{bmatrix} -B^T \hat{q}_k \\ \mathbf{1} \\ \mathbf{1} \end{bmatrix},$$

where ν_i, η_i are, respectively, Lagrange multipliers for the constraints (4.11a), (4.11b). Assuming convexity of the minimization in (4.9a), or equivalently

$$\begin{bmatrix} B^T E_i^T & G_i^T \end{bmatrix}_\perp (R + B^T \hat{P}_i B) \begin{bmatrix} E_i B \\ G_i \end{bmatrix}_\perp \succ 0$$

where the columns of $\begin{bmatrix} E_i B \\ G_i \end{bmatrix}_\perp$ span the kernel of $\begin{bmatrix} E_i B \\ G_i \end{bmatrix}$, then

$$\begin{bmatrix} u_i^*(x) \\ \nu_i^*(x) \\ \mu_i^*(x) \end{bmatrix} = L_i x + l_i,$$

for some matrix L_i and vector l_i.

The affine forms of $u_i^*(x)$ and $w_i^*(x,u)$ imply that the optimal cost $\check{J}_{N-i}^*(x)$ for $N-i$ stages to go is quadratic, justifying by induction the quadratic form assumed in (4.10). Furthermore, having determined M_i, m_i, L_i, l_i for each $i = 0, \ldots, N-1$ for a given active set, it is possible to express as functions of x_0 the sequences of minimizing control inputs and maximizing disturbance inputs

$$\mathbf{u}(x_0) = \{u_0^*, \ldots, u_{N-1}^*\}, \quad \mathbf{w}(x_0) = \{w_0^*, \ldots, w_{N-1}^*\},$$

and the corresponding multiplier sequences

$$\boldsymbol{\eta}(x_0) = \{\eta_0^*, \ldots, \eta_{N-1}^*\}, \quad \boldsymbol{\nu}(x_0) = \{\nu_0^*, \ldots, \nu_{N-1}^*\}, \quad \boldsymbol{\mu}(x_0) = \{\mu_0^*, \ldots, \mu_{N-1}^*\}.$$

If $\mathbf{u}(x_0)$ and $\mathbf{w}(x_0)$ satisfy the constraints of (4.9a–4.9e) and if the multipliers satisfy $\boldsymbol{\eta}(x_0) \geq 0$, $\boldsymbol{\nu}(x_0) \geq 0$, $\boldsymbol{\mu}(x_0) \geq 0$, then these sequences are optimal for the N-stage problem (4.8) provided the constraints in (4.9a–4.9e) are linearly dependent. The affine form of each of these functions implies that x_0 must lie in a convex polytopic set, which for a given active constraint set, denoted by $\mathcal{A}$, we denote as $\mathcal{R}(\mathcal{A})$.

Algorithm 4.2 (*Online active set DP*)
Offline: compute the controllable sets $\mathcal{X}^{(1)}, \mathcal{X}^{(2)}, \ldots, \mathcal{X}^{(N)}$ to $\mathcal{X}_T$.
Online, at each time instant $k = 0, 1, \ldots$:

(i) Set $x = x_k$ and initialize the solver with $x_0^{(0)}$ and an active set $\mathcal{A}^{(0)}$ such that $x_0^{(0)} \in \mathcal{R}(\mathcal{A}^{(0)})$. At each iteration $j = 0, 1, \ldots$:

(a) Determine M_i, m_i, L_i, l_i for $i = N-1, N-2, \ldots, 0$, and hence determine $\mathbf{u}(x_0)$, $\mathbf{w}(x_0)$, $\boldsymbol{\eta}(x_0)$, $\boldsymbol{\nu}(x_0)$, $\boldsymbol{\mu}(x_0)$.

(b) Perform the line search:

$$\alpha^{(j)} = \max_{\alpha \le 1} \ \alpha \ \text{ subject to } x_0^{(j)} + \alpha(x - x_0^{(j)}) \in \mathcal{R}(\mathcal{A}^{(0)}).$$

(c) If $\alpha^{(j)} < 1$, set $x_0^{(j+1)} = x_0^{(j)} + \alpha^{(j)}(x - x_0^{(j)})$ and use the active boundary of $\mathcal{R}(\mathcal{A}^{(j)})$ to determine $\mathcal{A}^{(j+1)}$ from $\mathcal{A}^{(j)}$.
Otherwise set $\mathcal{A}_k^* = \mathcal{A}^{(j)}$ and proceed to (ii).

(ii) Apply the control law $u_k = u_0^*(x)$. $\triangleleft$

If the constraints of (4.9a) are linearly dependent, the optimal multiplier sequences may be non-unique [12]. This situation can be handled by introducing into the problem additional equality constraints that enforce compatibility of the linearly dependent constraints. The first-order optimality conditions can then be used to relate the multipliers of these additional constraints to the free variables appearing in the solutions for the multipliers of linearly dependent constraints. Furthermore the degrees of freedom in the optimal multiplier sequences can be chosen so as to ensure that the multipliers are continuous when the active set changes. With this approach, the sequences of primal and dual variables are again determined uniquely as functions of x_0, and the continuity of these sequences at the boundaries of $\mathcal{R}(\mathcal{A})$ in x_0-space is preserved (see [11] for details).

The optimization in step (i) of Algorithm 4.2 can be initialized with a cold start by setting $x_0^{(0)} = 0$ and $\mathcal{A}^{(0)} = \emptyset$, since by assumption the origin lies inside the minimal RPI set for the unconstrained optimal feedback law $u = Kx$, and hence all constraints are inactive for this choice of x_0. Alternatively, if the previous optimal solution is known, then it can be warm started by setting $x_0^{(0)}$ at time k equal to the plant state x_{k-1} at the previous time step and setting $\mathcal{A}^{(0)}$ equal to the corresponding optimal active set $\mathcal{A}_{k-1}^*$. Convergence of the optimization in step (i) in a finite number of iterations is guaranteed since the active set $\mathcal{A}^{(j+1)}$ is uniquely defined at each iteration and since there are a finite number of possible active sets.

Finally it can be shown that the control law of Algorithm 4.2 is recursively feasible and robustly stabilizing for the system (4.1), for all initial conditions in the N-step controllability set, $\mathcal{X}^{(N)}$, to $\mathcal{X}_T$. Recursive feasibility follows from the constraints of (4.9a–4.9e) since these ensure that, at each time k, x_{k+1} will necessarily be steered into $\mathcal{X}^{(N-1)}$. Closed-loop stability for all initial conditions $x_0 \in \mathcal{X}^{(N)}$ can be demonstrated using an identical argument to the proof of Theorem 3.4 to show that the bound:

$$\sum_{k=0}^{n}(\|x_k\|_Q^2 + \|u_k\|_R^2) \leq \check{J}^*(x_0) + \gamma^2 \sum_{k=0}^{n} \|w_k\|^2 \tag{4.12}$$

holds for all $n \geq 0$ along closed-loop trajectories under Algorithm 4.2.

Example 4.3 For the triple integrator of Example 4.2, Figs. 4.4 and 4.5 show regions of state space within which the optimal control law for problem (4.8) is given by a single affine feedback law, for horizons of $N = 4$ and $N = 10$, respectively. Since this is a third-order system, figures show the intersection of these regions with the plane $\{x : [0\ 1\ 0]x = 0\}$. As expected, for any given active set $\mathcal{A}$, the region $\mathcal{R}(\mathcal{A})$ is a convex polytope, and the union of all regions covers the N-step controllable set. The terminal set $\mathcal{X}_T$ is contained in the region $\mathcal{R}(\emptyset)$, within which the solution of (4.8) coincides with the unconstrained optimal feedback law, $u = Kx$. Note that $\mathcal{R}(\emptyset)$ is not necessarily invariant under $u = Kx$, and hence $\mathcal{R}(\emptyset)$ extends beyond the boundaries of $\mathcal{X}_T$ even though $\mathcal{X}_T$ is the maximal RPI set for $u = Kx$. Comparing Figs. 4.4 and 4.5, it can be seen that the number of active set regions increases rapidly with N.

To illustrate how the online computational requirement of Algorithm 4.2 varies with horizon length for this system, the optimization in step (i) was solved for 50 values of the model state randomly selected on the boundary of $\mathcal{X}^{(N)}$ for $N = 1, 2, \ldots, 20$. Each optimization was cold started (initialized with $x_0^{(0)} = 0$ and $\mathcal{A}^{(0)} = \emptyset$). Figure 4.6 shows that the average, maximum and minimum execution times per iteration[1] for the chosen set of initial conditions depend approximately linearly on N. This is in agreement with the expectation that the computation per iteration should depend approximately linearly on horizon length. Despite exponential growth of the total number of active set regions with N, the number of iterations required is a polynomial function of N for this example. This is illustrated by Fig. 4.7, which shows that the average number of iterations grows approximately as $O(0.25N^2)$. For $N = 20$, the average total execution time was 0.58 s and the maximum execution time 2.23 s. ◊

[1]The execution times in Fig. 4.6 provide an indication of how computation scales with horizon length—Algorithm 4.2 was implemented in Matlab without being fully optimized for speed.

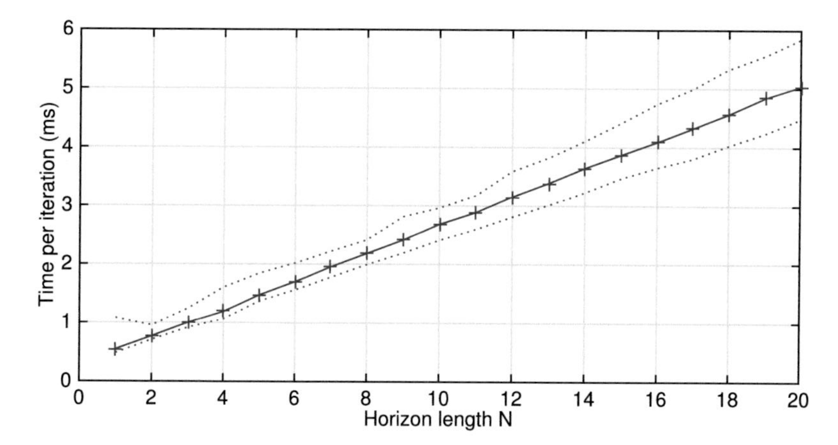

Fig. 4.6 Computation per iteration of the optimization in step (i) of Algorithm 4.2 for Example 4.3 with varying horizon N; average execution time (*solid line*), and minimum and maximum execution times (*dotted lines*)

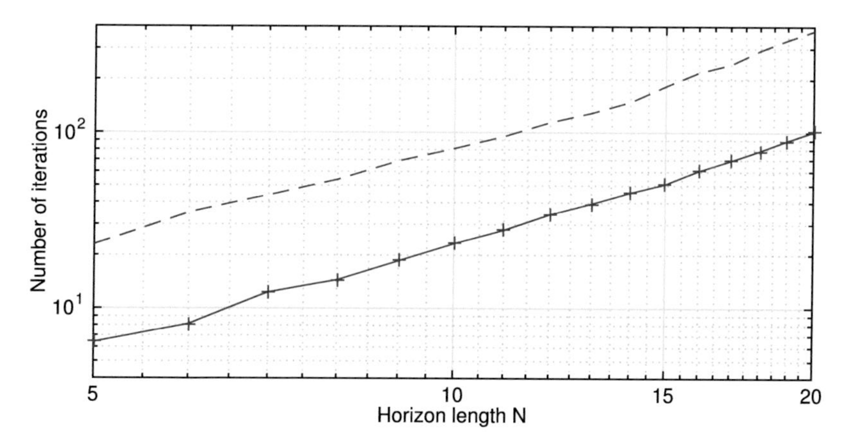

Fig. 4.7 Number of iterations of the optimization in step (i) of Algorithm 4.2 for Example 4.3 with varying horizon N; average (*solid line*) and maximum (*dashed line*)

4.1.2 MPC with General Feedback Laws

As stated previously, dynamic programming methods that provide global solutions are often computationally prohibitive, particularly for problems in which the optimal control is only required locally in a particular operating region of the state space. Although the approach of Sect. 4.1.1 reduces computation using active sets that apply locally within regions of state space, the approach requires prior knowledge of the controllable sets. However, even when performed offline, the computation of these controllable sets can be challenging, and this is a particular concern for problems with long horizons and for systems with many states and input variables.

A major advantage of MPC, and perhaps its defining feature, is that it determines an optimal control input locally, usually for a particular value of the system state, by propagating input and state sequences forwards in time. Since the constraints (4.4) are linear and the disturbance set $\mathcal{W}$ is polytopic, it should not be surprising that each vertex of the controllable set to a convex polytopic target set is determined by a particular maximizing sequence of vertices of $\mathcal{W}$ in (4.5). Therefore, to allow for the full generality of future feedback laws that is needed to achieve the largest possible set of feasible initial conditions, a predictive control strategy must assign a control sequence to each possible sequence of future disturbance inputs [13]. The approach, which is described in this section, leads to an optimization problem with computation that grows exponentially with horizon length. As a result it is computationally intractable for the vast majority of control applications and is mainly of interest from a conceptual point of view. However the approach provides important motivation for the computationally viable MPC strategies that are discussed in Sect. 4.2.

Consider a sequence of disturbance inputs, each of which is equal in value to one of the vertices $w^{(j)}$, $j = 1, \ldots, m$, of the disturbance set $\mathcal{W}$ in (4.3):

$$\{w^{(j_1)}, w^{(j_2)}, \ldots\}.$$

As before we assume that the plant state x_i is known to the controller at the ith time step and a causal control law is assumed; thus u_i depends on x_i but cannot depend on $x_{i+1}, x_{i+2}, \ldots$ If $\{w_0, \ldots, w_{i-1}\} = \{w^{(j_1)}, \ldots, w^{(j_i)}\}$, then it follows that u_i is a function of x_0 and $(j_1, \ldots, j_i)$. Hence we denote the control sequence as $\{u_0, u^{(j_1)}, u^{(j_1,j_2)}, \ldots\}$ and denote $\{x_0, x^{(j_1)}, x^{(j_1,j_2)}, \ldots\}$ as the corresponding sequence of states, which evolve according to the model (4.1):

$$x^{(j_1)} = Ax_0 + Bu_0 + Dw^{(j_1)} \tag{4.13a}$$

and for $i = 1, 2, \ldots$,

$$x^{(j_1,\ldots,j_i,j_{i+1})} = Ax^{(j_1,\ldots,j_i)} + Bu^{(j_1,\ldots,j_i)} + Dw^{(j_{i+1})}. \tag{4.13b}$$

In this prediction scheme, the m^i distinct disturbance sequences $\{w^{(j_1)}, \ldots w^{(j_i)}\}$, in which $j_r \in \{1, \ldots, m\}$ for $r = 1, \ldots, i$, generate m^i state and control sequences with the tree structure shown in Fig. 4.8.

The linearity of the dynamics (4.1) implies that the convex hull of $x^{(j_1,\ldots,j_{i+1})}$ for $j_r \in \{1, \ldots, m\}$, $r = 1, \ldots, i+1$ contains the model state x_{i+1} under any convex combination of control sequences $\{u_0, u^{(j_1)}, \ldots, u^{(j_1,\ldots,j_i)}\}$. To show this, let $X_0 = \{x_0\}$, $U_0 = \{u_0\}$, and for $i = 1, 2, \ldots$ define the sets

$$X_i = \mathrm{Co}\{x^{(j_1,\ldots,j_i)},\ (j_1, \ldots, j_i) \in \mathcal{L}_i\}, \tag{4.14a}$$

$$U_i = \mathrm{Co}\{u^{(j_1,\ldots,j_i)},\ (j_1, \ldots, j_i) \in \mathcal{L}_i\}, \tag{4.14b}$$

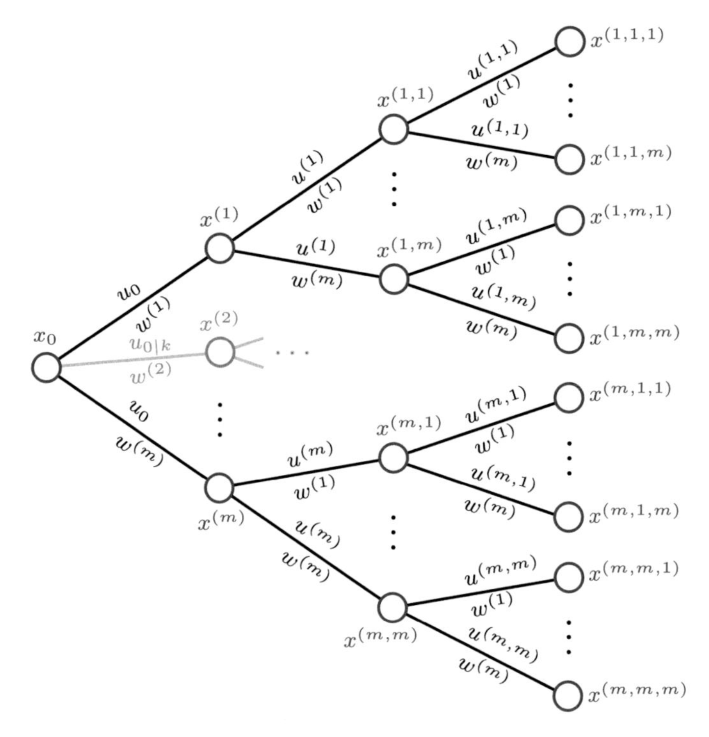

Fig. 4.8 The tree structure of predicted input and state sequences when each element of the disturbance sequence is equal to one of the m vertices of the disturbance set $\mathcal{W}$, shown here for a horizon of $N = 3$

where $\mathcal{L}_i$ denotes the set of all possible values of the sequence $\{j_1, \ldots, j_i\}$:

$$\mathcal{L}_i = \{(j_1, \ldots, j_i) : j_r \in \{1, \ldots, m\},\ r = 1, \ldots, i\},$$

for any positive integer i. Then, for any given $x_i \in X_i$, there necessarily exists $u_i \in U_i$ such that $x_{i+1} \in X_{i+1}$ for all $w_i \in \mathcal{W}$. In particular $x_i \in X_i$ implies $x_i = \sum_{(j_1,\ldots,j_i)\in\mathcal{L}_i} \lambda_{(j_1,\ldots,j_i)} x^{(j_1,\ldots,j_i)}$, where $\lambda_{(j_1,\ldots,j_i)}$ are non-negative scalars satisfying $\sum_{(j_1,\ldots,j_i)\in\mathcal{L}_i} \lambda_{(j_1,\ldots,j_i)} = 1$. Therefore setting

$$u_i = \sum_{(j_1,\ldots,j_i)\in\mathcal{L}_i} \lambda_{(j_1,\ldots,j_i)} u^{(j_1,\ldots,j_i)}$$

gives

$$Ax_i + Bu_i \in \mathrm{Co}\{Ax^{(j_1,\ldots,j_i)} + Bu^{(j_1,\ldots,j_i)},\ (j_1, \ldots, j_i) \in \mathcal{L}_i\},$$

but from (4.13a, 4.13b) and from the definition of X_i in (4.14a) we have

$$\mathrm{Co}\{Ax^{(j_1,\ldots,j_i)} + Bu^{(j_1,\ldots,j_i)},\ (j_1, \ldots, j_i) \in \mathcal{L}_i\} \oplus \mathcal{W} = X_{i+1}, \tag{4.15}$$

and it follows that the tubes $\{X_0, X_1, \ldots\}$ and $\{U_0, U_1, \ldots\}$ necessarily contain the state and control trajectories of (4.1) for any disturbance sequence $\{w_0, w_1, \ldots\}$ with $w_i \in \mathcal{W}$ for all $i = 0, 1, \ldots$.

Consider now a set of input and state trajectories predicted at time k in response to disturbance sequences, $\{w_{0|k}, \ldots, w_{i-1|k}\} = \{w^{(j_1)}, \ldots, w^{(j_i)}\}$, consisting entirely of vertices of $\mathcal{W}$. In order that the state and control trajectories $\{x_{0|k}, x_k^{(j_1)}, x_k^{(j_1,j_2)}, \ldots\}$ and $\{u_{0|k}, u_k^{(j_1)}, u_k^{(j_1,j_2)}, \ldots\}$ generated by the model:

$$x_k^{(j_1)} = Ax_k + Bu_{0|k} + Dw^{(j_1)} \tag{4.16a}$$

$$x_k^{(j_1,\ldots,j_k,j_{i+1})} = Ax_k^{(j_1,\ldots,j_i)} + Bu_k^{(j_1,\ldots,j_i)} + Dw^{(j_{i+1})}, \quad i = 1, 2, \ldots \tag{4.16b}$$

satisfy the constraints (4.4) for all $(j_1, \ldots, j_i) \in \mathcal{L}_i$, we require that

$$Fx_k + Gu_{0|k} \leq \mathbf{1} \tag{4.17a}$$

and, for $i = 1, 2, \ldots$,

$$Fx_k^{(j_1,\ldots,j_i)} + Gu_k^{(j_1,\ldots,j_i)} \leq \mathbf{1}, \text{ for all } (j_1, \ldots, j_i) \in \mathcal{L}_i. \tag{4.17b}$$

These constraints are imposed over an infinite future horizon if (4.17a, 4.17b) are invoked for $i = 1, \ldots, N-1$ together with a terminal constraint of the form

$$x_k^{(j_1,\ldots,j_N)} \in \mathcal{X}_T, \text{ for all } (j_1, \ldots, j_N) \in \mathcal{L}_N, \tag{4.18}$$

where $\mathcal{X}_T$ is a robustly positively invariant set for (4.1) and (4.4) under a particular feedback law. Assuming for convenience that this control law is linear, e.g. $u = Kx$, $\mathcal{X}_T$ can be defined as the maximal RPI set and determined using the method of Theorem 3.1.

Let $\mathcal{F}_N$ denote the set of feasible initial conditions x_k for (4.17a, 4.17b) and (4.18), i.e.

$$\mathcal{F}_N = \Big\{x_k : \exists\{u_{0|k}, u_k^{(j_1)}, \ldots, u_k^{(j_1,\ldots,j_{N-1})}\} \text{ for } (j_1, \ldots, j_{N-1}) \in \mathcal{L}_{N-1}$$
$$\text{such that } \{x_k, x_k^{(j_1)}, \ldots, x_k^{(j_1,\ldots,j_N)}\} \text{ satisfies}$$
$$\text{(4.16a, b), (4.17a, b) for } i = 1, \ldots, N-1, \text{ and (4.18)}\Big\}. \tag{4.19}$$

Theorem 4.1 *$\mathcal{F}_N$ is identical to $\mathcal{X}^{(N)}$, the N-step controllable set to $\mathcal{X}_T$ for the system (4.1) subject to the constraints (4.4), defined in Definition 4.1.*

Proof From (4.14a, 4.14b) and (4.15), constraints (4.17a, 4.17b) and (4.18) ensure that every point $x_0 \in \mathcal{F}_N$ belongs to $\mathcal{X}^{(N)}$. To show that $\mathcal{F}_N$ is in fact equal to $\mathcal{X}^{(N)}$, note that, if $\mathcal{X}_T = \{x : V_T x \leq \mathbf{1}\}$, then $\mathcal{X}^{(N)}$ can be expressed as $\{x : f_T(x) \leq \mathbf{1}\}$

where $f_T(\cdot)$ is the solution of a sequential min-max problem:

$$f_T(x_0) \doteq \min_{\substack{u_0 \\ Fx_0+Gu_0\le \mathbf{1}}} \max_{\substack{w_0 \\ w_0\in\mathcal{W}}} \cdots \min_{\substack{u_{N-1} \\ Fx_{N-1}+Gu_{N-1}\le \mathbf{1}}} \max_{\substack{w_{N-1} \\ w_{N-1}\in\mathcal{W}}} \max_{r\in\{1,\ldots,n_V\}} V_{T,r}\, x_N, \tag{4.20}$$

in which $V_{T,r}$ is the rth row of V_T. Each stage of this problem can be expressed as a linear program, and hence the sequence of maximizing disturbance inputs is $\{w^{(j_1)}, \ldots, w^{(j_N)}\}$ for some sequence $(j_1, \ldots, j_N)$ of vertices of $\mathcal{W}$. Therefore $x_0 \in \mathcal{X}^{(N)}$ if and only if the optimal control sequence for this problem is $\{u_0, u^{(j_1)}, \ldots, u^{(j_1,\ldots,j_{N-1})}\}$ such that $V_T x^{(j_1,\ldots,j_N)} \le \mathbf{1}$. Since $x_0 \in \mathcal{F}_N$ if V_T $x^{(j_1,\ldots,j_N)} \le \mathbf{1}$ for some sequence $\{u_0, u^{(j_1)}, \ldots, u^{(j_1,\ldots,j_{N-1})}\}$, for each $(j_1, \ldots, j_N)$ $\in \mathcal{L}_N$, it follows that every point $x_0 \in \mathcal{X}^{(N)}$ also belongs to $\mathcal{F}_N$. □

The following lemma shows that $\mathcal{F}_N$ is robustly positively invariant under the control law $u_k = u_{0|k}$.

Lemma 4.1 *For any $N > 0$, $\mathcal{F}_N$ is RPI for the dynamics (4.1), disturbance set (4.2) and constraints (4.4) if $u_k = u_{0|k}$.*

Proof Since $w_k \in \mathcal{W}$, there exist non-negative scalars λ_j, $j = 1, \ldots, m$, such that $\sum_{j=1}^m \lambda_j = 1$ and

$$w_k = \sum_{j=1}^m \lambda_j w^{(j)}.$$

Therefore, for any given $\{x_k, x_k^{(j_1)}, x_k^{(j_1,j_2)}, \ldots\}$ and $\{u_{0|k}, u_k^{(j_1)}, u_k^{(j_1,j_2)}, \ldots\}$ satisfying (4.16a, 4.16b), (4.17a, 4.17b) and (4.18), the trajectories defined by

$$u_{0|k+1} = \sum_{j=1}^m \lambda_j u_k^{(j)}, \quad u_{k+1}^{(j_1,\ldots,j_i)} = \sum_{j=1}^m \lambda_j u_k^{(j,j_1,\ldots,j_i)}, \ i = 1, \ldots, N-2,$$

$$x_{k+1}^{(j_1,\ldots,j_i)} = \sum_{j=1}^m \lambda_j x_k^{(j,j_1,\ldots,j_i)}, \ i = 1, \ldots, N-1,$$

for all $(j_1, \ldots, j_i) \in \mathcal{L}_i$, and

$$u_{k+1}^{(j_1,\ldots,j_{N-1})} = K x_{k+1}^{(j_1,\ldots,j_{N-1})}, \qquad \text{for all } (j_1, \ldots, j_{N-1}) \in \mathcal{L}_{N-1}$$

$$x_{k+1}^{(j_1,\ldots,j_{N-1},j_N)} = \Phi x_{k+1}^{(j_1,\ldots,j_{N-1})} + D w^{(j_N)}, \quad \text{for all } (j_1, \ldots, j_N) \in \mathcal{L}_N$$

satisfy, at time $k+1$, the constraints of (4.16a, 4.16b) (by linearity), (4.17a, 4.17b) (by convexity), and (4.18) since $x_{k+1}^{(j_1,\ldots,j_{N-1})} \in \mathcal{X}_T$ and $\mathcal{X}_T$ is RPI for (4.1) and (4.4) if $u = Kx$. □

Theorem 4.1 shows that x_0 lies in the N-step controllable set to a given target set if and only if a set of linear constraints is satisfied in the variables $\{u_0, u^{(j_1)}, \ldots, u^{(j_1,\ldots,j_{N-1})}\}$, $(j_1, \ldots, j_{N-1}) \in \mathcal{L}_{N-1}$. This feasibility problem could be combined with any chosen performance index to define an optimal control law with a set of feasible initial conditions equal to the controllable set for the given target set and horion. Furthermore, by Lemma 4.1, a receding horizon implementation of any such strategy would necessarily be recursively feasible, and hence could form basis of a robust MPC law. However, the feasible set $\mathcal{F}_N$ is defined in (4.19) in terms of the vertices of the predicted tubes for states and control inputs. In general, the optimal predictions with respect to a quadratic cost will be convex combinations of these vertices, and hence additional optimization variables would be needed in the MPC optimization if a quadratic performance index were used. With this in mind, a linear min-max cost is employed in [13], the cost index being defined as the sum of stage costs that depend linearly on the future state and control input. This choice of cost has the convenient property that the optimal control sequence is given by $\{u_{0|k}, u_k^{j_1}, \ldots, u_k^{j_1,\ldots,j_{N-1}}\}$ for some $(j_1, \ldots, j_{N-1}) \in \mathcal{L}_{N-1}$.

Linear or piecewise linear stage costs have the disadvantage that the unconstrained optimal is not straightforward to determine, and in [13] the terminal control law is restricted to one that enforces finite time convergence of the state (i.e. "deadbeat" control) in order that the cost is finite over an infinite prediction horizon. Instead we illustrate the MPC strategy here using a performance index similar to the nominal cost employed in Sect. 3.3. In this setting, we first reparameterize the predicted control input trajectories in terms of optimization variables $\mathbf{c}_k^{(l)} = \{c_{0|k}, c_k^{(j_1)}, \ldots, c_k^{(j_1,\ldots,j_{N-1})}\}$ for $l = (j_1, \ldots, j_{N-1}) \in \mathcal{L}_{N-1}$, so that

$$\begin{aligned} u_{0|k} &= Kx_k + c_{0|k} \\ u_k^{(j_1,\ldots,j_i)} &= \begin{cases} Kx_k^{(j_1,\ldots,j_i)} + c_k^{(j_1,\ldots,j_i)}, & i = 1, \ldots, N-1 \\ Kx_k^{(j_1,\ldots,j_i)}, & i = N, N+1, \ldots \end{cases} \end{aligned} \tag{4.21}$$

Next define a set of nominal state and control trajectories, one for each sequence $l = (j_1, \ldots, j_{N-1}) \in \mathcal{L}_{N-1}$, as follows

$$\begin{aligned} v_{i|k}^{(l)} &= Ks_{1|k}^{(l)} + c_{i|k}^{(l)}, & s_{i+1|k}^{(l)} &= \Phi s_{i|k}^{(l)} + Bc_{i|k}^{(l)}, & i &= 0, 1, \ldots, N-1 \\ v_{i|k}^{(l)} &= Ks_{1|k}^{(l)}, & s_{i+1|k}^{(l)} &= \Phi s_{i|k}^{(l)}, & i &= N, N+1, \ldots \end{aligned}$$

where $c_{i|k}^{(l)}$ is the ith element of the sequence $\mathbf{c}_k^{(l)}$ for each $l \in \mathcal{L}_{N-1}$, namely $c_{i|k}^{(l)} = c_{0|k}$ for $i = 0$ and $c_{i|k}^{(l)} = c_k^{(j_1,\ldots,j_i)}$ for $i = 1, \ldots, N-1$. Finally, we define the worst case quadratic cost over these nominal predicted sequences as

$$J\big(s_{0|k}, \{\mathbf{c}_k^{(l)}, l \in \mathcal{L}_{N-1}\}\big) = \max_{l \in \mathcal{L}_{N-1}} \sum_{i=0}^{\infty} \big(\|s_{i|k}^{(l)}\|_Q^2 + \|v_{i|k}^{(l)}\|_R^2\big) \tag{4.22}$$

Assuming for convenience that K is the unconstrained LQ-optimal feedback gain, Theorem 2.10 gives

$$J\big(s_{0|k}, \{\mathbf{c}_k^{(l)},\ l \in \mathcal{L}_{N-1}\}\big) = \|s_{0|k}\|_{W_x}^2 + \max_{l\in\mathcal{L}_{N-1}} \|\mathbf{c}_k^{(l)}\|_{W_c}^2,$$

where W_x is the solution of the Riccati equation (2.9), and where W_c is block diagonal, $W_c = \mathrm{diag}\{B^T W_x B + R, \ldots, B^T W_x B + R\}$. Hence the problem of minimizing $J(s_{0|k}, \{\mathbf{c}_k^{(l)}\})$ is equivalent to the minimization of $\|s_{0|k}\|_{W_x}^2 + \alpha^2$ subject to $\alpha^2 \geq \|\mathbf{c}_k^{(l)}\|_{W_c}^2$ for all $l \in \mathcal{L}_{N-1}$. This is a convex optimization problem that can be formulated as a second-order cone program (SOCP) [14].

By combining the cost of (4.22) with the linear constraints defining $\mathcal{F}_N$, we obtain the following MPC algorithm.

Algorithm 4.3 (*General feedback MPC*) At each time $k = 0, 1, \ldots$:

(i) Perform the optimization

$$\begin{aligned} &\underset{\{\mathbf{c}_k^{(l)},\, l\in\mathcal{L}_{N-1}\}}{\text{minimize}} && \max_{l\in\mathcal{L}_{N-1}} \|\mathbf{c}_k^{(l)}\|_{W_c}^2 \\ &\text{subject to} && (4.16\text{a, b}), (4.17\text{a, b}), i = 1, \ldots, N-1, \\ &&& (4.18) \text{ and } (4.21). \end{aligned} \tag{4.23}$$

(ii) Apply the control law $u_k = Kx_k + c_{0|k}^*$, where $\|\mathbf{c}_k^{l*}\|_{W_c}^2$ is the optimal value of the objective in (4.23) and $\mathbf{c}_k^{(l)*} = \{c_{0|k}^*, \ldots, c_k^{(j_1,\ldots,j_{N-1})*}\}$. ◁

The online optimization in step (i) can be formulated as a SOCP in $n_u(m^N - 1)/(m-1)$ variables, $n_c(m^N - 1)/(m-1) + n_T m^N$ linear inequality constraints and $m^N + 1$ second-order cone constraints, where n_T is the number of constraints defining $\mathcal{X}_T$. Furthermore Algorithm 4.3 is recursively feasible and enforces convergence to the minimal RPI set associated with $u = Kx$, as we now discuss.

The feasibility of (4.23) at all times given initial feasibility (i.e. $x_0 \in \mathcal{F}_N$) is implied by Lemma 4.1. Therefore, defining $c_{0|k+1}$ and $c_{k+1}^{(j_1,\ldots,j_i)}$ for $i = 1, \ldots, N-1$ analogously to the definitions of $u_{0|k+1}$ and $u_{k+1}^{(j_1,\ldots,j_i)}$ in the proof of Lemma 4.1 gives

$$\begin{aligned} c_{0|k+1} &= \sum_{j=1}^{m} \lambda_j c_k^{(j)*} \\ c_{k+1}^{(j_1,\ldots,j_i)} &= \sum_{j=1}^{m} \lambda_j c_k^{(j,j_1,\ldots,j_i)*}, \quad i = 1, \ldots, N-2 \\ c_{k+1}^{(j_1,\ldots,j_{N-1})} &= 0 \end{aligned}$$

for each $l = (j_1, \ldots, j_{N-1}) \in \mathcal{L}_{N-1}$, and hence by convexity we have $\|\mathbf{c}_{k+1}^{(l)}\|_{W_c}^2 \leq \|\mathbf{c}_k^{l*}\|_{W_c}^2 - \|c_{0|k}^*\|_{R+B^T W_x B}^2$ for all $l = (j_1, \ldots, j_{N-1}) \in \mathcal{L}_{N-1}$. The optimization of (4.23) at time $k+1$ therefore gives

$$\|\mathbf{c}_k^{(l)*}\|_{W_c}^2 - \|\mathbf{c}_{k+1}^{(l)*}\|_{W_c}^2 \geq \|c_{0|k}^*\|_{R+B^T W_x B}^2$$

which implies (by Lemma 3.1) that $c_{0|k}^* \to 0$ as $k \to \infty$. From Theorem 3.2, it follows that the l_2 gain from the disturbance input w to the state x is upper bounded by the l_2 bound for the unconstrained system under $u = Kx$, and furthermore $x_k \to \mathcal{X}^{\mathrm{mRPI}}$ as $k \to \infty$, where $\mathcal{X}^{\mathrm{mRPI}}$ is the minimal RPI for (4.1) under $u = Kx$.

Finally note that the control law of Algorithm 4.3 can also guarantee exponential convergence to an outer approximation, $\mathcal{S}$, of $\mathcal{X}^{\mathrm{mRPI}}$ if $s_{0|k}$ is retained as an optimization variable in (4.23). Specifically, replacing the objective of (4.23) with $J(s_{0|k}, \{\mathbf{c}_k^{(l)}, l \in \mathcal{L}_{N-1}\})$ and including $x_k - s_{0|k} \in \mathcal{S}$ as an additional constraint in (4.23) ensures, by an argument similar to the proof of Theorem 3.5, that the optimal value $J(s_{0|k}^*, \mathbf{c}_k^{(l)*})$ converges exponentially to zero and hence that $\mathcal{S}$ is exponentially stable with region of attraction equal to $\mathcal{F}_N$. This argument relies on the existence of constants a, b satisfying $a\|s_{0|k}^*\|^2 \leq J(s_{0|k}^*, \mathbf{c}_k^{(l)*}) \leq b\|s_{0|k}^*\|^2$, which is ensured by the continuity [14] of the optimal objective of (4.23) and by the fact that $J(s_{0|k}^*, \mathbf{c}_k^{(l)*}) \geq 0$, whereas $J(s_{0|k}^*, \mathbf{c}_k^{(l)*}) = 0$ if and only if $x_{0|k}$ lies in $\mathcal{S}$.

Example 4.4 Although the general feedback MPC law of Algorithm 4.3 and the active set DP approach of Algorithm 4.2 are both feasible for all initial conditions in the N-step controllable set $\mathcal{X}^{(N)}$, their computational requirements are very different. The exponential growth in the numbers of variables and constraints of the optimization in step (i) of Algorithm 4.3 implies that its computation grows very rapidly with N. For the system of Example 4.2, we have $n_u = 1$, $n_c = 4$, $m = 8$,

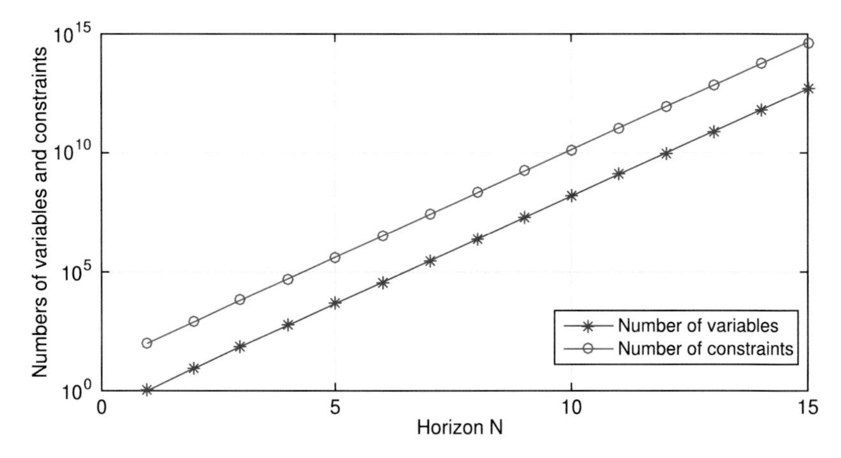

Fig. 4.9 Numbers of variables and linear inequality constraints for Algorithm 4.3 in Example 4.4

$n_T = 11$, and the number of optimization variables and linear inequality constraints grow with N as shown in Fig. 4.9. Even this simple third-order problem is limited to short horizons in order that the optimization (4.23) remains manageable; for example, if the number of optimization variables is required to be less than 1000, then N must be no greater than 4. $\diamondsuit$

4.2 Parameterized Feedback Strategies

Dynamic programming and MPC laws based on general feedback strategies have the definite advantages that they can provide optimal performance and the maximum achievable region of attraction. However, as discussed in Sect. 4.1.2 for the case of min-max robust control (and as also discussed in [15, 16] for stochastic problems), the computation of optimal control laws for these approaches often suffers from poor scalability with the problem size and the length of horizon. It is therefore perhaps inevitable that closed-loop optimization strategies with computational demands that grow less rapidly with problem size should be sought for robust MPC. These approaches reduce computation by restricting the class of closed-loop policies over which optimization is performed.

One such restriction is to the class of time-varying linear feedback plus feedforward control laws, where the linear feedback gains are parameters that are to be computed online. However, the dependence of the future state and input trajectories on these feedback gains is non-linear, and the optimization of predicted performance is non-convex. A way around this is offered by the Youla parameterization introduced into MPC in [17]. On account of the Bezout identity of (2.67), the transfer function matrices that map additive disturbances at the plant input to the plant output have an affine dependence on the Youla parameter. This property is exploited in [18] to devise a lower-triangular prediction structure in the degrees of freedom, leading to a convex online optimization. Later developments in this area, known as disturbance-affine MPC (DAMPC), are reported in [19, 20]. These proposals lead to an online optimization in a number of variables that grows quadratically with the length of the prediction horizon. This can be reduced to a linear growth if the lower-triangular structure is computed offline, but of course the resulting MPC algorithm is then no longer based on optimization of a closed-loop strategy.

An alternative triangular prediction structure to that of [18–20] was proposed in [21], which, like the approach of Sect. 4.1.2, parameterized predicted future feedback laws in terms of the vertices of the disturbance set. In this setting, the input at each prediction instant is known to lie in the convex hull of a linear combination of polytopic cross sections defined by the inputs associated with the disturbance set vertices. The approach thus implicitly employs a parameterization that defines tubes in which the predicted inputs and states will lie, and for this reason it is known as parameterized tube MPC (PTMPC). The disturbance-affine MPC strategy can be shown to be a special case of PTMPC (albeit with a restricted set of optimization variables) and both approaches have a number of optimization variables that grows quadratically with the prediction horizon. This can be reduced to a linear dependence if, instead of the triangular structure of PTMPC, a striped structure is employed [22]. This section discusses these three parameterized feedback MPC strategies.

4.2.1 Disturbance-Affine Robust MPC

A simple but highly restrictive feedback parameterization replaces the fixed linear feedback gain K in the open-loop strategy (3.6) with a linear time-varying state feedback law to give $u_{i|k} = K_k^{(i)} x_{i|k} + c_{i|k}$, where $K_k^{(i)}$ and $c_{i|k}$ for $i = 0, \ldots, N-1$ are to be optimized online at each time k. A less restrictive class of control laws is obtained by allowing the state dependence to be dynamic, in which case the time-varying state feedback is replaced by a convolutional sum, i.e. $u_{i|k} = \sum_{j=0}^{i} K_k^{(i-j)} x_{j|k} + c_{i|k}$. This class can be further widened if the state dependence of the control law is allowed to be both dynamic and time varying over the prediction horizon:

$$u_{i|k} = \sum_{j=0}^{i} K_{i|k}^{(j)} x_{j|k} + c_{i|k} \tag{4.24}$$

leading to prediction equations with a lower-triangular structure [18]

$$\begin{aligned} \mathbf{u}_k &= \bar{K}_k \mathbf{x}_k + \mathbf{c}_k \\ \bar{K}_k &= \begin{bmatrix} K_{0|k}^{(0)} & 0 & \cdots & 0 \\ K_{1|k}^{(0)} & K_{1|k}^{(1)} & \cdots & 0 \\ \vdots & \vdots & \ddots & \vdots \\ K_{N-1|k}^{(0)} & K_{N-1|k}^{(1)} & \cdots & K_{N-1|k}^{(N-1)} \end{bmatrix} \end{aligned} \tag{4.25}$$

Here $\mathbf{x}_k = (x_{0|k}, \ldots, x_{N|k})$ and $\mathbf{u}_k = (u_{0|k}, \ldots, u_{N-1|k})$ are vectors of predicted states and inputs for the model of (4.1), with $x_{0|k} = x_k$. Also $\mathbf{c}_k = (c_{0|k}, \ldots, c_{N-1|k})$ is a vector of feedforward parameters, and the subscript k is used as a reminder that $\bar{K}_k$ and $\mathbf{c}_k$ are to be computed online at each time k.

Using (4.1) the state predictions $\mathbf{x}_k$ can be written in terms of the future control and disturbance inputs, $\mathbf{u}_k$ and $\mathbf{w}_k = (w_{0|k}, \ldots, w_{N-1|k})$, as

$$\mathbf{x}_k = C_{xx} x_k + C_{xu} \mathbf{u}_k + C_{xw} \mathbf{w}_k \tag{4.26}$$

where C_{xx}, C_{xu} and C_{xw}, respectively, denote the convolution matrices from x, $\mathbf{u}$ and $\mathbf{w}$ to $\mathbf{x}$ given by

$$C_{xx} = \begin{bmatrix} I \\ A \\ \vdots \\ A^{N-1} \\ A^N \end{bmatrix}, \; C_{xu} = \begin{bmatrix} 0 & \cdots & 0 & 0 \\ B & \cdots & 0 & 0 \\ \vdots & \ddots & \vdots & \vdots \\ A^{N-2}B & \cdots & B & 0 \\ A^{N-1}B & \cdots & AB & B \end{bmatrix}, \; C_{xw} = \begin{bmatrix} 0 & \cdots & 0 & 0 \\ D & \cdots & 0 & 0 \\ \vdots & \ddots & \vdots & \vdots \\ A^{N-2}D & \cdots & D & 0 \\ A^{N-1}D & \cdots & AD & D \end{bmatrix}. \tag{4.27}$$

Substituting (4.26) into (4.25) and solving for $\mathbf{u}_k$ yields

$$\begin{aligned} \mathbf{u}_k &= (I - \bar{K}_k C_{xu})^{-1} \bar{K}_k C_{xx} x_k + (I - \bar{K}_k C_{xu})^{-1} \mathbf{c}_k \\ &\quad + (I - \bar{K}_k C_{xu})^{-1} \bar{K}_k C_{xw} \mathbf{w}_k & (4.28\text{a}) \\ \mathbf{x}_k &= \big[C_{xx} + C_{xu}(I - \bar{K}_k C_{xu})^{-1} \bar{K}_k C_{xx}\big] x_k + C_{xu}(I - \bar{K}_k C_{xu})^{-1} \mathbf{c}_k \\ &\quad + \big[C_{xw} + C_{xu}(I - \bar{K}_k C_{xu})^{-1} \bar{K}_k C_{xw}\big] \mathbf{w}_k & (4.28\text{b}) \end{aligned}$$

where $(I - \bar{K}_k C_{xu})$ is necessarily invertible since it is lower block diagonal with all its diagonal blocks equal to the identity. Thus, for given $\bar{K}_k$, the predicted state and control trajectories are given as affine functions of the current state and the vector of future disturbances. For any given value of $\bar{K}_k$, the worst case disturbance with respect to the constraints (4.4) could be determined by solving a sequence of linear programs similarly to the robust constraint handling approach of Sect. 3.2. However, for the closed-loop optimization strategy considered here, $\bar{K}_k$ is an optimization variable, and Eqs. (4.28a, 4.28b) depend nonlinearly on this variable. As a result, the implied optimization is non-convex and does not therefore lend itself to online implementation.

As mentioned in Sect. 2.10, the state and input predictions can be transformed into linear functions of the optimization variables through the use of a Youla parameter, and this is the route followed by [18]. Thus (4.28a) can be written equivalently as

$$\mathbf{u}_k = \bar{L}_k \mathbf{w}_k + \mathbf{v}_k, \tag{4.29}$$

with

$$\mathbf{v}_k = (I - \bar{K}_k C_{xu})^{-1} \bar{K}_k C_{xx} x_k + (I - \bar{K}_k C_{xu})^{-1} \mathbf{c}_k \tag{4.30a}$$

$$\bar{L}_k = (I - \bar{K}_k C_{xu})^{-1} \bar{K}_k C_{xw} = \begin{bmatrix} 0 & 0 & \cdots & 0 & 0 \\ L^{(1)}_{1|k} & 0 & \cdots & 0 & 0 \\ L^{(1)}_{2|k} & L^{(2)}_{2|k} & \cdots & 0 & 0 \\ \vdots & \vdots & \ddots & \vdots & \vdots \\ L^{(1)}_{N-1|k} & L^{(2)}_{N-1|k} & \cdots & L^{(N-1)}_{N-1|k} & 0 \end{bmatrix}. \tag{4.30b}$$

The transformation of (4.30a) is bijective, so for each $\mathbf{c}_k$ there exists a unique $\mathbf{v}_k$ and vice versa. Likewise $\bar{L}_k$ is uniquely defined by (4.30b) for arbitrary $\bar{K}_k$. On account of the lower-triangular structure of $\bar{L}_k$, the control policy of (4.29) is necessarily causal and realizable in that $u_{i|k}$ depends on disturbances $w_{0|k}, \ldots, w_{i-1|k}$ that will be known to the controller at the ith prediction time step. The corresponding vector of predicted states assumes the form

$$\mathbf{x}_k = C_{xx} x_k + C_{xu} \mathbf{v}_k + (C_{xw} + C_{xu} \bar{L}_k) \mathbf{w}_k. \tag{4.31}$$

The parameterization of (4.29) implies a disturbance-affine feedback law (e.g. [19, 23]) and is the basis of the disturbance-affine MPC laws proposed for example in [18, 20].

The predictions of (4.29) and (4.31) can be used to impose the constraints of (4.4) over an infinite future horizon by invoking

$$F x_{i|k} + G u_{i|k} \leq \mathbf{1}, \quad i = 0, \ldots, N-1, \tag{4.32a}$$

together with the terminal constraint, $x_{N|k} \in \mathcal{X}_T = \{x : V_T x \leq \mathbf{1}\}$:

$$V_T x_{N|k} \leq \mathbf{1}, \tag{4.32b}$$

where $\mathcal{X}_T$ is RPI for (4.1) and (4.4) under a feedback law $u = Kx$ with a fixed gain K. Here the linear dependence of the constraints (4.32a, 4.32b) on $x_{i|k}$ and $u_{i|k}$ means that they can be expressed in the form

$$\bar{F} \mathbf{x}_k + \bar{G} \mathbf{u}_k \leq \mathbf{1},$$

where $\bar{F}$ and $\bar{G}$ are block diagonal matrices:

$$\bar{F} = \begin{bmatrix} F & & \\ & \ddots & \\ & & F \\ & & & V_T \end{bmatrix}, \quad \bar{G} = \begin{bmatrix} G & & \\ & \ddots & \\ & & G \\ 0 & \cdots & 0 \end{bmatrix}.$$

Using (4.29) and (4.31), these constraints are equivalent to

$$\bar{F}_u \mathbf{v}_k + \max_{\mathbf{w} \in \mathcal{W} \times \cdots \times \mathcal{W}} (\bar{F}_w + \bar{F}_u \bar{L}_k) \mathbf{w} \leq \mathbf{1} - \bar{F}_x x_k,$$

where $\bar{F}_x = \bar{F} C_{xx}$, $\bar{F}_u = \bar{F} C_{xu} + \bar{G}$, $\bar{F}_w = \bar{F} C_{xw}$, and the maximization is performed element wise.

The vertex representation (4.2) of $\mathcal{W}$ allows these constraints to be expressed as a set of linear inequalities in $\mathbf{v}_k$. Thus, in the notation of Sect. 4.1.2, with $\mathbf{w}^{(l)}$ denoting the vector $(w^{(j_1)}, \ldots, w^{(j_N)})$ of vertices of $\mathcal{W}$ for each $l = (j_1, \ldots, j_N) \in \mathcal{L}_N$, the constraints (4.32a, 4.32b) are equivalent to linear constraints in $\mathbf{v}_k$ and $\bar{L}_k$:

$$\bar{F}_u \mathbf{v}_k + (\bar{F}_w + \bar{F}_u \bar{L}_k) \mathbf{w}^{(l)} \leq \mathbf{1} - \bar{F}_x x_k \quad \text{for all } l \in \mathcal{L}_N. \tag{4.33}$$

An alternative constraint formulation that preserves linearity while avoiding the exponential growth in the number of constraints with N that is implied by (4.33) uses convex programming duality to write these constraints equivalently as [20]:

$$H_k \mathbf{1} \leq \mathbf{1} - \bar{F}_x x_k - \bar{F}_u \mathbf{v}_k, \quad H_k \geq 0, \quad H_k \bar{V} = \bar{F}_w + \bar{F}_u \bar{L}_k. \tag{4.34}$$

Here $H_k \in \mathbb{R}^{(Nn_c+n_T)\times n_V}$ is a matrix of additional variables in the MPC optimization performed online at times $k = 0, 1, \ldots$ and $\bar{V}$ is a block diagonal matrix containing N diagonal blocks, each of which is equal to V. Therefore, (4.34) constitutes a set of linear constraints in decision variables $\mathbf{v}_k$, $\bar{L}_k$ and H_k, and the total number of these constraints grows linearly with N. The technique used to derive (4.34) from (4.33) and the equivalence of these sets of constraints is discussed in Chap. 5 (see Lemma 5.6).

We next consider the definition of the predicted cost that forms the objective of an MPC strategy employing the constraints (4.34). One possible choice is a nominal cost that assumes all future disturbance inputs to be zero, i.e. $w_{i|k} = 0, i = 0, 1, \ldots$. By combining a quadratic nominal cost with the constraints (4.34), the online MPC optimization can be formulated conveniently as a quadratic program. It can be shown that the resulting MPC law ensures a finite l_2 gain from the disturbance input to the state and control input (see [20] for details), but the implied l_2 gain could be arbitrarily large because the nominal cost contains no information on the feedback gain matrix $\bar{L}_k$. By including in the cost an additional quadratic penalty on elements of $\bar{L}_k$, it is possible to derive stronger stability results [24], in particular the state of the closed-loop system can be shown to converge asymptotically to the minimal RPI set under a specific known linear feedback law. However, this approach relies on using sufficiently large weights in the penalty on $\bar{L}_k$, and we therefore consider here a conceptually simpler min-max approach proposed in [25], which uses the worst case cost considered in Sect. 3.4.

The predicted cost is therefore defined as the maximum, with respect to disturbances $w \in \mathcal{W}$, of a quadratic cost over a horizon of N steps:

$$\check{J}(x_0, \{u_0, u_1, \ldots\}) = \max_{\substack{w_i \in \mathcal{W} \\ i=0,\ldots,N-1}} \sum_{i=0}^{N-1} \left(\|x_i\|_Q^2 + \|u_i\|_R^2 - \gamma^2 \|w_i\|^2\right) + \|x_N\|_{\check{W}_x}^2. \tag{4.35}$$

Here $\check{W}_x$ is the solution of the Riccati equation (3.42), and hence the cost (4.35) is equivalent to the maximum of an infinite horizon cost over $w_i \in \mathcal{W}$ for $i = 0, \ldots, N-1$ and over $w_i \in \mathbb{R}^{n_w}$ for $i \geq N$. We denote the predicted cost evaluated at time k along the trajectories of (4.1) with the feedback strategy of (4.29) as $\check{J}(x_k, \mathbf{v}_k, \bar{L}_k)$:

$$\check{J}(x_k, \mathbf{v}_k, \bar{L}_k) = \max_{\mathbf{w}_k \in \{\mathbf{w}:\bar{V}\mathbf{w} \leq \mathbf{1}\}} \Big\{ \|C_{xx} x_k + C_{xu} \mathbf{v}_k + (C_{xw} + C_{xu} \bar{L}_k)\mathbf{w}_k\|_{\bar{Q}}^2 + \|\bar{L}_k \mathbf{w}_k + \mathbf{v}_k\|_{\bar{R}}^2 - \gamma^2 \|\mathbf{w}_k\|^2 \Big\} \tag{4.36}$$

where

$$\bar{Q} = \begin{bmatrix} Q & & & \\ & \ddots & & \\ & & Q & \\ & & & \check{W}_x \end{bmatrix}, \quad \bar{R} = \begin{bmatrix} R & & \\ & \ddots & \\ & & R \end{bmatrix}.$$

Following [25] and using the approach of Sect. 3.4, the cost (4.36) can be expressed in terms of conditions that are convex in the variables $\mathbf{v}_k$ and $\bar{L}_k$, provided γ is sufficiently large so that $\check{J}(x_k, \mathbf{v}_k, \bar{L}_k)$ is concave in $\mathbf{w}_k$. This concavity condition is equivalent to the requirement that $\Delta \succ 0$, where

$$\Delta \doteq \gamma^2 I - \Big((C_{xw} + C_{xu}\bar{L}_k)^T \bar{Q}(C_{xw} + C_{xu}\bar{L}_k) + \bar{L}_k^T \bar{R}\bar{L}_k\Big).$$

Lemma 4.2 *If $\Delta \succ 0$, then $\check{J}(\mathbf{v}_k, \bar{L}_k) = \min_{\delta,\, \mu\in\{\mu:\mu\geq 0\}} \delta + \mathbf{1}^T \mu$ subject to the following LMI in $\mathbf{v}_k$, $\bar{L}_k$, μ and δ,*

$$\begin{bmatrix} \begin{bmatrix} \delta & \frac{1}{2}\mu^T \bar{V} \\ \bar{V}^T \mu & \gamma^2 I \end{bmatrix} & \begin{bmatrix} (C_{xx}x_k + C_{xu}\mathbf{v}_k)^T \bar{Q}^{1/2} & \mathbf{v}_k^T \bar{R}^{1/2} \\ (C_{xw} + C_{xu}\bar{L}_k)^T \bar{Q}^{1/2} & \bar{L}_k^T \bar{R}^{1/2} \end{bmatrix} \\ \star & I \end{bmatrix} \succeq 0, \tag{4.37}$$

where $\bar{Q}^{1/2}$ and $\bar{R}^{1/2}$ satisfy $(\bar{Q}^{1/2})^T \bar{Q}^{1/2} = \bar{Q}$ and $(\bar{R}^{1/2})^T \bar{R}^{1/2} = \bar{R}$.

Proof If $\Delta \succ 0$, then (4.37) can be shown (by considering Schur complements) to be equivalent to the condition

$$\begin{aligned} \delta \geq\ & \|C_{xx}x_k + C_{xu}\mathbf{v}_k\|_{\bar{Q}}^2 + \|\mathbf{v}_k\|_{\bar{R}}^2 \\ & + \|\tfrac{1}{2}\mu^T \bar{V} - (C_{xx}x_k + C_{xu}\mathbf{v}_k)^T \bar{Q}(C_{xw} + C_{xu}\bar{L}_k) - \mathbf{c}_k^T \bar{R}\bar{L}_k\|_{\Delta^{-1}}^2. \end{aligned} \tag{4.38}$$

Moreover if $\Delta \succ 0$, then the equivalence of the convex QP (3.52) and its dual (3.53) implies that $\check{J}(x_k, \mathbf{v}_k, \bar{L}_k)$ is equal to the minimum of $\delta + \mathbf{1}^T \mu$ over $\mu \geq 0$ and δ subject to (4.38). □

Lemma 4.2 allows the minimization of the worst case cost (4.36) subject to (4.32a, 4.32b) with the feedback strategy of (4.29) to be formulated as a semi-definite program in $O(N^2)$ variables. This is the online optimization that forms the basis of the following disturbance-affine MPC law.

Algorithm 4.4 (*DAMPC*) At each time instant $k = 0, 1, \ldots$:

(i) Perform the optimization

$$\underset{\mathbf{v}_k, \bar{L}_k, H_k, \delta_k, \mu_k}{\text{minimize}} \quad \delta_k + \mathbf{1}^T \mu_k \text{ subject to (4.34), (4.37) and } \mu_k \geq 0. \tag{4.39}$$

(ii) Apply the control law $u_k = v^*_{0|k}$, where $\mathbf{v}^*_k = (v^*_{0|k}, \ldots, v^*_{N-1|k})$ and $\bar{L}^*_k$ are the optimal values of $\mathbf{v}_k$, $\bar{L}_k$ in (4.39). ◁

The set of feasible states for the MPC optimization (4.39) is given by[2]

$$\mathcal{F}_N \doteq \{x_k : \exists(\mathbf{v}_k, \bar{L}_k) \text{ such that } \Delta \succ 0 \text{ and (4.34) holds for some } H_k \geq 0\}.$$

For any $N \geq 1$, $\mathcal{F}_N$ can be shown to be RPI under the control law of Algorithm 4.4 by constructing $\mathbf{v}_{k+1}$ and $\bar{L}_{k+1}$ so that conditions (4.32a, 4.32b) and $\Delta \succ 0$ are satisfied at time $k+1$. Specifically, let

$$\bar{L}_{k+1} = \begin{bmatrix} 0 & \cdots & 0 & 0 & 0 \\ L^{(2)*}_{2|k} & \cdots & 0 & 0 & 0 \\ \vdots & \ddots & \vdots & \vdots & \vdots \\ L^{(2)*}_{N-1|k} & \cdots & L^{(N-1)*}_{N-1|k} & 0 & 0 \\ KA^{N-2}D & \cdots & KAD & KD & 0 \end{bmatrix}, \quad \mathbf{v}_{k+1} = \begin{bmatrix} v^*_{1|k} + L^{(1)*}_{1|k} w_k \\ v^*_{2|k} + L^{(1)*}_{2|k} w_k \\ \vdots \\ v^*_{N-1|k} + L^{(1)*}_{N-1|k} w_k \\ v_{N|k+1} + KA^{N-1}Dw_k \end{bmatrix},$$

with $v_{N|k+1} = KA^N x_k + K[A^{N-1}B \ \cdots \ B]\mathbf{v}^*_k$. Then the sequence of control inputs given by $(u_{0|k+1}, \ldots, u_{N-1|k+1}) = \bar{L}_{k+1}\mathbf{w}_{k+1} + \mathbf{v}_{k+1}$ satisfies

$$u_{i|k+1} = \begin{cases} \tilde{u}_{i+1|k} & i = 0, 1, \ldots, N-2 \\ K\tilde{x}_{N|k} & i = N-1 \end{cases}$$

where $\tilde{u}_{i+1|k}$ and $\tilde{x}_{N|k}$ are elements of the optimal predicted input and state sequences at time k that would be obtained with $w_{0|k}$ equal to the actual disturbance realization, w_k, and which satisfy (4.32a, 4.32b) by construction. It can also be shown [25] that this choice of $\bar{L}_{k+1}$ satisfies

$$\gamma^2 I - (C_{xw} + C_{xu}\bar{L}_{k+1})^T \bar{Q}(C_{xw} + C_{xu}\bar{L}_{k+1}) + \bar{L}^T_{k+1}\bar{R}\bar{L}_{k+1} \succ 0,$$

whenever $\Delta \succ 0$. Therefore $x_{k+1} \in \mathcal{F}_N$ for all $w_k \in \mathcal{W}$, implying that Algorithm 4.4 is recursively feasible.

The MPC law of Algorithm 4.4 guarantees a bound, which depends on the parameter γ, on the l_2 gain from the disturbance input to the state and control input.

Corollary 4.1 *For any $x_0 \in \mathcal{F}_N$ and all $n \geq 0$, the closed-loop trajectories of (4.1) under Algorithm 4.4 satisfy the bound*

$$\sum_{k=0}^{n} \left(\|x_k\|^2_Q + \|u_k\|^2_R\right) \leq \check{J}(x_0, \mathbf{v}^*_0, \bar{L}^*_0) + \gamma^2 \sum_{k=0}^{n} \|w_k\|^2. \tag{4.40}$$

---

[2]There is no need to include the LMI (4.37) in the conditions defining the feasible set $\mathcal{F}_N$ because δ and $\mu \geq 0$ can always be found satisfying (4.38) (and hence also (4.37)) whenever $\Delta \succ 0$.

Proof This can be shown similarly to the proof of Theorem 3.4. □

Example 4.5 In this example we consider the set of feasible initial conditions and the optimal value of the objective of the DAMPC optimization in step (i) of Algorithm 4.4 for the system dynamics, constraints and terminal set of Example 4.2. The feasible initial condition sets $\mathcal{F}_N$ of DAMPC for $N = 4, 5, \ldots, 10$ are shown in Fig. 4.10 projected onto the plane $\{x : [0\ 1\ 0]x = 0\}$. This figure also shows the projections of $\mathcal{X}^{(N)}$, the N-step controllable sets to $\mathcal{X}_T$, onto the same plane, for the same values of N. Clearly $\mathcal{F}_N$ must be a subset of $\mathcal{X}^{(N)}$ since $\mathcal{X}^{(N)}$ is the largest feasible set for any controller parameterization with a prediction horizon of N. The Figure shows that $\mathcal{F}_N$ provides a good approximation of $\mathcal{X}^{(N)}$ for small values of N, but the approximation accuracy decreases as N increases, and $\mathcal{F}_N$ is much smaller than $\mathcal{X}^{(N)}$ for $N \geq 8$.

The degree of suboptimality of the DAMPC optimization at any given feasible point can be determined by comparing the value of the objective of the DAMPC optimization (4.39) with the optimal cost (4.8). Since the objective of (4.39) is equivalent to the min-max cost (4.35), the optimal value (4.8) is the minimum cost that can be obtained for this problem with any controller parameterization. Table 4.1 shows the percentage suboptimality of (4.39) relative to this ideal optimal cost. The average and maximum suboptimality are given for 50 values of the model state randomly selected on the boundary of $\mathcal{F}_N$ for $N = 10, 11, \ldots, 15$. For this example, the DAMPC opti-

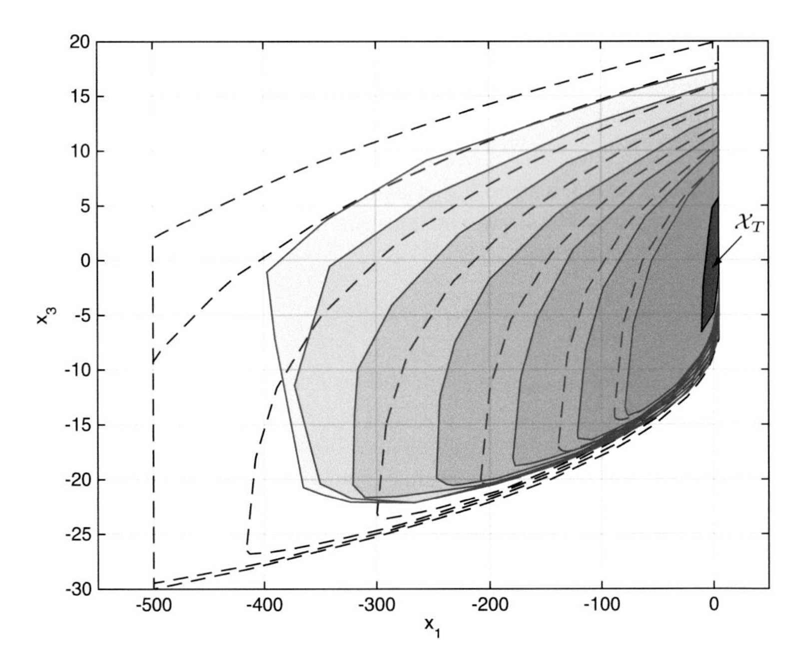

Fig. 4.10 The projections onto the (x_1, x_3)-plane of the feasible sets $\mathcal{F}_N$ for $N = 4, 5, \ldots, 10$ and the terminal set $\mathcal{X}_T$. The *dashed lines* show the projections of the N-step controllable sets to $\mathcal{X}_T$, for $N = 4, 5, \ldots, 10$

Table 4.1 The relative difference between the objective of the DAMPC optimization (4.39) and the ideal optimal cost (4.8) for varying N

Horizon N	Suboptimality	
	Average (%)	Maximum (%)
10	0.1	0.6
11	1.1	6.6
12	2.6	14.0
13	3.8	18.1
14	6.5	31.3
15	9.7	43.3

For each N the average and maximum percentage differences are given for 50 plant states randomly selected on the boundary of the feasible set $\mathcal{F}_N$

mization has negligible suboptimality for $N \leq 10$, but the degree of suboptimality rises quickly as N increases above 10.

To give an indication of the computation required by Algorithm 4.4, the average execution time[3] of (4.39) for $N = 10$ was 0.55 s and the maximum was 0.67 s, whereas for $N = 15$ the execution times were 2.1 s (average) and 2.6 s (maximum). For comparison, solving (4.8) by dynamic programming using Algorithm 4.2 with cold starting required 0.05 s (average) and 0.20 s (maximum) for $N = 10$, and 0.13 s (average) and 0.49 s (maximum) for $N = 15$. Although for this example the online computational requirement of the active set DP implemented by Algorithm 4.2 is considerably less than that of DAMPC, it has to be remembered that DAMPC has a much lower offline computational burden because it does not need the controllable sets $\mathcal{X}^{(1)}, \ldots, \mathcal{X}^{(N)}$ to be determined offline. ◊

We close this section by noting that the disturbance-affine structure of the feedback strategy (4.29) provides compensation for future disturbances that will be known to the controller when the control law is implemented. Clearly it could be advantageous to extend this compensation beyond the future horizon consisting of the first N prediction time steps, and this indeed is proposed in [26]. This approach is cast in the context of stochastic MPC and is described in Chap. 8 (see Sect. 8.2). An application of this idea to a closed-loop prediction strategy that addresses the robust MPC problem considered in this chapter is described in Sect. 4.2.3.

4.2.2 Parameterized Tube MPC

The disturbance-affine feedback strategy considered in Sect. 4.2.1 assumes predicted control inputs with the structure shown in Table 4.2. Although computationally convenient, this parameterization is more restrictive than the general feedback strategy

[3] Execution times are reported here to provide an indication of the computational load—Algorithm 4.4 was implemented using Matlab and Mosek v7.

Table 4.2 The feedback structure of DAMPC

Mode 1					Mode 2	
$v_{0\|k}$	$v_{1\|k}$	$v_{2\|k}$	$\cdots$	$v_{N-1\|k}$	$Kx_{N\|k}$	$\cdots$
	$L^{(1)}_{1\|k}w_{0\|k}$	$L^{(1)}_{2\|k}w_{0\|k}$	$\cdots$	$L^{(1)}_{N-1\|k}w_{0\|k}$		
		$L^{(2)}_{2\|k}w_{1\|k}$	$\cdots$	$L^{(2)}_{N-1\|k}w_{1\|k}$		
			$\ddots$	$\vdots$		
				$L^{(N-1)}_{N-1\|k}w_{N-2\|k}$		
$u_{0\|k}$	$u_{1\|k}$	$u_{2\|k}$	$\cdots$	$u_{N-1\|k}$	$u_{N\|k}$	$\cdots$

The i steps-ahead predicted control input at time k, $u_{i|k}$, is the sum of the entries that lie above the horizontal line in the $(i+1)$th column of the table (zero entries are left blank)

discussed in Sect. 4.1.2 because it forces the predicted control inputs to depend linearly on disturbance inputs. As a result, the i-steps-ahead predicted control input $u_{i|k}$ is determined for all disturbance sequences $(w_{0|k}, \ldots, w_{i-1|k}) \in \mathcal{W} \times \cdots \times \mathcal{W}$ by its values at only $in_w + 1$ vertices of the set $\mathcal{W} \times \cdots \times \mathcal{W}$. On the other hand, the general feedback policy of Sect. 4.1.2 allows $u_{i|k}$ to be chosen independently at each of the m^i vertices of $\mathcal{W} \times \cdots \times \mathcal{W}$ (recall that m is the number of vertices of the disturbance set $\mathcal{W} \subset \mathbb{R}^{n_w}$). Clearly the number of optimization variables required by the general feedback policy must therefore grow exponentially with horizon length (as illustrated for example by the tree structure of Fig. 4.8), making the approach intractable for many problems. However, it is often possible to achieve the same performance as the general feedback strategy, or at least a good approximation of it, using a much more parsimonious parameterization of predicted trajectories.

This is the motivation behind the parameterized tube MPC (PTMPC) formulation of [21], which allows for more general predicted feedback laws than the disturbance-affine strategy of Sect. 4.2.1 while requiring, like DAMPC, a number of optimization variables that grows quadratically with horizon length. PTMPC defines a predicted control tube in terms of the vertices of the sets that form the tube cross sections at each prediction time step over a horizon of N steps. The predicted control trajectories of PTMPC are taken to be convex combinations of these vertices, a subset of which are designated as optimization variables. As is the case for the general feedback strategy in Sect. 4.1.2, the linearity of the system model and constraints (4.1)–(4.4) imply that the conditions for robust constraint satisfaction depend on the vertices of the state and control tubes, but not on the interpolation parameters that define specific trajectories for a given set of vertices. Unlike the general feedback policy, however, PTMPC does not assign an optimization variable to every possible sequence of disturbance vertices over the N-step prediction horizson; instead the predicted state and control tubes are constructed from the Minkowski sum of tubes that model separately the effects of disturbances at individual future time instants.

The decomposition of predicted trajectories into responses to individual disturbances is key to the complexity reduction achieved by PTMPC relative to the general feedback strategy of Sect. 4.1.2, and it also explains why PTMPC cannot in general

Table 4.3 The control tube structure of PTMPC

Mode 1					Mode 2	
$u^{(0)}_{0\|k}$	$u^{(0)}_{1\|k}$	$u^{(0)}_{2\|k}$	$\cdots$	$u^{(0)}_{N-1\|k}$	$Kx^{(0)}_{N\|k}$	$\cdots$
	$U^{(1)}_{1\|k}$	$U^{(1)}_{2\|k}$	$\cdots$	$U^{(1)}_{N-1\|k}$	$KX^{(1)}_{N\|k}$	$\cdots$
		$U^{(2)}_{2\|k}$	$\cdots$	$U^{(2)}_{N-1\|k}$	$KX^{(2)}_{N\|k}$	$\cdots$
			$\ddots$	$\vdots$	$\vdots$	
				$U^{(N-1)}_{N-1\|k}$	$KX^{(N-1)}_{N\|k}$	$\cdots$
					$KX^{(N)}_{N\|k}$	$\cdots$
						$\ddots$
$U_{0\|k}$	$U_{1\|k}$	$U_{2\|k}$	$\cdots$	$U_{N-1\|k}$	$U_{N\|k}$	$\cdots$

The i steps-ahead predicted control input at time k, $u_{i|k}$, is contained in the set $U_{i|k}$, which is formed from the Minkowski sum of all the entries above the horizontal line in the $(i+1)$th column of the table

provide the same performance as control laws that are determined using dynamic programming. This decomposition is illustrated by the triangular tube structure in Table 4.3. For each $i < N$, the set $U_{i|k}$ that defines the tube cross section containing the predicted input $u_{i|k}$ is the (Minkowski) sum of a feedforward term $u^{(0)}_{i|k}$ and sets $U^{(l)}_{i|k}$, $l = 1, 2, \ldots i$:

$$u_{i|k} \in U_{i|k} = \{u^{(0)}_{i|k}\} \oplus U^{(1)}_{i|k} \oplus \cdots \oplus U^{(i)}_{i|k}.$$

Each set $U^{(l)}_{i|k}$ is a compact convex polytope with as many vertices as the disturbance set $\mathcal{W}$:

$$U^{(l)}_{i|k} = \text{Co}\{u^{(l,j)}_{i|k},\ j = 1, \ldots, m\},$$

where the vertices $u^{(l,j)}_{i|k}$, for each $i < N, l \le i$ and $j \le m$ are optimization variables. The tube cross section $U_{N|k}$ containing $u_{N|k}$ is similarly given by the sum of sets that are obtained by applying the fixed linear feedback law $u = Kx$ to each of a sequence of state tube cross sections.

The sets $X_{i|k}$ defining the cross sections of the tube containing the predicted state trajectories are decomposed similarly (Table 4.4) into the sum of a nominal predicted state $x^{(0)}_{i|k}$ and sets $\mathcal{W}$ and $X^{(l)}_{i|k}$ for $l = 1, \ldots, i-1$:

$$x_{i|k} \in X_{i|k} = \{x^{(0)}_{i|k}\} \oplus X^{(1)}_{i|k} \oplus \cdots \oplus X^{(i)}_{i|k}. \tag{4.41}$$

Table 4.4 The state tube structure of PTMPC

Mode 1					Mode 2		
$x^{(0)}_{0\|k}$	$x^{(0)}_{1\|k}$	$x^{(0)}_{2\|k}$	$\cdots$	$x^{(0)}_{N-1\|k}$	$x^{(0)}_{N\|k}$	$\Phi x^{(0)}_{N\|k}$	$\cdots$
	$D\mathcal{W}$	$X^{(1)}_{2\|k}$	$\cdots$	$X^{(1)}_{N-1\|k}$	$X^{(1)}_{N\|k}$	$\Phi X^{(1)}_{N\|k}$	$\cdots$
		$D\mathcal{W}$	$\cdots$	$X^{(2)}_{N-1\|k}$	$X^{(2)}_{N\|k}$	$\Phi X^{(2)}_{N\|k}$	$\cdots$
			$\ddots$	$\vdots$	$\vdots$	$\vdots$	
				$D\mathcal{W}$	$X^{(N-1)}_{N\|k}$	$\Phi X^{(N-1)}_{N\|k}$	$\cdots$
					$D\mathcal{W}$	$\Phi D\mathcal{W}$	$\cdots$
						$D\mathcal{W}$	$\cdots$
							$\ddots$
$X_{0\|k}$	$X_{1\|k}$	$X_{2\|k}$	$\cdots$	$X_{N-1\|k}$	$X_{N\|k}$	$X_{N+1\|k}$	$\cdots$

The i steps-ahead predicted state at time k, $x_{i|k}$, is contained in the set $X_{i|k}$, which is formed from the Minkowski sum of all the entries above the horizontal line in the $(i+1)$th column of the table; $\Phi = A + BK$ where K is the feedback gain appearing in Table 4.3

Each $X^{(l)}_{i|k}$ is a compact convex polytopic set defined by its vertices:

$$X^{(l)}_{i|k} = \mathrm{Co}\{x^{(l,j)}_{i|k},\ j = 1, \ldots, m\}.$$

The trajectory of $x^{(0)}_{i|k}$ is determined by the nominal model dynamics

$$x^{(0)}_{i+1|k} = Ax^{(0)}_{i|k} + Bu^{(0)}_{i|k},\quad i = 0, 1, \ldots, N-1, \tag{4.42a}$$

and the vertices $X^{(l)}_{i|k}$ are paired with those of $U^{(l)}_{i|k}$ so that the predicted state tube evolves as

$$x^{(l,j)}_{l|k} = Dw^{(j)}, \tag{4.42b}$$

$$x^{(l,j)}_{i+1|k} = Ax^{(l,j)}_{i|k} + Bu^{(l,j)}_{i|k},\quad i = l, l+1, \ldots, N-1, \tag{4.42c}$$

for $j = 1, \ldots, m$. Thus $X^{(i)}_{i|k} = D\mathcal{W}$ and the sets $D\mathcal{W}, X^{(l)}_{l+1|k}, X^{(l)}_{l+2|k}, \ldots$ in the $(l+1)$th row of Table 4.4 account for the effects of the disturbance input $w_{l|k}$ on predicted state trajectories. From (4.42a) to (4.42c), it follows that $x_{i+1|k} \in X_{i+1|k}$ for all $(x_{i|k}, u_{i|k}) \in X_{i|k} \times U_{i|k}$ and $w_{i|k} \in \mathcal{W}$, at each time-step $i = 0, 1, \ldots$.

The feedback policy of DAMPC is a special case of PTMPC since the disturbance affine feedback law of Table 4.2 is obtained if every vertex of the PTMPC control tube is defined as a linear function of the vertices of $\mathcal{W}$, namely if $u^{(l,j)}_{i|k} = L^{(l)}_{i|k} w^{(j)}$ for each i, j, l. In general, however, the feedback policy of PTMPC will be piecewise affine since the control input corresponding to any $x_{i|k}$ lying on the boundary of $X_{i|k}$ is given by a linear combination of vertices of $U_{i|k}$. At points on the boundary of its feasible set in state space therefore, the control law of PTMPC is, like the dynamic programming solution discussed in Sect. 4.1.1, a piecewise affine function of the plant state. However, except for some special cases (discussed in [21]), PTMPC is not equivalent to the general feedback policy of Sect. 4.1 since $u_{i|k}$ is specified by a linear combination of only $im + 1$ rather than m^i free variables.

Before considering the definition of a cost index for PTMPC, we first discuss how to impose the constraints (4.4) on the predicted state and control trajectories of PTMPC for all future disturbance realizations. An equivalent and computationally convenient formulation of the constraints $FX_{i|k} + GU_{i|k} \leq \mathbf{1}$ for $i = 0, 1, \ldots, N-1$ is given by

$$Fx_{0|k}^{(0)} + Gu_{0|k}^{(0)} \leq \mathbf{1} \tag{4.43a}$$

together with the following conditions for $i = 1 \ldots N-1$,

$$Fx_{i|k}^{(0)} + Gu_{i|k}^{(0)} + \sum_{l=1}^{i} f_{i|k}^{(l)} \leq \mathbf{1} \tag{4.43b}$$

$$Fx_{i|k}^{(l,j)} + Gu_{i|k}^{(l,j)} \leq f_{i|k}^{(l)}, \quad j = 1, \ldots, m \tag{4.43c}$$

where $f_{i|k}^{(l)}$ for $l = 1, \ldots, i$ are slack variables. The satisfaction of the condition $FX_{N|k} + GU_{N|k} \leq \mathbf{1}$ is ensured by the terminal constraint, $X_{N|k} \subseteq \mathcal{X}_T$, where $\mathcal{X}_T$ is a robustly positively invariant set for (4.1), (4.2) and (4.4) under $u = Kx$. Assuming that this terminal set is given by $\mathcal{X}_T = \{x : V_T x \leq \mathbf{1}\}$, the conditions for $X_{N|k} \subseteq \mathcal{X}_T$ are equivalent to

$$V_T x_{N|k}^{(0)} + \sum_{l=1}^{i} f_T^{(l)} \leq \mathbf{1} \tag{4.44a}$$

$$V_T x_{N|k}^{(l,j)} \leq f_T^{(l)}, \quad j = 1, \ldots, m, \tag{4.44b}$$

where $f_T^{(l)}$, $l = 1, \ldots, N$ are slack variables.

Let $\mathbf{u}_k$ denote the vector of optimization variables defining the predicted control tubes:

$$\mathbf{u}_k = \Big\{(u_{0|k}^{(0)}, \ldots, u_{N-1|k}^{(0)}),\ (u_{i|k}^{(l,j)},\ i = 1, \ldots, N-1,\ l = 1, \ldots, i,\ j = 1, \ldots, m)\Big\}.$$

Then the set $\mathcal{F}_N$ of feasible initial conditions for (4.43a–4.43c) and (4.44a, 4.44b) can be expressed

$$\mathcal{F}_N = \Big\{x_k : \exists\, \mathbf{u}_k \text{ such that for some } f_{i|k}^{(l)},\ i < N,\ l \leq i, \text{ and } f_T^{(l)},\ l \leq N,$$
$$\text{(4.42a–4.42c), (4.43a–4.43c) and (4.44a, 4.44b) hold with } x_{0|k}^{(0)} = x_k\Big\}. \tag{4.45}$$

Lemma 4.3 *For any $N > 0$, $\mathcal{F}_N$ is RPI for the dynamics (4.1), disturbance set (4.2) and constraints (4.4) if $u_k = u^{(0)}_{0|k}$.*

Proof This can be shown by constructing $\mathbf{u}_{k+1}$ such that the conditions defining $\mathcal{F}_N$ are satisfied at time $k+1$, given $x_k \in \mathcal{F}_N$ and $u_k = u^{(0)}_{0|k}$. Specifically, let $w_k = \sum_{j=1}^m \lambda_j w^{(j)}$ for scalars $\lambda_j \geq 0$ satisfying $\sum_{j=1}^m \lambda_j = 1$, and define

$$u^{(0)}_{i|k+1} = \begin{cases} u^{(0)}_{i+1|k} + \sum_{j=1}^{m} \lambda_j u^{(1,j)}_{i+1|k}, & i = 0, \ldots, N-2 \\ K x^{(0)}_{N|k} + \sum_{j=1}^{m} \lambda_j K x^{(1,j)}_{N|k}, & i = N-1 \end{cases} \tag{4.46a}$$

and

$$U^{(l)}_{i|k+1} = \begin{cases} U^{(l+1)}_{i+1|k}, & i = 1, \ldots, N-2, \ l = 1, \ldots, i \\ K X^{(l+1)}_{N|k}, & i = N-1, \ l = 1, \ldots, N-1 \end{cases} \tag{4.46b}$$

Then the sets $X_{i|k+1} = \{x^{(0)}_{i|k+1}\} \oplus X^{(1)}_{i|k+1} \oplus \cdots \oplus X^{(i-1)}_{i|k+1} \oplus D\mathcal{W}$ generated by the tube dynamics (4.42a, 4.42b) with $x^{(0)}_{0|k+1} = x_{k+1}$ satisfy $X_{i|k+1} \subseteq X_{i+1|k}$ for $i = 0, 1, \ldots, N-1$. Hence conditions (4.43a–4.43c) hold at time $k+1$ by convexity. Moreover (4.44a, 4.44b) hold at $k+1$ because $X_{N|k+1} = \Phi X_{N|k} + D\mathcal{W}$ and $X_{N|k} \in \mathcal{X}_T$ where $\mathcal{X}_T$ is by assumption RPI. □

The parameterized tubes of Tables 4.3 and 4.4 can be combined with various alternative performance indices in order to define recursively feasible receding horizon control laws. For example, a quadratic nominal cost similar to that of Sect. 3.3, which involves only the nominal predicted sequences $x^{(0)}_{i|k}$ and $u^{(0)}_{i|k}$ for $i = 0, 1, \ldots, N-1$ and the nominal terminal state $x^{(0)}_{N|k}$, results in a robustly stabilizing MPC law that ensures a finite l_2 gain from the disturbance input to the closed-loop state and control sequences. This approach is described in the context of a related parameterized tube MPC formulation in [22, 27] and also is discussed in Sect. 4.2.3. Here, however, we consider a performance index that ensures exponential convergence to a predefined target set.

The definition of the vertices of predicted state and control tubes as optimization variables in PTMPC motivates the use of a piecewise linear cost, since in this case the optimal state and control sequences are given by sequences of tube vertices. For example, a piecewise linear stage cost of the form

$$l(x, u) = \|Qx\|_\infty + \|Ru\|_\infty$$

is proposed in [21], where the weighting matrices Q and R are chosen so that the sets $\{x : \|Qx\|_\infty \leq 1\}$ and $\{u : \|Ru\|_\infty \leq 1\}$ are compact and contain the origin in their (non-empty) interiors. This stage cost is used in conjunction with a state decomposition in order to penalize the distance of predicted state trajectories from desired target set in [21]. However, for ease of exposition, we consider here a conceptually simpler approach that is discussed in [22], in which the distance of the predicted state from a target set is evaluated directly through a piecewise linear stage cost.

We denote the target set into which the controller is required to steer the state as $\mathcal{S}$, and we assume $\mathcal{S}$ is a convex polytopic subset of the terminal set $\mathcal{X}_T$ that contains the origin in its (non-empty) interior and has the representation

$$\mathcal{S} = \{x : Hx \leq \mathbf{1}\}.$$

We further assume that $\mathcal{S}$ is robustly positively invariant for (4.1) and (4.2) under the linear feedback law $u = Kx$, namely that

$$\Phi\mathcal{S} \oplus D\mathcal{W} \subseteq \mathcal{S}$$

and $Fx + GKx \leq \mathbf{1}$ for all $x \in \mathcal{S}$. A suitable target set is provided by the maximal RPI set (determined for example using Theorem 3.1), and as explained below it is convenient to choose $\mathcal{S}$ equal to the terminal set $\mathcal{X}_T$. We define a measure of the distance of a point $x \in \mathbb{R}^{n_x}$ from the set $\mathcal{S}$ as

$$|x|_{\mathcal{S}} \doteq \begin{cases} 0, & \text{if } x \in \mathcal{S} \\ \max\{Hx\} - 1, & \text{otherwise} \end{cases}$$

where $\max\{Hx\} \doteq \max\{H_1 x, H_2 x, \ldots, H_{n_H} x\}$ with H_i for $i = 1, \ldots, n_H$ denoting the rows of H. A straightforward extension of this definition provides a measure of the distance of a closed set $X \subset \mathbb{R}^{n_x}$ from $\mathcal{S}$ as

$$|X|_{\mathcal{S}} \doteq \max_{x \in X} |x|_{\mathcal{S}}.$$

Note that these measures of distance from a point x to $\mathcal{S}$ and from a set X to $\mathcal{S}$ are consistent with the requirements that $|x|_{\mathcal{S}} = 0$ if and only if $x \in \mathcal{S}$ and $|X|_{\mathcal{S}} = 0$ if and only if $X \subseteq \mathcal{S}$.

The PTMPC predicted cost can be defined in terms of the distances of the cross sections of the predicted state tube from $\mathcal{S}$ as

$$J(x_{0|k}^{(0)}, \mathbf{u}_k) = \sum_{i=0}^{N-1} |X_{i|k}|_{\mathcal{S}} + |X_{N|k}|_{\mathcal{S}_T}. \tag{4.47}$$

The terminal term $|X_{N|k}|_{\mathcal{S}_T}$ is assumed to be defined so that $|x|_{\mathcal{S}_T}$ satisfies, for all $x \in \mathcal{X}_T$,

$$|x|_{\mathcal{S}_T} \geq |\Phi x + Dw|_{\mathcal{S}_T} + |x|_{\mathcal{S}}.$$

This condition requires that $|x|_{\mathcal{S}_T}$ is a piecewise linear Lyapunov function for (4.1) under the feedback law $u = Kx$ for $x \in \mathcal{X}_T$; methods of computing such Lyapunov functions are discussed in [28]. For convenience and without loss of generality, we assume here that $\mathcal{S}$ is equal to the terminal set $\mathcal{X}_T$, in which case $\mathcal{S}_T = \mathcal{S}$ is a valid choice, and the cost of (4.47) reduces to

$$J(x_{0|k}^{(0)}, \mathbf{u}_k) = \sum_{i=0}^{N-1} |X_{i|k}|_{\mathcal{S}} \tag{4.48}$$

as a result of the terminal constraint $X_{N|k} \subseteq \mathcal{X}_T$. Note that terms of the form $|U_{i|k}|_{K\mathcal{S}}$ could also be included in the stage cost in order to place an explicit penalty on the deviation of predicted control sequences from the set $K\mathcal{S} = \{Kx : x \in \mathcal{S}\}$, but for simplicity we consider the cost of (4.47) without this modification.

From the expression for $X_{i|k}$ in (4.41), each term appearing in the cost (4.48) can be expressed as

$$|X_{i|k}|_{\mathcal{S}} = \max\{Hx_{i|k}^{(0)}\} + \sum_{l=1}^{i} \max_j\{Hx_{i|k}^{(l,j)}\} - 1$$

(or $|X_{i|k}|_{\mathcal{S}} = 0$ if this is negative), where $\max_j\{Hx^{(j)}\} \doteq \max_j \max_i\{H_i x^{(j)}\}$. For implementation purposes, each stage cost can therefore be equivalently replaced by a (tight) upper bound given in terms of a slack variable:

$$J(x_{0|k}^{(0)}, \mathbf{u}_k) = \sum_{i=0}^{N-1} d_{i|k}.$$

Here $d_{i|k}$ is a slack variable satisfying the linear constraints

$$d_{i|k} \geq h_{i|k}^{(0)} + \sum_{l=1}^{i} h_{i|k}^{(l)} - 1,$$
$$d_{i|k} \geq 0,$$

where $h_{i|k}^{(l)}$ for $i = 0, \ldots, N-1$ and $l = 0, \ldots, i$ are slack variables satisfying

$$Hx_{i|k}^{(0)} \leq h_{i|k}^{(0)}\mathbf{1}$$

for $i = 0, \ldots, N-1$, and

$$Hx_{i|k}^{(l,j)} \leq h_{i|k}^{(l)}\mathbf{1}, \quad j = 1, \ldots, m$$

for $i = 1, \ldots, N-1$ and $l = 1, \ldots, i$.

We can now state the PTMPC algorithm. This is based on an online optimization which is a linear program involving $O(\frac{1}{2}n_u mN^2)$ optimization variables and $O\big(\frac{1}{2}(n_c+1)N^2\big)$ slack variables, and which has $O\big(\frac{1}{2}(n_c+n_H)mN^2\big)$ linear inequality constraints.

Algorithm 4.5 (*PTMPC*) At each time $k = 0, 1, \ldots$:

(i) If $x_k \notin \mathcal{S}$:

(a) Perform the optimization

$$\begin{aligned} &\underset{\mathbf{u}_k}{\text{minimize}} && J(x_k, \mathbf{u}_k) \\ &\text{subject to} && \text{(4.42a–4.42c), (4.43a–4.43c), (4.44a, 4.44b).} \end{aligned} \tag{4.49}$$

(b) Apply the control law $u_k = u_{0|k}^{(0)*}$, where $\mathbf{u}_k^* = \big\{(u_{0|k}^{(0)*}, \ldots, u_{N-1|k}^{(0)*}), (u_{i|k}^{(l,j)*}, i < N, l \le i, j \le m)\big\}$ is the minimizing argument of (4.49).

(ii) Otherwise (i.e. if $x_k \in \mathcal{S}$), apply the control law $u_k = Kx_k$. ◁

Before discussing closed-loop stability of PTMPC, we first establish that (4.49) is recursively feasible and give a monoticity property of the optimal predicted cost, $J^*(x_k) \doteq J(x_k, \mathbf{u}_k^*)$.

Lemma 4.4 *For the system (4.1) with the control law of Algorithm 4.5 the feasible set $\mathcal{F}_N$ of (4.49) is RPI. Furthermore for any $x_0 \in \mathcal{F}_N$ the optimal objective satisfies, for all $k \ge 0$,*

$$J^*(x_{k+1}) \le J^*(x_k) - |x_k|_{\mathcal{S}}. \tag{4.50}$$

Proof Robust invariance of $\mathcal{F}_N$ under $u_k = u_{0|k}^{(0)*}$ is a direct consequence of Lemma 4.3. Hence the PTMPC optimization (4.49) is feasible at all times $k \ge 1$ if $x_0 \in \mathcal{F}_N$. For $x_k \in \mathcal{S}$, the assumption that $\mathcal{S}$ is RPI implies that (4.50) is trivially satisfied. To derive a bound on the optimal cost $J^*(x_{k+1})$ in terms of $J^*(x_k)$ when $x_k \notin \mathcal{S}$, consider the state tubes that are generated at time $k+1$ by the feasible but suboptimal control tubes given by (4.46a, 4.46b). From (4.42a to 4.42c) we obtain

$$\begin{aligned} x_{i|k+1}^{(0)} &= x_{i+1|k}^{(0)} + \sum_{j=1}^{m} \lambda_j x_{i+1|k}^{(1,j)}, && i = 0, \ldots, N-1 \\ X_{i|k+1}^{(l)} &= X_{i+1|k}^{(l+1)}, && i = 1, \ldots, N-1, \ \ l = 1, \ldots, i \end{aligned}$$

and the corresponding stage cost of (4.48) for $i = 0, \ldots, N-1$ is given by

$$|X_{i|k+1}|_{\mathcal{S}} = \max\Big\{H\big(x_{i+1|k}^{(0)} + \sum_{j=1}^{m} \lambda_j x_{i+1|k}^{(1,j)}\big)\Big\} + \sum_{l=1}^{i} \max_j \{Hx_{i+1|k}^{(l+1,j)}\} - 1$$

(or $|X_{i|k+1}|_{\mathcal{S}} = 0$ if this expression is negative). Moreover, since any pair of vectors a, b necessarily satisfies $\max\{a + b\} \le \max\{a\} + \max\{b\}$, and since $\lambda_j \ge 0$ and $\sum_{j=1}^{m} \lambda_j = 1$, we therefore obtain

$$|X_{i|k+1}|_{\mathcal{S}} \le \max\{Hx^{(0)}_{i+1|k}\} + \sum_{l=1}^{i+1} \max_j\{Hx^{(l,j)}_{i+1|k}\} - 1$$

(or $|X_{i|k+1}|_{\mathcal{S}} = 0$ if this bound is negative). Hence $|X_{i|k+1}|_{\mathcal{S}} \le |X_{i+1|k}|_{\mathcal{S}}$ for all $i = 0, \ldots, N-1$ and the optimal predicted cost at time $k+1$ satisfies $J^*(x_{k+1}) \le \sum_{i=1}^{N-1} |X_{i|k}|_{\mathcal{S}} = J^*(x_k) - |x_k|_{\mathcal{S}}$, where $X_{0|k} = \{x^{(0)}_{0|k}\} = \{x_k\}$ has been used. □

The following result gives the closed-loop stability property of PTMPC.

Theorem 4.2 *Under Algorithm 4.5 the set $\mathcal{S}$ is robustly exponentially stable for the system (4.1), (4.2) and (4.4), with a region of attraction equal to $\mathcal{F}_N$.*

Proof Since $\mathcal{S}$ is RPI for (4.1) under linear feedback $u = Kx$, the optimal value of the cost is given by $J^*(x) = 0$ for all $x \in \mathcal{S}$. Furthermore $J^*(x) \ge |x|_{\mathcal{S}}$ since the stage costs $|X_{i|k}|_{\mathcal{S}}$ for $i \ge 1$ are non-negative. In addition, the optimal cost $J^*(x)$ is the optimal value of a (right-hand side) parametric linear program, and $J^*(x)$ is therefore a continuous piecewise affine function of x on the feasible set $\mathcal{F}_N$ [29]. It follows that constants α, β with $\beta \ge \alpha \ge 1$ exist such that $J^*(x)$ is bounded for all $x \in \mathcal{F}_N$ by

$$\alpha|x|_{\mathcal{S}} \le J^*(x) \le \beta|x_k|_{\mathcal{S}}.$$

If $x_0 \in \mathcal{F}_N$, then from Lemma 4.4 we therefore obtain, for all $k \ge 1$

$$J^*(x_k) \le \left(1 - \frac{1}{\beta}\right)^k J^*(x_0)$$

where $1 - \dfrac{1}{\beta} \in (0, 1)$, and hence the bound

$$|x_k|_{\mathcal{S}} \le \frac{\beta}{\alpha}\left(1 - \frac{1}{\beta}\right)^k |x_0|_{\mathcal{S}} \tag{4.51}$$

holds for all $k \ge 1$. □

Note that the control law of Algorithm 4.5 is a dual mode control law in the strict sense, namely that it implements the linear feedback law $u_k = Kx_k$ whenever $x_k \in \mathcal{S}$. The implied switch to linear feedback is desirable because the minimizing argument of (4.49) is not uniquely defined for all $x_k \in \mathcal{S}$. Furthermore, although it has been designed with the objective of robustly stabilizing the target set $\mathcal{S}$, Algorithm 4.5 can also ensure convergence of the state to the minimal RPI set, $\mathcal{X}^{\mathrm{mRPI}}$, of (4.1) and (4.2) under $u_k = Kx_k$. Specifically, suppose that $\mathcal{S}$ (and hence also the terminal set $\mathcal{X}_T$) is chosen to be a proper subset of the maximal RPI set $\mathcal{X}^{\mathrm{MRPI}}$, and that

Algorithm 4.5 is modified so that the linear feedback law $u = Kx$ is implemented whenever $x_k \in \mathcal{X}^{\text{MRPI}}$. Then Theorem 4.2 implies that x_k converges to $\mathcal{X}^{\text{MRPI}}$ in a finite number of steps and subsequently converges exponentially to $\mathcal{X}^{\text{mRPI}}$.

The online optimization problem of Algorithm 4.5 is an efficiently solvable linear program. Furthermore the number of free variables (including the slack variables) and the number of inequality constraints both depend quadratically with the prediction horizon. Thus PTMPC avoids the exponential growth in computation experienced by a MPC law with the general feedback policy described in Sect. 4.1.2, and its computational requirement is comparable to that of DAMPC. However, the prediction formulation of PTMPC includes that of DAMPC as a special case, and it is therefore to be expected that PTMPC can achieve a larger region of attraction than DAMPC, for the same length of horizon and terminal set. Finally, we note that PTMPC is in general more restricted in the definition of its predicted control trajectories than the general feedback MPC strategy of Sect. 4.1.2 and the active set DP approach of Sect. 4.1.1, and its feasible set for a horizon of N steps is therefore smaller than the N-step controllable set to $\mathcal{X}_T$. The following numerical example illustrates these properties.

Example 4.6 For the system defined by the triple integrator dynamics and the disturbance set and constraints of Example 4.2, the feasible set $\mathcal{F}_N$ of the online optimization of Algorithm 4.5 is shown projected onto the (x_1, x_3)-plane for horizon lengths $4 \leq N \leq 10$ in Fig. 4.11. The linear feedback gain K and the terminal set

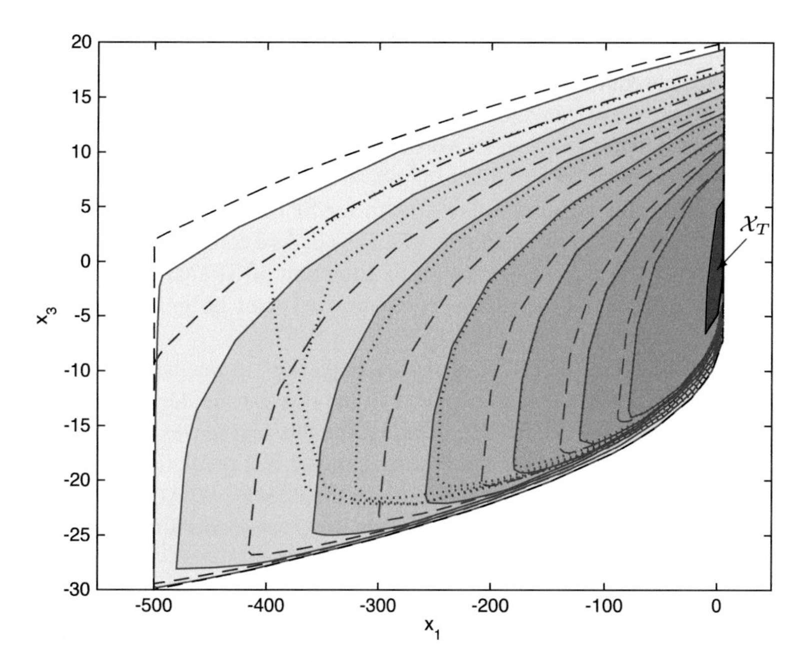

Fig. 4.11 The projections onto the (x_1, x_3)-plane of the feasible sets $\mathcal{F}_N$ for $N = 4, 5, \ldots, 10$ and the terminal set $\mathcal{X}_T$. The *dashed lines* show the projections of the N-step controllable sets to $\mathcal{X}_T$ and the *dotted lines* show the feasible sets for DAMPC

$\mathcal{X}_T$ are defined here as in Examples 4.2, 4.3 and 4.5, and $\mathcal{S}$ is taken to be equal to $\mathcal{X}_T$. It can be seen from this figure that the PTMPC feasible set is smaller than the N-step controllable set to $\mathcal{X}_T$ for each horizon length N in this range. As expected however, the projection of the PTMPC feasible set contains that of DAMPC for each N, and, although they are similar in size for $N \leq 6$, the PTMPC feasible set grows significantly faster than that of DAMPC for $N \geq 8$ in this example.

4.2.3 Parameterized Tube MPC Extension with Striped Structure

The online computational load of PTMPC can be reduced, as reported in [30], by designing some of the slack variables that appear in the constraints of (4.43) and (4.44) offline. Clearly this results in an open-loop strategy, however, and in general will therefore be conservative. An effective alternative is proposed in [22, 27], which uses the same tube parameterization as PTMPC but constrains the component $\{U^{(l)}_{l|k}, U^{(l)}_{l+1|k}, \ldots\}$ of the predicted control tube that appears in the $(l+1)$th row of Table 4.3 to be identical for each row $l = 1, 2, \ldots$. This strategy results in a striped (rather than triangular) structure for both state and control tubes, and for this reason the approach is referred to as Striped PTMPC (SPTMPC).

The key difference in this development is that the free variables defining predicted control tubes are allowed to affect directly the predicted response in both the mode 1 horizon consisting of the first N prediction time steps, and the mode 2 horizon containing the subsequent prediction instants. This can result in a relaxation of the terminal constraints and can thus allow regions of attraction that are larger than those obtained through the use of PTMPC. Furthermore, unlike PTMPC, for which the number of optimization variables and constraints grows quadratically with the prediction horizon, the corresponding growth for SPTMPC is linear, and this can result in significant reductions in computational load. As a consequence, it is possible to further increase the size of the region of attraction of SPTMPC using a longer horizon N while retaining an online computational load no greater than that of PTMPC.

The idea of allowing the effect of partial sequences to persist beyond mode 1 and into the mode 2 horizon was explored in the context the design of controlled invariant sets rather than MPC in [31]. Invariance in this setting was obtained through the use of a contraction variable, which was deployed in a prediction structure with the particular form illustrated in Table 4.5. As for the PTMPC state tubes in Table 4.4, the entries in the table indicate the components the cross sections of predicted state tubes; the corresponding input tubes have a similar structure.

Comparing Tables 4.5 and 4.4, the triangular prediction tube structure employed by PTMPC clearly introduces a greater number degrees of freedom for a given

Table 4.5 Striped prediction tube structure extending into the mode 2 horizon

Mode 1				Mode 2						
$X'_{0\|k}$	$X'_{1\|k}$	$\cdots$	$X'_{N-1\|k}$	$X'_{N\|k}$	$\alpha X'_{0\|k}$	$\cdots$	$\alpha X'_{N-1\|k}$	$\alpha X'_{N\|k}$	$\alpha^2 X'_{0\|k}$	$\cdots$
	$X'_{0\|k}$	$\cdots$	$X'_{N-2\|k}$	$X'_{N-1\|k}$	$X'_{N\|k}$	$\cdots$	$\alpha X'_{N-2\|k}$	$\alpha X'_{N-1\|k}$	$\alpha X'_{N\|k}$	$\cdots$
		$\ddots$	$\vdots$	$\vdots$	$\vdots$		$\vdots$	$\vdots$	$\vdots$	
			$X'_{0\|k}$	$X'_{1\|k}$	$X'_{2\|k}$	$\cdots$	$X'_{N\|k}$	$\alpha X'_{1\|k}$	$\alpha X'_{2\|k}$	$\cdots$
				$X'_{0\|k}$	$X'_{1\|k}$	$\cdots$	$X'_{N-1\|k}$	$X'_{N\|k}$	$\alpha X'_{1\|k}$	$\cdots$
					$\ddots$		$\vdots$	$\vdots$	$\vdots$	
$X_{0\|k}$	$X_{1\|k}$	$\cdots$	$X_{N-1\|k}$	$X_{N\|k}$	$X_{N+1\|k}$	$\cdots$	$X_{2N\|k}$	$X_{2N+1\|k}$	$X_{2N+2\|k}$	$\cdots$

The tube cross section $X_{i|k}$ containing $x_{i|k}$ is formed from the Minkowski sum of all the entries above the horizontal line in the $(i+1)$th column of the table. The contraction variable $\alpha \in (0, 1)$ is introduced for the purposes of controller invariance

horizon N, and these are available for expanding the region of attraction. However this is gained at the expense of additional online computation. Also it can be seen from the control input tubes in Table 4.3 that the PTMPC predicted control law in mode 2 assumes the form $u = Kx$, and hence PTMPC makes no degrees of freedom available for direct disturbance compensation over this part of the prediction horizon.

On the other hand, SPTMPC, like [31], allows the optimization variables to directly determine the cross sections of state and control tubes beyond the initial N-step horizon of mode 1. This is can be seen from the predicted state tube structure in Table 4.6 and the predicted control tube structure in Table 4.7. However, rather than repeating the sequence $X'_{1|k}, X'_{2|k}, \ldots, X'_{N|k}$, contracted by α as in [31], SPTPMC

Table 4.6 The state tube structure of SPTMPC which allows for the direct compensation to extend to mode 2

Mode 1					Mode 2		
$x^{(0)}_{0\|k}$	$x^{(0)}_{1\|k}$	$x^{(0)}_{2\|k}$	$\cdots$	$x^{(0)}_{N-1\|k}$	$x^{(0)}_{N\|k}$	$\Phi x^{(0)}_{N\|k}$	$\cdots$
	DW	$X'_{2\|k}$	$\cdots$	$X'_{N-1\|k}$	$X'_{N\|k}$	$\Phi X'_{N\|k}$	$\cdots$
		DW	$\cdots$	$X'_{N-2\|k}$	$X'_{N-1\|k}$	$X'_{N\|k}$	$\cdots$
			$\ddots$	$\vdots$	$\vdots$		
				DW	$X'_{2\|k}$	$X'_{3\|k}$	$\cdots$
					DW	$X'_{2\|k}$	$\cdots$
						$\ddots$	
$X_{0\|k}$	$X_{1\|k}$	$X_{2\|k}$	$\cdots$	$X_{N-1\|k}$	$X_{N\|k}$	$X_{N+1\|k}$	$\cdots$

The predicted state $x_{i|k}$ lies in the tube cross section $X_{i|k}$ formed from the Minkowski sum of all entries above the horizontal line in the $(i+1)$th column of the table

Table 4.7 The control tube structure of SPTMPC

Mode 1					Mode 2		
$u_{0\|k}^{(0)}$	$u_{1\|k}^{(0)}$	$u_{2\|k}^{(0)}$	$\cdots$	$u_{N-1\|k}^{(0)}$	$Kx_{N\|k}^{(0)}$	$K\Phi x_{N\|k}^{(0)}$	$\cdots$
	$U'_{1\|k}$	$U'_{2\|k}$	$\cdots$	$U'_{N-1\|k}$	$KX'_{N\|k}$	$K\Phi X'_{N\|k}$	$\cdots$
		$U'_{1\|k}$	$\cdots$	$U'_{N-2\|k}$	$U'_{N-1\|k}$	$KX'_{N\|k}$	$\cdots$
			$\ddots$	$\vdots$	$\vdots$		
				$U'_{1\|k}$	$U'_{2\|k}$	$U'_{3\|k}$	$\cdots$
					$U'_{1\|k}$	$U'_{2\|k}$	$\cdots$
						$\ddots$	
$U_{0\|k}$	$U_{1\|k}$	$U_{2\|k}$	$\cdots$	$U_{N-1\|k}$	$U_{N\|k}$	$U_{N+1\|k}$	$\cdots$

allows the predicted state and control tubes to decay in mode 2 through the closed-loop dynamics of (4.1) under $u = Kx$. Thus the state and control tube cross sections at prediction time i can be expressed

$$x_{i|k} \in X_{i|k} = x_{i|k}^{(0)} \oplus \bigoplus_{j=1}^{i} X'_{j|k}$$

$$u_{i|k} \in U_{i|k} = u_{i|k}^{(0)} \oplus \bigoplus_{j=1}^{i} U'_{j|k},$$

where $\mathcal{X}'_{1|k} = D\mathcal{W}$ and for $j = N, N+1, \ldots,$

$$X'_{j|k} = \Phi^{j-N} X'_{N|k}$$
$$U'_{j|k} = K\Phi^{j-N} X'_{N|k}.$$

State and input constraints are imposed on the predicted tubes of SPTMPC through the introduction of slack variables, similarly to the handling of constraints in PTMPC. However, since $X'_{1|k}, \ldots, X'_{N|k}$ appear at every prediction time step of the infinite mode 2 horizon, it is clear that a different terminal constraint is required in order to obtain a guarantee of recursive feasibility. In [22, 27], the set of feasible initial states is made invariant through the use of a supplementary horizon, N_2, within mode 2, over which system constraints are invoked under the mode 2 dynamics and a pair of terminal constraints.

For given N_2, the terminal conditions can be expressed

$$\Phi^{N_2} x_{N|k}^{(0)} \in \mathcal{X}_0 \tag{4.52}$$

and

$$\Phi^{N_2} X'_{N|k} \subseteq \mathcal{X}_1, \tag{4.53}$$

where $\mathcal{X}_0, \mathcal{X}_1$ are polytopic sets that contain the origin in their interior, and where $\mathcal{X}_0$ is RPI for the dynamics $x_{k+1} = \Phi x_k + \xi_k$ for all $\xi_k \in \Phi \mathcal{X}_1$, namely

$$\Phi \mathcal{X}_0 \oplus \Phi \mathcal{X}_1 \subseteq \mathcal{X}_0. \tag{4.54}$$

From (4.52) and (4.54), it follows that $\Phi^{N_2+i} x^{(0)}_{N|k} \subseteq \mathcal{X}_0$ for $i = 1, 2, \ldots$, and if (4.53) also holds, then (4.52) will be invariant for the system (4.1) and (4.2) under $u_k = u^{(0)}_{0|k}$ in the sense that, for all $w_k \in \mathcal{W}$,

$$\Phi^{N_2} x^{(0)}_{N|k+1} \in \mathcal{X}_0 \tag{4.55}$$

The reason (4.55) is implied by the conditions (4.52) and (4.53) and the property (4.54) is that (similarly to the proof of Lemma 4.3) a feasible nominal predicted state trajectory at time $k+1$ is given by the sum of the first row of Table 4.6 and a trajectory contained in the tube defined by the second row. This feasible nominal trajectory is given for $i = 0, 1, \ldots,$ by

$$x^{(0)}_{i|k+1} = \begin{cases} x^{(0)}_{i+1|k} + \sum_{j=1}^{m} \lambda_j x'^{(j)}_{i+1|k}, & i = 0, \ldots, N-1 \\ \Phi^{i+1-N}\Big(x^{(0)}_{N|k} + \sum_{j=1}^{m} \lambda_j x'^{(j)}_{N|k}\Big), & i = N, N+1, \ldots \end{cases}$$

where $\mathrm{Co}\{x'^{(1)}_{i|k}, \ldots, x'^{(m)}_{i|k}\} = X'_{i|k}$ and λ_j are non-negative scalars satisfying $w_k = \sum_{j=1}^{m} \lambda_j w^{(j)}$ and $\sum_{j=1}^{m} \lambda_j = 1$. Therefore

$$x^{(0)}_{N|k+1} \in \{\Phi x^{(0)}_{N|k}\} \oplus \Phi X'_{N|k}$$

so that (4.52) and (4.53) imply $\Phi^{N_2} x^{(0)}_{N|k+1} \in \Phi \mathcal{X}_0 \oplus \Phi \mathcal{X}_1$, and (4.55) then follows from (4.54).

With the conditions (4.52) and (4.53), it is now possible to formulate a condition that guarantees constraint satisfaction at all prediction times $i \geq N + N_2$, and which ensures recursive feasibility. To do this, we consider the components of the tube cross section X_{N+N_2+r}, for arbitrary $r \geq 0$, as specified by the terms appearing in the corresponding column of Table 4.6. Writing the entries in this column as a sequence:

$$\underbrace{\Phi^{N_2+r} x^{(0)}_{N|k}}_{\text{(A)}},\ \underbrace{\Phi^{N_2+r} X'_{N|k}, \ldots, \Phi^{N_2} X'_{N|k}}_{\text{(B)}},\ \underbrace{\Phi^{N_2-1} X'_{N|k}, \ldots, \Phi X'_{N|k}, X'_{N|k} \cdots X'_{1|k}}_{\text{(C)}}$$

we find that the first term (labelled A) lies in $\mathcal{X}_0$ by (4.52) and (4.54), while the block labelled C is common to the $(N + N_2 + r)$th column of the table for all $r \geq 0$. Thus the only challenge is presented by the block labelled B and to deal with this, we make use of the following bound

$$\bigoplus_{r=1}^{i} \Phi^{N_2+r} X'_{N|k} \subseteq \Omega_\infty(\mathcal{X}_1) \tag{4.56}$$

where $\Omega_\infty(\mathcal{X}_1)$ denotes a RPI set for the dynamics of $x_{k+1} = \Phi x_k + w_k$ with $w_k \in \mathcal{X}_1$. This is a direct consequence of (4.52), which implies that $\Phi^{N_2+r} X'_{N|k} \subset \Phi^r \mathcal{X}_1$. In this setting, it is convenient to represent the constraints $Fx + Gu \leq \mathbf{1}$, for $u = Kx$ in the format of a set inclusion:

$$x \in \mathcal{X} = \{x : (F + GK)x \leq \mathbf{1}\}. \tag{4.57}$$

Then, given the sets $\mathcal{X}_0$, $\mathcal{X}_1$ and $\Omega_\infty(\mathcal{X}_1)$, the condition

$$\mathcal{X}_0 \oplus \bigoplus_{i=1}^{N} X'_{i|k} \oplus \bigoplus_{i=1}^{N_2-1} \Phi^i X'_{N|k} \oplus \Omega_\infty(\mathcal{X}_1) \subseteq \mathcal{X} \tag{4.58}$$

ensures satisfaction of (4.57) at all prediction instants $i \geq N + N_2$ by construction. Furthermore condition (4.58) is invariant in the sense that it will be feasible at $k + 1$ if it is satisfied at time k.

In the following analysis, we make the assumption that the minimal robustly positively invariant set, $\mathcal{X}^{\text{mRPI}}$, for the system $x_{k+1} = Ax_k + Bu_k + w_k$ with $u_k = Kx_k$ and $w_k \in \mathcal{W}$ lies in the interior of $\mathcal{X}$. Note that $\mathcal{X}^{\text{mRPI}}$ must lie inside $\mathcal{X}$ in order that the linear feedback law $u = Kx$ is feasible in some neighbourhood of the origin; hence this is a mild assumption to make. In addition it ensures that (4.58) is necessarily feasible for sufficiently large N, N_2.

Lemma 4.5 *There exist N, N_2, $\mathcal{X}_0$, $\mathcal{X}_1$ such that (4.58) is feasible.*

Proof This result can be proved by construction. Thus let $\mathcal{X}_1 = \Phi^{N+N_2-1}\mathcal{W}$ and suppose that the control law in both mode 1 and mode 2 is chosen to be $u = Kx$. For this selection, (4.58) becomes

$$\mathcal{X}_0 \oplus \mathcal{X}^{\text{mRPI}} \subseteq \mathcal{X} \tag{4.59}$$

which, by assumption, will be feasible for sufficiently small $\mathcal{X}_0$. Such a choice for $\mathcal{X}_0$ is possible provided $N + N_2$ is chosen to be large enough. □

Note that the constraint (4.58) can be relaxed through the use of smaller $\mathcal{X}_0$, but at the same time the constraint (4.52) becomes more stringent. A sensible compromise between these two effects can be reached by introducing a tuning parameter α and defining $\mathcal{X}_0$ by

$$\mathcal{X}_0 = \alpha\tilde{\mathcal{X}} \oplus \tilde{\Omega}_\infty(\Phi\mathcal{X}_1) \tag{4.60}$$

where $\tilde{\mathcal{X}}$ is an invariant set for the dynamics $x_{k+1} = \Phi x_k$ and where $\tilde{\Omega}_\infty(\Phi\mathcal{X}_1)$ is the minimal invariant set (or an invariant outer approximation) for the dynamics $x_{k+1} = \Phi x_k + \xi_k$, $\xi_k \in \Phi\mathcal{X}_1$.

We are now able to construct a robust MPC strategy with the objective of steering the state to the minimal robust invariant set $\mathcal{X}^{\mathrm{mRPI}}$. This is achieved by replacing (4.58) with the condition

$$\mathcal{X}_0 \oplus \bigoplus_{i=1}^{N} X'_{i|k} \oplus \bigoplus_{i=1}^{N_2-1} \Phi^i X'_{N|k} \oplus \Omega_\infty(\mathcal{X}_1) \subseteq \mathcal{S} \tag{4.61}$$

where $\mathcal{S}$ is a robustly positively invariant polytopic set the system (4.1) and (4.2) and constraints under $u = Kx$.

We define the online objective function to penalize the distance of each of the tube cross sections, $X_{i|k}$, $i = 0, 1, \ldots$, from $\mathcal{S}$. Since (4.61) ensures $X_{i|k} \subseteq \mathcal{S}$ for all $i \geq N + N_2$, this cost has the form

$$J(x^{(0)}_{0|k}, \mathbf{u}_k) = \sum_{i=0}^{N+N_2-1} |X_{i|k}|_{\mathcal{S}}.$$

As in the case of the cost for the PTMPC algorithm discussed in Sect. 4.2.2, this cost is non-negative and its minimum over the optimization variables $\mathbf{u}_k = \{(u^{(0)}_{0|k}, \ldots, u^{(0)}_{N-1}), (U'_{1|k}, \ldots, U'_{N+N_2-1|k})\}$ is zero if and only if $x^{(0)}_{0|k} \in \mathcal{S}$. The minimization of this cost therefore forms the basis of a receding horizon strategy for steering the state of (4.1) into $\mathcal{S}$ while robustly satisfying constraints.

For $\mathcal{S}$ described by the inequalities $\mathcal{S} = \{x : Hx \leq \mathbf{1}\}$, the minimization of the cost $J(x, \mathbf{u}_k)$ can be performed by minimizing a sum of slack variables. Using again the approach of Sect. 4.2.2 we have

$$J(x^{(0)}_{0|k}, \mathbf{u}_k) = \sum_{i=0}^{N+N_2-1} d_{i|k}$$

where for $i = 0, \ldots, N + N_2 - 1$, the parameters $d_{i|k}$ satisfy

$$d_{i|k} \geq h^{(0)}_{i|k} + \sum_{l=1}^{i} h'_{i|k} - 1,$$

$$d_{i|k} \geq 0,$$

and $h^{(0)}_{i|k}$, $h'_{i|k}$ satisfy

$$h^{(0)}_{i|k}\mathbf{1} \geq \begin{cases} Hx^{(0)}_{i|k}, & i = 0, \ldots, N-1 \\ H\Phi^{i-N} x^{(0)}_{N|k}, & i = N, \ldots, N+N_2-1 \end{cases}$$

$$h'_{i|k}\mathbf{1} \geq \begin{cases} Hx'^{(j)}_{i|k}, & i = 1, \ldots, N-1, \;\; j = 1, \ldots, m \\ H\Phi^{i-N} x'^{(j)}_{N|k}, & i = N, \ldots, N+N_2-1, \;\; j = 1, \ldots, m \end{cases}$$

As discussed for the case of PTMPC in Sect. 4.2.2, the constraints of (4.4) at prediction times $i = 0, \ldots, N + N_2 - 1$ can be invoked through the use of slack variables. Likewise the constraints of (4.52), (4.53) and (4.61) constitute a set of linear inequalities that can be implemented using slack variables.

Algorithm 4.6 (*SPTMPC*) At each time $k = 0, 1, \ldots$:

(i) If $x_k \notin \mathcal{S}$:

(a) Perform the optimization

$$\begin{aligned} &\underset{\mathbf{u}_k}{\text{minimize}} \;\; J(x_k, \mathbf{u}_k) \\ &\text{subject to} \;\; FX_{i|k} + GU_{i|k} \leq \mathbf{1}, \; i = 0, \ldots, N + N_2 - 1 \\ &\qquad\qquad (4.52), (4.53) \text{ and } (4.61). \end{aligned} \tag{4.62}$$

(b) Apply the control law $u_k = u^{(0)*}_{0|k}$, where $\mathbf{u}^*_k = \big\{(u^{(0)*}_{0|k}, \ldots, u^{(0)*}_{N-1|k}), (U'^*_{1|k}, \ldots, U'^*_{N-1|k})\big\}$ is the minimizing argument of (4.62).

(ii) Otherwise (i.e. if $x_k \in \mathcal{S}$), apply the control law $u_k = Kx_k$. ◁

The closed-loop stability properties of Algorithm 4.6 can be stated in terms of the set of feasible states x_k for the online optimization (4.62), which we denote as $\mathcal{F}_{N,N_2}$.

Corollary 4.2 *For the system (4.1) and (4.2), constraints (4.4), and the control law of Algorithm 4.6, the feasible set $\mathcal{F}_{N,N_2}$ is RPI and $\mathcal{S}$ is exponentially stable with region of attraction equal to $\mathcal{F}_{N,N_2}$.*

Proof By construction the constraints of (4.62) are recursively feasible, namely feasibility at time $k = 0$ implies feasibility at all times $k = 1, 2, \ldots$. Exponential stability of $\mathcal{S}$ can be demonstrated using the same arguments as in the proofs of Lemma 4.4 and Theorem 4.2. In particular, the bound (4.50) holds along closed-loop trajectories and exponential convergence (4.51) therefore holds. □

Note that the state of (4.1) converges exponentially to the minimal RPI set $\mathcal{X}^{\text{mRPI}}$ under $u = Kx$ if $\mathcal{S}$ is chosen to be a proper subset of the RPI set on which the control law of Algorithm 4.6 switches to $u = Kx$.

The structure of SPTMPC indicated in Table 4.6 allows for only a single sequence $X'_{1|k}, \ldots, X'_{N|k}$ to be used in the definition of the predicted partial tubes. However, if it is desired to introduce more degrees of freedom into the SPTMPC algorithm, then it would be possible to implement a hybrid scheme. For example, this could be realized by introducing tubes $X^{(l)}_{1|k}, \ldots, X^{(l)}_{N|k}$, for $l = 1, \ldots, \nu$, with the full triangular structure of PTMPC into the upper rows of Table 4.6 before switching to a fixed sequence $X'_{1|k}, \ldots, X'_{N|k}$ to generate a striped tube structure in the remainder of the table. The implied algorithm can be shown to inherit the feasibility and stability properties of SPTMPC.

The advantage of SPTMPC over PTMPC is that it allows for a reduction of the online computational load and, at the same time, extends disturbance compensation into mode 2. Using the same prediction horizons for PTMPC and SPTMPC is likely to result in the constraints of PTMPC that apply to the mode 1 prediction horizon being less stringent than those of SPTMPC on account of the extra degrees of freedom introduced by PTMPC. For SPTMPC, however, the constraints of mode 2 are likely to be less stringent than for PTMPC, since SPTMPC allows for direct disturbance compensation in mode 2. In general, it is not possible to state which of the two methodologies will result in the larger region of attraction. We give next an illustrative example showing an instance of STMPC outperforming PTMPC.

Example 4.7 Consider the system defined by the model (4.1) with parameters

$$A = \begin{bmatrix} 0.787 & 1.02 \\ -0.93 & 1.03 \end{bmatrix}, \quad B = \begin{bmatrix} 0.331 \\ -1.01 \end{bmatrix}, \quad D = \begin{bmatrix} 1 & 0 \\ 0 & 1 \end{bmatrix}.$$

and disturbance set

$$\mathcal{W} = \left\{ w : \begin{bmatrix} -1 \\ -1 \end{bmatrix} \leq w \leq \begin{bmatrix} 1 \\ 1 \end{bmatrix} \right.$$

The constraints (4.4) are given by

$$\begin{aligned} \{(x, u) : \pm[-0.044 \;\; 0.092]x &\leq 1 \\ \pm[0.009 \;\; 0.093]x &\leq 1 \\ \pm u &\leq 1\} \end{aligned}$$

The set Ω_0 is constructed using (4.60), with

$$\Omega_0 = 0.01\tilde{X} \oplus \tilde{\Omega}_\infty(\Phi \mathcal{X}_1)$$

and to ensure satisfaction of (4.58), we take $N_2 = 5$.

The areas of the domains of attraction are given in Table 4.8 for three variant strategies: (i) the SPTMPC strategy of Algorithm 4.6; the SPTMPC strategy with a nominal cost, and with the constraint (4.61) replaced with (4.58); (iii) PTMPC. The table also gives the numbers of online optimization variables, numbers of equality constraints and numbers of inequality constraints for each algorithm.

Table 4.8 Areas of domains of attraction and numbers of variables, inequality constraints and inequality constraints for: (i) SPTMPC (Algorithm 4.6); (ii) SPTMPC (Algorithm 4.6 with nominal cost); (iii) PTMPC (Algorithm 4.5)

N	A_N			Variables			Inequalities			Equalities		
	(i)	(ii)	(iii)	(i)	(ii)	(iii)	(i)	(ii)	(iii)	(i)	(ii)	(iii)
1	2.38	2.38	2.38	93	67	24	296	186	61	2	2	10
2	3.01	2.87	3.01	120	89	71	347	216	158	12	12	28
3	3.57	3.57	3.66	147	111	144	398	246	303	22	22	54
4	4.44	4.53	4.19	174	133	243	449	276	496	32	32	88
5	5.02	5.14	4.59	201	155	368	500	306	737	42	42	130
6	5.25	5.39	4.88	228	177	519	551	336	1026	52	52	180
7	5.44	5.83	5.13	255	199	696	602	366	1363	62	62	238
8	5.57	5.87	5.34	282	221	899	653	396	1748	72	72	304
9	5.66	5.95	5.48	309	243	1128	704	426	2181	82	82	378
10	5.72	5.95	5.59	336	265	1383	755	456	2662	92	92	460

For the same value of N both variants of SPTMPC yield larger domains of attraction than PTMPC when $N \geq 4$. This is largely a result of the disturbance compensation that SPTMPC provides in mode 2. The full triangular structure of PTMPC is more general and can therefore outperform SPTMPC. However, the price of this triangular structure is that it implies a number of online optimization variables that grows quadratically with N, whereas the number of optimization variables required by SPTMPC grows only linearly with N. Thus SPTMPC can use longer horizons, thereby enlarging the size of the domain of attraction, at a computational cost which is still less than that required by PTMPC. For example, while for $N < 4$ both variants of SPTMPC have more variables and inequality constraints, for $N \geq 4$ the SPTMPC algorithms both provide a larger domain of attraction while using fewer variables and constraints. In particular, both variants of SPTMPC have fewer variables with $N = 10$ than PTMPC with $N = 5$.

The SPTMPC strategy with a nominal performance index does not constrain the predicted state to lie inside $\mathcal{S}$ at any instant and hence it achieves larger domains of attraction with fewer variables than Algorithm 4.6. However, the latter enjoys stronger stability properties which guarantee convergence to the minimal RPI set $\mathcal{X}^{\mathrm{mRPI}}$. ◊

References

1. H.S. Witsenhausen, A minimax control problem for sampled linear systems. IEEE Trans. Autom. Control **13**(1), 5–21 (1968)
2. M.C. Delfour, S.K. Mitter, Reachability of perturbed systems and min sup problems. SIAM J. Control **7**(4), 521–533 (1969)

3. J.D. Glover, F.C. Schweppe, Control of linear dynamic systems with set constrained disturbances. IEEE Trans. Autom. Control **16**(5), 411–423 (1971)
4. D.P. Bertsekas, I.B. Rhodes, On the minimax reachability of target sets and target tubes. Automatica **7**, 233–247 (1971)
5. R.E. Bellman, The theory of dynamic programming. Bull. Am. Math. Soc. **60**(6), 503–516 (1954)
6. D.P. Bertsekas, *Dynamic Programming and Optimal Control* (Academic Press, New York, 1976)
7. A. Bemporad, F. Borrelli, M. Morari, Min-max control of constrained uncertain discrete-time linear systems. IEEE Trans. Autom. Control **48**(9), 1600–1606 (2003)
8. H.J. Ferreau, H.G. Bock, M. Diehl, An online active set strategy to overcome the limitations of explicit MPC. Int. J. Robust Nonlinear Control **18**(8), 816–830 (2008)
9. M. Cannon, W. Liao, B. Kouvaritakis, Efficient MPC optimization using Pontryagin's minimum principle. Int. J. Robust Nonlinear Control **18**(8), 831–844 (2008)
10. J. Buerger, M. Cannon, B. Kouvaritakis, An active set solver for min-max robust control, in *Proceedings of the 2013 American Control Conference* (2013), pp. 4227–4233
11. J. Buerger, M. Cannon, B. Kouvaritakis, An active set solver for input-constrained robust receding horizon control. Automatica **1**, 155–161 (2014)
12. M.J. Best, *An Algorithm for the Solution of the Parametric Quadratic Programming Problem*. Applied Mathematics and Parallel Computing (Physica-Verlag, Heidelberg, 1996)
13. P.O.M. Scokaert, D.Q. Mayne, Min-max feedback model predictive control for constrained linear systems. IEEE Trans. Autom. Control **43**(8), 1136–1142 (1998)
14. J. Nocedal, S. Wright, *Numerical Optimization* (Springer, New York, 2006)
15. J.H. Lee, B.L. Cooley, Optimal feedback control strategies for state-space systems with stochastic parameters. IEEE Trans. Autom. Control **43**(10), 1469–1475 (1998)
16. I. Batina, A.A. Stoorvogel, S. Weiland, Optimal control of linear, stochastic systems with state and input constraints, in *Proceedings of the 41st IEEE Conference on Decision and Control*, Las Vegas, USA, vol. 2 (2002), pp. 1564–1569
17. B. Kouvaritakis, J.A. Rossiter, A.O.T. Chang, Stable generalised predictive control: an algorithm with guaranteed stability. IEE Proc. D Control Theory Appl. **139**(4), 349–362 (1992)
18. D.H. van Hessem, O.H. Bosgra, A conic reformulation of model predictive control including bounded and stochastic disturbances under state and input constraints, in *Proceedings of the 41st IEEE Conference on Decision and Control*, Las Vegas, USA, vol. 4 (2002), pp. 4643–4648
19. J. Löfberg, Approximations of closed-loop minimax MPC, in *Proceedings of the 42nd IEEE Conference on Decision and Control*, Maui, USA, vol. 2 (2003), pp. 1438–1442
20. P.J. Goulart, E.C. Kerrigan, J.M. Maciejowski, Optimization over state feedback policies for robust control with constraints. Automatica **42**(4), 523–533 (2006)
21. S.V. Rakovic, B. Kouvaritakis, M. Cannon, C. Panos, R. Findeisen, Parameterized tube model predictive control. IEEE Trans. Autom. Control **57**(11), 2746–2761 (2012)
22. D. Muñoz-Carpintero, B. Kouvaritakis, M. Cannon, Striped parameterized tube model predictive control. Automatica (2015), in press
23. A. Ben-Tal, A. Goryashko, E. Guslitzer, A. Nemirovski, Adjustable robust solutions of uncertain linear programs. Math. Program. **99**(2), 351–376 (2004)
24. C. Wang, C.J. Ong, M. Sim, Convergence properties of constrained linear system under MPC control law using affine disturbance feedback. Automatica **45**(7), 1715–1720 (2009)
25. P.J. Goulart, E.C. Kerrigan, T. Alamo, Control of constrained discrete-time systems with bounded l_2 gain. IEEE Trans. Autom. Control **54**(5), 1105–1111 (2009)
26. B. Kouvaritakis, M. Cannon, D. Muñoz-Carpintero, Efficient prediction strategies for disturbance compensation in stochastic MPC. Int. J. Syst. Sci. **44**(7), 1344–1353 (2013)
27. D. Muñoz-Carpintero, B. Kouvaritakis, M. Cannon, Striped parameterized tube model predictive control, in *Proceedings of the 19th IFAC World Congress*, Cape Town, South Africa (2014), pp. 11998–12003
28. F. Blanchini, S. Miani, *Set-Theoretic Methods in Control* (Birkhäuser, Boston, 2008)

29. T. Gal, *Postoptimal Analyses, Parametric Programming, and Related Topics*, 2nd edn. (De Gruyter, Berlin, 1995)
30. S.V. Raković, D. Muñoz-Carpintero, M. Cannon, B. Kouvaritakis, Offline tube design for efficient implementation of parameterized tube model predictive control, in *Proceedings of the 51st IEEE Conference on Decision and Control*, Maui, USA (2012), pp. 5176–5181
31. S.V. Raković, M. Barić, Parameterized robust control invariant sets for linear systems: theoretical advances and computational remarks. IEEE Trans. Autom. Control **55**(7), 1599–1614 (2010)

Chapter 5
Robust MPC for Multiplicative and Mixed Uncertainty

In this chapter, we consider constrained linear systems with imprecisely known parameters, namely systems that are subject to multiplicative uncertainty. Although of real concern in most applications, the unknown additive disturbances considered in Chaps. 3 and 4 are not the only form of uncertainty that may be present in a system model. Even if additive disturbances are not present, it is rarely the case that a linear time-invariant model is able to reproduce exactly the behaviour of a physical system. This may be a consequence of imprecise knowledge of model parameters, or it may result from inherent uncertainty in the system in the form of stochastic parameter variations. Model error can also arise through the use of reduced order models that neglect the high-order dynamics of the system. Moreover, the system dynamics may be linear but time-varying, or they may be nonlinear, or both time-varying and nonlinear.

In all such cases, it may be possible to capture the key features of system behaviour, albeit conservatively, using a linear model whose parameters are assumed to lie in a given uncertainty set. Thus for example in some instances it is possible to capture the behaviour of an uncertain nonlinear system using a linear difference inclusion (LDI). A linear difference inclusion is a discrete-time system consisting of a family of linear models, the parameters of which belong to the convex hull of a known set of vertices. An LDI model has the property that the trajectories of the actual system are contained in the convex hull of the trajectories that are generated by the linear models defined by these vertices (e.g. [1]).

Robust MPC techniques for systems subject to constraints and multiplicative model uncertainty are the focus of this chapter. As in earlier chapters, we examine receding horizon control methodologies that guarantee feasibility and stability while ensuring bounds on closed-loop performance. Also following earlier chapters, the emphasis is on computationally tractable approaches. Towards the end of the chapter, consideration is given to the situation most prevalent in practice, in which systems are subject to a mix of multiplicative uncertainty and unknown additive disturbances.

B. Kouvaritakis and M. Cannon, *Model Predictive Control*,
Advanced Textbooks in Control and Signal Processing,
DOI 10.1007/978-3-319-24853-0_5

5.1 Problem Formulation

Early work on MPC for systems with multiplicative uncertainty was of a heuristic nature and considered the case of a finite impulse response (FIR) model containing an uncertain scaling factor [2]. More general uncertain impulse response models in which the vector of impulse response coefficients is assumed to lie, at each instant, in a known polytopic set were considered in [3, 4]. These approaches posed the robust MPC problem in a min–max framework and hence minimized, over the trajectory of predicted future inputs, the maximum over all model uncertainty of a predicted cost. By defining the predicted cost in terms of the infinity-norms of tracking errors, the MPC optimization could be expressed as a linear program, with objective and constraints depending linearly on the optimization variables.

The computational requirement of the approach of [3] grows exponentially with the number of vertices used to model uncertainty. This growth can be reduced considerably by introducing slack variables, as proposed in [4], but the method remains restricted to FIR models. The convenience of FIR system descriptions is that, as mentioned in Chap. 2, they enable the implicit implementation of equality terminal conditions and for this reason FIRs have proved popular in the analysis and design of robust MPC methods for the case of multiplicative model uncertainty (e.g. [5, 6]).

To avoid excessive computational demands that make online optimization impracticable, an impulse response model must be truncated to a finite impulse response with a limited number of coefficients. Such truncation can render the model unrealistic, particularly when slow poles are present in the system dynamics. It is also clear that approaches based on FIR representations can only be applied to systems that are open-loop stable (although for the purposes of prediction alone this limitation may be overcome through the use of bicausal FIR models [7]). For these reasons, subsequent developments considered more general linear state-space models:

$$x_{k+1} = A_k x_k + B_k u_k \tag{5.1}$$

where $x_k \in \mathbb{R}^{n_x}$ and $u_k \in \mathbb{R}^{n_u}$ denote the system state and control input.

The parameters (A_k, B_k) of (5.1) are assumed to belong for all k to a convex compact polytopic set, Ω, defined by the convex hull of a known set of vertices

$$\Omega = \mathrm{Co}\{(A^{(1)}, B^{(1)}), \ldots, (A^{(m)}, B^{(m)})\}.$$

Therefore

$$(A_k, B_k) = \sum_{j=1}^{m} q_k^{(j)} \big(A^{(j)}, B^{(i)}\big) \tag{5.2}$$

for some scalars $q_k^{(1)}, \ldots, q_k^{(m)}$ that are unknown at time k and which satisfy

$$\sum_{j=1}^{m} q_k^{(j)} = 1 \;\text{ and }\; q_k^{(j)} \geq 0, \;\; j = 1, \ldots, m.$$

As in earlier chapters, the state and control inputs are assumed to be subject to linear constraints:

$$Fx_k + Gu_k \leq \mathbf{1}, \tag{5.3}$$

for given $F \in \mathbb{R}^{n_C \times n_x}$, $G \in \mathbb{R}^{n_C \times n_u}$, and we consider receding horizon strategies that aim to minimize predicted performance expressed as a sum of quadratic stage costs over a future horizon. The predicted cost may be defined in terms of either a nominal cost or a worst-case cost. A nominal cost is based on a nominal system description given by a known set of model parameters $(A^{(0)}, B^{(0)})$. Suitable nominal parameters could be defined, for example, by the expected value of the parameter set Ω or alternatively its centroid,

$$\big(A^{(0)}, B^{(0)}\big) = \frac{1}{m}\sum_{j=1}^{m}\big(A^{(j)}, B^{(j)}\big). \tag{5.4}$$

Conversely, worst-case performance is defined as the maximum value of the predicted cost over all model parameters in the uncertainty set Ω.

This chapter summarizes significant developments in robust MPC methodologies that provide guarantees of feasibility and stability over the entire uncertainty class Ω. We first discuss the approach of [8], which is based on Linear Matrix Inequalities (LMIs) and does not employ a mode 1 horizon but handles the cost and constraints implicitly through recursive quadratic bounds. Later work [9] reduced the conservativeness of this approach by constructing state tubes that take into account all possible predicted realizations of model uncertainty. This made it possible to introduce a mode 1 horizon over which the predicted cost and constraints are handled explicitly. However, [9] uses tubes that are minimal in the sense that the tube cross sections provide tight bounds on the predicted model states (i.e. every point in the tube cross section is a future predicted state for some realization of model uncertainty). This makes the computational complexity of the approach depend exponentially on the length of the mode 1 prediction horizon.

Computation can be greatly reduced, at the expense of a degree of conservativeness, if ellipsoidal bounds are used to invoke the constraints on predicted states and control inputs. This is the approach taken in [10, 11], where predicted state and input trajectories are generated by a dynamic feedback controller, and the degrees of freedom in predictions are incorporated as additional states of the prediction model. Related work [12] showed how to optimize the prediction system dynamics, similarly to the approach described in Sect. 2.9, so as to maximize ellipsoidal regions of attraction in robust MPC. An alternative form of optimized dynamics was introduced in [13] in order to improve robustness to multiplicative model uncertainty.

The conservativeness that is inherent in enforcing constraints using ellipsoidal tubes can be reduced or even avoided altogether using tubes with variable polytopic cross sections to bound predicted trajectories. Early work on this restricted tube cross sections to low-complexity polytopic sets (e.g. [14–17]). The approach was subsequently extended using an application of Farkas' Lemma to allow general polytopic sets [18–20].

After describing these developments, the chapter concludes by discussing MPC for systems that are subject to both multiplicative model uncertainty and additive disturbances [21–23]. In this case, the model of (5.1) and (5.2) becomes

$$x_{k+1} = A_k x_k + B_k u_k + D_k w_k \tag{5.5}$$

with polytopic uncertainty descriptions,

$$(A_k, B_k, D_k) \in \mathrm{Co}\{(A^{(1)}, B^{(1)}, D^{(1)}), \ldots, (A^{(m)}, B^{(m)}, D^{(m)})\}, \tag{5.6a}$$

$$w_k \in \mathrm{Co}\{w^{(1)}, \ldots, w^{(q)}\}, \tag{5.6b}$$

where the vertices $(A^{(j)}, B^{(j)}, D^{(j)})$ and $q^{(l)}$, $j = 1, \ldots, m$, $l = 1, \ldots, q$ are assumed to be known. As in earlier chapters, the nominal value of the additive disturbance is taken to be zero.

5.2 Linear Matrix Inequalities in Robust MPC

Linear matrix inequalities were encountered in earlier chapters in the context of ellipsoidal constraint approximations (Sects. 2.7 and 2.9) and worst-case quadratic performance indices (Sects. 3.4 and 4.2). We begin this section by considering analogous LMI conditions that ensure constraint satisfaction and performance bounds for systems subject to multiplicative model uncertainty. The simplest setting for this is the robust MPC strategy of [8], in which the predicted control trajectories are generated by a linear feedback law:

$$u_{i|k} = K_k x_{i|k}, \tag{5.7}$$

where $u_{i|k}$ and $x_{i|k}$ are the predictions at time k of the i steps ahead input and state variables. The corresponding predicted state trajectories satisfy

$$x_{i|k} = (A_{i|k} + B_{i|k} K_k) x_{i|k}, \quad (A_{i|k}, B_{i|k}) \in \Omega \tag{5.8}$$

for all $i \geq 0$.

The convexity property of linear matrix inequalities discussed in Sect. 2.7.3 is particularly useful in the context of robust MPC. Consider for example the function

$$M(x) \doteq M_0 + M_1 x_1 + \cdots + M_n x_n,$$

where $M_0, \ldots, M_n$ are given matrices, and suppose that x assumes values in a given convex compact polytope $\mathcal{X}$. Then $M(x) \succeq 0$ for all $x \in \mathcal{X}$ if and only if $M(x) \succeq 0$ holds at the vertices of $\mathcal{X}$, namely

$$M(x) \succ 0,\ \forall x \in \mathrm{Co}\{x^{(1)}, \ldots, x^{(m)}\} \iff M(x^{(j)}) \succ 0,\ j = 1, \ldots, m.$$

This property of LMIs makes it possible to express in terms of the vertices of the uncertainty set Ω the conditions under which an ellipsoidal set, defined for $P \succ 0$ by

$$\mathcal{E} \doteq \{x : \|x\|_P^2 \leq 1\},$$

is robustly positively invariant for the system (5.8). Clearly, $x_{i+1|k} \in \mathcal{E}$ for all $x_{i|k} \in \mathcal{E}$ if and only if $\|(A + BK_k)x\|_P^2 \leq \|x\|_P^2$ for all $(A, B) \in \Omega$ and all $x \in \mathbb{R}^{n_x}$, or equivalently

$$P - (A + BK_k)^T P (A + BK_k) \succeq 0,\ \forall (A, B) \in \Omega. \tag{5.9}$$

Although the dependence of (5.9) on the uncertain parameters (A, B) is quadratic, this condition can be rewritten in terms of linear conditions through the use of Schur complements, which are defined for general partitioned matrices in Sect. 2.7.3. Since $P \succ 0$, and hence also $P^{-1} \succ 0$, a necessary and sufficient condition for the quadratic inequality (5.9) is therefore

$$\begin{bmatrix} P & (A + BK_k)^T \\ A + BK_k & P^{-1} \end{bmatrix} \succeq 0,\ \ \forall (A, B) \in \Omega,$$

and the linear dependence of this condition on $A + BK_k$ implies that it is equivalent to the conditions

$$\begin{bmatrix} P & (A^{(j)} + B^{(j)} K_k)^T \\ A^{(j)} + B^{(j)} K_k & P^{-1} \end{bmatrix} \succeq 0,\ \ j = 1, \ldots, m \tag{5.10}$$

involving only the vertices of Ω.

We next consider conditions for robust satisfaction of the constraints (5.3). Following the approach of Sect. 2.7.1, a necessary and sufficient condition for the constraints (5.3) to hold under (5.7) for all $x_{i|k} \in \mathcal{E} = \{x : \|x\|_P^2 \leq 1\}$ is that there exists a symmetric matrix H satisfying

$$\begin{bmatrix} H & F + GK_k \\ (F + GK_k)^T & P \end{bmatrix} \succeq 0,\ \ e_i^T H e_i \leq 1,\ \ i = 1, 2, \ldots, n_C \tag{5.11}$$

where e_i is the ith column of the identity matrix. This can be shown using the argument that was used in the proof of Theorem 2.9, namely that, for $x \in \mathcal{E}$, the ith element of $(F + GK_k)x$ has upper bound

$$e_i^T (F + GK_k)x \leq \|(F + GK_k)^T e_i\|_{P^{-1}}, \tag{5.12}$$

and hence the Schur complements of (5.11) ensure that $(F + GK_k)x \leq \mathbf{1}$. Conversely, the bound (5.12) is satisfied with equality for some $x \in \mathcal{E}$, so if there exists no H satisfying (5.11), then $(F + GK_k)x \nleq \mathbf{1}$ for some $x \in \mathcal{E}$. If $\mathcal{E}$ is robustly invariant, then clearly (5.11) ensures that the constraints of (5.3) hold along all predicted trajectories of (5.7) and (5.8) starting from any initial state $x_{0|k}$ lying in $\mathcal{E}$.

A cost index forming the objective of a robust MPC law with guaranteed stability can be constructed from a suitable upper bound on the worst-case predicted cost, which is defined for given weights $Q \succ 0$ and $R \succ 0$ by

$$\check{J}(x_{0|k}, K_k) \doteq \max_{(A_{i|k}, B_{i|k}) \in \Omega,\, i=0,1,\dots} \sum_{i=0}^{\infty} \left(\|x_{i|k}\|_Q^2 + \|u_{i|k}\|_R^2 \right). \tag{5.13}$$

For this purpose [8] derived bounds on (5.13) using a quadratic function that, for computational convenience, was determined by strengthening the conditions defining the invariant set $\mathcal{E}$. Following a similar approach, if (5.9) is replaced by the condition

$$P - (A + BK_k)^T P (A + BK_k) \succeq \gamma^{-1}(Q + K_k^T R K_k), \quad \forall (A, B) \in \Omega \tag{5.14}$$

for some scalar $\gamma > 0$, then the cost of (5.13) necessarily satisfies the bound

$$\check{J}(x_{0|k}, K_k) \leq \gamma, \quad \forall x_{0|k} \in \mathcal{E} = \{x : \|x\|_P^2 \leq 1\}. \tag{5.15}$$

This is shown by the following lemma, which uses the variable transformations

$$P = S^{-1}, \quad K_k = Y S^{-1} \tag{5.16}$$

to express (5.14) equivalently in terms of LMIs in variables Y, S and γ.

Lemma 5.1 *If $Y \in \mathbb{R}^{n_u \times n_x}$, symmetric $S \in \mathbb{R}^{n_x \times n_x}$ and scalar γ satisfy*

$$\begin{bmatrix} \begin{bmatrix} S & (A^{(j)}S + B^{(j)}Y)^T \\ A^{(j)}S + B^{(j)}Y & S \end{bmatrix} & \begin{bmatrix} SQ^{1/2} & Y^T R^{1/2} \\ 0 & 0 \end{bmatrix} \\ \star & \gamma I \end{bmatrix} \succeq 0 \tag{5.17}$$

for $j = 1, \dots, m$, then the bound (5.15) holds with $P = S^{-1}$ and $K_k = YS^{-1}$.

Proof Condition (5.17) is linear in the parameters $A^{(j)}, B^{(j)}$. By considering all convex combinations of the matrix appearing on the LHS of (5.17), it can therefore be shown that (5.17) holds with $(A^{(j)}, B^{(j)})$ replaced by any $(A, B) \in \Omega$. Hence, by Schur complements, (5.17) implies $S \succ 0$, $\gamma > 0$ and

$$S - (AS + BY)^T S^{-1} (AS + BY) \succeq \gamma^{-1}(SQS + Y^T R Y), \quad \forall (A, B) \in \Omega.$$

Pre- and post-multiplying both sides of this inequality by P and using (5.16), we obtain (5.14), so from (5.7) it follows that

$$\|x_{i|k}\|_P^2 - \|Ax_{i|k} + Bu_{i|k}\|_P^2 \geq \gamma^{-1}\big(\|x_{i|k}\|_Q^2 + \|u_{i|k}\|_R^2\big), \; \forall (A, B) \in \Omega,$$

Summing over all $i \geq 0$ and making use of (5.8), we therefore obtain

$$\check{J}(x_{0|k}, K_k) \leq \gamma \|x_{0|k}\|_P^2,$$

which implies $\check{J}(x_{0|k}, K_k) \leq \gamma$ for all $x_{0|k} \in \mathcal{E}$. □

The cost bound provided by (5.15) could be conservative because it is constant on the boundary of the invariant set $\mathcal{E} = \{x : \|x\|_P^2 \leq 1\}$. Clearly, this may conflict with the requirement that it provides a tight bound on the sublevel sets of the cost (5.13) since the invariant set was constructed in order to approximate the set of feasible initial conditions. On the boundary of the feasible set in particular, (5.15) is likely to be a very conservative cost bound. To reduce the level of suboptimality therefore, the parameters (K_k, P, γ) are assigned as variables in the MPC optimization and recomputed online at each time step $k = 0, 1, \ldots$. The lemma below establishes the optimality conditions for the feedback gain K_k.

Lemma 5.2 *If the following optimization is feasible,*

$$(Y_k^*, S_k^*, \gamma_k^*) = \arg \min_{\substack{S=S^T \succ 0,\\ H=H^T \succ 0,\\ Y,\gamma}} \gamma \tag{5.18a}$$

subject to (5.17) and, for $x = x_k$ and some symmetric H, the conditions

$$\begin{bmatrix} 1 & x^T \\ x & S \end{bmatrix} \succeq 0, \tag{5.18b}$$

$$\begin{bmatrix} H & FS + GY \\ (FS + GY)^T & S \end{bmatrix} \succeq 0, \quad e_i^T H e_i \leq 1, \quad i = 1, \ldots, n_C \tag{5.18c}$$

then, with $K_k = Y_k^*(S_k^*)^{-1}$, the predicted trajectories of (5.7) and (5.8) satisfy $\check{J}(x_{0|k}, K_k) \leq \gamma_k^*$ and meet the constraints $Fx_{i|k} + Gu_{i|k} \leq \mathbf{1}$ for all $i \geq 0$.

Proof Using Schur complements, (5.18b) implies $x_k \in \mathcal{E}$, so by Lemma 5.1 the optimal objective in (5.18a) is an upper bound on the predicted cost $\check{J}(x_k, K_k)$. On the other hand, pre- and post-multiplying (5.18c) by the block diagonal matrix $\mathrm{diag}\{I, P\}$ (which is unitary since (5.18b) implies $P \succ 0$) yields (5.11), and hence (5.18c) ensures $(F + GK_k)x_{i|k} \leq \mathbf{1}$ for all $i \geq 0$. □

The optimization of Lemma 5.2 is the basis of the following robust MPC strategy, which requires the online solution of an SDP in $O(n_x^2)$ variables.

Algorithm 5.1 At each time instant $k = 0, 1, \ldots$:

(i) Perform the optimization of Lemma 5.2.
(ii) Apply the control law $u_k = K_k x_k$. ◁

Theorem 5.1 *For the system (5.1)–(5.3) and control law of Algorithm 5.1:*

(a) The optimization in step (i) is feasible for all times $k > 0$ if it is feasible at time $k = 0$.
(b) The origin of the state space of (5.1) is robustly asymptotically stable with region of attraction

$$\mathcal{F} \doteq \{x \in \mathbb{R}^{n_x} : (5.17), (5.18b), (5.18c) \text{ are feasible}\}, \tag{5.19}$$

and for $x_0 \in \mathcal{F}$ the trajectories of the closed-loop system satisfy the bound

$$\sum_{k=0}^{\infty} \left(\|x_k\|_Q^2 + \|u_k\|_R^2 \right) \le \gamma_0^*. \tag{5.20}$$

Proof We first demonstrate that, if the optimization of Lemma 5.2 is feasible at time k, then a feasible solution at time $k + 1$ is given by

$$(Y, S, \gamma) = (\alpha Y_k^*, \alpha S_k^*, \alpha \gamma_k^*) \tag{5.21}$$

where $\alpha = \|x_{k+1}\|_{P_k}^2$ for $P_k = (S_k^*)^{-1}$. Considering each of the constraints (5.17), (5.18b) and (5.18c) at time $k + 1$:

- With $S^{-1} = \alpha^{-1} P_k$, we obtain $\|x_{k+1}\|_{S^{-1}}^2 = \alpha^{-1} \|x_{k+1}\|_{P_k}^2 = 1$, and hence (5.18b) is necessarily satisfied with $x = x_{k+1}$.
- If $(Y_k^*, S_k^*, \gamma_k^*)$ is feasible for (5.17) then, since the inequality in (5.17) is unchanged if the matrix on the LHS is multiplied by $\alpha > 0$, the solution (Y, S, γ) in (5.21) must be feasible for (5.17), for any $\alpha > 0$.
- The feasibility of (5.17) and (5.18b) at time k implies $\|x_{k+1}\|_{P_k}^2 \le \|x_k\|_{P_k}^2$ and hence $\alpha \le 1$. Therefore, if (Y_k^*, S_k^*) satisfies (5.18c) with $H = H_k$, then (Y, S) in (5.21) must also satisfy (5.18c) with $H = \alpha H_k$.

It follows that the optimization in step (i) of Algorithm 5.1 is recursively feasible since if x_0 lies in the feasible set $\mathcal{F}$, then $x_k \in \mathcal{F}$ for all $k \ge 1$.

Turning next to closed-loop stability, from the optimality of the solution at time $k + 1$ and (5.17) we have

$$\begin{aligned} \gamma_{k+1}^* \le \alpha \gamma_k^* &= \gamma_k^* \|x_{k+1}\|_{P_k}^2 \\ &\le \gamma_k^* \|x_k\|_{P_k}^2 - \left(\|x_k\|_Q^2 + \|u_k\|_R^2 \right), \end{aligned}$$

but the constraint (5.18b) is necessarily active at the optimal solution, so $\|x_k\|_{P_k} = 1$ and

$$\gamma_{k+1}^* - \gamma_k^* \leq -\left(\|x_k\|_Q^2 + \|u_k\|_R^2\right).$$

This inequality implies that $x = 0$ is stable because $\gamma_k^* \geq \check{J}(x_k, K_k)$, while (5.13) implies $\check{J}(x, K)$ is positive definite in x since $Q \succ 0$. Summing both sides of this inequality over $k = 0, 1, \ldots$ yields the bound (5.20), which implies that $(\|x_k\|_Q^2 + \|u_k\|_R^2) \to 0$ as $k \to \infty$ for any $x_0 \in \mathcal{F}$, and hence $\lim_{k\to\infty}(x_k, u_k) = (0, 0)$, since Q and R are positive definite matrices. □

The online optimization posed in Algorithm 5.1 constitutes a convex program (see e.g. [24]) which can be solved efficiently (in polynomial time) using semidefinite programming solvers. However it should be noted that its computational burden increases considerably with the system dimension. For fast sampling applications (such as applications involving electromechanical systems), this algorithm is therefore viable only for small-scale models.

Example 5.1 An uncertain system is described by the model (5.1)–(5.2) with parameters

$$A^{(1)} = \begin{bmatrix} -0.7 & 0.15 \\ -0.35 & -0.6 \end{bmatrix}, \quad A^{(2)} = \begin{bmatrix} -0.75 & -0.1 \\ 0.15 & -0.65 \end{bmatrix}, \quad A^{(3)} = \begin{bmatrix} -0.65 & -0.35 \\ -0.1 & -0.55 \end{bmatrix}$$
$$B^{(1)} = \begin{bmatrix} 0.1 \\ 1 \end{bmatrix}, \quad B^{(2)} = \begin{bmatrix} 0.2 \\ 1.4 \end{bmatrix}, \quad B^{(3)} = \begin{bmatrix} 0.3 \\ 0.6 \end{bmatrix}.$$

The system is subject to state and input constraints

$$-10 \leq \begin{bmatrix} 0 & 1 \end{bmatrix} x_k \leq 10$$
$$-5 \leq u_k \leq 5$$

which are equivalent to (5.3) with

$$F = \begin{bmatrix} 0 & 0.1 \\ 0 & -0.1 \\ 0 & 0 \\ 0 & 0 \end{bmatrix}, \quad G = \begin{bmatrix} 0 \\ 0 \\ 0.2 \\ -0.2 \end{bmatrix}.$$

Figure 5.1 shows the feasible set, $\mathcal{F}$, for Algorithm 5.1. This set is formed from a union of ellipsoidal sets, and for this example is clearly not itself ellipsoidal. Figure 5.1 also shows that $\mathcal{F}$ contains points that lie outside the largest area ellipsoidal set that is robustly invariant under any given linear feedback law. This is to be expected since $\mathcal{F}$ contains every ellipsoidal set that is RPI for (5.1)–(5.3) under a linear feedback law.

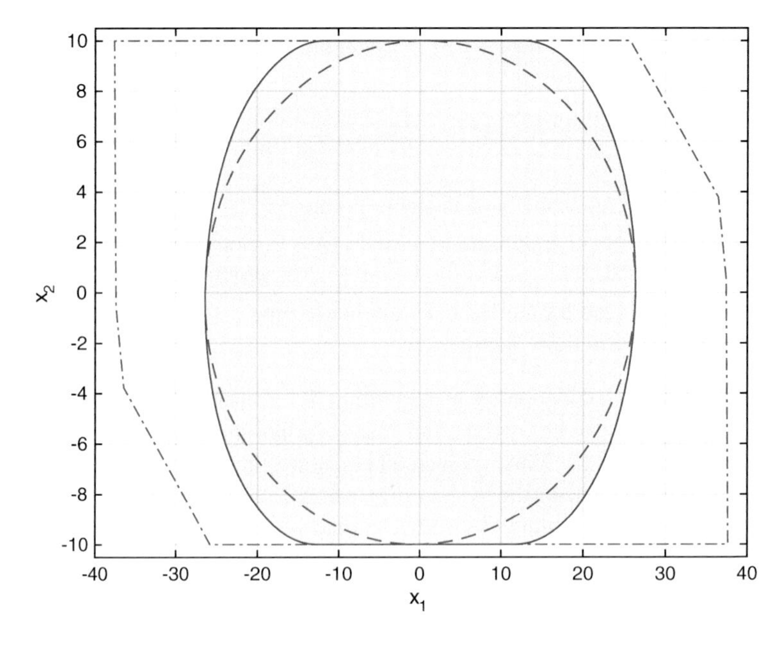

Fig. 5.1 The set of feasible initial states for Algorithm 5.1 applied to the uncertain system of Example 5.1 (*solid line*). For comparison, this figure also shows the maximal ellipsoidal RPI set under any linear feedback law (*dashed line*) and the maximal robustly controlled invariant set (*dash-dotted line*)

For 100 values of x evenly distributed on the dashed ellipsoid in Fig. 5.1, the average optimal value of the online MPC optimization (5.18) is 11,779 and the maximum is 69,505. However, the value of the minimum quadratic bound on the worst-case cost computed for the solution of (5.18) and averaged over this set of initial conditions is 939 and the corresponding maximum value is 1660. The discrepancy between these worst-case cost bounds is a result of the conservativeness of the cost bound that forms the objective of (5.18), which as previously discussed is constructed by scaling the ellipsoidal set $\mathcal{E}$ in order to ensure that the objective (5.18a) is an upper bound on the worst-case cost. ◊

5.2.1 Dual Mode Predictions

By the criteria of Chaps. 3 and 4, Algorithm 5.1 is a feedback MPC strategy since the predicted control trajectories depend on the realization of future uncertainty through a feedback gain computed online. Despite this, the parameterization of predicted trajectories in terms of a single linear feedback gain over the prediction horizon can be restrictive. Furthermore, for computational convenience the cost and constraints are approximated in Algorithm 5.1 using potentially conservative quadratic bounds.

A conceptually straightforward way to avoid these shortcomings is to extend Algorithm 5.1 by adjoining a mode 1 horizon containing additional degrees of freedom over which the cost and constraints are evaluated explicitly. This approach is proposed in [9], in which the predicted control trajectory is specified as

$$u_{i|k} = K_{i|k}x_{i|k} + c_{i|k} \quad \begin{cases} K_{i|k} = K_{k+i}, & i = 0, \ldots, N-1 \\ K_{i|k} = K_{N|k},\ c_{i|k} = 0, & i = N, N+1, \ldots \end{cases} \tag{5.22}$$

where $\mathbf{c}_k = (c_{0|k}, \ldots, c_{N-1|k})$ and $K_{N|k}$ are optimization variables at time k. In order to be able to make use of the conditions of Lemmas 5.1 and 5.2 for the computation of $K_{N|k}$ while retaining a convex online optimization, the feedback gains $K_k, \ldots, K_{N-1+k}$ are fixed at time k, their values being carried over from the optimization at a previous time instant.

A polytopic tube $\mathcal{X}_{0|k}, \ldots, \mathcal{X}_{N|k}$ containing the future state trajectories for all realizations of model uncertainty can be defined in terms of the vertices, $v_{i|k}^{(l)}$, of $\mathcal{X}_{i|k}$:

$$\mathcal{X}_{i|k} = \mathrm{Co}\{v_{i|k}^{(l)},\ l = 1, \ldots, m^i\},$$

where $\mathcal{X}_{0|k} = \{x_k\}$ and $x_{i|k} \in \mathcal{X}_{i|k}$. The tube cross sections $\mathcal{X}_{i|k}$ for $i = 1, \ldots, N$, are computed in [9] using the recursion

$$v_{i+1|k}^{(q)} = (A^{(j)} + B^{(j)}K_{k+i})v_{i|k}^{(l)} + B^{(j)}c_{i|k} \tag{5.23}$$

for $j = 1, \ldots, m$, $l = 1, \ldots, m^i$ and $q = 1, \ldots, m^{i+1}$. Clearly the number of vertices defining the tube cross sections increases exponentially with the length of the prediction horizon in this approach, limiting the approach to short horizons. The reason for this very rapid growth in complexity is that state tubes defined in this way are minimal in the sense that $\mathcal{X}_{i|k}$ is the smallest set containing the i-steps ahead predicted state $x_{i|k}$ for all possible realizations of the model uncertainty.

Let $\gamma_k \doteq (\gamma_{0|k}, \ldots, \gamma_{N|k})$ denote a sequence of upper bounds on the predicted stage costs of the worst-case predicted performance index (5.13):

$$\gamma_{i|k} \ge \|x_{i|k}\|_Q^2 + \|u_{i|k}\|_R^2.$$

By expressing these bounds as LMIs in the predicted states, they can be imposed for all $x_{i|k} \in \mathcal{X}_{i|k}$ through conditions on the vertices $v_{i|k}^{(l)}$, namely for $i = 1, \ldots, N-1$:

$$\begin{bmatrix} \gamma_{i|k} & \begin{bmatrix} v_{i|k}^{(l)\,T} Q^{1/2} & (K_{k+i}v_{i|k}^{(l)} + c_{i|k})^T R^{1/2} \end{bmatrix} \\ \star & I \end{bmatrix} \succeq 0, \quad l = 1, \ldots, m^i. \tag{5.24}$$

For all $i \ge N$, the predicted control inputs are defined in (5.22) by the feedback law $u_{i|k} = K_{N|k}x_{i|k}$, and an upper bound on the predicted cost over the mode 2 prediction horizon can be computed using the approach of Sect. 5.2. In the current context, however, the N-step ahead predicted state, $x_{N|k}$, is not known exactly, but instead is known to lie in the polytopic set $\mathcal{X}_{N|k}$. An upper bound on the cost to go over the mode 2 prediction horizon is therefore obtained by invoking the LMI of (5.18b) at each of the vertices defining $\mathcal{X}_{N|k}$:

$$\begin{bmatrix} 1 & v_{N|k}^{(l)\,T} \\ v_{N|k}^{(l)} & S \end{bmatrix} \succeq 0, \quad l = 1, \ldots, m^N. \tag{5.25}$$

Then, by an obvious extension of Lemma 5.1, the infinite horizon cost of (5.13) satisfies the bound

$$\check{J}(x_k, \mathbf{c}_k, K_{N|k}) \leq \sum_{i=0}^{N} \gamma_{i|k} \tag{5.26}$$

if (5.17) is invoked with $\gamma = \gamma_{N|k}$ and $K_{N|k} = YS^{-1}$.

Using the arguments of Lemma 5.2, it can be shown that (5.18c) ensures constraint satisfaction in mode 2, whereas in mode 1, given the linear nature of the constraints (5.3), a necessary and sufficient condition is that

$$(F + GK_{k+i})v_{i|k}^{(l)} + Gc_{i|k} \leq \mathbf{1}, \quad l = 1, \ldots, m^i, \quad i = 0, \ldots, N-1. \tag{5.27}$$

On the basis of this development, it is possible to state the following robust MPC algorithm, which requires the online solution of a SDP problem involving $O(n_x m^N)$ variables.

Algorithm 5.2 At each time instant $k = 0, 1, \ldots$:

(i) Perform the optimization:

$$\underset{\mathbf{c}_k, Y, S, \gamma_k}{\text{minimize}} \sum_{i=0}^{N} \gamma_{i|k} \text{ subject to (5.24), (5.25), (5.27),}$$
$$\text{(5.17) with } \gamma = \gamma_{N|k} \text{ and (5.18c)} \tag{5.28}$$

(ii) Apply the control law $u_k = K_k x_k + c_{0|k}^*$ and set $K_{k+N} = Y_k^*(S_k^*)^{-1}$, where $\mathbf{c}_k^* = (c_{0|k}^*, \ldots, c_{N-1|k}^*)$, S_k^* and Y_k^* are, respectively, the optimal values of $\mathbf{c}_k$, S and Y in (5.28). ◁

This algorithm must be initialized at $k = 0$ by computing the feedback gains $K_0, \ldots, K_{N-1}$ offline. Assuming knowledge of the initial state x_0 (or alternatively knowledge of a polytopic set $\mathrm{Co}\{v_{0|0}^{(1)}, \ldots, v_{0|0}^{(r)}\}$ containing x_0), this can be done by performing the optimization of Lemma 5.2 with $x = x_0$ in (5.18b) (or alternatively performing this optimization with

$$\begin{bmatrix} 1 & v_{0|0}^{(l)\,T} \\ v_{0|k}^{(l)} & S \end{bmatrix} \succeq 0, \quad l = 1, \ldots, m^N.$$

in place of (5.18b)) and by setting $K_i = K_0$ for all $i = 0, \ldots, N-1$.

Theorem 5.2 *Algorithm 5.2 is recursively feasible and robustly asymptotically stabilizes the origin of the state space of the system (5.1)–(5.3).*

Proof Assuming feasibility at time k, a feasible but suboptimal solution to (5.28) at time $k+1$ is given by

$$\mathbf{c}_{k+1} = (c_{1|k}^*, \ldots, c_{N-1|k}^*, 0),$$
$$\gamma_{k+1} = \bigl(\gamma_{1|k}^*, \ldots, \gamma_{N-1|k}^*, \max_{x \in \mathcal{X}_{N|k}} (\|x\|_Q^2 + \|K_{k+N}x\|_R^2), \alpha\gamma_{N|k}^*\bigr),$$
$$Y = \alpha Y_k^*, \quad S = \alpha S_k^*, \quad \alpha = \max_{x \in \mathcal{X}_{N|k}, j \in \{1,\ldots,m\}} \|(A^{(j)} + B^{(j)}K_{k+N})x\|_{P_k}^2$$

where $P_k = (S_k^*)^{-1}$. Feasibility of the constraints (5.24) and (5.27) follows directly from the inclusion property $\mathcal{X}_{i|k+1} \subseteq \mathcal{X}_{i+1|k}$, $i = 0, \ldots, N-1$, while feasibility of (5.25), (5.17) with $\gamma = \alpha\gamma_{N|k}^*$ and (5.18b) follows from the argument used in the proof of Theorem 5.1 and the property that $\mathcal{X}_{N|k+1} \subseteq \mathrm{Co}\{(A^{(j)} + B^{(j)}K_{k+N})\mathcal{X}_{N|k}, j = 1, \ldots, m\}$, as a consequence of $u_{N|k+1} = K_{k+N}x_{N|k+1}$ and $\mathcal{X}_{N-1|k+1} \subseteq \mathcal{X}_{N|k}$. The proof of Theorem 5.1 also shows that $\alpha\gamma_{N|k}^* \le \gamma_{N|k}^* - (\|x\|_Q^2 + \|K_{k+N}x\|_R^2)$ for all $x \in \mathcal{X}_{N|k}$. Hence the sum of the elements of γ_{k+1} is no greater than $\sum_{i=1}^N \gamma_{i|k}^*$, and optimality at time $k+1$ therefore implies

$$\sum_{i=0}^{N} \gamma_{i|k+1}^* \le \sum_{i=1}^{N} \gamma_{i|k}^* = \sum_{i=0}^{N} \gamma_{i|k}^* - (\|x_k\|_Q^2 + \|u_k\|_R^2).$$

Since the bound (5.26) and $Q \succ 0$ and $R \succ 0$ imply that $\sum_{i=0}^N \gamma_{i|k}^*$ is positive definite in x_k, it follows that $x = 0$ is asymptotically stable. □

5.3 Prediction Dynamics in Robust MPC

In the presence of multiplicative uncertainty, robust MPC algorithms employing minimal tubes to bound predicted trajectories suffer from the same disadvantage as Algorithm 5.2, namely that the number of constraints, and hence also the computational demand, grows rapidly (in general, exponentially) with the prediction horizon N. Therefore the use of minimal tubes in this context is generally impractical for anything other than short horizons and descriptions of model uncertainty with small numbers of vertices. To avoid this problem, it is necessary to bound predicted state and control trajectories using non-minimal tubes with lower complexity cross sections.

An approach that is extremely computationally efficient, though somewhat conservative, is based on ellipsoidal sets used in conjunction with autonomous prediction dynamics [10, 11]. As in the nominal case considered in Sect. 2.7, an ellipsoidal invariant set can be computed offline for a prediction system that incorporates the degrees of freedom in predicted trajectories into its state vector. This section extends the methods of Sects. 2.7 and 2.9 to the case of robust MPC, and considers the design of the prediction dynamics in order to enlarge feasible sets and to reduce the sensitivity of predicted performance to multiplicative uncertainty.

Following the approach of Sect. 2.7, the predicted control trajectory at time k is defined for all $i \geq 0$ by

$$u_{i|k} = Kx_{i|k} + c_{i|k},$$

where $\mathbf{c}_k = (c_{0|k}, \ldots, c_{N-1|k})$ is a vector of variables in the MPC optimization at time k and $c_{i|k} = 0$ for $i \geq N$. The feedback gain K is fixed and is assumed to be the unconstrained LQ optimal feedback gain associated with the nominal cost

$$J(s_{0|k}, \mathbf{c}_k) = \sum_{i=0}^{\infty} \left(\|s_{i|k}\|_Q^2 + \|v_{i|k}\|_R^2 \right), \tag{5.29}$$

which is evaluated along state and control trajectories of the nominal model:

$$\begin{aligned} s_{i+1|k} &= \Phi^{(0)} s_{i|k} + B^{(0)} c_{i|k} \\ v_{i|k} &= K s_{i|k} + c_{i|k} \end{aligned}$$

with $\Phi^{(0)} = A^{(0)} + B^{(0)} K$. We further assume that $u = Kx$ robustly quadratically stabilizes the uncertain system (5.1) and (5.2) in the absence of constraints. This assumption (which will be removed in Sect. 5.3.2) requires that a quadratic Lyapunov function exists for the unconstrained system (5.1) and (5.2) under $u = Kx$, namely that there exists $P \succ 0$ satisfying

$$P - \Phi^{(j)T} P \Phi^{(j)} \succ 0, \quad j = 1, \ldots, m \tag{5.30}$$

where $\Phi^{(j)} = A^{(j)} + B^{(j)} K$.

A prediction system incorporating the uncertain model (5.1) and (5.2) is described for all $i \geq 0$ by

$$z_{i+1|k} = \Psi_{i|k} z_{i|k}, \quad \Psi_{i|k} \in \mathrm{Co}\{\Psi^{(1)}, \ldots, \Psi^{(m)}\}, \tag{5.31a}$$

where the vertices of the parameter uncertainty set are given by

$$\Psi^{(j)} = \begin{bmatrix} \Phi^{(j)} & B^{(j)} E \\ 0 & M \end{bmatrix}, \quad \Phi^{(j)} = A^{(j)} + B^{(j)} K \tag{5.31b}$$

for $j = 1, \dots, m$. The initial prediction system state and the predicted state and input trajectories are defined, as in Sect. 2.7, by

$$z_{0|k} = \begin{bmatrix} x_k \\ \mathbf{c}_k \end{bmatrix}, \quad \begin{aligned} u_{i|k} &= \begin{bmatrix} K & E \end{bmatrix} z_{i|k} \\ x_{i|k} &= \begin{bmatrix} I & 0 \end{bmatrix} z_{i|k} \end{aligned} \tag{5.32}$$

and matrices E and M are as given in (2.26b), so that $E\mathbf{c}_k = c_{0|k}$ and $M\mathbf{c}_k = (c_{1|k}, \dots, c_{N-1|k}, 0)$.

The predicted state and input trajectories generated by (5.31a, 5.31b) and (5.32) necessarily satisfy the constraints (5.3) at all prediction times $i \geq 0$ if the initial prediction system state, $z_{0|k}$, is constrained to lie in a set that is robustly invariant for (5.31a, 5.31b) and feasible with respect to (5.3). This is demonstrated by the following extension of Theorem 2.9.

Corollary 5.1 *The ellipsoidal set* $\mathcal{E}_z \doteq \{z : z^T P_z z \leq 1\}$, *with* $P_z \succ 0$, *is robustly positively invariant for the dynamics (5.31a, 5.31b) and constraints* $\begin{bmatrix} F + GK & GE \end{bmatrix} z \leq \mathbf{1}$ *if and only if* P_z *satisfies*

$$P_z - \Psi^{(j)T} P_z \Psi^{(j)} \succeq 0, \quad j = 1, \dots, m \tag{5.33}$$

and

$$\begin{bmatrix} H & \begin{bmatrix} F + GK & GE \end{bmatrix} \\ \begin{bmatrix} (F + GK)^T \\ (GE)^T \end{bmatrix} & P_z \end{bmatrix} \succeq 0, \quad e_i^T H e_i \leq 1, \; i = 1, \dots, n_C \tag{5.34}$$

for some symmetric matrix H, *where* e_i *is the* i*th column of the identity matrix.*

Proof Using Schur complements, (5.33) is equivalent to

$$\begin{bmatrix} P_z & \Psi^{(j)T} \\ \Psi^{(j)} & P_z^{-1} \end{bmatrix} \succeq 0, \quad j = 1, \dots, m.$$

Since this is an LMI in $\Psi^{(j)}$, it is equivalent, again using Schur complements, to $P_z - \Psi^T P_z \Psi \succeq 0$ for all $\Psi \in \mathrm{Co}\{\Psi^{(1)}, \dots, \Psi^{(m)}\}$. Therefore, the sufficiency and necessity of (5.33) and (5.34) can be shown using the same argument as the proof of Theorem 2.9. □

Adopting the nominal cost of (5.29) as the objective of the MPC online optimization, we have, from Lemma 2.1 and Theorem 2.10,

$$J(x_k, \mathbf{c}_k) = \|z_{0|k}\|_W^2 = \|x_k\|_{W_x}^2 + \|\mathbf{c}_k\|_{W_c}^2, \tag{5.35a}$$

where W is the solution of the Lyapunov equation

$$W - \Psi^{(0)^T} W \Psi^{(0)} = \hat{Q}, \quad \hat{Q} = \begin{bmatrix} Q + K^T R K & K^T R E \\ E^T R K & E^T R E \end{bmatrix}. \tag{5.35b}$$

The block diagonal structure of $W = \mathrm{diag}\{W_x, W_c\}$ in (5.35a, 5.35b) follows from the definition of K as the unconstrained LQ optimal feedback gain. Furthermore, W_x is the solution of the Riccati equation (2.9) with $(A, B) = (A^{(0)}, B^{(0)})$, and W_c satisfies

$$W_c - M^T W_c M = E^T \big(B^{(0)^T} W_x B^{(0)} + R\big) E. \tag{5.36}$$

Hence, by Theorem 2.10, W_c is block diagonal with diagonal blocks equal to $(B^{(0)})^T W_x B^{(0)} + R$. Clearly, the problem of minimizing $J(x_k, \mathbf{c}_k)$ for given x_k is equivalent to that of minimizing $\|\mathbf{c}_k\|^2_{W_c}$. Therefore the MPC algorithm can be stated in terms of W_c and an ellipsoidal set $\mathcal{E}_z$ satisfying the conditions of Corollary 5.1 as follows.

Algorithm 5.3 At each time instant $k = 0, 1, \ldots$:

(i) Perform the optimization:

$$\underset{\mathbf{c}_k}{\text{minimize}} \ \|\mathbf{c}_k\|^2_{W_c} \ \text{subject to} \ \begin{bmatrix} x_k \\ \mathbf{c}_k \end{bmatrix} \in \mathcal{E}_z. \tag{5.37}$$

(ii) Apply the control law $u_k = K x_k + c^*_{0|k}$, where $\mathbf{c}^*_k = (c^*_{0|k}, \ldots, c^*_{N-1|k})$ is the optimal value of $\mathbf{c}_k$ in (5.37). ◁

Along the trajectories of the closed-loop system,

$$x_{k+1} = \Phi_k x_k + B_k c^*_{0|k}, \quad (\Phi_k, B_k) \in \mathrm{Co}\big\{(\Phi^{(1)}, B^{(1)}), \ldots, (\Phi^{(m)}, B^{(m)})\big\}, \tag{5.38}$$

the optimal value of the objective in (5.37) is not necessarily non-increasing at successive time instants since the predicted cost in (5.29) is computed assuming that model parameters are equal to their nominal values. However, Algorithm 5.3 can be shown to be robustly stabilizing using a method similar to the analysis in Sect. 3.3 of robust MPC based on a nominal cost in the presence of additive disturbances. We first give an l_2 stability property of the closed-loop system (5.38), based on the assumption that $u = Kx$ quadratically stabilizes the model (5.1) and (5.2).

Lemma 5.3 *If the quadratic stability condition (5.30) holds, then the state of the closed-loop system (5.38) satisfies the quadratic bound*

$$\sum_{k=0}^{\infty} \|x_k\|^2 \le \|x_0\|^2_P + \gamma^2 \sum_{k=0}^{\infty} \|c^*_{0|k}\|^2 \tag{5.39}$$

for some matrix $P \succ 0$ and a scalar γ.

Proof Suppose that $P_0 - \Phi^{(j)T} P_0 \Phi^{(j)} \succ 0$ for some $P_0 \succ 0$ and $j = 1, \ldots, m$. Then $P_0 - \Phi^{(j)T} P_0 \Phi^{(j)} \succ \epsilon I_{n_x}$ for some $\epsilon > 0$ and hence $P - \Phi^{(j)T} P \Phi^{(j)} \succ I_{n_x}$ for $j = 1, \ldots, m$, where $P = \epsilon^{-1} P_0 \succ 0$. Using Schur complements, this implies that there must exist $\gamma > 0$ satisfying

$$\begin{bmatrix} P & \Phi^{(j)T} P \\ P\Phi^{(j)} & P \end{bmatrix} \succ \begin{bmatrix} I_{n_x} & 0 \\ 0 & \gamma^{-2} P B^{(j)} B^{(j)T} P \end{bmatrix},$$

and using Schur complements again, this is equivalent to the condition

$$\begin{bmatrix} P & \Phi^{(j)T} P & 0 \\ \star & P & P B^{(j)} \\ \star & \star & \gamma^2 I_{n_u} \end{bmatrix} \succ \begin{bmatrix} I_{n_x} & 0 & 0 \\ 0 & 0 & 0 \\ 0 & 0 & 0 \end{bmatrix}.$$

Since this condition is an LMI in the parameters $\Phi^{(j)}$ and $B^{(j)}$, it must hold with $(\Phi^{(j)}, B^{(j)})$ replaced by any $(\Phi, B) \in \mathrm{Co}\{(\Phi^{(1)}, B^{(1)}), \ldots, (\Phi^{(m)}, B^{(m)})\}$. Using Schur complements once more, we therefore have

$$\begin{bmatrix} P & 0 \\ 0 & \gamma^2 I_{n_u} \end{bmatrix} - \begin{bmatrix} \Phi^T \\ B^T \end{bmatrix} P \begin{bmatrix} \Phi & B \end{bmatrix} \succeq \begin{bmatrix} I_{n_x} & 0 \\ 0 & 0 \end{bmatrix},$$

for all $(\Phi, B) \in \mathrm{Co}\{(\Phi^{(j)}, B^{(j)}), j = 1, \ldots, m\}$. Pre- and post-multiplying both sides of this inequality by $z_k = (x_k, c^*_{0|k})$ and using (5.38) gives

$$\|x_k\|_P^2 + \gamma^2 \|c^*_{0|k}\|^2 - \|x_{k+1}\|_P^2 \geq \|x_k\|^2$$

and summing both sides of this inequality over $k \geq 0$ yields the bound (5.39). □

Theorem 5.3 *For the system (5.1)–(5.3) with the control law of Algorithm 5.3, the optimization (5.37) is recursively feasible and the origin of state space is robustly asymptotically stable with region of attraction equal to the feasible set* $\mathcal{F} = \{x : \exists \mathbf{c}$ *such that* $(x, \mathbf{c}) \in \mathcal{E}_z\}$.

Proof If x_k lies in the feasible set $\mathcal{F}$, then a feasible but suboptimal solution of (5.37) at time $k+1$ is given by $\mathbf{c}_{k+1} = M\mathbf{c}^*_k$ since the robust positive invariance of $\mathcal{E}_z$ implies that $\Psi[x_k^T \; (\mathbf{c}^*_k)^T]^T \in \mathcal{E}_z$ for all $\Psi \in \mathrm{Co}\{\Psi^{(1)}, \ldots, \Psi^{(m)}\}$. Hence (5.36) implies that the optimal solution of (5.37) at time $k+1$ satisfies

$$\|\mathbf{c}^*_{k+1}\|_{W_c}^2 \leq \|M\mathbf{c}^*_k\|_{W_c}^2 \leq \|\mathbf{c}^*_k\|_{W_c}^2 - \|c^*_{0|k}\|^2_{R + B^{(0)T} W_x B^{(0)}}.$$

Summing this inequality over all $k \geq 0$ yields the bound

$$\sum_{k=0}^{\infty} \|c^*_{0|k}\|^2 \leq \frac{1}{\underline{\lambda}(R)} \|\mathbf{c}^*_0\|_{W_c}^2$$

where $\underline{\lambda}(R)$ is the smallest eigenvalue of R. Therefore Lemma 5.3 implies

$$\sum_{k=0}^{\infty} \|x_k\|^2 \leq \|x_0\|_P^2 + \frac{\gamma^2}{\underline{\lambda}(R)} \|\mathbf{c}_0^*\|_{W_c}^2 \tag{5.40}$$

and hence $x_k \to 0$ as $k \to \infty$. Finally, we note that $x = 0$ is Lyapunov stable since Algorithm 5.3 coincides with the feedback law $u_k = Kx_k$ (which is robustly stabilizing by assumption) for all x_k in the set $\{x : (x, 0) \in \mathcal{E}_z\}$, and this set necessarily contains $x = 0$ in its interior since $P_z \succ 0$. □

To maximize the volume of the feasible set, $\mathcal{F} = \{x : \exists \mathbf{c} \text{ such that } (x, \mathbf{c}) \in \mathcal{E}_z\}$, for Algorithm 5.3, the matrix P_z defining $\mathcal{E}_z$ can be designed offline analogously to the case of nominal MPC considered in Sect. 2.7.3. Thus, rewriting (5.33) and (5.34) in terms of $S = P_z^{-1}$ gives the equivalent conditions

$$\begin{bmatrix} S & \Psi^{(j)} S \\ S\Psi^{(j)T} & S \end{bmatrix} \succeq 0, \quad j = 1, \ldots, m \tag{5.41a}$$

$$\begin{bmatrix} H & \begin{bmatrix} F + GK & GE \end{bmatrix} S \\ S \begin{bmatrix} (F + GK)^T \\ (GE)^T \end{bmatrix} & S \end{bmatrix} \succeq 0, \quad e_i^T H e_i \leq 1, \quad i = 1, \ldots, n_C \tag{5.41b}$$

which are LMIs in the variables S and H. Hence the volume of $\mathcal{F}$ is maximized if $P_z = S^{-1}$ where S is the solution of the SDP problem

$$\underset{S,H}{\text{maximize}} \ \log\det\left(\begin{bmatrix} I_{n_x} & 0 \end{bmatrix} S \begin{bmatrix} I_{n_x} \\ 0 \end{bmatrix} \right) \ \text{subject to (5.41a, 5.41b).} \tag{5.42}$$

In concluding this section, we note that the online computation required by Algorithm 5.3 is identical to that for the nominal MPC law of Algorithm 2.2. Therefore the computational advantages of Algorithm 2.2 also apply to Algorithm 5.3. In particular, the online minimization of Algorithm 5.3 can be performed extremely efficiently by solving for the unique negative real root of a well-behaved polynomial using the Newton–Raphson iteration described in Sect. 2.8.

5.3.1 Prediction Dynamics Optimized to Maximize the Feasible Set

The system (5.31a, 5.31b) and (5.32) that generates the predicted state and control trajectories underpinning Algorithm 5.3 can be interpreted in terms of a dynamic feedback law applied to the uncertain model (5.1) and (5.2). The initial state of this dynamic controller is defined by the vector, $\mathbf{c}_k$, of degrees of freedom in the state and input predictions at time k. In order to maximize the ellipsoidal region of attraction of Algorithm 5.3, it is possible to optimize the predicted controller dynamics simultaneously with the invariant set $\mathcal{E}_z$ by solving a convex optimization problem [12]. This is achieved through an extension of the method of optimizing prediction dynamics described in Sect. 2.9 to the case of systems with uncertain dynamics.

Let the vertices $\Psi^{(j)}$ of the uncertainty set in (5.31a) be redefined as

$$\Psi^{(j)} = \begin{bmatrix} \Phi^{(j)} & B^{(j)}C_c \\ 0 & A_c \end{bmatrix}, \quad j = 1, \ldots, m \tag{5.43}$$

where $A_c \in \mathbb{R}^{\nu_c \times \nu_c}$ and $C_c \in \mathbb{R}^{n_u \times \nu_c}$ are to be designed offline together with the matrix P_z defining the ellipsoid $\mathcal{E}_z = \{z \in \mathbb{R}^{n_x+\nu_c} : z^T P_z z \le 1\}$. The conditions for robust invariance of $\mathcal{E}_z$ can then be obtained by restating Corollary 5.1 in terms of these uncertainty set vertices. However the resulting conditions are nonconvex when A_c, C_c and P_z are treated as variables. We therefore use the transformation (2.60) to reformulate them as equivalent convex conditions in terms of variables X, Y, U, V parameterizing P_z and variables Ξ, Γ parameterizing A_c, C_c. Using the approach of Sect. 2.9, we then obtain, analogously to (2.62a, 2.62b), the LMI conditions:

$$\begin{bmatrix} \begin{bmatrix} Y & X \\ X & X \end{bmatrix} & \begin{bmatrix} \Phi^{(j)}Y + B^{(j)}\Gamma & \Phi^{(j)}X \\ \Xi + \Phi^{(j)}Y + B^{(j)}\Gamma & \Phi^{(j)}X \end{bmatrix} \\ \star & \begin{bmatrix} Y & X \\ X & X \end{bmatrix} \end{bmatrix} \succeq 0 \tag{5.44a}$$

$$\begin{bmatrix} H & \begin{bmatrix} (F + GK)Y + G\Gamma & (F + GK)X \end{bmatrix} \\ \star & \begin{bmatrix} Y & X \\ X & X \end{bmatrix} \end{bmatrix} \succeq 0, \quad e_i^T H e_i \le 1, \quad i = 1, \ldots, n_C \tag{5.44b}$$

for $j = 1, \ldots, m$. Therefore, matrices A_c, C_c, P_z satisfying (5.33), (5.34) and (5.43) exist only if the LMIs (5.44a, 5.44b) hold for some X, Y, Ξ, Γ. Furthermore, the conditions (5.44a, 5.44b) are sufficient as well as necessary if the dimension of $\mathbf{c}$ is equal to that of x, namely if $\nu_c = n_x$, since in this case, by the argument of Sect. 2.9, the inverse transformation

$$P_z = \begin{bmatrix} X^{-1} & X^{-1}U \\ U^T X^{-1} & -U^T X^{-1} Y V^{-T} \end{bmatrix}, \quad A_c = U^{-1} \Xi V^{-T}, \quad C_c = \Gamma V^{-T} \tag{5.45}$$

where $UV^T = X - Y$, necessarily exists and defines A_c, C_c and P_z uniquely.

The set $\mathcal{F}$ of feasible initial conditions for Algorithm 5.3 is the projection of $\mathcal{E}_z$ onto the x-subspace (i.e. $\mathcal{F} = \{x : \exists \mathbf{c} \text{ such that } (x, \mathbf{c}) \in \mathcal{E}_z\}$). Therefore $\mathcal{F}$ can be maximized offline by solving the SDP problem:

$$\underset{\Xi,\Gamma,X,Y}{\text{maximize}} \ \log\det(Y) \ \text{ subject to (5.44a, 5.44b)},$$

and then determining A_c, C_c, P_z using (5.45). With these parameters the nominal cost $J(x_k, \mathbf{c}_k)$ is given by (5.35a), where W_x is the solution of the Riccati equation (2.9) with $(A, B) = (A^{(0)}, B^{(0)})$ and W_c is the solution of the Lyapunov equation

$$W_c - A_c^T W_c A_c = C_c^T \big(B^{(0)^T} W_x B^{(0)} + R\big) C_c.$$

A robust MPC law can then be defined analogously to Algorithm 5.3 with the difference that the control law is given by

$$u_k = K x_k + C_c \mathbf{c}_k^*$$

where $\mathbf{c}_k^*$ is the optimal solution of the online MPC optimization. Through a straightforward extension of Theorem 5.3, it can be shown that the closed-loop system is robustly asymptotically stable with region of attraction $\mathcal{F}$.

If there is no model uncertainty, then, as discussed in Sect. 2.9, the offline optimization of the prediction dynamics results in the remarkable property that the feasible set $\mathcal{F}$ (namely the projection of $\mathcal{E}_z$ onto the x-subspace) is equal to the maximal ellipsoidal invariant set under any linear feedback law subject to the constraints (5.3). Since this property holds regardless of the choice of feedback gain K, the MPC law is able to deploy a highly tuned linear feedback law close to the origin without reducing the set of initial conditions that are feasible for the online optimization. This result does not carry over to the case of polytopic model uncertainty if a single matrix A_c is employed in (5.43). It does, however, hold for the robust case if A_c is allowed to assume values in the convex hull of a set of vertices, each vertex being associated with one of the vertices of the model parameter uncertainty set and optimized offline [12]. Thus we let

$$A_c \in \mathrm{Co}\{A_c^{(1)}, \ldots, A_c^{(m)}\}, \ \ \Psi^{(j)} = \begin{bmatrix} \Phi^{(j)} & B^{(j)} C_c \\ 0 & A_c^{(j)} \end{bmatrix}, \ \ j = 1, \ldots, m \tag{5.46a}$$

and

$$\Xi^{(j)} = U A_c^{(j)} V^T, \ \ j = 1, \ldots, m. \tag{5.46b}$$

Then, replacing (5.44a) with the condition

$$\begin{bmatrix} \begin{bmatrix} Y & X \\ X & X \end{bmatrix} & \begin{bmatrix} \Phi^{(j)} Y + B^{(j)} \Gamma & \Phi^{(j)} X \\ \Xi^{(j)} + \Phi^{(j)} Y + B^{(j)} \Gamma & \Phi^{(j)} X \end{bmatrix} \\ \star & \begin{bmatrix} Y & X \\ X & X \end{bmatrix} \end{bmatrix} \succeq 0, \tag{5.47}$$

the projection of $\mathcal{E}_z$ onto the x-subspace is maximized subject to (5.33) and (5.34) by solving the SDP problem:

$$\underset{\Xi^{(1)},\ldots,\Xi^{(m)},\Gamma,X,Y}{\text{maximize}} \quad \log\det(Y) \ \text{ subject to (5.47) and (5.44b)} \tag{5.48}$$

and applying the inverse transformation (5.45) with $A_c^{(j)} = U^{-1}\Xi^{(j)}V^{-T}$ for $j = 1,\dots,m$. By considering the conditions for feasibility of (5.47) and (5.44b), it can be shown that the solution of (5.48) defines an ellipsoidal set $\mathcal{E}_z$ whose projection onto the x-subspace is equal to the maximal ellipsoidal invariant set for the uncertain model (5.1)–(5.3) under any linear feedback law. For a proof of this result, we refer the interested reader to [12].

From $z_{i+1|k} \in \mathrm{Co}\{\Psi^{(1)}z_{i|k},\dots,\Psi^{(m)}z_{i|k}\}$ and (5.46a), we obtain

$$\mathbf{c}_{i+1|k} \in \mathrm{Co}\{A_c^{(1)}\mathbf{c}_{i|k},\dots,A_c^{(m)}\mathbf{c}_{i|k}\}$$

along predicted trajectories. Hence, the controller dynamics are subject to polytopic uncertainty and the value of A_c is unknown at each prediction time step since it depends, implicitly through (5.46a), on the realization of the uncertain model parameters. A robust MPC law based on these predictions falls into the category of feedback MPC strategies since the evolution of the predicted control trajectory depends on the realization of future model uncertainty. Despite the future evolution of the controller state being uncertain, the implied MPC law is implementable since only the value of $\mathbf{c}_k$ need be known in order to evaluate $u_k = Kx_k + C_c\mathbf{c}_k$.

The objective of the MPC online optimization can be defined as in (5.35a), but to ensure closed-loop stability we require that the weighting matrix W_c satisfies

$$W_c - A_c^{(j)^T} W_c A_c^{(j)} \succeq C_c^T\big(B^{(0)^T} W_x B^{(0)} + R\big)C_c, \quad j = 1,\dots,m \tag{5.49}$$

where W_x is the solution of (2.9) with $(A,B) = (A^{(0)},B^{(0)})$. Thus W_c can be computed by minimizing $\mathrm{trace}(W_c)$ subject to (5.49). The MPC algorithm, which requires the same online computation as Algorithm 5.3, is stated next.

Algorithm 5.4 At each time instant $k = 0,1,\dots$:

(i) Perform the optimization:

$$\underset{\mathbf{c}_k}{\text{minimize}}\ \|\mathbf{c}_k\|_{W_c}^2 \ \text{ subject to } \begin{bmatrix} x_k \\ \mathbf{c}_k \end{bmatrix} \in \mathcal{E}_z. \tag{5.50}$$

(ii) Apply the control law $u_k = Kx_k + C_c\mathbf{c}_k^*$, where $\mathbf{c}_k^*$ is the optimal value of $\mathbf{c}_k$ in (5.50). ◁

Theorem 5.4 *For the system (5.1)–(5.3) under the control law of Algorithm 5.4, the optimization (5.50) is recursively feasible and $x = 0$ is asymptotically stable with region of attraction $\mathcal{F} = \{x : \exists\mathbf{c}$ such that $(x,\mathbf{c}) \in \mathcal{E}_z\}$.*

Proof Recursive feasibility is a consequence of the robust invariance of $\mathcal{E}_z$, which implies that a feasible solution to (5.50) at time $k+1$ is given by $\mathbf{c}_{k+1} = A_{c,k}\mathbf{c}_k$, for some $A_{c,k} \in \mathrm{Co}\{A_c^{(1)},\dots,A_c^{(m)}\}$. Asymptotic stability can be shown using a similar argument to the proof of Theorem 5.3. In particular, (5.30) implies that the bound

$$\sum_{k=0}^{\infty} \|x_k\|^2 \leq \|x_0\|_P^2 + \gamma^2 \sum_{k=0}^{\infty} \|C_c \mathbf{c}_k^*\|^2 \tag{5.51}$$

holds for the closed-loop system $x_{k+1} = \Phi_k x_k + B_k C_c \mathbf{c}_k^*$, for some $P \succ 0$ and scalar γ. Also, from (5.49) and feasibility of $\mathbf{c}_{k+1} = A_{c,k}\mathbf{c}_k^*$ the optimal solution at time $k+1$ necessarily satisfies

$$\|\mathbf{c}_{k+1}^*\|_{W_c}^2 \leq \|\mathbf{c}_k^*\|_{W_c}^2 - \|C_c \mathbf{c}_k^*\|_{R+B^{(0)T} W_x B^{(0)}}^2.$$

Summing this inequality over all $k \geq 0$ and using (5.51) gives the asymptotic bound (5.40), which implies $\lim_{k\to\infty} x_k = 0$ for all $x_0 \in \mathcal{F}$. Stability of $x = 0$ follows from the fact that the control law of Algorithm 5.4 is equal to $u_k = Kx_k$ for all $x_k \in \{x : (x, 0) \in \mathcal{E}_z\}$, and this feedback law is robustly stabilizing by (5.30). $\square$

In order to minimize a worst-case predicted cost instead of the cost of Algorithm 5.4, the optimization (5.50) can be replaced with

$$\underset{\mathbf{c}_k}{\text{minimize}} \ \|z_k\|_{\check{W}}^2 \ \text{subject to} \ z_k = \begin{bmatrix} x_k \\ \mathbf{c}_k \end{bmatrix} \in \mathcal{E}_z$$

where $\check{W}$ satisfies the following LMIs for $j = 1, \ldots, m$

$$\check{W} - \Psi^{(j)^T} \check{W} \Psi^{(j)} \succeq \begin{bmatrix} I & K^T \\ 0 & C_c^T \end{bmatrix} \begin{bmatrix} Q & 0 \\ 0 & R \end{bmatrix} \begin{bmatrix} I & 0 \\ K & C_c \end{bmatrix}. \tag{5.52}$$

With this modification, Algorithm 5.4 minimizes the upper bound on predicted performance:

$$\|z_k\|_{\check{W}}^2 \geq \max_{\substack{\Psi_{i|k} \in \text{Co}\{\Psi^{(1)},\ldots,\Psi^{(m)}\} \\ i=0,1,\ldots}} \sum_{i=0}^{\infty} \left(\|x_{i|k}\|_Q^2 + \|u_{i|k}\|_R^2\right),$$

and the closed-loop system has the properties given in Theorem 5.4.

Example 5.2 For the uncertain system and constraints defined in Example 5.1, the unconstrained optimal feedback law for the nominal model (5.4) and cost weights $Q = I$ and $R = 1$ is $u = Kx$, $K = [0.19\ 0.34]$, and the offline optimization (5.48) yields

$$A_c^{(1)} = \begin{bmatrix} -0.69 & 0.20 \\ -0.27 & -0.14 \end{bmatrix}, \quad A_c^{(2)} = \begin{bmatrix} -0.74 & -0.01 \\ 0.26 & -0.01 \end{bmatrix}, \quad A_c^{(3)} = \begin{bmatrix} -0.63 & -0.21 \\ -0.05 & -0.27 \end{bmatrix},$$

and $C_c = [0.12\ -0.12]$.

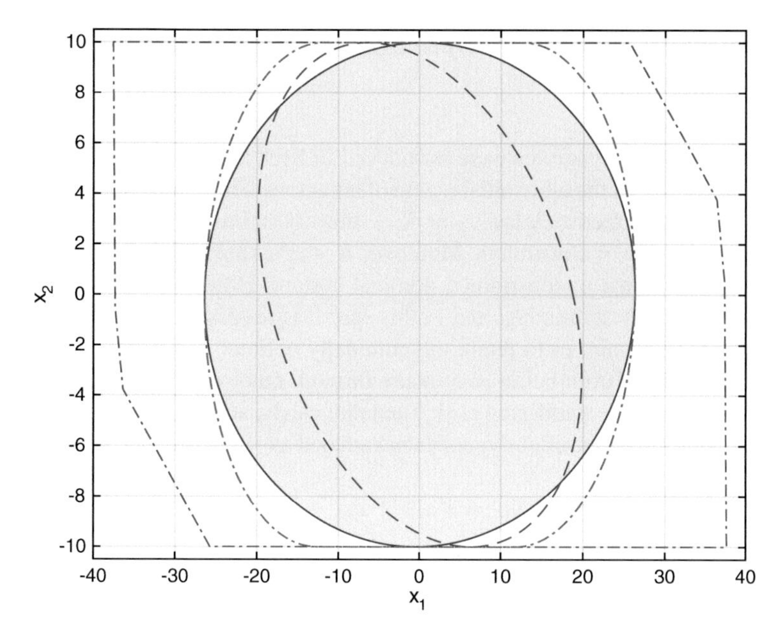

Fig. 5.2 The set of feasible initial states for Algorithm 5.4 applied to the system of Example 5.1. Also shown are the maximal ellipsoidal RPI set under the LQ-optimal feedback law $u = Kx$ (*dashed line*), the feasible set for Algorithm 5.1 (*inner dash-dotted line*) and the maximal robustly controlled invariant set (*outer dash-dotted line*)

Figure 5.2 shows the set, $\mathcal{F}$, of feasible initial conditions for Algorithm 5.4. Comparing Figs. 5.2 and 5.1, it can be seen that, as expected, this feasible set is identical to the maximum area robustly invariant ellipsoidal set under any linear feedback law. The average quadratic bound on the predicted worst-case cost computed for the solution of (5.50) at 100 initial conditions evenly distributed on the boundary of $\mathcal{F}$ is 1002, while the maximum worst-case cost bound for these states is 1823. These bounds are on average 4 % higher than the worst-case cost bounds obtained using Algorithm 5.1 with the same set of initial conditions (discussed in Example 5.1); however, neither approach results in cost bounds that are consistently lower for all initial conditions. The conservativeness in this case results from the offline computation of the parameters K, C_c and $A_c^{(1)}, \ldots, A_c^{(m)}$ defining the predicted control law. On the other hand, the online computation of Algorithm 5.4 is typically orders of magnitude lower than that of Algorithm 5.1[1] ◊

[1]For this example, the time required to solve (5.50) using the Newton–Raphson method is between one and two orders of magnitude less than the computation time for (5.18) using the Mosek.

5.3.2 Prediction Dynamics Optimized to Improve Worst-Case Performance

Polytopic controller dynamics were introduced in Sect. 5.3.1 with the aim of maximizing the volume of the ellipsoidal region of attraction. This was achieved under the assumption that the feedback law $u_k = Kx_k$ robustly stabilizes the system (5.1) and (5.2) in the absence of constraints. Moreover K was assumed to be the LQ optimal feedback gain for the unconstrained nominal system. However this feedback law may not be robustly stabilizing, and in this case it is necessary either to resort to a different feedback gain, or to retain the nominally optimal unconstrained feedback law in the problem formulation and ensure through other means that the predicted control law is robustly stabilizing [13]. The latter can be achieved through the use of predicted polytopic controller dynamics introduced as

$$u_{i|k} = Kx_{i|k} + v_{i|k} + c_{i|k} \tag{5.53a}$$

where $c_{i|k} = 0$ for all $i \geq N$ and

$$v_{i+1|k} \in \mathrm{Co}\{L^{(j)}x_{i|k} + N^{(j)}v_{i|k},\ j = 1, \dots, m\} \tag{5.53b}$$

for all $i \geq 0$.

The perturbations $c_{i|k}$ serve the same purpose here as in earlier sections of this chapter, namely to ensure satisfaction of the constraints (5.3), while the polytopic dynamics of (5.53b) are designed with the aim of improving robustness and reducing the sensitivity of closed-loop performance to model uncertainty. This is done by invoking the following conditions for $W \succ 0$:

$$W - \begin{bmatrix} \Phi^{(j)} & B^{(j)} \\ L^{(j)} & N^{(j)} \end{bmatrix}^T W \begin{bmatrix} \Phi^{(j)} & B^{(j)} \\ L^{(j)} & N^{(j)} \end{bmatrix} \succeq \begin{bmatrix} Q + K^T RK & K^T R \\ RK & R \end{bmatrix}, \quad j = 1, \dots, m \tag{5.54}$$

where $\Phi^{(j)} = A^{(j)} + B^{(j)}K$, and choosing $L^{(j)}, N^{(j)}, j = 1, \dots, m$ and W by solving

$$(L^{(1)}, N^{(1)}, \dots, L^{(m)}, N^{(m)}, W) = \arg \min_{L^{(1)}, N^{(1)}, \dots, L^{(m)}, N^{(m)}, W} \bar{\lambda}(W_x - W_{xv} W_v^{-1} W_{vx}) \tag{5.55}$$

subject to (5.54), where $\bar{\lambda}(P)$ denotes the maximum eigenvalue of the matrix P, and where W_x, W_{xv} and W_v are the blocks of the partition

$$W = \begin{bmatrix} W_x & W_{xv} \\ W_{xv}^T & W_v \end{bmatrix}.$$

Although the optimization is nonconvex as stated in (5.55), its solution can be determined by solving the following equivalent (convex) SDP problem:

$$(Y^{(1)}, \dots, Y^{(m)}, S) = \arg \max_{Y^{(1)},\dots,Y^{(m)},S,\lambda} \lambda \quad \text{subject to}$$

$$\begin{bmatrix} I_{n_x} & 0 \end{bmatrix} S \begin{bmatrix} I_{n_x} \\ 0 \end{bmatrix} \succeq \lambda I_{n_x}$$

$$\begin{bmatrix} S & \begin{bmatrix} S \begin{bmatrix} \Phi^{(j)T} \\ B^{(j)T} \end{bmatrix} & Y^{(j)T} \end{bmatrix} & S \begin{bmatrix} Q^{1/2} \\ 0 \end{bmatrix} & S \begin{bmatrix} K^T R^{1/2} \\ R^{1/2} \end{bmatrix} \\ \star & S & 0 & 0 \\ \star & \star & I & 0 \\ \star & \star & \star & I \end{bmatrix} \succeq 0, \quad j = 1, \dots, m \tag{5.56}$$

and then computing $W = S^{-1}$ and $\begin{bmatrix} L^{(j)} & N^{(j)} \end{bmatrix} = Y^{(j)} W$, for $j = 1, \dots, m$.

The rationale behind this strategy is that, if $L^{(j)}, N^{(j)}$ for $j = 1, \dots, m$ satisfy (5.54) for some $W \succ 0$, then the control law $u_{i|k} = K x_{i|k} + v_{i|k}$ robustly stabilizes the system

$$\begin{aligned} x_{i+1|k} &= A_{i|k} x_{i|k} + B_{i|k} u_{i|k}, \\ v_{i+1|k} &= L_{i|k} x_{i|k} + N_{i|k} v_{i|k}, \end{aligned} \quad \begin{bmatrix} A_{i|k} & B_{i|k} \\ L_{i|k} & N_{i|k} \end{bmatrix} \in \mathrm{Co}\left\{ \begin{bmatrix} \Phi^{(j)} & B^{(j)} \\ L^{(j)} & N^{(j)} \end{bmatrix}, \ j = 1, \dots, m \right\}$$

Furthermore, (5.54) implies that the predicted cost for this system under the control law $u_{i|k} = K x_{i|k} + v_{i|k}$ satisfies the bound

$$\sum_{i=0}^{\infty} \left(\|x_{i|k}\|_Q^2 + \|u_{i|k}\|_R^2 \right) \le \begin{bmatrix} x_{0|k} \\ v_{0|k} \end{bmatrix}^T W \begin{bmatrix} x_{0|k} \\ v_{0|k} \end{bmatrix} \tag{5.57}$$

for all admissible realizations of model uncertainty. If the constraints (5.3) are inactive, then the minimum, over all $v_{0|k} \in \mathbb{R}^{n_u}$, of this bound can be shown to be $x_{0|k}^T (W_x - W_{xv} W_v^{-1} W_{xv}^T) x_{0|k}$. Therefore the optimization (5.55) chooses $(L^{(1)}, N^{(1)}, \dots, L^{(m)}, N^{(m)})$ so as to minimize the maximum, over all $x_{0|k}$ in the ball $\{x : \|x\| \le r\}$, of the minimum value of the bound (5.57) in the absence of constraints.

Clearly, $v_{0|k}$ provides degrees of freedom for minimizing the predicted cost in the online MPC optimization subject to constraints. However the system constraints (5.3) must be satisfied and for this reason we introduce the vector of perturbations $\mathbf{c}_k = (c_{0|k}, \dots, c_{N-1|k})$ as additional degrees of freedom. Under the control law of (5.53a, 5.53b), the predicted state and control trajectories are generated by the prediction system

$$z_{i+1|k} = \Psi_{i|k} z_{i|k}, \quad \Psi_{i|k} = \mathrm{Co}\{\Psi^{(1)}, \dots, \Psi^{(m)}\} \tag{5.58}$$

where

$$z_{0|k} = \begin{bmatrix} x_{0|k} \\ v_{0|k} \\ \mathbf{c}_k \end{bmatrix}, \quad \Psi^{(j)} = \begin{bmatrix} \Phi^{(j)} & B^{(j)} & B^{(j)}E \\ L^{(j)} & N^{(j)} & 0 \\ 0 & 0 & M \end{bmatrix} \tag{5.59}$$

with E and M as defined in (5.31a, 5.31b). As before, the constraints of (5.3) can be imposed on predictions in a computationally efficient (though somewhat conservative) way by constraining the prediction system state $z_{0|k}$ to lie in a robustly invariant ellipsoidal set $\mathcal{E}_z \doteq \{z : \; z^T P_z z \le 1\}$. Analogously to Corollary 5.1, the conditions for invariance and constraint satisfaction are given by

$$P_z - \Psi^{(j)T} P_z \Psi^{(j)} \succeq 0, \quad j = 1, \ldots, m \tag{5.60}$$

and

$$\begin{bmatrix} H & \begin{bmatrix} F + GK & G & GE \end{bmatrix} \\ \begin{bmatrix} F + GK & G & GE \end{bmatrix}^T & P_z \end{bmatrix} \succeq 0, \quad e_i^T H e_i \le 1, \;\; i = 1, \ldots, n_C \tag{5.61}$$

for some symmetric matrix H, where e_i is the ith column of the identity matrix.

Additionally, if $\check{W}$ satisfies the condition

$$\check{W} - \Psi^{(j)T} \check{W} \Psi^{(j)} \succeq \begin{bmatrix} Q + K^T RK & K^T R & K^T RE \\ RK & R & RE \\ E^T RK & E^T R & E^T RE \end{bmatrix}, \quad j = 1, \ldots, m, \tag{5.62}$$

then the worst-case cost along predicted trajectories of (5.1) and (5.2) under the control law (5.53a, 5.53b)

$$\check{J}(x_{0|k}, v_{0|k}, \mathbf{c}_k) \doteq \max_{(A_{i|k}, B_{i|k}) \in \Omega, \, i=0,1,\ldots} \sum_{i=0}^{\infty} \left(\|x_{i|k}\|_Q^2 + \|u_{i|k}\|_R^2 \right),$$

satisfies the bound $\check{J}(x, v, \mathbf{c}) \le \|(x, v, \mathbf{c})\|_{\check{W}}^2$. On the basis of this predicted performance bound, we can state the following min–max robust MPC algorithm.

Algorithm 5.5 At each time instant $k = 0, 1, \ldots$:

(i) Perform the optimization:

$$\underset{v_k, \mathbf{c}_k}{\text{minimize}} \;\; \|(x_k, v_k, \mathbf{c}_k)\|_{\check{W}}^2 \;\; \text{subject to} \;\; \begin{bmatrix} x_k \\ v_k \\ \mathbf{c}_k \end{bmatrix} \in \mathcal{E}_z \tag{5.63}$$

(ii) Apply the control law $u_k = Kx_k + v_k^* + c_{0|k}^*$, where $\mathbf{c}_k^* = (c_{0|k}^*, \ldots, c_{N-1|k}^*)$ and $(v_k^*, \mathbf{c}_k^*)$ is the optimal solution of (5.63). ◁

Theorem 5.5 *For the system (5.1)–(5.3) and control law of Algorithm 5.4, the optimization (5.63) is recursively feasible and* $x = 0$ *is asymptotically stable with region of attraction* $\mathcal{F} = \{x : \exists(v, \mathbf{c}) \text{ such that } (x, v, \mathbf{c}) \in \mathcal{E}_z\}$.

Proof Recursive feasibility of (5.63) follows from (5.60) and (5.61), which imply that $\Psi z_k \in \mathcal{E}_z$ for some $\Psi \in \text{Co}\{\Psi^{(1)}, \ldots, \Psi^{(m)}\}$ if $z_k \in \mathcal{E}_z$, and hence $(v_{k+1}, \mathbf{c}_{k+1}) = (Lx_k + Nv_k^*, Mc_k^*)$ is a feasible solution to (5.63) at time $k+1$ for some $(L, N) \in \text{Co}\{L^{(1)}, N^{(1)}, \ldots, L^{(m)}, N^{(m)}\}$ if $x_k \in \mathcal{F}$. From (5.62), we therefore have

$$\|z_{k+1}\|^2_{\check{W}} \leq \|z_k\|^2_{\check{W}} - \left(\|x_k\|^2_Q + \|u_k\|^2_R\right),$$

where $z_k = (x_k, v_k^*, \mathbf{c}_k^*)$, and this implies that $x = 0$ is asymptotically stable since $\|z_k\|^2_{\check{W}} \geq \check{J}(x_k, v_k^*, \mathbf{c}_k^*)$ and $Q, R \succ 0$. □

Theorem 5.5 implies that the closed-loop system necessarily converges to a region of state space on which the trajectories of the nominal system model satisfy the constraints (5.3) at all future times under the unconstrained nominal LQ optimal feedback law. To allow the predicted control trajectories (5.53a, 5.53b) to realize this feedback law, the additional constraint that $(L^{(0)}, N^{(0)}) = (0, 0)$ must be included in the optimization (5.55) defining $(L^{(j)}, N^{(j)})$, where

$$(L^{(0)}, N^{(0)}) = \sum_{j=1}^{m} \mu^{(j)}(L^{(j)}, N^{(j)}) \tag{5.64}$$

and where $\mu^{(1)}, \ldots, \mu^{(m)}$ are scalar constants that define the nominal model parameters $(A^{(0)}, B^{(0)})$ via

$$(A^{(0)}, B^{(0)}) = \sum_{j=1}^{m} \mu^{(j)}(A^{(j)}, B^{(j)}).$$

The condition (5.64) can be imposed in the optimization (5.56) through a constraint which is linear in $Y^{(1)}, \ldots, Y^{(m)}$, namely that $\sum_{j=1}^{m} \mu^{(j)} Y^{(j)} = 0$.

To ensure that the MPC law recovers the nominal LQ optimal feedback law whenever it is feasible, a nominal cost computed analogously to (5.35b) can be used in place of the worst-case cost of Algorithm 5.5. In this case, however, the nominal cost does not have the block diagonal structure of (5.35a), and hence the predicted cost may not be monotonically non-increasing. Closed-loop stability is therefore ensured in [13] by tightening condition (5.60) by replacing the RHS of the LMI with ρP_z for some $\rho \in (0, 1)$, and by replacing the constraint in the online optimization (5.63) with the condition $(x_k, v_k, \mathbf{c}_k) \in \rho^k \mathcal{E}_z$, thus guaranteeing exponential stability of $x = 0$.

Robustness can alternatively be addressed by introducing controller dynamics through the Youla parameter [25]. Moreover, the approach can be recast in terms of state-space models, allowing constraints to be imposed in a computationally efficient way through the use of robustly invariant ellipsoidal sets. A nominal cost can be adopted for the definition of an MPC law and the free Youla parameters can be optimized so as to minimize a sensitivity transfer function, thus reducing the sensitivity of predicted trajectories to the model uncertainty and making a nominal cost more representative of system performance [13]. This is an indirect way of minimizing cost sensitivity to uncertainty, a problem that remains open, both for MPC and for constrained optimal control in general.

5.4 Low-Complexity Polytopes in Robust MPC

Although computationally convenient for robust MPC involving multiplicative model uncertainty, the ellipsoidal invariant sets discussed in Sect. 5.3 necessarily result in conservative feasible sets as a result of their implicit handling of state and control constraints. On the other hand, the exact tubes considered in Sect. 5.2.1, which enable constraints to be imposed on predicted states and inputs non-conservatively, are in general impractical because their computational requirements grow exponentially with the length of prediction horizon. To avoid these difficulties, this section considers the application of tubes with low-complexity polytopic cross sections (which were discussed in Sect. 3.6.2 in the context of robust MPC with additive model uncertainty) to the case of multiplicative model uncertainty.

5.4.1 Robust Invariant Low-Complexity Polytopic Sets

Tubes defined in terms of low-complexity polytopic sets are the basis of a family of computationally efficient methods of imposing constraints on predicted trajectories in the presence of multiplicative uncertainty [16–18, 26]. We discuss terminal constraints in this section and consider the uncertain system (5.1)–(5.3) under a given terminal feedback law $u = Kx$:

$$x_{k+1} = \Phi_k x_k, \quad \Phi_k \in \Omega_K \doteq \mathrm{Co}\{\Phi^{(1)}, \ldots, \Phi^{(m)}\} \tag{5.65}$$

where $\Phi^{(j)} = A^{(j)} + B^{(j)}K$, $j = 1, \ldots, m$. Recall that a low-complexity polytope, denoted here by $\Pi(V, \alpha)$, is defined for a non-singular matrix $V \in \mathbb{R}^{n_x \times n_x}$ and a positive vector $\alpha \in \mathbb{R}^{n_x}$ by

$$\Pi(V, \alpha) \doteq \{x : |Vx| \le \alpha\} \tag{5.66}$$

where the extraction of absolute values and the inequality sign apply on an element-by-element basis. The conditions under which $\Pi(V, \alpha)$ is a suitable terminal set, namely invariance under the dynamics of (5.65) for a given terminal control law

$u = Kx$ and feasibility with respect to the system constraints (5.3), can be stated as follows.

Theorem 5.6 *The low-complexity polytope $\Pi(V, \alpha) = \{x : |Vx| \leq \alpha\}$ is robustly invariant for the dynamics (5.65) if and only if*

$$|V\Phi^{(j)}W|\alpha \leq \alpha, \quad j = 1, \ldots, m \tag{5.67}$$

where $W = V^{-1}$. Furthermore, under $u = Kx$ the constraints (5.3) are satisfied for all $x \in \Pi(V, \alpha)$ if and only if

$$|(F + GK)W|\alpha \leq \mathbf{1}. \tag{5.68}$$

Proof The set $\Pi(V, \alpha)$ is robustly invariant if and only if $\Phi x \in \Pi(V, \alpha)$ for all $x \in \Pi(V, \alpha)$ and all $\Phi \in \text{Co}\{\Phi^{(j)}, \ j = 1, \ldots, m\}$. Equivalently, for all x such that $|Vx| \leq \alpha$, we require $|V\Phi x| \leq \alpha$. Therefore, given that

$$|V\Phi x| = |V\Phi W V x| \leq |V\Phi W||Vx| \leq |V\Phi W|\alpha$$

for all $x \in \Pi(V, \alpha)$, a sufficient condition for invariance is

$$|V\Phi W|\alpha \leq \alpha.$$

This condition can be expressed in terms of inequalities that depend linearly on Φ, and therefore it needs to be enforced only at the vertices of the model uncertainty set, as is done in (5.67). The conditions of (5.67) are necessary as well as sufficient because, for each $i = 1, \ldots, n_x$, the ith element of $|V\Phi^{(j)}x|$ satisfies $|V_i\Phi^{(j)}x| = |V_i\Phi^{(j)}WVx| = |V_i\Phi^{(j)}W|\alpha$ for x equal to one of the vertices of $\Pi(V, \alpha)$, where V_i is the ith row of V. The necessity and sufficiency of (5.68) follows similarly from the inequalities

$$(F + GK)x \leq |(F + GK)x| = |(F + GK)WVx| \leq |(F + GK)W|\alpha$$

for all $x \in \Pi(V, \alpha)$, where, for each i, $(F + GK)_i x = |(F + GK)_i W|\alpha$ for some x such that $|Vx| = \alpha$, with $(F + GK)_i$ denoting the ith row of $F + GK$. □

A non-symmetric low-complexity polytope is defined for $V \in \mathbb{R}^{n_x \times n_x}$ and $\underline{\alpha}, \overline{\alpha} \in \mathbb{R}^{n_x}$, with $\underline{\alpha} < 0$ and $\overline{\alpha} > 0$, as the set

$$\tilde{\Pi}(V, \underline{\alpha}, \overline{\alpha}) \doteq \{x : \underline{\alpha} \leq Vx \leq \overline{\alpha}\}. \tag{5.69}$$

These sets share many of the computational advantages of symmetric low-complexity polytopic sets, but can provide larger invariant sets if the system constraints (5.3) are non-symmetric. The conditions of Theorem 5.6 for robust invariance and feasibility with respect to constraints extend to non-symmetric low-complexity polytopes in an obvious way, as we show next.

Corollary 5.2 *Let $A^+ \doteq \max\{A, 0\}$ and $A^- \doteq \max\{-A, 0\}$ for any real matrix A. Under the feedback law $u = Kx$, the non-symmetric low-complexity polytope $\tilde{\Pi}(V, \underline{\alpha}, \overline{\alpha})$ is robustly invariant for the dynamics (5.65) and the constraint (5.3) is satisfied for all $x \in \tilde{\Pi}(V, \underline{\alpha}, \overline{\alpha})$ if and only if*

$$\begin{bmatrix} (V\Phi^{(j)}W)^+ & (V\Phi^{(j)}W)^- \\ (V\Phi^{(j)}W)^- & (V\Phi^{(j)}W)^+ \end{bmatrix} \begin{bmatrix} \overline{\alpha} \\ -\underline{\alpha} \end{bmatrix} \leq \begin{bmatrix} \overline{\alpha} \\ -\underline{\alpha} \end{bmatrix}, \quad j = 1, \ldots, m \tag{5.70}$$

$$\begin{bmatrix} \big((F+GK)W\big)^+ & \big((F+GK)W\big)^- \end{bmatrix} \begin{bmatrix} \overline{\alpha} \\ -\underline{\alpha} \end{bmatrix} \leq \mathbf{1}. \tag{5.71}$$

where $W = V^{-1}$.

Proof This follows from Lemma 3.6, which implies that the following bounds hold for all x such that $\underline{\alpha} \leq Vx \leq \overline{\alpha}$:

$$\begin{aligned} V\Phi x &\leq (V\Phi W)^+\overline{\alpha} + (V\Phi W)^-(-\underline{\alpha}) \\ V\Phi x &\geq (V\Phi W)^+\underline{\alpha} + (V\Phi W)^-(-\overline{\alpha}) \\ (F+GK)x &\leq \big((F+GK)W\big)^+\overline{\alpha} + \big((F+GK)W\big)^-(-\underline{\alpha}). \end{aligned}$$

Lemma 3.6 also shows that each inequality in these conditions must hold with equality for some x such that $\underline{\alpha} \leq Vx \leq \overline{\alpha}$. A similar argument to the proof of Theorem 5.6 therefore implies that (5.70) and (5.71) are necessary and sufficient for robust invariance of $\tilde{\Pi}(V, \underline{\alpha}, \overline{\alpha})$ for the system (5.65) and for satisfaction of the constraints (5.3) under $u = Kx$ at all points in this set. □

To make it possible to compute offline a low-complexity polytopic terminal set for use in an online MPC optimization, we require that the conditions of Theorem 5.6 (or Corollary 5.2) hold for some V and $\alpha > 0$ (or V, $\underline{\alpha} < 0$ and $\overline{\alpha} > 0$). The crucial condition to be satisfied is the invariance condition (5.67), since, by linearity of the dynamics (5.65), the existence of V and $\alpha > 0$ satisfying (5.67) is necessary as well as sufficient for (5.70) to hold for some V, $\underline{\alpha} < 0$ and $\overline{\alpha} > 0$. Likewise the feasibility conditions (5.68) and (5.70) can be satisfied simply by scaling α whenever (5.67) holds.

Necessary and sufficient conditions for existence of an ellipsoidal invariant set for the uncertain system (5.65) follow directly from the discussion of Sect. 5.2, namely the system admits an ellipsoidal invariant set if and only if it is quadratically stable. Moreover, quadratic stability is relatively easy to check numerically by determining whether a set of LMIs is feasible. Likewise a polytopic invariant set defined by a finite (but arbitrary) number of vertices exists whenever the system (5.65) is exponentially stable [27]. This condition is equivalent to the requirement that the joint spectral radius defined by

$$\rho \doteq \lim_{k\to\infty} \max_{\Phi_i \in \Omega_K,\, i=1,2,\ldots} \|\Phi_1 \cdots \Phi_k\|^{1/k} \tag{5.72}$$

satisfies $\rho < 1$.[2] For the special case of low-complexity polytopic sets, nonconservative conditions for the existence of low-complexity invariant sets are not available for general systems of the form (5.65); however, the following result provides a useful sufficient condition.

Lemma 5.4 *For given V, define $\bar{\Phi}$ as the matrix with (i, k)th element*

$$[\bar{\Phi}]_{ik} = \max_{j\in\{1,\dots,m\}} |V_i \Phi^{(j)} W_k|, \quad i = 1, \dots, n_x, \quad k = 1, \dots, n_x,$$

where V_i and W_k are, respectively, the ith row of V and kth column of $W = V^{-1}$. Then α exists such that $\Pi(V, \alpha)$ is robustly invariant for the dynamics (5.65) if $\lambda_{PF}(\bar{\Phi}) \leq 1$ where $\lambda_{PF}(\bar{\Phi})$ is the Perron–Frobenius eigenvalue of $\bar{\Phi}$.

Proof For each $i = 1, \dots, n_x$ and $j = 1, \dots, m$ we have $|V_i \Phi^{(j)} W|\alpha \leq \bar{\Phi}_i \alpha$ where $\bar{\Phi}_i$ is the ith row of $\bar{\Phi}$. Thus, if α is chosen as the Perron–Frobenius eigenvector of $\bar{\Phi}$, then $|V \Phi W|\alpha \leq \bar{\Phi}\alpha = \lambda_{PF}(\bar{\Phi})\alpha \leq \alpha$ which implies that (5.67) admits at least one feasible α. □

The volume in $\mathbb{R}^{n_x}$ of a low-complexity polytope $\Pi(V, \alpha)$, with $\alpha = (\alpha_1, \dots, \alpha_{n_x})$, is given by

$$C_{n_x} |\det(V^{-1})| \prod_{i=1}^{n_x} \alpha_i$$

where C_{n_x} is independent of V and α. If $\Pi(V, \alpha)$ exists satisfying the conditions of Theorem 5.6, then the maximum volume invariant set is therefore given by $\Pi(V, \mathbf{1})$ where V is the maximizing argument of

$$\underset{V,W}{\text{maximize}}\ |\det(W)| \ \text{ subject to (5.67), (5.68) with } \alpha = \mathbf{1}, \text{ and } V = W^{-1}.$$

Although this problem is nonconvex, its constraints can be reformulated in terms of equivalent bilinear constraints:

$$\begin{aligned} &\underset{W,H^{(1)},\dots,H^{(m)}}{\text{maximize}} && |\det(W)| \\ &\text{subject to} && W H^{(j)} = \Phi^{(j)} W, \qquad |H^{(j)}|\mathbf{1} \leq \mathbf{1}, \qquad j = 1, \dots, m \\ &&& |(F + GK)W|\mathbf{1} \leq \mathbf{1} \end{aligned} \tag{5.73}$$

Expressed in this form, the problem can be solved approximately by solving a sequence of convex programs [26, 29].

To avoid the inherent nonconvexity (offline) of (5.73), the volume of $\Pi(V, \alpha)$ can be maximized instead over $\alpha > 0$ for a given fixed V by solving

[2] In general, it is not possible to compute ρ exactly. However, upper bounds on ρ can be computed to any desired accuracy, for example, using sum of squares programming [28].

$$\underset{\alpha=(\alpha_1,\ldots,\alpha_{n_x})}{\text{maximize}} \prod_{i=1}^{n_x} \alpha_i \quad \text{subject to (5.67) and (5.68).} \tag{5.74}$$

The constraints of this problem are linear in α and the objective can be expressed as the determinant of a symmetric positive-definite matrix (namely $\text{diag}(\alpha_1, \ldots, \alpha_{n_x})$). Therefore the optimization can be performed by solving an equivalent (convex) semidefinite program (see e.g. [30]). An obvious modification of this problem allows the volume of the non-symmetric low-complexity polytope $\{x : \underline{\alpha} \le Vx \le \overline{\alpha}\}$ to be maximized over $\underline{\alpha}, \overline{\alpha}$ for given V by solving a similar semidefinite program.

If α is chosen so as to maximize the volume of $\Pi(V, \alpha)$ for fixed V using (5.74), then clearly V must be designed so as to ensure that (5.67) is feasible for some α. For example, robust invariance of $\Pi(V, \alpha)$ under (5.65) can be ensured for some α by choosing V on the basis of a robustly invariant ellipsoidal set $\mathcal{E} = \{x : x^T P x \le 1\}$, computed using semidefinite programming. If P is a symmetric positive-definite matrix satisfying

$$\Phi^{(j)T} P \Phi^{(j)} \preceq P/n_x, \quad j = 1, \ldots, m,$$

then the choice $V = P^{1/2}$ ensures that (5.67) is feasible since the bounds $\|x\|_\infty \le \|x\|_2 \le \sqrt{n_x}\|x\|_\infty$ (which hold for all $x \in \mathbb{R}^{n_x}$) then imply

$$\|V\Phi^{(j)}x\|_\infty \le \|V\Phi^{(j)}x\|_2 \le \frac{1}{\sqrt{n_x}}\|Vx\|_2 \le \|Vx\|_\infty, \quad \forall x \in \mathbb{R}^{n_x}, \quad j = 1, \ldots, m. \tag{5.75}$$

It follows that $\Pi(V, \mathbf{1})$ is robustly invariant and the constraints of (5.74) are therefore necessarily feasible. Note that this approach can only be used if the dynamics (5.65) satisfy the strengthened quadratic stability condition (5.75), and of course this may not be the case if K is designed as the unconstrained LQ optimal feedback gain for the nominal dynamics.

Alternatively, V could be chosen on the basis of the nominal model parameters. In particular, in the absence of uncertainty (i.e. for $\Phi = \Phi^{(0)}$), and for the case that the eigenvalues of $\Phi^{(0)}$ are real, an obvious choice for V is the inverse of the (right) eigenvector matrix of $\Phi^{(0)}$ since this gives $V\Phi^{(0)}W = \Lambda$, where Λ is the eigenvalue matrix of $\Phi^{(0)}$ and W is the corresponding eigenvector matrix. Given that the feedback gain K is stabilizing by assumption, the elements of Λ must be no greater than 1 in absolute value, and hence (5.67) holds for any chosen $\alpha > 0$ in this case. Similarly, if $\Phi^{(0)}$ has complex eigenvalues, then V and $W = V^{-1}$ can be chosen to be real matrices such that $V\Phi^{(0)}W$ is in (real) Jordan normal form. In this case, $V\Phi^{(0)}W$ is block diagonal—for example if the eigenvalues of $\Phi^{(0)}$ are distinct and equal to

$$\lambda_1, \ldots, \lambda_p, \sigma_1 \pm j\omega_1, \ldots, \sigma_q \pm j\omega_q,$$

where $p + 2q = n_x$, then

$$V\Phi^{(0)}W = \begin{bmatrix} \lambda_1 & & & & & \\ & \ddots & & & & \\ & & \lambda_r & & & \\ & & & \begin{bmatrix} \sigma_1 & \omega_1 \\ -\omega_1 & \sigma_1 \end{bmatrix} & & \\ & & & & \ddots & \\ & & & & & \begin{bmatrix} \sigma_q & \omega_q \\ -\omega_q & \sigma_q \end{bmatrix} \end{bmatrix}.$$

For this choice of V, the structure of $V\Phi^{(0)}W$ implies that a necessary and sufficient condition for existence of $\alpha > 0$ satisfying $|V\Phi^{(0)}W|\alpha \leq \alpha$ is that the eigenvalues of $\Phi^{(0)}$ should lie in the box in the complex plane with vertices at $\pm 1 \pm j$ shown in Fig. 3.9. For the case considered here, in which the model (5.1) and (5.2) has uncertain parameters A, B, the same approach can be used to define V on the basis of the nominal model pair $A^{(0)}$, $B^{(0)}$. If the uncertainty is sufficiently small and the eigenvalues of $\Phi^{(0)} = A^{(0)} + B^{(0)}K$ are sufficiently contractive, then the invariance condition (5.67) will hold for all $j = 1, \ldots, m$ for some $\alpha > 0$, and (5.74) could be used to maximize $\Pi(V, \alpha)$ for this choice of V.

Example 5.3 For the uncertain system (5.1)–(5.3) with model parameters defined in Example 5.1, define V as the inverse of the (right) eigenvector matrix of $\Phi^{(0)} = A^{(0)} + B^{(0)}K$ for $K = [0.19\ 0.34]$. For this choice of V, the Perron–Frobenius eigenvalue of $\tilde{\Phi}$ defined in Lemma 5.4 is 0.89, so Lemma 5.4 indicates that the low-complexity polytopic set $\Pi(V, \alpha)$ is robustly positively invariant for some $\alpha > 0$ for (5.1)–(5.3) under $u = Kx$. Choosing α so as to maximize the area of $\Pi(V, \alpha)$ by solving the convex optimization (5.74) then gives

$$\Pi(V, \alpha) = \left\{ x : \left| \begin{bmatrix} -0.997 & -0.084 \\ -0.229 & -0.973 \end{bmatrix} x \right| \leq \begin{bmatrix} 15.80 \\ 5.90 \end{bmatrix} \right\} \tag{5.76}$$

This set is shown in Fig. 5.3. ◊

5.4.2 *Recursive State Bounding and Low-Complexity Polytopic Tubes*

In order to account for the effects of model uncertainty on predicted state and control trajectories, this section considers the construction, through a recursive bounding procedure, of tubes $\{\mathcal{X}_{0|k}, \mathcal{X}_{1|k}, \ldots\}$ such that $x_{i|k} \in \mathcal{X}_{i|k}$ for all $i = 0, 1, \ldots$. For convenience, and assuming the terminal constraint to be defined by a low-complexity

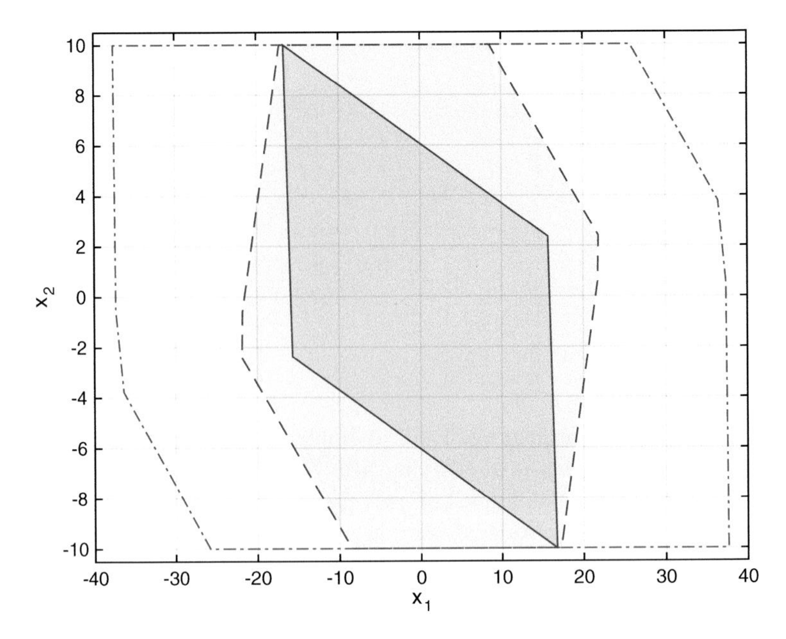

Fig. 5.3 The low-complexity polytopic set $\Pi(V, \alpha)$ defined by (5.76) (*solid line*) and the maximal RPI set (*dashed line*) under the LQ-optimal feedback law $u = Kx$ for the system of Example 5.1. Also shown is the maximal robustly controlled invariant set for this system (*dash-dotted line*)

polytopic invariant set $\Pi(V, \alpha)$, we define each tube cross section as a low-complexity polytope of the form

$$\mathcal{X}_{i|k} = \{x : \underline{\alpha}_{i|k} \leq Vx \leq \overline{\alpha}_{i|k}\}.$$

Here the matrix $V \in \mathbb{R}^{n_x \times n_x}$ is assumed to be determined offline so that $\Pi(V, \alpha)$ satisfies the robust invariance conditions of Theorem 5.6 for some α and for a given linear feedback gain K, and the parameters $\underline{\alpha}_{i|k}, \overline{\alpha}_{i|k}$ are variables in the online MPC optimization. We also assume that the predicted control input is given by the open-loop strategy

$$u_{i|k} = Kx_{i|k} + c_{i|k}, \quad i = 0, 1 \ldots \tag{5.77}$$

with $c_{i|k} = 0$ for all $i \geq N$.

The tube cross sections $\mathcal{X}_{i|k}$ are computed using a sequence of one-step ahead bounds on future model states. For the transformed state variable

$$\xi_{i|k} \doteq Vx_{i|k}$$

and the predicted control law (5.77), we obtain the dynamics

$$\xi_{i+1|k} = \tilde{\Phi}\xi_{i|k} + \tilde{B}c_{i|k}, \quad (\tilde{\Phi}, \tilde{B}) \in \text{Co}\big\{(\tilde{\Phi}^{(1)}, \tilde{B}^{(1)}), \ldots, (\tilde{\Phi}^{(m)}, \tilde{B}^{(m)})\big\}$$

where $\tilde{\Phi}^{(j)} = V\Phi^{(j)}W$, $W = V^{-1}$, and $\tilde{B}^{(j)} = VB^{(j)}$ for $j = 1, \ldots, m$. Using these transformed dynamics, it is easy to show that $\underline{\alpha}_{i+1|k} \le \xi_{i+1|k} \le \overline{\alpha}_{i+1|k}$ for all $\xi_{i|k}$ satisfying $\underline{\alpha}_{i|k} \le \xi_{i|k} \le \overline{\alpha}_{i|k}$ if and only if

$$\underline{\alpha}_{i+1|k} \le \tilde{\Phi}^{+}\underline{\alpha}_{i|k} + \tilde{\Phi}^{-}(-\overline{\alpha}_{i|k}) + \tilde{B}c_{i|k}$$
$$\overline{\alpha}_{i+1|k} \ge \tilde{\Phi}^{+}\overline{\alpha}_{i|k} + \tilde{\Phi}^{-}(-\underline{\alpha}_{i|k}) + \tilde{B}c_{i|k}$$

where, as in Sect. 5.4.1, $A^{+} = \max\{A, 0\}$ and $A^{-} = \max\{-A, 0\}$ denote, respectively, the absolute values of the positive and negative elements of a real matrix A. Invoking these conditions for all $(\tilde{\Phi}, \tilde{B})$ in the uncertainty set associated with the model parameters yields, by linearity and convexity, a finite set of conditions:

$$\begin{aligned}\underline{\alpha}_{i+1|k} &\le (\tilde{\Phi}^{(j)})^{+}\underline{\alpha}_{i|k} - (\tilde{\Phi}^{(j)})^{-}\overline{\alpha}_{i|k} + \tilde{B}^{(j)}c_{i|k}\\ \overline{\alpha}_{i+1|k} &\ge (\tilde{\Phi}^{(j)})^{+}\overline{\alpha}_{i|k} - (\tilde{\Phi}^{(j)})^{-}\underline{\alpha}_{i|k} + \tilde{B}^{(j)}c_{i|k}\end{aligned} \tag{5.78}$$

for $j = 1, \ldots, m$ and $i = 0, \ldots, N-1$. Thus the linear constraints (5.78) ensure that the predicted model trajectories satisfy $x_{i|k} \in \mathcal{X}_{i|k}$ for $i = 1, \ldots, N$, with $\mathcal{X}_{i|k} = \tilde{\Pi}(V, \underline{\alpha}_{i|k}, \overline{\alpha}_{i|k})$.

Given a tube $\{\mathcal{X}_{0|k}, \ldots, \mathcal{X}_{N-1|k}\}$ bounding predicted state trajectories and the predicted control law (5.77), the constraints $Fx_{i|k} + Gu_{i|k} \le \mathbf{1}$ at prediction times $i = 0, \ldots, N-1$ can be imposed through the conditions

$$(\tilde{F} + G\tilde{K})^{+}\overline{\alpha}_{i|k} - (\tilde{F} + G\tilde{K})^{-}\underline{\alpha}_{i|k} + Gc_{i|k} \le 1, \tag{5.79}$$

for $i = 0, \ldots, N-1$, where $\tilde{F} = FW$ and $\tilde{K} = KW$. Likewise, the initial and terminal conditions

$$\underline{\alpha}_{0|k} \le Vx_k, \qquad \overline{\alpha}_{0|k} \ge Vx_k \tag{5.80}$$

$$\underline{\alpha}_{N|k} = -\alpha, \qquad \overline{\alpha}_{N|k} = \alpha \tag{5.81}$$

enforce the constraints that $x_k \in \mathcal{X}_{0|k}$ and $\mathcal{X}_{N|k} \subseteq \Pi(V, \alpha)$ (Fig. 5.4).

Consider next the definition of the MPC performance index. Assuming a nominal (rather than a worst case) approach, the predicted cost is defined by (5.29) with $\mathbf{c}_k = (c_{0|k}, \ldots, c_{N-1|k})$. Thus if K is the unconstrained LQ optimal feedback gain for the nominal model, then by Theorem 2.10 we have

$$J(s_{0|k}, \mathbf{c}_k) = \|s_{0|k}\|^2_{W_x} + \|\mathbf{c}_k\|^2_{W_c} \tag{5.82}$$

where $W_c = \mathrm{diag}\{B^{(0)T}W_xB^{(0)} + R, \ldots, B^{(0)T}W_xB^{(0)} + R\}$. For the case that K is not LQ optimal for (5.29), the cost is a quadratic function of $(s_{0|k}, \mathbf{c}_k)$,

$$J(s_{0|k}, \mathbf{c}_k) = \begin{bmatrix} s_{0|k} \\ \mathbf{c}_k \end{bmatrix}^T W \begin{bmatrix} s_{0|k} \\ \mathbf{c}_k \end{bmatrix}$$

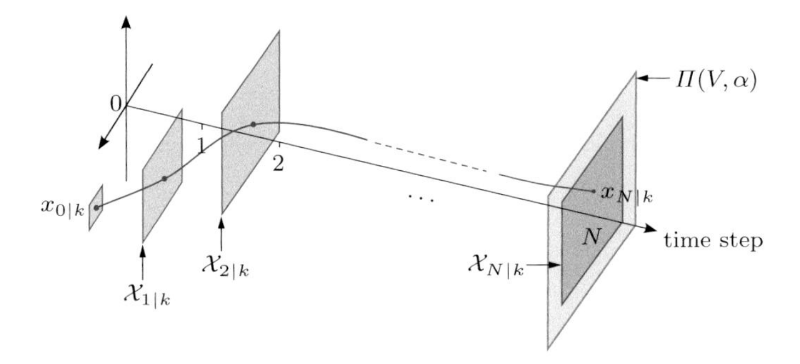

Fig. 5.4 A low-complexity polytopic tube and the predicted evolution of the state for a single realization of the model parameters

where W is the solution of the Lyapunov equation (2.34). Rather than set the initial state $s_{0|k}$ of the nominal cost index equal to x_k as was done in Sect. 5.3, we give here a more general formulation of the algorithm in which $s_{0|k}$ is an optimization variable.

The robust MPC algorithm, which requires the online solution of a QP in $O\big(N(2n_x + n_u)\big)$ variables and $O\big(N(2n_x + n_C)\big)$ inequality constraints, can be stated as follows.

Algorithm 5.6 At each time instant $k = 0, 1, \ldots$:

(i) Perform the optimization:

$$\underset{\substack{s_{0|k},\mathbf{c}_k\\ \underline{\alpha}_{0|k},\ldots,\underline{\alpha}_{N|k}\\ \overline{\alpha}_{0|k},\ldots,\overline{\alpha}_{N|k}}}{\text{minimize}}\ J(s_{0|k}, \mathbf{c}_k) \text{ subject to (5.78)–(5.81) and } \underline{\alpha}_{0|k} \le V s_{0|k} \le \overline{\alpha}_{0|k} \tag{5.83}$$

(ii) Apply the control law $u_k = K x_k + c^*_{0|k}$, where $\mathbf{c}^*_k = (c^*_{0|k}, \ldots, c^*_{N-1|k})$ is the optimal value of $\mathbf{c}_k$ in (5.83). ◁

The construction of the constraints (5.78)–(5.81) ensures recursive feasibility of the online MPC optimization (as demonstrated by Theorem 5.7). Therefore, if the dynamics (5.65) that govern predicted states for $i \ge N$ satisfy the quadratic stability condition (5.30), then the closed-loop system under Algorithm 5.6 can be analysed using Lemma 5.3 and the argument of Theorem 5.3. This approach (which assumes that (5.83) incorporates the additional constraint $s_{0|k} = x_k$) shows that the control law of Algorithm 5.6 asymptotically stabilizes the origin of the state space of (5.1)–(5.3).

However there is no intrinsic requirement in Algorithm 5.6 that the mode 2 prediction dynamics (5.65) should be quadratically stable. Instead it is required that $\Pi(V, \alpha)$ is robustly invariant for these dynamics. This is equivalent to the requirement that $\|V_\alpha x\|_\infty$, where $V_\alpha \doteq (\text{diag}\{\xi_1, \ldots, \xi_{n_x}\})^{-1} V$, is a (piecewise-linear) Lyapunov function for (5.65). Consequently, the additional assumption of quadratic stability may be over-restrictive here, and we therefore use a different approach to analyse stability that does not require an assumption of quadratic stability but is based instead on the robust invariance property of $\Pi(V, \alpha)$. Thus we assume that $\Pi(V, \alpha)$ is a λ-contractive set [27] for (5.65), namely that $\lambda \in [0, 1)$ exists such that

$$\Phi \Pi(V, \alpha) \subseteq \lambda \Pi(V, \alpha), \quad \forall \Phi \in \Omega_K, \tag{5.84}$$

or equivalently

$$\|V_\alpha \Phi^{(j)} x\|_\infty \leq \lambda \|V_\alpha x\|_\infty, \quad j = 1, \ldots, m \tag{5.85}$$

for all $x \in \Pi(V, \alpha) = \{x : \|V_\alpha x\|_\infty \leq 1\}$. Under this assumption, the trajectories of the closed-loop system

$$x_{k+1} = \Phi_k x_k + B_k c^*_{0|k}, \quad (\Phi_k, B_k) \in \mathrm{Co}\big\{(\Phi^{(1)}, B^{(1)}), \ldots, (\Phi^{(m)}, B^{(m)})\big\}, \tag{5.86}$$

satisfy the following bound.

Lemma 5.5 *If (5.85) holds for $\lambda \in [0, 1)$, then the bound*

$$\limsup_{k \to \infty} \|V_\alpha x_k\|_\infty \leq \frac{1}{(1 - \lambda)} \sup_{k \geq 0} \max_j \|V_\alpha B^{(j)} c^*_{0|k}\|_\infty \tag{5.87}$$

holds along trajectories (5.86).

Proof Using the triangle inequality and (5.85), we have, for all $k \geq 0$ along trajectories of (5.86),

$$\|V_\alpha x_{k+1}\|_\infty \leq \lambda \|V_\alpha x_k\|_\infty + \max_j \|V_\alpha B^{(j)} c^*_{0|k}\|_\infty.$$

It follows that

$$\|V_\alpha x_k\|_\infty \leq \lambda^k \|V_\alpha x_0\|_\infty + \sum_{i=0}^{k-1} \lambda^{k-i-1} \max_j \|V_\alpha B^{(j)} c^*_{0|i}\|_\infty,$$

and the bound (5.87) is obtained from this inequality in the limit as $k \to \infty$ since $\max_j \|V_\alpha B^{(j)} c^*_{0|k}\|_\infty \leq \sup_{k \geq 0} \max_j \|V_\alpha B^{(j)} c^*_{0|k}\|_\infty$. □

Theorem 5.7 *The optimization (5.83) is recursively feasible and if $\Pi(V, \alpha)$ satisfies (5.85) for $\lambda \in [0, 1)$, then $x = 0$ is an asymptotically stable equilibrium of the closed-loop system (5.1)–(5.3) under the control law of Algorithm 5.6, with region of attraction equal to the feasible set:*

$$\mathcal{F}_N = \big\{x_k : \exists(\mathbf{c}_k, \underline{\alpha}_{0|k}, \ldots, \underline{\alpha}_{N|k}, \overline{\alpha}_{0|k}, \ldots, \overline{\alpha}_{N|k}) \text{ satisfying (5.78)–(5.81)}\big\}.$$

Proof Given feasibility at time k, a feasible solution at $k+1$ is obtained with $\mathcal{X}_{i|k+1} = \mathcal{X}_{i+1|k}, i = 0, \ldots, N-2, \mathcal{X}_{N-1|k+1} = \mathcal{X}_{N|k+1} = \Pi(V, \alpha)$, and

$$\begin{aligned} s_{0|k+1} &= \Phi^{(0)} s^*_{0|k} + B^{(0)} c^*_{0|k} \\ \mathbf{c}_{k+1} &= (c^*_{1|k}, \ldots, c^*_{N-1|k}, 0) \\ \underline{\alpha}_{i|k+1} &= \underline{\alpha}^*_{i+1|k}, \ i = 0, \ldots, N-1, \quad \underline{\alpha}_{N|k+1} = -\alpha, \\ \overline{\alpha}_{i|k+1} &= \overline{\alpha}^*_{i+1|k}, \ i = 0, \ldots, N-1, \quad \overline{\alpha}_{N|k+1} = \alpha. \end{aligned}$$

The feasibility of this solution follows from the constraints at time k and robust invariance of $\Pi(V, \alpha)$ (which imply that constraints (5.78), (5.79) and (5.81) are necessarily satisfied), and from $x_{1|k} \in \mathcal{X}_{1|k} = \mathcal{X}_{0|k+1}$ (which ensures that (5.80) and $s_{0|k+1} \in \mathcal{X}_{0|k+1}$ are satisfied).

The optimality of the solution of (5.83) at time $k+1$ therefore ensures that the optimal cost satisfies

$$J(s^*_{0|k+1}, \mathbf{c}^*_{k+1}) \leq J(s^*_{0|k}, \mathbf{c}^*_k) - (\|s^*_{0|k}\|^2_Q + \|K s^*_{0|k} + c^*_{0|k}\|^2_R),$$

for all $k \geq 0$, and hence

$$\sum_{k=0}^{\infty} (\|s^*_{0|k}\|^2_Q + \|K s^*_{0|k} + c^*_{0|k}\|^2_R) \leq J(s^*_{0|k}, \mathbf{c}^*_0).$$

Since $Q, R \succ 0$, it follows that $s^*_{0|k} \to 0$ and $c^*_{0|k} \to 0$ as $k \to \infty$, and, for any $\epsilon > 0$ there necessarily exists a finite n such that

$$\|c^*_{0|k}\|_\infty \leq \epsilon \ \ \forall k \geq n.$$

Therefore Lemma 5.5 implies

$$\limsup_{m \to \infty} \|V_\alpha x_{n+m}\|_\infty \leq \frac{\epsilon}{(1-\lambda)} \max_j \|V_\alpha B^{(j)}\|_\infty$$

and since this bound can be made arbitrarily small by choosing sufficiently small ϵ, it follows that $x_k \to 0$ as $k \to \infty$ for all $x_0 \in \mathcal{F}_N$. To complete the proof, we note that Algorithm 5.6 gives $u = Kx$ for all $x \in \Pi(V, \alpha)$ (since (5.82) and the robust invariance of $\Pi(V, \alpha)$ imply that $\mathbf{c} = 0$ is necessarily optimal for (5.83) in this case), and from (5.85) this feedback law is locally exponentially stabilizing for $x = 0$. □

It may be desirable to detune the state feedback gain K with a view to enlarging the terminal set and hence also the size of the region of attraction, and in such cases the justification for the nominal cost of (5.82) will no longer be valid. It is, however, possible to construct a worst-case cost [17] using the l_1-norm of the bounds $\underline{\alpha}_{i|k}, \overline{\alpha}_{i|k}$ $i = 0, \ldots, N$ together with a terminal penalty term that is designed to preserve the monotonic non-increasing property of the optimized cost. The resulting robust MPC law has the same closed-loop properties as Algorithm 5.6 and its online optimization is a linear program as a result of the linear dependence of the cost on the degrees of freedom.

5.5 Tubes with General Complexity Polytopic Cross Sections

The low-complexity polytopic tubes discussed in Sect. 5.4 are convenient from the perspective of online computation but could be unduly conservative. In order to obtain tighter bounds on predicted trajectories, it is possible to remove the restriction to low-complexity polytopes and instead require that the predicted states lie in polytopic tubes with cross sections described by arbitrary but fixed numbers of linear inequalities [18–20, 23]. This modification can be beneficial in terms of the size of the feasible set of initial conditions and closed-loop performance, and it can also simplify offline computation by removing the need for a low-complexity robustly invariant terminal set.

In this section, we consider tubes $\{\mathcal{X}_{0|k}, \mathcal{X}_{1|k}, \ldots\}$ with polytopic cross sections defined in terms of linear inequalities:

$$\mathcal{X}_{i|k} = \{x : Vx \leq \alpha_{i|k}\} \tag{5.88}$$

where $V \in \mathbb{R}^{n_V \times n_x}$ is a full-rank matrix with a number of rows, n_V, that is typically greater than the number, $2n_x$, required to define a low-complexity polytope. The matrix V is to be chosen offline, whereas the parameter $\alpha_{i|k} \in \mathbb{R}^{n_V}$ is retained as a variable online MPC optimization. Since it is expressed as an intersection of half-spaces, $\mathcal{X}_{i|k}$ is necessarily convex; however, unlike a low-complexity polytope, there is no requirement here for $\mathcal{X}_{i|k}$ to be bounded.

As in Sect. 5.4, we assume an open-loop prediction strategy, and hence predicted states and inputs of the models (5.1) and (5.2) evolve according to

$$u_{i|k} = Kx_{i|k} + c_{i|k}, \tag{5.89a}$$

$$x_{i+1|k} = \Phi_{i|k}x_{i|k} + B_{i|k}c_{i|k}, \ (\Phi_{i|k}, B_{i|k}) \in \mathrm{Co}\{(\Phi^{(j)}, B^{(j)}), j = 1, \ldots, m\} \tag{5.89b}$$

for $i = 0, 1, \ldots$ with $c_{i|k} = 0$ for all $i \geq N$. Although the associated predicted state and control trajectories are the same as in Sect. 5.4, the recursive bounding approach of Sect. 5.4 is no longer applicable. Essentially, this is because closed-form expressions for the extreme points of $\mathcal{X}_{i|k}$ are not generally available when the set is parameterized, as in (5.88), by an intersection of half-spaces. Instead we use the following result based on Farkas' Lemma [31, 32], to express polytopic set inclusion conditions in terms of algebraic conditions.

Lemma 5.6 *Let $\mathcal{S}_i \doteq \{x : F_i x \leq b_i\}$, $i = 1, 2$, be non-empty subsets of $\mathbb{R}^{n_x}$. Then $\mathcal{S}_1 \subseteq \mathcal{S}_2$ if and only if there exists a nonnegative matrix H satisfying*

$$HF_1 = F_2 \tag{5.90a}$$

$$Hb_1 \leq b_2 \tag{5.90b}$$

Proof To show that the conditions (5.90a, 5.90b) are sufficient for $\mathcal{S}_1 \subseteq \mathcal{S}_2$, suppose that x satisfies $F_1 x \le b_1$. Then $HF_1 x \le Hb_1$ since $H \ge 0$, so (5.90a, 5.90b) imply $F_2 x \le b_2$ and it follows that $x \in \mathcal{S}_2$ for all $x \in \mathcal{S}_1$.

To show the necessity of (5.90a, 5.90b), assume $\mathcal{S}_1 \subseteq \mathcal{S}_2$. Then we must have $\mu \le b_2$ where, for each i, the ith element of μ is defined by

$$\mu_i \doteq \max_{x \in \mathbb{R}^{n_x}} (F_2)_i x \ \text{ subject to } \ F_1 x \le b_1$$

with $(F_2)_i$ denoting the ith row of F_2. The optimal value of this linear program (which is feasible by assumption) is equal to the optimal value of the dual problem [33]:

$$\mu_i = \min_{h \in \mathbb{R}^{n_1}} b_1^T h \ \text{ subject to } \ h^T F_1 = (F_2)_i \text{ and } h \ge 0 \tag{5.91}$$

where n_1 is the number of rows of F_1. Let h_i^* be the optimal solution of this linear program and let H be the matrix with ith row equal to h_i^{*T}. Then (5.91) implies that $H \ge 0$ exists satisfying (5.90a, 5.90b) whenever $\mathcal{S}_1 \subseteq \mathcal{S}_2$. □

From (5.89b) and Lemma 5.6, we have $x_{i+1|k} \in \mathcal{X}_{i+1|k}$ for all $x_{i|k} \in \mathcal{X}_{i|k}$ and $i = 0, \ldots, N-1$ if there exist matrices $H^{(j)} \ge 0$, $j = 1, \ldots, m$ satisfying

$$\alpha_{i+1|k} \ge H^{(j)} \alpha_{i|k} + V B^{(j)} c_{i|k}, \ \ i = 0, \ldots, N-1 \tag{5.92a}$$

$$H^{(j)} V = V \Phi^{(j)} \tag{5.92b}$$

for $j = 1, \ldots, m$. Similarly, (5.89a) and Lemma 5.6 imply that the constraint $Fx_{i|k} + Gu_{i|k} \le \mathbf{1}$ is satisfied for all $x_{i|k} \in \mathcal{X}_{i|k}$, $i = 0, \ldots, N-1$, if there exists a matrix $H_c \ge 0$ satisfying

$$H_c \alpha_{i|k} + G c_{i|k} \le \mathbf{1}, \ \ i = 0, \ldots, N-1 \tag{5.93a}$$

$$H_c V = F + GK. \tag{5.93b}$$

The constraints (5.92a) and (5.93a) are nonlinear if $H^{(j)}$ and H_c are treated as variables concurrently with $\alpha_{i|k}$. However these conditions become linear constraints in the online MPC optimization if $H^{(j)}$ and H_c are determined offline. This has the effect of making (5.92a) and (5.93a) only sufficient (not necessary) to ensure $x_{i+1|k} \in \mathcal{X}_{i+1|k}$ and $Fx_{i|k} + Gc_{i|k} \le \mathbf{1}$ for all $x_{i|k} \in \mathcal{X}_{i|k}$. Therefore, to relax the constraints on $\alpha_{i|k}$, a convenient criterion for the offline design of $H^{(j)}$ and H_c is to minimize the sum of the elements in each row of these matrices subject to (5.92b) and (5.93b). This suggests computing the rows, $(H_c)_i$, of H_c for $i = 1, \ldots, n_C$ offline by solving the linear programs

$$h_i^* = \arg\min_{h \in \mathbb{R}^{n_V}} \mathbf{1}^T h \ \text{ subject to } \ h^T V = (F + GK)_i \text{ and } h \ge 0 \tag{5.94}$$

and setting $(H_c)_i = h_i^{*T}$, where $(H_c)_i$ and $(F + GK)_i$ denote the ith rows of H_c and $F + GK$, respectively. Similarly, each $H^{(j)}$, for $j = 1, \ldots, m$, can be determined offline by solving the linear programs

$$h_i^{(j)*} = \arg \min_{h \in \mathbb{R}^{n_V}} \mathbf{1}^T h \text{ subject to } h^T V = V_i \Phi^{(j)} \text{ and } h \geq 0 \tag{5.95}$$

for $i = 1, \ldots, n_V$ and setting $H_i^{(j)} = h_i^{(j)*^T}$, where $H_i^{(j)}$ and V_i denote, respectively, the ith rows of $H^{(j)}$ and V.[3]

The constraints $Fx_{i|k} + Gc_{i|k} \leq \mathbf{1}$ can be imposed for $i \geq N$ through terminal constraints on $\alpha_{N|k}$. From (5.92a) and (5.93a) with $c_{i|k} = 0$, the required conditions are given by

$$\alpha_{i+1|k} \geq H^{(j)} \alpha_{i|k}, \tag{5.96a}$$

$$H_c \alpha_{i|k} \leq \mathbf{1} \tag{5.96b}$$

for $j = 1, \ldots, m$ and $i = N, N+1, \ldots$. Although these conditions are expressed in terms of inequalities, the fact that H_c and $H^{(j)}$ are nonnegative matrices implies that (5.96a, 5.96b) are feasible for all $i = N, N+1, \ldots$ if and only if they are feasible for the unique trajectory $\{\alpha_{N|k}, \alpha_{N+1|k}, \ldots\}$ defined by the piecewise-linear dynamics

$$(\alpha_{i+1|k})_l = \max_{j \in \{1, \ldots, m\}} H_l^{(j)} \alpha_{i|k}, \quad l = 1, \ldots, n_V, \tag{5.97}$$

where $(\alpha_{i|k})_l$ denotes the lth element of $\alpha_{i|k}$. The following result shows that these dynamics are stable if V is chosen so that the set $\{x : Vx \leq \mathbf{1}\}$ is contractive for the mode 2 prediction dynamics defined in (5.65).

Lemma 5.7 *If $\{x : Vx \leq \mathbf{1}\}$ is λ-contractive for some $\lambda \in [0, 1)$ under the dynamics (5.65), then (5.97) satisfies $\|\alpha_{i+1|k}\|_\infty \leq \lambda \|\alpha_{i|k}\|_\infty$, for all i.*

Proof If $\mathcal{S} = \{x : Vx \leq \mathbf{1}\}$ is λ-contractive, namely if $\Phi^{(j)} \mathcal{S} \subseteq \lambda \mathcal{S}$ for all $j = 1, \ldots, m$, then Lemma 5.6 implies $\|H^{(j)}\|_\infty \leq \lambda$. □

---

[3] With H_c and $H^{(j)}$, $j = 1, \ldots, m$ chosen so as to minimize sum of elements in each of their rows, the conditions (5.92) and (5.93) include the corresponding conditions that were derived in Sect. 5.4.2 for low-complexity polytopes as a special case. Thus, expressing the low-complexity polytopic set $\{x : \underline{\alpha} \leq V_0 x \leq \overline{\alpha}\}$ equivalently as $\{x : Vx \leq \alpha\}$ with $V = [V_0^T \; -V_0^T]^T$ and $\alpha = [\overline{\alpha}^T \; -\underline{\alpha}^T]^T$, the solutions of (5.94) and (5.95) can be obtained in closed form as

$$H_c = [(\tilde{F} + G\tilde{K})^+ \; (\tilde{F} + G\tilde{K})^-] \text{ and } H^{(j)} = \begin{bmatrix} (\tilde{\Phi}^{(j)})^+ & (\tilde{\Phi}^{(j)})^- \\ (\tilde{\Phi}^{(j)})^- & (\tilde{\Phi}^{(j)})^+ \end{bmatrix}, \; j = 1, \ldots, m.$$

Therefore conditions (5.78) and (5.79) are identical to (5.92) and (5.93) for this case.

To meet the requirement in Lemma 5.7 that the set $\{x : Vx \leq \mathbf{1}\}$ is λ-contractive for $\lambda \in [0, 1)$, V can be chosen, for example, so that this set is the maximal robust invariant set for the system $x_{k+1} = (1/\lambda)\Phi x_k$, $\Phi \in \Omega_K$ subject to $(F + GK)x_k \leq \mathbf{1}$. For any λ in the interval $(\rho, 1)$, where ρ is the joint spectral radius defined in (5.72), it can be shown that this maximal robust invariant set is defined by a set of linear inequalities of the form

$$\begin{aligned}\mathcal{X}^{(n)} = \{x : (F + GK)x \leq \mathbf{1}, (F + GK)\Phi^{(j_1)}x \\ \leq \lambda\mathbf{1}, \ldots, (F + GK)\Phi^{(j_1)} \cdots \Phi^{(j_n)}x \leq \lambda^n\mathbf{1}, \ j_i = 1, \ldots, m, \ i = 1, \ldots, n\}\end{aligned} \tag{5.98}$$

where n is necessarily finite if the pair $\big(\Phi, (F + GK)\big)$ is observable for some $\Phi \in \Omega_K$. Rather than prove this result (details of which can be found in [34]), we consider next the related problem of determining the maximal positively invariant set for (5.97) contained in the set on which $H_c\alpha \leq \mathbf{1}$.

Under the assumption that the set $\{x : Vx \leq \mathbf{1}\}$ is λ-contractive, which by Lemma 5.7 implies that the system describing the evolution of $\alpha_{i|k}$ in (5.97) is asymptotically stable, the constraint that $H_c\alpha_{i|k} \leq \mathbf{1}$ for all $i \geq N$ can be ensured by imposing a finite set of linear constraints on $\alpha_{N|k}$. To see this, let $\mathcal{A}^{(n)}$ denote the set

$$\begin{aligned}\mathcal{A}^{(n)} \doteq \big\{\alpha : H_c\alpha \leq \mathbf{1}, \ H_cH^{(j_1)}\alpha \leq \mathbf{1}, \ \ldots, \ H_cH^{(j_1)} \cdots H^{(j_n)}\alpha \leq \mathbf{1}, \\ j_i = 1, \ldots, m, \ i = 1, \ldots, n\big\},\end{aligned}$$

then the maximal positively invariant set defined by

$$\mathcal{A}^{\mathrm{MPI}} \doteq \{\alpha_{N|k} : \text{ (5.96b), and (5.97) hold for } i = N, N + 1, \ldots\},$$

is given by $\mathcal{A}^{\mathrm{MPI}} = \lim_{n\to\infty} \mathcal{A}^{(n)}$. The following result, which is an extension of Theorems 2.3 and 3.1, gives a characterisation of $\mathcal{A}^{\mathrm{MPI}}$ in terms of a finite number of linear conditions.

Corollary 5.3 *The maximal positively invariant set for the system (5.97) and constraints (5.96b) is given by*

$$\mathcal{A}^{\mathrm{MPI}} = \mathcal{A}^{(\nu)}$$

where ν is the smallest integer such that $\mathcal{A}^{(\nu)} \subseteq \mathcal{A}^{(\nu+1)}$. If $\mathcal{A}^{\mathrm{MPI}}$ is bounded and $\{x : Vx \leq \mathbf{1}\}$ *is λ-contractive for* $\lambda \in [0, 1)$ *under (5.65), then n is necessarily finite.*

Proof If $\mathcal{A}^{(\nu)} \subseteq \mathcal{A}^{(\nu+1)}$, then $\mathcal{A}^{(\nu)}$ is necessarily invariant under (5.97) and is therefore a subset of $\mathcal{A}^{\mathrm{MPI}}$ in this case. But $\mathcal{A}^{\mathrm{MPI}}$ is by definition a subset of $\mathcal{A}^{(n)}$ for all n and it follows that $\mathcal{A}^{(\nu)} = \mathcal{A}^{\mathrm{MPI}}$.

Furthermore, if $\mathcal{A}^{\mathrm{MPI}}$ is bounded, then $\mathcal{A}^{(n)}$ must also be bounded for some n. From the definition of $\mathcal{A}^{(n)}$ and Lemma 5.7, we have

$$\begin{aligned} \mathcal{A}^{(n+1)} &= \mathcal{A}^{(n)} \cap \big\{\alpha : H_c H^{(j_1)} \cdots H^{(j_{n+1})}\alpha \leq \mathbf{1},\ j_i = 1, \ldots, m, \\ &\qquad\qquad\qquad\qquad\qquad\qquad\qquad\qquad i = 1, \ldots, n+1\big\} \\ &\supseteq \mathcal{A}^{(n)} \cap \{\alpha : \|H_c\|_\infty \lambda^{n+1} \|\alpha\|_\infty \leq 1\}, \end{aligned}$$

and $\lambda \in [0, 1)$ therefore implies $\mathcal{A}^{(n+1)} \supseteq \mathcal{A}^{(n)}$ for some finite n. □

Note that the assertion that $\mathcal{A}^{\mathrm{MPI}}$ is bounded, which is used in Corollary 5.3 to ensure that the maximal positively invariant set for (5.97) and (5.96b) is finitely determined, can always be assumed to hold in practice. This is because unboundedness of $\mathcal{A}^{\mathrm{MPI}}$ would indicate that some of the rows of V are redundant. Therefore, these rows could be removed without affecting the set $\{x : Vx \leq \alpha\}$ for $\alpha \in \mathcal{A}^{\mathrm{MPI}}$.

Lemma 5.7 implies that a small value of λ will make the trajectories of (5.97) converge more rapidly, but choosing λ close to the joint spectral radius ρ could require a large number of rows in V such that $\{x : Vx \leq \mathbf{1}\}$ is λ-contractive. Conversely, larger values of λ typically result in smaller values of n_V but at the same time allow the tube cross sections to grow more rapidly along predicted trajectories, and this can cause the set of feasible initial conditions of an associated MPC law to increase more slowly with the prediction horizon N. The design of λ is discussed further in Example 5.4.

From the preceding discussion, it follows immediately that the predicted trajectories of (5.89a, 5.89b) necessarily satisfy the constraints $Fx_{i|k} + Gu_{i|k} \leq \mathbf{1}$ for all $i \geq 0$ and for all realizations of model uncertainty if $\{\alpha_{0|k}, \ldots, \alpha_{N|k}\}$ satisfy the initial and terminal conditions

$$Vx_k \leq \alpha_{0|k} \tag{5.99}$$

$$\alpha_{N|k} \in \mathcal{A}^{\mathrm{MPI}} \tag{5.100}$$

in addition to the conditions for inclusion (5.92a) and feasibility (5.93a) for $i = 0, \ldots, N-1$. On the basis of these constraints and using the nominal cost (5.82) discussed in Sect. 5.4.2, a robust MPC algorithm can be stated as follows. The associated online optimization is a QP with $O(N(n_x + n_u))$ variables and $O(N(n_x + n_C) + m^\nu)$ constraints.

Algorithm 5.7 At each time instant $k = 0, 1, \ldots$:

(i) Perform the optimization:

$$\underset{\substack{s_{0|k}, \mathbf{c}_k \\ \alpha_{0|k}, \ldots, \alpha_{N|k}}}{\text{minimize}}\ J(s_{0|k}, \mathbf{c}_k) \ \text{subject to} \ \ (5.92a), (5.93a), (5.99), (5.100),$$

$$\text{and } Vs_{0|k} \leq \alpha_{0|k} \tag{5.101}$$

(ii) Apply the control law $u_k = Kx_k + c^*_{0|k}$, where $\mathbf{c}^*_k = (c^*_{0|k}, \ldots, c^*_{N-1|k})$ is the optimal value of $\mathbf{c}_k$ in (5.101). ◁

Theorem 5.8 *If $\{x : Vx \leq \mathbf{1}\}$ is λ-contractive for some $\lambda \in [0, 1)$, then the optimization (5.101) is recursively feasible and the control law of Algorithm 5.7 asymptotically stabilizes the origin of the system (5.1)–(5.3), with a region of attraction equal to the feasible set:*

$$\mathcal{F}_N = \big\{x_k : \exists(\mathbf{c}_k, \alpha_{0|k}, \ldots, \alpha_{N|k}) \text{ satisfying (5.92a), (5.93a), (5.99), (5.100)}\big\}.$$

Proof The proof of this result closely follows that of Theorem 5.7, and we therefore provide only a sketch of the argument here. The recursive feasibility of (5.101) follows from the feasibility of the solution at time $k+1$ given by

$$\begin{aligned} s_{0|k+1} &= \Phi^{(0)} s^*_{0|k} + B^{(0)} c^*_{0|k} \\ \mathbf{c}_{k+1} &= (c^*_{1|k}, \ldots, c^*_{N-1|k}, 0) \\ \alpha_{i|k+1} &= \alpha^*_{i+1|k}, \quad i = 0, \ldots, N-1, \\ (\alpha_{N|k+1})_l &= \max_{j \in \{1,\ldots,m\}} H^{(j)}_l \alpha^*_{N|k}, \quad l = 1, \ldots, n_V \end{aligned}$$

since with these parameters we obtain $\mathcal{X}_{i|k+1} = \mathcal{X}_{i+1|k}$ for $i = 0, \ldots, N-1$ and $\alpha_{N|k+1} \in \mathcal{A}^{\text{MPI}}$ (since $\alpha^*_{N|k} \in \mathcal{A}^{\text{MPI}}$). Thus at time $k+1$, (5.92a) and (5.93a) hold for $i = 0, \ldots, N-1$ and likewise (5.100) holds, whereas (5.99) and $V s_{0|k+1} \leq \mathbf{1}$ are satisfied at time $k+1$ because $x_{k+1} \in \mathcal{X}_{1|k}$ for all realizations of model uncertainty at time k.

Asymptotic convergence can be shown using the bound

$$\limsup_{k\to\infty} \max\{V x_k\} \leq \frac{1}{(1-\lambda)} \sup_{k\geq 0} \max_j \max\{V B^{(j)} c^*_{0|k}\} \tag{5.102}$$

(where $\max\{\cdot\}$ indicates the maximum element of a vector). This follows, analogously to Lemma 5.5, from the assumption that $\{x : Vx \leq \mathbf{1}\}$ is λ-contractive for $\lambda \in [0, 1)$, and hence

$$\begin{aligned} \max\{V x_{k+1}\} &\leq \max\{V \Phi_k x_k\} + \max\{V B_k c^*_{0|k}\} \\ &\leq \lambda \max\{V x_k\} + \max\{V B_k c^*_{0|k}\} \end{aligned}$$

along the trajectories of the closed-loop system. Asymptotic convergence, $x_k \to 0$, then follows from the argument of the proof of Theorem 5.7 applied to (5.102), whereas stability of $x = 0$ follows from local stability under $u = Kx$ and the property that $\mathbf{c}^*_k = 0$ if x_k is sufficiently close to zero. □

Computing offline the matrices $H^{(j)}$ and H_c that appear in the constraints of the online MPC optimization (5.101) as opposed to retaining these matrices as variables in the optimization of predicted performance at each time step could make the handling of constraints conservative. The following corollary of Lemma 5.6

gives conditions under which computing $H^{(j)}$ and H_c offline incurs no conservativeness.

Corollary 5.4 *For given $\mathcal{F}_i \in \mathbb{R}^{n_i \times n_x}$ and $b_i \in \mathbb{R}^{n_i}$, let $\mathcal{S}_i = \{x : F_i x \leq b_i\}$, $i = 1, 2$. If there exists $H \geq 0$ satisfying $HF_1 = F_2$ such that each row of H has only one non-zero element, then $\mathcal{S}_1 \subseteq \mathcal{S}_2$ if and only if $Hb_1 \leq b_2$.*

Proof A sufficient condition for $\mathcal{S}_1 \subseteq \mathcal{S}_2$ is $Hb_1 \leq b_2$ since then $H \geq 0$ and $HF_1 = F_2$ imply that $F_2 x \leq b_2$ whenever $F_1 x \leq b_1$. To show that $Hb_1 \leq b_2$ is also necessary if each row of H contains only one non-zero element, we first note that $\{x : F_1 x \leq b_1\}$ can be assumed without loss of generality to be an irreducible representation of $\mathcal{S}_1$, i.e. for each $i \in \{1, \ldots, n_1\}$ there exists $x \in \mathcal{S}_1$ such that $(F_1)_i x = (b_1)_i$. Suppose that the jth element of the ith row of H is non-zero and choose $x \in \mathcal{S}_1$ so that $(F_1)_j x = (b_1)_j$. Then $HF_1 = F_2$ implies $(F_2)_i x = H_i b_1$, where H_i is the ith row of H. Hence $H_i b_1 \leq (b_2)_i$ is needed in order that $x \in \mathcal{S}_2$; repeating this argument for each $i = 1, \ldots, n_2$ shows that $Hb_1 \leq b_2$ is necessary for $\mathcal{S}_1 \subseteq \mathcal{S}_2$. □

If V is chosen so that $\{x : Vx \leq \mathbf{1}\}$ is the maximal RPI set for the dynamics $x_{k+1} = (1/\lambda)\Phi x_k$, $\Phi \in \Omega_K$, with $(F + GK)x_k \leq \mathbf{1}$, then from (5.98), V has the form $V = [(F + GK)^T \; V'^T]^T$ for some V'. Therefore it is always possible to choose $H_c \geq 0$ satisfying (5.93b) so that $H_c = [I_{n_C} \; 0]$. Hence, by Corollary 5.4, the conditions (5.93a) in the online optimization (5.101) are necessary as well as sufficient for satisfaction of the constraints $Fx_{i|k} + Gc_{i|k} \leq \mathbf{1}$ for all $x_{i|k} \in \mathcal{X}_{i|k}, i = 0, \ldots, N-1$ in this case.

The linear dependence of conditions (5.92a), (5.93a), (5.99) and (5.100) on the variables $\{\alpha_{0|k}, \ldots, \alpha_{N|k}\}$ allows the MPC optimization (5.101) to be formulated as a QP problem. However defining the terminal condition in terms of the maximal MPI set $\mathcal{A}^{\text{MPI}}$ may introduce a large number of constraints into the online optimization. At worst this terminal constraint contributes $O(m^\nu)$ linear constraints to (5.101), and although removing redundant constraints reduces this number substantially (see for example the related approach of [35]), the possibility of rapid growth of the number of constraints with m may limit the usefulness of $\mathcal{A}^{\text{MPI}}$ as a terminal set if either the number, m, of vertices of the uncertain model or the value of ν in Corollary 5.3 is large.

To reduce the number of constraints in the online MPC optimization (5.101), it is possible to define the terminal constraint using an invariant set for (5.97) and (5.93b) that is not necessarily maximal. Provided this terminal set is invariant for the dynamics (5.97), the guarantee of recursive feasibility of the MPC optimization in Theorem 5.8 will not be affected. A simple alternative is to replace (5.100) with the terminal constraints

$$\alpha_{N|k} \geq H^{(j)} \alpha_{N|k}, \quad j = 1, \ldots, m \tag{5.103a}$$

$$H_c \alpha_{N|k} \leq \mathbf{1}. \tag{5.103b}$$

These constraints are necessarily feasible for some $\alpha_{N|k}$ because of the conditions on H_c and $H^{(j)}$ in (5.92b) and (5.93b) and the asymptotic stability property of the system (5.97). To lessen the impact of this modification, it is possible to invoke the terminal constraint on $\alpha_{T|k}$ by introducing a supplementary horizon of $T - N$ and invoking the constraints (5.96a, 5.96b) for $i = N, \ldots, T - N - 1$ [19, 20]. For further details, see Question 9 on p. 237.

Finally we note that Algorithm 5.7 is based on a nominal predicted cost, but it could be reformulated as a min–max robust MPC algorithm using the worst-case cost discussed in Sect. 5.3.1. For example, if $\check{W}$ satisfies the LMIs, for $j = 1, \ldots, m$,

$$\check{W} - \Psi^{(j)T} \check{W} \Psi^{(j)} \succeq \begin{bmatrix} I & K^T \\ 0 & E^T \end{bmatrix} \begin{bmatrix} Q & 0 \\ 0 & R \end{bmatrix} \begin{bmatrix} I & 0 \\ K & E \end{bmatrix}, \quad \Psi^{(j)} = \begin{bmatrix} \Phi^{(j)} & B^{(j)}E \\ 0 & M \end{bmatrix},$$

then $\|(x_k, \mathbf{c}_k)\|^2_{\check{W}}$ is an upper bound on the worst-case cost:

$$\left\| \begin{bmatrix} x_k \\ \mathbf{c}_k \end{bmatrix} \right\|^2_{\check{W}} \geq \check{J}(x_k, \mathbf{c}_k) \doteq \max_{(A_{i|k}, B_{i|k}) \in \Omega,\, i=0,1,\ldots} \sum_{i=0}^{\infty} \bigl(\|x_{i|k}\|^2_Q + \|u_{i|k}\|^2_R \bigr), \quad (5.104)$$

and replacing the objective of (5.101) with $\|(x_k, \mathbf{c}_k)\|^2_{\check{W}}$ converts Algorithm 5.7 to a min–max robust MPC requiring a QP optimization online. The implied control law can be shown to be asymptotically stabilizing by demonstrating, similarly to the proof of Theorem 5.5, that $\|(x_k, \mathbf{c}_k)\|^2_{\check{W}}$ is a Lyapunov function for the closed-loop system.

Example 5.4 For the system (5.1)–(5.3) with the model parameters given in Example 5.1 and $K = [0.19\ 0.34]$, the joint spectral radius of the system $x_{k+1} = \Phi x_k$, $\Phi \in \Omega_K$, is $\rho = 0.7415$. For values of λ in the interval $(\rho, 1)$, Table 5.1 gives numbers of rows, n_V, of the matrix V such that $\{x : Vx \leq \mathbf{1}\}$ is the maximal RPI set for the dynamics $x_{k+1} = (1/\lambda)\Phi x_k$, $\Phi \in \Omega_K$ and constraints $(F + GK)x_k \leq \mathbf{1}$. The table also shows the number of terminal constraints associated with either $\mathcal{A}^{\text{MPI}}$ or the alternative conditions (5.103a, 5.103b) when the tube cross sections are given by $\mathcal{X}_{i|k} = \{x : Vx \leq \alpha_{i|k}\}$. For all values of λ in this range and with either of these definitions of the terminal constraints, the terminal set (namely the set $x_{N|k}$ such that there exists $\alpha_{N|k}$ satisfying the terminal constraints and $Vx_{N|k} \leq \alpha_{N|k}$) is equal to the maximal RPI set under $u = Kx$ (Fig. 5.5).

From Table 5.1, it can be seen that, as expected, n_V increases as λ is reduced. This causes the required number of variables and constraints in the MPC online

Table 5.1 Number of facets of tube cross sections and number of terminal constraints for varying λ

λ	Tube cross section n_V	Terminal constraints $\mathcal{A}^{\text{MPI}}$	Terminal constraints (5.103a, 5.103b)
0.999	6	12	22
0.9	8	20	28
0.8	12	60	40
0.742	28	808	88

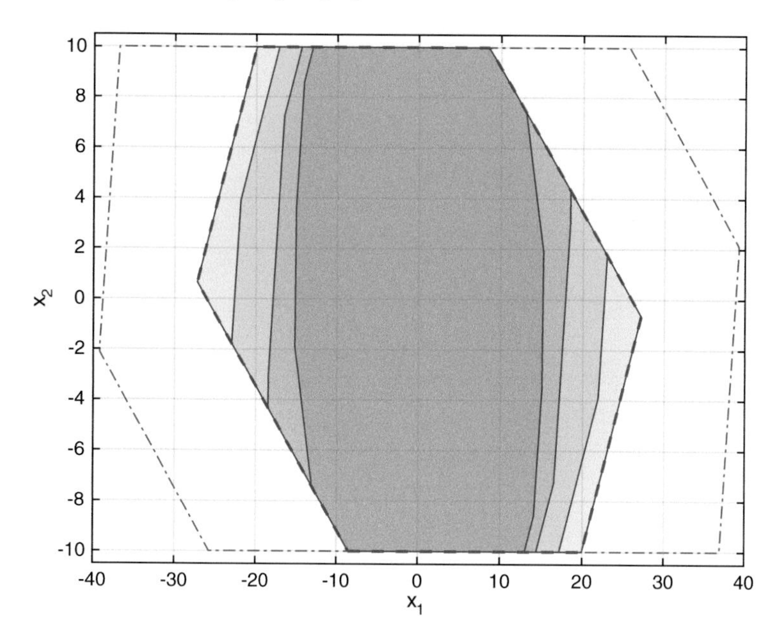

Fig. 5.5 The λ-contractive sets $\{x : Vx \leq \mathbf{1}\}$ for $\lambda = 0.742, 0.8, 0.9, 0.999$ (*solid lines*) and the corresponding terminal sets defined either by $\mathcal{A}^{\text{MPI}}$ or (5.103a, 5.103b) (*dashed line*). The maximal robustly controlled invariant set is also shown (*dash-dotted line*)

optimization (5.101) to increase, and in order to reduce online computation it is therefore desirable to choose λ so as to avoid large values of n_V. But for values of λ close to unity the set of feasible initial conditions for Algorithm 5.7 grows slowly with N, and for this example a good compromise is obtained with $\lambda = 0.9$.

The sets of feasible initial conditions, $\mathcal{F}_N$, are shown in Fig. 5.6 for Algorithm 5.7 with $\lambda = 0.9$ and with terminal constraints defined by $\mathcal{A}^{\text{MPI}}$. The feasible set $\mathcal{F}_4$ is equal to the maximal robustly controlled positively invariant (CPI) set for this example, which has an area of 1415. Hence, there can be no further increase in $\mathcal{F}_N$ for $N > 4$. For comparison, Fig. 5.7 shows the feasible sets for Algorithm 5.6 with tube cross sections and a terminal set defined by the low-complexity polytope given in Example 5.3. Using low-complexity tubes results in smaller numbers of variables and constraints in the online MPC optimization (Table 5.2), but for this example there is no increase in the feasible set for Algorithm 5.6 for $N > 5$, which implies that Algorithm 5.6 remains infeasible irrespective of the value of N for some initial conditions in the robustly CPI set.

The use of low-complexity polytopic tubes has a small but appreciable effect on performance. For a set of initial conditions consisting of 34 points lying on the boundary of $\mathcal{F}_5$ for Algorithm 5.6, the worst-case predicted cost of Algorithm 5.6 with $N = 5$ is on average 0.8 % (and at most 1.6 %) greater than the worst-case predicted cost of Algorithm 5.7 with $N = 4$. ◊

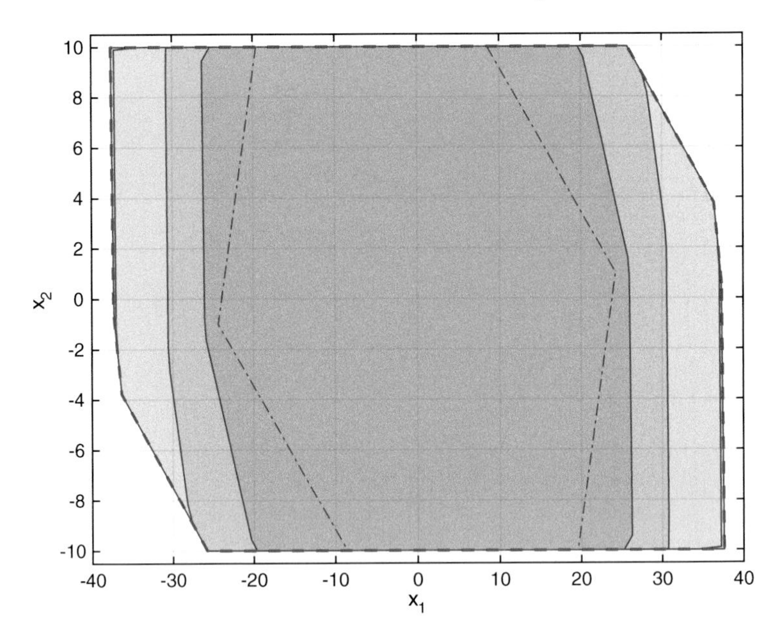

Fig. 5.6 The feasible sets, $\mathcal{F}_N$ for Algorithm 5.7 with $N = 1, 2, 3, 4$ (*solid lines*), and the maximal robustly controlled invariant set (*dashed line*). The terminal set is shown by the *dash-dotted lines*

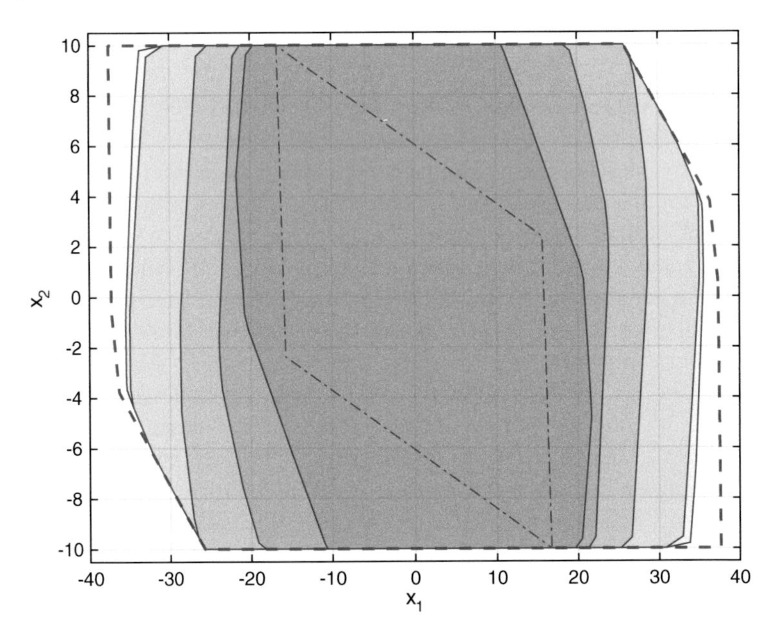

Fig. 5.7 The feasible sets, $\mathcal{F}_N$ for Algorithm 5.6 with $N = 1, 2, 3, 4, 5$ (*solid lines*), and the maximal robustly controlled invariant set (*dashed line*). The terminal set is indicated by the *dash-dotted lines*

Table 5.2 Number of variables, constraints and area of feasible region for the general complexity and low-complexity polytopic tube MPC strategies of Algorithms 5.6 and 5.7

N	General complexity tube			Low-complexity tube		
	# variables	# constraints	Area of $\mathcal{F}_N$	# variables	# constraints	Area of $\mathcal{F}_N$
1	17	56	989	9	24	747
2	26	84	1198	14	40	902
3	35	112	1407	19	56	1110
4	44	140	1415	24	72	1324
5				29	88	1345

5.6 Mixed Additive and Multiplicative Uncertainty

The results presented in Sects. 5.4 and 5.5 allow for the definition of polytopic tubes that contain the predicted state and control trajectories of an uncertain system for all future realizations of model uncertainty. Such tubes provide a systematic means of handling constraints and, crucially, the complexity of their cross sections can be controlled by the designer and does not depend on the length of prediction horizon. Sections 5.4 and 5.5 considered multiplicative model uncertainty only but the method can easily be extended to cater for the more general case of mixed multiplicative and additive uncertainty. This section considers the model (5.5) with dynamics $x_{k+1} = A_k x_k + B_k u_k + D_k w_k$ containing both multiplicative parametric uncertainty and unknown additive disturbances. Here (A_k, B_k, D_k) belong for all k to a compact polytopic parameter set $\tilde{\Omega}$ and w_k is confined for all k to a compact polytopic set $\mathcal{W}$ containing $w = 0$:

$$(A_k, B_k, D_k) = \sum_{j=1}^{m} q_k^{(j)} (A^{(j)}, B^{(j)}, D^{(j)}), \quad w_k = \sum_{l=1}^{q} r_k^{(l)} w^{(l)} \tag{5.105a}$$

$$q_k^{(j)}, r_k^{(l)} \geq 0, \quad \sum_{j=1}^{m} q_k^{(j)} = 1, \quad \sum_{l=1}^{q} r_k^{(l)} = 1 \tag{5.105b}$$

where the superscripts j and l are used to denote, respectively, multiplicative and additive uncertainty vertices.

The approach of Sect. 5.5 can be extended to the case of mixed model uncertainty by modifying the conditions (5.92a) and (5.96a) defining the evolution of tubes that bound the predicted states over the mode 1 and mode 2 prediction horizons, respectively. However we discuss here a more general approach that uses optimized controller dynamics to design the mode 2 prediction dynamics for the case of mixed uncertainty [21–23]. The use of general polytopic tubes in this context results in an approach that subsumes the method of Sect. 5.5 as a special case and allows for a terminal set larger than the maximal robustly positively invariant set under any given linear feedback law.

The optimization of the mode 2 prediction dynamics is aimed at maximizing the volume of an invariant terminal set, and we therefore introduce further degrees of freedom over the mode 1 prediction horizon for the purposes of improving performance and increasing the region of attraction. Hence the predicted control strategy is of the form

$$u_{i|k} = Kx_{i|k} + C_c\mathbf{c}_{i|k} + f_{i|k}, \tag{5.106a}$$

$$\mathbf{c}_{i+1|k} \in \text{Co}\{A_c^{(j)}\mathbf{c}_{i|k} + C_w^{(j)}w_{i|k},\ j = 1, \dots, m\} \tag{5.106b}$$

with $f_{i|k} = 0$ for all $i \geq N$. At each time instant $k \geq 0$, $\mathbf{c}_k \doteq \mathbf{c}_{0|k} \in \mathbb{R}^{n_x}$ and $\mathbf{f}_k \doteq (f_{0|k}, \dots, f_{N-1|k}) \in \mathbb{R}^{Nn_u}$ are variables in the online MPC optimization. We note that the disturbance affine term in (5.106b) provides feedback from the future disturbances which, although unknown when the predictions are optimized at time k, will be available to the controller at time $k+i$, and (5.106a, 5.106b) therefore constitutes a closed-loop prediction strategy.

In this setting, the optimized controller dynamics are introduced at every time step of the entire prediction horizon, with C_c, $A_c^{(j)}$, $C_w^{(j)}$ chosen so as to maximize the volume of a robustly invariant ellipsoidal terminal set. For $i \geq N$, we use the following description of the mode 2 prediction dynamics corresponding to (5.5) under the terminal control law $u_{i|k} = Kx_{i|k} + C_c\mathbf{c}_{i|k}$:

$$z_{i+1|k} = \Psi_{i|k}z_{i|k} + \tilde{D}_{i|k}w_{i|k}, \quad z_{N|k} = \begin{bmatrix} x_{N|k} \\ \mathbf{c}_{N|k} \end{bmatrix} \tag{5.107}$$

where

$$(\Psi_{i|k}, \tilde{D}_{i|k}) \in \text{Co}\big\{(\Psi^{(j)}, \tilde{D}^{(j)}),\ j = 1, \dots, m\big\}$$

and, for $j = 1, \dots, m$,

$$\Psi^{(j)} = \begin{bmatrix} \Phi^{(j)} & B^{(j)}C_c \\ 0 & A_c^{(j)} \end{bmatrix}, \quad \tilde{D}^{(j)} = \begin{bmatrix} D^{(j)} \\ C_w^{(j)} \end{bmatrix}.$$

Theorem 5.9 *The ellipsoidal set $\mathcal{E}_z = \{z : z^T P_z z \leq 1\}$ is robustly invariant for the dynamics (5.107) with the constraint (5.3) if and only if there exists a positive definite matrix P_z and a nonnegative scalar σ such that*

$$\begin{bmatrix} P_z & P_z\Psi^{(j)} & P_z\tilde{D}^{(j)}w^{(l)} \\ \star & \sigma P_z & 0 \\ \star & \star & 1-\sigma \end{bmatrix} \succeq 0, \quad j = 1, \dots, m, \quad l = 1, \dots, q. \tag{5.108}$$

Proof The conditions for invariance require $z_{i+1|k}^T P_z z_{i+1|k} \leq 1$ for all $z_{i|k}$ such that $z_{i|k}^T P_z z_{i|k} \leq 1$. From (5.107), this is equivalent to the conditions that, for all $z \in \mathbb{R}^{2n_x}$, and for $j = 1, \dots, m$ and $l = 1, \dots, q$,

$$1 - \big(\Psi^{(j)}z + \tilde{D}^{(j)}w^{(l)}\big)^T P_z \big(\Psi^{(j)}z_{k|k} + \tilde{D}^{(j)}w^{(l)}\big) \le \sigma_{jl}(1 - z^T P_z z).$$

These conditions can be expressed in terms of σ, defined as the minimum of σ_{jl} over j and l, as

$$\begin{bmatrix} z \\ 1 \end{bmatrix}^T \begin{bmatrix} \sigma P_z - \Psi^{(j)^T} P_z \Psi^{(j)} & \Psi^{(j)^T} P_z \tilde{D}^{(j)} w^{(l)} \\ \star & 1 - \sigma - w^{(l)^T} \tilde{D}^{(j)^T} P_z \tilde{D}^{(j)} w^{(l)} \end{bmatrix} \begin{bmatrix} z \\ 1 \end{bmatrix} \ge 0 \quad (5.109)$$

for all z, j and l, which can be shown, using Schur complements, to be equivalent to (5.108). From the linear dependence of (5.108) on $w^{(l)}$ for $l = 1, \dots, q$, it follows that (5.108) is necessary and sufficient for invariance over the entire class of additive uncertainty. The same applies in respect of the multiplicative uncertainty class since (5.108) depends linearly on $(\Psi^{(j)}, \tilde{D}^{(j)})$ for $j = 1, \dots, m$. This argument makes the implicit assumption that $\sigma P_z - \Psi^{(j)T} P_z \Psi^{(j)}$ is strictly positive definite for each j, but we note that, if this matrix is only positive semidefinite, a similar argument applies with the matrix inverse replaced by the relevant Moore–Penrose pseudoinverse [24]. □

Given the invariance property of Theorem 5.9, the constraints (5.3) can be imposed over the mode 2 prediction horizon by ensuring that $Fx + Gu \le \mathbf{1}$ holds for all $z \in \mathcal{E}_z = \{z : z^T P_z z \le 1\}$. This is equivalent to the conditions

$$\begin{bmatrix} H & \big[F + GK \;\; GC_c\big] P_z^{-1} \\ \star & P_z^{-1} \end{bmatrix} \succeq 0, \quad e_i^T H e_i \le 1, \quad i = 1, \dots, n_C, \qquad (5.110)$$

for some symmetric matrix H, where e_i is the ith column of the identity matrix. Using the convexification technique of Sect. 5.3.1, it is possible to express (5.108) and (5.110) in terms of the equivalent LMI conditions:

$$\begin{bmatrix} \begin{bmatrix} Y & X \\ X & X \end{bmatrix} & \begin{bmatrix} \Phi^{(j)}Y + B^{(j)}\Gamma & \Phi^{(j)}X \\ \Xi^{(j)} + \Phi^{(j)}Y + B^{(j)}\Gamma & \Phi^{(j)}X \end{bmatrix} & \begin{bmatrix} D^{(j)} \\ D^{(j)} + \Gamma_w^{(j)} \end{bmatrix} \\ \star & \begin{bmatrix} Y & X \\ X & X \end{bmatrix} & 0 \\ \star & \star & 1 - \sigma \end{bmatrix} \succeq 0 \qquad (5.111)$$

for $j = 1, \dots, m$ and (5.44b). Here the transformed variables $X, Y, \Xi^{(j)}, \Gamma$ are as defined in (5.45) and (5.46b), and $\Gamma_w^{(j)}$ is an additional variable satisfying $C_w^{(j)} = \Gamma_w^{(j)} U^{-1}$ for $j = 1, \dots, m$.

The parameters P_z, C_c and $A_c^{(j)}$, $C_w^{(j)}$, $j = 1, \dots, m$ that maximize the volume of the x-subspace projection of $\mathcal{E}_z$ for the mode 2 prediction dynamics of (5.107) can be computed offline by solving a semidefinite program. In particular, the projection of $\mathcal{E}_z$ onto the x-subspace is given by $\{x : x^T Y^{-1} x \le 1\}$, and the volume of this ellipsoidal set is maximized through the maximization of $\log\det(Y)$ subject to (5.111) and (5.44b).

As pointed out in Sect. 5.3.1, using the ellipsoid $\mathcal{E}_z$ to impose linear constraints on predicted trajectories implies a degree of conservativeness in the handling of constraints. Here instead we use the design of $\mathcal{E}_z$ simply as a means of optimizing the mode 2 prediction dynamics, and impose the constraints $Fx_{i|k} + Gu_{i|k} \leq \mathbf{1}$ at all prediction times using polytopic tubes. This can be done by modifying the approach of Sect. 5.5 to account for the predicted values of the controller state $\mathbf{c}_{i|k}$ and the additive uncertainty $w_{i|k}$. For $i = 0, \ldots, N-1$ the predicted trajectories are governed by

$$z_{i+1|k} = \Psi_{i|k} z_{i|k} + \tilde{B}_{i|k} f_{i|k} + \tilde{D}_{i|k} w_{i|k}, \quad z_{0|k} = \begin{bmatrix} x_k \\ \mathbf{c}_k \end{bmatrix}$$
$$(\Psi_{i|k}, \tilde{B}_{i|k}, \tilde{D}_{i|k}) \in \mathrm{Co}\big\{(\Psi^{(j)}, \tilde{B}^{(j)}, \tilde{D}^{(j)}),\ j = 1, \ldots, m\big\}$$
$$w_{i|k} \in \mathrm{Co}\{w^{(l)},\ l = 1, \ldots, q\}$$

where

$$\tilde{B}_{i|k} = \begin{bmatrix} B_{i|k} \\ 0 \end{bmatrix}, \quad \tilde{B}^{(j)} = \begin{bmatrix} B^{(j)} \\ 0 \end{bmatrix}.$$

Define the polytopic set

$$\mathcal{Z}_{i|k} = \{z : Vz \leq \alpha_{i|k}\},$$

where $V \in \mathbb{R}^{n_V \times 2n_x}$ is to be designed offline and $\alpha_{i|k}$ for $i = 0, \ldots, N$ are variables in the online MPC optimization performed at each time step k. Then, by Lemma 5.6, the conditions $z_{i|k} \in \mathcal{Z}_{i|k}$, for $i = 1 \ldots, N$ are enforced by the following constraints, for some $H^{(j)} \geq 0$,

$$\alpha_{i+1|k} \geq H^{(j)} \alpha_{i|k} + V\tilde{B}^{(j)} f_{i|k} + V\tilde{D}^{(j)} w^{(l)}, \quad i = 0, \ldots, N-1 \tag{5.112a}$$
$$H^{(j)} V = V\Psi^{(j)} \tag{5.112b}$$

for $j = 1, \ldots, m, l = 1, \ldots, q$. The constraints $Fx_{i|k} + Gu_{i|k} \leq \mathbf{1}$ are likewise satisfied for all $i = 0, \ldots, N-1$ if, for some $H_c \geq 0$ the following constraints hold:

$$H_c \alpha_{i|k} + G f_{i|k} \leq \mathbf{1}, \quad i = 0, \ldots, N-1 \tag{5.113a}$$
$$H_c V = \begin{bmatrix} F + GK & GC_c \end{bmatrix}. \tag{5.113b}$$

Based on the discussion of terminal conditions in Sect. 5.5, we introduce the terminal conditions

$$H^{(j)} \alpha_{N|k} + V\tilde{D}^{(j)} w^{(l)} \leq \alpha_{N|k} \tag{5.114a}$$
$$H_c \alpha_{N|k} \leq \mathbf{1} \tag{5.114b}$$

for $j = 1, \ldots, m$, $l = 1, \ldots, q$. By Lemma 5.6, these conditions are sufficient to ensure $Fx_{i|k} + Gu_{i|k} \leq \mathbf{1}$ for all $i \geq 0$ if $\alpha_{0|k}$ satisfies the initial condition

$$\alpha_{0|k} \geq V \begin{bmatrix} x_k \\ \mathbf{c}_k \end{bmatrix}. \tag{5.115}$$

As in Sect. 5.5, the matrix V is assumed to be chosen offline so that $\{z : Vz \leq \mathbf{1}\}$ is λ-contractive, for some $\lambda \in [0, 1)$, under the mode 2 prediction dynamics (5.107). The nonnegative matrices H_c are computed offline by solving the linear programs

$$h_i^* = \arg \min_{h \in \mathbb{R}^{n_V}} \mathbf{1}^T h \;\text{ subject to }\; h^T V = \big(\big[F + GK \;\; GC_c\big]\big)_i \text{ and } h \geq 0$$

and setting $(H_c)_i = h_i^{*T}$ for $i = 1, \ldots, n_C$, where $(H_c)_i$ and $([F + GK \;\; GC_c])_i$ denote the ith rows of H_c and $[F + GK \;\; GC_c]$. Likewise $H^{(j)}$, $j = 1, \ldots, m$, are computed by solving

$$h_i^{(j)*} = \arg \min_{h \in \mathbb{R}^{n_V}} \mathbf{1}^T h \;\text{ subject to }\; h^T V = V_i \Psi^{(j)} \text{ and } h \geq 0$$

and setting $H_i^{(j)} = h_i^{(j)*T}$ for $i = 1, \ldots, n_V$ and $j = 1, \ldots, m$, where $H_i^{(j)}$ and V_i denote the ith rows of $H^{(j)}$ and V, respectively.

To construct a robust MPC algorithm using this formulation of constraints, we next consider the form of the MPC cost index. Here we consider a worst-case cost with respect to the model uncertainty. To evaluate this cost, the prediction dynamics are first expressed in compact form as

$$\xi_{i+1|k} = \Theta_{i|k}\xi_{i|k} + \hat{D}w_{i|k}, \quad \xi_{0|k} = \begin{bmatrix} x_k \\ \mathbf{c}_k \\ \mathbf{f}_k \end{bmatrix} \tag{5.116}$$

with $\mathbf{f}_k = (f_{0|k}, \ldots, f_{N-1|k})$ and

$$(\Theta_{i|k}, \hat{D}_{i|k}) = \mathrm{Co}\big\{(\Theta^{(j)}, D^{(j)}),\; j = 1, \ldots, m\big\},$$

$$\Theta^{(j)} = \begin{bmatrix} \Phi^{(j)} & B^{(j)}C_c & B^{(j)}E \\ 0 & A_c^{(j)} & 0 \\ 0 & 0 & M \end{bmatrix}, \quad \hat{D}^{(j)} = \begin{bmatrix} D^{(j)} \\ C_w^{(j)} \\ 0 \end{bmatrix}$$

with the matrices E and M defined as in (2.26b). For the predicted cost defined by

$$\check{J}(x_k, \mathbf{c}_k, \mathbf{f}_k) = \max_{\substack{(A_{i|k}, B_{i|k}, D_{i|k}) \in \tilde{\Omega}, \\ w_{i|k},\, i=0,1,\ldots}} \sum_{i=0}^{\infty} \|x_{i|k}\|_Q^2 + \|u_{i|k}\|_R^2 - \gamma^2 \|w_{i|k}\|^2 \tag{5.117}$$

the following result provides conditions allowing an upper bound on $\check{J}(x_k, \mathbf{c}_k, \mathbf{f}_k)$ to be determined.

Lemma 5.8 *For given γ, the predicted cost (5.117) is bounded from above as*

$$\check{J}(x_k, \mathbf{c}_k, \mathbf{f}_k) \leq \xi_{0|k}^T \check{W} \xi_{0|k}$$

for all $(x_k, \mathbf{c}_k, \mathbf{f}_k)$ if and only if $\check{W} \succ 0$ satisfies the LMIs, for $j = 1, \ldots, m$:

$$\check{W} - \begin{bmatrix} \Theta^{(j)T} & 0 \\ \hat{D}^{(j)T} & 1 \end{bmatrix} \check{W} \begin{bmatrix} \Theta^{(j)} & \hat{D}^{(j)} \\ 0 & 1 \end{bmatrix} \succeq \begin{bmatrix} \hat{Q} & 0 \\ 0 & -\gamma^2 I \end{bmatrix} \tag{5.118}$$

where

$$\hat{Q} = \begin{bmatrix} Q + K^T R K & K^T R C_c & K^T R E \\ \star & C_c^T R C_c & C_c^T R E \\ \star & \star & E^T R E \end{bmatrix}.$$

Proof The bound on $\check{J}(x_k, \mathbf{c}_k, \mathbf{f}_k)$ is obtained by summing over all $i \geq 0$ the bound $\|\xi_{i|k}\|^2_{\check{W}} - \|\xi_{i+1|k}\|^2_{\check{W}} \geq \|x_{i|k}\|^2_Q + \|u_{i|k}\|^2_R - \gamma^2\|w_{i|k}\|^2$, which by (5.116) can be expressed as $\|\xi\|^2_{\check{W}} - \|\Theta\xi + \hat{D}w\|^2_{\check{W}} \geq \|\xi\|^2_{\hat{Q}} - \gamma^2\|w\|^2$ or equivalently as

$$\begin{bmatrix} \xi \\ w \end{bmatrix}^T \begin{bmatrix} \check{W} - \Theta^T \check{W} \Theta & \Theta \check{W} \hat{D} \\ \star & \gamma^2 - \hat{D}^T \check{W} \hat{D} \end{bmatrix} \begin{bmatrix} \xi \\ w \end{bmatrix} \geq 0$$

Since the disturbance set $\mathcal{W}$ contains $w = 0$ by assumption, the matrix appearing in this expression must be positive semidefinite, and by rearranging terms this condition is equivalent to

$$\check{W} - \begin{bmatrix} \Theta^T & 0 \\ \hat{D}^T & 1 \end{bmatrix} \check{W} \begin{bmatrix} \Theta & \hat{D} \\ 0 & 1 \end{bmatrix} \succeq \begin{bmatrix} \hat{Q} & 0 \\ 0 & -\gamma^2 I \end{bmatrix}.$$

Using Schur complements, this can be written as an LMI in Θ and $\hat{D}$, which by convexity is satisfied for all $(\Theta, \hat{D}) \in \text{Co}\{(\Theta^{(j)}, \hat{D}^{(j)}, \ j = 1, \ldots, m\}$ if and only (5.118) holds. □

A unique value of $\check{W}$ corresponding to a tight bound on $J(x_k, \mathbf{c}_k, \mathbf{f}_k)$ can be obtained by solving the semidefinite program:

$$\underset{\check{W}}{\text{minimize}} \ \text{tr}(\check{W}) \ \text{subject to (5.118).}$$

Algorithm 5.8 At each time instant $k = 0, 1, \ldots$:

(i) Perform the optimization:

$$\begin{aligned} &\underset{\substack{\mathbf{c}_k, \mathbf{f}_k \\ \alpha_{0|k}, \ldots, \alpha_{N|k}}}{\text{minimize}} \; \|(x_k, \mathbf{c}_k, \mathbf{f}_k)\|^2_{\check{W}} \\ &\text{subject to } (5.112a), (5.113a), (5.114a,b), (5.115) \end{aligned} \tag{5.119}$$

(ii) Apply the control law $u_k = Kx_k + C_c\mathbf{c}_k^* + f_{0|k}^*$, where $\mathbf{f}_k^* = (f_{0|k}^*, \ldots, f_{N-1|k}^*)$ and $\mathbf{c}_k^*$, $\mathbf{f}_k^*$ are the optimal values of $\mathbf{c}_k$ and $\mathbf{f}_k$ in (5.119). ◁

Theorem 5.10 *The optimization (5.119) is recursively feasible, and for all initial conditions x_0 in the feasible set*

$$\mathcal{F}_N = \{x_k : \exists(\mathbf{c}_k, \mathbf{f}_k) \textit{ satisfying } (5.112a), (5.113a), (5.114a, 5.114b), (5.115)\},$$

the trajectories of the system (5.1)–(5.3) under the control law of Algorithm 5.8 satisfy the l_2 bound:

$$\sum_{k=0}^{\infty}(\|x_k\|_Q^2 + \|u_k\|_R^2) \le \gamma^2 \sum_{k=0}^{\infty} \|w\|_k^2 + \|(x_0, \mathbf{c}_0^*, \mathbf{f}_0^*)\|^2_{\check{W}}. \tag{5.120}$$

Proof Assume $x_k \in \mathcal{F}_N$ and consider the solution at time $k+1$ given by

$$\begin{aligned} \mathbf{c}_{k+1} &= A_{c,k}\mathbf{c}_k^* + C_{w,k}w_k \\ \mathbf{f}_{k+1} &= (f_{1|k}^*, \ldots, f_{N-1|k}^*, 0) \\ \alpha_{i|k+1} &= \alpha_{i+1|k}^*, \quad i = 0, \ldots, N-1, \\ (\alpha_{N|k+1})_l &= \max_{j \in \{1,\ldots,m\}} H_l^{(j)}\alpha_{N|k}^*, \quad l = 1, \ldots, n_V \end{aligned}$$

for some $(A_{c,k}, C_{w,k}) \in \text{Co}\{(A_c^{(1)}, C_w^{(1)}), \ldots, (A_c^{(m)}, C_w^{(m)})\}$. This choice of variables gives $\mathcal{Z}_{i|k+1} = \mathcal{Z}_{i+1|k}$ for $i = 0, \ldots, N-1$ and hence (5.112a), (5.113a) and (5.114a) are necessarily satisfied at $k+1$. Also the definition of $\alpha_{N|k+1}$ satisfies (5.114b) and $z_{1|k} \in \mathcal{Z}_{1|k}$ implies that (5.115) is satisfied by $(x_{k+1}, \mathbf{c}_{k+1})$, which demonstrates that (5.119) is recursively feasible.

From (5.108) and optimality of the solution of (5.119) at time $k+1$, it follows that

$$\begin{aligned} \|(x_{k+1}, \mathbf{c}_{k+1}^*, \mathbf{f}_{k+1}^*)\|^2_{\check{W}} &\le \|(x_{k+1}, \mathbf{c}_{k+1}, \mathbf{f}_{k+1})\|^2_{\check{W}} \\ &\le \|(x_k, \mathbf{c}_k^*, \mathbf{f}_k^*)\|^2_{\check{W}} - \|x_k\|_Q^2 + \|u_k\|_R^2 - \gamma^2\|w_k\|^2 \end{aligned}$$

Summing both sides of this inequality over all $i \ge 0$ yields the bound (5.120). □

To conclude this section, we note that, if the sequence $\{w_k,\ k = 0, 1, \ldots\}$ is square summable, then under the control law of Algorithm 5.8 the origin of the state-space will be asymptotically stable since in this case the bound (5.120) implies that the stage cost converges to zero. We also note that, as explained in Chap. 3, the definition of the MPC cost as an upper bound on the min–max cost (5.117) ensures that γ provides a bound on the induced l_2 norm from the disturbance to the closed-loop system state. In this sense, one may wish to choose γ to be small. However, the smaller the γ is, the larger the trace of $\check{W}$ will have to be in order that $\check{W}$ can satisfy (5.118). Thus a compromise between the tightness of the cost bound and the disturbance rejection ability of Algorithm 5.8 is needed.

Example 5.5 Consider the system (5.5) with parametric uncertainty and unknown disturbances contained in the polytopic sets defined in (5.105a, 5.105b). The parameters $A^{(j)}$, $B^{(j)}$ for $j = 1, 2, 3$ are as given in Example 5.1. Also $D^{(j)} = I$ for $j = 1, 2, 3$ and the disturbance set $\mathcal{W} \subset \mathbb{R}^2$ is given by

$$\mathcal{W} = \mathrm{Co}\left\{\begin{bmatrix}1\\1\end{bmatrix}, \begin{bmatrix}1\\-1\end{bmatrix}, \begin{bmatrix}-1\\1\end{bmatrix}, \begin{bmatrix}-1\\-1\end{bmatrix}\right\}.$$

The system is subject to the constraints of Example 5.1, namely the state constraints $-10 \le [0\ 1]x_k \le 10$ and input constraints $-5 \le u_k \le 5$.

The feedback gain K is again given by $K = [0.19\ 0.34]$, and optimizing the prediction dynamics subject to (5.109) and (5.110) yields $\sigma = 0.898$ and

$$A_c^{(1)} = \begin{bmatrix}-0.69 & 0.20\\-0.28 & -0.12\end{bmatrix}, \quad A_c^{(2)} = \begin{bmatrix}-0.74 & -0.004\\0.25 & 0.02\end{bmatrix}, \quad A_c^{(3)} = \begin{bmatrix}-0.63 & -0.21\\-0.06 & -0.26\end{bmatrix}$$

with $C_w^{(j)} = -I$ (to 2 decimal places) for $j = 1, 2, 3$ and

$$C_c = \begin{bmatrix}0.13 & -0.14\end{bmatrix}.$$

The matrix V defining the polytopic tube cross sections $\mathcal{Z}_{i|k}$ is chosen so that the set $\{z : Vz \le \mathbf{1}\}$ is λ-contractive, with $\lambda = 0.9$. Thus $\{z : Vz \le \mathbf{1}\}$ is defined as the maximal RPI set for the system $z_{k+1} = (1/\lambda)(\Psi z_k + \tilde{D}w_k)$, $(\Psi, \tilde{D}) \in \mathrm{Co}\{(\Psi^{(j)}, \tilde{D}^{(j)}),\ j = 1, \ldots, m\}$, $w_k \in \mathcal{W}$, which yields a V with 22 rows. The corresponding terminal conditions (5.113) consist of 70 constraints and the terminal set is shown in Fig. 5.8. From Fig. 5.8, it can be seen that this terminal set contains and extends beyond the maximal RPI set under the linear feedback law $u = Kx$.

For comparison, if the terminal control law is chosen as $u = Kx$ and if the state tube cross sections are defined in the space of x (rather than z) in terms of the matrix V_x such that $\{x : V_x x \le \mathbf{1}\}$ is the maximal RPI set for $x_{k+1} = (1/\lambda)(\Phi x_k + Dw_k)$, $(\Phi, D) \in \mathrm{Co}\{(\Phi^{(j)}, D^{(j)}),\ j = 1, \ldots, m\}$, $w_k \in \mathcal{W}$, then V_x has 8 rows, the terminal conditions are defined by defined by 28 constraints and the corresponding terminal set coincides with the maximal RPI set under $u = Kx$. The larger terminal set that is obtained with the optimized prediction dynamics enables the feasible set

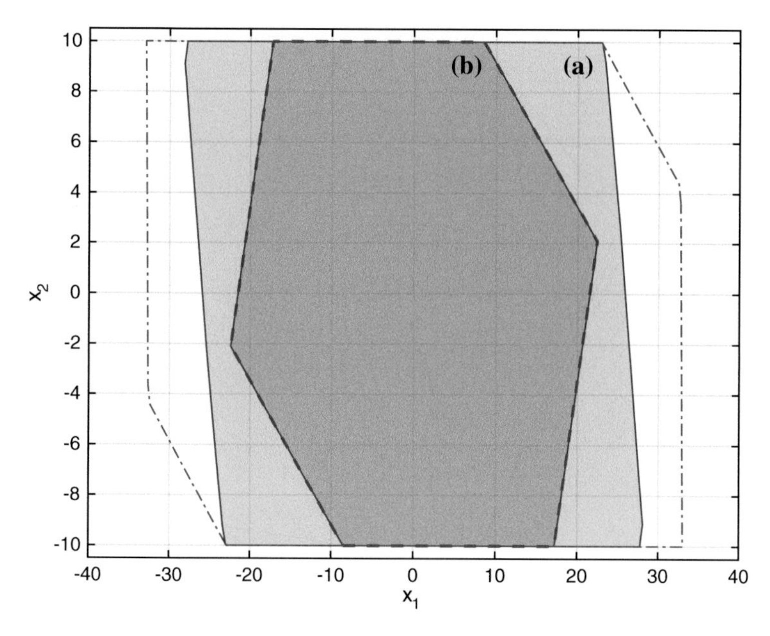

Fig. 5.8 The terminal sets for the case that the terminal controller is defined using **a** optimized prediction dynamics, and **b** a fixed linear controller (*solid lines*). Also shown are the maximal robustly controlled invariant set (*dash-dotted line*) and the maximal RPI set under the linear feedback law (*dashed line*)

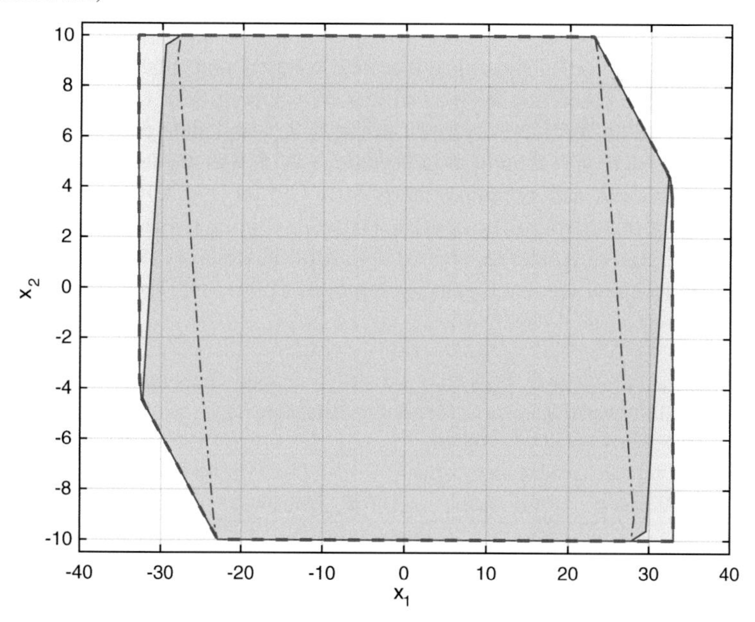

Fig. 5.9 The feasible initial condition sets $\mathcal{F}_N$ for Algorithm 5.8, for $N = 1, 2$ (*solid lines*). The terminal set (*dash-dotted line*) and the maximal robustly controlled invariant set (*dashed line*) are also shown

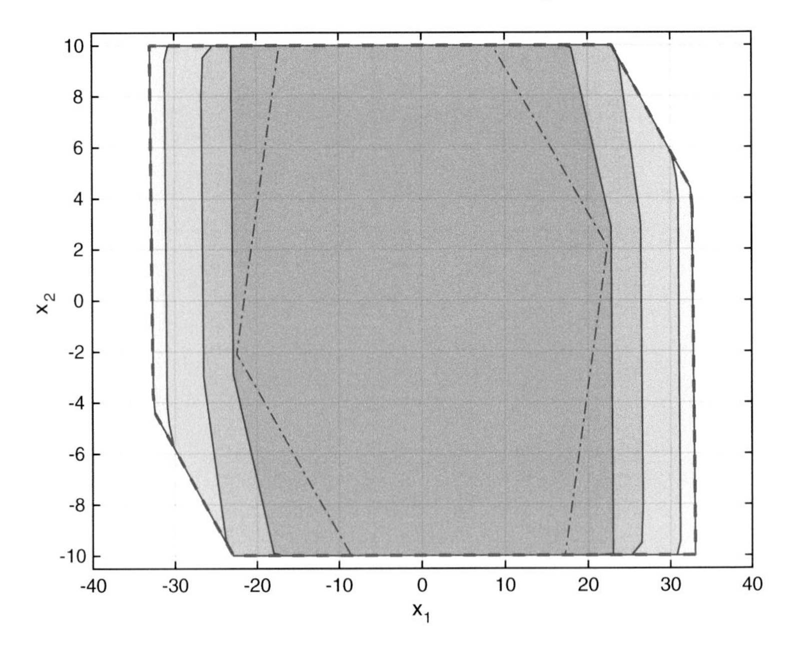

Fig. 5.10 The feasible initial condition sets $\mathcal{F}_N$ for Algorithm 5.8 with a fixed linear terminal controller, for $N = 1, 2, 3, 4$ (*solid lines*). The terminal set (*dash-dotted line*) and the maximal robustly controlled invariant set (*dashed line*) are also shown

for Algorithm 5.8 to cover the entire maximal robustly controlled invariant set for this system with a prediction horizon of just $N = 2$ (Fig. 5.9). On the other hand, with the fixed linear terminal feedback law $u = Kx$ and state tube cross sections defined in the space of x, a prediction horizon of $N = 4$ is needed in order that the feasible set coincides with the maximal robustly CPI set (Fig. 5.10).

However, for this example the approach based on optimized prediction dynamics with $N = 2$ requires a greater number of optimization variables and constraints than when a fixed terminal feedback gain is employed (Table 5.3). The extra degrees of

Table 5.3 Number of variables, constraints and area of feasible region for the MPC strategy of Algorithm 5.8 with terminal control law defined by optimized prediction dynamics and with terminal control law defined by fixed linear feedback

N	Optimized prediction dynamics			Fixed terminal feedback gain		
	Variables	Constraints	Area of $\mathcal{F}_N$	Variables	Constraints	Area of $\mathcal{F}_N$
1	47	162	1200	17	64	880
2	70	332	1255	26	92	1038
3				35	120	1202
4				44	148	1255

freedom associated with the optimized prediction dynamics translates into improved predicted performance. For example, the worst-case predicted cost, evaluated at the vertices of the maximal CPI set, is on average 3.1 % (at most 4.6 %) higher if a fixed linear terminal feedback gain is used than if the optimized prediction dynamics are employed. ◊

5.7 Exercises

1 A symmetric, real-valued $n \times n$ matrix P is positive definite ($P \succ 0$) if and only if $v^T P v > 0$ for all non-zero real vectors v. Use this property to prove the following statements.

(a) The linear matrix inequality $M(x) \succ 0$ is convex in $x = (x_1, \ldots, x_n)$ where

$$M(x) = M_0 + x_1 M_1 + \cdots + x_n M_n$$

for given symmetric matrices $M_0, \ldots, M_n$.

(b) The condition

$$\begin{bmatrix} P & Q \\ Q^T & R \end{bmatrix} \succ 0$$

for symmetric P and R is equivalent to the Schur complements

$$R \succ 0, \quad P - QR^{-1}Q^T \succ 0.$$

(c) If $P = S^{-1} \succ 0$, then the condition $S - ASA^T \succ 0$ is equivalent to $P - A^T PA \succ 0$.

2 Let $\mathcal{E}$ be the ellipsoidal set defined by $\mathcal{E} = \{x : x^T Px \leq 1\}$, for some symmetric matrix $P \succ 0$.

(a) Show that $\mathcal{E} \subseteq \mathcal{X}$, where $\mathcal{X}$ is the polytope $\mathcal{X} = \{x : Vx \leq \mathbf{1}\}$ for a given matrix $V \in \mathbb{R}^{n_V \times n_x}$, if and only if

$$V_i P^{-1} V_i^T \leq 1, \quad i = 1, \ldots, n_V$$

where V_i for $i = 1, \ldots, n_V$ are the rows of V.

(b) Hence, show that $Fx + Gu \leq \mathbf{1}$ holds for all $x \in \mathcal{E}$, where $u = Kx$, if and only if the following LMI conditions in variables S, Y hold

$$\begin{bmatrix} 1 & F_i S + G_i Y \\ (F_i S + G_i Y)^T & S \end{bmatrix} \succeq 0, \quad i = 1, \ldots, n_C$$

where F_i and G_i, $i = 1, \ldots, n_C$ are the rows of F and G, and where $(S, Y) = (P^{-1}, KP^{-1})$.

3 A system is described by the model $x_{k+1} = A_k x_k + B_k u_k$ with

$$(A_k, B_k) \in \text{Co}\{(A^{(1)}, B^{(1)}), (A^{(2)}, B^{(2)})\}$$
$$= \text{Co}\left\{\left(\begin{bmatrix} 0.7 & -0.6 \\ -0.7 & -1.8 \end{bmatrix}, \begin{bmatrix} 0.3 \\ -0.5 \end{bmatrix}\right), \left(\begin{bmatrix} 0.5 & -0.8 \\ -0.6 & -1.8 \end{bmatrix}, \begin{bmatrix} 0.1 \\ -0.4 \end{bmatrix}\right)\right\}$$

at each time step $k = 0, 1, \ldots$. The control input is subject to the constraints $-1 \le u_k \le 1$ for all k. A robust MPC law $u_k = K_k x_k$ is to be designed for this system with the aim of minimizing, at each time k, a quadratic bound on the worst-case predicted cost:

$$\check{J}(x_k, K_k) = \max_{\{(A_k, B_k), (A_{k+1}, B_{k+1}), \ldots\}} \sum_{i=0}^{\infty} \left(\|x_{i|k}\|_Q^2 + \|u_{i|k}\|_R^2\right)$$

subject to $-1 \le u_{i|k} \le 1$, $i = 0, 1, \ldots$.

(a) Show that the worst-case cost satisfies the bound $\check{J}(x_k, K_k) \le \gamma_k$ if

$$P - (A^{(j)} + B^{(j)} K_k)^T P (A^{(j)} + B^{(j)} K_k) \succeq \gamma_k^{-1}(Q + K_k^T R K_k), \quad j = 1, 2$$
$$(F_i + G_i K_k) P^{-1} (F_i + G_i K_k)^T \le 1, \quad i = 1, \ldots, n_C$$
$$x_k^T P x_k \le 1$$

for some matrix $P = P^T \succ 0$, where F_i, G_i for $i = 1, \ldots, n_C$ are the rows of F and G.

(b) Suggest a suitable transformation of optimization variables to enable the value of K_k that minimizes γ_k subject to the constraints of (a) to be computed using semidefinite programming. Hence, verify that for $x_0 = (4, -1)$ this gives $K_0 = \begin{bmatrix} -0.962 & -3.678 \end{bmatrix}$ and $\gamma_0 = 152.4$.

(c) Explain why a better approximation of the worst-case cost is given by the bound $J(x_k, K_k) \le x_k^T \Theta_k^* x_k$, where

$$\Theta_k^* = \arg\min_{\Theta} \; x_k^T \Theta x_k \;\text{ subject to}$$
$$\Theta - (A^{(j)} + B^{(j)} K_k)^T \Theta (A^{(j)} + B^{(j)} K_k) \succeq Q + K_k^T R K_k, \quad j = 1, 2$$

Hence, verify that for $x_0 = (4, -1)$ and $K_0 = \begin{bmatrix} -0.962 & -3.678 \end{bmatrix}$ the worst-case cost satisfies the upper bound $J(x_0, K_0) \le 69.0$.

(d) Comment on the suggestion that a better control strategy could be constructed by choosing K_k so as to minimize, at each time k, the value of $x_k^T \Theta x_k$ subject to

$$\Theta - (A^{(j)} + B^{(j)} K_k)^T \Theta (A^{(j)} + B^{(j)} K_k) \succeq Q + K_k^T R K_k, \quad j = 1, 2$$
$$P - (A^{(j)} + B^{(j)} K_k)^T P (A^{(j)} + B^{(j)} K_k) \succeq 0, \quad j = 1, 2$$

$$(F_i + G_i K_k)P^{-1}(F_i + G_i K_k)^T \preceq 1, \qquad i = 1, \ldots, n_C$$
$$x_k^T P x_k \le 1.$$

4 (a) Let $\mathcal{E} = \{x : x^T P x \le 1\}$ for given $P \succ 0$ and let the vector $x \in \mathbb{R}^n$ be partitioned according to $x = (u, v)$, $u \in \mathbb{R}^m$, $v \in \mathbb{R}^{n-m}$. Show that the u-subspace projection of $\mathcal{E}$ (i.e. the set $\{u : \exists v$ such that $(u, v) \in \mathcal{E}\}$) is given by

$$\mathcal{E}_u = \{u : u^T P_u u \le 1\}$$

where

$$P_u^{-1} = \begin{bmatrix} I_m & 0 \end{bmatrix} P^{-1} \begin{bmatrix} I_m \\ 0 \end{bmatrix}$$

and $\begin{bmatrix} I_m & 0 \end{bmatrix} x = u$.

(b) Using the system model and constraints of Question 3, calculate the robustly invariant ellipsoidal set $\mathcal{E}_z$ for the uncertain dynamics

$$z_{i+1|k} \in \mathrm{Co}\{\Psi^{(1)} z_{i|k}, \Psi^{(2)} z_{i|k}\}$$

and constraints $-1 \le \begin{bmatrix} K & E \end{bmatrix} z_{i|k} \le 1$, with $N = 12$ (i.e. $z_{i|k} \in \mathbb{R}^{14}$) and

$$\Psi^{(j)} = \begin{bmatrix} A^{(j)} + B^{(j)} K & B^{(j)} E \\ 0 & M \end{bmatrix}, \quad j = 1, 2,$$

$$K = \begin{bmatrix} -1.078 & -3.523 \end{bmatrix}, \quad E = \begin{bmatrix} 1 & 0 & \cdots & 0 \end{bmatrix}, \quad M = \begin{bmatrix} 0 & 1 & \cdots & 0 \\ \vdots & \vdots & \ddots & \vdots \\ 0 & 0 & \cdots & 1 \\ 0 & 0 & \cdots & 0 \end{bmatrix},$$

such that the area of its projection onto the state space of the system is maximized. Verify that the maximum area projection is given by $\{x : x^T P_x x \le 1\}$ with $\det(P_x) = 0.783$.

(c) For the system in Question 4, with nominal model parameters defined by $(A^{(0)}, B^{(0)}) = \frac{1}{2}\big((A^{(1)}, B^{(1)}) + (A^{(2)}, B^{(2)})\big)$, the unconstrained optimal feedback gain for the nominal cost with weights $Q = I$, $R = 1$ is $K = \begin{bmatrix} -1.078 & -3.523 \end{bmatrix}$ and the corresponding solution of the Riccati equation (2.9) is

$$W_x = \begin{bmatrix} 2.691 & 3.668 \\ 3.668 & 21.94 \end{bmatrix}.$$

Taking $N = 12$, determine the matrix W in the expression for the nominal predicted cost:

$$J(x_k, \mathbf{c}_k) = \sum_{i=0}^{\infty} \big(\|x_{i|k}\|^2 + u_{i|k}^2 \big) = \begin{bmatrix} x_k \\ \mathbf{c}_k \end{bmatrix}^T W \begin{bmatrix} x_k \\ \mathbf{c}_k \end{bmatrix},$$

where the nominal predicted state and control trajectories evolve according to

$$x_{i+1|k} = A^{(0)} x_{i|k} + B^{(0)} u_{i|k}, \quad u_{i|k} = K x_{i|k} + c_{i|k},$$

with $\mathbf{c}_k = (c_{0|k}, \ldots, c_{N-1|k})$ and $c_{i|k} = 0$ for $i \geq N$.

Show that the minimum value of this cost at $k = 0$ with $x_0 = (4, -1)$ subject to the constraint $(x_0, \mathbf{c}_0) \in \mathcal{E}_z$, where $\mathcal{E}_z$ is the robustly invariant ellipsoid determined in part (b), is $J^*(x_0) = 39.9$.

5 It is suggested that the value of the predicted cost in Question 4 could be reduced using a univariate search:

$$\alpha_k^* = \min_{\alpha_k \in [0,1]} \alpha_k \text{ subject to } -1 \leq K x_k + \alpha_k \mathbf{c}_k^* \leq 1,$$

$$\Psi^{(j)} \begin{bmatrix} x_k \\ \alpha_k \mathbf{c}_k^* \end{bmatrix} \in \mathcal{E}_z, \quad j = 1, 2$$

where $\mathbf{c}_k^*$ is the solution of the minimization in Question 4(c) at time k.

(a) Explain the purpose of each the constraints in this line search.

(b) For the system of Question 4, find the smallest value of σ for which there exists $\Theta \succ 0$ satisfying

$$\begin{bmatrix} \Theta - I & (A^{(j)} + B^{(j)} K)^T \Theta & 0 \\ \star & \Theta & \Theta B^{(j)} \\ \star & \star & \sigma^2 I_{n_u} \end{bmatrix} \succeq 0, \quad j = 1, 2$$

What does this imply about the state of the closed-loop system under a control law of the form $u_k = K x_k + c_{0|k}$?

(c) Suppose that, at each time step, $k = 0, 1, \ldots$, the optimization of Question 4(c) and the line search in (a) is performed and the solution is used to define an MPC law $u_k = K x_k + \alpha_k^* \mathbf{c}_k^*$. Will the closed-loop system be stable?

6 (a) For the system of Question 3 with $K = \begin{bmatrix} -1.078 & -3.523 \end{bmatrix}$, solve (5.48) to determine the prediction dynamics that give the robustly invariant ellipsoid $\mathcal{E}_z$ with the largest area projection onto the model state space, and confirm that the solution gives

$$A_c^{(1)} = \begin{bmatrix} 0.835 & -1.539 \\ -0.035 & -0.612 \end{bmatrix}, \quad A_c^{(1)} = \begin{bmatrix} 0.692 & -1.113 \\ 0.026 & -0.821 \end{bmatrix}$$

$$C_c = \begin{bmatrix} -0.176 & -0.394 \end{bmatrix}.$$

(b) Compute the matrix $W_c \succ 0$ with smallest trace satisfying

$$W_c - A_c^{(j)^T} W_c A_c^{(j)} \succeq B^{(0)T} W_x B^{(0)} + R, \quad j = 1, 2$$

Hence, verify that, with $x_0 = (4, -1)$, the minimum over $\mathbf{c}_0$ of the cost $J(x_0, \mathbf{c}_0) = \|x_0\|_{W_x}^2 + \|\mathbf{c}_0\|_{W_c}^2$ subject to $(x_0, \mathbf{c}_0) \in \mathcal{E}_z$, where $\mathcal{E}_z$ is the ellipsoidal set determined in (a), is $J^*(x_0) = 44.7$.

(c) Explain how the inequality defining W_c in (b) ensures the stability of the closed-loop system under the control law $u_k = Kx_k + c^*_{0|k}$, where at each time $k = 0, 1, \ldots, c^*_{0|k}$ is the first element of the optimal $\mathbf{c}^*_k$ for the minimization of the cost $J(x_k, \mathbf{c}_k)$ subject to $(x_k, \mathbf{c}_k) \in \mathcal{E}_z$.

(d) Determine the maximum scaling σ such that σx_0, with $x_0 = (4, -1)$, lies in the feasible set for the minimization of $J(x_0, \mathbf{c}_0)$ in (b) (i.e. Algorithm 5.4). Compare this with the maximum value of σ such that σx_0 is feasible for the minimization in Question 4(c) (i.e. Algorithm 5.3) and that of Question 3(b) (i.e. Algorithm 5.1). What conclusions can be drawn from this comparison?

7 Explain why the problem of maximizing the volume of the low-complexity polytopic set, $\Pi(V, \alpha) \doteq \{x : |Vx| \leq \alpha\}$, where $V \in \mathbb{R}^{n_x \times n_x}$ is an invertible matrix, is in general nonconvex when V and α are both considered to be optimization variables. Show that the volume maximization can be formulated as a convex problem if V is fixed.

8 This question considers how to construct a low-complexity polytopic set for the system and constraints of Question 3 under the control law $u = Kx$, $K = \begin{bmatrix}-1.078 & -3.523\end{bmatrix}$.

(a) Show that if $V = W^{-1}$, where W is the (right) eigenvector matrix of $\Phi^{(0)} = A^{(0)} + B^{(0)}K$, i.e.

$$W = \begin{bmatrix} 0.982 & 0.851 \\ -0.187 & 0.525 \end{bmatrix}, \quad V = \begin{bmatrix} 0.778 & -1.262 \\ 0.277 & 1.456 \end{bmatrix},$$

then there necessarily exists a vector α such that the low-complexity polytope $\{x : |Vz| \leq \alpha\}$ is invariant for the uncertain dynamics $x_{k+1} \in \mathrm{Co}\{\Phi^{(1)}x_k, \Phi^{(2)}x_k\}$ where $\Phi^{(j)} = A^{(j)} + B^{(j)}K$, $j = 1, 2$.

(b) Formulate and solve a convex optimization to determine α so that the volume of the set $\Pi(V, \alpha)$ is maximized, where V is fixed at the value specified in (a).

9 The maximal robustly invariant polytopic set for the system and constraints of Question 3 under $u = Kx$ is given by

$$\{x : Vx \leq \mathbf{1}\}, \quad V = \begin{bmatrix} -1.078 & -3.523 \\ 1.078 & 3.523 \\ 0.172 & 2.622 \\ -0.172 & -2.622 \end{bmatrix}$$

(a) Determine nonnegative matrices $H^{(1)}$, $H^{(2)}$ and H_c satisfying

$$\begin{aligned} H^{(j)}V &= V\Phi^{(j)}, \quad j = 1,2 \\ H_c V &= GK \end{aligned}$$

where $G = \begin{bmatrix} 1 & -1 \end{bmatrix}^T$, such that the sum of the elements in each row of each of $H^{(1)}$, $H^{(2)}$ and H_c is minimized.

(b) Show that, for any given mode 1 horizon N, if there exists a pair of sequences $\mathbf{c}_k = (c_{0|k}, \ldots, c_{N-1|k})$ and $\boldsymbol{\alpha}_k = (\alpha_{0|k}, \ldots, \alpha_{N|k})$ satisfying the constraints defined at time k by

$$\begin{aligned} Vx_k &\le \alpha_{0|k} \\ H^{(j)}\alpha_{i|k} + VB^{(j)}c_{i|k} &\le \alpha_{i+1|k}, \quad j = 1,2, && i = 0,\ldots,N-1 \\ H_c\alpha_{i|k} + Gc_{i|k} &\le \mathbf{1}, && i = 0,\ldots,N-1 \\ H^{(j)}\alpha_{N|k} &\le \alpha_{N|k}, \quad j = 1,2 \\ H_c\alpha_{N|k} &\le \mathbf{1} \end{aligned}$$

then there must exist $\mathbf{c}_{k+1}$ and $\boldsymbol{\alpha}_{k+1}$ satisfying the corresponding constraints at time $k+1$ if $u_k = Kx_k + c_{0|k}$.

(c) For a mode 1 horizon of $N = 8$ and $x_0 = (4, -1)$, determine the maximum scalar σ such that σx_0 is feasible for the constraints in part (b).

(d) At $k = 0$ with $x_0 = (4, -1)$, solve the MPC optimization:

$$\underset{s_{0|k},\mathbf{c}_k,\boldsymbol{\alpha}_k}{\text{minimize}} \ J(s_{0|k}, \mathbf{c}_k)$$

subject to $Vs_{0|k} \le \alpha_{0|k}$ and the constraints of part (b) with $N = 8$, where the nominal cost is defined by

$$J(s_{0|k}, \mathbf{c}_k) = \|s_{0|k}\|^2_{W_x} + \|\mathbf{c}_k\|^2_{W_c}.$$

Confirm that the optimal predicted cost for this initial condition is $J(s^*_{0|k}, \mathbf{c}^*_0) = 37.9$.

References

1. R.W. Liu, Convergent systems. IEEE Trans. Autom. Control **13**(4), 384–391 (1968)
2. J. Richalet, A. Rault, J.L. Testud, J. Papon, Model predictive heuristic control: applications to industrial processes. Automatica **14**(5), 413–428 (1978)
3. P.J. Campo, M. Morari, Robust model predictive control, in *Proceedings of the 1987 American Control Conference*, Minneapolis, USA. vol. 2 (1987), pp. 1021–1026
4. J.C. Allwright, G.C. Papavasiliou, On linear programming and robust model-predictive control using impulse-responses. Syst. Control Lett. **18**(2), 159–164 (1992)

5. A. Zheng, M. Morari, Robust stability of constrained model predictive control, in *Proceedings of the 1993 American Control Conference*, San Francisco, USA (1993), pp. 379–383
6. H. Genceli, M. Nikolaou, Robust stability analysis of constrained l_1-norm model predictive control. AIChE J. **39**(12), 1954–1965 (1993)
7. B. Kouvaritakis, J.A. Rossiter, Use of bicausal weighting sequences in least squares identification of open-loop unstable dynamic systems. Control Theory Appl. IEE Proc. D **139**(3), 328–336 (1992)
8. M.V. Kothare, V. Balakrishnan, M. Morari, Robust constrained model predictive control using linear matrix inequalities. Automatica **32**(10), 136–1379 (1996)
9. J. Schuurmans, J.A. Rossiter, Robust predictive control using tight sets of predicted states. Control Theory Appl., IEE Proc. **147**(1), 13–18 (2000)
10. B. Kouvaritakis, J.A. Rossiter, J. Schuurmans, Efficient robust predictive control. IEEE Trans. Autom. Control **45**(8), 1545–1549 (2000)
11. B. Kouvaritakis, M. Cannon, J.A. Rossiter, Who needs QP for linear MPC anyway? Automatica **38**(5), 879–884 (2002)
12. M. Cannon, B. Kouvaritakis, Optimizing prediction dynamics for robust MPC. IEEE Trans. Autom. Control **50**(11), 1892–1897 (2005)
13. Q. Cheng, M. Cannon, B. Kouvaritakis, The design of dynamics in the prediction structure of robust MPC. Int. J. Control **86**(11), 2096–2103 (2013)
14. Y.I. Lee, B. Kouvaritakis, Stabilizable regions of receding horizon predictive control with input constraints. Syst. Control Lett. **38**(1), 13–20 (1999)
15. Y.I. Lee, B. Kouvaritakis, Robust receding horizon predictive control for systems with uncertain dynamics and input saturation. Automatica **36**(10), 1497–1504 (2000)
16. Y.I. Lee, B. Kouvaritakis, A linear programming approach to constrained robust predictive control. IEEE Trans. Autom. Control **45**(9), 1765–1770 (2000)
17. Y.I. Lee, B. Kouvaritakis, Linear matrix inequalities and polyhedral invariant sets in constrained robust predictive control. Int. J. Robust Nonlinear Control **10**(13), 1079–1090 (2000)
18. M. Evans, M. Cannon, B. Kouvaritakis, Robust MPC for linear systems with bounded multiplicative uncertainty, in *Proceedings of the 51st IEEE Conference on Decision and Control*, Maui, USA (2012), pp. 248–253
19. J. Fleming, B. Kouvaritakis, M. Cannon, Regions of attraction and recursive feasibility in robust MPC, in *Proceedings of the 21st Mediterranean Conference on Control and Automation*, Chania, Greece (2013), pp. 801–806
20. J. Fleming, B. Kouvaritakis, M. Cannon, Robust tube MPC for linear systems with multiplicative uncertainty. IEEE Trans. Autom. Control **60**(4), 1087–1092 (2015)
21. Y.C. Gautam, A. Chu, Y.C. Soh, Optimized dynamic policy for receding horizon control of linear time-varying systems with bounded disturbances. IEEE Trans. Autom. Control **57**(4), 973–998 (2012)
22. D. Muñoz-Carpintero, M. Cannon, B. Kouvaritakis, Recursively feasible robust MPC for linear systems with additive and multiplicative uncertainty using optimized polytopic dynamics, in *Proceedings of the 52nd IEEE Conference on Decision and Control*, Florence, Italy (2013),pp. 1101–1106
23. D. Muñoz-Carpintero, M. Cannon, B. Kouvaritakis, Robust MPC strategy with optimized polytopic dynamics for linear systems with additive and multiplicative uncertainty. Syst. Control Lett. (2015), in press
24. S.P. Boyd, L. El Ghaoui, E. Feron, V. Balakrishnan, *Linear Matrix Inequalities in System and Control Theory* (Society for Industrial and Applied Mathematics, Philadelphia, 1994)
25. B. Kouvaritakis, J.A. Rossiter, A.O.T. Chang, Stable generalised predictive control: an algorithm with guaranteed stability. Control Theory Appl., IEE Proc. D **139**(4), 349–362 (1992)
26. T. Barjas Blanco, M. Cannon, B. De Moor, On efficient computation of low-complexity controlled invariant sets for uncertain linear systems. Int. J. Control **83**(7), 1339–1346 (2010)
27. F. Blanchini, Ultimate boundedness control for uncertain discrete-time systems via set-induced Lyapunov functions. IEEE Trans. Autom. Control **39**(2), 428–433 (1994)

28. P.A. Parrilo, A. Jadbabaie, Approximation of the joint spectral radius using sum of squares. Linear Algebra Appl. **428**(10), 2385–2402 (2008)
29. M. Cannon, V. Deshmukh, B. Kouvaritakis, Nonlinear model predictive control with polytopic invariant sets. Automatica **39**(8), 1487–1494 (2003)
30. Y. Nesterov, A. Nemirovsky, *Interior-Point Polynomial Algorithms in Convex Programming* (SIAM, Philadelphia, 1994)
31. G. Bitsoris, On the positive invariance of polyhedral sets for discrete-time systems. Syst. Control Lett. **11**(3), 243–248 (1988)
32. A. Benzaouia, C. Burgat, Regulator problem for linear discrete-time systems with non-symmetrical constrained control. Int. J. Control **48**(6), 2441–2451 (1988)
33. R. Fletcher, *Practical Methods of Optimization*, 2nd edn. (Wiley, New York, 1987)
34. F. Blanchini, S. Miani, *Set-Theoretic Methods in Control* (Birkhäuser, Boston, 2008)
35. B. Pluymers, J.A. Rossiter, J.A.K. Suykens, B. De Moor, The efficient computation of polyhedral invariant sets for linear systems with polytopic uncertainty, in *Proceedings of the 2005 American Control Conference* (2005), pp. 804–809

Part III
Stochastic MPC

Chapter 6
Introduction to Stochastic MPC

Uncertainty forms an integral part of most control problems and earlier chapters discussed how MPC algorithms can be constructed in order to treat model uncertainty in a robust sense. One of the key features of robust MPC is that it requires constraints to be satisfied for all possible realizations of uncertainty. Thus each element of the set of values that can be assumed by an uncertain model parameter or disturbance input is treated with equal importance, and robust MPC does not discriminate between alternative realizations on the basis of their respective likelihood.

However, in practical applications it is often the case that some realizations of model uncertainty, for example parameter realizations that lie close to the nominal value of that parameter, are more likely than others, such as parameter realizations that lie on the boundary of the uncertainty set. In fact model uncertainty is often stochastic with a probability distribution that is known, either as a result of statistical analysis performed during model identification, or because of physical principles underlying the model. Clearly the distribution of model uncertainty is useful information that should be taken into account in the design of MPC algorithms.

An obvious way to address this is through the definition of the cost. Thus, rather than being defined as a nominal or worst-case value, the MPC cost can be chosen to be the expected value, over the distribution of uncertainty, of the usual quadratic predicted cost. An additional and often more crucial use of distribution information corresponds to the case in which some or all of the system constraints are probabilistic in nature. In this case, constraint violations are permitted provided the frequency of constraint violations (or more generally the number of violations in a given time interval) is below a predefined limit.

An example of a control problem involving probabilistic constraints concerns the optimal allocation of resources for sustainable development. Consider, for example, investment in electricity generating technology, where it is important to minimize (among other objectives) the cost of energy for the consumer while meeting (among other constraints) limits on emissions of greenhouse gases. Taking the horizon of

B. Kouvaritakis and M. Cannon, *Model Predictive Control*,
Advanced Textbooks in Control and Signal Processing,
DOI 10.1007/978-3-319-24853-0_6

interest to be the 30 years separating generations, it does not make sense to expect the accumulated emissions of CO_2 over 30 years, say y, to be less than a given amount, say A, because of the various unknown factors associated with economic and technological development over a horizon of this length. In fact, given that such factors are stochastic in nature, it follows that y itself will be a random variable, so that a constraint that $y \leq A$ is meaningless. If, however, information is available on the probability distribution of y, then an aspirational constraint such as $\Pr\{y \leq A\} \geq p$, namely that the probability that y is less that a target bound A should exceed a given level p, is eminently sensible [1, 2].

Many other examples where probabilistic constraints are more natural than deterministic constraints can be found in diverse areas of engineering and related fields such as process control [3, 4], financial engineering [5, 6], electricity generation, distribution and pricing [7, 8], building climate control [9] and telecommunications network traffic control [10]. For the purposes of motivation, we briefly describe here another example taken from a problem that concerns the operation of wind turbines. Large wind turbines for electricity generation are typically designed for a given service life, which is typically around 20 years, but this lifespan may be compromised as a result of fatigue damage caused, for example, by the fore-aft movement of the tower. Controlling this movement so as to control the rate of accumulation of fatigue damage in the wind turbine forms one of the objectives of a supervisory controller for variable speed wind turbines (see e.g. [11]). However, in order to maximize electrical power output, it is possible to allow violations of the constraints on the tower oscillations, provided these do not happen more often than some pre-specified limit. Such violations occur as a result of fluctuations in wind speed, the variability of which can be modelled by suitable probability distributions (e.g. [12]). This approach therefore imposes constraints on stochastic variables, and the implied constraints are naturally stated in a probabilistic manner.

Exacting performance requirements often cause constrained variables to reach their limits and in this sense it is vital that the definition of the system constraints takes into account the stochastic nature of the given application. Even if information on the probability distribution of model uncertainty is available, it may of course be possible to act cautiously and enforce constraints robustly, namely to impose conditions that require constraints to be satisfied with certainty. But in cases where constraint violation is allowed (up to a specified probability) it should of course be evident that such a strategy is conservative, and will result in poorer performance and smaller regions of attraction.

It is the purpose of Stochastic MPC (SMPC) to address these issues. In particular, SMPC is concerned with the repetitive optimization of an appropriate predicted cost for systems with stochastic uncertainty. This optimization is to be performed subject to constraints, some of which are probabilistic in nature. Constraint satisfaction requires the determination of probability distributions for predicted variables, something which is relatively easy in the case of additive uncertainty only, given the linear dependence of predictions on disturbances. Stochastic uncertainty that appears multiplicatively in the system model, for example as a result of stochastic model parameters, is often more challenging. This is because the predicted future model states are random variables, and these are multiplied by stochastic model parameters to generate the successor states, compounding the problem of determining the distributions of predicted states over multiple prediction time steps.

As with classical and robust MPC, there is a concern here with recursive feasibility. In fact it is easy to see (as will be discussed in detail Chap. 7) that, except in certain special cases, strict recursive feasibility can only be guaranteed for the case that the uncertainty in predicted states and inputs has a finitely supported distribution. The reason for this is that whereas the current state may be such that a feasible predicted trajectory exists, uncertainty with unbounded support could, albeit perhaps with low probability, result in a successor state for which it is impossible to guarantee the existence of a feasible trajectory. This particular difficulty has been a feature of earlier SMPC formulations, which almost exclusively considered uncertainty with Gaussian distributions. This choice was natural given the mathematical convenience of the Gaussian assumption. However, in addition to preventing the statement of recursive feasibility results, this assumption is often not consistent with practice, since for many physical systems the probability of an uncertainty realization exceeding an arbitrarily large threshold is zero rather than arbitrarily small. SMPC is also concerned with the definition of a suitable cost that enables the statement of stability.

This chapter introduces the SMPC problem formulation, describes earlier work in this area and discusses quadratic expected value costs. Probabilistic constraints are introduced here in the context of uncertain moving average models, for which recursive feasibility can be ensured even when the probability distribution of model uncertainty is not finitely supported. The discussion of probabilistic constraints for general classes of linear system and model uncertainty is deferred to Chap. 7, where the requirements for a guarantee of recursive feasibility and methods of handling general probability distributions are considered. Chapter 7 also extends the framework for analysing closed-loop stability to the case of expected value costs. Chapter 8 combines these techniques with tube-based methods of constraint handling in order to construct stochastic MPC algorithms with guaranteed closed-loop properties for general classes of model uncertainty and probabilistic constraints.

6.1 Problem Formulation

There are several ways to obtain mathematical descriptions of systems that are subject to stochastic uncertainty. Perhaps the most convenient is through black box identification of auto-regressive moving average (ARMA) relationships between the input variable u and output variable y:

$$\begin{aligned} y_k + a_1 y_{k-1} + \cdots + a_n y_{k-n} \\ = b_0 u_{k-d} + \cdots + b_n u_{k-d-n} + e_k + c_1 e_{k-1} + \cdots + c_n e_{k-n}, \end{aligned} \tag{6.1}$$

where e denotes a zero-mean disturbance and the positive integer d accounts for the delay in the system. This type of model is general enough to include systems with multiple inputs and outputs since y, u, e can be vector-valued and a, b, c matrices of conformal dimensions. Descriptions of this kind (which include moving average (MA) models as a special case when the parameters $a_1, \ldots, a_n$ are identically equal to zero) have been used in diverse fields including, for example, process control and econometrics [13–15]. A useful by-product of black box identification is that it provides information on the statistical properties of the model parameters [16]. This could be obtained from a single experiment on the basis of information on the noise affecting the system output measurements or by repeated experiments capturing different realizations of the uncertain parameters.

To use the constraint handling machinery of invariant sets and state and control tubes, it is convenient to convert the ARMA model (6.1) into a state-space form. Given the linearity of (6.1), this can be expressed as

$$x_{k+1} = A_k x_k + B_k u_k + D w_k \tag{6.2}$$

where $x_k \in \mathbb{R}^{n_x}$ and $u_k \in \mathbb{R}^{n_u}$ are the state and control inputs at time k. In setting up the stochastic MPC problem, we make the assumption that the matrices A_k and B_k containing multiplicative model parameters and the additive disturbance input $w_k \in \mathbb{R}^{n_w}$ can be expressed in terms of a linear expansion over a known basis set:

$$(A_k, B_k, w_k) = (A^{(0)}, B^{(0)}, 0) + \sum_{j=1}^{\rho} (A^{(j)}, B^{(j)}, w^{(j)}) q_k^{(j)}. \tag{6.3a}$$

Here $q^{(j)}$ is a scalar random variable. The realization of $q^{(j)}$ at time k, denoted $q_k^{(j)}$, is unknown at time k but has a known probability distribution.[1] The vector $q_k = (q_k^{(1)}, \ldots, q_k^{(\rho)})$ may be time varying (in the sense that it has a different realization at each time instant k) but q_k is assumed to be identically distributed for each k.

We assume that q_k has a mean value of zero and covariance matrix equal to the identity matrix, namely

$$\mathbb{E}(q_k) = 0, \quad \mathbb{E}(q_k q_k^T) = I \tag{6.3b}$$

where $\mathbb{E}(\cdot)$ denotes the expectation operator. These assumptions do not imply any loss of generality because of the linearity of the dynamics (6.2) and the expansion (6.3a). In particular, the expected value of the additive disturbance is taken to be zero since a state translation can be used to account for any non-zero components. Thus a non-zero value of $\mathbb{E}(q_k)$ and a corresponding non-zero expected value

[1] Unless explicitly stated otherwise, all random variables that will be encountered in the discussion of stochastic MPC are functions of $q^{(j)}$ appearing in (6.3a) and every random event of interest is related in a straightforward way to the realizations of $q_k^{(j)}, j = 1, \ldots, \rho$ at time instant k. With this understanding, the underlying probability space is well defined and we are able to avoid cumbersome measure-theoretic notation.

of the disturbance, $\mathbb{E}(w_k) = w^{(0)}$, can be absorbed by transforming the state and disturbance variables of (6.2) according to $x_k - x^{(0)}$ and $w_k - w^{(0)}$, respectively, where $x^{(0)} = (I - A^{(0)})^{-1}Dw^{(0)}$, and by replacing q_k with $q_k - \mathbb{E}(q_k)$ in (6.3a) and appropriately redefining $A^{(0)}$ and $B^{(0)}$.

Likewise, the assumption that $\mathbb{E}(q_k q_k^T) = I$ is justified by a linear transformation since (6.3a) can be expressed equivalently as

$$\begin{aligned}\text{vec}(A_k, B_k, w_k) &= \text{vec}(A_0, B_0, 0) \\ &\quad + \left[\text{vec}(A^{(1)}, B^{(1)}, w^{(1)}) \cdots \text{vec}(A^{(m)}, B^{(m)}, w^{(m)})\right] q_k\end{aligned}$$

where vec$(\cdot)$ is a vectorization operation that rearranges the elements of a matrix into a column vector. Therefore, if the covariance matrix of q_k has the eigenvalue decomposition

$$\mathbb{E}(q_k q_k^T) = W_\Sigma \Lambda_\Sigma W_\Sigma^T$$

then it is possible to define triples $(\tilde{A}^{(j)}, \tilde{B}^{(j)}, \tilde{w}^{(j)})$ for $j = 1, \ldots, m$ and a vector $\tilde{q}$ by

$$\begin{aligned}\text{vec}(\tilde{A}^{(j)}, \tilde{B}^{(j)}, \tilde{w}^{(j)}) &= \text{vec}(A^{(j)}, B^{(j)}, w^{(j)}) W_\Sigma \Lambda_\Sigma^{1/2} \\ \tilde{q}_k &= \Lambda_\Sigma^{-1/2} W_\Sigma^T q_k.\end{aligned}$$

This transformation leaves the parameterization of (6.3a) unaffected because

$$\begin{aligned}\sum_{j=1}^{m} (\tilde{A}^{(j)}, \tilde{B}^{(j)}, \tilde{w}^{(j)}) \tilde{q}_k^{(j)} &= \left[\text{vec}(A^{(1)}, B^{(1)}, w^{(1)}) \cdots \text{vec}(A^{(m)}, B^{(m)}, w^{(m)})\right] \tilde{q}_k \\ &= \sum_{j=1}^{m} (A^{(j)}, B^{(j)}, w^{(j)}) q_k^{(j)},\end{aligned}$$

but at the same time the covariance matrix of the transformed vector of coefficients is given by $\mathbb{E}(\tilde{q}_k \tilde{q}_k^T) = I$.

We make the further assumption that q_k and q_i are statistically independent for $k \neq i$. This assumption is not necessarily restrictive in respect of the additive disturbance because a linear filter can be introduced into the state-space model (6.2) in order to generate temporally correlated disturbances if required. But the same approach cannot be used to conveniently introduce temporal correlation between multiplicative parameters in the model (6.2) since the dynamics are assumed to be linear. However, the assumption of independence of q_k and q_i is used here only to simplify the computation of predicted costs based on the expected value of sums of quadratic stage costs, and to simplify the analysis of stability based on this cost. The methods of handling constraints that are discussed here and in the following chapters do not rely on this assumption.

Like the nominal and robust MPC strategies considered in earlier chapters, the predicted performance cost of stochastic MPC is often taken to be a quadratic function of the degrees of freedom in predicted state and control trajectories. For example, for an N-step horizon we can define

$$\hat{J}(x_k, \mathbf{u}_k, \mathbf{q}_k) = \sum_{i=0}^{N-1} \big(\|x_{i|k}\|_Q^2 + \|u_{i|k}\|_R^2\big) + \|x_{N|k}\|_{W_T}^2$$

where $x_{i|k}, u_{i|k}$ are predicted values of x_{k+i}, u_{k+i} at time k with $x_k = x_{0|k}$ and $\mathbf{u}_k = \{u_{0|k}, \ldots, u_{N-1|k}\}$, where $\mathbf{q}_k = \{q_{0|k}, \ldots, q_{N-1|k}\}$ is a realization of the sequence $\{q_k, \ldots, q_{N-1}\}$ of uncertain model parameters, and where matrices $Q \succeq 0$ and $R \succ 0$ are cost weights with W_T a terminal weighting matrix. On account of the stochastic uncertainty in $\mathbf{q}_k$, the cost $\hat{J}(x_k, \mathbf{u}_k, \mathbf{q}_k)$ is stochastic and the cost index, $J(x_k, \mathbf{u}_k)$, of stochastic MPC must therefore be constructed under specific assumptions on $\mathbf{q}_k$. Thus it is possible to adopt the nominal cost

$$J(x_k, \mathbf{u}_k) \doteq \hat{J}(x_k, \mathbf{u}_k, 0),$$

or, if the model uncertainty is subject to bounds $q_k \in \mathcal{Q}$ for some compact set $\mathcal{Q}$, then the worst-case cost can be employed,

$$J(x_k, \mathbf{u}_k) \doteq \max_{\mathbf{q}_k \in \mathcal{Q} \times \cdots \times \mathcal{Q}} \hat{J}(x_k, \mathbf{u}_k, \mathbf{q}_k).$$

Given knowledge of the distribution of q_k, it is more common, however, to use a predicted cost that takes into account the stochastic nature of the problem through the expectation of a quadratic cost. Therefore we focus the discussion of stochastic MPC on an expected cost of the form

$$\begin{aligned} J(x_k, \mathbf{u}_k) &\doteq \mathbb{E}_k\big(\hat{J}(x_k, \mathbf{u}_k, \mathbf{q}_k)\big) \\ &= \sum_{i=0}^{N-1} \mathbb{E}_k\big(\|x_{i|k}\|_Q^2 + \|u_{i|k}\|_R^2\big) + \mathbb{E}_k\big(\|x_{N|k}\|_{W_T}^2\big). \end{aligned} \tag{6.4}$$

Here, the notation $\mathbb{E}_k(\cdot)$ has been introduced[2] to indicate that the expectation is conditional on information available to the controller at time k, namely the current plant state x_k (for the case that x_k is measured directly), and is therefore dependent on the distribution of the model uncertainty sequence $\mathbf{q}_k$. Variations on this cost will also be discussed, such as quadratic costs based on probabilistic bounds on predicted variables or on combinations of their means and variances.

[2]We use the simpler notation $\mathbb{E}(\cdot)$ for the expectation of a random variable that depends on a single realization of model uncertainty, e.g. $\mathbb{E}(A_k) = \mathbb{E}_k(A_k) = A^{(0)}$.

The aim of stochastic MPC is to obtain, through the repetitive minimization of a predicted cost, an approximation of the optimal control law, namely the optimal argument of (6.4) subject to constraints on the system states and/or control inputs. This is to be done while providing stability guarantees for all initial conditions in some region of state space. A precondition for such guarantees is that the expected value of the predicted stage cost in (6.4) converges to a finite limit as $i \to \infty$ under some control law. This in turn requires that the pair (A_k, B_k) is mean-square stabilizable [17], which is ensured by the following assumption.

Assumption 6.1 There exist matrices K and P such that $P = P^T \succ 0$ and

$$P - \mathbb{E}\Big((A_k + B_k K)^T P (A_k + B_k K)\Big) \succ 0. \tag{6.5}$$

It can be determined whether the system of (6.2) and (6.3a, 6.3b) satisfies Assumption 6.1 simply by checking the feasibility of a linear matrix inequality. Thus, using (6.3a, 6.3b) condition (6.5) can be expressed equivalently as

$$P - \sum_{j=0}^{m} (A^{(j)} + B^{(j)}K)^T P (A^{(j)} + B^{(j)}K) \succ 0.$$

Introducing convexifying transformations $S = P^{-1}$, and $Y = KP^{-1}$ and using Schur complements this condition can be written as

$$\begin{bmatrix} S & (A^{(0)}S + B^{(0)}Y)^T & (A^{(1)}S + B^{(1)}Y)^T & \cdots & (A^{(m)}S + B^{(m)}Y)^T \\ \star & S & 0 & \cdots & 0 \\ \star & \star & S & \cdots & 0 \\ \vdots & \vdots & \vdots & \ddots & \vdots \\ \star & \star & \star & \cdots & S \end{bmatrix} \succ 0 \tag{6.6}$$

(with $\star$ indicating a block of a symmetric matrix). Thus the conditions of Assumption 6.1 are satisfied by $P = S^{-1}$ and $K = YS^{-1}$ if and only if matrices $S = S^T \succ 0$ and Y exist satisfying (6.6).

For the case of multiplicative uncertainty alone and in the absence of constraints, Assumption 6.1 implies that the control law $u_k = Kx_k$ ensures that the variance of the state of (6.2) converges asymptotically to zero. This guarantees the existence of a predicted control sequence $\{u_{i|k},\ i = 0, 1, \ldots\}$ that causes the predicted state $x_{i|k}$ to converge (with probability 1) to zero as $i \to \infty$ [17]. Clearly, no feedback law can make the state converge identically to zero in the presence of persistent additive disturbances. In this case, Assumption 6.1 ensures, in the absence of constraints, that the expected value of the stage cost of (6.4) tends to a finite limit, which is discussed in Sect. 6.2.

Analogously to the formulations of classical MPC and robust MPC, a dual-mode prediction strategy can be used in stochastic MPC to define predicted state and control trajectories over an infinite prediction horizon while retaining only a finite number of free variables in their parameterization. This makes it possible to extend the cost of (6.4) over an infinite horizon by suitably choosing the weighting matrix W. We therefore assume that the predicted control law over the prediction horizon of mode 2 is defined by a predetermined feedback law: $u_{i|k} = Kx_{i|k}$ for all $i \geq N$.

Several different forms of state and input constraints have been proposed for stochastic MPC in the context of a model such as (6.2) and (6.3a, 6.3b). For example, constraints may be stated in terms of expected values as

$$\mathbb{E}_k(Fx_{i|k} + Gu_{i|k}) \leq \mathbf{1}, \quad i = 1, 2, \ldots \tag{6.7}$$

(e.g. [6, 10]). Alternatively, pointwise in time probabilistic constraints can be stated as

$$\Pr_k(Fx_{1|k} + Gu_{1|k} \leq \mathbf{1}) \geq p \tag{6.8}$$

for some specified probability p. An alternative form of probabilistic constraint can be imposed over an interval of T time steps:

$$\Pr_k(Fx_{i|k} + Gu_{i|k} \leq \mathbf{1},\ i = 1, \ldots, T) \geq p \tag{6.9}$$

for some given probability p and horizon T. In each case, the variable $Fx + Gu$ may be vector-valued; thus for example (6.8) requires that the probability of any element of $Fx_{k+1|k} + Gu_{k+1|k}$ exceeding a threshold of 1 should be less than $1 - p$.

An equivalent statement of (6.9) is that the expected number of times that any element of $Fx + Gu$ exceeds 1 over an interval of T predicted time steps should be less than $(1-p)T$. Clearly, it is possible to construct constraint sets involving various different probabilities and probabilistic conditions by combining constraints of the form of (6.8) and (6.9). This also applies to the special case of $p = 1$, thus allowing for problems that involve a mixture of probabilistic ($p < 1$) and robust ($p = 1$) constraints.

The notation $\Pr_k(\mathcal{A})$ in (6.8) and (6.9) refers to the probability of an event $\mathcal{A}$ that depends[3] on the sequence $\mathbf{q}_k = \{q_k, \ldots, q_{k+N-1}\}$, given that the initial prediction model state is x_k. Hence, $\Pr_k(Fx_{1|k} + Gu_{1|k} \leq \mathbf{1})$ is the probability that the one-step ahead predicted state and input satisfy $Fx_{1|k} + Gu_{1|k} \leq \mathbf{1}$, which is therefore a function of x_k and the probability distribution of the random variable q_k. Similarly, $\Pr_k(Fx_{i|k} + Gu_{i|k} \leq \mathbf{1})$ for $i \geq 1$ depends on x_k and on the probability distribution of $\{q_k, \ldots, q_{k+i-1}\}$. Our treatment of stochastic MPC concentrates on problem formulations for which recursive feasibility can be guaranteed, and we consider mainly the constraints of (6.8) and (6.9). A detailed discussion of recursive feasibility is given in Sect. 7.1.

[3]We also use $\Pr(\mathcal{A})$ for the probability of an event $\mathcal{A}$ that depends on a single realization of model uncertainty when it is obvious from the context which random variable $\mathcal{A}$ depends on, e.g. $\Pr(w_k \leq 0) = \Pr_k(w_k \leq 0) = \Pr(\sum_{j=1}^{m} w^{(j)} q_k^{(j)} \leq 0)$.

6.2 Predicted Cost and Unconstrained Optimal Control Law

Early stochastic MPC strategies, such as those discussed in Sect. 6.4, ensure closed-loop stability by imposing equality terminal constraints on the predicted input and output trajectories of the plant model. In the context of a moving average model, this can be done simply by setting the predicted input equal to a constant value at the end of the mode 1 prediction horizon. After a subsequent interval of n prediction time steps, where n is the order of the model, the expected value of the predicted output necessarily reaches its desired steady-state value. However equality terminal constraints can be overly stringent and may result in limited regions of attraction. A way to avoid this is to replace equality with inequality terminal constraints (as discussed in Chap. 2) and in this setting, recursive feasibility and closed-loop stability can be ensured through the use of invariant terminal sets defined in the plant model state space.

The construction of suitable terminal sets and the handling of constraints are key ingredients in the development of a stochastic MPC strategy, and will be considered in more detail in Chaps. 7 and 8. Before discussing these, however, we turn our attention to another two basic components of SMPC. The first of these concerns the definition and evaluation of a predicted cost, while the second concerns the unconstrained optimal feedback law that achieves the minimum of the predicted cost in the absence of constraints. Such unconstrained control laws are obvious candidates for use as terminal control laws.

We begin with the observation that the mean-square stability property of Assumption 6.1 implies that, when no constraints are present, the expected value of the state x_k of the model (6.2) and (6.3a, 6.3b) under the linear feedback law $u_k = Kx_k$ converges asymptotically to zero whenever K satisfies (6.5). However, the variance of x_k (and hence also the expected stage cost in (6.4)) converges to a non-zero value because of the presence of additive model uncertainty, as the following lemma shows.

Lemma 6.1 *[18] In the absence of constraints, the state of (6.2) and (6.3a, 6.3b) under* $u_k = Kx_k$ *satisfies the asymptotic conditions* $\lim_{k\to\infty} \mathbb{E}_0(x_k) = 0$ *and* $\lim_{k\to\infty} \mathbb{E}_0(x_k x_k^T) = \Theta$*, where* Θ *is the solution of*

$$\Theta - \mathbb{E}\Big((A_k + B_k K)\Theta(A_k + B_k K)^T\Big) = D\mathbb{E}(w_k w_k^T)D^T, \tag{6.10}$$

if and only if (6.5) holds for some $P \succ 0$.

Proof Since the system (6.2) is linear, its state can be decomposed for all $k = 0, 1, \ldots$ as $x_k = \zeta_k + \xi_k$, with

$$\zeta_{k+1} = \Phi_k \zeta_k, \qquad \zeta_0 = x_0 \tag{6.11a}$$

$$\xi_{k+1} = \Phi_k \xi_k + Dw_k, \qquad \xi_0 = 0 \tag{6.11b}$$

where $\Phi_k = A_k + B_k K$ and A_k, B_k take values from the uncertainty class of (6.3a, 6.3b). Existence of $P \succ 0$ satisfying (6.5) is necessary and sufficient for mean-square stability of (6.11a). To prove sufficiency, let $Z_k \doteq \mathbb{E}_0(\zeta_k \zeta_k^T)$, then, since ζ_k and Φ_k are by assumption independent, (6.11a) gives

$$\operatorname{tr}(PZ_{k+1}) = \mathbb{E}\Big(\operatorname{tr}(P\Phi_k Z_k \Phi_k^T)\Big) = \operatorname{tr}\Big(\mathbb{E}(\Phi_k^T P \Phi_k) Z_k\Big).$$

Hence (6.5) and $Z_k \succeq 0$ imply that $\operatorname{tr}(PZ_{k+1}) < \operatorname{tr}(PZ_k)$ whenever $Z_k \neq 0$, and since $P \succ 0$ implies $\operatorname{tr}(PZ_k) > 0$ for all $Z_k \neq 0$, it follows that

$$\lim_{k\to\infty} Z_k = 0, \tag{6.12}$$

which implies that (6.11a) is mean-square stable. To show the necessity of (6.5), let $P_k \doteq \sum_{i=0}^{k} S_i$ where $S_{k+1} = \mathbb{E}(\Phi_k S_k \Phi_k)$ for arbitrary $S_0 \succ 0$. Then $P_{k+1} = \mathbb{E}(\Phi_k P_k \Phi_k^T) + S_0$ and the mean-square stability of (6.11a) implies that P_k converges to a finite limit as $k \to \infty$. Defining this limit as P, we can conclude from $S_0 \succ 0$ that $P \succ 0$ exists satisfying (6.5) whenever (6.11a) is mean-square stable.

From the mean-square stability property (6.12), it follows that $\zeta_k \to 0$ as $k \to \infty$ with probability 1 [17]. On the other hand, (6.11b) implies $\mathbb{E}_0(\xi_k) = 0$ for all k, and since $x_k = \zeta_k + \xi_k$ we can therefore conclude that $\lim_{k\to\infty} \mathbb{E}_0(x_k) = 0$.

Using (6.11b) and the zero-mean property of ξ_k, and noting that ξ_k and Φ_k are independent by assumption, we obtain

$$\mathbb{E}_0(\xi_{k+1}\xi_{k+1}^T) = \mathbb{E}\Big(\Phi_k \mathbb{E}_0(\xi_k \xi_k^T)\Phi_k^T\Big) + D\mathbb{E}(w_k w_k^T)D^T.$$

Combining this relationship with (6.10) and defining $\hat{\Theta}_k \doteq \mathbb{E}_0(\xi_k \xi_k^T) - \Theta$, it follows that

$$\hat{\Theta}_{k+1} = \mathbb{E}(\Phi_k \hat{\Theta}_k \Phi_k^T).$$

Therefore, $\lim_{k\to\infty} \hat{\Theta}_k = 0$ if and only if (6.11a) is mean-square stable, in which case we have $\lim_{k\to\infty} \mathbb{E}(\xi_k \xi_k^T) = \Theta$. This completes the proof because it then follows that $\lim_{k\to\infty} \mathbb{E}(x_k x_k^T) = \Theta$ since $\zeta_k \to 0$ with probability 1 as $k \to \infty$. □

Note that the parameterization of model uncertainty in (6.3a, 6.3b) allows the expectation in (6.10) to be evaluated explicitly. Hence (6.10) is equivalent to a set of linear conditions on the elements of Θ:

$$\Theta - \sum_{j=0}^{m} (A^{(j)} + B^{(j)}K)\Theta(A^{(j)} + B^{(j)}K)^T = D\sum_{j=1}^{m} w^{(j)} w^{(j)\,T} D^T,$$

which can be shown to yield a unique positive definite solution for Θ whenever Assumption 6.1 is satisfied.

Many stochastic control problems are formulated in terms of an expected infinite horizon quadratic performance index of the form

$$\sum_{i=0}^{\infty} \mathbb{E}_0\big(\|x_k\|_Q^2 + \|u_k\|_R^2\big). \tag{6.13}$$

However, in the context of stochastic MPC two difficulties with this cost are immediately apparent. First, the cost is evaluated over an infinite sequence of control inputs, and in order to formulate an objective that can be minimized numerically subject to input and state constraints, this infinite control sequence must therefore be parameterized in terms of a finite number of optimization variables. Second, a consequence of Lemma 6.1 is that the cost (6.13) is in general infinite for the system (6.2) and (6.3a, 6.3b) under a linear feedback law. Moreover, in the absence of constraints the optimal controller is a linear-state feedback law (as we show later in this section), and therefore the minimum value of cost (6.13) is in general also infinite.

The first of these issues can be handled by introducing a dual-mode prediction scheme as was done for nominal and robust MPC in Sects. 2.3 and 3.1. Therefore we define the predicted control sequence at time k as

$$u_{i|k} = Kx_{i|k} + c_{i|k}, \quad i = 0, 1, \ldots \tag{6.14}$$

where $\{c_{0|k}, \ldots, c_{N-1|k}\}$ are decision variables at time k and $c_{i|k} = 0$ for all prediction times $i \geq N$. The gain K is assumed to satisfy the mean-square stability condition (6.5), and ideally K should be chosen as the optimal feedback gain in the absence of constraints. To tackle the second issue, we ensure that the minimum value of the predicted cost associated with this predicted input sequence is finite by subtracting a constant from each stage of (6.13). Lemma 6.1 implies that the expected value of the stage cost for the system (6.2) and (6.3a, 6.3b) under $u_{i|k} = Kx_{i|k}$ converges to a steady-state value, which we denote as l_{ss}:

$$l_{ss} \doteq \lim_{i\to\infty} \mathbb{E}_k\big(\|x_{i|k}\|_Q^2 + \|u_{i|k}\|_R^2\big) = \operatorname{tr}\Big(\Theta(Q + K^T RK)\Big).$$

We therefore define the predicted cost as

$$J(x_k, \mathbf{c}_k) = \sum_{i=0}^{\infty} \mathbb{E}_k\big(\|x_{i|k}\|_Q^2 + \|u_{i|k}\|_R^2 - l_{ss}\big) \tag{6.15}$$

where $\mathbf{c}_k \in \mathbb{R}^{Nn_u}$ is defined by $\mathbf{c}_k \doteq (c_{0|k}, \ldots, c_{N-1|k})$.

From its definition in (6.15), it is clear that the cost $J(x_k, \mathbf{c}_k)$ is a quadratic function of the optimization variable $\mathbf{c}_k$. The most convenient way to compute this cost function is to express the predicted dynamics using the lifted autonomous formulation of Sect. 2.7. For the case of the model (6.2), this has the following form:

$$z_{i+1|k} = \Psi_{k+i} z_{i|k} + \bar{D} w_{k+i}, \quad z_{0|k} = \begin{bmatrix} x_k \\ \mathbf{c}_k \end{bmatrix} \tag{6.16}$$

with state variable $z_{i|k} \in \mathbb{R}^{n_z}$, $n_z = n_x + Nn_u$. Here

$$\Psi_k = \begin{bmatrix} \Phi_k & B_k E \\ 0 & M \end{bmatrix}, \quad \Phi_k = A_k + B_k K, \quad \bar{D} = \begin{bmatrix} D \\ 0 \end{bmatrix}$$

and the matrices E and M are defined as in (2.26b) so that $E\mathbf{c}_k = c_{0|k}$ and $M\mathbf{c}_k = (c_{1|k}, \ldots, c_{N-1|k}, 0)$. This autonomous formulation is the basis of the following result for evaluating the predicted cost.

Theorem 6.1 *[18] The predicted cost of (6.15), computed for the model (6.2) under the control law (6.14) is given by*

$$J(x_k, \mathbf{c}_k) = \begin{bmatrix} z_k \\ 1 \end{bmatrix}^T \begin{bmatrix} W_z & w_{z1} \\ w_{z1}^T & w_1 \end{bmatrix} \begin{bmatrix} z_k \\ 1 \end{bmatrix}, \tag{6.17}$$

where $W_z = W_z^T \in \mathbb{R}^{n_z \times n_z}$, $w_{z1} \in \mathbb{R}^{n_z}$ *and* $w_1 \in \mathbb{R}$ *are defined by*

$$W_z - \mathbb{E}(\Psi_k^T W_z \Psi_k) = \hat{Q} \tag{6.18a}$$

$$w_{z1}^T \Big(I - \mathbb{E}(\Psi^{(0)})\Big) = \mathbb{E}\big(w_k^T \bar{D}^T W_z \Psi_k\big) \tag{6.18b}$$

$$w_1 = -\mathrm{tr}(\Theta W_x) \tag{6.18c}$$

with $W_x = \begin{bmatrix} I_{n_x} & 0 \end{bmatrix} W_z \begin{bmatrix} I_{n_x} \\ 0 \end{bmatrix}$ *and* $\hat{Q} = \begin{bmatrix} Q + K^T RK & K^T RE \\ E^T RK & E^T RE \end{bmatrix}$.

Proof Let $V_{i|k} \doteq \|z_{i|k}\|_{W_z}^2 + 2w_{z1}^T z_{i|k} + w_1$ for all $i \geq 0$. Then, since $z_{i|k}$ is by assumption independent of (Ψ_{k+i}, w_{k+i}), (6.16) implies

$$\begin{aligned} \mathbb{E}_k(V_{i|k}) - \mathbb{E}_k(V_{i+1|k}) = {} & \mathbb{E}_k\Big(z_{i|k}^T \big(W_z - \mathbb{E}(\Psi_{k+i}^T W_z \Psi_{k+i})\big) z_{i|k}\Big) \\ & + 2\Big(w_{z1}^T \big(I - \mathbb{E}(\Psi_{k+i})\big) - \mathbb{E}\big(w_{k+i}^T \bar{D}^T W_z \Psi_{k+i}\big)\Big)\mathbb{E}_k(z_{i|k}) \\ & - \mathbb{E}\left(w_{k+i}^T D^T W_x D w_{k+i}\right). \end{aligned}$$

From (6.18a, 6.18b), the sum of the first two terms on the RHS of this equation is equal to $\mathbb{E}_k(z_{i|k}^T \hat{Q} z_{i|k})$. Furthermore the last term is equal to $-l_{ss}$ since post-multiplying (6.10) by W_x and extracting the trace gives

$$\begin{aligned} \mathrm{tr}\Big(\Theta W_x - \mathbb{E}(\Phi_k \Theta \Phi_k^T) W_x\Big) &= \mathrm{tr}\Big(\Theta\big(W_x - \mathbb{E}(\Phi_k^T W_x \Phi_k)\big)\Big) \\ &= \mathrm{tr}\big(D\mathbb{E}(w_k w_k^T) D^T W_x\big) = \mathbb{E}(w_k^T D^T W_x D w_k), \end{aligned}$$

and hence, noting that (6.18a) implies $W_x - \mathbb{E}(\Phi_k^T W_x \Phi_k) = Q + K^T RK$, we have

$$\mathbb{E}(w_k^T D^T W_x D w_k) = \mathrm{tr}\big(\Theta(Q + K^T RK)\big) = l_{ss}.$$

Therefore $\mathbb{E}_k(V_{i|k}) - \mathbb{E}_k(V_{i+1|k}) = \mathbb{E}_k(\|x_{i|k}\|_Q^2 + \|u_{i|k}\|_R^2) - l_{ss}$, and by summing both sides of this equation over $i = 0, 1, \ldots$, we can conclude that

$$V_{0|k} - \lim_{i\to\infty} \mathbb{E}_k(V_{i|k}) = \sum_{i=0}^{\infty} \mathbb{E}_k\big(\|x_{i|k}\|_Q^2 + \|u_{i|k}\|_R^2 - l_{ss}\big). \tag{6.19}$$

Finally, we note that (6.18c) ensures that $\lim_{i\to\infty} \mathbb{E}_k(V_{i|k}) = 0$, since the definition of $V_{i|k}$ implies

$$\mathbb{E}_k(V_{i|k}) = \mathbb{E}_k(z_{i|k}^T W_z z_{i|k}) + 2w_{z1}^T \mathbb{E}_k(z_{i|k}) - \operatorname{tr}(\Theta W_x)$$

where, by Lemma 6.1, $\lim_{i\to\infty} \mathbb{E}_k(z_{i|k}) = 0$ and $\lim_{i\to\infty} \mathbb{E}_k(z_{i|k}^T W_z z_{i|k}) = \operatorname{tr}(\Theta W_x)$. From (6.19) it then follows that $V_{0|k} = J(x_k, \mathbf{c}_k)$. □

In the absence of constraints, the expression for the predicted cost provided by Theorem 6.1 allows the optimal value of $\mathbf{c}_k$ to be determined analytically for any horizon N. Using a similar argument to that of Sect. 2.7.2, it is possible to deduce from this the unconstrained optimal value of the linear feedback gain K. First, we partition W_z and w_{z1} into blocks conformal with the dimensions of x_k and $\mathbf{c}_k$:

$$W_z = \begin{bmatrix} W_x & W_{xc} \\ W_{cx} & W_c \end{bmatrix}, \quad w_{z1} = \begin{bmatrix} w_{x1} \\ w_{c1} \end{bmatrix}.$$

Next note that mean-square stability of the dynamics $x_{k+1} = \Phi_k x_k$ in Assumption 6.1 implies that $z_{k+1} = \Psi_k z_k$ is also mean-square stable. From (6.18a), it follows that $W_z \succ 0$, and this in turn implies that W_c is positive definite. Therefore the $\mathbf{c}_k$ that achieves the minimum of (6.17) in the absence of constraints is given by

$$\arg\min_{\mathbf{c}_k} J(x_k, \mathbf{c}_k) = -W_c^{-1} W_{cx} x_k - W_c^{-1} w_{c1}. \tag{6.20}$$

The constant term $-W_c^{-1} w_{c1}$ appearing in this expression indicates that the optimal control law is in general an affine rather than a linear function of the model state. In fact, w_{c1} can be determined from (6.18b) using

$$w_{c1}^T = \Big(\mathbb{E}(w_k^T D^T W_x B_k) + w_{x1}^T B^{(0)}\Big) \big[I_{n_u} \ \cdots \ I_{n_u}\big] \tag{6.21a}$$

$$w_{x1}^T = \mathbb{E}(w_k^T D^T W_x \Phi_k)(I - \Phi^{(0)})^{-1}. \tag{6.21b}$$

These expressions imply that the second term on the RHS of (6.20) applies a constant perturbation to the predicted control law (6.14), which is independent of the prediction time step and independent of the horizon N.

From (6.21a, 6.21b), it can be seen that w_{c1} is non-zero unless the additive disturbance term w_k and the multiplicative uncertainty in the model parameters A_k, B_k are statistically uncorrelated. However, even in the more general case in which the optimal control law for the cost (6.13) is affine rather than linear-state feedback, it is still possible to determine the unconstrained optimal linear feedback gain since this corresponds to the case in which the minimizing argument of (6.17) is independent of x_k. From (6.20), this requires $W_{cx} = 0$, and from (6.18a) we therefore require that K satisfies

$$K = -\big(R + \mathbb{E}(B_k^T W_x B_k)\big)^{-1} \mathbb{E}(B_k^T W_x A_k) \tag{6.22}$$

where W_x is the solution of

$$W_x - \mathbb{E}\big((A_k + B_k K)^T W_x (A_k + B_k K)\big) = Q + K^T R K. \tag{6.23}$$

The corresponding solution for W_c can be determined from (6.18a) as

$$W_c = \operatorname{diag}\big\{R + \mathbb{E}(B_k^T W_x B_k), \ldots, R + \mathbb{E}(B_k^T W_x B_k)\big\}. \tag{6.24}$$

The preceding results concerning the unconstrained optimal linear feedback gain and the predicted cost are summarized as follows.

Corollary 6.1 *The linear feedback law that minimizes the performance index (6.13) for the dynamics of (6.2) and (6.3) is $u_k = K x_k$ where K is given by (6.22) and (6.23). If the predicted control sequence (6.14) is defined in terms of this gain K, then the predicted cost of (6.15) is given by*

$$J(x_k, \mathbf{c}_k) = x_k^T W_x x_k + \mathbf{c}_k^T W_c \mathbf{c}_k + 2 w_{x1}^T x_k + 2 w_{c1}^T \mathbf{c}_k + w_1$$

where W_x, W_c and w_{x1}, w_{c1} are given by (6.23), (6.24) and (6.21a, 6.21b).

We conclude this section by considering how to compute K and W_x satisfying (6.22) and (6.23). It is possible to derive a set of algebraic conditions on W_x by using (6.22) to eliminate K from (6.23) and evaluating expectations using the parameterization of model uncertainty in (6.3a, 6.3b). However, the resulting algebraic Riccati equation is nonlinear and multivariate, and a computationally more convenient approach is to consider (W_x, K) as an extremal point of the feasible set of a particular LMI, as we now briefly discuss. For any given mean-square stabilizing $\tilde{K}$, let $W_x' \succ 0$ satisfy the equality

$$W_x' - \mathbb{E}\big((A_k + B_k \tilde{K})^T W_x' (A_k + B_k \tilde{K})\big) = Q + \tilde{K}^T R \tilde{K}$$

and let $\tilde{W}_x \succ 0$ satisfy the inequality

$$\tilde{W}_x - \mathbb{E}\big((A_k + B_k \tilde{K})^T \tilde{W}_x (A_k + B_k \tilde{K})\big) \succeq Q + \tilde{K}^T R \tilde{K} \tag{6.25}$$

Then, subtracting (6.23) we have

$$(\tilde{W}_x - W_x') - \mathbb{E}\big((A_k + B_k\tilde{K})^T(\tilde{W}_x - W_x')(A_k + B_k\tilde{K})\big) \succeq 0,$$

from which it follows that $\tilde{W}_x \succeq W_x'$ since $A_k + B_k\tilde{K}$ is mean-square stable by assumption. But $\tilde{W}_x \succeq W_x'$ implies $\mathrm{tr}(\tilde{W}_x) \geq \mathrm{tr}(W_x')$ and W_x' is therefore equal to the solution of the problem of minimizing $\mathrm{tr}(\tilde{W}_x)$ over $\tilde{W}_x$ subject to (6.25) for a given fixed value of $\tilde{K}$. Since K in (6.22) and (6.23) is the feedback gain that minimizes the cost (6.15), the value of $\mathrm{tr}(W_x)$ satisfying (6.23) is equal to the minimum of $\mathrm{tr}(\tilde{W}_x)$ over the set of all $\tilde{W}_x$ satisfying (6.25) for variable $\tilde{K}$. In other words, the pair (W_x, K) satisfying (6.22) and (6.23) is the optimal argument of the problem

$$\underset{\tilde{W}_x, \tilde{K}}{\text{minimize}} \quad \mathrm{tr}(\tilde{W}_x) \quad \text{subject to (6.25)}.$$

This problem is nonconvex, but introducing transformed variables $S = \tilde{W}_x^{-1}$ and $Y = \tilde{K}\tilde{W}_x^{-1}$, (6.25) can be rewritten as

$$S - \mathbb{E}\big((A_kS + B_kY)^T S^{-1}(A_kS + B_kY)\big) \succeq SQS + Y^T RY.$$

Using the model parameterization (6.3a, 6.3b), this inequality can be expressed as the following LMI in $S \succ 0$, and Y:

$$\begin{bmatrix} S & (A^{(0)}S + B^{(0)}Y)^T & \cdots & (A^{(m)}S + B^{(m)}Y)^T & \begin{bmatrix} SQ^{1/2} & Y^T R^{1/2} \end{bmatrix} \\ \star & S & \cdots & 0 & 0 \\ \vdots & \vdots & \ddots & \vdots & \vdots \\ \star & \star & \cdots & S & 0 \\ \star & \star & \cdots & \star & I \end{bmatrix} \succeq 0 \qquad (6.26)$$

Thus W_x and K satisfying (6.22) and (6.23) can be computed by solving the semi-definite program:

$$(W_x, S, Y) = \arg \min_{W_x, S, Y} \ \mathrm{tr}(W_x) \ \text{subject to (6.26) and} \begin{bmatrix} W_x & I \\ I & S \end{bmatrix} \succeq 0$$

and setting $K = YW_x$.

6.3 Mean-Variance Predicted Cost

The cost considered in Sect. 6.2 gives a measure of the second moments of predicted states and control inputs. Thus (6.15) can be written for $\kappa = 1$ as

$$J(x_k, \mathbf{c}_k) = \sum_{i=0}^{\infty} \big(\|x_{i|k}^{(0)}\|_Q^2 + \|u_{i|k}^{(0)}\|_R^2\big)$$
$$+ \kappa^2 \sum_{i=0}^{\infty} \mathbb{E}_k\big(\|x_{i|k} - x_{i|k}^{(0)}\|_Q^2 + \|u_{i|k} - u_{i|k}^{(0)}\|_R^2 - l_{ss}\big) \tag{6.27}$$

where $x_{i|k}^{(0)} = \mathbb{E}(x_{i|k})$ and $u_{i|k}^{(0)} = \mathbb{E}(u_{i|k})$ denote the nominal values of the predicted states and control inputs and $l_{ss} = \mathrm{tr}\big(\Theta(Q + K^T RK)\big)$. These nominal sequences are governed by the nominal prediction dynamics:

$$x_{i+1|k}^{(0)} = A^{(0)} x_{i|k}^{(0)} + B^{(0)} u_{i|k}^{(0)}. \tag{6.28}$$

Expressed this way, the cost $J(x_k, \mathbf{c}_k)$ can be seen to evaluate a particular linear mix of two cost indices: one based on the mean and the other on the variance of predicted states and inputs.

In applications such as the sustainable development problem considered in Sect. 6.5 and those involving, for example, portfolio selection [19, 20], a typical control objective is to minimize a cost based on probabilistic bounds within which the future predicted states and control inputs will lie. In this case different affine mixes, corresponding to $\kappa \neq 1$ in (6.27), may be appropriate. For example, the cost proposed in [21] is concerned with minimizing the width of the probabilistic band defined by

$$\begin{aligned} &\Pr_k(y_{i|k} \le \bar{y}_{k+i}) \ge p \\ &\Pr_k(y_{i|k} \ge \underline{y}_{k+i}) \ge p \end{aligned} \tag{6.29}$$

Then, under the assumption that $y_{i|k}$ is normally distributed, an appropriate stage cost would be

$$\frac{1}{2}(\underline{y}_{k+i}^2 + \bar{y}_{k+i}^2) = (y_{i|k}^{(0)})^2 + \kappa^2 \mathbb{E}_k\big((y_{i|k} - y_{i|k}^{(0)})^2\big) \tag{6.30}$$

where κ is the argument of the standard cumulative normal distribution function corresponding to probability p, i.e. $\Pr(X \le \kappa) = p$ for a normally distributed random variable X with zero mean and unit variance.

For such applications, the cost of (6.27) with $\kappa^2 \neq 1$ is more appropriate and results in a predictive control strategy known as mean-variance SMPC [22]. Note that this cost can also be written as

$$J(x_k, \mathbf{c}_k) = (1 - \kappa^2) \sum_{i=0}^{\infty} \big(\|x_{i|k}^{(0)}\|_Q^2 + \|u_{i|k}^{(0)}\|_R^2\big)$$
$$+ \kappa^2 \sum_{i=0}^{\infty} \mathbb{E}_k\big(\|x_{i|k}\|_Q^2 + \|u_{i|k}\|_R^2 - l_{ss}\big). \tag{6.31}$$

From this expression it can be seen that $J(x_k, \mathbf{c}_k)$ reduces to the expected quadratic cost of (6.15) for $\kappa = 1$, gives the nominal cost for $\kappa = 0$ and tends to the minimum variance cost as $\kappa \to \infty$.

The cost of (6.31), when evaluated for the model (6.2) and (6.3a, 6.3b) with the predicted control law (6.14), is a quadratic function of the degrees of freedom, $\mathbf{c}_k$. Using Theorem 2.10 to compute the nominal cost in the first term on the RHS of (6.31) and using the conditions of Theorem 6.1 to compute the second term, this function is given by

$$J(x_k, \mathbf{c}_k) = \begin{bmatrix} z_k \\ 1 \end{bmatrix}^T \begin{bmatrix} W_z & w_{z1} \\ w_{z1}^T & w_1 \end{bmatrix} \begin{bmatrix} z_k \\ 1 \end{bmatrix}$$

where $z_k = (x_k, \mathbf{c}_k)$ and W_z, w_{z1}, w_1 are defined by

$$W_z = (1 - \kappa^2)\bar{W}_z + \kappa^2 \hat{W}_z \qquad \begin{cases} \bar{W}_z - \Psi^{(0)\,T} \bar{W}_z \Psi^{(0)} = \hat{Q} \\ \hat{W}_z - \mathbb{E}(\Psi_k^T \hat{W}_z \Psi_k) = \hat{Q} \end{cases} \tag{6.32a}$$

and

$$w_{z1}^T (I - \Psi^{(0)}) = \mathbb{E}\Big(w_k^T \begin{bmatrix} D^T & 0 \end{bmatrix} \hat{W}_z \Psi_k \Big) \tag{6.32b}$$

$$w_1 = -\mathrm{tr}(\Theta \hat{W}_x), \tag{6.32c}$$

with $\hat{W}_x$ and $\Psi^{(0)}$ defined as

$$\hat{W}_x = \begin{bmatrix} I_{n_x} & 0 \end{bmatrix} \hat{W}_z \begin{bmatrix} I_{n_x} \\ 0 \end{bmatrix}, \qquad \Psi^{(0)} = \begin{bmatrix} \Phi^{(0)} & B^{(0)}E \\ 0 & M \end{bmatrix}.$$

The unconstrained optimal linear feedback gain can be determined by an analysis similar to that of Sect. 6.2, but in this case the gain K is given by the solution of a pair of coupled algebraic Riccati equations:

$$\bar{W}_x - (A^{(0)} + B^{(0)}K)^T \bar{W}_x (A^{(0)} + B^{(0)}K) = Q + K^T R K \tag{6.33a}$$

$$\hat{W}_x - \mathbb{E}\big((A_k + B_k K)^T \hat{W}_x (A_k + B_k K)\big) = Q + K^T R K. \tag{6.33b}$$

with

$$K = -\big(R + \mathbb{E}(B_k^T \hat{W}_x B_k)\big)^{-1} \big((1 - \kappa^2) B^{(0)\,T} \bar{W}_x A^{(0)} + \kappa^2 \mathbb{E}(B_k^T \hat{W}_x A_k)\big). \tag{6.33c}$$

The details of the derivation of this expression and an iterative method of computing the solution K can be found in [23].

6.4 Early Stochastic MPC Algorithms

Optimization methods for problems involving probabilistic constraints have been used in applications of Operations Research since the 1950s (see e.g. [24]). In the context of predictive control, one of the first proposals to use dynamic models containing stochastic uncertainty was reported in [25], where systems with constant but unknown parameters were considered. In this strategy, unknown model parameters are identified online and used to update a control law, and for this reason it is known as self-tuning control. The aim of the approach, however, is to minimize the expected value of a predicted cost that is computed using a dynamic model, and the optimal predicted control input is implemented as a receding horizon control law. Therefore self-tuning control can be viewed as a type of stochastic MPC strategy. This section briefly reviews self-tuning control strategies before considering the formulation of probabilistic constraints for moving average models.

6.4.1 Auto-Regressive Moving Average Models

In its simplest form, self-tuning control is based on a single-input single-output (SISO) ARMA model, as defined in (6.1). In [25], the model parameters a_i, b_i are assumed to be unknown constants and the additive noise process, $\{e_0, e_1, \ldots\}$, is assumed to be an independent and identically distributed (i.i.d.) sequence in which e_k is normally distributed with zero mean. In compact form, using z-transform operators, the system dynamics may be written as

$$A(z)y_k = B(z)u_{k-d} + C(z)e_k \tag{6.34}$$

where

$$\begin{aligned}
A(z) &= 1 + a_1 z^{-1} + \cdots + a_n z^{-n} \\
B(z) &= b_0 + b_1 z^{-1} + \cdots + b_n z^{-n} \\
C(z) &= 1 + c_1 z^{-1} + \cdots + c_n z^{-n}
\end{aligned}$$

with z^{-1} representing the backward shift operator. The system is assumed to be unconstrained, and hence no constraints act on y_k and u_k.

The predicted cost is taken to be

$$J(x_k, \mathbf{u}_k) = \mathbb{E}_k(y_{d|k}^2) \tag{6.35}$$

where $\mathbf{u}_k = (u_{0|k}, \ldots, u_{d-1|k})$ and $y_{i|k}$ is the predicted value of the output y_{k+i} at time k. Because of this objective the strategy is sometimes described as a minimum variance (MV) control law. We note that on account of the delay d of the model (6.34), the d steps-ahead output, $y_{d|k}$, is the first output variable that can be influenced by the control input $u_{0|k}$.

In the absence of constraints, the minimization of the cost (6.35) can be performed analytically. To see this, define polynomials $F(z)$ and $G(z)$ by

$$z^d C(z) = A(z)F(z) + G(z), \tag{6.36}$$

where

$$\begin{aligned} F(z) &= z^d + f_1 z^{d-1} + \cdots + f_d \\ G(z) &= g_1 z^{-1} + g_2 z^{-2} + \cdots + g_n z^{-n}. \end{aligned}$$

This identity allows the dependence of $y_{d|k}$ on the additive disturbance sequence to be split into terms that depend on the values of $e_{k+d}, \ldots, e_k$, which are unknown at time k, and terms involving only the past values $e_{k-1}, e_{k-2}, \ldots$, which have already been realized at time k and can therefore be determined from the observed values of the control input and system output. In particular, from the model (6.34) and the identity (6.36) we obtain

$$\begin{aligned} y_{k+d} &= \frac{B(z)}{A(z)} u_k + \frac{z^d C(z)}{A(z)} e_k \\ &= \left[\frac{B(z)}{A(z)} u_k + \frac{G(z)}{A(z)} e_k \right] + F(z) e_k \end{aligned} \tag{6.37}$$

where the square-bracketed quantity on the right-hand side of (6.37) is known at time k, whereas the last term depends on the unknown noise sequence $e_{k+d}, \ldots, e_k$, which is a random variable at time k. Given the i.i.d. and zero-mean assumptions on e_k, it follows that the control law that minimizes the cost of (6.35) necessarily satisfies

$$\frac{B(z)}{A(z)} u_k + \frac{G(z)}{A(z)} e_k = 0. \tag{6.38}$$

The dependence of e_k on present and past outputs and past inputs can be deduced from (6.34), which implies

$$\begin{aligned} e_k &= \frac{A(z)}{C(z)} y_k - \frac{B(z)}{C(z)} u_{k-d} \\ &= \frac{A(z)}{C(z)} y_k - \frac{B(z)}{z^d C(z)} u_k. \end{aligned}$$

In conjunction with (6.38), this results in the condition

$$z^d C(z) B(z) u_k - G(z) B(z) u_k + z^d G(z) A(z) y_k = 0. \tag{6.39}$$

Using the identity (6.36) to replace $z^d C(z)$ by $A(z)F(z) + G(z)$ then leads to the optimal control solution

$$u_k = -\frac{z^d G(z)}{B(z)F(z)} y_k. \tag{6.40}$$

Substituting the control law (6.40) into the system model (6.34) gives the closed-loop characteristic equation

$$\frac{A(z)F(z) + G(z)}{F(z)} y_k = \frac{C(z)}{F(z)} y_{k+d} = 0, \tag{6.41}$$

which therefore identifies the roots of $C(z)$ as the closed-loop poles of the prediction dynamics. Hence a necessary condition for this strategy to stabilize the system (6.34) is that every root of $C(z)$ should lie within the unit circle centred at the origin. In addition, by considering (6.40) it is easy to show that, for internal stability, the roots of $B(z)$ must also lie inside the unit circle centred at the origin. Therefore this strategy is restricted to minimum phase plants.

The restriction to minimum phase systems was subsequently removed by the generalized minimum variance (GMV) strategy [26], which modifies the predicted cost so as to include a term that penalizes control activity. However, neither the MV nor the GMV control strategy can provide an *a priori* guarantee of stability for the case of unknown model parameters (although stability can be checked *a posteriori*, namely after the specification of the problem parameters and the derivation of the optimal control law). Moreover, these approaches do not take into account constraints on outputs and control inputs.

Constraints in stochastic MPC in the context of ARMA models were not introduced until much later. For example, [27] considered SISO systems described by the model (6.34), but removed the assumption of independent additive disturbances and imposed hard and probabilistic constraints of the form

$$\underline{u} \leq u_k \leq \bar{u} \tag{6.42}$$

$$\Pr_k(\underline{y}_{k+i} \leq y_{i|k} \leq \bar{y}_{k+i}) \geq p. \tag{6.43}$$

Without the assumption of i.i.d. disturbances, the minimization subject to these constraints of the predicted cost, which in [27] was defined so as to penalize control increments

$$J(\mathbf{u}_k) = (u_{0|k} - u_{k-1})^2 + \sum_{i=1}^{N-1} (u_{i|k} - u_{i-1|k})^2, \tag{6.44}$$

becomes a computationally demanding problem that has to be solved online, and for which there is in general no guarantee of convergence to the global optimum. In addition, the use of a finite horizon cost precludes the possibility of providing *a priori* stability guarantees. Furthermore the presence in the system model of random variables with distributions that are not finitely supported makes it impossible to ensure the recursive feasibility of this approach.

6.4.2 Moving Average Models

Constraints of the form (6.7) and (6.9) present two major difficulties for stochastic MPC algorithms: how to impose these constraints on predicted control sequences given the distribution of model uncertainty, and how to ensure that the optimization problem to be solved online is recursively feasible. Both of these difficulties are encountered with the ARMA model (6.34) with stochastic coefficients (and likewise with the state-space model of (6.2) and (6.3a, 6.3b)) since the future evolution of the model state and output trajectories depends on earlier realizations of the model uncertainty. In particular, if the uncertainty does not have bounded support, then clearly it is not possible in general to ensure that constraints will be satisfied at future time instants.

However it is relatively straightforward to obtain a guarantee of recursive feasibility for a stochastic MPC strategy based on a moving average (MA) model. This form of model arises when the system dynamics are described by a discrete-time finite impulse response model of the form

$$y_k = \sum_{j=1}^{n} H_j u_{k-j} + d_k \tag{6.45}$$

where $u \in \mathbb{R}^{n_u}$, $y \in \mathbb{R}^{n_y}$ are the control input and system output, $d \in \mathbb{R}^{n_y}$ is a stochastic additive disturbance, and $H_1, \ldots, H_n$ are the elements of the uncertain system impulse response after truncation to n terms. The probability distributions of d and $H_j, j = 1, \ldots, n$ are assumed to be known.

Knowledge of the distributions of uncertain model parameters makes it possible to transform probabilistic constraints into deterministic constraints on predicted input sequences. For the case of the model (6.45) with normally distributed model parameters, this is easy to do for the constraints of (6.8) because the output y_k in (6.45) depends linearly on the model parameters, and hence $y_{i|k}$ is also a normally distributed random variable. In this case, it is possible to invoke the chance-constrained optimization framework [24, 28] to convert probabilistic system constraints of the form of (6.8) into second-order cone (SOC) constraints as was done for example in [3].

Thus let $y_{i|k,l}$ denote the lth element of the predicted value at time k of the i steps-ahead output vector y_{k+i}. Then from (6.45), it follows that

$$y_{i|k,l} = h_l^T \begin{bmatrix} u_i^f \\ u_i^p \end{bmatrix} + d_{i,l}$$

where $u_i^f = (u_{i-1|k}, \ldots, u_{0|k})$ and $u_i^p = (u_{k-1}, \ldots, u_{k+i-n})$ denote, respectively, the predicted future and the past components of the input sequence, also $h_l^T = e_l^T \begin{bmatrix} H_1 \ldots H_n \end{bmatrix}$, with e_l denoting the lth column of the identity matrix, and $d_{i,l}$ is the lth element of d_{k+i}. Then, under the assumption that $(h_l, d_{i,l})$ is normally distributed with mean $(\bar{h}, 0)$ and covariance matrix Σ, the probabilistic constraint

$$\Pr_k(y_{i|k,l} \leq a_l) \geq p \tag{6.46}$$

can be written as a deterministic second-order cone constraint of the form

$$a_l - \bar{h}^T \begin{bmatrix} u_i^f \\ u_i^p \end{bmatrix} \geq \kappa \left\| (u_i^f, u_i^p, 1) \right\|_\Sigma. \tag{6.47}$$

The constant κ in (6.47) is defined by

$$\Pr(X \leq \kappa) = p,$$

for a normally distributed scalar random variable X with mean 0 and variance 1. For p in the interval $[0.5, 1)$, the value of κ is nonnegative and, since the condition $a^T x - b \geq \kappa \|c + Dx\|$ is convex in x if $\kappa \geq 0$ for any given constants a, b, c, D, the constraint (6.47) is therefore convex in this case. Hence, for $p \geq 0.5$, the minimization of a convex quadratic predicted cost subject to (6.47) is a second-order cone programming problem (SOCP) that can be solved efficiently.

This is the approach used in [3], where a finite horizon cost is used to formulate a SMPC algorithm. However this formulation lacked a guarantee of closed-loop stability (on account of the finite horizon cost). In the presence of hard input constraints in addition to probabilistic output constraints, feasibility may also become problematic and requires constraint softening this context (see for example [29]).

A further disadvantage of the use of MA models is that they are limited to open-loop stable systems; the impulse responses of open-loop unstable systems cannot be truncated to give finite-order MA models. These aspects of closed-loop stability and feasibility, as well as the extension of the use of MA to the case of open-loop unstable systems have been addressed in [1, 2], where stochastic MPC is introduced as a tool for the assessment of sustainable development policies.

6.5 Application to a Sustainable Development Problem

Sustainable development addresses the problem of balancing the needs for economic, technological and industrial development of the current generation with those of future generations [30]. Human generations can be considered to be separated by about 30 years, and this defines a natural prediction horizon for assessing the likely effects of policy decisions on sustainable development. The treatment of the problem has been mostly discursive but it can be posed in a formal mathematical manner. Thus [31] considers the question of assessing policy in budget allocation between alternative technologies for electricity generation and develops a strongly stochastic model. This model describes the effect of adjusting a number of inputs ("instruments"), which form the elements of an input vector $u \in \mathbb{R}^{n_u}$, and which include for example measures of investment in combined cycle gas turbine technology and investment in renewable energy such as wind turbines, on a number of outputs ("indicators"). Amongst these indicators is included a measure of *benefit*, say y_1, related to the cost of the energy produced, and another, y_2 that measures *cost*, related to accumulated CO_2 emissions. These are measured at the end of the 30-year horizon and are therefore strongly stochastic on account of the vagaries of world economy. Clearly, it is desirable to maximize benefit while respecting constraints on cost, but since both y_1 and y_2 are random variables, a more appropriate stochastic optimization problem is given by

$$\begin{aligned} &\underset{u, A_1}{\text{maximize}} \ A_1 \\ &\text{subject to} \ \Pr(y_1 \geq A_1) \geq p_1 \\ &\qquad\qquad \Pr(y_2 \leq A_2) \geq p_2 \\ &\qquad\qquad \mathbf{1}^T u \leq b, \quad u \geq 0 \end{aligned} \tag{6.48}$$

which is given in terms of target bounds, A_1 and A_2, rather than directly in terms of the variables y_1, y_2 which, as stated earlier, are stochastic. The hard constraint in (6.48), that the sum of the inputs should not exceed a specified value b, expresses budgetary limitations.

It possible to introduce (6.48) into a SMPC framework by performing the optimization repetitively in a receding horizon manner, for example at the beginning of the kth year, for $k = 0, 1, \ldots$. Furthermore, rather than consider a single input adjustment and its effect on the output variables at the end of the prediction horizon, one can allow input adjustments to occur at each step of the prediction horizon and measure their effect on the output variables, again, over the entire horizon. This introduces an explicit time dependence into the input and output variables and requires the use of dynamic models. Accordingly, we denote the vector of budget adjustments and the vector of indicators in year k as u_k and $y_k \doteq (y_{k,1}, y_{k,2})$ respectively.

In [1], MA models of the form of (6.45) are used to describe the input–output dependence. Such models are convenient for the derivation of the parameter distributions from identification of experiments (performed using world economy models) since they avoid the difficulty of ARMA models in which stochastic model parameters multiply disturbance values which are also random variables. The limitation of MA models to open-loop stable systems can be overcome through the artifice of bicausality, according to which causal relationships such as given in (6.45) can be used to account for the stable dynamics, whereas unstable dynamics can be modelled using anti-causal regressions. As an illustration of this, consider a system described by a transfer function with a single pole which is unstable, namely $1/(1 - \lambda z^{-1})$

where $|\lambda| > 1$. This transfer function can also be written as $-\lambda^{-1}z/(1-\lambda^{-1}z)$ and, after suitable truncation, leads to the anti-causal regression

$$y_k = -\frac{1}{\lambda}\left(u_k + \frac{1}{\lambda}u_{k+1} + \cdots + \left(\frac{1}{\lambda}\right)^{n-1} u_{k+n-1}\right) \tag{6.49}$$

Superposition can be deployed to extend this treatment to the case of complex conjugate unstable poles and/or multiple poles.

The overall stochastic MPC strategy can be split into two phases, of which the first defines a suitable setpoint, and the second is concerned with tracking this setpoint. In the interests of maximizing the value of the objective steady state, Phase 1 considers a predicted input trajectory which assumes a constant steady-state value, u_{ss}, after N_u time steps. If N_u satisfies $N_u + n - 1 < N = 30$, then from (6.45) the outputs necessarily reach their steady-state values, $y_{ss,1}$ and $y_{ss,2}$, within the 30-year horizon. Hence, to maximize benefit, the setpoints for u and y_1 can be defined as

$$r \doteq \mathbb{E}(h_1^T)\mathbf{u}_{ss}$$

where $\mathbf{u}_{ss} = (u_{ss}, \ldots, u_{ss})$ and u_{ss} is defined, for given values of the probabilities $p_1, p_2 \in [0.5, 1)$, the threshold A_2 and the overall budget B for an N-step horizon, as the solution of the convex optimization:

$$\begin{aligned} \underset{u_{ss}, A_1}{\text{maximize}} \quad & A_1 \\ \text{subject to} \quad & \Pr(y_{ss,1} \geq A_1) \geq p_1 \\ & \Pr(y_{ss,2} \leq A_2) \geq p_2 \\ & \sum_{j=0}^{N-1} \rho^j \mathbf{1}^T u_{ss} \leq B, \quad u_{ss} \geq 0. \end{aligned} \tag{6.50}$$

The parameter ρ is a discounting factor chosen to lie in the interval $[0, 1)$, which is used in recognition of the fact that the value of expenditure decreases with the advance of time. The implication of the optimization (6.50) that defines this first phase of the algorithm is that the undershoot in the predicted steady-state benefit, $r - y_{ss,1}$, will with probability p_1 be no greater the threshold value $t_{ss} \doteq r - A_1$.

It is now possible to measure performance over the N-step prediction horizon using probabilistic thresholds:

$$\Pr_k(r - y_{i|k,1} \leq t_{i|k}) \geq p_1, \quad i = 1, \ldots, N.$$

The second phase of the algorithm aims at the minimization of these thresholds. However, given the stochastic nature of the problem and the desire to secure a monotonically decreasing property for the predicted cost, rather than penalize in the cost all $t_{i|k}$, only those that exceed t_{ss} will be taken into account. This suggests the following online implementation of stochastic MPC.

Algorithm 6.1 At each time instant $k = 0, 1, \ldots$:

(i) Perform the optimization:

$$\underset{\substack{u_{0|k},\ldots,u_{N_u-1|k},u_{ss} \\ s_{i|k},\, t_{i|k},\, i=1,\ldots,N}}{\text{minimize}} \quad \sum_{i=1}^{N} s_{i|k}^2 \tag{6.51a}$$

subject to

$$s_{i|k} \geq t_{i|k} - t_{ss}, \quad s_{i|k} \geq 0 \tag{6.51b}$$

$$\Pr_k(r - y_{i|k,1} \geq t_{i|k}) \geq p_1 \tag{6.51c}$$

$$\Pr_k(y_{i|k,2} \leq A_2) \geq p_2 \tag{6.51d}$$

$$\sum_{j=l}^{N_u-1} \rho^{j-l} \mathbf{1}^T u_{j|k} + \Big(\frac{\rho^{N_u-l} - \rho^N}{1-\rho}\Big) \mathbf{1}^T u_{ss} \leq B \tag{6.51e}$$

$$u_{l|k} \geq 0, \quad u_{ss} \geq 0 \tag{6.51f}$$

for $i = 1, \ldots, N$ and $l = 0, \ldots, N_u - 1$.

(ii) Apply the control law $u_k = u^*_{0|k}$, where $(u^*_{0|k}, \ldots, u^*_{N_u-1|k}, u^*_{ss})$ is the minimizing input sequence in (6.51). ◁

For $l = 0$, (6.51e) applies the discounted budgetary constraint

$$\sum_{j=0}^{N-1} \rho^j \mathbf{1}^T u_{j|k} \leq B$$

to the predicted input sequence $(u_{0|k}, u_{1|k}, \ldots)$ with $u_{j|k} = u_{ss}$ for $j \geq N_u$. However (6.51e) is also applied for each $l = 1, \ldots, N_u - 1$ to ensure that the time-shifted input sequences, $(u_{l|k}, u_{l+1|k}, \ldots)$, with $u_{j|k} = u_{ss}$ for $j \geq N_u$, satisfy the corresponding discounted budgetary constraint:

$$\sum_{j=l}^{N-1+l} \rho^{j-l} \mathbf{1}^T u_{j|k} \leq B.$$

Given that the constraints of (6.50) are satisfied for the given t_{ss}, A_2 and B for some u_{ss}, this is all that is needed to ensure recursive feasibility of Algorithm 6.1. In addition, if $s^*_{i|k}$, $i = 1, \ldots, N$ is optimal for (6.51) at time k, then the sequence defined by $s_{i|k+1} = s^*_{i+1|k}$ for $i = 1, \ldots, N-1$ and $s_{N|k+1} = 0$ is by construction feasible at time $k+1$. Therefore the optimal objective of (6.51), denoted $J^*_k \doteq \sum_{i=1}^{N} s^{*\,2}_{i|k}$, necessarily satisfies

$$J^*_{k+1} \leq J^*_k - s^{*\,2}_{1|k} \tag{6.52}$$

for all k. Summing both sides of this inequality over all $k \geq 0$ gives

$$\sum_{k=0}^{\infty} s_{1|k}^{*\,2} \leq J_0^*$$

which implies that $s_{1|k}^* \to 0$ as $k \to \infty$. These results are summarized below.

Theorem 6.2 *For the closed-loop system (6.45) under the control law of Algorithm 6.1, the optimization (6.51) is feasible at all times* $k = 0, 1, \ldots$ *and* $r - y_{1|k,1} \leq t_{ss}$ *with probability* p_1 *as* $k \to \infty$.

Assuming the parameters of the model (6.45) to be normally distributed, the probabilistic constraints in (6.50), as well as those of (6.51c, 6.51d), can be converted into second-order cone constraints using the method by which (6.46) was transformed into the deterministic constraint (6.47). Therefore, the Phase 1 optimization (6.50) is convex and can be performed for example by solving a sequence of SOCPs. Similarly, the online optimization of Algorithm 6.2 can be expressed as a single SOCP problem.

References

1. B. Kouvaritakis, M. Cannon, P. Couchman, MPC as a tool for sustainable development integrated policy assessment. IEEE Trans. Autom. Control **51**(1), 145–149 (2006)
2. B. Kouvaritakis, M. Cannon, V. Tsachouridis, Recent developments in stochastic MPC and sustainable development. Annu. Rev. Control **28**(1), 23–35 (2004)
3. A.T. Schwarm, M. Nikolaou, Chance constrained model predictive control. AIChE J. **45**(8), 1743–1752 (1999)
4. D.H. van Hessem, O.H. Bosgra, Closed-loop stochastic model predictive control in a receding horizon implementation on a continuous polymerization reactor example, in *Proceedings of the 2004 American Control Conference*, Boston, USA (2004), pp. 914–919
5. F. Herzog, S. Keel, G. Dondi, L.M. Schumann, H.P. Geering, Model predictive control for portfolio selection, in *Proceedings of the 2006 American Control Conference*, Minneapolis, USA (2006), pp. 1252–1259
6. J.A. Primbs, Dynamic hedging of basket options under proportional transaction costs using receding horizon control. Int. J. Control **82**(10), 1841–1855 (2009)
7. P. Couchman, B. Kouvaritakis, M. Cannon, F. Prashad, Gaming strategy for electric power with random demand. IEEE Trans. Power Syst. **20**(3), 1283–1292 (2005)
8. P. Patrinos, S. Trimboli, A. Bemporad, Stochastic MPC for real-time market-based optimal power dispatch, in *Proceedings of the 50th Conference on Decision and Control*, Orlando, USA (2011), pp. 7111–7116
9. F. Oldewurtel, A. Parisio, C.N. Jones, D. Gyalistras, M. Gwerder, V. Stauch, B. Lehmann, M. Morari, Use of model predictive control and weather forecasts for energy efficient building climate control. Energy Build. **45**, 15–27 (2012)
10. J. Yan, R.R. Bitmead, Incorporating state estimation into model predictive control and its application to network traffic control. Automatica **41**(4), 595–604 (2005)
11. M. Evans, M. Cannon, B. Kouvaritakis, Robust MPC tower damping for variable speed wind turbines. IEEE Trans. Control Syst. Technol. **23**(1), 290–296 (2014)
12. J.F. Manwell, J.G. McGowan, A.L. Rogers, *Wind Energy Explained: Theory, Design and Application* (Wiley, New York, 2002)

13. J. Richalet, S. Abu el Ata-Doss, C. Arber, H.B. Kuntze, A. Jacubash, W. Schill, Predictive functional control: application to fast and accurate robots, in *Proceedings of the 10th IFAC World Congress*, Munich, (1987), pp. 251–258
14. G.E.P. Box, G.M. Jenkins, *Time Series Analysis: Forecasting and Control* (Holden-Day, San Francisco, 1976)
15. P.J. Brockwell, R.A. Davis, *Introduction to Time Series and Forecasting* (Springer, New York, 2002)
16. L. Ljung, *System Identification: Theory for the User*, 2nd edn. (Prentice Hall, New Jersey, 1999)
17. H.J. Kushner, *Introduction to Stochastic Control* (Holt, Rinehart and Winston, New York, 1971)
18. M. Cannon, B. Kouvaritakis, X. Wu, Probabilistic constrained MPC for multiplicative and additive stochastic uncertainty. IEEE Trans. Autom. Control **54**(7), 1626–1632 (2009)
19. D. Li, W.L. Ng, Optimal dynamic portfolio selection: multiperiod mean-variance formulation. Math. Financ. **10**(3), 387–406 (2000)
20. S.-S. Zhu, D. Li, S.-Y. Wang, Risk control over bankruptcy in dynamic portfolio selection: a generalized mean-variance formulation. IEEE Trans. Autom. Control **49**(3), 447–457 (2004)
21. P. Couchman, M. Cannon, B. Kouvaritakis, Stochastic MPC with inequality stability constraints. Automatica **42**(12), 2169–2174 (2006)
22. P. Couchman, B. Kouvaritakis, M. Cannon, MPC on state space models with stochastic input map, in *Proceedings of the 45th Conference on Decision and Control*, San Diego, USA (2006), pp. 3216–3221
23. M. Cannon, B. Kouvaritakis, P. Couchman, Mean-variance receding horizon control for discrete-time linear stochastic systems in *Proceedings of the 17th IFAC World Congress*, Seoul, Korea (2008), pp. 15321–15326
24. A. Charnes, W.W. Cooper, Chanced-constrained programming. Manag. Sci. **6**(1), 73–79 (1959)
25. K.J. Astrom, B. Wittenmark, On self tuning regulators. Automatica **9**(2), 185–199 (1973)
26. D.W. Clarke, P.J. Gawthrop, Self-tuning controller. Proc. Inst. Electr. Eng. **122**(9), 929–934 (1975)
27. P. Li, M. Wendt, G. Wozny, A probabilistically constrained model predictive controller. Automatica **38**(7), 1171–1176 (2002)
28. A. Charnes, W.W. Cooper, Deterministic equivalents for optimizing and satisficing under chance constraints. Oper. Res. **11**(1), 18–39 (1963)
29. E. Zafiriou, H.W. Chiou, Output constraint softening for SISO model predictive control *Proceedings of the 1993 American Control Conference*, San Francisco, USA (1993), pp. 372–376
30. United Nations, Report of the world commission on environment and development. Technical report, General Assembly Resolution 42/187, December 1987
31. N. Kouvaritakis, Methodologies for integrating impact assessment in the field of sustainable development (MINIMA-SUD). Technical report, European Commission Project EVG1-CT-2002-00082 (2004)

Chapter 7
Feasibility, Stability, Convergence and Markov Chains

This chapter considers the closed-loop properties of stochastic MPC strategies based on the predicted costs and probabilistic constraints formulated in Chap. 6. To make the analysis of closed-loop stability and performance possible, it must first be ensured that the MPC law is well-defined at all times and the most natural way to approach this is to ensure that the associated receding horizon optimization problem remains feasible whenever it is initially feasible. We therefore begin by discussing the conditions for recursive feasibility.

The requirement for future feasibility of probabilistic constraints induces constraints on the model state that must be satisfied for all realizations of model uncertainty. Although this introduces robust constraints into the problem, we show that these constraints provide the least restrictive means of ensuring recursive feasibility, and hence they are less restrictive than the conservative robust counterparts of the probabilistic constraints. This suggests a general framework for stochastic MPC that combines robust constraints for recursive feasibility with the probabilistic or expected value constraints of the control problem.

Closed-loop stability and convergence are discussed next in the context of a prototype stochastic MPC algorithm. We present an analysis of asymptotic mean-square bounds on the closed-loop state and control trajectories that are derived from bounds on the optimal value of the predicted cost. This is the basis of the stability analysis of the various stochastic MPC strategies considered in this chapter and in Chap. 8. We also briefly discuss an interesting and seldom-used alternative based on supermartingale convergence theory, which provides additional insight into the behaviour of the closed-loop system.

The conditions for recursive feasibility of pointwise-in-time probabilistic constraints require the predicted state and control trajectories to lie in a tube that ensures robust feasibility and satisfaction of the probabilistic constraints. Similar conditions apply to the case of probabilistic constraints that are imposed jointly at more than a single future time step, and we conclude the chapter by considering a strategy for

B. Kouvaritakis and M. Cannon, *Model Predictive Control*,
Advanced Textbooks in Control and Signal Processing,
DOI 10.1007/978-3-319-24853-0_7

this case that makes use of probabilistic bounds on the uncertain model parameters. The approach is first introduced using the concept of probabilistic invariance and then extended to a more general framework using Markov chains.

Throughout this chapter, we consider systems described by the uncertain model introduced in Chap. 6:

$$x_{k+1} = A_k x_k + B_k u_k + D w_k \tag{7.1a}$$

where A_k, B_k, w_k are functions of a stochastic parameter vector q_k which is unknown at time k. Thus $(A_k, B_k, w_k) = \big(A(q_k), B(q_k), w(q_k)\big)$ with

$$\big(A(q), B(q), w(q)\big) = (A^{(0)}, B^{(0)}, 0) + \sum_{j=1}^{m} (A^{(j)}, B^{(j)}, w^{(j)}) q^{(j)} \tag{7.1b}$$

and the probability distribution of $q_k = (q_k^{(1)}, \ldots, q_k^{(m)})$ is assumed to be known and to satisfy $\mathbb{E}(q_k) = 0$ and $\mathbb{E}(q_k q_k^T) = I$. Furthermore, q_k and q_i are assumed to be independent for all $k \neq i$.

7.1 Recursive Feasibility

This section examines conditions under which the online optimization of predicted performance in stochastic MPC can be guaranteed to remain feasible at all future sampling instants if it is initially feasible. Probabilistic or expectation constraints such as (6.7)–(6.9) are usually regarded as "soft" constraints since they are not required to hold for every possible realization of model uncertainty. However, in order that a problem is feasible, we require that all conditions of the problem are met, whether these are invoked for a predefined subset or for all model uncertainty realizations. Probabilistic or expectation constraints are, in general, only feasible if the system state belongs to a particular subset of state space, and the conditions for their feasibility thus impose additional constraints on states and control inputs.

Recursive feasibility of stochastic MPC algorithms can be handled in one of two ways. Either conditions to ensure robust feasibility are imposed as explicit constraints in the online optimization or the optimization is allowed to become infeasible whenever necessary. The latter approach typically includes a penalty on constraint violation in the MPC cost index [1, 2], or else directly minimizes a measure of the distance of the state from the feasible set whenever the problem is infeasible [3]. Without a guarantee of feasibility, however, it is generally impossible to make a definite statement about the degree to which the closed-loop system under a receding horizon controller satisfies constraints. Moreover, the closed-loop system may not satisfy the constraints of the problem, even if these are feasible at initial time.

In this section, we focus on the robust feasibility of stochastic MPC optimization problems subject to pointwise in time probabilistic constraints (including, by extension, mixtures of probabilistic and robust constraints that hold with probability 1). However we note that the same principles and analogous conditions for ensuring feasibility apply to other constraint formulations such as the expectation constraints and the joint probabilistic constraints of (6.7) and (6.9). To place this discussion in a general context, we consider constraints applied to an uncertain output variable:

$$\Pr_k(z_{0|k} \leq 0) \geq p \tag{7.2}$$

where $z_{i|k}$ is the predicted value at time k of the i-steps ahead output z_{k+i}, and where z_k is a function of the state x_k and control input u_k of (7.1a):

$$z_k \doteq f(x_k, u_k, v_k). \tag{7.3}$$

Here v is a random variable that is defined in terms of a stochastic model parameter $r_k = (r_k^{(1)}, \ldots, r_k^{(m)})$, which is unknown at time k but which has a known probability distribution, and a known set of vectors $\{v^{(1)}, \ldots, v^{(m)}\}$:

$$v_k = \sum_{j=1}^{m} v^{(j)} r_k^{(j)}. \tag{7.4}$$

In this set-up, r_k is not assumed to be independent of the stochastic parameter q_k appearing in (7.1a). Therefore the state constraints $\Pr_k(Fx_{1|k} \leq \mathbf{1}) \geq p$ are a special case of (7.2) with $z_k = f(x_k, u_k, v_k) = FAx_k + FBu_k + FDv_k - \mathbf{1}$ and $v_k = w_k$. Likewise, the mixed state and input constraints of (6.8) are included in (7.2) through a change of the definition of z_k:

$$z_k = f(x_k, u_k, u_{k+1}, v_k) = FA_k x_k + FB_k u_k + Gu_{k+1} + FDv_k - \mathbf{1}$$

and $v_k = w_k$.

Example 7.1 This example uses a simple system model to motivate the derivation of recursively feasible probabilistic constraints. The state x_k, control input u_k and output z_k at time k are governed by the first order dynamics:

$$\begin{aligned} x_{k+1} &= x_k + u_k + w_k \\ z_k &= x_k + v_k \end{aligned}$$

where, at time k, x_k is known and w_k, v_k are unknown discrete random variables with

$$\Pr(w_k = j) = \Pr(v_k = j) = \tfrac{1}{3}, \quad j = -1, 0, 1.$$

The system is subject to a probabilistic constraint, which is required to hold for all $k \geq 0$:

$$\Pr_k(z_{0|k} \leq 0) \geq \tfrac{1}{2}.$$

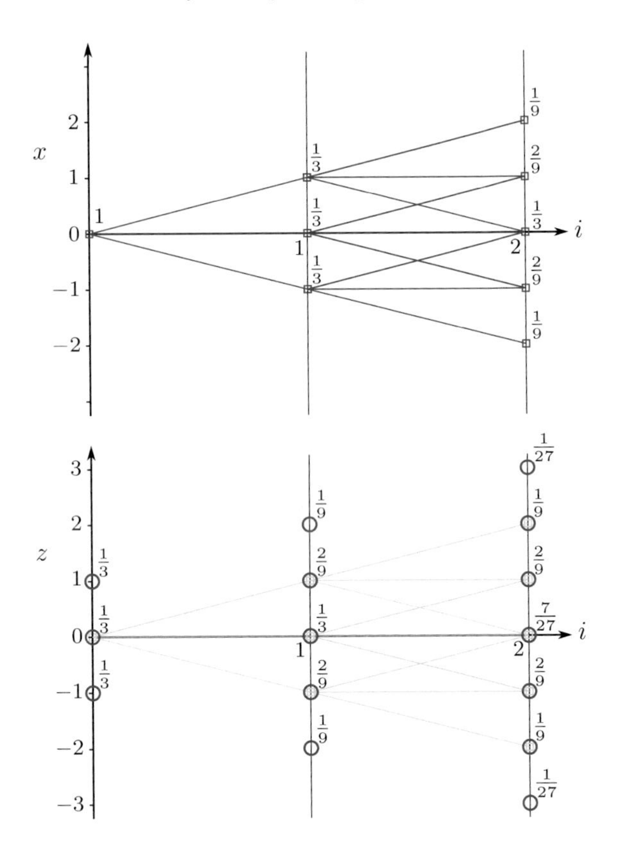

Fig. 7.1 The probability distributions of the predicted states, $\Pr_0(x_{i|0})$ (*upper plot*), and outputs, $\Pr_0(z_{i|0})$ (*lower plot*), of Example 7.1 for $i = 0, 1, 2$ with $u_{0|0} = u_{1|0} = 0$ and $x_0 = 0$

For the initial condition $x_0 = 0$, Fig. 7.1 shows the distributions of $x_{i|0}$ and $z_{i|0}$ that are obtained with the predicted input sequence $u_{0|0} = u_{1|0} = 0$. From this it can be seen that

$$\Pr_0(z_{0|0} \le 0) = \tfrac{2}{3}, \quad \Pr_0(z_{1|0} \le 0) = \tfrac{2}{3}, \quad \Pr_0(z_{2|0} \le 0) = \tfrac{17}{27},$$

which implies that $\Pr_0(z_{i|0} \le 0) \ge \frac{1}{2}$ holds for $i = 0, 1, 2$ (in fact it is easy to show that $\Pr_0(z_{i|0} \le 0) \ge \frac{1}{2}$ holds for all $i \ge 0$) if $u_{i|0} = 0$ for all $i \ge 0$.

However, somewhat counterintuitively, the existence of a predicted control sequence such that $\Pr_0(z_{i|0} \le 0) \ge \frac{1}{2}$ holds for given i does not ensure that the constraint $\Pr_k(z_{0|k} \le 0) \ge \frac{1}{2}$ will be feasible at time $k = i$. For example, if $u_0 = 0$, then at time $k = 1$ the condition $\Pr_1(z_{0|1} \le 0) \ge \frac{1}{2}$ may be violated since if $w_0 = 1$, then $x_1 = 1$ so that $z_1 = 1 + v_1$ and hence

$$\Pr_1(z_{0|1} \le 0) = \Pr(v_1 = -1) = \tfrac{1}{3}$$

in this case.

From $z_k = x_k + v_k$ and the probability distribution of v_k, it is easy to see that the condition $\Pr_k(z_{0|k} \le 0) \ge \frac{1}{2}$ is satisfied if and only if $x_k \le 0$. Given that $x_0 = 0$ and w_k lies in the interval $[-1, 1]$, we must therefore have $u_0 \le -1$ in order that $\Pr_1(z_{0|1} \le 0) \ge \frac{1}{2}$ is feasible. By this reasoning, a control law that ensures satisfaction of the probabilistic constraint for all k is given by

$$u_0 = -1, \quad u_k = -w_{k-1}, \quad k = 1, 2, \ldots$$

since this control strategy ensures $x_k \le 0$ for all k.

Returning to the problem of determining a feasible predicted control sequence at time $k = 0$, in order to ensure the future feasibility of the constraint $\Pr_i(z_{0|i} \le 0) \ge \frac{1}{2}$, the predicted control sequence must meet the condition $x_{i|0} \le 0$ for all possible realizations of the random sequence $\{w_0, \ldots, w_{i-1}\}$, for each $i > 0$. This can be formulated as the problem of determining $\{u_{0|0}, u_{1|0}, \ldots\}$ satisfying, for each $i > 0$,

$$\Pr_i\Bigl(\max_{w_0,\ldots,w_{i-1}\in[-1,1]} z_{i|0} \le 0\Bigr) \ge \tfrac{1}{2}$$

or equivalently

$$\Pr_i\Bigl(i + \sum_{j=0}^{i-1} u_{j|0} + v_i \le 0\Bigr) \ge \tfrac{1}{2}.$$

In order to impose the probabilistic constraint (7.2) in a way that guarantees recursive feasibility, it is therefore necessary to consider for each $i > 0$ the worst case realization of $\{w_0, \ldots, w_{i-1}\}$ whereas v_i can be treated as a stochastic variable. Note also that a robust reformulation of the constraint, i.e. $\Pr_k(z_{0|k} \le 0) = 1$, would have to consider worst case realizations of both w and v, and hence would require that $x_k \le -1$. $\diamondsuit$

Given the probability distribution of v, the function f defines a set of states for which there exists a control input such that the constraint (7.2) holds. Let

$$\mathcal{X} \doteq \bigl\{x : \exists u \ \text{ such that } \ \Pr\bigl(f(x, u, v) \le 0\bigr) \ge p\bigr\}, \tag{7.5}$$

then the future feasibility of (7.2) can be ensured by requiring that the predicted state sequence at time k satisfies $x_{i|k} \in \mathcal{X}$ for all possible realizations of the uncertainty sequence $\{q_k, \ldots, q_{k+i-1}\}$, for each $i > 0$. Denoting the support of q (namely the set of all possible realizations q_k) as $\mathcal{Q} \subseteq \mathbb{R}^m$, so that

$$\Pr(q_k \in \mathcal{Q}) = 1 \ \text{ and } \ \Pr(q_k \notin \mathcal{Q}) = 0, \tag{7.6}$$

the implied constraints on the control sequence predicted at time k are then

$$\Pr_k(z_{0|k} \leq 0) \geq p \tag{7.7a}$$

$$\Pr_{k+i}\Big(\max_{q_k,\ldots,q_{k+i-1}\in\mathcal{Q}} z_{i|k} \leq 0\Big) \geq p, \quad i = 1, 2, \ldots \tag{7.7b}$$

Thus the constraints on the i steps ahead output variable $z_{i|k}$ have been made robust with respect to $\{q_k, \ldots, q_{k+i-1}\}$ but remain stochastic with respect to v_{k+i}.

To verify these constraints are recursively feasible, suppose $\{u_{0|k}, u_{1|k}, \ldots\}$ satisfies (7.7a, 7.7b) for some k and consider the sequence defined at time $k+1$ by $u_{i|k+1} = u_{i+1|k}$ for $i = 0, 1, \ldots$ (or by $u_{i|k+1}(x) = u_{i+1|k}(x)$ if optimization is performed over feedback laws rather than open-loop control sequences). Then, for any $q_k \in \mathcal{Q}$, the condition (7.7b) with $i = 1$ implies $x_{k+1} \in \mathcal{X}$ and

$$\Pr_{k+1}(z_{0|k+1} \leq 0) \geq p.$$

Likewise, the condition (7.7b), when invoked for $i = j + 1$, ensures that $x_{j|k+1} \in \mathcal{X}$ for all $\{q_k, \ldots, q_{k+j}\} \in \mathcal{Q} \times \cdots \times \mathcal{Q}$ and

$$\Pr_{k+1+j}\Big(\max_{q_{k+1},\ldots,q_{k+j}\in\mathcal{Q}} z_{j|k+1} \leq 0\Big) \geq p, \quad j = 1, 2, \ldots$$

Hence there exists a predicted control sequence at time $k+1$ such that the conditions of (7.7a and 7.7b) hold with k replaced by $k+1$.

This argument demonstrates that the conditions of (7.7a and 7.7b) provide a recursively feasible set of constraints ensuring satisfaction of (7.2). It is important to note that these conditions are necessary as well as sufficient for recursive feasibility. In particular, if the predicted control sequence is optimized over arbitrary feedback laws with no restriction on the controller parameterization, then infeasibility of (7.7a, 7.7b) implies that, for some $i \geq 0$, it is not possible to satisfy $\Pr_{k+j}(z_{0|k+j} \leq 0) \geq p$ for all $j = 0, \ldots, i$ under any control law.

A consequence of the constraints (7.7b) involving a maximization over a subset of the uncertain model parameters is that feasibility cannot generally be guaranteed for models containing random variables with infinite support. Thus if q has infinite support, then a predicted control sequence or feedback law satisfying (7.2) can only exist if the model state x is unobservable from the constrained output variable z, and furthermore the unbounded model uncertainty associated with q must only affect the components of x that are unobservable from z. In general, this rules out problems in which q is normally distributed since such models allow disturbances to have an arbitrarily large effect on the model state, albeit with vanishingly small probability.

In most applications the restriction to finitely supported model uncertainty is not limiting since control systems are rarely subject to unbounded uncertainty in practice. However it does affect both the modelling of disturbances and the numerical methods required to handle probabilistic constraints. On the other hand, the parameter v that appears directly in the constrained output z is treated as a stochastic variable and is not required to have finite support. One consequence of this is that problems based on moving average dynamic models, such as those considered in Chap. 6, are not restricted to finitely supported random variables since in this case all of the model uncertainty is contained in the output map.

To conclude this section, we discuss how the conditions of (7.2), consisting of an infinite number of constraints over an infinite prediction horizon, can be reduced to a finite number of constraints. As in the case of robust MPC, it is possible to impose the constraints (7.2) on predicted state and control trajectories through a finite set of conditions, which nevertheless are recursively feasible over an infinite horizon, by using a dual-mode prediction strategy and an appropriate terminal constraint. In this context the terminal constraint $x_{N|k} \in \mathcal{X}_T$ is required to hold for all realizations of the uncertain sequence $\{q_k, \ldots, q_{k+N-1}\}$ over the initial N-step prediction horizon. Moreover recursive feasibility requires that $\mathcal{X}_T$ is a robustly invariant subset of the feasible set $\mathcal{X}$ in (7.5) and hence the probabilistic constraint (7.2) must hold for all $x \in \mathcal{X}_T$ under the terminal feedback law.

Assuming a linear terminal control law, $u_k = Kx_k$, we therefore require that $\mathcal{X}_T$ satisfies, for all $x \in \mathcal{X}_T$, the conditions

$$\big(A(q) + B(q)K\big)x + Dw(q) \in \mathcal{X}_T \quad \forall q \in \mathcal{Q} \tag{7.8a}$$

and

$$\Pr\big(f(x, Kx, v) \leq 0\big) \geq p. \tag{7.8b}$$

If (7.8b) can be invoked through an equivalent algebraic condition, then the maximal robustly positively invariant set satisfying (7.8a, 7.8b) (or a convex inner approximation of this set—see e.g. [4]) can be determined by a conceptually straightforward extension of Theorem 3.1, as we now briefly discuss. Defining the sequence of sets $\{\mathcal{S}_0, \mathcal{S}_1, \ldots\}$ by

$$\mathcal{S}_0 = \big\{x : \Pr\big(f(x, Kx, v) \leq 0\big) \geq p\big\}$$

and, for $k = 1, 2, \ldots$

$$\mathcal{S}_k = \big\{x : \big(A(q) + B(q)K\big)x + Dw(q) \in \mathcal{S}_{k-1} \ \forall q \in \mathcal{Q}\big\},$$

the MRPI set is given by

$$\mathcal{X}^{\text{MRPI}} \doteq \bigcap_{k=0}^{\infty} \mathcal{S}_k = \bigcap_{k=0}^{\nu} \mathcal{S}_k$$

where ν satisfies $\bigcap_{k=0}^{\nu} \mathcal{S}_k \subseteq \mathcal{S}_{\nu+1}$. In many cases of practical interest (see e.g. [5]), it is not easy to determine an algebraic equivalent of (7.8b) and it may therefore be necessary to resort to a conservative approximation, for example based on random sampling methods.

7.2 Prototype SMPC Algorithm: Stability and Convergence

This section proposes a general formulation of stochastic MPC for the system (7.1a, 7.1b). The algorithms presented here are conceptual in the sense that we do not consider how to solve the implied online MPC optimization problem (this is discussed in detail in later sections of this chapter and in Chap. 8). Instead, the focus of this section is on analysing closed-loop behaviour using the optimal value of the predicted cost. We show that the closed-loop system inherits a quadratic stability property when the MPC objective function is a quadratic predicted cost.

Two alternatives are considered for the MPC cost: the expected value predicted cost of Sect. 6.2 and the nominal predicted cost of Sect. 3.3. The system is subject to the pointwise-in-time probabilistic constraints of (6.8) and the constraints of the MPC optimization are constructed to ensure recursive feasibility as described in Sect. 7.1. Using a dual-mode prediction scheme, the predicted control sequence at time k is parameterized as

$$u_{i|k} = Kx_{i|k} + c_{i|k}, \quad i = 0, 1, \ldots \tag{7.9}$$

where $\mathbf{c}_k \doteq (c_{0|k}, \ldots, c_{N-1|k})$ is a vector of optimization variables at time k and $c_{i|k} = 0$ for all prediction times $i \geq N$, and where K satisfies the mean-square stability condition (6.5). It is assumed that a terminal set $\mathcal{X}_T$ is known, where $\mathcal{X}_T$ satisfies the condition (7.8) for robust invariance and the probabilistic constraint

$$\Pr\Big(\tilde{F}\big(\Phi(q)x + Dw(q)\big) \leq \mathbf{1}\Big) \geq p$$

for all $x \in \mathcal{X}_T$, where $\tilde{F} = F + GK$ and $\Phi(q) = A(q) + B(q)K$.

7.2.1 Expectation Cost

Define the predicted cost at time k as the expectation cost (6.15) of Sect. 6.2:

$$J(x_k, \mathbf{c}_k) = \sum_{i=0}^{\infty} \mathbb{E}_k\big(\|x_{i|k}\|_Q^2 + \|u_{i|k}\|_R^2 - l_{ss}\big) \tag{7.10}$$

where $l_{ss} = \mathrm{tr}\big(\Theta(Q + K^T RK)\big)$ and Θ is the solution of (6.10), and let K be the optimal linear feedback gain for this cost given by (6.22) and (6.23). Then, by Corollary 6.1, $J(x, \mathbf{c})$ has the quadratic form:

$$J(x, \mathbf{c}) = x^T W_x x + \mathbf{c}^T W_c \mathbf{c} + 2w_{x1}^T x + 2w_{c1}^T \mathbf{c} + w_1$$

where $W_x \succ 0$, $W_c \succ 0$ and w_{x1}, w_{c1} are given by (6.23), (6.24) and (6.21a), (6.21b). A recursively feasible MPC algorithm that minimizes this cost subject to the constraint $\Pr_k(Fx_{1|k} + Gu_{1|k} \leq \mathbf{1}) \geq p$ can be stated as follows.

Algorithm 7.1 At each time instant $k = 0, 1, \ldots$:

(i) Perform the optimization:

$$\underset{\mathbf{c}_k}{\text{minimize}} \quad J(x_k, \mathbf{c}_k) \tag{7.11a}$$

subject to

$$\Pr_k\big(\tilde{F}x_{1|k} + Gc_{1|k} \leq \mathbf{1}\big) \geq p \tag{7.11b}$$

$$\Pr_{k+i}\Bigg(\max_{q_k,\ldots,q_{k+i-1}\in\mathcal{Q}} \tilde{F}x_{i+1|k} + Gc_{i+1|k} \leq \mathbf{1}\Bigg) \geq p, \qquad i = 1,\ldots,N-2 \tag{7.11c}$$

$$\Pr_{k+N-1}\Bigg(\max_{q_k,\ldots,q_{k+N-2}\in\mathcal{Q}} \tilde{F}x_{N|k} \leq \mathbf{1}\Bigg) \geq p \tag{7.11d}$$

$$\text{and } x_{N|k} \in \mathcal{X}_T \quad \text{for all} \quad \{q_k,\ldots,q_{k+N-1}\} \in \mathcal{Q}\times\cdots\times\mathcal{Q} \tag{7.11e}$$

(ii) Apply the control law $u_k = Kx_k + c^*_{0|k}$, where $\mathbf{c}^*_k = (c^*_{0|k},\ldots,c^*_{N-1|k})$ is the optimal argument of (7.11). ◁

Theorem 7.1 *For the system (7.1a, 7.1b) under the control law of Algorithm 7.1, if the optimization (7.11) is feasible at $k = 0$, then it remains feasible for all $k > 0$. Also the closed-loop system satisfies the quadratic stability condition*

$$\lim_{r\to\infty}\frac{1}{r}\sum_{k=0}^{r}\mathbb{E}_0\big(\|x_k\|_Q^2 + \|u_k\|_R^2\big) \leq l_{ss} \tag{7.12}$$

and $\Pr_k(Fx_{1|k} + Gu_{1|k} \leq \mathbf{1}) \geq p$ *holds for all $k > 0$.*

Proof The argument of Sect. 7.1 shows that the constraints (7.11b–7.11e) are recursively feasible since if the optimization of (7.11) is feasible at time k, then

$$\mathbf{c}_{k+1} = (c^*_{1|k},\ldots,c^*_{N-1|k},0)$$

necessarily satisfies (7.11b–7.11e) at time $k+1$. Feasibility of (7.11b), therefore, implies satisfaction of the probabilistic constraint $\Pr_k(Fx_{1|k} + Gu_{1|k} \leq \mathbf{1}) \geq p$ for all $k \geq 0$.

From the definition of the predicted cost, we have

$$\mathbb{E}_k\big(J(x_{k+1}, \mathbf{c}_{k+1})\big) \le J^*(x_k) - \big(\|x_k\|_Q^2 + \|u_k\|_R^2 - l_{ss}\big)$$

where $J^*(x_k) = J(x_k, \mathbf{c}_k^*)$, and since optimality at time $k+1$ implies, for any realization of $q_k \in \mathcal{Q}$, that $J^*(x_{k+1}) \le J(x_{k+1}, \mathbf{c}_{k+1})$ it follows that

$$\mathbb{E}_k\big(J^*(x_{k+1})\big) \le J^*(x_k) - \big(\|x_k\|_Q^2 + \|u_k\|_R^2 - l_{ss}\big). \tag{7.13}$$

Taking expectations conditional on x_0 of both sides of this inequality and noting that $\mathbb{E}_0\big(\mathbb{E}_k\big(J^*(x_{k+1})\big)\big) = \mathbb{E}_0\big(J^*(x_{k+1})\big)$, we obtain, for all $r > 0$:

$$\frac{1}{r}\sum_{k=0}^{r-1} \mathbb{E}_0\big(\|x_k\|_Q^2 + \|u_k\|_R^2\big) \le l_{ss} + \frac{1}{r}\Big(J^*(x_0) - \mathbb{E}_0\big(J^*(x_r)\big)\Big).$$

Here $J^*(x_0)$ is by assumption finite whereas $J^*(x)$ has a finite lower bound since W_x and W_c are positive definite matrices; therefore, the second term on the RHS of this inequality vanishes as $r \to \infty$. □

The optimal value of the predicted cost $J(x, \mathbf{c})$ is not necessarily non-negative and the convergence analysis in (7.1) therefore relies on the existence of a finite lower bound on the predicted cost. A consequence of this is that an asymptotic value of the expected stage cost lower than l_{ss} may be achievable using an affine rather than linear state feedback law whenever the additive and multiplicative uncertainty in the model (7.1a, 7.1b) are correlated. From the expression for the unconstrained minimizer of (7.11) given by (6.20) and (6.21a) it can be seen that the control law of Algorithm 7.1 will in fact be equal to this affine feedback law if constraints are inactive, and this is the explanation for the inequality in the asymptotic bound of (7.12).

Two special cases are of particular interest in this convergence analysis. If the additive and multiplicative uncertain model parameters in (7.1b) are uncorrelated, then the discussion of Sect. 6.2 shows that l_{ss} is the minimum expected value of the stage cost that can be achieved under any control law. Therefore in this case the bound (7.12) implies that the control law of Algorithm 7.1 converges asymptotically to $u_k = Kx_k$, and it can then be shown that x_k converges with probability 1 to the minimal RPI set under this feedback law as $k \to \infty$.

A second special case is that in which the system (7.1a, 7.1b) is only affected by multiplicative uncertainty (i.e., $w_k = 0$ for all k). In this case, $l_{ss} = 0$ and it follows that $J^*(x)$ is a positive definite function of x. Therefore (7.13) implies that the closed-loop system is quadratically stable and hence $x_k \to 0$ as $k \to \infty$ with probability 1.

7.2.2 Mean-Variance Cost

A quadratic stability result also applies if the performance objective of Algorithm 7.1 is replaced by the mean-variance predicted cost (7.14) of Sect. 6.3:

$$J(x_k, \mathbf{c}_k) = \sum_{i=0}^{\infty} \big(\|x^{(0)}_{i|k}\|^2_Q + \|u^{(0)}_{i|k}\|^2_R\big)$$
$$+ \kappa^2 \sum_{i=0}^{\infty} \mathbb{E}_k\big(\|x_{i|k} - x^{(0)}_{i|k}\|^2_Q + \|u_{i|k} - u^{(0)}_{i|k}\|^2_R - l_{ss}\big). \qquad (7.14)$$

with $l_{ss} = \mathrm{tr}\big(\Theta(Q + K^T RK)\big)$ as before, and where κ is a given constant. However the optimal value of this predicted cost does not necessarily satisfy a condition such as (7.13), and as a result the available bounds on the mean-square value of the closed-loop system state are, in general, weaker than the bound given by Theorem 7.1. To analyse stability, we take an indirect approach similar to the analysis in Sect. 3.3 of robust stability of nominal MPC. This is based on the fact that the optimal linear feedback gain for the problem of minimizing (7.14) in the absence of constraints necessarily satisfies the condition (6.5) for mean-square stability, and it therefore induces a finite l_2 gain between the disturbance input and the closed-loop system state under an associated MPC law.

Assuming K to be the unconstrained optimal linear feedback gain defined by (6.33a–6.33c), the matrix W_z in (6.32a) is block-diagonal and the cost (7.14) can be expressed

$$J(x_k, \mathbf{c}_k) = x_k^T W_x x_k + \mathbf{c}_k^T W_c \mathbf{c}_k + 2w_{x1}^T x_k + 2w_{c1}^T \mathbf{c}_k + w_1.$$

Furthermore the structure of Ψ_k and $\Psi^{(0)}$ in (6.32a, 6.32b) implies that W_c and w_{c1} have the block structure:

$$W_c = \mathrm{diag}\{S, \ldots, S\}, \qquad w_{c1}^T = \begin{bmatrix} v^T & \cdots & v^T \end{bmatrix}$$

where $S \in \mathbb{R}^{n_u \times n_u}$ and $v \in \mathbb{R}^{n_u}$, with $S \succ 0$. The statement of a stochastic MPC algorithm based on the minimization of this cost subject to the constraints of (7.11) is as follows.

Algorithm 7.2 At each time instant $k = 0, 1, \ldots$:

(i) Perform the optimization:

$$\underset{\mathbf{c}_k}{\text{minimize}} \ \ \|\mathbf{c}_k\|^2_{W_c} + 2w_{c1}^T \mathbf{c}_k \ \ \text{subject to (7.11b–7.11e)} \qquad (7.15)$$

(ii) Apply the control law $u_k = Kx_k + c^*_{0|k}$, where $\mathbf{c}^*_k = (c^*_{0|k}, \ldots, c^*_{N-1|k})$ is the optimal argument of (7.15). ◁

Theorem 7.2 *For the control law of Algorithm 7.2 applied to the system (7.1a, 7.1b): if the optimization (7.15) is feasible at $k = 0$, then it is feasible for all $k > 0$, the closed-loop system satisfies the quadratic stability condition*

$$\lim_{r\to\infty} \frac{1}{r} \sum_{k=0}^{r} \mathbb{E}_0\big(\|x_k\|^2\big) \leq \gamma^2 \mathbb{E}\big(\|w_k\|^2\big) \tag{7.16}$$

for some finite scalar γ, and $\Pr_k(Fx_{1|k} + Gu_{1|k} \leq \mathbf{1}) \geq p$ *holds for all $k > 0$.*

The proof of the bound in (7.16) relies on the following result.

Lemma 7.1 *There exist scalars β, γ and a matrix $P \succ 0$ such that, under the control law of Algorithm 7.2, the following bound holds*

$$\mathbb{E}_k\big(\|x_{k+1}\|_P^2\big) \leq \|x_k\|_P^2 - \|x_k\|^2 + \beta^2\Big(\|c^*_{0|k}\|_S^2 + 2v^T c^*_{0|k}\Big) + \gamma^2 \mathbb{E}\big(\|w_k\|^2\big). \tag{7.17}$$

Proof Since K satisfies (6.33b) with $\hat{W}_x \succ 0$, there exists $P \succ 0$ satisfying $P - \mathbb{E}(\Phi_k^T P \Phi_k) \succ I$, and this ensures that β, γ exist so that $H_1 \succ 0$, where

$$H_1 \doteq \begin{bmatrix} P - I & 0 & 0 \\ 0 & \beta^2 S & \beta^2 v \\ 0 & \beta^2 v^T & \gamma^2 \mathbb{E}\big(\|w_k\|^2\big) \end{bmatrix} - \mathbb{E}\left(\begin{bmatrix} \Phi_k^T \\ B_k^T \\ w_k^T D^T \end{bmatrix} P \begin{bmatrix} \Phi_k & B_k & Dw_k \end{bmatrix}\right).$$

The bound in (7.17) is then obtained by pre- and post-multiplying H_1 by the vector $(x_k, c^*_{0|k}, 1)$ and using the system dynamics (7.1a). To show that $H_1 \succ 0$, suppose that $P - \mathbb{E}(\Phi_k^T P \Phi_k) \succeq (1 + \epsilon)I$ for some $\epsilon > 0$, then using Schur complements we find that $H_1 \succ 0$ if $H_2 \succ 0$, where

$$\begin{aligned} H_2 \doteq & \begin{bmatrix} \beta^2 S & \beta^2 v \\ \beta^2 v^T & \gamma^2 \mathbb{E}\big(\|w_k\|^2\big) \end{bmatrix} - \begin{bmatrix} \mathbb{E}(B_k^T P B_k) & \mathbb{E}(B_k^T P D w_k) \\ \mathbb{E}(w_k^T D^T P B_k) & \mathbb{E}\big(\|w_k\|_P^2\big) \end{bmatrix} \\ & + \epsilon^{-1} \begin{bmatrix} \mathbb{E}(B_k^T P \Phi_k) \\ \mathbb{E}(w_k^T D^T P \Phi_k) \end{bmatrix} \begin{bmatrix} \mathbb{E}(\Phi_k^T P B_k) & \mathbb{E}(\Phi_k^T P D w_k) \end{bmatrix}. \end{aligned}$$

The bottom right element of H_2 has the lower bound

$$\Big[\gamma^2 - \bar{\lambda}(P)\Big(1 + \epsilon^{-1} \sum_{j=1}^{m} \|P^{1/2} \Phi^{(j)}\|^2\Big)\Big] \mathbb{E}\big(\|w_k\|^2\big)$$

and is therefore positive for all non-zero $\mathbb{E}(\|w_k\|^2)$ if γ is sufficiently large. Since $S \succ 0$ it follows that $H_2 \succ 0$ for any given P, $\epsilon > 0$ and $\mathbb{E}(\|w_k\|^2)$ whenever the values of β and γ are sufficiently large. □

We can now give the proof of Theorem 7.2.

Proof (*of Theorem* 7.2) The constraints of (7.15) are identical to those of the MPC optimization in Algorithm 7.1. Therefore recursive feasibility and satisfaction of the probabilistic constraint therefore follow from feasibility of

$$\mathbf{c}_{k+1} = (c^*_{1|k}, \ldots, c^*_{N-1|k}, 0)$$

in (7.15) at time $k+1$. Let $V^*(x_k)$ denote the optimal value of the objective of (7.15) at time k, then the optimality of the MPC optimization at $k+1$ implies that $V^*(x_{k+1}) \le \|\mathbf{c}_{k+1}\|^2_{W_c} + 2w_{c1}^T \mathbf{c}_k$ for every realization $q_k \in \mathcal{Q}$. From the block structure of W_c and w_{c1}, we therefore have

$$\mathbb{E}_k\big(V^*(x_{k+1})\big) \le \|\mathbf{c}_{k+1}\|^2_{W_c} + 2w_{c1}^T \mathbf{c}_k = V^*(x_k) - \big(\|c^*_{0|k}\|^2_S + 2v^T c^*_{0|k}\big),$$

and hence from (7.17)

$$\mathbb{E}_k\big(\|x_{k+1}\|^2_P\big) \le \|x_k\|^2_P - \|x_k\|^2 + \beta^2\Big(V^*(x_k) - \mathbb{E}_k\big(V^*(x_{k+1})\big)\Big) + \gamma^2\mathbb{E}\big(\|w_k\|^2\big).$$

Taking expectations and summing over $k = 0, \ldots, r-1$, we have

$$\frac{1}{r}\sum_{k=0}^{r-1} \mathbb{E}_0\big(\|x_k\|^2\big) \le \gamma^2\mathbb{E}\big(\|w\|^2\big) + \frac{1}{r}\Big(\|x_0\|_P - \mathbb{E}_0\big(\|x_r\|^2_P\big)\Big) + \frac{\beta^2}{r}\Big(V^*(x_0) - \mathbb{E}_k\big(V^*(x_r)\big)\Big).$$

In the limit as $r \to \infty$, this implies the bound (7.16) since the second and third terms on the RHS of this inequality are necessarily bounded from above. □

Theorem 7.2 demonstrates the existence of a finite upper bound on the gain between the mean-square value of the additive disturbance and that of the closed-loop system state. But this result gives no indication of how the gain bound depends on the distribution of multiplicative model uncertainty, and hence it is a weaker result than Theorem 7.1. It is possible, however, to generalize the result of Theorem 7.1 for the case of the cost (7.14) with $\kappa > 1$, since then, using (6.32a–6.32c) and the expression (6.31) for the predicted cost, it can be shown that

$$\mathbb{E}_k\big(J^*(x_{k+1})\big) \le J^*(x_k) - \big(\|x_k\|^2_Q + \|u_k\|^2_R - \kappa^2 l_{ss}\big) \tag{7.18}$$

where $J^*(x_k) = J(x_k, \mathbf{c}^*_k)$ is the value of the cost (7.14) at the solution of the MPC optimization (7.15). By the argument that is used in the proof of Theorem 7.1, it follows from (7.18) that the closed-loop system under Algorithm 7.2 satisfies the quadratic stability condition

$$\lim_{r\to\infty} \frac{1}{r} \sum_{k=0}^{r} \mathbb{E}_0\big(\|x_k\|_Q^2 + \|u_k\|_R^2\big) \le \kappa^2 l_{ss} \tag{7.19}$$

whenever $\kappa > 1$.

7.2.3 Supermartingale Convergence Analysis

The quadratic bounds of Theorems 7.1 and 7.2 provide an indication of the asymptotic behaviour of the mean-square value of the closed-loop system state under Algorithms 7.1 and 7.2. However, except for the special case in which the model (7.1a, 7.1b) contains no additive disturbance, these results do not demonstrate asymptotic convergence of the state to a particular neighbourhood of the origin. Yet on the basis of the bounds on the evolution of the optimal value of the cost in (7.13) and (7.18), it is possible to state a convergence result for the state of the closed-loop system.

In order to do this, we define the ellipsoidal set

$$\Omega_\kappa \doteq \big\{x : x^T Q x \le \kappa^2 l_{ss}\big\}$$

and, given a sequence of states $\{x_0, x_1, \ldots\}$, we define the sequence $\{\hat{x}_0, \hat{x}_1, \ldots\}$ by $\hat{x}_0 = x_0$ and

$$\hat{x}_k = \begin{cases} x_k & \text{if } x_i \notin \Omega_\kappa \text{ for all } i < k \\ \hat{x}_{k-1} & \text{if } x_i \in \Omega_\kappa \text{ for some } i < k \end{cases} \tag{7.20}$$

for all $k > 0$. If x_k satisfies (7.13) for $\kappa = 1$ or (7.18) for $\kappa > 1$, then the sequence $\{J^*(\hat{x}_0), J^*(\hat{x}_1), \ldots\}$ is a supermartingale, namely a sequence of random variables with the property that $\mathbb{E}_k\big(J^*(\hat{x}_{k+1})\big) \le J^*(\hat{x}_k)$ for all $k \ge 0$ [6]. This follows from the fact that (7.13) or (7.18) imply

$$\mathbb{E}_k\big(J^*(\hat{x}_{k+1})\big) \le J^*(\hat{x}_k) - \big(\|\hat{x}_k\|_Q^2 - \kappa^2 l_{ss}\big) \le J^*(\hat{x}_k)$$

if $x_i \notin \Omega_\kappa$ for all $i \le k$ whereas $J^*(\hat{x}_{k+1}) = J^*(\hat{x}_k)$ if $x_i \in \Omega_\kappa$ for any $i \le k$, by (7.20). The stochastic convergence properties of supermartingales are well known, see e.g., [7], and in particular the following result (adapted from [8]) is useful in the current context.

Theorem 7.3 *Under Algorithm 7.1 with* $\kappa = 1$ *or Algorithm 7.2 with* $\kappa > 1$*, the state of the closed-loop system satisfies* $x_k \in \Omega_\kappa$ *for some k with probability* 1.

Proof Define the function $l(x)$ by

$$l(x) = \begin{cases} \|x\|_Q^2 - \kappa^2 l_{ss} & \text{if } x \notin \Omega_\kappa \\ 0 & \text{if } x \in \Omega_\kappa \end{cases}$$

and note that $l(x) > 0$ if and only if $x \notin \Omega_\kappa$. Then from (7.13), (7.18) and (7.20) we have, for all $k \geq 0$,

$$\mathbb{E}_k\big(J^*(\hat{x}_{k+1})\big) - J^*(\hat{x}_k) \leq -l(\hat{x}_k), \tag{7.21}$$

and summing over all $k < r$ yields, for any $r > 0$,

$$\sum_{k=0}^{r-1} \mathbb{E}_0\big(l(\hat{x}_k)\big) \leq J^*(x_0) - \mathbb{E}_0\big(J^*(\hat{x}_r)\big).$$

The RHS of this inequality has a finite upper bound because $J^*(x)$ is bounded from below for all x; it follows (by the Borel–Cantelli Lemma—see [6]) that $l(\hat{x}_k) \to 0$ with probability 1 and hence $\hat{x}_k \to \Omega_\kappa$ with probability 1. □

Theorem 7.3 implies that every state trajectory of the closed-loop system converges to the set Ω_κ. Although subsequently the state may not remain in Ω_κ, successive applications of Theorem 7.3 show that it must continually return to Ω_κ. The convergence of $\hat{x}$ to Ω_κ with probability 1 is equivalent to convergence in probability [8] since $\Pr\big(l(\hat{x}_k) \geq \epsilon\big) \to 0$ as $k \to \infty$ for all $\epsilon > 0$.

Analogous stability and convergence results can also be obtained for problems incorporating soft constraints that may be violated as often as required. For example, in applications that involve model uncertainty with unbounded support and for which the satisfaction of probabilistic constraints cannot be guaranteed, feasibility can be maintained by performing the optimization (7.11) (or (7.15) for $\kappa > 1$) whenever this is feasible, and otherwise minimizing the worst-case constraint violation subject to

$$J(x_k, c_k) \leq J\big(x_k, (c^*_{1|k-1}, \ldots, c^*_{N-1|k-1}, 0)\big).$$

Clearly this approach cannot guarantee that the closed-loop system will satisfy constraints of the form (6.7) or (6.8) and (6.9) at the required level of probability. However, it can impose a supermartingale-like condition such as (7.13) (or (7.18)), thus ensuring the quadratic stability condition of (7.12) (or (7.16)) and the convergence property of Theorem 7.3.

7.3 Probabilistically Invariant Ellipsoids

We next consider a generalized form of the probabilistic constraints (6.9). Rather than constraining the probability of the variable $Fx_k + Gu_k$ exceeding some threshold, we consider constraints on n_ψ output variables that are defined as the elements of the vector [9]

$$\psi_k = F_k x_k + G_k u_k + \eta_k. \tag{7.22}$$

The parameters $F_k \in \mathbb{R}^{n_\psi \times n_x}$, $G_k \in \mathbb{R}^{n_\psi \times n_u}$ and the noise process $\eta_k \in \mathbb{R}^{n_\psi}$ are subject to stochastic uncertainty which is described in terms of the random variable $q_k = (q_k^{(1)}, \ldots, q_k^{(m)})$ appearing in the model (7.1a, 7.1b) as $(F_k, G_k, \eta_k) = \big(F(q_k), G(q_k), \eta(q_k)\big)$, where

$$\big(F(q), G(q), \eta(q)\big) = (F^{(0)}, G^{(0)}, 0) + \sum_{j=1}^{m} (F^{(j)}, G^{(j)}, \eta^{(j)}) q^{(j)}. \tag{7.23}$$

Furthermore we consider a form of probabilistic constraint in which the output variable ψ_k is allowed to lie outside a prescribed interval $I_\psi = [\underline{\psi}, \bar{\psi}]$ provided the average probability of this happening over a given horizon N_c does not exceed a given limit:

$$\frac{1}{N_c} \sum_{i=0}^{N_c-1} \Pr_k(\psi_{i|k} \notin I_\psi) \le \frac{N_{\max}}{N_c}. \tag{7.24}$$

Here, $N_{\max}$ is a limit on the expected number of times the output variable can lie outside the prescribed interval over a horizon of N_c steps.

The form of constraint in (7.24) is applicable to situations in which it is not realistic to invoke a pointwise in time probabilistic constraint of the form of (6.8) at every time instant, but where it is desirable to constrain the average rate of violations. For example in the design of supervisory controllers for large wind turbines with the aim of damping structural vibrations (such as fore-aft tower oscillations [10]) so as to reduce fatigue damage, it may not be possible to constrain the material stresses to a given range with a pre-specified probability at each sampling instant. But to limit potential fatigue damage, it is essential that the expected rate of such violations does not exceed a given threshold.

Propagating stochastic uncertainty over a prediction horizon can present considerable computational challenges, so to provide an efficient method of invoking the constraints (7.24), here we make use of probabilistic bounds on the model parameters. Thus it is assumed that with a probability of at least p, the parameter q lies in a known set $\mathcal{Q}_p$. In the following, we assume $\mathcal{Q}_p$ to be a compact convex polytope defined in terms of its vertices, namely

$$\Pr(q \in \mathcal{Q}_p) \ge p, \qquad \mathcal{Q}_p \doteq \mathrm{Co}\{q_{p,1}, \ldots, q_{p,\nu}\}. \tag{7.25}$$

Here $\mathcal{Q}_p$ defines a confidence region for the uncertain parameters which is non-unique in general, and this can be a source of suboptimality. For example, a probabilistic constraint such as $\Pr\big(f(c,q) \le 0\big) \ge p$, where c is a decision variable and f a given function, is implied by the condition $f(c,q) \le 0$ for all $q \in \mathcal{Q}_p$. However, replacing the original probabilistic constraint in an optimization problem by the condition $f(c,q) \le 0$ for all $q \in \mathcal{Q}_p$ will, in general, result in a suboptimal solution for c if $\mathcal{Q}_p$ is taken to be a fixed set rather than an optimization variable.

The confidence region $\mathcal{Q}_p$ of (7.25) is computationally convenient in that it allows a straightforward algebraic reformulation of probabilistic constraints and it can be computed offline. Thus, if q were normally distributed with zero mean and identity covariance matrix, then $q^T q$ would have a chi-squared distribution and hence the radius of a sphere centred at the origin and containing q with probability p could be computed using the chi-squared distribution with m degrees of freedom. Consequently $\mathcal{Q}_p$ could be defined as any polytopic set that over-bounds this sphere. Normal distributions do not of course have finite support, but this approach can be used to approximate $\mathcal{Q}_p$ if q has a truncated normal distribution. The implication of (7.25) is that

$$\begin{aligned}
&\Pr\Big(\Phi_k \in \Phi^{(0)} + \mathrm{Co}\big\{\Phi(q_{p,1}), \ldots, \Phi(q_{p,\nu})\big\}\Big) \ge p \\
&\Pr\Big(B_k \in B^{(0)} + \mathrm{Co}\big\{B(q_{p,1}), \ldots, B(q_{p,\nu})\big\}\Big) \ge p \\
&\Pr\Big(w_k \in \mathrm{Co}\big\{w(q_{p,1}), \ldots, w(q_{p,\nu})\big\}\Big) \ge p
\end{aligned} \tag{7.26}$$

where $\Phi_k = A_k + B_k K$ and $\Phi(q) = \Phi^{(0)} + \sum_{j=1}^m \Phi^{(j)} q^{(j)}$.

The concept of probabilistic invariance is defined as follows.

Definition 7.1 A set $\mathcal{S} \subset \mathbb{R}^{n_x}$ is invariant with probability p for a system with state x_k if $\Pr_k(x_{1|k} \in \mathcal{S}) \ge p$ for all $x_k \in \mathcal{S}$.

To determine conditions under which an ellipsoidal set is invariant with probability p, we make use of the lifted autonomous state-space formulation of (6.16) in Sect. 6.2:

$$z_{i+1|k} = \Psi_{k+i} z_{i|k} + \bar{D} w_{k+i}, \quad z_{0|k} = \begin{bmatrix} x_k \\ \mathbf{c}_k \end{bmatrix}, \tag{7.27}$$

where $z_{i|k} \in \mathbb{R}^{n_z}$, $n_z = n_x + N n_u$, and $(\Psi_k, w_k) = \big(\Psi(q_k), w(q_k)\big)$ with

$$(\Psi(q), w(q)) = (\Psi^{(0)}, 0) + \sum_{j=1}^m (\Psi^{(j)}, w^{(j)}) q^{(j)},$$

$$\Psi^{(j)} = \begin{bmatrix} \Phi^{(j)} & B^{(j)} \\ 0 & M \end{bmatrix}, \quad \Phi^{(j)} = A^{(j)} + B^{(j)} K, \quad \bar{D} = \begin{bmatrix} D \\ 0 \end{bmatrix}.$$

The implied predicted control law is given by

$$u_{i|k} = Kx_{i|k} + c_{i|k}$$

with $c_{i|k} = 0$ for all $i \geq N$. As before, $\mathbf{c}_k = (c_{0|k}, \ldots, c_{N-1|k})$ is a decision variable at time k. We consider ellipsoidal sets $\mathcal{E}_z \subset \mathbb{R}^{n_z}$ defined in terms of $P_z \succ 0$, and their projections onto the x-subspace:

$$\mathcal{E}_z \doteq \{z : z^T P_z z \leq 1\}, \qquad \mathcal{E}_x \doteq \{x : x^T P_x x \leq 1\},$$

where, as discussed in Chap. 2 (Sect. 2.7.2), $P_x = \left(\begin{bmatrix} I_{n_x} & 0 \end{bmatrix} P_z^{-1} \begin{bmatrix} I_{n_x} \\ 0 \end{bmatrix}^T \right)^{-1}$.

Theorem 7.4 *The set $\mathcal{E}_z$ is invariant with probability p for the system (7.27) if a scalar λ exists such that, for $j = 1, \ldots, \nu$,*

$$\begin{bmatrix} P_z^{-1} & \Psi(q_{p,j})P_z^{-1} & \bar{D}w(q_{p,j}) \\ \star & \lambda P_z^{-1} & 0 \\ \star & \star & 1 - \lambda \end{bmatrix} \succeq 0. \tag{7.28}$$

Proof A sufficient condition for invariance of $\mathcal{E}_z$ with probability p is that $\|z_{1|k}\|^2_{P_z} \leq 1$ whenever $\|z_{0|k}\|^2_{P_z} \leq 1$, for all $q \in \mathcal{Q}_p$. Using (7.27) to express $z_{1|k}$ in terms of $z_{0|k}$ and applying the S-procedure [11] to these two inequalities, we obtain the equivalent condition that $\lambda > 0$ should exist such that

$$1 - \big(\Psi(q)z + \bar{D}w(q)\big)^T P_z \big(\Psi(q)z + \bar{D}w(q)\big) \geq \lambda(1 - z^T P_z z)$$

for all $z \in \mathbb{R}^{n_z}$ and all $q \in \mathcal{Q}_p$. An equivalent condition is that

$$\begin{bmatrix} \lambda P_z & 0 \\ 0 & 1 - \lambda \end{bmatrix} - \begin{bmatrix} \Psi^T(q) \\ w^T(q)\bar{D}^T \end{bmatrix} P_z \begin{bmatrix} \Phi(q) & \bar{D}w(q) \end{bmatrix} \succeq 0$$

for all q in $\mathcal{Q}_p$. Using Schur complements, it can be shown that this is equivalent to an LMI in $\Psi(q)$ and $w(q)$, which, when invoked for all q in the polytope $\mathcal{Q}_p$, is by linearity and convexity equivalent to the LMIs (7.28) corresponding to the vertices $q_{p,j}$, $j = 1, \ldots, \nu$ of $\mathcal{Q}_p$. □

The theorem gives conditions under which the state of (7.27) returns, at the next time instant, to $\mathcal{E}_z$ with probability p. A second confidence polytope, $\mathcal{Q}_{\tilde{p}}$, corresponding to a confidence level of $\tilde{p}$, can be used to state conditions such that $\psi_{0|k} \in I_\psi$ with probability $\tilde{p}$ for all $z_{0|k} \in \mathcal{E}_z$. To do this, ψ is first expressed as a function of z and q:

$$\psi_{0|k} = C(q_k)z_{0|k} + \eta(q_k)$$
$$C(q) = \begin{bmatrix} F(q) + G(q)K & G(q)E \end{bmatrix}.$$

Corollary 7.1 *Let $q_{\tilde{p},j}$ for $j = 1, \ldots, \nu$ denote the vertices of $\mathcal{Q}_{\tilde{p}}$. Then* $\Pr(\psi_{0|k} \in I_\psi \mid z_{0|k} \in \mathcal{E}_z) \geq \tilde{p}$ *if*

$$\underline{\psi} \leq \eta(q_{\tilde{p},j}) \leq \bar{\psi} \tag{7.29a}$$

and

$$[C(q_{\tilde{p},j})P_z^{-1}C(q_{\tilde{p},j})^T]_{l,l} \leq [\bar{\psi} - \eta(q_{\tilde{p},j})]_l^2 \tag{7.29b}$$

$$[C(q_{\tilde{p},j})P_z^{-1}C(q_{\tilde{p},j})^T]_{l,l} \leq [\eta(q_{\tilde{p},j}) - \underline{\psi}]_l^2 \tag{7.29c}$$

for $j = 1, \ldots, \nu$ and $l = 1, \ldots, n_\psi$, where $[\cdot]_{l,l}$ and $[\cdot]_l$ denote, respectively, the lth diagonal element of the matrix $[\cdot]$ and the lth element of the vector $[\cdot]$.

Proof For any given q, the maximum absolute value of the lth element of $C(q)z_{0|k}$ over all $z_{0|k} \in \mathcal{E}_z$ is equal to $[C(q)P_z^{-1}C(q)^T]_{l,l}^{1/2}$. It follows that $\Pr(\psi_{0|k} \in I_\psi \mid z_{0|k} \in \mathcal{E}_z) \geq \tilde{p}$ if (7.29a) holds and

$$[C(q)P_z^{-1}C(q)^T]_{l,l}^{1/2} \leq [\bar{\psi} - \eta(q)]_l$$

$$[C(q)P_z^{-1}C(q)^T]_{l,l}^{1/2} \leq [\eta(q) - \underline{\psi}]_l$$

for all $q \in \mathcal{Q}_{\tilde{p}}$ and $l = 1, \ldots, n_\psi$. Given the affine dependence of C and η on q, these conditions are convex in q and it follows that they are equivalent to the conditions (7.29b, 7.29c), which correspond to the vertices $q_{\tilde{p},1}, \ldots, q_{\tilde{p},\nu}$ of $\mathcal{Q}_{\tilde{p}}$. □

Theorem 7.4 and Corollary 7.1 can be used to invoke the constraint (7.24) by deploying a Markov chain model. To illustrate this, we consider a pair of ellipsoidal sets $\mathcal{E}_1, \mathcal{E}_2 \subset \mathbb{R}^{n_x}$ where $\mathcal{E}_1 \subset \mathcal{E}_2$, and $\mathcal{E}_1$ is probabilistically invariant with probability $p_{1,1}$ whereas $\mathcal{E}_2$ is robustly invariant (i.e. invariant with probability 1). Although two sets are considered here, the approach is also applicable to a larger number of nested sets. Define $\mathcal{S}_1 \doteq \mathcal{E}_1$ and $\mathcal{S}_2 \doteq \mathcal{E}_2 - \mathcal{E}_1$, and assume that the predicted state x is steered by (7.27) from $\mathcal{S}_l$ to $\mathcal{S}_j$ in a single time step with probability $p_{j,l}$ for $j = 1, 2, l = 1, 2$, i.e.

$$\Pr_{k+i}(x_{i+1|k} \in \mathcal{S}_j \mid x_{i|k} \in \mathcal{S}_l) = p_{j,l}.$$

Also let the probability that $\psi_{i|k} \notin I_\psi$ given that $x_{i|k} \in \mathcal{S}_j$ be p_j, for $j = 1, 2$, i.e.

$$\Pr_{k+i}(\psi_{i|k} \notin I_\psi \mid x_{i|k} \in \mathcal{S}_j) \leq p_j.$$

Then it follows that

$$\Pr_{k+i}\{\psi_{i|k} \notin I_\psi\} \leq \begin{bmatrix} p_1 & p_2 \end{bmatrix} \Pi^i e \tag{7.30}$$

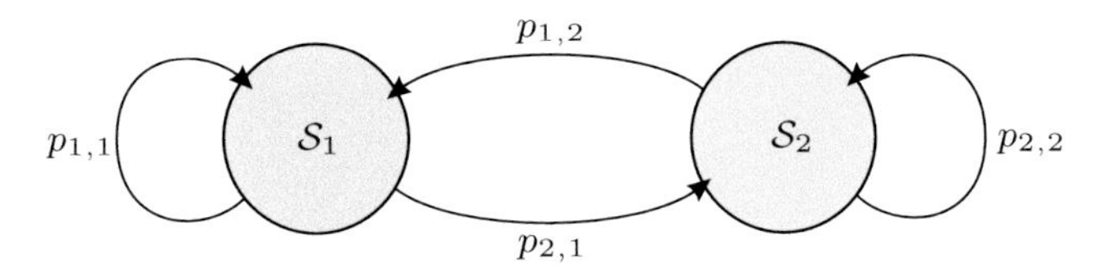

Fig. 7.2 Markov chain model with two discrete states

where

$$\Pi = \begin{bmatrix} p_{1,1} & p_{1,2} \\ p_{2,1} & p_{2,2} \end{bmatrix} \quad \text{and} \quad e = \begin{cases} [1\ 0]^T & \text{if } x_k \in \mathcal{S}_1 \\ [0\ 1]^T & \text{if } x_k \in \mathcal{S}_2 \end{cases}$$

By definition Π is the transition matrix of a Markov chain [12] and therefore satisfies $\begin{bmatrix}1 & 1\end{bmatrix}\Pi = \begin{bmatrix}1 & 1\end{bmatrix}$ since the successor state must belong to either $\mathcal{S}_1$ or $\mathcal{S}_2$ (Fig. 7.2). This implies that $\begin{bmatrix}1 & 1\end{bmatrix}$ is a left eigenvector of Π with 1 as the corresponding eigenvalue. Thus the eigenvector decomposition of Π can be written as

$$\Pi = \begin{bmatrix} w_1 & w_2 \end{bmatrix} \begin{bmatrix} 1 & 0 \\ 0 & \lambda_2 \end{bmatrix} \begin{bmatrix} v_1^T \\ v_2^T \end{bmatrix}, \quad 0 \le \lambda_2 \le 1. \tag{7.31}$$

Together with (7.30), this implies that constraint (7.24) will be satisfied if

$$\begin{bmatrix} p_1 & p_2 \end{bmatrix} w_1 v_1^T e_j + \frac{(\lambda_2 - \lambda_2^{N_c})}{N_c(1-\lambda_2)} \begin{bmatrix} p_1 & p_2 \end{bmatrix} w_2 v_2^T e_j \le \frac{N_{\max}}{N_c}, \quad j = 1, 2 \tag{7.32}$$

However, we must have $N_{\max} = \mu N_c$ for some $\mu \in (0, 1)$, and hence this condition can always be satisfied for sufficiently large N_c so long as

$$\begin{bmatrix} p_1 & p_2 \end{bmatrix} w_1 v_1^T e_j \le \mu, \quad j = 1, 2. \tag{7.33}$$

In the current setting, we can choose $\mathcal{E}_1$ as $\mathcal{E}_x$, the x-subspace projection of an ellipsoidal set $\mathcal{E}_z = \{z : z^T P_z z \le 1\}$ that is constrained to be invariant with a specified probability, $p_{1,1}^0$, through the conditions Theorem 7.4. Similarly, $\mathcal{E}_2$ can be defined as the x-subspace projection of a robustly invariant ellipsoidal set. Theorem 7.4 then implies that $p_{1,1} \ge p_{1,1}^0$ and $p_{2,1} \le p_{2,1}^0 \doteq 1 - p_{1,1}^0$. However, since the outer ellipsoid $\mathcal{E}_2$ contains all states of the system for which this MPC strategy can be applied, it is reasonable to assume that the probability, p_2, of $\psi \notin I_\psi$ given $x \in \mathcal{E}_2$ will be higher than, p_1, the corresponding probability given $x \in \mathcal{E}_1$. In this case, replacing $p_{1,1}$ and $p_{2,1}$ in Π with their conservative bounds, $p_{1,1}^0$ and $p_{2,1}^0$, has the effect of reinforcing the inequality (7.30) and thus the condition in (7.32) for satisfaction of the constraint (7.24) will continue to hold.

For $x_k \in \mathcal{E}_1$, the MPC cost can be defined, for example, as the expected quadratic cost $J(x_k, \mathbf{c}_k)$ of Sect. 7.2.1. However, whenever $x_k \notin \mathcal{E}_1$, rather than minimize this cost, a sensible strategy is to steer the nominal successor state towards $\mathcal{E}_1$ so as to increase the value of $p_{1,2}$. This can be done by minimizing an objective function such as $\|\Psi^{(0)} z_k\|^2_{P_z}$. The implied MPC algorithm can be stated as follows.

Algorithm 7.3 At times $k = 0, 1, \ldots$:

(i) If $x_k \in \mathcal{E}_1$, compute

$$\mathbf{c}_k^* = \arg\min_{\mathbf{c}_k} J(x_k, \mathbf{c}_k) \text{ subject to } z_k^T P_z z_k \leq 1 \tag{7.34}$$

(ii) Otherwise (i.e. if $x_k \notin \mathcal{E}_1$), compute

$$\mathbf{c}_k^* = \arg\min_{\mathbf{c}_k} z_k^T {\Psi^{(0)}}^T P_z \Psi^{(0)} z_k \text{ subject to } J(x_k, \mathbf{c}_k) \leq J(x_k, M\mathbf{c}_{k-1}^*) \tag{7.35}$$

(iii) Apply the control law $u_k = K x_k + c_{0|k}^*$ where $\mathbf{c}_k^* = (c_{0|k}^*, \ldots, c_{N-1|k}^*)$. ◁

The algorithm requires the minimization of a quadratic objective function subject to a single quadratic constraint in the optimization problems in each of steps (i) and (ii). The online MPC optimization is thus convex and can be solved efficiently. The constraint in (7.35) ensures that the time-average of the expected value of $\|x_k\|^2_Q + \|u_k\|^2_R$ satisfies an asymptotic bound, as shown by the following result.

Theorem 7.5 *For the system (7.1a, 7.1b) under the control law of Algorithm 7.3 with $J(x_k \mathbf{c}_k)$ defined by (7.10), the online optimization (7.34) or (7.35) is recursively feasible and the closed-loop state and control trajectories satisfy the asymptotic mean-square bound*

$$\lim_{r\to\infty} \frac{1}{r} \sum_{k=0}^{r-1} \mathbb{E}_0\big(\|x_k\|^2_Q + \|u_k\|^2_R\big) \leq l_{ss}. \tag{7.36}$$

Proof By construction, $\mathcal{E}_2$ is robustly invariant under Algorithm 7.3 and if $x_k \in \mathcal{E}_2$, then $\mathbf{c}_{k+1} = M\mathbf{c}_k^*$ is necessarily feasible for (7.34) and (7.35) at time $k+1$, thus establishing recursive feasibility. To demonstrate the bound (7.36), let $J^*(x_k) = J(x_k, \mathbf{c}_k^*)$, then

$$\begin{aligned}\mathbb{E}_k\big(J^*(x_{k+1})\big) &= \mathbb{E}_k\big(J^*(x_{k+1}) \mid x_{k+1} \in \mathcal{E}_1\big)\Pr_k(x_{k+1} \in \mathcal{E}_1) \\ &\quad + \mathbb{E}_k(J^*(x_{k+1}) \mid x_{k+1} \notin \mathcal{E}_1\big)\Pr_k(x_{k+1} \notin \mathcal{E}_1)\end{aligned}$$

where the feasibility of $\mathbf{c}_{k+1} = M\mathbf{c}_k^*$ implies that $J^*(x_{k+1}) \leq J(x_{k+1}, M\mathbf{c}_k^*)$ for all realizations of q_k at time k (this follows from the objective in (7.34) and the constraint in (7.35)). Therefore,

$$\mathbb{E}_k\big(J^*(x_{k+1})\big) \leq J^*(x_k) - \big(\|x_k\|^2_Q + \|u_k\|^2_R - l_{ss}\big)$$

and the bound (7.36) follows from the argument of Theorem 7.1. □

Algorithm 7.3 needs to be initialized through the offline design of the parameters $p_{1,1}, p_{1,2}, p_1, p_2$ and P_z. A possible procedure for this is as follows. Specify the initial values, $p^0_{1,1}$ and $p^0_{1,2}$, of $p_{1,1}$ and $p_{1,2}$ (and hence also the values of $p^0_{2,1} = 1 - p^0_{1,1}$ and $p^0_{2,2} = 1 - p^0_{1,2}$). Then, to make the constraints on $\mathcal{E}_1$ and $\mathcal{E}_2$ as unrestrictive as possible, set $p_2 = 1$ and choose p_1 as the maximum allowable value for p_1 according to (7.32). Next, construct the confidence polytopes $\mathcal{Q}_p$ (with $p = p^0_{1,1}$) and $\mathcal{Q}_{\tilde{p}}$ (with $\tilde{p} = p_1$) and maximize the volume of $\mathcal{E}_1$ by solving the optimization problem

$$\underset{P_z^{-1},\lambda}{\text{maximize}} \ \det(P_x^{-1}) \ \text{ subject to (7.28) and (7.29b, 7.29c)} \tag{7.37}$$

to determine P_z. This problem is convex for fixed values of λ in (7.28) and it can therefore be solved via a univariate search over $\lambda \in (0, 1)$. The optimization (7.37) can also be used to determine a robustly invariant ellipsoid $\mathcal{E}_z$ with maximum volume x-subspace projection, although the vertices of $\mathcal{Q}$ such that $q_k \in \mathcal{Q}$ with probability 1 must be used in place of those of $\mathcal{Q}_p$ in the constraint (7.28), and the constraints (7.29b, 7.29c) are not needed in (7.37) given that $p_2 = 1$ is assumed.

For given $\mathcal{E}_1$ and $\mathcal{E}_2$, the actual value of $p_{1,2}$ can be determined through Monte Carlo simulations (for example, by searching over the boundary of $\mathcal{E}_2$ for the minimum probability of inclusion of the successor state in $\mathcal{E}_1$). In order to ensure satisfaction of (7.32) and hence of the constraint (7.24), we require that $p_{1,2} \geq p^0_{1,2}$. If this condition is not satisfied, the procedure must be repeated with reduced values for the initial guesses $p^0_{1,1}$ and $p^0_{1,2}$.

Problems involving constraints on the rate of accumulation of fatigue damage typically place higher weighting on larger amplitude fluctuations of the constrained output ψ. This suggests using a number of intervals for ψ and modifying the constraint (7.24) in order to define an upper bound on a weighted sum of expected constraint violations over an interval. Further improvements in the accuracy with which the closed-loop system satisfies the probabilistic constraints of the problem can be obtained by using a larger number of sets $\mathcal{S}_j$, since the accuracy of the Markov chain model in predicting constraint violations improves as a finer discretization of the state space is employed. These modifications are considered in Sect. 7.4, which also introduces a mode 1 prediction horizon and tubes in order to obtain more accurate approximations of predicted probability distributions based on Markov chains.

7.4 Markov Chain Models Based on Tubes with Polytopic Cross Sections

This section uses a Markov chain model to approximate the evolution of the probability distribution of the states of the model (7.1a, 7.1b) over a prediction horizon. In this setting, the Markov chain imposes a discretization of the state space based on a sequence of nested tubes. The approach determines offline bounds on the transition probabilities between the sets that form the tube cross sections at successive time steps so as to meet probabilistic constraints. Confidence bounds on the model parameters can be used to apply these probabilistic bounds (together with the hard constraints of the problem) to predicted state and control trajectories which are constructed during the online MPC optimization.

Tubes with polytopic cross sections are considered here. These are taken to be low-complexity polytopes such as the sets considered in Sects. 3.6.2 and 5.4. Thus the shapes of the sets defining the tube cross sections are fixed but their centres and scalings along a set of fixed directions can be adjusted online. At each prediction time step $i = 0, 1, \ldots$, we define μ polytopic cross sections:

$$\mathcal{X}_{i|k}^{(j)} \doteq \{z : \underline{z}_{i|k}^{(j)} \leq z \leq \bar{z}_{i|k}^{(j)}\}, \quad j = 1, \ldots, \mu, \tag{7.38}$$

where z denotes the transformed state $z = Vx$ for a non-singular matrix $V \in \mathbb{R}^{n_x \times n_x}$. The dynamics of z are given by

$$z_{i+1|k} = \tilde{\Phi}_{k+i} z_{i|k} + \tilde{B}_{k+i} c_{i|k} + \tilde{D} w_{i|k}, \quad i = 0, 1, \ldots$$

where $\tilde{\Phi}_{k+i} = V \Phi_{k+i} V^{-1}$, $\tilde{B}_{k+i} = V B_{k+i}$, $\tilde{D} = VD$.

The methodology of this section is not limited to low-complexity polytopic tubes, and general complexity polytopic tube cross sections such as those considered in Sect. 5.5 could be used in place of the sets defined in (7.38). In this more general case, the set inclusion conditions that are developed in this section could be imposed using an approach based on Farkas Lemma such as that of Lemma 5.6. For simplicity however we present the ideas here using low-complexity tubes.

We assume that V is fixed and designed offline, while $\underline{z}_{i|k}^{(j)}$ and $\bar{z}_{i|k}^{(j)}$, for $i = 0, \ldots, N$, $j = 1, \ldots, \mu$, are online optimization variables that are computed simultaneously with $\mathbf{c}_k = (c_{0|k}, \ldots, c_{N-1|k})$ at each time k. The tube cross sections defined in (7.38) are constrained to be nested (Fig. 7.3):

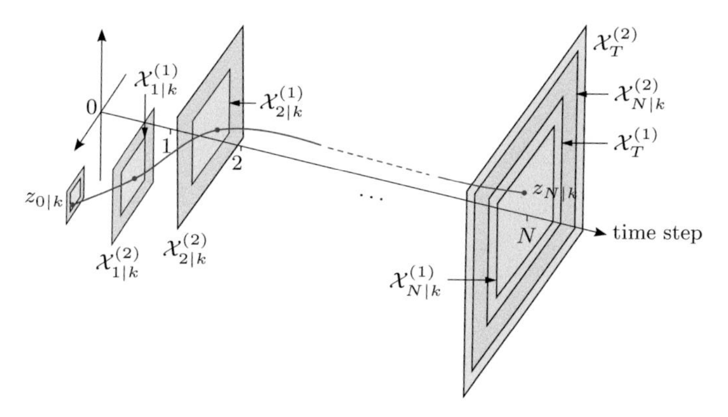

Fig. 7.3 Low-complexity polytopic tubes $\{\mathcal{X}_{0|k}^{(j)}, \ldots, \mathcal{X}_{N|k}^{(j)}\}$ and terminal sets $\mathcal{X}_T^{(j)}$ for $j = 1, 2$, for the case of a 2-dimensional state space

$$\mathcal{X}_{i|k}^{(1)} \subseteq \mathcal{X}_{i|k}^{(2)} \subseteq \cdots \subseteq \mathcal{X}_{i|k}^{(\mu)}, \quad i = 0, \ldots, N. \tag{7.39}$$

At the start of the prediction horizon, we require $z_{0|k} = Vx_k$ to lie $\mathcal{X}_{0|k}^{(j)}$ for some $l = 1, \ldots, \mu$, and terminal condition involving a polytopic terminal set is imposed on each tube layer,

$$\mathcal{X}_{N|k}^{(j)} \subseteq \mathcal{X}_{T}^{(j)} = \{z : |z| \leq z_T^{(j)}\}, \quad j = 1, \ldots, \mu, \tag{7.40}$$

where the terminal sets are also nested:

$$\mathcal{X}_{T}^{(1)} \subseteq \mathcal{X}_{T}^{(2)} \subseteq \cdots \subseteq \mathcal{X}_{T}^{(\mu)}. \tag{7.41}$$

The nested conditions of (7.39) and (7.41) can be invoked simply by imposing the constraints

$$\underline{z}_{i|k}^{(1)} \geq \underline{z}_{i|k}^{(2)} \geq \cdots \geq \underline{z}_{i|k}^{(\mu)}, \tag{7.42a}$$

$$\bar{z}_{i|k}^{(1)} \leq \bar{z}_{i|k}^{(2)} \leq \cdots \leq \bar{z}_{i|k}^{(\mu)}, \tag{7.42b}$$

$$z_T^{(1)} \leq z_T^{(2)} \leq \cdots \leq z_T^{(\mu)}. \tag{7.42c}$$

The constraints of the problem are taken to be the same as those considered in Sect. 7.3 (i.e. in (7.24)), but here we separate the constraints into probabilistic and hard constraints through the definition of two output vectors, ψ^p and ψ^h. These are given in terms of $z_{i|k}$ and $c_{i|k}$ as

$$\psi_{i|k}^p = F_p z_{i|k} + G_p c_{i|k} \tag{7.43a}$$

$$\psi_{i|k}^h = F_h z_{i|k} + G_h c_{i|k} \tag{7.43b}$$

For simplicity, the parameters F_p and G_p are assumed to be deterministic, although constraints such as (7.22)–(7.23) can be handled by a simple extension of the same approach. We consider the constraints:

$$\frac{1}{N_c}\sum_{i=1}^{N_c} \Pr_k(\psi_{i|k}^p > \mathbf{1}) \leq \frac{N_{\max}}{N_c} \tag{7.44a}$$

$$\psi_{i|k}^h \leq \mathbf{1}. \tag{7.44b}$$

It is straightforward to show that with an appropriate choice of the parameters F_p, G_p, (7.44a) has the same form as (7.24).

The strategy for handling the probabilistic constraints (7.44a) resembles that of Sect. 7.3 in that conditions are imposed on the probability with which the predicted state makes a transition between different tube layers and on the probability of the

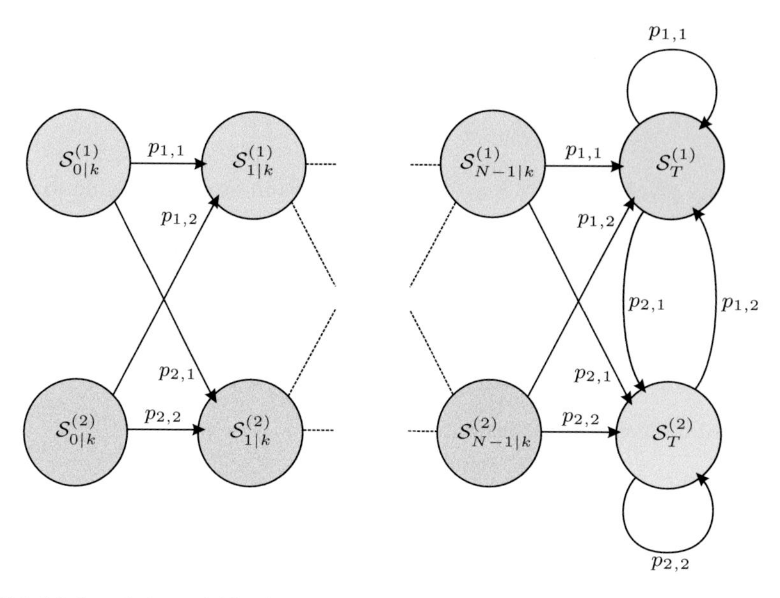

Fig. 7.4 Markov chain model for the case of $\mu = 2$

constrained output variables in (7.43a, 7.43b) exceeding threshold values within each tube layer. Define sets $\mathcal{S}^{(j)}_{i|k}$ for $i = 0, \ldots, N$ by

$$\mathcal{S}^{(j)}_{i|k} = \begin{cases} \mathcal{X}^{(1)}_{i|k} & j = 1 \\ \mathcal{X}^{(j)}_{i|k} \setminus \mathcal{X}^{(j-1)}_{i|k} & j = 2, \ldots, \mu \end{cases}$$

and let $p_{j,m}$ be the probability that $z_{i+1|k}$ lies in $\mathcal{S}^{(j)}_{i+1|k}$ given that $z_{i|k}$ lies in $\mathcal{S}^{(m)}_{i|k}$ (Fig. 7.4). Then

$$\begin{bmatrix} \Pr_k\big(z_{i+1|k} \in \mathcal{S}^{(1)}_{i+1|k}\big) \\ \vdots \\ \Pr_k\big(z_{i+1|k} \in \mathcal{S}^{(\mu)}_{i+1|k}\big) \end{bmatrix} = \Pi \begin{bmatrix} \Pr_k\big(z_{i|k} \in \mathcal{S}^{(1)}_{i|k}\big) \\ \vdots \\ \Pr_k\big(z_{i|k} \in \mathcal{S}^{(\mu)}_{i|k}\big) \end{bmatrix}, \quad \Pi \doteq \begin{bmatrix} p_{1,1} & \cdots & p_{1,\mu} \\ \vdots & \ddots & \vdots \\ p_{\mu,1} & \cdots & p_{\mu,\mu} \end{bmatrix} \tag{7.45}$$

and, for $z_{0|k} \in \mathcal{S}^{(j)}_{0|k}$, we have

$$\begin{bmatrix} \Pr_k\big(z_{i|k} \in \mathcal{S}^{(1)}_{i|k}\big) \\ \vdots \\ \Pr_k\big(z_{i|k} \in \mathcal{S}^{(\mu)}_{i|k}\big) \end{bmatrix} = \Pi^i e_j \tag{7.46}$$

where e_j is the jth column of the identity matrix. If, in addition, the probability that $\psi^p_{i+1|k} > 1$ given $z_{i|k} \in \mathcal{S}^{(m)}_{i|k}$ is no greater than p_m for each $m = 1, \ldots, \mu$, then the bound

$$\Pr_k(\psi^p_{i+1|k} > \mathbf{1}) \leq \begin{bmatrix} p_1 & \cdots & p_\mu \end{bmatrix} \Pi^i e_j$$

holds for any given i whenever $z_{0|k} \in \mathcal{S}^{(j)}_{0|k}$, and this in turn can be used to ensure that (7.44a) is satisfied.

In order to obtain a computationally convenient set of constraints for the online MPC optimization, we formulate the constraints in terms of the probabilities $\tilde{p}_{j,m}$ of transition from $\mathcal{X}^{(m)}_{i|k}$ to $\mathcal{X}^{(j)}_{i+1|k}$ rather than transition probabilities between $\mathcal{S}^{(m)}_{i|k}$ and $\mathcal{S}^{(j)}_{i+1|k}$. Furthermore these conditions are imposed through inequalities rather than equality constraints. The required set of constraints is as follows.

(i) Transition probability constraints, for $j, m = 1, \ldots, \mu$

$$\Pr_{k+i}\big(z_{i+1|k} \in \mathcal{X}^{(j)}_{i+1|k} \mid z_{i|k} \in \mathcal{X}^{(m)}_{i|k}\big) \geq \tilde{p}_{j,m}, \quad i = 0, \ldots, N-1 \qquad (7.47a)$$

$$\Pr_{k+i}\big(z_{i+1|k} \in \mathcal{X}^{(j)}_{T} \mid z_{i|k} \in \mathcal{X}^{(m)}_{T}\big) \geq \tilde{p}_{j,m}, \quad i \geq N. \qquad (7.47b)$$

(ii) Probabilistic output constraints, for $j = 1, \ldots, \mu$

$$\Pr_{k+i}\big(\psi^p_{i+1|k} > \mathbf{1} \mid z_{i|k} \in \mathcal{X}^{(j)}_{i|k}\big) \leq p_j, \quad i = 0, \ldots, N-1 \qquad (7.48a)$$

$$\Pr_{k+i}\big(\psi^p_{i+1|k} > \mathbf{1} \mid z_{i|k} \in \mathcal{X}^{(j)}_{T}\big) \leq p_j, \quad i \geq N. \qquad (7.48b)$$

(iii) Robust output constraints

$$\psi^h_{i+1|k} \leq \mathbf{1} \;\; \forall\, z_{i|k} \in \mathcal{X}^{(\mu)}_{i|k}, \quad i = 0, \ldots, N-1 \qquad (7.49a)$$

$$\psi^h_{i+1|k} \leq \mathbf{1} \;\; \forall\, z_{i|k} \in \mathcal{X}^{(\mu)}_{T}, \quad i \geq N. \qquad (7.49b)$$

(iv) Initial and terminal constraints, for $j = 1, \ldots, \mu$

$$V x_k \in \mathcal{X}^{(\mu)}_{0|k} \qquad (7.50a)$$

$$\mathcal{X}^{(j)}_{N|k} \subseteq \mathcal{X}^{(j)}_{T}. \qquad (7.50b)$$

The transition probability constraints are invoked through inequalities in (7.47a, 7.47b), which implies that the transition probabilities of the Markov chain model (7.45) will not, in general, hold with equality. However it is still possible to ensure satisfaction of the constraints (7.44a, 7.44b) if we make the following assumption on the probabilities $\tilde{p}_{j,m}$, p_j chosen by the designer.

Assumption 7.1 The probabilities $\tilde{p}_{j,m}$ and p_j satisfy:

$$\tilde{p}_{j,m} \geq \tilde{p}_{j,m+1}, \quad j, m = 1, \dots, \mu - 1 \tag{7.51a}$$
$$\tilde{p}_{\mu,m} = 1, \quad m = 1, \dots, \mu \tag{7.51b}$$
$$p_{j+1} \geq p_j, \quad j = 1, \dots, \mu. \tag{7.51c}$$

These conditions correspond to a unimodality assumption on the distributions of probabilistically constrained outputs. In particular, the conditions of (7.51c) require that the probability with which the one-step ahead output satisfies $\psi^p \leq \mathbf{1}$ should be smaller at points further from the centre of the tube. Likewise, (7.51a) requires that the probability of transition to any given layer should decrease away from the tube centre, while (7.51b) ensures that the outer tube layer bounds robustly the predicted state trajectories for all possible uncertainty realizations.

We next use (7.47)–(7.50) to derive an upper bound on the probability $\Pr_k(\psi^p_{i+1|k} > 1)$ for any $i \geq 0$.

Lemma 7.2 *Under conditions (7.47)–(7.50) and Assumption 7.1, the probability that $\psi^p_{i+1|k} > \mathbf{1}$ given that $z_{0|k} \in \mathcal{S}^{(j)}_{0|k}$ is bounded by*

$$\Pr_k\big(\psi^p_{i+1|k} > \mathbf{1}\big) \leq \big[p_1 \ \cdots \ p_\mu\big](T\tilde{\Pi})^i e_j \tag{7.52}$$

for any $i \geq 0$, where $\tilde{\Pi}$ and T are defined by

$$\tilde{\Pi} = \begin{bmatrix} \tilde{p}_{1,1} & \cdots & \tilde{p}_{1,\mu} \\ \vdots & \ddots & \vdots \\ \tilde{p}_{\mu,1} & \cdots & \tilde{p}_{\mu,\mu} \end{bmatrix}, \quad T = \begin{bmatrix} 1 & 0 & 0 & \cdots & 0 & 0 \\ -1 & 1 & 0 & \cdots & 0 & 0 \\ 0 & -1 & 1 & \cdots & 0 & 0 \\ \vdots & \vdots & \vdots & \ddots & \vdots & \vdots \\ 0 & 0 & 0 & \cdots & 1 & 0 \\ 0 & 0 & 0 & \cdots & -1 & 1 \end{bmatrix}. \tag{7.53}$$

Proof First, let $\tilde{v}_{i|k}$ and $v_{i|k}$ denote the vectors

$$\tilde{v}_{i|k} = \Big[\Pr_k\big(z_{i|k} \in \mathcal{X}^{(1)}_{i|k}\big) \ \cdots \ \Pr_k\big(z_{i|k} \in \mathcal{X}^{(\mu)}_{i|k}\big)\Big]^T$$
$$v_{i|k} = \Big[\Pr_k\big(z_{i|k} \in \mathcal{S}^{(1)}_{i|k}\big) \ \cdots \ \Pr_k\big(z_{i|k} \in \mathcal{S}^{(\mu)}_{i|k}\big)\Big]^T$$

and note that the nested property (7.41) and the definition of $\mathcal{S}^{(j)}_{i|k}$ as the set difference $\mathcal{X}^{(j)}_{i|k} \setminus \mathcal{X}^{(l-1)}_{i|k}$ implies that

$$\Pr_k\big(z_{i|k} \in \mathcal{X}^{(j)}_{i|k}\big) = \Pr_k\big(z_{i|k} \in \mathcal{X}^{(l-1)}_{i|k}\big) + \Pr_k\big(z_{i|k} \in \mathcal{S}^{(j)}_{i|k}\big), \quad l = 2, \dots, \mu$$

and hence $v_{i|k} = T\tilde{v}_{i|k}$. Next we show by induction that if $z_{0|k} \in \mathcal{X}^{(j)}_{0|k}$, then

$$\tilde{v}_{i|k} \geq (\tilde{\Pi} T)^{i-1} \tilde{\Pi} e_j. \tag{7.54}$$

Specifically, conditions (7.47a, 7.47b) imply $\Pr_{k+i}\big(z_{i+1|k} \in \mathcal{X}^{(j)}_{i+1|k} \mid z_{i|k} \in \mathcal{S}^{(m)}_{i|k}\big) \geq \tilde{p}_{j,m}$, and since $\Pr\big(z_{0|k} \in \mathcal{S}^{(j)}_{0|k}\big) = 1$, we therefore have $\tilde{v}_{1|k} \geq \tilde{\Pi} e_j \geq 0$. Furthermore condition (7.51a) of Assumption 7.1 ensures that the elements of $\tilde{\Pi} T$ are non-negative and it follows that

$$\tilde{v}_{i+1|k} \geq \tilde{\Pi} v_{i|k} = \tilde{\Pi} T \tilde{v}_{i|k}$$

whenever $\tilde{v}_{i|k} \geq 0$. Therefore (7.54) holds for all $i > 0$.

Finally we obtain a bound on the probability that $\psi^{p}_{i+1|k} > \mathbf{1}$ using (7.54) and

$$\Pr_k\big(\psi^{p}_{i+1|k} > \mathbf{1}\big) \geq \big[p_1 \ \cdots \ p_\mu\big] v_{i|k} = \big[p_1 \ \cdots \ p_\mu\big] T \tilde{v}_{i|k}. \tag{7.55}$$

Here the first $\mu - 1$ elements of the row vector $\big[p_1 \ \cdots \ p_\mu\big] T$ are non-positive because of (7.51c), whereas every element of $\tilde{v}_{i|k}$ except the last (which by (7.51b) is equal to 1) is over-estimated by (7.54). Hence replacing $\tilde{v}_{i|k}$ in (7.55) by the RHS of (7.54) yields an upper bound on $\Pr_k\big(\psi^{p}_{i+1|k} > \mathbf{1}\big)$ and the bound in (7.52) then follows since $T(\tilde{\Pi} T)^{i-1} \tilde{\Pi} = (T\tilde{\Pi})^i$ for $i > 0$. □

The bounds of Lemma 7.2 provide sufficient conditions for the satisfaction of the probabilistic constraint (7.44a). Summing (7.52) over $i = 0, \ldots, N_c - 1$ yields directly the result that (7.44a) necessarily holds if the probabilities $\tilde{p}_{j,m}$ and p_j, for $j, m = 1, \ldots, \mu$ are chosen so as to satisfy

$$\frac{1}{N_c} \sum_{i=0}^{N_c-1} \big[p_1 \ \cdots \ p_\mu\big] (T\tilde{\Pi})^i e_j \leq \frac{N_{\max}}{N_c} \tag{7.56}$$

for all $j = 1, \ldots, \mu$.

Before stating the stochastic MPC algorithm, we first show that the constraints (7.47)–(7.50) can be expressed as linear inequalities in the optimization variables $\mathbf{c}_k$, $\underline{z}^{(j)}_{i|k}$ and $\bar{z}^{(j)}_{i|k}$, $i = 0, \ldots, N$, $j = 1, \ldots, \mu$. There is little advantage in including the parameters $z^{(j)}_T$ of the terminal sets in the list of degrees of freedom; instead, it is suggested that these parameters are chosen offline in order to maximize the volume of each terminal set, which, given that $\mathcal{X}^{(j)}_T$, $j = 1, \ldots, \mu$ are orthotopes, can be taken to be the product of the elements of $z^{(j)}_T$. Clearly, such a maximization has to be subject to conditions (7.47b)–(7.49b) or the equivalent inequalities to be presented below. The implied optimization can be solved using the techniques described in Sect. 5.4.1.

Each of the constraints (7.47)–(7.50) is conditioned on either $z_{i|k} \in \mathcal{X}_{i|k}^{(j)}$ or $z_{i|k} \in \mathcal{X}_T^{(j)}$. Moreover, if these constraints are imposed using convex confidence polytopes for the uncertain model parameters, then as we show below, they each depend linearly on $z_{i|k}$. Therefore, (7.47)–(7.50) are equivalent to the corresponding constraints invoked only at the vertices of $\mathcal{X}_{i|k}^{(j)}$ and $\mathcal{X}_T^{(j)}$. Furthermore, these conditions can be invoked using the recursive bounding technique of Sect. 5.4.2 without the need to evaluate the vertices of the tube cross sections.

Conditions (7.47)–(7.48) are probabilistic and can be converted into linear inequalities by invoking them at the vertices of confidence polytopes for the uncertain model parameters. However these polytopes must be defined separately for each of the probabilities $\tilde{p}_{j,m}$ involved in (7.47)–(7.48). Therefore we define $\mathcal{Q}(\tilde{p}_{j,m})$ as a confidence polytope such that $\Pr\big(q \in \mathcal{Q}(\tilde{p}_{j,m})\big) \geq \tilde{p}_{j,m}$, and denote its vertices by $q^{(s)}(\tilde{p}_{j,m})$, $s = 1, \ldots, \nu(\tilde{p}_{j,m})$. With these definitions, the constraints of (7.47)–(7.50) can be expressed as the following linear inequalities.

(i) Transition probability constraints, for $j, m = 1, \ldots, \mu$:

$$\tilde{\Phi}^+\big(q^{(s)}(\tilde{p}_{j,m})\big)\bar{z}_{i|k}^{(m)} - \tilde{\Phi}^-\big(q^{(s)}(\tilde{p}_{j,m})\big)\underline{z}_{i|k}^{(m)} + \tilde{B}\big(q^{(s)}(\tilde{p}_{j,m})\big)c_{i|k} + \tilde{D}w\big(q^{(s)}(\tilde{p}_{j,m})\big) \leq \bar{z}_{i+1|k}^{(j)}, \quad i = 0, \ldots, N-1 \tag{7.57a}$$

$$\tilde{\Phi}^+\big(q^{(s)}(\tilde{p}_{j,m})\big)\underline{z}_{i|k}^{(m)} - \tilde{\Phi}^-\big(q^{(s)}(\tilde{p}_{j,m})\big)\bar{z}_{i|k}^{(m)} + \tilde{B}\big(q^{(s)}(\tilde{p}_{j,m})\big)c_{i|k} + \tilde{D}w\big(q^{(s)}(\tilde{p}_{j,m})\big) \geq \underline{z}_{i+1|k}^{(j)}, \quad i = 0, \ldots, N-1 \tag{7.57b}$$

$$\big|\tilde{\Phi}\big(q^{(s)}(\tilde{p}_{j,m})\big)\big|\bar{z}_T^{(m)} + \tilde{D}w\big(q^{(s)}(\tilde{p}_{j,m})\big) \leq \bar{z}_T^{(j)} \tag{7.57c}$$

(ii) Probabilistic output constraints, for $j = 1, \ldots, \mu$:

$$\Big(F_p\tilde{\Phi}\big(q^{(s)}(1-p_j)\big)\Big)^+\bar{z}_{i|k}^{(j)} - \Big(F_p\tilde{\Phi}\big(q^{(s)}(1-p_j)\big)\Big)^-\underline{z}_{i|k}^{(j)} + F_p\tilde{B}\big(q^{(s)}(1-p_j)\big)c_{i|k} + F_p\tilde{D}w\big(q^{(s)}(1-p_j)\big) + G_p c_{i|k} \leq \mathbf{1}, \quad i = 0, \ldots, N-1 \tag{7.58a}$$

$$\big|F_p\tilde{\Phi}\big(q^{(s)}(1-p_j)\big)\big|\bar{z}_T^{(j)} + F_p\tilde{D}w\big(q^{(s)}(1-p_j)\big) \leq \mathbf{1} \tag{7.58b}$$

(iii) Robust output constraints

$$F_h^+\bar{z}_{i|k}^{(\mu)} - F_h^-\underline{z}_{i|k}^{(\mu)} + G_h c_{i|k} \leq \mathbf{1}, \quad i = 0, \ldots, N-1 \tag{7.59a}$$

$$|F_h|\bar{z}_T^{(\mu)} \leq \mathbf{1} \tag{7.59b}$$

(iv) Initial and terminal constraints, for $j = 1, \ldots, \mu$

$$\underline{z}_{0|k}^{(\mu)} \leq V x_k \leq \bar{z}_{0|k}^{(\mu)} \tag{7.60a}$$

$$|\bar{z}_{N|k}^{(j)}| \leq z_T^{(j)} \tag{7.60b}$$

$$|\underline{z}_{N|k}^{(j)}| \leq z_T^{(j)} \tag{7.60c}$$

Here $A^+ \doteq \max\{A, 0\}$ and $A^- \doteq \max\{-A, 0\}$ denote the absolute values of the positive and negative elements of a matrix A.

Having defined the constraints of the online optimization problem, we now formulate a stochastic MPC algorithm with the objective of minimizing the expected quadratic cost $J(x_k, \mathbf{c}_k)$ of Sect. 7.2.1. This requires the online solution of a quadratic program.

Algorithm 7.4 At each time instant $k = 0, 1, \ldots$:

1. Solve the optimization

$$\underset{\substack{\mathbf{c}_k, \\ (\bar{z}_{0|k}^{(1)}, \ldots, \bar{z}_{0|k}^{(\mu)}) \cdots (\bar{z}_{N|k}^{(1)}, \ldots, \bar{z}_{N|k}^{(\mu)}) \\ (\underline{z}_{0|k}^{(1)}, \ldots, \underline{z}_{0|k}^{(\mu)}) \cdots (\underline{z}_{N|k}^{(1)}, \ldots, \underline{z}_{N|k}^{(\mu)})}}{\text{minimize}} \quad J(x_k, \mathbf{c}_k) \tag{7.61}$$

subject to (7.42a, 7.42b), (7.57a, 7.57b), (7.58a), (7.59a) and (7.60a–7.60c)

2. Implement the control law $u_k = K x_k + c_{0|k}^*$ where $\mathbf{c}_k^* = (c_{0|k}^*, \ldots, c_{N-1|k}^*)$. ◁

By construction, this algorithm is recursively feasible and hence Theorem 7.1 demonstrates that the closed-loop system satisfies a quadratic stability condition. These properties can be summarized as follows.

Corollary 7.2 *For the system (7.1a, 7.1b) with the control law of Algorithm 7.4, if* $\tilde{p}_{j,m}$ *and* p_j, $j, m = 1, \ldots, \nu$ *satisfy (7.56), then the optimization (7.61) is recursively feasible and the closed-loop system satisfies the constraints (7.44) and the asymptotic mean-square bound*

$$\lim_{r \to \infty} \frac{1}{r} \sum_{k=0}^{r-1} \mathbb{E}_0 \big(\|x_k\|_Q^2 + \|u_k\|_R^2 \big) \leq l_{ss}. \tag{7.62}$$

We conclude this section by noting that, in the interests of optimality, K should ideally be chosen as the unconstrained optimal feedback gain discussed in Sect. 6.2. However it may be necessary to detune this feedback gain in order to use the methods of Sect. 5.4.1 in the design of V and the terminal set parameters $\bar{z}_T^{(j)}$. It should also be noted that the number of optimization variables can be reduced by using tube cross sections that are parameterized by scalar variables, for example by redefining $\mathcal{X}_{i|k}^{(j)}$ as the set $\{z : |z - z_{i|k}^{(0)}| \leq \alpha_{i|k}^{(j)} \bar{z}_T^{(j)}\}$, where the scalar $\alpha_{i|k}^{(j)}$ and the vector $z_{i|k}^{(0)}$ are decision variables.

References

1. D. Bernardini, A. Bemporad, Stabilizing model predictive control of stochastic constrained linear systems. IEEE Trans. Autom. Control **57**(6), 1468–1480 (2012)
2. G.C. Calafiore, L. Fagiano, Stochastic model predictive control of LPV systems via scenario optimization. Automatica **49**(6), 1861–1866 (2013)
3. M. Cannon, B. Kouvaritakis, P. Couchman, Mean-variance receding horizon control for discrete-time linear stochastic systems, In *Proceedings of the 17th IFAC World Congress, Seoul, Korea*, (2008), pp. 15321–15326
4. C. Wang, C.J. Ong, M. Sim, Linear systems with chance constraints: Constraint-admissible set and applications in predictive control, *Proceedings of the 48th IEEE Conference on Decision and Control and 28th Chinese Control Conference*, Shanghai, China, (2009), pp. 2875–2880
5. A. Prékopa, *Stochastic Programming, Mathematics and Its Applications*, vol. 324 (Kluwer Academic Publishers, Dordrecht, 1995)
6. W. Feller, *Introduction to Probability Theory and Its Applications*, vol. 2 (Wiley, New York, 1971)
7. J.L. Doob, *Stochastic Processes* (Wiley, New York, 1953)
8. H.J. Kushner, *Introduction to Stochastic Control* (Holt, Rinehart and Winston, New York, 1971)
9. M. Cannon, B. Kouvaritakis, X. Wu, Probabilistic constrained MPC for multiplicative and additive stochastic uncertainty. IEEE Trans. Autom. Control **54**(7), 1626–1632 (2009)
10. M. Evans, M. Cannon, B. Kouvaritakis, Robust MPC tower damping for variable speed wind turbines. IEEE Trans. Control Syst. Technol. **23**(1), 290–296 (2014)
11. S.P. Boyd, L. El Ghaoui, E. Feron, V. Balakrishnan, *Linear Matrix Inequalities in System and Control Theory*. Society for Industrial and Applied Mathematics (1994)
12. W. Feller, *Introduction to Probability Theory and Its Applications*, vol. 1 (Wiley, New York, 1968)

Chapter 8
Explicit Use of Probability Distributions in SMPC

The previous chapter introduced the use of tubes with ellipsoidal or polytopic cross sections in stochastic MPC. However the probabilistic constraints on predicted states and control inputs were handled using confidence regions for stochastic model parameters, namely sets determined offline that contain the uncertain parameters of the model with a specified probability. This provides a computationally convenient but indirect way to exploit the available information on the probability distribution of uncertainty.

In this chapter the probability distributions of model parameters are directly employed in the formulation of constraints in the online MPC optimization. We do this first in the context of stochastic additive model disturbances. Then, towards the end of this chapter, we consider stochastic multiplicative model uncertainty.

Sections 8.1 and 8.2 consider pointwise-in-time probabilistic constraints that apply to individual scalar random variables for problems with additive disturbances. This makes it possible to transfer the majority of the computational burden of handling probabilistic constraints to offline calculations, and thus allows highly efficient online implementations. Section 8.3 describes a method of adapting constraints according to the number of past constraint violations with the aim of maintaining a specified average rate of violations in closed-loop operation. Still considering the case of additive model uncertainty, Sect. 8.4 deals with joint probabilistic constraints that apply simultaneously to several random variables by constructing stochastic tubes containing the uncertain component of the predicted model state. Section 8.5 is likewise concerned with the design of stochastic tube MPC strategies, but here polytopic tubes are constructed online for the case in which both multiplicative and additive model uncertainty are present.

In Sects. 8.1 and 8.2, we avoid discussing the details of the computational methods that are needed to determine probabilistic bounds on random variables. This is possible because these sections deal with simple cases of scalar random variables, which can be handled for example by numerical integration or random sampling

B. Kouvaritakis and M. Cannon, *Model Predictive Control*,
Advanced Textbooks in Control and Signal Processing,
DOI 10.1007/978-3-319-24853-0_8

performed offline. Numerical methods of computing probabilistic bounds are discussed in more detail in Sects. 8.4 and 8.5. In particular, the approach of Sect. 8.4 lends itself to methods of computing probabilistic bounds that are based on numerical integration, whereas that of Sect. 8.5 is naturally suited to an online optimization based on random sampling.

8.1 Polytopic Tubes for Additive Disturbances

We begin by considering linear systems that are subject only to additive model uncertainty. As in Chaps. 3 and 4, the system model is given by

$$x_{k+1} = Ax_k + Bu_k + Dw_k, \tag{8.1}$$

with state $x_k \in \mathbb{R}^{n_x}$, which is assumed to be known to the controller at time k, control input $u_k \in \mathbb{R}^{n_u}$, and disturbance input $w_k \in \mathbb{R}^{n_w}$ which is unknown to the controller at time k. The disturbance w_k is taken to be the realization at time k of a bounded i.i.d. random variable. Unlike the model employed in Chaps. 3 and 4, the probability distribution of w_k is assumed to be known. We further assume that the distribution of w_k is finitely supported with $w_k \in \mathcal{W}$ for all k, where $\mathcal{W}$ is a compact convex polytopic set that contains the origin.

The aim of the stochastic MPC strategy is to minimize a predicted cost, which is designed to ensure that the closed-loop system is stable in a suitable sense, subject to a pointwise-in-time probabilistic constraint of the form (6.8). However, in this section and in Sects. 8.2 and 8.3, we treat (6.8) as a collection of n_C individual probabilistic constraints, each of which is defined by a row of F and G, namely

$$\Pr_k(F_j x_{1|k} + G_j u_{1|k} \leq 1) \geq p, \quad j = 1, \ldots, n_C \tag{8.2}$$

where F_j, G_j denote the jth rows of F, G, and p is a specified probability. This differs from the pointwise-in-time probabilistic constraints considered in Chaps. 6 and 7, as well as those to be considered in Sects. 8.4 and 8.5, which are treated as constraints on the probability that all elements of a random vector should not exceed a given threshold.

Hard constraints can be included in the problem formulation by setting the probability p in (8.2) equal to 1. Also the treatment of intersections of constraint sets corresponding to different probabilities presents no particular challenges. As discussed in Sect. 7.1, the assumption of a compact bounding set $\mathcal{W}$ for w_k is needed in order to establish a guarantee of recursive feasibility.

The control strategy considered in this section minimizes the infinite horizon expected quadratic cost discussed in Sect. 6.2, subject to the constraint (8.2), with the aim of asymptotically steering the state to a neighbourhood of the origin. This asymptotic target set can be taken to be the minimal robust invariant set of Definition 3.4 under the control law $u = Kx$, with K defined as the unconstrained optimal state feedback gain discussed in Sect. 6.2. We thus formulate the stochastic MPC objective as the minimization of the predicted cost of (6.15):

$$\sum_{i=0}^{\infty} \mathbb{E}_k\big(\|x_{i|k}\|_Q^2 + \|u_{i|k}\|_R^2 - l_{ss}\big) \tag{8.3}$$

where $l_{ss} = \mathrm{tr}\big(\Theta(Q + K^T RK)\big)$ and Θ is the solution of the Lyapunov equation (6.10). Note that, for the case considered here, the expectation operator on the LHS of (6.10) is not needed because the model (8.1) is not subject to multiplicative uncertainty. For the dual prediction mode strategy, namely

$$u_{i|k} = Kx_{i|k} + c_{i|k} \tag{8.4}$$

with $c_{i|k} = 0$ for all $i \geq N$, Theorem 6.1 shows that the cost (8.3) can be expressed as a quadratic function, $J(x_k, \mathbf{c}_k)$, of the vector of degrees of freedom $\mathbf{c}_k = (c_{0|k}, \ldots, c_{N-1|k})$. Furthermore, given that K is the unconstrained optimal feedback gain, Corollary 6.1 gives

$$J(x_k, \mathbf{c}_k) = \sum_{i=0}^{\infty} \mathbb{E}_k\big(\|x_{i|k}\|_Q^2 + \|u_{i|k}\|_R^2 - l_{ss}\big) = x_k^T W_x x_k + \mathbf{c}_k^T W_c \mathbf{c}_k + w_1$$

where $w_1 = -\mathrm{tr}(W_x\Theta)$ and W_x, W_c are given by (6.23)–(6.24), which reduce to the certainty equivalent conditions of Theorem 2.10 since no multiplicative uncertainty is included in the model (8.1). The minimization of the predicted cost is to be performed subject to the constraint (8.2), which is to be invoked in a manner that allows a guarantee of recursive feasibility of the online MPC optimization problem. This results in a set of constraints that apply at each time step of the initial N-step prediction horizon (Mode 1) and a terminal constraint requiring $x_{N|k}$ to lie in a set that is robustly invariant under the terminal control law $u = Kx$.

When considering constraints, it is convenient to use the decomposition of prediction dynamics introduced in Sect. 3.2:

$$s_{i+1|k} = \Phi s_{i|k} + Bc_{i|k} \tag{8.5a}$$

$$e_{i+1|k} = \Phi e_{i|k} + Dw_{i|k} \tag{8.5b}$$

where $\Phi = A + BK$ and $x_{i|k} = s_{i|k} + e_{i|k}$, with $s_{0|k} = x_k$ and $e_{0|k} = 0$. The convenience of this decomposition is due to the fact that the nominal system in (8.5a) is deterministic and the effects of uncertainty are treated by (8.5b). Furthermore, the uncertain component, $e_{i|k}$, of the predicted state is independent of x_k since the initial condition for (8.5b) is taken to be $e_{0|k} = 0$ and $\{w_0, w_1, \ldots\}$ is a stationary process. Hence the effects of model uncertainty on the constraints of the problem can be computed offline. As explained in this section, which is based on the approach of

[1, 2], the explicit use of such information makes it possible to construct constraints that are as tight as possible for the open-loop dual-mode prediction strategy of (8.4).

In order to formulate the constraints of the stochastic MPC algorithm, we first consider the conditions on the vector $\mathbf{c}_k = (c_{0|k}, \dots, c_{N-1|k})$ of decision variables under which the constraints (8.2) are satisfied at each future time step of an infinite prediction horizon.

Lemma 8.1 *The predictions generated by (8.4) and (8.5a, 8.5b) satisfy the constraints* $\Pr_k(F_j x_{i|k} + G_j u_{i|k} \le \mathbf{1}) \ge p$ *for* $j = 1, \dots, n_C$ *and* $i = 1, 2, \dots$ *if and only if*

$$\tilde{F} s_{i|k} + G c_{i|k} \le \mathbf{1} - \gamma_i, \quad i = 1, 2, \dots \tag{8.6}$$

where $\tilde{F} = F + GK$ *and where the elements of* $\gamma_i = (\gamma_{i,1}, \dots, \gamma_{i,n_C})$ *are defined by*

$$\gamma_{i,j} \doteq \min_{\gamma_{i,j}} \quad \gamma_{i,j} \qquad \text{subject to} \quad \Pr\Big(\tilde{F}_j(\Phi^{i-1} D w_i + \cdots + D w_1) \le \gamma_{i,j}\Big) \ge p \tag{8.7}$$

for $j = 1, \dots, n_C$ *and* $i = 1, 2, \dots$.

Proof This is a consequence of the predicted control sequence (8.4) and the decomposition $x_{i|k} = s_{i|k} + e_{i|k}$, according to which $F x_{i|k} + G u_{i|k} = \tilde{F} s_{i|k} + G c_{i|k} + \tilde{F} e_{i|k}$, whereas (8.5b) with $e_{0|k} = 0$ implies

$$e_{i|k} = \Phi^{i-1} D w_k + \cdots + D w_{k+i-1}. \tag{8.8}$$

Hence the definition (8.7) implies that the jth element of γ_i has the minimum value such that $\tilde{F}_j e_{i|k} \le \gamma_{i,j}$ with probability p. It follows that the conditions $\Pr_k(F_j x_{i|k} + G_j u_{i|k} \le \mathbf{1}) \ge p$ are satisfied if and only if (8.6) holds. □

The constraints in (8.6) are analogous to constraints that were derived in Sect. 3.2 for systems with additive model uncertainty, and which were based on tubes bounding the uncertain components of predicted state and control trajectories. The formulation here likewise uses polytopic uncertainty tubes and hence results in linear constraints on nominal predictions.

A key observation concerns the computation of $\gamma_{i,j}$ in (8.7), which requires knowledge of the probability distribution of $\tilde{F}_j e_{i|k}$. However $e_{i|k}$ does not depend on the system state at time k but is instead a function of additive disturbance realizations whose distributions are known a priori (this explains the absence of the time index k in (8.7)), and hence $\gamma_{i,j}$ can be computed for each given i and j offline. In practice the computation of $\gamma_{i,j}$ in (8.7) has to be performed approximately, for example by numerically approximating the associated convolution integrals or by random sampling methods (see e.g. [3, 4]).

Although the conditions of Lemma 8.1 impose the probabilistic constraints in (8.2) on the predicted state and control trajectories over an infinite prediction horizon, these conditions do not ensure existence of a feasible decision variable $\mathbf{c}_{k+1}$ at time $k+1$ given feasibility at time k. However it is possible to guarantee recursive feasibility by imposing the constraint at time k that $\mathbf{c}_{k+1} = M\mathbf{c}_k$ will be feasible at time $k+1$ for all realizations of w_k, where M is the shift matrix defined in (2.26b). As discussed in Sect. 7.1, this approach requires that the disturbances $w_k, \ldots, w_{k+i-1}$ are handled robustly (by considering their worst-case values) in the probabilistic constraints imposed at time k on the predicted state and control input i steps ahead, whereas w_{k+i} is treated probabilistically in these i steps ahead constraints. The resulting constraint set is defined as follows.

Theorem 8.1 *If $\mathbf{c}_k$ satisfies the constraints defined at time k by*

$$\tilde{F}s_{i|k} + Gc_{i|k} \le \mathbf{1} - \beta_i, \quad i = 1, 2, \ldots \tag{8.9}$$

where β_i and α_i are defined for all $i \ge 1$ by

$$\beta_i \doteq \gamma_1 + \sum_{j=1}^{i-1} a_j \tag{8.10a}$$

$$a_i \doteq \max_{w \in \mathcal{W}} \tilde{F}\Phi^i D w, \tag{8.10b}$$

then the constraints (8.2) hold, and, at time $k+1$, $\mathbf{c}_{k+1} = M\mathbf{c}_k$ will necessarily satisfy $\tilde{F}s_{i|k+1} + Gc_{i|k+1} \le 1 - \beta_i$ for $i = 1, 2, \ldots$.

Proof First note that, in order that the constraints (8.2) hold for $\mathbf{c}_k = M^k\mathbf{c}_0$ at all times $k \ge 0$ it is sufficient (and also necessary) that: (i) the i steps ahead constraints $\Pr_0(F_j x_{i|0} + G_j u_{i|0} \le \mathbf{1}) \ge p, j = 1, \ldots, n_C$ hold for all $i \ge 1$ at time $k = 0$, and (ii) for each $i \ge 1$ these constraints are feasible with $\mathbf{c}_k = M^k\mathbf{c}_0$ at times $k = 1, \ldots, i-1$. To prove the theorem, we show that both (i) and (ii) are ensured by the constraints of (8.9) time $k = 0$.

Consider the constraints $\Pr_0(F_j x_{i|0} + G_j u_{i|0} \le \mathbf{1}) \ge p, j = 1, \ldots, n_C$, which apply to the i steps ahead state and control input predicted at time $k = 0$ for some particular i. By Lemma 8.1, these constraints are satisfied if (8.6) holds at time $k = 0$.

To ensure that at time $k = 1$ the predictions generated by $\mathbf{c}_1 = M\mathbf{c}_0$ will satisfy the $i-1$ steps ahead constraints, we also require that

$$\Pr_1\Bigl(\max_{w_0 \in \mathcal{W}} (F_j x_{i|0} + G_j u_{i|0}) \le \mathbf{1}\Bigr) \ge p, \quad j = 1, \ldots, n_C$$

holds at time $k = 0$. Using $Fx_{i|k} + Gu_{i|k} = \tilde{F}s_{i|k} + Gc_{i|k} + \tilde{F}e_{i|k}$ and (8.8), this condition can be expressed as the constraint

$$\begin{aligned}\tilde{F}s_{i|0} + Gc_{i|0} &\leq \mathbf{1} - \gamma_{i-1} - \max_{w_0\in\mathcal{W}} \tilde{F}\Phi^{i-1}Dw_0 \\ &= \mathbf{1} - \gamma_{i-1} - a_{i-1}.\end{aligned}$$

To ensure that the predictions generated at time $k = 2$ by $\mathbf{c}_2 = M^2\mathbf{c}_0$ will satisfy the $i-2$ steps ahead constraints, we require that

$$\Pr_2\Big(\max_{\{w_0,w_1\}\in\mathcal{W}\times\mathcal{W}} (F_j x_{i|0} + G_j u_{i|0}) \leq \mathbf{1}\Big) \geq p, \quad j = 1, \ldots, n_C$$

holds at time $k = 0$. From (8.8) this is equivalent to

$$\begin{aligned}\tilde{F}s_{i|0} + Gc_{i|0} &\leq \mathbf{1} - \gamma_{i-2} + \max_{w_0\in\mathcal{W}} \tilde{F}\Phi^{i-1}Dw_0 - \max_{w_1\in\mathcal{W}} \tilde{F}\Phi^{i-2}Dw_1 \\ &= \mathbf{1} - \gamma_{i-2} - a_{i-2} - a_{i-1}.\end{aligned}$$

Repeating this argument for the predictions generated by $\mathbf{c}_k = M^k\mathbf{c}_0$ at times $k = 3, \ldots, i-1$, we obtain the conditions:

$$\tilde{F}s_{i|0} + Gc_{i|0} \leq \mathbf{1} - \max\big\{\gamma_i,\ (\gamma_{i-1} + a_{i-1}),\ \ldots,\ (\gamma_1 + a_1 + \cdots + a_{i-1})\big\}. \tag{8.11}$$

But $\gamma_i \leq \gamma_{i-1} + a_{i-1}$ for all $i > 1$ since (8.7) and (8.10b) imply that

$$\Pr\Big(\tilde{F}_j(\Phi^{i-1}Dw_i + \cdots + Dw_1) \leq \gamma_{i-1,j} + a_{i-1,j}\Big) \geq p, \quad j = 1, \ldots, n_C,$$

and it follows that (8.11) is equivalent to the constraint $\tilde{F}s_{i|0} + Gc_{i|0} \leq 1 - \beta_i$. These conditions, when invoked for all $i \geq 1$ and any given $k \geq 0$, are equivalent to (8.9). □

The constraints of (8.9) consist of an infinite number of inequalities which correspond to constraints applied to the predicted trajectories of the model (8.1) over an infinite prediction horizon. However these conditions can be expressed equivalently in terms of a finite number of inequalities using the approach of Sect. 3.2.1. Of course, these conditions are only meaningful if the constraints (8.9) are feasible for some x_k and $\mathbf{c}_k$, and, since the prediction model is by assumption stable, we therefore require that β_i is strictly less than 1 for all $i \geq 1$. The following lemma shows that the limit $\bar{\beta} \doteq \lim_{i\to\infty}\beta_i$ exists and provides bounds on its value.

Lemma 8.2 *The sequence $\{\beta_1, \beta_2, \ldots\}$ is monotonically non-decreasing and converges to a limit $\bar{\beta} \doteq \lim_{i\to\infty}\beta_i$, where the lth element of $\bar{\beta}$ is bounded by*

$$\bar{\beta}_l \leq \gamma_{1,l} + \sum_{j=1}^{\rho-1} a_{j,l} + \frac{\lambda^\rho}{1-\lambda}\|\tilde{F}_l^T\|_S, \quad l = 1, \ldots, n_C \tag{8.12}$$

for any non-negative integer ρ, and for $S \succ 0$ and λ satisfying the conditions

$$\max_{w \in \mathcal{W}} \|Dw\|_{S^{-1}} \le 1 \tag{8.13a}$$

$$\Phi S \Phi^T \preceq \lambda^2 S, \quad \lambda \in [0, 1). \tag{8.13b}$$

Proof The non-decreasing property of the sequence $\beta_1, \beta_2, \ldots$ follows from the definition (8.10a) and from the fact that $a_i \ge 0$ for all i, which follows from (8.10b). Also (8.10a) implies

$$\bar{\beta} = \lim_{i \to \infty} \beta_i = \gamma_1 + \sum_{i=1}^{\infty} a_i \tag{8.14}$$

and, by condition (8.13a), we have

$$a_{i,l} = \max_{w \in \mathcal{W}} \tilde{F}_l \Phi^i D w \le \max_{\|v\|_{S^{-1}} \le 1} \tilde{F}_l \Phi^i v \le \left\| {\Phi^i}^T \tilde{F}_l^T \right\|_S \tag{8.15}$$

for $l = 1, \ldots, n_C$. However, (8.13b) implies that

$$\left\| {\Phi^i}^T \tilde{F}_l^T \right\|_S \le \lambda \left\| {\Phi^{i-1}}^T \tilde{F}_l^T \right\|_S,$$

which, combined with (8.15), gives $a_{i,l} \le \lambda^i \|\tilde{F}_l^T\|_S$ for $l = 1, \ldots, n_C$. Replacing $a_{i,l}$ in (8.14) with this bound for all $i \ge \rho$ gives the bound in (8.12). □

By assumption Φ is strictly stable and hence (8.13b) will necessarily have solutions for S and $\lambda \in [0, 1)$. These can be scaled so that (8.13a) will be met. Furthermore, it follows from (8.14) and the non-negative property of a_i that the maximum error in the bound in (8.12) on the lth element of $\bar{\beta}$ can be no greater than $\lambda^\rho \|\tilde{F}_l^T\|_S/(1 - \lambda)$, which can be made as small as desired by using a sufficiently large value of ρ. To ensure the existence of feasible x_k and $\mathbf{c}_k$ satisfying the conditions of Theorem 8.1, we therefore assume

$$\bar{\beta} < \mathbf{1}. \tag{8.16}$$

The constraints of (8.9) are more convenient to handle (both in terms of computation and notation) using the lifted prediction dynamics introduced in Sect. 2.7 and extended to the case of additive model uncertainty in Sect. 3.2:

$$\bar{F} \Psi^i z_{0|k} \le \mathbf{1} - \beta_i, \quad i = 1, 2, \ldots$$

where

$$z_{0|k} = \begin{bmatrix} x_k \\ \mathbf{c}_k \end{bmatrix}, \quad \Psi = \begin{bmatrix} \Phi & BE \\ 0 & M \end{bmatrix}, \quad \bar{F} = \begin{bmatrix} \tilde{F} & GE \end{bmatrix}$$

with E and M as defined in (2.26b). An equivalent formulation of (8.9) in terms of a finite number of inequalities is given by the following result.

Theorem 8.2 *If (8.16) holds, then for any given z, $\bar{F}\Psi^i z \leq \mathbf{1} - \beta_i$ is satisfied for all $i \geq 1$ if and only if*

$$\bar{F}\Psi^i z \leq \mathbf{1} - \beta_i, \quad i = 1, 2, \ldots, \nu \tag{8.17}$$

where ν is the smallest integer such that $\bar{F}\Psi^{\nu+1} z \leq \mathbf{1} - \beta_{\nu+1}$ holds for all z satisfying (8.17).

Proof The necessity of (8.17) is obvious and we therefore prove sufficiency. For ν satisfying the conditions of the theorem, define $\mathcal{Z}^{(\nu)}$ as the set

$$\mathcal{Z}^{(\nu)} \doteq \{z : \bar{F}\Psi^i z \leq 1 - \beta_i, \ i = 1, 2, \ldots, \nu\}$$

then for all $z \in \mathcal{Z}^{(\nu)}$we have:

$$\bar{F}\Psi z \leq \mathbf{1} - \beta_1 \tag{8.18a}$$

$$\bar{F}\Psi^{i+1} z \leq \mathbf{1} - \beta_{i+1}, \quad i = 1, \ldots, \nu \tag{8.18b}$$

and since $\beta_{i+1} = \beta_i + a_i \geq \beta_i + \tilde{F}\Phi^i D w$ for all $w \in \mathcal{W}$, (8.18b) implies, for all $w \in \mathcal{W}$

$$\bar{F}\Psi^i(\Psi z + \bar{D}w) \leq \mathbf{1} - \beta_i, \quad i = 1, \ldots, \nu$$

where $\bar{D} = [D^T \ 0]^T$. Therefore $\mathcal{Z}^{(\nu)}$ is a robustly invariant set for the system with dynamics $z_{k+1} = \Psi z_k + \bar{D}w_k$, $w_k \in \mathcal{W}$ and constraints $\bar{F}\Psi z_k \leq \mathbf{1} - \beta_1$. This property and the equivalence of (8.18a) for $z = (x_k, \mathbf{c}_k)$ with the constraints of (8.2) imply that $\bar{F}\Psi^i z \leq \mathbf{1} - \beta_i$ holds for all $i \geq 1$ whenever $z \in \mathcal{Z}^{(\nu)}$. □

There necessarily exists a finite integer ν satisfying the conditions of Theorem 8.2 whenever $(\Psi, \bar{F})$ is observable. This can be shown using an argument identical to the one used in the proof of Theorem 3.1. It also follows from the argument of Theorem 3.1 that $\mathcal{Z}^{(\nu)}$ is the maximal RPI set for the dynamics $z_{k+1} = \Psi z_k + \bar{D}w_k$, $w_k \in \mathcal{W}$ and constraints $\bar{F}\Psi z_k \leq \mathbf{1} - \beta_1$, and from this it can be concluded that every feasible pair $(x_k, \mathbf{c}_k)$ for the conditions of (8.9) must lie in the set $\mathcal{Z}^{(\nu)}$. The value of ν can be determined by solving a sequence of linear programs to determine for $i = 0, 1, \ldots$ the maximum of $\bar{F}\Psi^{i+1} z$ over $z \in \mathcal{Z}^{(i)}$ in order to check whether $\bar{F}\Psi^{i+1} z \leq \mathbf{1} - \beta_{i+1}$ for all $z \in \mathcal{Z}^{(i)}$.

A stochastic MPC algorithm based on the constraint formulation of this section and the quadratic cost $J(x_k, \mathbf{c}_k)$ of (8.3) can be stated as follows. The online optimization in step (i) requires the solution of a QP at each sampling instant.

Algorithm 8.1 At each time instant $k = 0, 1, \ldots$

(i) Perform the optimization:

$$\begin{aligned} &\underset{\mathbf{c}_k}{\text{minimize}} \quad J(x_k, \mathbf{c}_k) \\ &\text{subject to} \quad \bar{F}\Psi^i \begin{bmatrix} x_k \\ \mathbf{c}_k \end{bmatrix} \leq \mathbf{1} - \beta_i, \quad i = 1, \ldots, \nu \end{aligned} \tag{8.19}$$

(ii) Implement the control law $u_k = Kx_k + c^*_{0|k}$ where $\mathbf{c}^*_k = (c^*_{0|k}, \ldots, c^*_{N|k})$ is the minimizing argument of (8.19). ◁

The recursive feasibility of the optimization (8.19), is guaranteed by Theorem 8.1. Also the definition of MPC cost $J(x_k, \mathbf{c}_k)$ in 8.3 implies, by Theorem 7.1, that the closed-loop system satisfies a quadratic stability condition. For completeness, these properties are summarized in the following corollary.

Corollary 8.1 *Algorithm 8.1 applied to the system (8.1) is feasible at all times $k = 1, 2, \ldots$ if it is feasible at $k = 0$. The constraints of (8.2) hold for all $k \geq 0$ and the mean-square bound*

$$\lim_{r\to\infty} \frac{1}{r} \sum_{k=0}^{r-1} \mathbb{E}_0\big(\|x_k\|_Q^2 + \|u_k\|_R^2\big) \leq l_{ss} \tag{8.20}$$

where $l_{ss} = \lim_{k\to\infty}(\|x_k\|_Q^2 + \|u_k\|_R^2)$, is satisfied along trajectories of the closed-loop system. If $u = Kx$ is the optimal feedback law for the cost (8.3) in the absence of constraints, then x_k converges with probability 1 to the minimal RPI set for the dynamics (8.1) under this feedback law as $k \to \infty$.

8.2 Striped Prediction Structure with Disturbance Compensation in Mode 2

In this section, we consider the system description of (8.1) but the matrix D defining the disturbance input map is, for simplicity, taken to be equal to the identity matrix I. The disturbance input is again stochastic with a known distribution. For convenience, we express the constraints in terms of an output vector $\psi_k \in \mathbb{R}^{n_C}$:

$$\psi_k = \tilde{G}x_{k+1} + \tilde{F}u_k \tag{8.21}$$

and hence the constraints take the form

$$\Pr_k(\psi_k \leq h) \geq p. \tag{8.22}$$

As in the case of the constraints assumed in Sect. 8.4, the constrained output variable here depends on the unknown additive disturbance at time k and represents a mixture of state and input constraints. The formulation of constraints in terms of (8.22) rather than (8.2) simplifies the presentation of this section but is no less general than (8.2).

The approach employed in Sect. 8.1 does not make use of disturbance feedback. However, feedback can be used to attenuate the effects of the future disturbances w_{k+j} for $j = 0, \dots, i-1$ on the i steps ahead predicted state and control input $x_{i|k}$ and $u_{i|k}$ at time k. These predicted disturbance values are not known a priori at time k but will be available to the controller at time $k+i$. For this reason, when considering the i steps ahead state and control input we refer in this section to the predictions, $w_{j|k}$, of w_{k+j}, for $j = 0, \dots, i-1$ at time k as known future disturbances (KFD), whereas $w_{j|k}$ for all $j \geq i$ are referred to as unknown future disturbances (UFD). This section describes the approach of [5], which introduces disturbance feedback with a striped structure into the predicted control trajectories of a stochastic MPC strategy.

To cater for KFD, the predicted control inputs and the resulting predicted dynamics are decomposed as

$$u_{i|k} = Kx_{i|k} + c_{i|k} + v_{i|k}, \quad i = 0, 1, \dots \tag{8.23}$$

where $c_{i|k} = 0$ for all $i \geq N$, $\mathbf{c}_k = (c_{0|k}, \dots, c_{N-1|k})$ is an optimization variable at time k and $v_{i|k}$ depends on the KFD linearly, as explained later in this section. We again use the decomposition of prediction dynamics introduced in Sect. 3.2 into nominal ($s_{i|k}$) and uncertain ($e_{i|k}$) components

$$\begin{aligned} s_{i+1|k} &= \Phi s_{i|k} + Bc_{i|k} \\ e_{i+1|k} &= \Phi e_{i|k} + Bv_{i|k} + w_{i|k} \end{aligned}$$

where

$$x_{i|k} = s_{i|k} + e_{i|k} \tag{8.24}$$

with $s_{0|k} = x_k$ and $e_{0|k} = 0$.

Like the striped parameterized tube MPC approach of Sect. 4.2.3 (and unlike the disturbance affine of Sect. 4.2.1), the component v of the predicted control input is applied throughout the infinite prediction horizon, and thereby enables disturbance compensation to extend to the predicted control law of Mode 2. This allows constraints to be relaxed and therefore leads to larger sets of feasible initial conditions. The scheme of (8.24) generates predictions for the constrained output according to

$$\psi_{i|k} = (Gs_{i|k} + Fc_{i|k}) + (Ge_{i|k} + Fv_{i|k}) + \tilde{G}w_{i|k}$$

where

$$G = \tilde{G}\Phi + \tilde{F}K, \quad F = \tilde{G}B + \tilde{F}$$

This leads to the overall vector of constraint output predictions

$$\psi_k = \zeta_k + \epsilon_k + \eta_k \tag{8.25}$$

where

$$\begin{aligned} \zeta_k &= C_3 s_k + C_{1,N}\mathbf{c}_k \\ \epsilon_k &= C_1 \mathbf{v}_k + C_2 \mathbf{w}_k \\ \eta_k &= \mathrm{diag}\{\tilde{G}, \tilde{G}, \tilde{G}, \ldots\}\mathbf{w}_k \end{aligned} \tag{8.26}$$

with bold symbols being used to indicate the entire sequence of predictions over an infinite future horizon; of these only $\mathbf{c}_k = (c_{0|k}, \ldots, c_{N|k})$ is finite-dimensional. Here $C_{1,N}$ denotes the matrix consisting of the first N block-columns of matrix C_1 and the matrices C_1, C_2, C_3 are defined as

$$C_1 = \begin{bmatrix} F & 0 & 0 & \cdots \\ GB & F & 0 & \cdots \\ G\Phi B & GB & F & \cdots \\ \vdots & \vdots & \vdots & \ddots \end{bmatrix}, \quad C_2 = \begin{bmatrix} 0 & 0 & 0 & \cdots \\ G & 0 & 0 & \cdots \\ G\Phi & G & 0 & \cdots \\ \vdots & \vdots & \vdots & \ddots \end{bmatrix}, \quad C_3 = \begin{bmatrix} G \\ G\Phi \\ G\Phi^2 \\ \vdots \end{bmatrix}.$$

Equation (8.25) separates the sequence of predicted outputs into three components:

(i) ζ_k, generated by the nominal (disturbance free) dynamics;
(ii) ϵ_k, associated with the KFD;
(iii) η_k, associated with the UFD.

The first of these is to be controlled by the variables $c_{i|k}, i = 0, \ldots, N-1$, the second can be compensated for through the use of $v_{i|k}, i = 1, 2, \ldots$, while the third is beyond control since it depends on unknown future disturbances.

In the special case of the number of constraints being equal to the number of inputs and where the dynamics defined by the state space model (Φ, B, G, F) are minimum-phase, the effects of the KFD can be completely eliminated by setting $\epsilon_k = 0$, which gives

$$\mathbf{v}_k = -C_1^{-1}C_2\mathbf{w}_k = L\mathbf{w}_k, \tag{8.27}$$

where L is a lower block triangular matrix whose first block row is zero. In this case, the predictions of the component v take the form:

$$\begin{aligned} v_{0|k} &= 0 \\ v_{i|k} &= L_{i,1}w_{0|k} + L_{i,2}w_{1|k} + \cdots + L_{i,j}w_{i-1|k}. \end{aligned} \tag{8.28}$$

In general, however, complete cancellation of the KFD is not be possible and instead the parameters $L_{i,j}$ for $j = 1, \ldots, i$, $j = 1, \ldots, N-1$ could be computed online. This is the approach that is employed by the disturbance affine MPC algorithms proposed by [6, 7] for the robust case, and by [8] for the stochastic case. Clearly the number of degrees of freedom in such algorithms grows quadratically with the prediction horizon, and this could result in unacceptably high online computational loads for long horizons. It was seen in Sect. 4.2.3 that in the case of robust MPC there may be no loss of performance in using a striped prediction structure, particularly if disturbance compensation is allowed to extend into the Mode 2 prediction horizon, in which case SPTMPC can outperform PTMPC.

To develop the idea, consider first the case with no disturbance compensation. As was seen in the previous section, for this case recursive feasibility is achieved through the satisfaction of the constraint

$$C_3 x_k + C_{1,N}\mathbf{c}_k \leq 1 - \beta \tag{8.29}$$

where the ith block elements of β are given by the sum of: (i) γ_1 and (ii) $\sum_{j=0}^{i-1} \max_{w \in \mathcal{W}} G\Phi^j w$. Thus γ_1, the jth element of which is defined here as the minimum value of $\gamma_{1,j}$ satisfying the condition $\Pr(\tilde{G}_j w < \gamma_{1,j}) = p$, for $j = 1, \ldots, n_C$, accounts for the term η_k in (8.25) which is associated with the UFD and is treated probabilistically. On the other hand, (ii) accounts for of the term ϵ_k in (8.25) which is associated with the KFD and therefore has to be treated robustly. It is noted that although (8.29) involves an infinite number of inequalities, as explained in Sect. 8.1 it is only necessary to consider the inequalities implied by the ith block for $i = 1, \ldots, \nu$, for some finite integer ν which is independent of x_k and can be determined offline. The implied inequality is here denoted by

$$\hat{C}_3 x_k + \hat{C}_{1,N}\mathbf{c}_k \leq 1 - \hat{\beta}. \tag{8.30}$$

Allowing now for disturbance compensation, a non-zero vector $\mathbf{v}_k$ can be used to reduce the amount of constraint tightening, β, in (8.29) needed to account for ϵ_k. Adopting a striped structure, $\mathbf{v}_k$ is written as

$$\mathbf{v}_k = L\mathbf{w}_k \tag{8.31}$$

where

$$L = \begin{bmatrix} 0 & 0 & \cdots & 0 & 0 & \cdots \\ L_1 & 0 & \cdots & 0 & 0 & \cdots \\ L_2 & L_1 & \cdots & 0 & 0 & \cdots \\ \vdots & \vdots & \ddots & \vdots & \vdots & \\ L_{N-1} & L_{N-2} & \cdots & L_1 & 0 & \cdots \\ 0 & L_{N-1} & \cdots & L_2 & L_1 & \cdots \\ \vdots & \vdots & \ddots & \vdots & \vdots & \ddots \end{bmatrix} \tag{8.32}$$

The aim is to choose L so as to reduce the constraint tightening parameter $\hat{\beta}$. In particular, by substituting (8.31) into (8.26), $\tilde{\beta}$ must satisfy

$$(\tilde{C}_1 L + \tilde{C}_2)\mathbf{w}_k \leq \tilde{\beta} - \hat{\gamma} \leq \hat{\beta} - \hat{\gamma} \tag{8.33}$$

where $\hat{\gamma}$ is a vector of commensurate dimensions whose block elements are equal to the corresponding blocks of γ.

In order to keep the computational load low, and in fact to make it comparable to classical MPC, we begin by selecting L offline. This is done in a sequential manner, designing L_1, then L_2, and so on. Not only is such a design computationally convenient but it favours the design of the L parameters that apply at the beginning of the prediction horizon at the expense of those applying at later prediction times, which is consistent with the expectation that constraints will be more stringent during the initial transients and are less likely to be active as the predicted state and control trajectories tend towards the steady state.

Before stating an algorithm for the design of the elements of L, note that $\hat{C}_1 L + \hat{C}_2$ has ν row blocks and that all the column blocks of this matrix beyond the νth are zero. Thus the infinite-dimensional $\mathbf{w}_k$ appearing in (8.33) can be replaced by the vector, denoted $\hat{\mathbf{w}}_k$, that contains only the first ν blocks of $\mathbf{w}_k$. For the same reason, L is replaced by $\hat{L}$, which comprises only the first ν row blocks and column blocks of L. Then let $\hat{\mathcal{W}}$ denote the set $\{\hat{\mathbf{w}}_k : w_{i|k} \in \mathcal{W},\ i = 0, \ldots, \nu\}$ and let E_i be the matrix which is such that $E_i M$ gives the ith block of M.

Algorithm 8.2

(i) Solve the minimization

$$\left(L_1^{(1)}, L_2^{(1)}, \ldots, L_{N-1}^{(1)}, r^{(1)}\right) = \arg \min_{L_1,\ldots,L_{N-1},r} r$$

subject to

$$\begin{aligned} E_1(\hat{C}_1\hat{L} + \hat{C}_2)\hat{\mathbf{w}} &\leq r\mathbf{1}, & \forall\, \hat{\mathbf{w}} \in \hat{\mathcal{W}} \\ (\hat{C}_1\hat{L} + \hat{C}_2)\hat{\mathbf{w}} &\leq r\hat{\beta} - \hat{\gamma}, & \forall\, \hat{\mathbf{w}} \in \hat{\mathcal{W}} \end{aligned}$$

(ii) Then, for each $i = 2, \ldots, N-1$, set $(L_1, \ldots, L_{i-1}) = \left(L_1^{(1)}, \ldots, L_{i-1}^{(i-1)}\right)$ and solve the minimization

$$\left(L_i^{(i)}, L_{i+1}^{(i)}, \ldots, L_{N-1}^{(i)}, r^{(i)}\right) = \arg \min_{L_1,\ldots,L_{N-1},r} r$$

subject to

$$\begin{aligned} E_i(\hat{C}_1\hat{L} + \hat{C}_2)\hat{\mathbf{w}} &\leq r\mathbf{1}, & \forall\, \hat{\mathbf{w}} \in \hat{\mathcal{W}} \\ (\hat{C}_1\hat{L} + \hat{C}_2)\hat{\mathbf{w}} &\leq r\hat{\beta} - \hat{\gamma}, & \forall\, \hat{\mathbf{w}} \in \hat{\mathcal{W}} \end{aligned}$$

◁

Since the constraints of the algorithm depend linearly on $\hat{\mathbf{w}}$, they need only be invoked at the vertices of $\hat{\mathcal{W}}$, which therefore implies that the minimizations in steps (i) and (ii) of the algorithm are linear programming problems. It is also noted that Algorithm 8.2 is necessarily feasible under the assumption that $\lim_{i\to\infty} \beta_i < \mathbf{1}$, since then $L_i = 0$, $i = 1, \ldots, N-1$ gives at least one feasible solution.

The reduced tightening parameters derived by Algorithm 8.2 are given as

$$\tilde{\beta}_i = E_i|C_1 L + C_2|\boldsymbol{\alpha} + \gamma, \quad i = 1, \ldots, N+m \tag{8.34}$$

where $\boldsymbol{\alpha}$ is an infinite-dimensional vector whose blocks are α. By definition these tightening parameters satisfy the condition $\tilde{\beta}_i \le \hat{\beta}_i$, and they can be shown by simple algebra to share the monotonically increasing property of $\hat{\beta}_i$ and to tend to a limit $\tilde{\beta}_\infty \le \beta_\infty$.

Using the approach described in this section, disturbance compensation can be introduced into the stochastic MPC strategy of Sect. 8.1 simply by replacing the constraint tightening parameters $\hat{\beta}_i = \beta_i$ with $\tilde{\beta}_i$. However two further issues must also be addressed. The first concerns the definition of a finite constraint set, which may lead to a different value of ν, say $\tilde{\nu}$. The definition of this is exactly the same as for ν, namely through the conditions (8.17) of Theorem 8.2, but with β_i replaced by $\tilde{\beta}_i$ (and ν replaced by $\tilde{\nu}$). The second issue concerns the extension of disturbance compensation into Mode 2, which causes the steady-state value of the expectation of the predicted stage cost, say $\tilde{l}_{ss}$, to differ from the value $l_{ss} = \mathrm{tr}\big(\Theta(Q + K^T RK)\big)$ that is subtracted from the stage cost of (6.15). This steady state expected stage cost can be computed, given the feedback gain matrix L, through a straightforward modification of the method of computing l_{ss} in Sect. 6.2.

8.3 SMPC with Bounds on Average Numbers of Constraint Violations

The probabilistic constraints considered thus far allow constraint violations to occur but require that the expected frequency of violations remains at all times below a threshold of $1 - p$. Maintaining the frequency of constraint violations below a given threshold is essential in some applications. For example, the specification of a wind turbine control problem might require that the power captured from the wind is maximized while the frequency with which the material stresses in the turbine blades and tower violate certain thresholds is kept below a required rate defined by fatigue damage considerations. In such a scenario, formulating constraints in the probabilistic manner presented in Sects. 8.1 and 8.2 can be conservative because it does not account for the fact that, during certain periods of the past, the average number of constraint violations may fortuitously have been low (for example because of periods of low turbulence). In such circumstances, the controller is able to be more aggressive, causing a higher number of violations while still maintaining the overall average of constraint violations within acceptable limits.

To take advantage of such situations a stochastic MPC strategy is proposed in [9] that controls the average violation of a given constraint

$$g^T x_{k+1} \leq h$$

where for convenience we consider here only a single constraint, i.e. $g \in \mathbb{R}^{n_x}$. This average is defined as $v_k = L_k/s_k$ where L_k is the accumulated loss (weighted by a forgetting factor $\gamma \in [0, 1]$):

$$L_k = \sum_{i=0}^{k} \gamma^{k-i} l(g^T x_i - h)$$

and s_k is the normalizing factor $s_k = \sum_{i=0}^{k} \gamma^{k-i}$. The function $l(\cdot)$ is a non-decreasing (and lower semi-continuous) loss function; it could for example be the indicator function that is equal to 1 when $g^T x > h$ and 0 otherwise. The strategy adopted is to try and keep v_k below a given threshold ξ. More precisely, if the current average is below ξ then the constraint

$$\mathbb{E}_k\big(L_{k+1}/s_{k+1}\big) \leq \xi \tag{8.35}$$

is employed. Otherwise, namely if $v_k > \xi$, then the aim is to return v_k, with probability 1, to a value below ξ.

Given that $L_{k+1} = \gamma L_k + l(g^T x_{k+1} - h)$ it follows that (8.35) can be written as

$$\mathbb{E}_k\Big(l(g^T x_{k+1} - h)\Big) \leq \gamma(\xi s_k - L_k) + \xi$$

and to allow for a relaxation of this constraint at times when L_k is small, this condition is replaced by

$$\mathbb{E}_k\Big(l(g^T x_{k+1} - h)\Big) \leq \beta_k$$
$$\beta_k = \max\big\{\gamma(\xi s_k - L_k) + \xi, \alpha\big\}, \quad \alpha \leq \xi$$

Thus at times of low average violation (i.e. when $L_k/s_k \leq \xi$), we have $\beta_k > \alpha$ thereby resulting in a relaxed constraint which still meets the requirement of (8.35). At all other times, the expected 1-step-ahead loss $\mathbb{E}_k\big(l(g^T x_{k+1} - h)\big)$ is forced to be less than or equal to α thereby ensuring that the average loss will, at some point in the future, be no greater than ξ.

To ensure recursive feasibility, the successor state is constrained to lie at each time instant in a robust controlled invariant set, $\mathcal{S}$, which satisfies the condition:

$$\forall x \in \mathcal{S}, \ \exists u \in \mathcal{U} \ \text{such that} \ Ax + Bu + Dw \in \mathcal{S}, \quad \forall w \in \mathcal{W}, \tag{8.36a}$$

$$\mathbb{E}_k\Big(l\big(g^T(Ax + Bu + Dw) - h\big)\Big) \le \alpha \tag{8.36b}$$

Here $\mathcal{U}$ denotes a set of allowable inputs and $\mathcal{W}$ defines a set of admissible disturbances, with $\mathcal{U}$ and $\mathcal{W}$ assumed to be convex and polytopic. At times of low average constraint violation, restricting the successor state to lie in $\mathcal{S}$ is unnecessarily conservative because, as argued above, it is possible to relax the constraint of the expected loss. Instead it is possible to define a set of nested reachable sets according to the recursion

$$\mathcal{S}_{j+1} = \{x : \exists u \in \mathcal{U} \ \text{such that} \ Ax + Bu + Dw \in \mathcal{S}_j \ \forall w \in \mathcal{W}\}, \quad \mathcal{S}_1 = \mathcal{S},$$

so that the successor state can be allowed to lie in a set, $\mathcal{S}_{j(k)}$, larger than $\mathcal{S}$ without affecting recursive feasibility. The choice of the largest index $j(k)$ which retains recursive feasibility depends on the current value of accumulated constraint violations, as discussed in [9]. The overall constraint set then becomes

$$u \in \mathcal{U} \tag{8.37a}$$

$$Ax + Bu + Dw \in \mathcal{S}_{j(k)} \quad \forall w \in \mathcal{W} \tag{8.37b}$$

$$\mathbb{E}_k\Big(l\big(g^T(Ax + Bu + Dw) - h\big)\Big) \le \beta_k \tag{8.37c}$$

and this can be grafted into a stochastic MPC algorithm with guaranteed recursive feasibility.

In Sects. 8.1 and 8.2 probabilistic constraints involving affine functions of future disturbances were converted in to affine constraints on the degrees of freedom through numerical integration techniques. Similarly here, given the assumptions on the loss function $l(\cdot)$, it is possible to use numerical integration to convert (8.36b) into an affine inequality in u. In particular, (8.36b) can be written as

$$g^T(Ax + Bu) \le h + q(\alpha)$$

where

$$q(\alpha) = \sup\left\{\mu : \int_{-\infty}^{\infty} l(\mu + y) f_{g^T Dw}(y)\, dy \le \alpha\right\}$$

with $f_{g^T Dw}(y)$ denoting the probability density function of $g^T Dw$. The same reasoning applies to the constraint (8.37c), with $q(\alpha)$ replaced by $q(\beta_k)$, however in this case β_k is not known a priori. To avoid computing $q(\beta_k)$ online, it is possible to compute $q(\hat{\beta}_j)$ for a large number of predefined points $\hat{\beta}_j$ in the range of possible values for β_k and select the value of $\hat{\beta}_{j(\beta_k)}$ that is nearest to β_k and such that $\hat{\beta}_{j(\beta_k)} \le \beta_k$. Note that, since (8.36b) can be converted into an inequality which is affine in x and u, it follows that $\mathcal{S}$ and hence $\mathcal{S}_j$ are polyhedra. These can be determined as the maximal such sets or as inner approximations of the maximal sets; the latter offering a compromise between the demands of computation of these sets and the relaxation of (8.37b).

We conclude the section by noting that a stochastic MPC algorithm which minimizes an appropriate predicted cost subject to the constraints

$$u \in \mathcal{U}$$
$$Ax + Bu + Dw \in \mathcal{S}_{j(k)} \quad \forall w \in \mathcal{W}$$
$$g^T(Ax + Bu + Dw) \le h + q\big(\hat{\beta}_{j(\beta_k)}\big)$$

will preserve feasibility at all future times given feasibility at initial time, and in closed-loop operation will meet the average violation constraints, either instantaneously or with probability 1 at some time in the future.

8.4 Stochastic Quadratic Bounds for Additive Disturbances

Each of the probabilistic constraints considered in Sects. 8.1–8.3 is specified in terms of a bound on the probability of a scalar linear function exceeding a given threshold. This leads to computationally efficient algorithms in which the bulk of the computation that is needed to convert probabilistic constraints on random variables into deterministic conditions on optimization variables is performed offline. We now return to the case of multiple linear constraints that are required to jointly hold with a specified probability. Using the approach of [10], this can be achieved by constructing tubes with ellipsoidal cross sections that are defined on the basis of information on the distribution of a quadratic function of the unknown model disturbance input, w_k. The approach is computationally convenient since it characterizes the effects of model uncertainty on the predicted state and control trajectories in terms of a scalar stochastic variable that defines the scaling of the tube cross sections.

We consider again the model of (8.1), which is subject to an i.i.d. additive disturbance sequence $\{w_0, w_1, \ldots\}$, where the probability distribution of w_k is known and assumed to be finitely supported. The system is assumed to be subject to the pointwise-in-time probabilistic constraints of (6.8), defined in terms of given matrices F and G and a specified probability $p \in (0, 1]$ by

$$\Pr_k(Fx_{1|k} + Gu_{1|k} \le \mathbf{1}) \ge p, \tag{8.38}$$

where, as before, $x_{i|k}$ and $u_{i|k}$ denote the i steps ahead predictions at time k of the system state x_{k+i} and control input u_{k+i}. Hard constraints can be handled in this framework by setting the probability p equal to 1.

In the following development, we consider ellipsoidal sets defined for given $V_w \succ 0$ by

$$\mathcal{E}_w(\alpha) = \{w : \; w^T V_w w \le \alpha\},$$

where α is a random variable whose realization at time k is defined by $\alpha_k \doteq w_k^T V_w w_k$. We assume that it is possible to determine the distribution of α from knowledge of the distribution of w, and in particular we assume that the cumulative distribution function

$$\mathcal{F}_\alpha(a) = \Pr(\alpha_k \le a)$$

can be computed, for example by numerical integration. The assumption that w_k is bounded implies that $w_k \in \mathcal{W}$ with probability 1 for some bounded set $\mathcal{W}$. On account of this bound it is possible to determine an upper bound $\bar{\alpha}$ for α_k; this is formally stated in the following assumption.

Assumption 8.1 $\mathcal{F}_\alpha(\bar{\alpha}) = 1$, where $\bar{\alpha} = \max_{w \in \mathcal{W}} w^T V_w w$.

Using the dual-mode predicted control strategy of (8.4), the i steps ahead predicted state, $x_{i|k}$, at time k is decomposed into nominal ($s_{i|k}$) and uncertain ($e_{i|k}$) components according to (8.5a, 8.5b). We begin by considering the constraints of the Mode 1 prediction horizon. For each $i > 0$ let $\mathcal{E}_e(\beta_i)$ denote an ellipsoidal set,

$$\mathcal{E}_e(\beta_i) \doteq \{e : e^T V_e e \le \beta_i\},$$

that contains $e_{i|k}$. Then, since $e_{i|k}$ is a stochastic variable, the minimum β_i such that $e_{i|k} \in \mathcal{E}_e(\beta_i)$ is likewise stochastic. Given the distribution $\mathcal{F}_\alpha$ of the parameter α_k that determines the scaling of the set $\mathcal{E}_w(\alpha_k)$, it is possible to compute a probability distribution for β_i. Moreover a value b_i such that $e_{i|k} \in \mathcal{E}_e(b_i)$ with a probability of at least p can be computed on the basis of this distribution. Then the constraint $\Pr_k(Fx_{i|k} + Gu_{i|k} \le \mathbf{1}) \ge p$ can be transformed, using (8.4) and (8.5a, 8.5b), into the linear deterministic constraint

$$\tilde{F}s_{i|k} + Gc_{i|k} \le \mathbf{1} - b_i^{1/2} h,$$

where $\tilde{F} = F + GK$ and the vector $h = (h_1, \ldots, h_{n_C}) \in \mathbb{R}^{n_C}$ is defined by

$$h_j = (\tilde{F}_j V_e^{-1} \tilde{F}_j^T)^{1/2}, \quad j = 1, \ldots, n_C,$$

with $\tilde{F}_j$ denoting the jth row of $\tilde{F}$. The justification for this transformation is that $b_i^{1/2} h_j$ is the attainable upper bound for $\tilde{F}_j e_{i|k}$ over all $e_{i|k} \in \mathcal{E}_e(b_i)$. To simplify notation we use β_i and b_i in preference to $\beta_{i|k}$ and $b_{i|k}$; these variables do not depend on k because the probability distribution of $e_{i|k}$ is independent of k. For the same reason the notation α_i is used here instead of $\alpha_{i|k}$.

To obtain the probability distributions of β_i, $i = 1, 2, \ldots$ we need to determine how $\mathcal{E}_e(\beta_i)$ evolves over the prediction horizon. A recurrence relation governing β_i can be deduced from the requirement that the sequence $\beta_0, \beta_1, \ldots$ must satisfy the 1-step-ahead inclusion condition

$$\max_{e \in \mathcal{E}_e(\beta_i),\, w \in \mathcal{E}_w(\alpha_i)} (\Phi e + Dw)^T V_e (\Phi e + Dw) \le \beta_{i+1} \tag{8.39}$$

in order to ensure that $e_{i+1|k} \in \mathcal{E}_e(\beta_{i+1})$ whenever $e_{i|k} \in \mathcal{E}_e(\beta_i)$. However, the problem of determining the minimum β_{i+1} satisfying (8.39) is nonconvex (in fact it is NP-complete [11]), and instead we make use of the following sufficient condition.

Theorem 8.3 *The* 1*-step-ahead inclusion condition, that* $e_{i+1|k} \in \mathcal{E}_e(\beta_{i+1})$ *whenever* $e_{i|k} \in \mathcal{E}_e(\beta_i)$*, holds if*

$$\beta_{i+1} = \lambda\beta_i + \alpha_i \tag{8.40a}$$

$$V_e^{-1} - \frac{1}{\lambda}\Phi V_e^{-1}\Phi^T \succeq D V_w^{-1} D^T \tag{8.40b}$$

for some $V_e \succ 0$ *and* $\lambda > 0$*. Furthermore there exist* $V_e \succ 0$ *and* $\lambda \in (0, 1)$ *satisfying (8.40b) if* Φ *is strictly stable.*

Proof Using the S-procedure [11], sufficient conditions for (8.39) are given by

$$\beta_{i+1} \geq \lambda\beta_i + \mu\alpha_i \tag{8.41a}$$

$$\begin{bmatrix} \Phi^T \\ D^T \end{bmatrix} V_e \begin{bmatrix} \Phi & D \end{bmatrix} \preceq \lambda \begin{bmatrix} I \\ 0 \end{bmatrix} V_e \begin{bmatrix} I & 0 \end{bmatrix} + \mu \begin{bmatrix} 0 \\ I \end{bmatrix} V_w \begin{bmatrix} 0 & I \end{bmatrix} \tag{8.41b}$$

for some scalars $\lambda, \mu > 0$. Scaling β_i, β_{i+1} and V_e by μ^{-1} removes μ from these conditions. The equivalence of the scaled version of (8.41b) with (8.40b) can be shown using Schur complements, whereas (8.41a) implies (8.40a) since we are interested in the minimum value of β_{i+1} satisfying (8.39). If all eigenvalues of Φ are no greater than ρ in absolute value and $\rho < 1$, then (8.41a) has a solution $V_e \succ 0$ whenever $\lambda \in (\rho^2, 1)$ since, for any $S = S^T \succ 0$ and λ in this interval, the Lyapunov matrix equation $V - \frac{1}{\lambda}\Phi V \Phi^T = S$ has a solution $V \succ 0$. □

Theorem 8.3 makes it possible to propagate the distribution of β_i over the prediction horizon given the distribution of β_0. Before doing this we state the following corollary to Theorem 8.1.

Corollary 8.2 *If* $\lambda \in (0, 1)$*, then* β_i *lies in the interval* $\beta_i \in [0, \bar{\beta}_i]$ *for all i, where* $\bar{\beta}_0 = 0$ *and*

$$\bar{\beta}_{i+1} = \lambda\bar{\beta}_i + \bar{\alpha}$$

Furthermore $\bar{\beta}_i \leq \bar{\beta}$ *for all i, where* $\bar{\beta} \doteq \bar{\alpha}/(1 - \lambda)$.

Proof If $\beta_i \in [0, \bar{\beta}_i]$ for some i, then from (8.40a) and Assumption 8.1 we obtain $\beta_{i+1} \in [0, \lambda\bar{\beta}_i + \bar{\alpha}]$, and, since $e_{0|k} = 0$ implies $\beta_0 = \bar{\beta}_0 = 0$, it follows that $\beta_i \in [0, \bar{\beta}_i]$ for all i. For any given $\lambda \in (0, 1)$ the asymptotic bound $\lim_{i\to\infty} \bar{\beta}_i = \bar{\beta} = \bar{\alpha}/(1 - \lambda)$ therefore holds, and from $\bar{\beta} - \bar{\beta}_{i+1} = \lambda(\bar{\beta} - \bar{\beta}_i)$ we have $\bar{\beta}_i \leq \bar{\beta}_{i+1} \leq \bar{\beta}$ for all i. □

The recursion in (8.40a) expresses β_{i+1} as the sum of two random variables, namely $\lambda\beta_i$ and α_i. Therefore the distribution function for β_{i+1} is given by a convolution integral (see e.g. [12]),

$$\mathcal{F}_{\beta_{i+1}}(\tau) = \lambda \int_0^{\bar{\beta}} \mathcal{F}_{\beta_i}(\theta) f_\alpha(\tau - \lambda\theta) d\theta, \tag{8.42}$$

where $f_\alpha(\cdot)$ is the probability density function of α and $\mathcal{F}_{\beta_i}(\cdot)$ is the cumulative distribution function of β_i. In general it is not possible to perform this integration analytically but it can be approximated using a Markov chain similar to those introduced in Sects. 7.3 and 7.4.

Consider, for example, subdividing the interval $[0, \bar{\beta}]$ into r intervals $[\tau_i, \tau_{i+1})$, $i = 0, \ldots, r-1$, where

$$0 = \tau_0 < \tau_1 < \cdots < \tau_r = \bar{\beta},$$

and approximating the distribution function $\mathcal{F}_{\beta_i}(\tau)$ by a piecewise constant function $\hat{\mathcal{F}}_{\beta_i}$, defined by

$$\hat{\mathcal{F}}_{\beta_i}(\tau) = \begin{cases} \pi_{i,j} & \tau \in [\tau_j, \tau_{j+1}) \\ \pi_{i,r} = 1 & \tau \geq \tau_r. \end{cases}$$

Under mild assumptions on the continuity of f_α (see e.g. [10]), it can be shown that a generic numerical quadrature approximation of (8.42) provides uniform convergence of the approximation error, $\mathcal{F}_{\beta_i} - \hat{\mathcal{F}}_{\beta_i} \to 0$ for given i, as $\max_j(\tau_{j+1} - \tau_j) \to 0$. Let π_i denote the vector $\pi_i = (\pi_{i,0}, \ldots, \pi_{i,r})$. Then numerical integration applied to the convolution integral (8.42) results in a linear relationship defining π_{i+1} in terms of π_i:

$$\pi_{i+1} = P\pi_i. \tag{8.43}$$

where $\pi_0 = \mathbf{1}$ since $\beta_0 = 0$ implies $\mathcal{F}_{\beta_0}(\tau) = 1$ for all $\tau \geq 0$.

The transition matrix P in (8.43) has as elements the probabilities $p_{l,m}$ that β_{i+1} lies in the interval $[0, \tau_l)$ given that β_i lies in the interval $[0, \tau_m)$. Thus P in (8.43) differs from the transition matrix Π of (7.45) in Sect. 7.4 in that it relates cumulative probabilities, but it can be converted into analogous form by pre-multiplying (8.43) by the matrix T of (7.53) and writing

$$T\pi_{i+1} = \Pi(T\pi_i), \quad \Pi = TPT^{-1}. \tag{8.44}$$

Therefore it may be concluded (as was done in Sects. 7.3 and 7.4) that TPT^{-1}, and hence also P, has one eigenvalue equal to 1, while all other eigenvalues of P are less than 1 in absolute value. The implication of this is that π_{ss}, the eigenvector of P that corresponds to the eigenvalue at 1 describes the steady-state behaviour of the approximation π_i of $\mathcal{F}_{\beta_i}$ as $i \to \infty$.

Now, by construction P is such that the elements $\pi_{i,j}$ of π_i satisfy the inequality $\pi_{i,j} \leq \pi_{i,j+1}$ and $\pi_{i,r} = 1$ so that for any given probability p it is possible to determine the smallest j such that $\beta_i \leq \tau_j$ with probability at least p. Formally this can be achieved through the use of the function $b(\pi_i, p)$ defined by

$$\begin{aligned}\text{ind}(\pi_i, p) &= \min\{j : \pi_{i,j} \geq p\},\\ b(\pi_i, b) &= \tau_j, \quad j = \text{ind}(\pi_i, p).\end{aligned}$$

With this definition, we can state that the i steps ahead prediction $e_{i|k}$ lies with a probability of at least p in the ellipsoid $\mathcal{E}_e\big(b(\pi_i, p)\big)$, i.e.

$$\Pr_k\Big(e_{i|k} \in \mathcal{E}_e\big(b(\pi_i, p)\big)\Big) \geq p. \tag{8.45}$$

The probabilistic inclusion condition in (8.45) defines a stochastic tube with ellipsoidal cross sections $\mathcal{E}_e\big(b(\pi_i, p)\big)$ containing the uncertain component $e_{i|k}$ of the predicted state with probability at least p. Equivalently, it defines a stochastic tube with cross sections $\{s_{i|k}\} \oplus \mathcal{E}_e\big(b(\pi_i, p)\big)$ that will contain the predicted state $x_{i|k}$ with the same probability. It is important to note that these tubes can be computed offline and hence the dimension r of π_i can be taken to be as large as desired. This allows the error in the approximation of the integral in (8.43) to be made insignificant without increasing the online computational load. We next use these tubes to derive linear inequalities that ensure the state predictions satisfy constraints (8.38).

Lemma 8.3 *The constraint* $\Pr_k(Fx_{i|k} + Gu_{i|k} \leq \mathbf{1}) \geq p$ *is satisfied by the predictions of the model (8.5) for given i if*

$$\tilde{F}s_{i|k} + Gc_{i|k} \leq \mathbf{1} - \big(b(\pi_i, p)\big)^{1/2} h \tag{8.46}$$

where $h_j = (\tilde{F}_j V_e^{-1} \tilde{F}_j^T)^{1/2}$ *for* $j = 1, \ldots, n_C$, *and* $\tilde{F} = [\,\tilde{F}_1^T \cdots \tilde{F}_{n_C}^T\,]^T$.

Proof From the state decomposition (8.5a, 8.5b), we have that $Fx_{i|k} + Gu_{i|k} \leq \mathbf{1}$ whenever $\tilde{F}e_{i|k} \leq \mathbf{1} - (\tilde{F}s_{i|k} + Gc_{i|k})$. But $e_{i|k}$ lies in $\mathcal{E}_e(b)$ with probability p, where $b = b(\pi_i, p)$, and hence $Fx_{i|k} + Gu_{i|k} \leq \mathbf{1}$ with probability p if the maximum of each element of $\tilde{F}e$ over all e in the ellipsoid $\mathcal{E}_e(b)$ is no greater than the corresponding element of $\mathbf{1} - (\tilde{F}s_{i|k} + Gc_{i|k})$. This condition is ensured by the inequality of (8.46) since $\max_{e \in \mathcal{E}_e(b)} \tilde{F}_j e = b^{1/2}(\tilde{F}_j V_e^{-1} \tilde{F}_j^T)^{1/2}$. □

Lemma 8.3 allows the probabilistic constraint $\Pr_k(Fx_{i|k} + Gu_{i|k} \leq \mathbf{1}) \geq p$ on the i steps ahead predicted state and control input to be imposed for $i = 1, 2, \ldots$ through linear constraints on the online optimization variable $\mathbf{c}_k$. However these constraints are not necessarily recursively feasible and hence they cannot ensure the future feasibility of the probabilistic constraint in (8.38). This can be explained using the reasoning of Sect. 7.1 as follows. Suppose $\mathbf{c}_k$ is such that (8.46) is satisfied at time k, thus ensuring that the i steps ahead constraint $\Pr_k(Fx_{i|k} + Gu_{i|k} \leq \mathbf{1}) \geq p$ holds for given i. Then, to ensure feasibility of the corresponding $i - 1$ steps ahead probabilistic constraint at time $k+1$, we require that $\mathbf{c}_{k+1} = (c_{1|k}, \ldots, c_{N-1|k}, 0)$ satisfies (8.46), with k and i replaced by $k+1$ and $i-1$, respectively. But at time $k+1$, the disturbance input w_{k+1} has already been realized and therefore it cannot be treated as a stochastic variable; hence the stochastic tube that was used to formulate the probabilistic constraint in (8.46) at time k is no loger valid.

In order to ensure that $\mathbf{c}_{k+1}$ is feasible for the $i-1$ steps ahead probabilistic constraint at time $k+1$, it is necessary at time k to take into account the effect of the worst-case value of w_{k+1} on the i-step-ahead probabilistic constraint. This can be done by considering the stochastic tube $\{\mathcal{E}_e(\bar{\beta}_1), \mathcal{E}_e(\beta_2), \ldots, \mathcal{E}_e(\beta_i)\}$. For this tube the distribution functions $\mathcal{F}_{\beta_j}$ for $j \geq 2$ are again governed by (8.42), but with the initial condition

$$\mathcal{F}_{\beta_1}(\tau) = \begin{cases} 0 & \tau < \bar{\beta}_1 \\ 1 & \tau \geq \bar{\beta}_1 \end{cases}$$

which corresponds to $\beta_1 = \bar{\beta}_1$, and the approximation of $\mathcal{F}_{\beta_j}$ therefore evolves according to (8.43), but with an initial condition π_1 corresponding to $\beta_1 = \bar{\beta}_1$. The constraint that $\Pr_{k+1}(Fx_{i-1|k+1} + Gu_{i-1|k+1} \leq \mathbf{1}) \geq p$ should hold for all realizations of w_{k+1} then has the same form as (8.46), but with the RHS of the inequality adjusted to account for the new value of π_1. The feasibility of this constraint would then be ensured at $k+1$, but we need also to guarantee that it remains feasible at times $k+2, \ldots, k+i-1$, which requires that the worst-case bounds on β_i for $i = 2, \ldots, i-1$ must similarly be taken into account in the constraints imposed at time k.

To simplify the analysis, we introduce the notation $\pi_{i|j}$ to denote the approximation of the distribution of β_i when β_j, for some given $j \leq i$, assumes its maximum value of $\bar{\beta}_j$. We therefore define

$$\pi_{i|j} = P^{i-j}\pi_{j|j}, \qquad \pi_{j|j} = \mathbf{u}(\bar{\beta}_j), \tag{8.47}$$

where the bound $\bar{\beta}_j$ is given by Corollary 8.2 as

$$\bar{\beta}_j = \frac{1-\lambda^j}{1-\lambda}\bar{\alpha} \tag{8.48}$$

and where $\mathbf{u}(\bar{\beta}_j)$ is the vector of 0s and 1s, the lth element of which is equal to 1 if $\tau_{l+1} < \bar{\beta}_j$ and is equal to 0 otherwise (i.e. if $\tau_{l+1} \geq \bar{\beta}_j$). Then ensuring that constraint (8.46) is feasible j steps ahead implies that it must also hold with $b(\pi_{i|0}, p)$ replaced by $b(\pi_{i|j}, p)$. Invoking this argument for all $j = 0, 1, \ldots, i-1$ results in the constraint

$$\tilde{F}s_{i|k} + Gc_{i|k} \leq \mathbf{1} - \big(\max\{b(\pi_{i|0}, p), \ldots, b(\pi_{i|i-1}, p)\}\big)^{1/2} h.$$

This constraint can be simplified using the following result, which is consistent with the intuition that the maximum of the bounds $b(\cdot, p)$ appearing on the RHS corresponds to the case in which β_{i-1} assumes its worst-case value and α_{i-1} is treated as a stochastic variable.

Lemma 8.4 *For all $i \geq 1$ and $0 \leq j < i$ we have $b(\pi_{i|j}, p) \leq b(\pi_{i|i-1}, p)$.*

Proof Given that $\bar{\beta}_{i-1}$ defines an upper bound on β_{i-1} for all uncertainty realizations, we must have $\bar{\beta}_{i-1} \geq b(\pi_{i-1|j}, 1)$ for all $i \geq 1$ and $0 \leq j < i$, and hence $\pi_{i-1|i-1} = \mathbf{u}(\bar{\beta}_{i-1}) \leq \pi_{i-1|j}$. It follows that $\pi_{i|i-1} \leq \pi_{i|j}$ since the elements of P are non-negative, and this implies that $b(\pi_{i|j}, p) \leq b(\pi_{i|i-1}, p)$ for any given $p \in (0, 1]$. □

Applying Lemma 8.4 to the constraints of this section results in conditions that are equivalent to the constraints of (7.7a, 7.7b), formulated for the general case in Sect. 7.1. These constraints and their recursive feasibility property can be summarized as follows.

Theorem 8.4 *The constraints defined at time k by*

$$\tilde{F} s_{i|k} + G c_{i|k} \leq \mathbf{1} - \big(b(\pi_{i|i-1}, p)\big)^{1/2} h, \quad i = 1, 2, \ldots \tag{8.49}$$

ensure that (8.38) holds. Furthermore if $\mathbf{c}_k$ satisfies (8.49) at time k, then $\mathbf{c}_{k+1} = M\mathbf{c}_k$ will be feasible at time $k+1$ for (8.49) with k replaced by $k+1$, where M is the shift matrix defined in (2.26b).

The conditions of (8.49) impose an infinite number of constraints on the predicted state and control trajectories over an infinite prediction horizon. Using the approach of Sect. 3.2.1 however, it is possible to impose these constraints through a finite number of inequalities. To demonstrate this, we reintroduce the lifted autonomous prediction dynamics of Sect. 3.2 and thus write

$$z_{i+1|k} = \Psi z_{i|k}, \quad z_{0|k} = \begin{bmatrix} x_k \\ \mathbf{c}_k \end{bmatrix}, \quad \Psi = \begin{bmatrix} \Phi & BE \\ 0 & M \end{bmatrix}$$

where E and M are defined in (2.26b). This allows the predicted trajectories of the state and control input to be generated as $x_{i|k} = [I\ 0] z_{i|k} + e_{i|k}$ and $u_{i|k} = [K\ E] z_{i|k} + K e_{i|k}$, and hence the constraints (8.49) can be expressed equivalently (and more conveniently) as

$$\bar{F} \Psi^i z_{0|k} \leq \mathbf{1} - \big(b(\pi_{i|i-1}, p)\big)^{1/2} h, \quad i = 1, 2, \ldots \tag{8.50}$$

where $\bar{F} = [F + GK\ \ GE]$.

Given that the prediction system is stable, the constraints in (8.50) are necessarily satisfied for some x_k and $\mathbf{c}_k$ if and only if the RHS of the inequality is non-negative for all $i \geq 1$. This condition is satisfied if

$$\bar{\beta}^{1/2} h < \mathbf{1} \tag{8.51}$$

since the definition of $\bar{\beta}_i$ as a bound on β_i implies that $b(\pi_{i|i-1}, p) \leq \bar{\beta}_i$, and by Corollary 8.2 the sequence $\{\bar{\beta}_0, \bar{\beta}_1, \ldots\}$ is monotonically non-decreasing with $\lim_{i\to\infty} \bar{\beta}_i = \bar{\beta}$. Under the assumption that (8.51) holds, the following result, which is closely related to Theorem 3.1, shows that the infinite set of inequalities in (8.50) is equivalent to a finite number of inequality constraints.

Theorem 8.5 *If (8.51) holds, then* $\bar{F}\Psi^i z \leq \mathbf{1} - \big(b(\pi_{i|i-1}, p)\big)^{1/2} h$ *is satisfied for all* $i \geq 1$ *if and only if*

$$\bar{F}\Psi^i z \leq \mathbf{1} - \big(b(\pi_{i|i-1}, p)\big)^{1/2} h, \quad i = 1, 2, \ldots, \nu \tag{8.52}$$

where ν *is the smallest integer such that* $\bar{F}\Psi^{\nu+1} z \leq \mathbf{1} - \big(b(\pi_{\nu+1|\nu}, p)\big)^{1/2} h$ *holds for all* z *satisfying (8.52).*

Proof Clearly (8.52) must hold in order that $\bar{F}\Psi^i z \leq \mathbf{1} - \big(b(\pi_{i|i-1}, p)\big)^{1/2} h$ holds for all $i \geq 1$. To show that (8.52) is also sufficient, let $\mathcal{Z}^{(\nu)}$ denote the set

$$\mathcal{Z}^{(\nu)} = \big\{z : \bar{F}\Psi^i z \leq \mathbf{1} - \big(b(\pi_{i|i-1}, p)\big)^{1/2} h,\ i = 1, 2, \ldots, \nu\big\}.$$

Then, under the conditions of the theorem, $z \in \mathcal{Z}^{(\nu)}$ implies

$$\bar{F}\Psi z \leq \mathbf{1} - \big(b(\pi_{1|0}, p)\big)^{1/2} h \tag{8.53a}$$

$$\bar{F}\Psi^i z \leq \mathbf{1} - \big(b(\pi_{i|i-1}, p)\big)^{1/2} h, \quad i = 2, \ldots, \nu + 1 \tag{8.53b}$$

and, by Theorem 8.4, the constraints in (8.53b) imply that the conditions

$$\bar{F}\Psi^i (\Psi z + \bar{D} w) \leq \mathbf{1} - \big(b(\pi_{i|i-1}, p)\big)^{1/2} h, \quad i = 1, \ldots, \nu$$

hold for all $w \in \mathcal{W}$, where $\bar{D} = [D^T\ 0]^T$. Thus $\mathcal{Z}^{(\nu)}$ is robustly invariant for the system

$$z_{k+1} = \Psi z_k + \bar{D} w_k, \quad w_k \in \mathcal{W}, \tag{8.54}$$

while (8.53a) implies that $\Pr_k(F x_{1|k} + G u_{1|k} \leq \mathbf{1}) \geq p$ is satisfied for all $z_k \in \mathcal{Z}^{(\nu)}$ and it follows that $\bar{F}\Psi^i z \leq \mathbf{1} - (b(\pi_{i|i-1}, p))^{1/2} h$ holds for all $i \geq 1$ whenever $z \in \mathcal{Z}^{(\nu)}$. □

A straightforward extension of the arguments of Theorem 3.1 can be used to show that there exists a finite ν satisfying the conditions of Theorem 8.5 if $(\Psi, \bar{F})$ is observable, and also that $\mathcal{Z}^{(\nu)}$ is the maximal RPI set for the system (8.54) and the constraints of (8.53a).

We are now in a position to state the stochastic MPC algorithm. Since the constraints (8.52) are linear in the optimization variable $\mathbf{c}_k$, the online MPC optimization problem requires the solution of a QP at each sampling instant.

Algorithm 8.3 At each time instant $k = 0, 1, \ldots$

(i) Perform the optimization:

$$\begin{aligned} &\underset{\mathbf{c}_k}{\text{minimize}}\ J(x_k, \mathbf{c}_k) \\ &\text{subject to}\ \bar{F}\Psi^i \begin{bmatrix} x_k \\ \mathbf{c}_k \end{bmatrix} \leq \mathbf{1} - \bigl(b(\pi_{i|i-1}, p)\bigr)^{1/2} h, \quad i = 1, \ldots, \nu \end{aligned} \tag{8.55}$$

(ii) Implement the control law $u_k = Kx_k + c^*_{0|k}$ where $\mathbf{c}^*_k = (c^*_{0|k}, \ldots, c^*_{N|k})$ is the minimizing argument of (8.55). ◁

The statement of the algorithm presupposes that the parameters V_e and λ have been designed offline. These parameters can be determined for example by solving the optimization problem

$$(V_e^{-1}, \lambda) = \arg \min_{V_e^{-1},\, \lambda \in (0,1)} \frac{\bar{\alpha}}{1-\lambda} \max_j (\tilde{F}_j V_e^{-1} \tilde{F}_j^T) \ \text{ subject to (8.40b)}$$

The rationale behind this optimization is that it minimizes the steady-state effect of the uncertainty on constraints of (8.55) by minimizing the maximum element of the LHS of (8.51). The optimization can be performed by combining a univariate search over $\lambda \in (0, 1)$ with semidefinite programming to compute the optimal V_e for fixed λ.

Algorithm 8.3 has the guarantee of recursive feasibility provided by Theorem 8.4, and, by Theorem 7.1, the use of the cost $J(x_k, \mathbf{c}_k)$ therefore ensures that the closed-loop system satisfies a quadratic stability condition. These properties are summarized as follows.

Corollary 8.3 *Algorithm 8.3 applied to the system (8.1) is feasible at all times $k = 1, 2, \ldots$ if it is feasible at $k = 0$. The closed-loop system satisfies the constraints of (8.38) for all $k \geq 0$ as well as the mean-square bound*

$$\lim_{r\to\infty} \frac{1}{r} \sum_{k=0}^{r-1} \mathbb{E}_0\bigl(\|x_k\|_Q^2 + \|u_k\|_R^2\bigr) \leq l_{ss} \tag{8.56}$$

where $l_{ss} = \lim_{k\to\infty}(\|x_k\|_Q^2 + \|u_k\|_R^2)$. If $u = Kx$ is optimal in the absence of constraints for the cost (8.3), then x_k converges with probability 1 to the minimal RPI set for the dynamics (8.1) under this feedback law as $k \to \infty$.

8.5 Polytopic Tubes for Additive and Multiplicative Uncertainty

The system dynamics considered in Sects. 8.1–8.4 of this chapter are subject only to additive disturbances. We now turn to the case of linear models that are subject to stochastic multiplicative uncertainty as well as additive uncertainty. We consider systems described by the model of (6.2)–(6.3):

$$x_{k+1} = A_k x_k + B_k u_k + D w_k \tag{8.57}$$

where the additive disturbance $w_k \in \mathbb{R}^{n_w}$ and the matrices A_k, B_k that contain the model parameters depend linearly on a set of zero-mean random variables $q_k^{(j)}, j = 1, \ldots, m$ so that $(A_k, B_k, w_k) = \big(A(q_k), B(q_k), w(q_k)\big)$ at time k, with

$$\big(A(q), B(q), w(q)\big) = (A^{(0)}, B^{(0)}, 0) + \sum_{j=1}^{m} (A^{(j)}, B^{(j)}, w^{(j)}) q^{(j)}, \tag{8.58}$$

and $(A^{(j)}, B^{(j)}, w^{(j)}), j = 0, 1, \ldots, m$, are known, constant parameters. We assume that the probability distribution of $q_k = (q_k^{(1)}, \ldots, q_k^{(m)})$ is known and time invariant, and that q_k and q_j are statistically independent for all $k \neq j$.

The MPC strategy discussed in this section carries a guarantee of recursive feasibility and is designed for problems with mixed hard and probabilistic constraints. As discussed in Sect. 7.1, this requires knowledge of a bounding set, $\mathcal{Q}$, such that $q_k \in \mathcal{Q}$ with probability 1 for all k. For convenience, we assume that $\mathcal{Q}$ is a compact, convex polytope, with known vertices $q_v^{(1)}, \ldots, q_v^{(\nu)}$. Corresponding to each vertex $q_v^{(l)}$ of $\mathcal{Q}$ is a vertex of the uncertainty set for the model parameters (A, B, w), which we denote as $(A_v^{(l)}, B_v^{(l)}, w_v^{(l)})$, so that

$$\big(A_v^{(l)}, B_v^{(l)}, w_v^{(l)}\big) = \big(A(q_v^{(l)}), B(q_v^{(l)}), w(q_v^{(l)})\big)$$

for $l = 1, \ldots, \nu$.

The system of (8.57) is considered to be subject at all times $k = 0, 1, \ldots$ to a mixture of hard constraints:

$$F_h x_k + G_h u_k \leq \mathbf{1}, \tag{8.59}$$

and probabilistic constraints:

$$\Pr_k(F_p x_{1|k} + G_p u_{1|k} \leq \mathbf{1}) \geq p, \tag{8.60}$$

for a given set of matrices F_h, G_h, F_p, G_p and a given probability $p \in (0, 1]$. Here (8.60) requires, similarly to the pointwise probabilistic constraint (6.8), that the joint probability of all elements of the vector $F_p x_{1|k} + G_p u_{1|k}$ being less than or equal to unity should be no less than p, where $F_p x_{1|k} + G_p u_{1|k}$ is the 1 step ahead prediction of $F_p x + G_p u$ at time k. As in Sect. 8.1 it is straightforward to extend the approach of this section to intersections of probabilistic constraints (8.60) with different probabilities p.

Using an open-loop prediction strategy, the predicted control sequence is parameterized in terms of a perturbed linear feedback law:

$$u_{i|k} = Kx_{i|k} + c_{i|k}, \quad i = 0, 1, \ldots$$

with $c_{i|k} = 0, i = N, N+1, \ldots$. The dynamics governing predicted state trajectories therefore assume the form

$$x_{i+1|k} = \Phi_{k+i} x_{i|k} + B_{k+i} c_{i|k} + D w_{k+i}$$

where $\Phi_k = A_k + B_k K$. The feedback gain K is assumed to be stabilizing in the sense that $x_{k+1} = \Phi_k x_k$ is mean-square stable in the absence of constraints. Hence the quadratic predicted cost, which is taked to be (6.15), can be evaluated using the approach of Sect. 6.2.

It was mentioned in Chap. 7 that propagating the effects of multiplicative uncertainty over a prediction horizon can cause computational difficulties because both Φ_{k+i} and $x_{i|k}$ are uncertain. Instead, similarly to the robust MPC strategies considered in Chap. 5, we construct tubes with polytopic cross sections [13, 14], defined by

$$\mathcal{X}_{i|k} = \{x_{i|k} : Vx_{i|k} \leq \alpha_{i|k}\}, \tag{8.61}$$

that contain the predicted state trajectories. The vectors $\alpha_{i|k}$ for $i = 0, \ldots, N$ are treated as variables in the online MPC optimization whereas the matrix $V \in \mathbb{R}^{n_V \times n_x}$ is determined offline and remains fixed online. The choice of V is based on the considerations detailed in Sects. 5.5 and 5.6, summarized by the following assumption.

Assumption 8.2 V is chosen so that the set $\mathcal{X} = \{x : Vx \leq \mathbf{1}\}$ is λ-contractive for the dynamics $x_{k+1} = \Phi_k x_k + Dw_k$, for some $\lambda < 1$.

On account of the requirement for a recursively feasible MPC strategy (and also satisfaction of hard constraints, when these are present), the constraint that the predicted state $x_{i|k}$ should lie in the tube cross section $\mathcal{X}_{i|k}$ must be handled robustly. Through the application of Farkas' Lemma discussed in Chap. 5, Lemma 5.6 shows that this is achieved by the conditions, for $H^{(l)} \geq 0$ and $i = 0, 1, \ldots,$

$$\alpha_{i+1|k} \geq H^{(l)} \alpha_{i|k} + VB_v^{(l)} c_{i|k} + VDw_v^{(l)} \tag{8.62a}$$

$$H^{(l)} V = V\Phi_v^{(l)} \tag{8.62b}$$

for $l = 1, \ldots, \nu$, where $\Phi_v^{(l)} \doteq A_v^{(l)} + B_v^{(l)} K$. Recursive feasibility with respect to satisfaction of the hard constraints is then guaranteed by the existence of a matrix $H_h \geq 0$ such that, for $i = 0, 1, \ldots,$

$$H_h\alpha_{i|k} + G_h c_{i|k} \leq \mathbf{1} \tag{8.63a}$$

$$H_h V = \tilde{F}_h. \tag{8.63b}$$

where $\tilde{F}_h \doteq F_h + G_h K$.

In order to handle the probabilistic constraints however, we require a probabilistic extension of Farkas' Lemma. This is provided by the following result.

Theorem 8.6 ([14]) *Let*

$$\mathcal{X}_1 = \{x : V_1 x \leq b_1\}, \quad \mathcal{X}_2 = \{x : \Pr(V_2 x \leq b_2) \geq p\}$$

where V_2 and b_2 are random variables. Then $\mathcal{X}_1 \subseteq \mathcal{X}_2$ (i.e. $\Pr(V_2 x \leq b_2) \geq p$ for all x such that $V_1 x \leq b_1$) if and only if there exists a random variable $H \geq 0$ satisfying

$$HV_1 = V_2 \tag{8.64a}$$

$$\Pr(Hb_1 \leq b_2) \geq p. \tag{8.64b}$$

Proof Sufficiency follows from the fact that, by (8.64a), for $x \in \mathcal{X}_1$ we can write

$$V_2 x = HV_1 x \leq Hb_1$$

which, from (8.64b), implies that

$$\Pr(V_2 x \leq b_2) \geq p.$$

Thus every $x \in \mathcal{X}_1$ also belongs to $\mathcal{X}_2$, thereby implying that $\mathcal{X}_1 \subseteq \mathcal{X}_2$.

To prove necessity, assume that $\mathcal{X}_1 \subseteq \mathcal{X}_2$ holds. Then

$$\Pr(\mu_i \leq b_{2,i}) \geq p$$

with

$$\mu_i = \max_x \{V_{2,i} x : V_1 x \leq b_1\} \tag{8.65}$$

where $V_{2,i}$ and $b_{2,i}$ denote the ith row and ith element of V_2 and b_2, respectively. By strong duality, the dual of the linear program (8.65) for a given realization of V_2 gives

$$\mu_i = \min_h \{h^T b_1 : h^T V_1 = V_{2,i}, \ h \geq 0\}. \tag{8.66}$$

Let h_i^* be the minimizing argument of this dual LP and define H as the matrix with ith row equal to h_i^*. Note that h_i^* is a continuous piecewise affine function of the random variable $V_{2,i}$ since (8.66) has the form of a (right-hand side) parametric linear program in the parameter $V_{2,i}$. Hence h_i^* is itself a random variable [15]. The constraints of (8.66) imply that $H \geq 0$ and that H satisfies (8.64a). From the objective of (8.66) it follows that H also satisfies (8.64b). □

Theorem 8.6 can be used to derive conditions which ensure that the predicted states $x_{i|k}$ satisfy the probabilistic constraints of (8.60) for all $x_{i|k} \in \mathcal{X}_{i|k}$, $i = 0, 1, \ldots$, namely that

$$\begin{aligned}\Pr_{k+i}&(F_p x_{i+1|k} + G_p u_{i+1|k}) \\ &= \Pr_{k+i}\big(\tilde{F}_p \Phi_{k+i} x_{i|k} + \tilde{F}_p(B_{k+i} c_{i|k} + D w_{k+i}) + G_p c_{i+1|k} \le \mathbf{1}\big) \ge p\end{aligned}$$

where $\tilde{F}_p = F_p + G_p K$. From (8.64a, 8.64b), these conditions are equivalent to the requirement that there exists $H_p \ge 0$ satisfying, for $i = 0, 1, \ldots$

$$\Pr\big(H_p \alpha_{i|k} + \tilde{F}_p(B c_{i|k} + D w) + G_p c_{i+1|k} \le \mathbf{1}\big) \ge p \tag{8.67a}$$

$$H_p V = \tilde{F}_p \Phi. \tag{8.67b}$$

Since Φ is a function of the random variable q, in general H_p satisfying (8.67b) will also be a random variable. This means the method of handling (8.67a, 8.67b) differs from that of (8.62a, 8.62b) and (8.63a, 8.63b); however given knowledge of the distribution of q it is possible to construct a computationally tractable online optimization, as discussed below.

In summary therefore, a recursively feasible set of conditions that impose the constraints of (8.59–8.60) are:

(i) Tube inclusion constraints—(8.62a, 8.62b);
(ii) Hard constraints—(8.63a, 8.63b);
(iii) Probabilistic constraints—(8.67a, 8.67b).

The degree of conservativeness of the conditions of (8.62), (8.63) and (8.67) would be minimized if V, $H^{(l)}$, H_h and H_p were computed online for each prediction instant, $i = 0, 1, \ldots$. However this strategy is unlikely to be implementable as it would require the solution of a large nonconvex optimization problem online. Instead we discuss how to design these matrices offline. Thus V is to be chosen, as described in Assumption 8.2, so as to define a λ-contractive set for the dynamics $x_{k+1} = \Phi_k x_k$, whereas each of the matrices $H^{(l)}$, H_h and H_p is designed to have minimum row sum with the aim of relaxing the associated constraints. In particular the ith rows of these matrices are selected according to

$$H_i^{(l)\,T} = \arg\min_h \ \mathbf{1}^T h \text{ subject to } h^T V = V_i \Phi_v^{(l)} \text{ and } h \ge 0, \quad l = 1, \ldots, \nu \tag{8.68a}$$

$$H_{h,i}^T = \arg\min_h \ \mathbf{1}^T h \text{ subject to } h^T V = \tilde{F}_{h,i} \text{ and } h \ge 0 \tag{8.68b}$$

$$H_{p,i}^T = \arg\min_h \ \mathbf{1}^T h \text{ subject to } h^T V = \tilde{F}_{p,i} \Phi \text{ and } h \ge 0. \tag{8.68c}$$

The values of $H^{(l)}$ and H_h in (8.68a) and (8.68b) are fixed for given V, $\Phi_v^{(l)}$ and $\tilde{F}_h$ as the deterministic solutions of a set of linear programs. However (8.68c) specifies H_p as a random variable, the probability distribution of which is defined by the parametric solutions of a set of linear programs. In particular, from (8.58) the linear program (8.68c) is equivalent to

$$\begin{aligned} h^*(q) = \arg\min_{h} \quad & \mathbf{1}^T h \\ \text{subject to} \quad & h^T V = \tilde{F}_{p,i}(A^{(0)} + B^{(0)}K) + \sum_{j=1}^{m} \tilde{F}_{p,i}(A^{(j)} + B^{(j)}K)q^{(j)} \\ & h \geq 0. \end{aligned} \tag{8.69}$$

Since the constraints of this problem depend linearly on both h and the random variable $q = (q^{(1)}, \ldots, q^{(m)})$, it can be shown that the solution $h^*(q)$ is a continuous, piecewise affine function of q [16]. Thus H_p is given by (8.68c) as a continuous and piecewise affine function of q. For most problems of practical interest it is unlikely to be computationally feasible to determine the probability distribution of H_p by using multiparametric linear programming to solve (8.68c) for H_p as an explicit function of q. Instead (8.68c) can be used as a means of generating random samples of the distribution of H_p given samples of q; the approach is discussed further at the end of this section.

The matrices $H^{(l)}$, H_h and H_p are necessarily sparse in the sense that each of their rows can have at most n_x non-zero elements. This follows from the fact that the problems posed in (8.68a–8.68c) for given q, have $n_V - n_x$ active constraints so that $n_V - n_x$ of the elements of the optimizing h must in each case be zero. This affords computational advantages.

Using the argument of Sect. 7.1, conditions (8.62a), (8.63a) and (8.67a) would ensure recursive feasibility if they were applied over an infinite prediction horizon, but this would of course require an infinite number of constraints and is clearly not implementable. This difficulty can be avoided through the use of terminal conditions, as described in Sect. 5.5. For example, let the sequence of parameters $\{\alpha_{0|k}, \ldots, \alpha_{N|k}\}$ satisfy the constraints of (8.62a), (8.63a) and (8.67a) for $i = 0, \ldots, N-1$ and impose terminal constraints on $\alpha_{N|k}$:

$$H^{(l)}\alpha_{N|k} + VDw_v^{(l)} \leq \alpha_{N|k}, \quad l = 1, \ldots, \nu \tag{8.70a}$$

$$H_h \alpha_{N|k} \leq \mathbf{1} \tag{8.70b}$$

$$\Pr(H_p \alpha_{N|k} \leq \mathbf{1}) \geq p. \tag{8.70c}$$

The following lemma shows that, under Assumption 8.2, the matrices $H^{(l)}$ satisfy the condition $\|H^{(l)}\|_\infty \leq 1$, which is necessary for feasibility of (8.70a). With these constraints we are able to state the following result.

Lemma 8.5 *Under Assumption 8.2, the definition of* $H^{(l)}$ *in (8.68a) implies that* $H^{(l)}\mathbf{1} + VDw_v^{(l)} \leq \lambda\mathbf{1}$ *for all* $l = 1, \ldots, \nu$.

Proof Since the set $\{x : Vx \leq \mathbf{1}\}$ is λ-contractive for $x_{k+1} = \Phi_k x_k + Dw_k$, we have $V\Phi_v^{(l)}x + VDw_v^{(l)} \leq \lambda\mathbf{1}$ for all $l = 1, \ldots, \nu$ and x such that $Vx \leq \mathbf{1}$, so the bound $H^{(l)}\mathbf{1} + VDw_v^{(l)} \leq \lambda\mathbf{1}$ follows from the constraints of (8.68a). □

Theorem 8.7 *If V is chosen according to Assumption 8.2, then the constraints of (8.62a), (8.63a) and (8.67a), invoked for $i = 0, \ldots, N-1$, and the terminal and initial constraints, (8.70a–8.70c) and $Vx_k \leq \alpha_{0|k}$, are jointly recursively feasible for the system (8.57–8.58) under the control law $u_k = Kx_k + c_{0|k}$.*

Proof Suppose that $\mathbf{c}_k = (c_{0|k}, \ldots, c_{N-1|k})$ and $\{\alpha_{0|k}, \ldots, \alpha_{N|k}\}$ satisfy the constraints of the theorem at time k. Then a feasible set of parameters at time $k+1$ is given by

$$\mathbf{c}_{k+1} = (c_{1|k}, \ldots, c_{N-1|k}, 0)$$
$$\alpha_{i|k+1} = \alpha_{i+1|k}, \quad i = 0, \ldots, N-1$$
$$\alpha_{N|k+1} = \alpha_{N|k}$$

since these parameters give $\mathcal{X}_{i|k+1} = \mathcal{X}_{i+1|k}$ for $i = 0, 1, \ldots, N-1$ and $\mathcal{X}_{N|k+1} = \mathcal{X}_{N|k}$. It follows that (8.62a), (8.63a) and (8.67a), with k replaced by $k+1$, hold for $i = 0, \ldots, N-1$. Also the conditions (8.70a–8.70c) are trivially satisfied when k is replaced by $k+1$ if $\alpha_{N|k+1} = \alpha_{N|k}$. Furthermore, we have $Vx_{k+1} \leq \alpha_{0|k+1}$ since $x_{k+1} \in \mathcal{X}_{1|k}$ for all realizations of model uncertainty at time k. □

We can now formulate the stochastic MPC algorithm. For simplicity the objective function to be minimized online is chosen here as the quadratic predicted cost of (6.15) in Sect. 6.2:

$$J(x_k, \mathbf{c}_k) = \sum_{i=0}^{\infty} \mathbb{E}\big(\|x_{i|k}\|_Q^2 + \|u_{i|k}\|_R^2 - l_{ss}\big). \tag{8.71}$$

By Theorem 6.1, this cost is a quadratic function of the vector, $\mathbf{c}_k = (c_{0|k}, \ldots, c_{N-1|k})$, of free variables in the predicted control sequence.

Algorithm 8.4 At each time instant $k = 0, 1, \ldots$

(i) Perform the optimization:

$$\begin{aligned} &\underset{\substack{\mathbf{c}_k \\ \alpha_{0|k}, \ldots, \alpha_{N|k}}}{\text{minimize}} \ J(x_k, \mathbf{c}_k) \\ &\text{subject to } (8.62a), (8.63a), (8.67a) \text{ for } i = 0, \ldots, N-1, \\ &\qquad\qquad (8.70a\text{-c}) \text{ and } Vx_k \leq \alpha_{0|k}. \end{aligned} \tag{8.72}$$

(ii) Implement the control law $u_k = Kx_k + c_{0|k}^*$ where $\mathbf{c}_k^* = (c_{0|k}^*, \ldots, c_{N|k}^*)$ is the minimizing argument of (8.72). ◁

The constraints of Algorithm 8.4 are recursively feasible by Theorem 8.7. Furthermore, Theorem 7.1 of Chap. 7.2.1 demonstrates that the closed-loop system satisfies the constraints (8.59–8.60) for all $k = 0, 1, \ldots$ and the optimal value of the cost, $J^*(x_k)$ satisfies

$$\mathbb{E}_k\big(J^*(x_{k+1})\big) \leq J^*(x_k) - \big(\|x_k\|_Q^2 + \|u_k\|_R^2 - l_{ss}\big).$$

Therefore the quadratic stability condition holds for the closed-loop system:

$$\lim_{r\to\infty} \frac{1}{r} \sum_{k=0}^{r} \mathbb{E}_0\big(\|x_k\|_Q^2 + \|u_k\|_R^2\big) \leq l_{ss}.$$

The online MPC optimization in step (i) of Algorithm 8.4 is not stated in form that can be implemented directly. This is because the constraints (8.67a) and (8.70c) involve products of optimization variables $\alpha_{i|k}$ and $c_{i|k}$ with the random variables H_p and B [19], and because the probability distribution of H_p is implicitly defined by (8.68c). A way to circumvent these difficulties is to use methods for imposing probabilistic constraints based on random sampling [3, 4, 20].

Let $q^{[j]}$, $j = 1, \ldots, n_s$ denote a set of n_s independent samples drawn from the known probability distribution for q. Given these samples, the corresponding samples of B and w:

$$B^{[j]} = B(q^{[j]}), \quad w^{[j]} = w(q^{[j]}), \quad j = 1, \ldots, n_s$$

are generated by (8.58). Likewise samples of H_p are obtained by defining the ith row of $H_p^{[j]}$ using (8.69) as

$$H_{p,i}^{[j]} = \big(h^*(q^{[j]})\big)^T, \quad j = 1, \ldots, n_s.$$

Using this set of samples, the probabilistic constraints (8.67a) and (8.70c) in the online MPC optimization can be approximated using sampled constraints defined by

$$H_p^{[j]}\alpha_{i|k} + \tilde{F}_p(B^{[j]}c_{i|k} + Dw^{[j]}) + G_p c_{i+1|k} + s_{i|k}^{[j]} = \mathbf{1}, \quad i = 0, \ldots, N-1 \tag{8.73a}$$

$$H_p^{[j]}\alpha_{N|k} + s_{N|k}^{[j]} = \mathbf{1} \tag{8.73b}$$

$$s_{i|k}^{[j]} \geq 0, \quad \forall j \in \mathcal{I}_k \subseteq \{1, \ldots, n_s\}, \quad |\mathcal{I}_k| \geq rn_s. \tag{8.73c}$$

Here $|\mathcal{I}_k|$ denotes the number of elements in the set $\mathcal{I}_k$. Thus (8.73c) ensures that the conditions

$$H_p^{[j]}\alpha_{i|k} + \tilde{F}_p(B^{[j]}c_{i|k} + Dw^{[j]}) + G_p c_{i+1|k} \leq \mathbf{1}, \quad i = 0, \ldots, N-1$$
$$H_p^{[j]}\alpha_{N|k} \leq \mathbf{1}$$

are imposed for all $j \in \mathcal{I}_k$, where $\mathcal{I}_k$ in an index set containing no fewer than $\lceil rn_s \rceil$ of the samples $j \in \{1, \ldots, n_s\}$. The remaining $\lfloor (1-r)n_s \rfloor$ samples are discarded since the corresponding slack variables $s_{i|k}^{[j]}$ are not constrained to be non-negative in (8.73a, 8.73b). Since samples are selected randomly, the constraints (8.73a–8.73c) are not equivalent to (8.67a) and (8.70c) for finite n_s. However it is possible to derive bounds on the probability, which is dependent on n_s and r, such that a solution to the MPC optimization (8.72) with (8.67a) and (8.70c) replaced by (8.73a–8.73c) satisfies the probabilistic constraints of (8.67a) and (8.70c). For details we refer the reader to [4, 17].

This approach therefore approximates the probabilistic constraints by using samples to empirically approximate the distributions of the stochastic variables appearing in (8.72). Since the index set $\mathcal{I}_k$ is an optimization variable, the resulting optimization has the form of a mixed integer quadratic program (MIQP). The effects of varying the number of samples on the confidence of constraint satisfaction are discussed in the following example.

Example 8.1 This example provides a simple illustration of the use of sampling to approximate a probabilistically constrained optimization problem. Consider the minimization

$$\begin{aligned} &\underset{x}{\text{minimize}} && f(x) \\ &\text{subject to} && g_i(x) \leq 0, \quad \forall i \in \mathcal{I} \subseteq \{1, \ldots, n_s\}, \quad |\mathcal{I}| \geq rn_s \end{aligned} \tag{8.74}$$

where the functions $f(x)$ and $g_i(x) = g(x, q^{[i]})$ are convex in the optimization variable x, and where $q^{[i]}$, $i = 1, \ldots, n_s$ are independent samples of a random variable q. For suitable choices of n_s and r, the constraints of (8.74) provide an approximation of the probabilistic constraint

$$\Pr\big(g(x, q) \leq 0\big) \geq p. \tag{8.75}$$

Let $F_{n,m}(p)$ denote the binomial distribution function giving the probability of m or fewer successes in n independent trials, each of which has a probability p of success:

$$F_{n,m}(p) \doteq \sum_{i=0}^{m} \binom{n}{i} p^i (1-p)^{n-i}.$$

Then a lower bound on the probability that the solution of the convex program (8.74) satisfies the probabilistic constraint (8.75) is given in [4, 17] as $1 - \epsilon$, where the parameter ϵ satisfies

$$\epsilon \leq \binom{\lfloor n_s(1-r) \rfloor + \rho - 1}{\lfloor n_s(1-r) \rfloor} F_{n_s, \lceil rn_s \rceil - \rho}(1-p)$$

Here ρ is the number of support constraints of the problem (8.74), which is essentially the number of constraints of the form $g_i(x) \leq 0$ that can be active at the solution of (8.74) (where removal of an active constraint causes a reduction in the value of the objective). For a feasible problem we must clearly have $\rho \leq n_x$ [17]. Similarly, the probability that a solution of (8.74) is feasible for (8.75) has an upper bound β, where

$$\beta \leq F_{n_s, \lceil rn_s \rceil - 1}(p).$$

The variation with r of these confidence bounds is shown for $p = 0.9$ and various sample sizes n_s, with $\rho = 1$ and $\rho = 2$ in Figs. 8.1 and 8.2. ◊

Example 8.2 This example illustrates the use of random sampling in approximating the probabilistically constrained online MPC optimization of Algorithm 8.4. We consider a system model of the form (8.57) containing multiple independent sources of stochastic uncertainty. The expected values of the system matrices are given by

$$A^{(0)} = \begin{bmatrix} -1.9 & -1.4 \\ 0.7 & 0.5 \end{bmatrix}, \quad B^{(0)} = \begin{bmatrix} 1 \\ -0.25 \end{bmatrix}$$

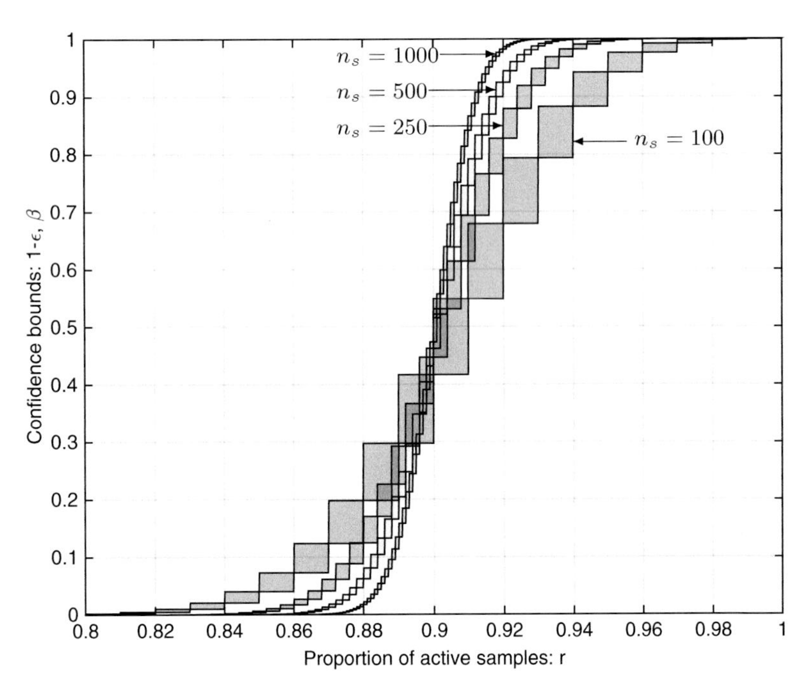

Fig. 8.1 *Upper* and *lower* bounds on the probability that a solution of the sampled program (8.74) satisfies the probabilistic constraint (8.75) for the case of $\rho = 1$

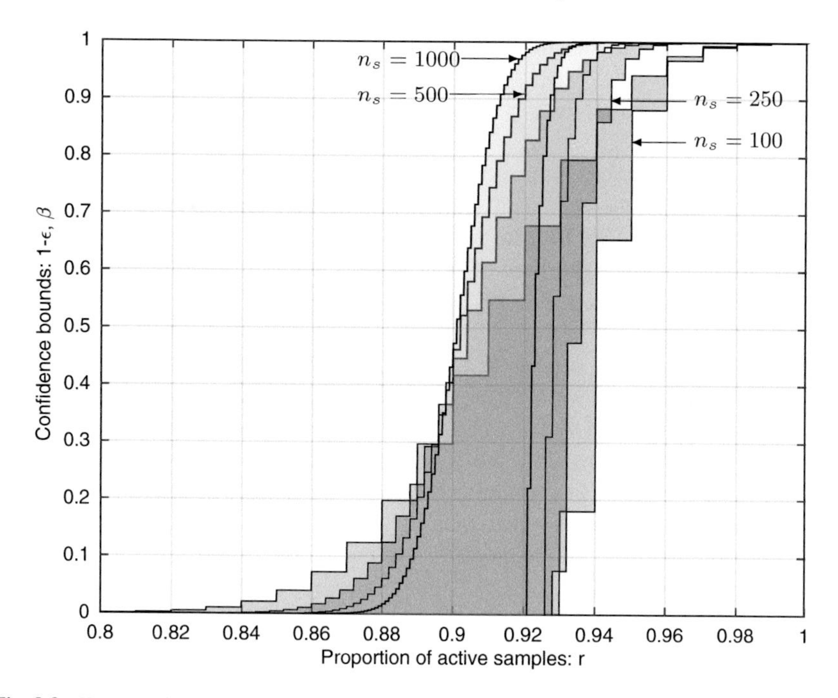

Fig. 8.2 *Upper* and *lower* bounds on the probability that a solution of the sampled program (8.74) satisfies the probabilistic constraint (8.75) with $\rho = 2$

and $D = I$. The realizations of A_k, B_k and w_k are given by

$$A_k = A^{(0)} + A^{(1)} q_k^{(1)} + A^{(2)} q_k^{(2)} + A^{(3)} q_k^{(3)}$$
$$B_k = B^{(0)} + B^{(4)} q_k^{(4)} + B^{(5)} q_k^{(5)}$$
$$w_k = w^{(6)} q_k^{(6)} + w^{(7)} q_k^{(7)}$$

where $q^{(j)}$ is a scalar random variable, uniformly distributed on the interval $[-0.5, 0.5]$ for $j = 1, \dots, 7$, and $q_k = (q_k^{(1)}, \dots, q_k^{(7)})$ satisfies $\mathbb{E}(q_k q_k^T) = \frac{1}{24} I$ and $\mathbb{E}(q_k q_i^T) = 0$ for all $i \neq k$. The remaining model parameters are

$$A^{(1)} = \begin{bmatrix} 0.03 & 0.15 \\ -0.15 & -0.03 \end{bmatrix}, \quad A^{(2)} = \begin{bmatrix} -0.03 & -0.15 \\ 0 & -0.03 \end{bmatrix}, \quad A^{(3)} = \begin{bmatrix} 0 & 0 \\ 0.15 & 0.06 \end{bmatrix},$$
$$B^{(1)} = \begin{bmatrix} 0.036 \\ -0.024 \end{bmatrix}, \quad B^{(2)} = \begin{bmatrix} -0.036 \\ 0.024 \end{bmatrix}, \quad w^{(1)} = \begin{bmatrix} 0.04 \\ -0.04 \end{bmatrix}, \quad w^{(2)} = \begin{bmatrix} -0.04 \\ 0.04 \end{bmatrix}.$$

The system is subject to the constraint

$$\Pr_k(Fx_{1|k} \leq 1) \geq 0.9, \quad F = \begin{bmatrix} -5 & 10 \end{bmatrix},$$

and the weighting matrices in the cost (8.71) are defined by $Q = I$ and $R = 1$.

In order to apply the stochastic MPC law of Algorithm 8.4, the probabilistic constraints (8.67a) and (8.70c) in the online optimization (8.72) are replaced by (8.73a–8.73c). A mode 1 horizon of $N = 4$ steps and 250 samples are employed at each prediction time step. For this problem the number of support constraints at each time step is at most $\rho = 1$, and the confidence bounds of Fig. 8.1 with $n_s = 250$ can therefore be used to determine the fraction r of the samples that should be activated in order to achieve a given confidence of feasibility with respect to the probabilistic constraints [18]. For a confidence level of 90 % we need $r = 0.93$, and hence $\lceil rn_s \rceil = 233$ samples activated at each prediction time step.

The MPC optimization incorporating (8.73a–8.73c) is solved approximately using a greedy algorithm. This attempts to identify the optimal samples to be discarded at each prediction time step by successively solving the QP problem that corresponds to a fixed set of discarded samples, and then discarding the samples that correspond to the constraints (8.73a, 8.73b) that have the largest associated multipliers. Note that the implementation of (8.73a–8.73c) in terms of slack variables has the advantage that not all constraints in the online optimization need to be recomputed at each iteration.

We next compare the performance of the stochastic MPC law with that of its robust counterpart, which is obtained by invoking the robust constraints $Fx_{i|k} \leq \mathbf{1}$, $i = 0, 1, \ldots,$ for all realizations of model uncertainty. For problems involving multiple independent sources of uncertainty, the robust MPC approach is likely to be very conservative. In particular, although each uncertain component of the model is uniformly distributed, the model uncertainty combines to give a one step-ahead probability density function for the model state that is heavily centre-weighted and quickly drops to a negligible value a short distance from its centroid. A probabilistically constrained stochastic MPC algorithm can explicitly account for this effect whereas robust MPC must take into account the worst-case value of each source of uncertainty.

The high degree of conservativeness of robust MPC can be seen in Fig. 8.3, which shows how the optimal predicted cost for robust MPC compares with that of stochastic MPC for the initial condition $x_0 = (0.4, 0.4)$ as p varies: clearly a small reduction in p causes a relatively large reduction in predicted cost. The state trajectories of the closed-loop system under SMPC with $p = 0.9$ and RMPC are shown in Figs. 8.4 and 8.5 for 100 model uncertainty sequences. Again it is clear that using robust MPC in this example results in conservative closed-loop responses.

Table 8.1 compares the closed-loop performance of the robust MPC and stochastic MPC algorithms using 500 realizations of model uncertainty. Here the mean cost computed along closed-loop system trajectories is 16 % lower in the stochastic case than the robust, and this is achieved with approximately double the computation

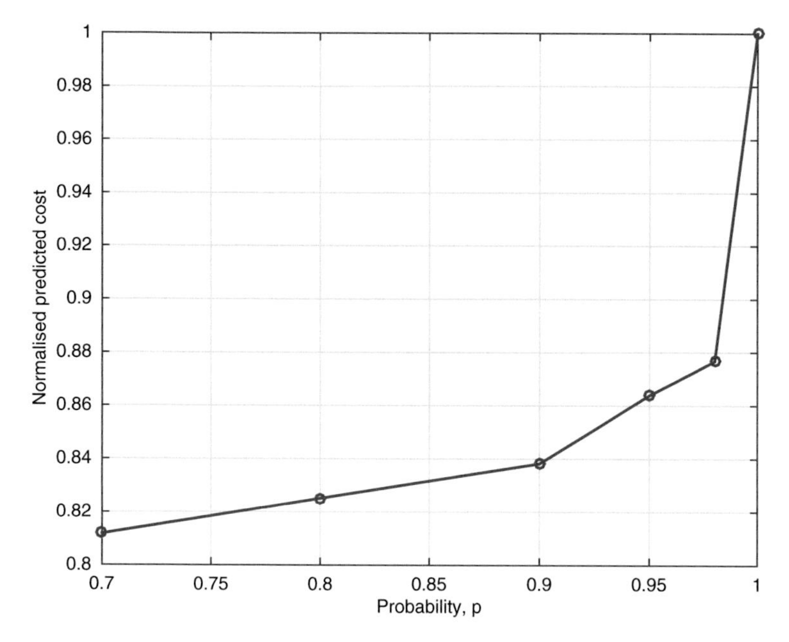

Fig. 8.3 Variation with p of the optimal predicted cost of stochastic MPC for a fixed initial condition, relative to the optimal predicted cost for $p = 1$

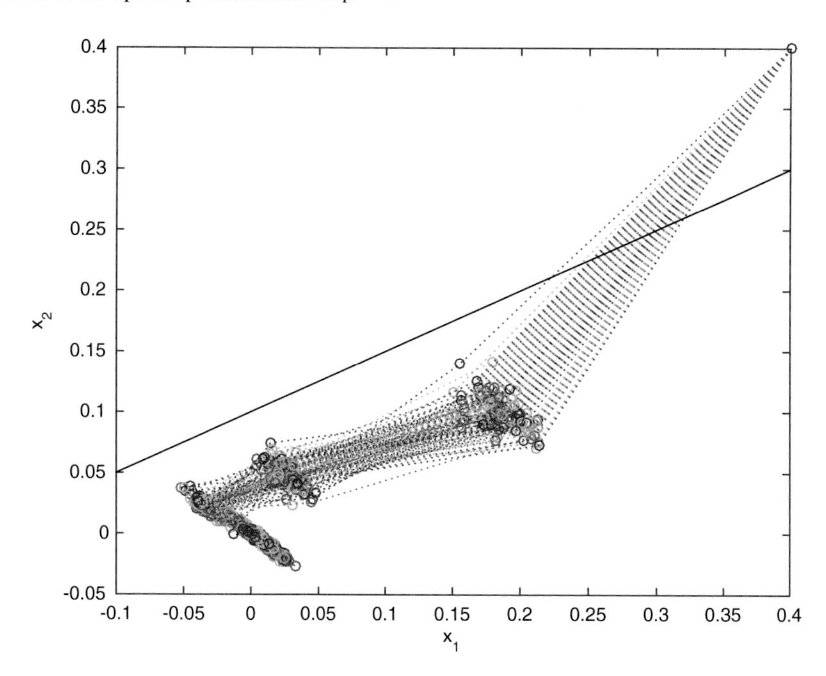

Fig. 8.4 Robust MPC: closed-loop state trajectories for 100 uncertainty sequences

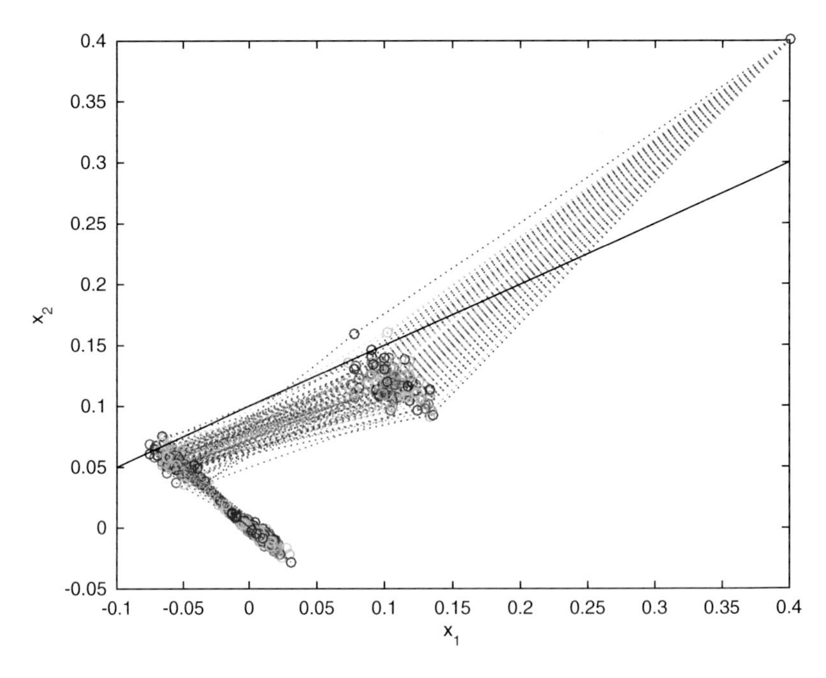

Fig. 8.5 Stochastic MPC: closed-loop state trajectories, 100 uncertainty sequences

Table 8.1 Closed-loop comparison of robust and stochastic MPC: average costs, constraints and computation times for 500 sequences of model uncertainty

	Robust	Stochastic
Mean closed-loop cost	25.7	21.5
Proportion of realizations satisfying constraints	100	93.8
Mean computation time (ms)	26	60

time required for the stochastic algorithm. The proportion of trajectories satisfying constraints (93.8 %) implies a degree of conservativism in the stochastic algorithm. However this is expected from the confidence parameter of 90 % and the relatively small sample size, $n_s = 250$. $\diamond$

References

1. B. Kouvaritakis, M. Cannon, S.V. Raković, Q. Cheng, Explicit use of probabilistic distributions in linear predictive control. Automatica **46**(10), 1719–1724 (2010)
2. M. Cannon, Q. Cheng, B. Kouvaritakis, S.V. Rakovic, Stochastic tube MPC with state estimation. Automatica **48**(3), 536–541 (2012)
3. G. Calafiore, M.C. Campi, Uncertain convex programs: randomized solutions and confidence levels. Math. Progr. **102**(1), 25–46 (2005)

4. M.C. Campi, S. Garatti, A sampling-and-discarding approach to chance-constrained optimization: feasibility and optimality. J. Optim. Theory Appl. **148**, 257–280 (2011)
5. B. Kouvaritakis, M. Cannon, D. Muñoz-Carpintero, Efficient prediction strategies for disturbance compensation in stochastic MPC. Int. J. Syst. Sci. **44**(7), 1344–1353 (2013)
6. J. Löfberg, Approximations of closed-loop minimax MPC, in *Proceedings of the 42nd IEEE Conference on Decision and Control*, Maui, USA, vol. 2 (2003), pp. 1438–1442
7. P.J. Goulart, E.C. Kerrigan, J.M. Maciejowski, Optimization over state feedback policies for robust control with constraints. Automatica **42**(4), 523–533 (2006)
8. D.H. van Hessem, O.H. Bosgra, A conic reformulation of model predictive control including bounded and stochastic disturbances under state and input constraints, in *Proceedings of the 41st IEEE Conference on Decision and Control*, Las Vegas, USA, vol. 4 (2002), pp. 4643–4648
9. M. Korda, R. Gondhalekar, F. Oldewurtel, C.N. Jones, Stochastic MPC framework for controlling the average constraint violation. IEEE Trans. Autom. Control **59**(7), 1706–1721 (2014)
10. M. Cannon, B. Kouvaritakis, S.V. Raković, Q. Cheng, Stochastic tubes in model predictive control with probabilistic constraints. IEEE Trans. Autom. Control **56**(1), 194–200 (2011)
11. S.P. Boyd, L. El Ghaoui, E. Feron, V. Balakrishnan, Linear Matrix Inequalities in System and Control Theory. Society for Industrial and Applied Mathematics (1994)
12. W. Feller, *Introduction to Probability Theory and Its Applications*, vol. 2 (John Wiley, New York, 1971)
13. J. Fleming, B. Kouvaritakis, M. Cannon, Robust tube MPC for linear systems with multiplicative uncertainty. IEEE Trans. Autom. Control **60**(4), 1087–1092 (2015)
14. J. Fleming, M. Cannon, and B. Kouvaritakis, Stochastic tube MPC for LPV systems with probabilistic set inclusion conditions, in *Proceedings of the 53rd IEEE Conference on Decision and Control, Los Angeles, USA* (2014)
15. M. Capinski, P.E. Kopp, *Measure, Integral and Probability* (Springer, London, 2004)
16. T. Gal, *Postoptimal Analyses, Parametric Programming, and Related Topics*, 2nd edn. (De Gruyter, Berlin, 1995)
17. G.C. Calafiore, Random convex programs. SIAM J. Optim. **20**(6), 3427–3464 (2010)
18. G. Schildbach, L. Fagiano, M. Morari, Randomized solutions to convex programs with multiple chance constraints. SIAM J. Optim. **23**(4), 2479–2501 (2013)
19. P. Kall, S.W. Wallace, *Stochastic Programming* (John Wiley, Chichester, 1994)
20. J. Luedtke, S. Ahmed, A sample approximation approach for optimization with probabilistic constraints. SIAM J. Optim. **19**(2), 674–699 (2008)

Chapter 9
Conclusions

The aim of this final chapter is to give a short discursive summary of some of the key results presented in this book. We also speculate on extensions that could, in our opinion, be pursued in future.

It is perhaps difficult to pinpoint precisely the origins of predictive control but it appears that the early development of the subject ignored the presence of constraints. The perception of the subject has changed considerably over the last few decades and now the justification and success of predictive control is almost exclusively attributed to its ability to provide near optimal solutions that account for constraints. This feature alone makes MPC a particularly useful tool for the solution of real life problems where typically limits in actuation and safety considerations imply the presence of constraints.

This development brought with it two difficulties, one of which is theoretical and the other practical, namely the guarantee of the stability of the control system and the implementation within the inter-sample interval. The practicability of implementation implied the need to turn what, in essence, was an infinite-dimensional optimization problem into a finite-dimensional problem and this was made possible through the split of the prediction horizon into the near horizon (mode 1), where the control moves are considered to be degrees of freedom, and the far horizon (mode 2), where the control moves are dictated by a prescribed control law. To guarantee feasibility within mode 2, use was made of the concept of invariance and the implied terminal constraints. Thereafter, closed-loop stability could be established by applying a Lyapunov-like analysis to the closed-loop behaviour of the optimal predicted cost. Clearly, feasibility of the MPC optimization was required at each time step, and in general this was guaranteed recursively by ensuring that a specific predicted trajectory satisfied the constraints of the problem at successive time instants.

There has been a proliferation of MPC strategies proposed in the literature over the last few decades and their relative success has been judged mostly on the basis of their ease of computation on the one hand and comparing the size of their respective

B. Kouvaritakis and M. Cannon, *Model Predictive Control*,
Advanced Textbooks in Control and Signal Processing,
DOI 10.1007/978-3-319-24853-0_9

regions of attraction on the other. Often one may also wish to see a comparison of performance, but in general this is example- and initial condition-dependent, thereby rendering such an exercise meaningless. The overall aim has nevertheless been to strive for the best possible balance between ease of computation and optimality. Several attempts, some quite effective, have been proposed and are described in this book. Thus one of the ways of achieving efficiency in computation is to reduce the number of degrees of freedom over which the online repetitive optimization of MPC is to be performed. This can be done for example by input blocking, or interpolating between given trajectories or using the homotopy-based active constraint approach. Alternatively, for low-dimensional systems, one may use a multi-parametric approach which identifies regions in which the optimal predicted control is known and given by affine relationships to the state. The lifted autonomous formulation of the predicted dynamics provides yet another way in which the online computational load can be reduced significantly through the replacement of the optimization by a well-behaved Newton–Raphson procedure. Arguably, there is no unqualified best amongst all these, and an array of other approaches exist, which have not been mentioned in this book. The designer has to choose the approach that best meets the demands (in terms of degree of optimality, efficiency of computation and size of region of attraction) of the particular problem to be addressed. It is to be hoped that more original ideas will come about in future years and that some of the existing approaches will be developed further.

As mentioned earlier, over and above computation, one needs to consider the size of the region of attraction of a particular MPC algorithm. To improve on this, one can use as large a terminal set as possible and for a given terminal control law that leads naturally to the employment of the maximal invariant set. Further improvements are possible through the use of longer prediction horizons but this carries the penalty of increased computational load. An alternative to longer horizons is the introduction of controller dynamics whose action extends across an infinite prediction horizon. Such dynamics can be optimized to give the largest ellipsoidal region of attraction that can be attained over all terminal linear feedback laws. Yet this benefit is attained regardless of the choice of the terminal control law which can be chosen to be the unconstrained optimal.

The body of ideas of classical MPC carry over to robust MPC, but catering for uncertainty clearly requires more intensive online optimization. Low complexity tubes provide a convenient (albeit potentially conservative) way to define sets that contain predictions for all possible realizations of uncertainty. On the basis of these, constraints can be invoked robustly, and, coupled with a monotonically non-increasing property of a cost based on the nominal predictions or on perturbations to an unconstrained optimal control law, this leads to algorithms with guaranteed closed-loop stability. It is also possible to use general (rather than low complexity) tubes and in particular for additive uncertainty only one can employ rigid or homothetic tubes with the attendant inclusion conditions, whereas in the case of multiplicative (and also additive) uncertainty one can construct general tubes through inclusion conditions that are based on the use Farkas' Lemma.

For the case of additive disturbances only, improved results can be achieved through a re-parameterization of predictions that is affine in future disturbances and has a lower triangular structure. The drawback of this approach however is that, through the lower triangular structure, it introduces a greater number of degrees of freedom, which is in fact of the order of N^2, where N is the mode 1 prediction horizon. For the same order of magnitude of degrees of freedom, it is possible to use a more general lower-triangular tube parameterization which is piecewise affine rather than affine in the future disturbances. Both the disturbance affine MPC and PTMPC assume that the uncertainty set is polytopic but the former works with the facets of the set whereas the latter assumes a set description in terms of its vertices. The numbers of facets and vertices could differ significantly and this in turn implies a significant difference in the number of inequalities in the online optimization of the two approaches. This difference could be removed if an extension were found that enabled the methodology of PTMPC to deploy uncertainty facets rather than vertices.

An alternative that reduces computational complexity considerably replaces the lower triangular structure of PTMPC by a striped lower triangular structure, thus leading to a striped PTMPC (or SPTMPC) algorithm. Despite the reduction of the number of degrees of freedom (which are of order N for SPTMPC), this modification allows disturbance compensation to extend to mode 2. On account of this it can potentially outperform the parameterized tube MPC in terms of the size of its region of attraction.

In all of these endeavours, the goal is to get as close to the dynamic programming (DP) solution for the optimal feedback law without restrictive assumptions on controller parameterization, but to do so with a computational load that is tractable. PTMPC has narrowed the gap between available algorithms and the DP solution and indeed produces optimal results for several special cases. However, for fast sampling applications, the computational requirement of PTMPC could be excessive, while the degree of sub-optimality in SPTMPC could be more than desired. The field is wide open for researchers to come up with ideas that sit somewhere between PTMPC and SPTMPC in respect of the balance between optimality and computability. The field is also wide open in respect of re-parameterizations of tube MPC for the case of multiplicative uncertainty.

Robust MPC is clearly not the answer to controlling systems that are subject to random uncertainty with known probability distributions and that are subject to constraints, some of which could be probabilistic. The answer to this problem is provided by stochastic MPC, which has received considerable attention over the last decade. There were significant developments in this field, especially on the control theoretic front, leading to recursive feasibility (through a combined robust and probabilistic treatment of constraints) and stability guarantees. In general this is only possible for model uncertainty with finitely supported distributions. Such distributions are perhaps not as convenient as the Gaussian but accord well with the physical world where variations in the uncertain parameters are almost never unbounded.

These developments considered first the case of additive uncertainty and only more recently have been extended to the case of more general multiplicative uncertainty models. The particular difficulty here is that the multiplication of predicted states, which are random variables, by parameters which themselves are stochastic, makes it difficult to determine the distributions of predictions. The combined use of Farkas' Lemma with sampling circumvents this difficulty in terms of practical implementation and it is anticipated that further advances in this area will carry on being proposed in the near future. It is perhaps to be expected that some re-parameterization of stochastic MPC (e.g. along the lines of PTMPC) might be available for the case of additive model uncertainty, and possibly also the multiplicative uncertainty case.

Another area that may attract attention in future concerns the definition of predicted performance costs that preserve as much of the probabilistic nature of the cost as viable computation allows. Costs expressed in terms of nominal, expected values or worst-case values tend to conceal much of the stochastic nature of the control problem. An attempt at overcoming this difficulty was proposed in the solution to the sustainable problem discussed in Chap. 6 through the definition of a cost on the basis of probabilistic bands, but certainly there will be alternatives which are yet to be worked on. A topic related to the stochastic nature of the problem is the possibility of relaxation of future constraints on the basis of past realizations of uncertainty. Preliminary results in this area have been reported in the last chapter of the book but this area deserves further development.

In conclusion, classical MPC is now mature enough to suggest that further future developments, though still possible, will be few. The same, is not true of MPC applied to uncertain systems, especially for cases in which uncertainty is stochastic.

Solutions to Exercises

Solutions to Exercises for Chap. 2

1 (a) The predicted state and control sequences at time k with $N = 2$ are

$$\mathbf{x} = \begin{bmatrix} x_{0|k} \\ x_{1|k} \\ x_{2|k} \end{bmatrix}, \quad \mathbf{u} = \begin{bmatrix} u_{0|k} \\ u_{1|k} \end{bmatrix}, \quad \mathbf{x} = \begin{bmatrix} 1 \\ 1.5 \\ 2.25 \end{bmatrix} x + \begin{bmatrix} 0 & 0 \\ 1 & 0 \\ 1.5 & 1 \end{bmatrix} \mathbf{u}.$$

Hence the predicted cost for $q = 1$ is $J(x, \mathbf{u}) = \mathbf{x}^T \mathbf{x} + 10\mathbf{u}^T \mathbf{u}$

$$\begin{aligned} J(x, \mathbf{u}) &= \mathbf{u}^T \left(\begin{bmatrix} 10 & 0 \\ 0 & 10 \end{bmatrix} + \begin{bmatrix} 0 & 1 & 1.5 \\ 0 & 0 & 1 \end{bmatrix} \begin{bmatrix} 0 & 0 \\ 1 & 0 \\ 1.5 & 1 \end{bmatrix} \right) \mathbf{u} \\ &\quad + 2x \begin{bmatrix} 1 & 1.5 & 2.25 \end{bmatrix} \begin{bmatrix} 0 & 0 \\ 1 & 0 \\ 1.5 & 1 \end{bmatrix} \mathbf{u} + x^2 \begin{bmatrix} 1 & 1.5 & 2.25 \end{bmatrix} \begin{bmatrix} 1 \\ 1.5 \\ 2.25 \end{bmatrix} \\ &= \mathbf{u}^T H \mathbf{u} + 2x F^T \mathbf{u} + x^2 G, \end{aligned}$$

Since $\mathbf{u} = -H^{-1} F x$, we get the unconstrained MPC law

$$u_k = -\left(\frac{11(1.5 + 1.5^3) - 1.5^3}{11(11 + 1.5^2) - 1.5^2} \right) x_k = -0.350 x_k.$$

(b) Let $u_{i|k} = K x_{i|k}$ for all $i \geq 2$, with $K = -0.88$ (the LQ optimal feedback gain). If q satisfies $q - (A + BK)^2 q = 1 + 10K^2$, i.e. if

$$q = \frac{1 + 10K^2}{1 - (A + BK)^2} = \frac{1 + 10(0.88)^2}{1 - (1.5 - 0.88)^2} = 14.20,$$

B. Kouvaritakis and M. Cannon, *Model Predictive Control*,
Advanced Textbooks in Control and Signal Processing,
DOI 10.1007/978-3-319-24853-0

then the predicted cost satisfies

$$J = \sum_{i=0}^{N-1} (x_{i|k}^2 + 10u_{i|k}^2) + qx_{2|k}^2 = \sum_{i=0}^{\infty} (x_{i|k}^2 + 10u_{i|k}^2)$$

so the unconstrained MPC law is identical to LQ optimal control.

(c) Constraints: $-0.5 \le u_{i|k} \le 1$ for $i = 0, 1, \ldots$ imply constraints on the predicted input sequence:

$$-0.5 \le u_{i|k} \le 1, \quad i = 0, 1, \ldots, N + \nu$$

where $N = 2$ and ν must be large enough so that

$$-0.88(1.5 - 0.88)^{\nu+1} x \in [-0.5, 1] \text{ for all } x \text{ such that}$$
$$-0.88(1.5 - 0.88)^{i} x \in [-0.5, 1], \; i = 0, \ldots, \nu$$

Here $1.5 - 0.88 = 0.62$, so $\nu = 0$ is sufficient.

2 (a) The dynamics are stable if and only if $|\alpha| < 1$, which is therefore a requirement for $|y_k| \le 1$ for all $k \ge 0$. Also

$$\begin{aligned} y_0 &= \begin{bmatrix} 1 & 0 \end{bmatrix} x_0, & &\text{so } |y_0| \le 1 \iff -1 \le \begin{bmatrix} 1 & 0 \end{bmatrix} x_0 \le 1 \\ y_1 &= \begin{bmatrix} 0 & 1 \end{bmatrix} x_0, & &\text{so } |y_1| \le 1 \iff -1 \le \begin{bmatrix} 0 & 1 \end{bmatrix} x_0 \le 1 \\ y_2 &= \alpha \begin{bmatrix} 0 & 1 \end{bmatrix} x_0, & &\text{so } |y_2| \le 1 \Longleftarrow \begin{bmatrix} -1 \\ -1 \end{bmatrix} \le x_0 \le \begin{bmatrix} 1 \\ 1 \end{bmatrix} \end{aligned}$$

so we can conclude that $|y_k| \le 1$ for all $k \ge 0$ if and only if each element of x_0 is less than or equal to 1 in absolute value.

The same result can also be deduced from $y_i = \alpha^{i-1} \begin{bmatrix} 1 & 0 \end{bmatrix} x_0$ for $i \ge 1$.

(b) If $u_{i|k} = \begin{bmatrix} -\beta & 0 \end{bmatrix} x_{i|k}$ for all $i \ge N$, then

$$\sum_{i=N}^{\infty} (y_{i|k}^2 + u_{i|k}^2) = x_{N|k}^T \begin{bmatrix} p_1 & p_{12} \\ p_{12} & p_2 \end{bmatrix} x_{N|k}$$

where

$$\begin{bmatrix} p_1 & p_{12} \\ p_{12} & p_2 \end{bmatrix} - \begin{bmatrix} 0 & 1 \\ 0 & \alpha \end{bmatrix}^T \begin{bmatrix} p_1 & p_{12} \\ p_{12} & p_2 \end{bmatrix} \begin{bmatrix} 0 & 1 \\ 0 & \alpha \end{bmatrix} = \begin{bmatrix} 1 + \beta^2 & 0 \\ 0 & 0 \end{bmatrix} \implies \begin{cases} p_1 = 1 + \beta^2 \\ p_{12} = 0 \\ p_2 = \dfrac{p_1}{1 - \alpha^2} \end{cases}$$

which proves (i). To demonstrate (ii) we can use the result from part (a) by replacing the constraint $|y_k| \leq 1$ with $|u_k| \leq 1$ and noting that $u_{i|k} = -\beta y_{i|k}$ for all $i \geq N$.

(c) Although the terminal equality constraint $x_{N|k} = 0$ would ensure recursive feasibility closed loop stability, it would severely restrict the operating region of the controller. In particular the second element of the state is uncontrollable so this terminal constraint would require the second element of the state to be equal to zero at all points in the operating region.

3 (a) If $u_k = Kx_k$ and $y_k = Cx_k$, with $C = \frac{1}{\sqrt{2}}\begin{bmatrix}1 & 1\end{bmatrix}$ and $K = \frac{1}{\sqrt{2}}C$, then

$$I - (A + BK)^T(A + BK) = \frac{1}{2}(C^TC + K^TK) = \frac{3}{4}C^TC = \frac{3}{8}\begin{bmatrix}1 & 1\\ 1 & 1\end{bmatrix}.$$

Hence the solution of $P - (A + BK)^T P(A + BK) = \frac{1}{2}(C^TC + K^TK)$ is $P = I$, which implies $\sum_{k=0}^{\infty} \frac{1}{2}(y_k^2 + u_k^2) = x_0^T x_0$.

(b) From part (a), the cost function is equal to $\sum_{i=0}^{\infty} \frac{1}{2}(y_{i|k}^2 + u_{i|k}^2)$ and the control input is $u_{i|k} = \frac{1}{\sqrt{2}} y_{i|k}$ for all $i \geq N$. Let $J^*(x_k)$ be the minimum value of this cost over $u_{0|k}, \ldots, u_{N-1|k}$, at time k. Then, at time $k+1$, the predicted input sequence $u_{i|k+1} = u_{i+1|k}$, $i = 0, 1, \ldots$ gives

$$J(x_{k+1}) = \sum_{i=1}^{\infty} \frac{1}{2}(y_{i|k}^2 + u_{i|k}^2) = J^*(x_k) - \frac{1}{2}(y_k^2 + u_k^2).$$

and since the optimal cost at time $k+1$ satisfies $J^*(x_{k+1}) \leq J(x_{k+1})$, we can conclude that $J^*(x_{k+1}) \leq J^*(x_k) - \frac{1}{2}(y_k^2 + u_k^2)$. This implies closed loop stability because $J^*(x_k)$ is positive definite in x_k since (A, C) is observable.

(c) The closed loop system will be stable if the predicted trajectories satisfy $-1 \leq y_{i|k} \leq 1$ for all $i \geq 0$. The constraints give $-1 \leq y_{i|k} \leq 1$ for $i = 0, 1, \ldots, N-1$ and

$$-1 \leq y_{N+i|k} = C(A + BK)^i x_{N|k} \leq 1, \quad i = 0, 1.$$

Here $C = \frac{1}{\sqrt{2}}\begin{bmatrix}1 & 1\end{bmatrix}$, $C(A + BK) = \frac{1}{\sqrt{2}}\begin{bmatrix}-1 & 1\end{bmatrix}$ and $(A + BK)^2 = -\frac{1}{2}I$. Therefore

$$\left.\begin{aligned} -1 \leq Cx \leq 1 \\ -1 \leq C(A + BK)x \leq 1 \end{aligned}\right\} \implies -1 \leq C(A + BK)^2 x \leq 1$$

Hence $-1 \leq C(A + BK)^i x \leq 1$ for all $i \geq 0$ which implies $-1 \leq y_{N+i|k} \leq 1$ for all $i \geq 0$ if $x = x_{N|k}$.

4 (a) The largest invariant set compatible with the constraints is given by

$$S_\nu = \{x : F\Phi^i x \leq \mathbf{1},\ i = 0, \ldots \nu\},\ F = \begin{bmatrix} 1 & -1 \\ -1 & 1 \end{bmatrix},\ \Phi = \begin{bmatrix} 0.42 & -0.025 \\ -0.16 & -0.35 \end{bmatrix}$$

where ν is such that $F\Phi^{\nu+1}x \leq \mathbf{1}$ for all $x \in S_\nu$. Since S_ν is symmetric about $x = 0$ this condition can be checked by solving the linear program: $\mu = \max_{x \in S_\nu} \begin{bmatrix} 1 & -1 \end{bmatrix} \Phi^{\nu+1} x$, and determining whether $\mu \leq 1$.

(b) To check $\nu = 1$:

$$\mu = \max_x \begin{bmatrix} 0.193 & -0.127 \end{bmatrix} x \text{ subject to } \begin{bmatrix} 1 & -1 \\ -1 & 1 \\ 0.578 & 0.324 \\ -0.578 & -0.324 \end{bmatrix} x \leq \begin{bmatrix} 1 \\ 1 \\ 1 \\ 1 \end{bmatrix}$$

gives $\mu = 0.224$, so $\mu \leq 1$ as required.

5 (a) Solving $W - \Phi^T W \Phi = I + K^T K$ for W with $K = \begin{bmatrix} 0.244 & 1.751 \end{bmatrix}$ gives

$$W = \begin{bmatrix} 1.33 & 0.58 \\ 0.58 & 4.64 \end{bmatrix}$$

and hence $-(B^T W B + 1)^{-1} B^T W A = \begin{bmatrix} 0.244 & 1.751 \end{bmatrix}$, which confirms that K is the LQ-optimal feedback gain.

(b) By construction $\begin{bmatrix} F & 0 \end{bmatrix} \Psi^i z_k \leq \mathbf{1}$ implies $F x_{i|k} \leq \mathbf{1}$ for $i = 0, \ldots, N + 1$. But $F x_{N|k} \leq \mathbf{1}$ and $F\Phi x_{N|k} \leq \mathbf{1}$ implies that $x_{N|k}$ lies in the invariant set of Question 4(b) and hence $F\Phi^i x_{N|k} \leq \mathbf{1}$ for all $i \geq 0$.

(c) The quadratic form of the cost, with $\rho = B^T W B + 1 = 6.56$, follows from Theorem 2.10.

(d) Since $u = Kx$ is the feedback law that minimizes the MPC cost index for the case of no constraints, and since the MPC cost is evaluated over an infinite horizon, there cannot be any reduction in the predicted cost when N is increased above the minimum value, say $\bar{N}$, for which the terminal constraints are inactive, i.e. $J_N^*(x_0) = J_{\bar{N}}^*(x_0)$ for all $N > \bar{N}$. This is likely to be the case for this initial condition with $\bar{N} = 9$ since the cost seems to have converged, with $J_9(x_0) = J_{10}(x_0)$.

If the terminal constraints are inactive, then the optimal predicted control sequence is optimal for an infinite mode 1 horizon and hence it must be equal to the closed loop control sequence generated by the receding horizon control law.

6 (a) Solving the SDP defining the invariant ellipsoid $\mathcal{E}_z = \{z : z^T P_z z \leq 1\}$ gives

$$P_z = \begin{bmatrix} 1.09 & -1.16 & -0.07 & 0.45 \\ -1.16 & 4.20 & -3.06 & 1.07 \\ -0.07 & -3.06 & 4.20 & -3.10 \\ 0.45 & 1.07 & -3.10 & 4.20 \end{bmatrix}.$$

The projection onto the x-subspace is $\mathcal{E}_x = \{x : x^T P_x x \leq 1\}$, where

$$P_x = \left(\begin{bmatrix} I & 0 \end{bmatrix} P_z^{-1} \begin{bmatrix} I \\ 0 \end{bmatrix} \right)^{-1} = \begin{bmatrix} 1.01 & -0.96 \\ -0.96 & 1.24 \end{bmatrix}$$

and the maximum value of α is $\alpha = 1/(v^T P_x v)^{1/2} = 1.79$, $v = \begin{bmatrix} 1 \\ 1 \end{bmatrix}$.

(b) Solving the linear program

$$\max_{\alpha, \mathbf{c}} \ \alpha \ \text{subject to} \ \begin{bmatrix} F & 0 \end{bmatrix} \Psi^i \begin{bmatrix} \alpha v \\ \mathbf{c} \end{bmatrix} \leq \mathbf{1}, \ i = 0, \ldots, N+1$$

with $v = \begin{bmatrix} 1 \\ 1 \end{bmatrix}$ gives $\alpha = 2.41$.

This value is necessarily greater than the value of α in (a) because the set $\mathcal{Z} = \{z : \begin{bmatrix} F & 0 \end{bmatrix} \Psi^i z \leq \mathbf{1}, \ i = 0, \ldots, N+1\}$ is the maximal invariant set for the dynamics $z_{k+1} = \Psi z_k$ and constraints $\begin{bmatrix} F & 0 \end{bmatrix} z_k \leq \mathbf{1}$, so it must contain $\mathcal{E}_z = \{z : z^T P_z z \leq 1\}$ as a subset, and therefore the projection of $\mathcal{E}_z$ onto the x-subspace must be a subset of the projection of $\mathcal{Z}$ onto the x-subspace.

(c) Solving the SDP for the invariant ellipsoidal set $\hat{\mathcal{E}}_z = \{z : z^T \hat{P}_z z \leq 1\}$ and the optimized prediction dynamics gives C_c and A_c as stated in the question and

$$\hat{P}_z = \begin{bmatrix} 2.43 & -1.31 & 1.36 & -1.31 \\ -1.31 & 3.12 & 1.21 & 3.12 \\ 1.36 & 1.21 & 2.45 & 1.21 \\ -1.31 & 3.12 & 1.21 & 8.88 \end{bmatrix}.$$

The projection onto the x-subspace is $\hat{\mathcal{E}}_x = \{x : x^T \hat{P}_x x \leq 1\}$, where

$$\hat{P}_x = \left(\begin{bmatrix} I & 0 \end{bmatrix} \hat{P}_z^{-1} \begin{bmatrix} I \\ 0 \end{bmatrix} \right)^{-1} = \begin{bmatrix} 1.19 & -1.38 \\ -1.38 & 1.75 \end{bmatrix}$$

and hence the maximum value of α is $\alpha = 1/(v^T P_x v)^{1/2} = 2.32$.

(d) With $\hat{\Psi}$ defined on the basis of the optimized prediction dynamics, the maximal invariant set for the dynamics $z_{k+1} = \hat{\Psi} z_k$ and constraints $\begin{bmatrix} F & 0 \end{bmatrix} z_k \leq \mathbf{1}$ is $\{z : \begin{bmatrix} F & 0 \end{bmatrix} \hat{\Psi}^i z_k \leq \mathbf{1}, \ i = 0, \ldots, 5\}$. Solving the LP

$$\max_{\alpha,\mathbf{c}} \ \alpha \ \text{subject to} \ \begin{bmatrix} F & 0 \end{bmatrix} \hat{\Psi}^i \begin{bmatrix} \alpha v \\ \mathbf{c} \end{bmatrix} \leq \mathbf{1}, \ i = 0, \ldots, 5$$

with $v = \begin{bmatrix} 1 \\ 1 \end{bmatrix}$ gives $\alpha = 3.82$.

(e) Computing the MPC cost for the optimized prediction dynamics by solving the Lyapunov equation (2.57) gives

$$J(x_k, \mathbf{c}_k) = \|x_k\|_W^2 + \mathbf{c}_k^T \begin{bmatrix} 106 & 32.8 \\ 32.8 & 10.2 \end{bmatrix} \mathbf{c}_k.$$

Solving $\min_{\mathbf{c}} J(x, \mathbf{c})$ subject to $\begin{bmatrix} F & 0 \end{bmatrix} \hat{\Psi}^i (x, \mathbf{c}) \leq \mathbf{1}$, $i = 0, \ldots, 5$ gives the optimal predicted cost for $x = (3.8, 3.8)$ as 1686. This is larger than (in fact more than double) the optimal predicted cost in Question 5(d) for $N = 9$, which is the minimum that can be obtained by any control sequence. The advantage of the optimized prediction dynamics is that the associated set of feasible initial conditions is almost identical to that of the MPC law in Question 5(d) despite using only 2 degrees of freedom rather than 9 degrees of freedom.

7 (a) In this case $\begin{bmatrix} 1 & 1 \end{bmatrix} (A + BK)x = 0$ for all $x \in \mathbb{R}^2$, and the eigenvalues of $A + BK$ lie inside the unit circle. It follows that there exists a stabilizing control law (namely $u = Kx$) such that, starting from the initial condition $x_0 = (\alpha, -\alpha)$ for arbitrarily large $|\alpha|$, the constraints $|\begin{bmatrix} 1 & 1 \end{bmatrix} (A + BK)^k x_0| \leq 1$ are satisfied for all k. Hence the maximal CPI set is unbounded (in fact it is equal to $\{x : |\begin{bmatrix} 1 & 1 \end{bmatrix} x| \leq 1\}$), and the maximal feasible initial condition set of an MPC law will increase monotonically with N.

(b) The transfer function from u_k to the constrained output $y_k = \begin{bmatrix} 1 & -1 \end{bmatrix} x_k$ is nonminimum-phase (its zero lies outside the unit circle at 1.33). Hence there is no stabilizing control law under which the constraints $|\begin{bmatrix} 1 & -1 \end{bmatrix} x_k| \leq \mathbf{1}$ are satisfied for all k when $\|x_0\|$ is arbitrarily large, in other words the maximal CPI set is bounded.

8 The predicted cost is

$$J_k = \mathbf{u}_k^T (\hat{R} + C_u^T \hat{Q} C_u) \mathbf{u}_k + 2\mathbf{u}_k C_u^T \hat{Q} C_x x_k + x_k^T C_x \hat{Q} C_x x_k,$$

and the optimal control sequence $\mathbf{u}_k^* = -(\hat{R} + C_u^T \hat{Q} C_u)^{-1} C_u^T \hat{Q} C_x x_k$ for the case of no constraints can be obtained by differentiation. The MPC law is given by the first element of this sequence, and is therefore a feedback law of the form $u_k = K_{(N,N_u)} x_k$.

For the given (A, B, C), computing the spectral radius of $A + BK_{(N,N_u)}$ for $N = 1, 2, \ldots$ and for $1 \leq N_u \leq N$ shows that $N = 9$ is the smallest output horizon for which stability can be achieved for any input horizon $N_u \leq N$. In fact for $N_u = 1$, stability is achieved only if $N \geq 10$.

The poles of the open loop system are at 0.693, 0.997, and thus both lie within the unit circle, whereas the system zero is at 1.16. This non-minimum phase zero implies that the predicted output sequence initially sets off in the wrong direction and this effect will be exacerbated at the next time step, when larger inputs will needed in order to return the output to the correct steady state. This indicates the tendency towards instability. For a sufficiently large output horizon (in this case for $N \geq 9$) the predicted cost can be shown to be monotonically non-increasing along closed loop system trajectories, indicating closed loop stability.

9 (a) The numerator and denominator polynomials of the system transfer function are given by

$$B(z^{-1}) = B_1 z^{-1} + B_0 = -0.6527z^{-1} + 0.5647$$
$$A(z^{-1}) = A_2 z^{-1} + A_1 z^{-1} + A_0 = 0.6908z^{-2} - 1.69z^{-1} + 1$$

Hence for the given $\tilde{X}(z^{-1})$ and $\tilde{Y}(z^{-1})$ polynomials we obtain $\tilde{Y}(z^{-1})A(z^{-1}) + z^{-1}\tilde{X}(z^{-1})B(z^{-1}) = 1$.

(b) For the single input single output case we have $\tilde{A}(z^{-1}) = A(z^{-1})$, $\tilde{B}(z^{-1}) = B(z^{-1})$, $\tilde{X}(z^{-1}) = X(z^{-1})$, $\tilde{Y}(z^{-1}) = Y(z^{-1})$ so that

$$C_{z^{-1}\tilde{X}} = \begin{bmatrix} 0 & 0 & 0 & 0 \\ X_0 & 0 & 0 & 0 \\ X_1 & X_0 & 0 & 0 \\ 0 & X_1 & X_0 & 0 \end{bmatrix} \qquad C_{\tilde{Y}} = \begin{bmatrix} Y_0 & 0 & 0 & 0 \\ Y_1 & Y_0 & 0 & 0 \\ 0 & Y_1 & Y_0 & 0 \\ 0 & 0 & Y_1 & Y_0 \end{bmatrix}$$

$$C_{\tilde{A}} = \begin{bmatrix} A_0 & 0 & 0 & 0 \\ A_1 & A_0 & 0 & 0 \\ A_2 & A_1 & A_0 & 0 \\ 0 & A_2 & A_1 & A_0 \end{bmatrix} \qquad C_{\tilde{B}} = \begin{bmatrix} B_0 & 0 & 0 & 0 \\ B_1 & B_0 & 0 & 0 \\ 0 & B_1 & B_0 & 0 \\ 0 & 0 & B_1 & B_0 \end{bmatrix}$$

where $X_0 = -32.2308$, $X_1 = 21.0529$, $Y_0 = 1$, $Y_1 = 19.89$. Inverting the 16×16 matrix consisting of these four blocks gives the matrix with blocks $C_A, C_Y, C_B, -C_{z^{-1}X}$, as required. This inverse can then be used to obtain the predicted output and control sequences from (2.72).

(c) From the predicted output and control sequences it is obvious that $y_{i+4|k} = 0$ and $u_{i+3|k} = 0$ for all $i \geq 1$. This implies that SGPC invokes (implicitly) a terminal equality constraint, and therefore the optimal predicted cost is monotonically non-increasing, from which it can be deduced that SGPC guarantees closed loop stability. The property that the predicted output and control sequences reach their steady state values of zero after $\nu + n_A$ prediction steps, where ν is the length of $\mathbf{c}_k$ and n_A the system order (i.e. here $\nu = n_A = 2$) is generally true given

Table A.1 Frequency responses of $S_a(\omega T)$ and $S_b(\omega T)$

ωT	0°	18°	36°	54°	72°	90°	108°	126°	144°	162°	180°
$\lvert S_a \rvert$	0.009	1.74	7.3	19.4	39.3	66.1	97.3	128.7	155.6	173.7	180.1
$\lvert S_b \rvert$	0.009	1.72	6.7	15.7	26.2	33.5	34.3	29.9	27.5	31.7	33.8

the structure of the convolution and Hankel matrices in the expression for the predicted output and control sequences. Hence, in the absence of constraints, SGPC ensures closed loop stability for any initial condition.

10 Denoting by S_a and S_b the transfer functions $K(z^{-1})/\big(1 + G(z^{-1})K(z^{-1})\big)$ corresponding to (a) $Q = 0$ and (b) $Q(z^{-1}) = -11.7z^{-1} + 43$, respectively, and denoting the sampling interval as T, we obtain the transfer function moduli given in Table A.1. These indicate that $|S_b(\omega T)| < |S_a(\omega T)|$ at all frequencies, and the ratio $|S_a(\omega T)|/|S_b(\omega T)|$ becomes larger at high frequencies ($|S_a(\omega T)|/|S_b(\omega T)| > 5$ for $\omega T > 140°$). Thus for $Q(z^{-1})$ as given in (b) the closed loop system will have enhanced robustness to additive uncertainty in the open loop system transfer function.

Solutions to Exercises for Chap. 3

1 (a) Two advantages of receding horizon control for this application:

- The receding horizon optimization is repeated at each time step, thus providing feedback (since the optimal predicted input sequence at k depends on the state x_k) and reducing the effect of the uncertainty in w_k.
- The optimization has to be performed over a finite number of free variables because of the presence of constraints. Using a receding horizon optimization reduces the degree of suboptimality with respect to the infinite horizon optimal control problem.

(b) With $u_{i|k} = \hat{w} - (x_{i|k} - x^0) + c_{i|k}$ we get $x_{i+1|k} - x^0 = \hat{w} - w_{k+i} + c_{i|k}$, so setting the disturbance equal to its nominal value, $w_{k+i} = \hat{w}$, gives $x_{i+1|k} - x^0 = s_{i+1|k} = c_{i|k}$ for all $i \geq 0$, and hence the nominal cost is $J(x_k, \mathbf{c}_k) = s_{0|k}^2 + \|\mathbf{c}_k\|^2$.

(c) Setting $s_{i|k} + e_{i|k} = x_{i|k} - x^0$ gives $e_{0|k} = 0$ and $e_{i+1|k} = \hat{w} - w_{k+i}$ for all $i \geq 0$. Hence $e_{i|k}$ lies in the interval $[\hat{w} - W, \hat{w}]$ for all $i \geq 0$. Using these bounds and $x_{i|k} = s_{i|k} + x^0 + e_{i|k}$, $u_{i|k} = \hat{w} - s_{i|k} - e_{i|k} + c_{i|k}$ we obtain

$$x_{i|k} \in [0, X] \iff \begin{cases} c_{i-1|k} + x^0 + \hat{w} \in [W, X], & 1 \leq i \leq N \\ W \leq X, & i > N \end{cases}$$

$$u_{i|k} \in [0, U] \iff \begin{cases} c_{0|k} - x_k + x^0 + \hat{w} \in [0, U - W], & i = 0 \\ c_{i|k} - c_{i-1|k} \in [0, U - W], & 1 \le i \le N - 1 \\ -c_{N-1|k} \in [0, U - W], & i = N \\ W \le U, & i > N \end{cases}$$

(d) By construction, $\mathbf{c}_{k+1} = (c^*_{1|k}, \ldots, c^*_{N-1|k}, 0)$ is feasible at time $k + 1$ if $\mathbf{c}^*_k$ is optimal at time k, so the problem is recursively feasible. Convergence of $c^*_{0|k}$ to zero as $k \to \infty$ then follows from the property that $\|c^*_{k+1}\|^2 \le \|c^*_k\|^2 - (c^*_{0|k})^2$, which implies that the l_2 norm of the sequence $\{c^*_{0|0}, c^*_{0|1}, \ldots\}$ is finite.

(e) The constraint $u_{i|k} \ge 0$ for $i = 1, \ldots, N$ requires $c_{i|k} \ge c_{i-1|k}$ and $c_{N-1} \le 0$. Hence $c_{0|k} \le 0$ so $u_{0|k} \ge 0$ requires $x_k \le x^0 + \hat{w}$. To relax this condition we need to use a less aggressive auxiliary control law, e.g. $u_{i|k} = \hat{w} - \alpha(x_{i|k} - x^0) + c_{i|k}$ for $0 < \alpha < 1$.

2 The structure of Ψ implies that, for any integer q,

$$\Psi^q = \begin{bmatrix} \Phi^q & \Gamma_q \\ 0 & M^q \end{bmatrix}$$

for some matrix Γ_q. Since $\Phi^n = 0$ and by construction $M^N = 0$, it follows that

$$\Psi^{n+N} = \Psi^n \Psi^N = \begin{bmatrix} \Gamma_n \\ M^n \end{bmatrix} \begin{bmatrix} 0 & I \end{bmatrix} \begin{bmatrix} I \\ 0 \end{bmatrix} \begin{bmatrix} \Phi^N & \Gamma_N \end{bmatrix} = 0.$$

(a) Since $\Phi^n = 0$ the minimal RPI set (3.23) is given by

$$\mathcal{X}^{\mathrm{mRPI}} = D\mathcal{W} \oplus \cdots \oplus \Phi^{n-1} D\mathcal{W}.$$

(b) Let $\bar{F} = \begin{bmatrix} F_1 & F_2 \end{bmatrix}$ and define $h_0 = 0$ and h_i for $i \ge 1$ as in (3.13):

$$h_i = \sum_{j=0}^{i-1} \max_{w_j \in \mathcal{W}} F_1 \Phi^j D w_j,$$

Since $h_i = h_n$ for all $i > n$ and $\Psi^{n+N} = 0$, the MRPI set (3.16) is

$$\mathcal{Z}^{\mathrm{MRPI}} = \{z : \bar{F}\Psi^i z \le \mathbf{1} - h_i, \ \ i = 0, \ldots, n + N\}.$$

(c) The MRPI set must contain the origin, while from (a) the maximum of $F_1 e$ over $e \in \mathcal{X}^{\mathrm{mRPI}}$ is h_n. Hence from (b) the MRPI set is non-empty if and only if

$$h_n \le \mathbf{1}.$$

3 For the given A, B and K:

$$\Phi = A + BK = 0.5 \begin{bmatrix} -1 & 1 \\ -1 & 1 \end{bmatrix}$$

and $\Phi^2 = 0$, which verifies that Φ is nilpotent. The constraint tightening parameters h_i that bound the effects of future disturbances on the constraints can be determined either by solving linear programs or by using the vertices of the disturbance set $\mathcal{W}$. In this example $\mathcal{W}$ has 4 vertices:

$$\mathcal{W} \doteq \text{Co}\left\{\begin{bmatrix} 0 \\ \sigma \end{bmatrix}, \begin{bmatrix} 0 \\ -\sigma \end{bmatrix}, \begin{bmatrix} \sigma \\ 0 \end{bmatrix}, \begin{bmatrix} -\sigma \\ 0 \end{bmatrix}\right\} = \text{Co}\{w^{(j)},\ j = 1, 2, 3, 4\}$$

Hence $h_0 = 0$ and

$$h_1 = \max_j Fw^{(j)} = \sigma\mathbf{1},$$
$$h_2 = h_1 + \max_j F(A + BK)w^{(j)} = 1.5\sigma\mathbf{1},$$

with $h_i = h_2$ for all $i > 2$. For $N = 2$ the MRPI set is therefore given by

$$\begin{aligned} \mathcal{Z}^{\text{MRPI}}(\sigma) = \{z :& \bar{F}z \le \mathbf{1}, \\ & \bar{F}\Psi z \le (1-\sigma)\mathbf{1}, \\ & \bar{F}\Psi^i z \le (1 - 1.5\sigma)\mathbf{1},\ \ i = 2, 3\} \end{aligned}$$

and $\mathcal{Z}^{\text{MRPI}}(\sigma)$ is non-empty if $1 - 1.5\sigma \ge 0$, i.e. $\sigma \le \frac{2}{3}$.

4 (a) Robust invariance of $\mathcal{Z}^{\text{MRPI}}(\sigma)$ implies that $(x_{k+1}, M\mathbf{c}_k) \in \mathcal{Z}^{\text{MRPI}}(\sigma)$ holds for all $w_k \in \sigma\mathcal{W}_0$ if $(x_k, \mathbf{c}_k^*) \in \mathcal{Z}^{\text{MRPI}}(\sigma)$. Hence if the MPC optimization is feasible at time k, then $\mathbf{c}_{k+1} = M\mathbf{c}_k^*$ is feasible at time $k + 1$. By optimality therefore $\|\mathbf{c}_{k+1}^*\|^2 \le \|\mathbf{c}_k^*\|^2 - (c_{0|k}^*)^2$, which implies that $\sum_{k=0}^{\infty}(c_{0|k}^*)^2 \le \|\mathbf{c}_0^*\|^2$ and hence $c_{0|k}^* \to 0$ as $k \to \infty$.
Lemma 3.2 implies that the state x_k of the closed loop system satisfies the quadratic bound (3.35) since $A + BK$ is stable, and since the sequence $\{c_{0|0}^*, c_{0|1}^*, \ldots\}$ is square-summable, the argument of Theorem 3.2 implies that x_k converges to the minimal RPI set:

$$\begin{aligned} \mathcal{X}^{\text{mRPI}}(\sigma) &= \sigma\mathcal{W}_0 \oplus (A + BK)\mathcal{W}_0 \\ &= \sigma\text{Co}\left\{\begin{bmatrix} 1 \\ 0 \end{bmatrix}, \begin{bmatrix} -1 \\ 0 \end{bmatrix}, \begin{bmatrix} 0 \\ 1 \end{bmatrix}, \begin{bmatrix} 0 \\ -1 \end{bmatrix}\right\} \oplus \text{Co}\left\{\begin{bmatrix} 0.5 \\ 0.5 \end{bmatrix}, \begin{bmatrix} -0.5 \\ -0.5 \end{bmatrix}\right\} \\ &= \sigma\text{Co}\left\{\begin{bmatrix} 0.5 \\ 1.5 \end{bmatrix}, \begin{bmatrix} 1.5 \\ 0.5 \end{bmatrix}, \begin{bmatrix} -0.5 \\ -1.5 \end{bmatrix}, \begin{bmatrix} -1.5 \\ -0.5 \end{bmatrix}\right\} \end{aligned}$$

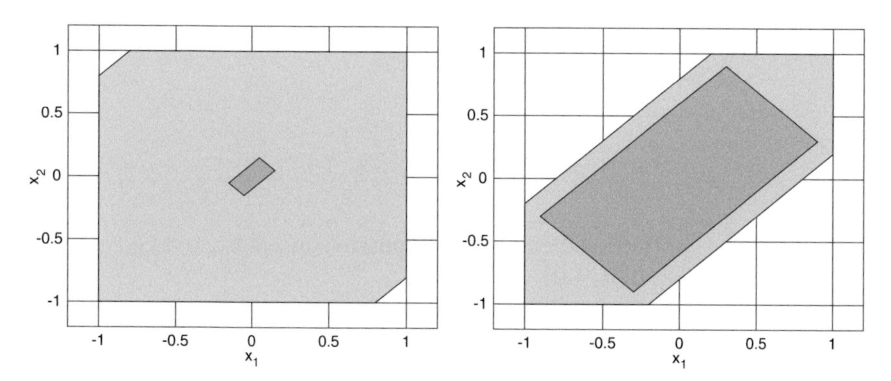

Fig. A.1 The set of feasible initial conditions for the MPC law of Question 4(a) (i.e. the projection $\mathcal{Z}^{\text{MRPI}}(\sigma)$ onto the x-subspace) and the minimal RPI set $\mathcal{X}^{\text{mRPI}}(\sigma)$ for $\sigma = 0.1$ (*left*) and $\sigma = 0.6$ (*right*)

The feasible initial condition sets and the minimal RPI sets for two values of σ are shown in Fig. A.1.

(b) Although the objective of the suggested MPC optimization problem is equal to the nominal predicted value of the cost, this is not a good suggestion since closed loop stability cannot, in general, be guaranteed with this combination of predicted cost and constraints. This is because the choice $\mathbf{c}_{k+1} = M\mathbf{c}_k^*$ does not ensure that the optimal predicted cost is monotonically non-increasing here due to the unknown disturbance that acts on the terms of the cost that depend on x_k. Note also that because K is not equal to the unconstrained optimal feedback gain, the predicted cost cannot be separated into terms that depend only on x_k and $\mathbf{c}_k$, as was done in the stability analysis of Sect. 3.3.

5 (a) Since $\Phi^r \mathcal{W} \subseteq \rho\mathcal{W}$ we have, for any $i \geq r$,

$$\max_{w\in\mathcal{W}} F\Phi^i w = \max_{w\in\Phi^r\mathcal{W}} F\Phi^{i-r} w \leq \max_{w\in\rho\mathcal{W}} F\Phi^{i-r} w = \rho \max_{w\in\mathcal{W}} F\Phi^{i-r} w,$$

which implies

$$\begin{aligned} h_\infty &= \sum_{j=0}^{\infty} \max_{w_j\in\mathcal{W}} F\Phi^j w_j \\ &\leq \sum_{j=0}^{r-1} \max_{w_j\in\mathcal{W}} F\Phi^j w_j + \rho \sum_{j=0}^{r-1} \max_{w_j\in\mathcal{W}} F\Phi^j w_j + \rho^2 \sum_{j=0}^{r-1} \max_{w_j\in\mathcal{W}} F\Phi^j w_j + \cdots \\ &= \frac{1}{1-\rho} \sum_{j=0}^{r-1} \max_{w_j\in\mathcal{W}} F\Phi^j w_j = \hat{h}_\infty. \end{aligned}$$

Also $h_\infty \geq h_r = \sum_{j=0}^{r-1} \max_{w_j \in \mathcal{W}} F\Phi^j w_j$ implies

$$\frac{\hat{h}_\infty - h_\infty}{h_\infty} \leq \frac{\hat{h}_\infty - h_r}{h_\infty} \leq \frac{\hat{h}_\infty - h_r}{h_r} = \frac{\rho}{1-\rho}.$$

(b) In order that the fractional error in the approximation of h_∞ is no greater than 0.01 we need $\rho \leq 0.01/1.01 = 0.0099$.
Two alternative methods of finding ρ such that $\Phi^r \mathcal{W} \subseteq \rho\mathcal{W}$ for given r: (i) Using the representation $\mathcal{W} = \{w : Vw \leq \mathbf{1}\}$, ρ is given by the maximum element of $\max_{w\in\mathcal{W}} V\Phi^r w$. (ii) Using the vertex representation, $\mathcal{W} = \mathrm{Co}\{w^{(1)}, \ldots, w^{(4)}\}$, the value of ρ is the maximum element of $\max_{j\in\{1,\ldots,4\}} V\Phi^r w^{(j)}$.
For the system of Question 4 with $K = \begin{bmatrix}0.479 & 0.108\end{bmatrix}$ we need $r = 7$ for $\rho \leq 0.0099$, and this gives $\rho = 0.0055$ and

$$\hat{h}_\infty = \begin{bmatrix}0.175 & 0.199 & 0.175 & 0.199\end{bmatrix}^T.$$

(c) An approximation to the minimal RPI set $\mathcal{X}^{\mathrm{mRPI}}$ is given by

$$\hat{\mathcal{X}}^{\mathrm{mRPI}} = \frac{1}{1-\rho} \bigoplus_{j=0}^{r-1} \Phi^j \mathcal{W}.$$

The discussion in (a) implies that $\hat{\mathcal{X}}^{\mathrm{mRPI}}$ contains $\mathcal{X}^{\mathrm{mRPI}}$ and, for any vector v, the support function $\max_e\{v^T e$ subject to $e \in \mathcal{X}^{\mathrm{mRPI}}\}$ is approximated by $\max_e\{v^T e$ subject to $e \in \hat{\mathcal{X}}^{\mathrm{mRPI}}\}$ with a fractional error no greater than $\rho/(1-\rho)$.

6 (a) For the given system parameters with $K = \begin{bmatrix}0.479 & 0.108\end{bmatrix}$ and $N = 1$,

$$\Psi = \begin{bmatrix} -0.521 & 0.308 & 1 \\ -0.489 & 0.596 & -0.5 \\ 0 & 0 & 0 \end{bmatrix}, \quad \bar{D} = \begin{bmatrix} 1 & 0 \\ 0 & 1 \\ 0 & 0 \end{bmatrix}, \quad \bar{F} = \begin{bmatrix} 1 & 0 & 0 \\ 0 & 1 & 0 \\ -1 & 0 & 0 \\ 0 & -1 & 0 \end{bmatrix}.$$

The maximal RPI set for the dynamics $z_{i+1|k} = \Psi z_{i|k} + \bar{D} w_{k+i}$ and constraint $\bar{F} z_{i|k} \leq \mathbf{1}$ is given by

$$\mathcal{Z}^{\mathrm{MRPI}} = \{z : \bar{F}\Psi^i z \leq \mathbf{1} - h_i,\ i = 0, 1, 2\},$$

$$\{h_0, h_1, h_2\} = \left\{ \begin{bmatrix}0\\0\\0\\0\end{bmatrix}, \begin{bmatrix}0.5\\0.5\\0.5\\0.5\end{bmatrix}, \begin{bmatrix}0.761\\0.798\\0.761\\0.798\end{bmatrix} \right\}.$$

since $\bar{F}\Psi^3 z \leq \mathbf{1} - h_3$ holds for all $z \in \mathcal{Z}^{\text{MRPI}}$. Since this is the maximal RPI set for the prediction dynamics, its projection onto the x-subspace:

$$\mathcal{F}_1 = \{x : \exists \mathbf{c}_k \text{ such that } (x_k, \mathbf{c}_k) \in \mathcal{Z}^{\text{MRPI}}\}$$

must be equal to the maximal set of feasible model states x_k.

(b) Solving $W_z - \Psi^T W_z \Psi = \bar{Q}$ for W_z, where $\bar{Q} = \text{diag}\{I, 0\} + \begin{bmatrix} K & 1 \end{bmatrix}^T \begin{bmatrix} K & 1 \end{bmatrix}$:

$$W_z = \begin{bmatrix} 1.87 & -0.48 & 0.00 \\ -0.48 & 1.57 & 0.00 \\ 0.00 & 0.00 & 3.74 \end{bmatrix}.$$

With $x_0 = (0, 1)$ the optimal solution of the QP,

$$\underset{\mathbf{c}_0}{\text{minimize}} \quad \left\| \begin{bmatrix} x_0 \\ \mathbf{c}_0 \end{bmatrix} \right\|_{W_z}^2 \quad \text{subject to } \bar{F}\Psi^i \begin{bmatrix} x_0 \\ \mathbf{c}_0 \end{bmatrix} \leq \mathbf{1} - h_i, \ i = 0, 1, 2$$

is $\mathbf{c}_0^* = 0.192$ and the corresponding cost value is $z_0^T W_z z_0 = 1.707$.

(c) With the disturbance sequence as given in the question, the closed loop state and control sequences are

$$\{x_0, x_1, x_2, x_3, \ldots\} = \left\{ \begin{bmatrix} 0 \\ 1 \end{bmatrix}, \begin{bmatrix} 1 \\ 0.5 \end{bmatrix}, \begin{bmatrix} -0.867 \\ -0.192 \end{bmatrix}, \begin{bmatrix} 0.393 \\ 0.810 \end{bmatrix} \right\}$$
$$\{u_0, u_1, u_2, u_3, \ldots\} = \{0.3, \ 0.533, \ -0.436, \ 0.276\}$$

and hence $\sum_{k=0}^{3} \big(\|x_k\|^2 + u_k^2\big) = 4.489$.

7 (a) The solution $\check{W}_x$ of the Riccati equation (3.42) and the optimal gain K given in the question can be computed by using semidefinite programming to minimize $\text{tr}(\check{W}_x)$ subject to (3.43) and $\left[\begin{smallmatrix} \check{W}_x & I \\ I & S \end{smallmatrix}\right] \succeq 0$, with the value of γ^2 fixed at 3.3. Lemma 3.3 then implies that $\check{J}(x_k, \mathbf{c}_k) = \|x_k\|_{W_x}^2 + \|\mathbf{c}_k\|_{W_c}^2$ where

$$W_x = \check{W}_x = \begin{bmatrix} 2.336 & -0.904 \\ -0.904 & 2.103 \end{bmatrix},$$
$$W_c = B^T \big(\check{W}_x + \check{W}_x(\gamma^2 I - \check{W}_x)^{-1} \check{W}_x\big) B + 1 = 72.78.$$

(b) Repeating the procedure of Question 6(a) with the new value of K we get $\nu = 2$, so the online MPC optimization becomes the QP

$$\mathbf{c}_k^* = \arg \min_{\mathbf{c}_k} \ \|\mathbf{c}_k\|_{W_c}^2 \quad \text{subject to } \bar{F}\Psi^i \begin{bmatrix} x_k \\ \mathbf{c}_k \end{bmatrix} \leq \mathbf{1} - h_i, \ i = 0, 1, 2$$

with $\bar{F}$ as defined in Question 6 and

$$\Psi = \begin{bmatrix} -0.460 & 0.449 & 1 \\ -0.520 & 0.526 & -0.5 \\ 0 & 0 & 0 \end{bmatrix}, \quad \{h_0, h_1, h_2\} = \left\{ \begin{bmatrix} 0 \\ 0 \\ 0 \\ 0 \end{bmatrix}, \begin{bmatrix} 0.5 \\ 0.5 \\ 0.5 \\ 0.5 \end{bmatrix}, \begin{bmatrix} 0.730 \\ 0.763 \\ 0.730 \\ 0.763 \end{bmatrix} \right\}.$$

Solving this QP with $x_0 = (0, 1)$ is $\mathbf{c}_0^* = 0.051$ and $\check{J}^*(x_0) = 2.29$.

(c) From Lemma 3.4 we get

$$\Delta = \begin{bmatrix} 0.964 & 0.904 \\ 0.904 & 1.197 \end{bmatrix},$$

$$W_{\mu z} = \begin{bmatrix} -1.29 & 1.25 & 4.17 \\ 1.29 & -1.25 & -4.17 \\ -0.01 & -0.61 & 56.6 \\ 0.01 & 0.61 & -56.6 \end{bmatrix}, \quad W_{\mu\mu} = \begin{bmatrix} 4.19 & -4.19 & 2.78 & -2.78 \\ -4.19 & 4.19 & -2.78 & 2.78 \\ 2.78 & -2.78 & 47.2 & -47.2 \\ -2.78 & 2.78 & -47.2 & 47.2 \end{bmatrix},$$

and hence the predicted cost is given by

$$\check{J}(x_k, \mathbf{c}_k) = \min_{\mu_k \geq 0} \begin{bmatrix} x_k \\ \mathbf{c}_k \\ \mu_k \end{bmatrix}^T \begin{bmatrix} \begin{bmatrix} W_x & 0 \\ 0 & W_c \end{bmatrix} & -W_{\mu z}^T \\ -W_{\mu z} & W_{\mu\mu} \end{bmatrix} \begin{bmatrix} x_k \\ \mathbf{c}_k \\ \mu_k \end{bmatrix} + 2\mu^T \mathbf{1}.$$

With this cost and $x_0 = (0, 1)$ the solution of the online MPC optimization is $\mathbf{c}_0^* = 0.051$ and $\check{J}^*(x_0) = 2.22$.

(d) The cost in (c) is the maximum over $\{w_k, w_{k+1}, \ldots\}$ subject to the disturbance bounds $w_{k+i} \in \mathcal{W}$ for $i = 0, \ldots, N-1$. Therefore its minimum value over $\mathbf{c}_k$ is no greater than that of the cost in (b) for any given x_k. For this example $x_0 = (0, 1)$ lies on the boundary of the feasible set and this is why the optimal $\mathbf{c}_0$ is the same for both costs.
By taking into account the disturbance bounds over the first N predicted times steps, the cost of (c) provides a tighter worst case bound and hence is more representative of the worst case predicted performance than (b). On the other hand, the cost of (c) requires $5N$ optimization variables rather than N for (b). Also (b) ensures convergence to a limit set that is relatively easy to compute, namely the minimal RPI set for $e_{k+1} = \Phi e_k + w_k$, $w_k \in \mathcal{W}$, whereas this is not necessarily the case for (c). In this example however, the closed loop state sequences under the MPC laws for (b) and (c) are identical. This is illustrated in Fig. A.2, which compares the closed loop evolution of the state under the two MPC laws for a random disturbance sequence.

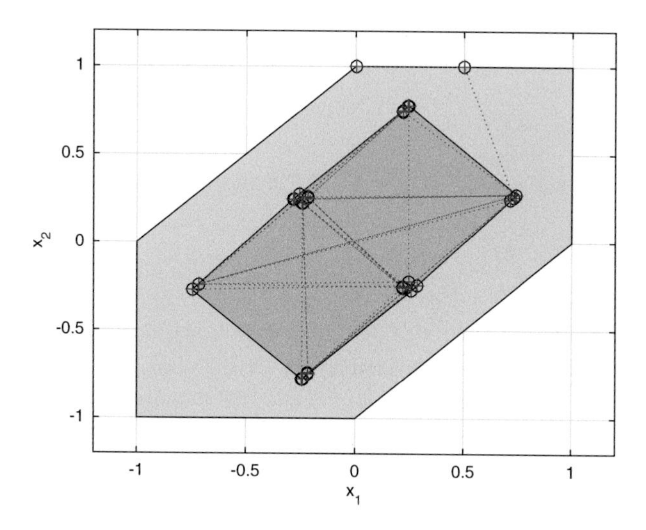

Fig. A.2 The evolution of the closed loop system state for the MPC laws in Question 7 part (b) (*marked* ∘) and part (c) (*marked* +) for a random sequence of disturbances $\{w_0, w_1, \ldots\}$ in which w_k is equal to a vertex of $\mathcal{W}$ for all k. Also shown are the set of feasible states and the minimal RPI set for $e_{k+1} = \Phi e_k + w_k$, $w_k \in \mathcal{W}$

8 (a) The minimum ρ such that $\Phi^2 \mathcal{W} \subseteq \rho \mathcal{W}$ is equal to the largest element of $\max_{w \in \mathcal{W}} V \Phi^2 w$. Using the disturbance set representation

$$\mathcal{W} = \{w : Vw \leq \mathbf{1}\}, \quad V = \begin{bmatrix} 2.5 & 2.5 \\ -2.5 & -2.5 \\ -2.5 & 2.5 \\ 2.5 & -2.5 \end{bmatrix}$$

we can compute ρ by solving a linear program:

$$\rho = \min_{t,w} t \quad \text{subject to} \quad \begin{aligned} V\Phi^2 w &\leq t\mathbf{1} \\ Vw &\leq \mathbf{1} \end{aligned}$$

Hence $\rho = 0.228$ and $\mathcal{S} = 1.295(\mathcal{W} \oplus \Phi \mathcal{W})$. With this expression for $\mathcal{S}$ we can obtain $h_{\mathcal{S}} = \max_{e \in \mathcal{S}} Fe$ by solving a set of linear programs (one pair of linear programs for each element of $h_{\mathcal{S}}$):

$$h_{\mathcal{S}} = 1.295 \Big(\max_{w : Vw \leq \mathbf{1}} Fw + \max_{w : Vw \leq \mathbf{1}} F\Phi w \Big) = \begin{bmatrix} 0.788 \\ 0.827 \\ 0.788 \\ 0.827 \end{bmatrix}.$$

(b) With the numerical values of Ψ and $\bar{F}$ given in the solution of Question 6(a), we get

$$\max_{z}\left\{\bar{F}\Psi^3 z \text{ subject to } \bar{F}\Psi^i z \leq \mathbf{1} - h_{\mathcal{S}},\ i = 0, 1, 2\right\} = \begin{bmatrix} 0.030 \\ 0.041 \\ 0.030 \\ 0.041 \end{bmatrix} \leq \mathbf{1} - h_{\mathcal{S}}$$

which implies that $\{z : \bar{F}\Psi^i z \leq \mathbf{1} - h_{\mathcal{S}},\ i = 0, 1, 2\}$ is the maximal invariant set for the nominal prediction system $z_{k+1} = \Psi z_k$ and constraints $\bar{F} z_k \leq \mathbf{1} - h_{\mathcal{S}}$.

(c) The predicted state is decomposed as $x_{i|k} = s_{i|k} + e_{i|k}$ where $e_{i|k} \in \mathcal{S}$ for all $i \geq 0$ and $s_{i|k}$ evolves according to the nominal dynamics (in which $w_{k+i} = 0$ for all $i \geq 0$ so that $s_{i|k} = \begin{bmatrix} I & 0 \end{bmatrix} z_{i|k}$ where $z_{i+1|k} = \Psi z_{i|k}$). This decomposition allows the constraints $F x_{i|k} \leq \mathbf{1}$ to be imposed robustly through the conditions $\bar{F}\Psi^i z_k \leq \mathbf{1} - h_{\mathcal{S}},\ i = 0, 1, 2$ with $z_k = (s_{0|k}, \mathbf{c}_k)$. Thus $s_{0|k}$ is the initial state of the nominal prediction system, the predicted trajectories of which are constrained so that $F(s_{i|k} + e_{i|k}) \leq \mathbf{1}$ for all $i \geq 0$.
To ensure that $e_{i|k} \in \mathcal{S}$ for all $i \geq 0$, we require that $s_{0|k}$ satisfies $e_{0|k} = x_k - s_{0|k} \in \mathcal{S}$. Using the hyperplane description of $\mathcal{W}$:

$$\mathcal{W} = \text{Co}\{w^{(j)},\ j = 1, \ldots, 4\} = \text{Co}\left\{\pm\begin{bmatrix} 0.4 \\ 0 \end{bmatrix}, \pm\begin{bmatrix} 0 \\ 0.4 \end{bmatrix}\right\}$$

it is possible to compute $\mathcal{S}$ as the convex hull of the 16 vertices that are formed from $w^{(i)} + \Phi w^{(j)}$ for all $i, j = 1, \ldots, 4$. This convex hull has 8 vertices:

$$\mathcal{S} = \text{Co}\left\{\pm\begin{bmatrix} 0.788 \\ 0.254 \end{bmatrix}, \pm\begin{bmatrix} 0.270 \\ 0.772 \end{bmatrix}, \pm\begin{bmatrix} 0.159 \\ 0.827 \end{bmatrix}, \pm\begin{bmatrix} -0.359 \\ 0.309 \end{bmatrix}\right\}$$

In order to invoke the constraint $x_k - s_{0|k} \in \mathcal{S}$ in a manner that avoids introducing additional optimization variables, we need to compute the hyperplane representation:

$$\mathcal{S} = \{e : V_{\mathcal{S}} e \leq \mathbf{1}\}, \quad V_{\mathcal{S}} = \begin{bmatrix} -0.960 & -0.960 \\ -0.551 & -1.103 \\ 1.498 & -1.498 \\ 1.681 & -1.284 \\ 0.960 & 0.960 \\ 0.551 & 1.103 \\ -1.498 & 1.498 \\ -1.681 & 1.284 \end{bmatrix}.$$

For $\mathcal{S} \subset \mathbb{R}^2$ with a small number of vertices this can done simply by plotting the vertices to determine which lie on each facet of $\mathcal{S}$, then computing the corresponding hyperplanes. For more complicated sets in higher dimensions, dedicated software (for example lrslib [1], cddlib [2]) can be used to convert between vertex and hyperplane representations.

(d) The objective function of the MPC optimization is $\|(s_{0|k}, \mathbf{c}_k)\|^2_{W_z}$, where $W_z = \text{diag}\{W_x, W_c\}$ is as given in the solution to Question 6(c). The MPC optimization is then

$$\underset{s_{0|k},\mathbf{c}_k}{\text{minimize}} \quad \left\| \begin{bmatrix} s_{0|k} \\ \mathbf{c}_k \end{bmatrix} \right\|^2_{W_z} \quad \text{subject to} \quad \begin{aligned} \bar{F}\Psi^i \begin{bmatrix} s_{0|k} \\ \mathbf{c}_k \end{bmatrix} &\leq \mathbf{1} - h_{\mathcal{S}}, \ i = 0, 1, 2 \\ V_{\mathcal{S}}(x_k - s_{0|k}) &\leq \mathbf{1} \end{aligned}$$

For $x_0 = (0, 1)$ the optimal solution is given by $s^*_{0|0} = (-0.159, 0.173)$, $\mathbf{c}^*_0 = 0.0163$ and $\|s^*_{0|0}\|^2_{W_x} + \|\mathbf{c}^*_0\|^2_{W_c} = 0.122$.

(e) Rigid tubes provide conservative bounds on the effects of disturbances on predicted trajectories, thus reducing the size of the feasible set of states relative to e.g. the robust MPC strategy of Question 6. This effect can be countered (but not eliminated, in general) by choosing K_e to improve the disturbance rejection properties of the system $e_{k+1} = (A + BK_e)e_k + w_k$, thus reducing the size of $\mathcal{S}$ and allowing tighter bounds on the effects of disturbances on constrained variables.

9 (a) From $\Phi\mathcal{S} \oplus \mathcal{W} \subseteq \mathcal{S}$, where $\mathcal{S} = \{s : V_{\mathcal{S}}s \leq \mathbf{1}\}$, we have

$$V_{\mathcal{S}}\Phi e + V_{\mathcal{S}}w \leq V_{\mathcal{S}}e \leq \mathbf{1}, \quad \forall e \in \mathcal{S}, \ \forall w \in \mathcal{W}.$$

This condition is equivalent to $\bar{e} + \bar{w} \leq \mathbf{1}$.

(b) Let

$$\mathcal{Z}^{(\nu)} = \big\{(z, \boldsymbol{\alpha}) : \bar{F}\Psi^i z \leq \mathbf{1} - \alpha_i h_{\mathcal{S}}, \ \alpha_i \bar{e} + \bar{w} \leq \alpha_{i+1}, \ i = 0, \ldots, \nu\big\}$$

where $\boldsymbol{\alpha} = (\alpha_0, \ldots, \alpha_{N-1})$ and $\alpha_i = 1$ for all $i \geq N$. If $\nu \geq N - 1$ satisfies $\bar{F}\Psi^{\nu+1} z \leq h_{\mathcal{S}}$ for all $(z, \boldsymbol{\alpha}) \in \mathcal{Z}^{(\nu)}$, then from Theorem 2.3 it follows that $\mathcal{Z}^{(\nu)}$ is an invariant set (in fact it is the MPI set) for the dynamics $z^+ = \Psi z$ and $\boldsymbol{\alpha}^+ = (\alpha_1, \ldots, \alpha_{N-1}, 1)$, and the constraints $\bar{F}z \leq \mathbf{1} - \alpha_0 h_{\mathcal{S}}$.
From the robust invariance property of $\mathcal{S}$ and the definition of $\mathcal{Z}^{(\nu)}$, if $x_k - s_{0|k} \in \alpha_{0|k}\mathcal{S}$ and $(z_k, \boldsymbol{\alpha}_k) \in \mathcal{Z}^{(\nu)}$ where $z_k = (s_{0|k}, \mathbf{c}_k)$, then by construction $x_{k+1} - s_{1|k} \in \alpha_{1|k}\mathcal{S}$ for all $w_k \in \mathcal{W}$. Furthermore the invariance of $\mathcal{Z}^{(\nu)}$ implies $\big(\Psi z_k, (\alpha_{1|k}, \ldots, \alpha_{N-1|k}, 1)\big) \in \mathcal{Z}^{(\nu)}$. Hence

$$\begin{bmatrix} s_{0|k+1} \\ \mathbf{c}_{k+1} \end{bmatrix} = \Psi \begin{bmatrix} s_{0|k} \\ \mathbf{c}_k \end{bmatrix}$$
$$\boldsymbol{\alpha}_k + 1 = (\alpha_{1|k}, \ldots, \alpha_{N-1|k}, 1)$$

satisfy $x_{k+1} - s_{0|k+1} \in \alpha_{0|k+1}\mathcal{S}$ and $(z_{k+1}, \boldsymbol{\alpha}_k + 1) \in \mathcal{Z}^{(\nu)}$ at time $k + 1$.

(c) From part (b), feasibility at $k = 0$ implies feasibility at all times $k > 0$. Therefore the definition of $h_{\mathcal{S}}$ and the constraints $x_k - s_{0|k} \in \alpha_{0|k}\mathcal{S}$ and $\bar{F}z_k = Fs_{0|k} \leq \mathbf{1} - \alpha_{0|k}h_{\mathcal{S}}$ imply that $Fx_k \leq \mathbf{1}$ for all $k \geq 0$.
From the feasible solution in part (b) we obtain the cost bound:

$$
\begin{aligned}
J(s^*_{0|k+1}, \mathbf{c}^*_{k+1}, {}^*_{k+1}) &\leq \begin{bmatrix} s^*_{0|k} \\ \mathbf{c}^*_k \end{bmatrix}^T \Psi^T W_z \Psi \begin{bmatrix} s^*_{0|k} \\ \mathbf{c}^*_k \end{bmatrix} + \sum_{i=1}^{N-1} q_\alpha(\alpha^*_{i|k} - 1)^2 \\
&= \begin{bmatrix} s^*_{0|k} \\ \mathbf{c}^*_k \end{bmatrix}^T \left(W_z - \begin{bmatrix} Q & 0 \\ 0 & 0 \end{bmatrix} - \begin{bmatrix} K^T \\ E^T \end{bmatrix} R \begin{bmatrix} K & E \end{bmatrix} \right) \begin{bmatrix} s^*_{0|k} \\ \mathbf{c}^*_k \end{bmatrix} \\
&\quad + \sum_{i=1}^{N-1} q_\alpha(\alpha^*_{i|k} - 1)^2 \\
&= J(s^*_{0|k}, \mathbf{c}^*_k, {}^*_k) - (\|s_{0|k}\|^2_Q + \|v_{0|k}\|^2_R) - q_\alpha(\alpha^*_{0|k} - 1)^2.
\end{aligned}
$$

Using, for example, the argument of the proof of Theorem 3.6, it then follows that $x_k \to \mathcal{S}$ asymptotically as $k \to \infty$ (in fact the minimum distance from x_k to any point in $\mathcal{S}$ decays exponentially with k).

10 (a) The solution of Question 8 gives $\rho = 0.228$ and $\mathcal{S} = 1.295(\mathcal{W} \oplus \Phi\mathcal{W})$, thus allowing $V_{\mathcal{S}}$ to be determined so that $\mathcal{S} = \{s : V_{\mathcal{S}}s \leq \mathbf{1}\}$. Hence by solving a set of linear programs (one LP for each element of $\bar{w}$ and $\bar{e}$) we obtain

$$
\bar{e} = \max_{e: V_{\mathcal{S}}e \leq \mathbf{1}} V_{\mathcal{S}}\Phi e = \begin{bmatrix} 0.616 & 0.552 & 0.365 & 0.259 & 0.616 & 0.552 & 0.365 & 0.259 \end{bmatrix}^T
$$

$$
\bar{w} = \max_{w: Vw \leq \mathbf{1}} V_{\mathcal{S}}w = \begin{bmatrix} 0.384 & 0.441 & 0.599 & 0.673 & 0.384 & 0.441 & 0.599 & 0.673 \end{bmatrix}^T
$$

For $N = 1$ and the numerical values of Ψ, $\bar{F}$ given in the solution of Question 6(a) we obtain

$$
\max_{z,} \left\{ \bar{F}\Psi^3 z \ \text{ subject to } (z,) \in \mathcal{Z}^{(2)} \right\} = \begin{bmatrix} 0.030 \\ 0.041 \\ 0.030 \\ 0.041 \end{bmatrix} \leq \mathbf{1} - h_{\mathcal{S}}
$$

and hence $\nu = 2$ satisfies the conditions for invariance of $\mathcal{Z}^{(\nu)}$.

(b) Solving the QP that defines the MPC optimization at $k = 0$ with $x_0 = (0, 1)$ we get

$$
s^*_{0|0} = (-0.159, 0.173), \quad \mathbf{c}^*_0 = 0.0163, \quad {}^*_0 = 1.0,
$$

and hence $\mathbf{c}^*_0$ is equal to the optimal solution for rigid tube MPC for the same initial condition. The explanation for this is that the HTMPC online optimization places a penalty on $|1 - \alpha_{i|k}|$ and hence its optimal solution will be equal to that of rigid tube MPC if it is feasible.

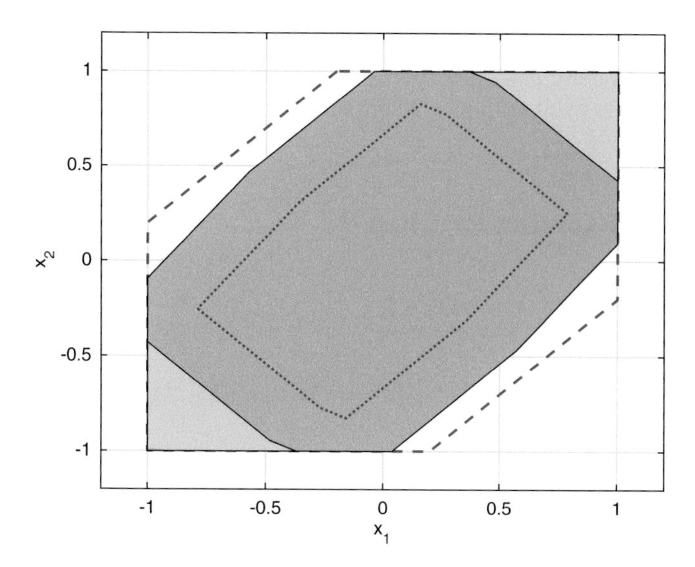

Fig. A.3 The feasible sets for the rigid and homothetic tube MPC laws in Questions 8 and 9, *shaded dark* and *light blue* respectively. Also shown are the set of feasible states for the MPC law of Question 6 (*outer, red dashed line*) and the minimal RPI set for $e_{k+1} = \Phi e_k + w_k$, $w_k \in \mathcal{W}$ (*inner, red dotted line*)

(c) Comments on the size of the feasible sets for these algorithms:

- The feasible initial condition set of the rigid tube MPC strategy is necessarily a subset of that for HTMPC, since if $\alpha_{i|k} = 1$ for all i, then the constraints of HTMPC are identical to those of rigid tube MPC. Since it is able to scale the uncertainty set $\mathcal{S}$, HTMPC may also be feasible for initial conditions for which the rigid tube MPC is infeasible.
- The feasible initial condition set of HTMPC is itself a subset of the robust MPC strategy of Question 6 (namely Algorithm 3.1), since this approach employs the tightest available bounds on the unknown future disturbances appearing in the constraints on predicted variables, whereas the corresponding bounds in HTMPC are computed using outer (possibly conservative) bounds based on $\alpha_{i|k}\mathcal{S}$.

  This nested property can be seen in Fig. A.3, which shows the feasible sets for the numerical examples in Questions 6, 8 and 9.
  Comments on performance:

- As mentioned in the solution to (b), the performance of the HTMPC and rigid tube MPC algorithms of Questions 8 and 9 are identical for every initial condition such that rigid tube MPC is feasible.
- The predicted cost for HTMPC must be greater than or equal to that of the robust MPC strategy of Question 6 since the overbounding of the constraints in HTMPC implies that any feasible $\mathbf{c}_k$ for HTMPC is also feasible for the strategy of Question 6. A possible advantage of the algorithms of Questions 7 and 8 over Question 6

is that they ensure exponential convergence to an outer approximation of the mRPI set, while the control law Question 6 ensures convergence of the state to the mRPI set, but not necessarily exponential convergence.

Solutions to Exercises for Chap. 5

1 (a) Each element of $M(x) = M_0 + x_1 M_1 + \cdots + x_n M_n$ is an affine function of x. For any $y, z \in \mathbb{R}^n$ and any scalar λ we therefore have

$$M\big(\lambda y + (1-\lambda)z\big) = \lambda M(y) + (1-\lambda)M(z).$$

Suppose that $M(y) \succ 0$ and $M(z) \succ 0$, so that $v^T M(y) v > 0$ and $v^T M(z) v > 0$ for all vectors $v \neq 0$. Then for all $0 \leq \lambda \leq 1$ we have

$$v^T M\big(\lambda y + (1-\lambda)z\big)v = \lambda v^T M(y)v + (1-\lambda)v^T M(z)v > 0$$

for all $v \neq 0$. This implies that, if $x = \lambda y + (1-\lambda)z$, then $M(x) \succ 0$ for all $0 \leq \lambda \leq 1$ and thus demonstrates that $M(x) \succ 0$ is a convex condition on x.

(b) The matrix $\begin{bmatrix} P & Q \\ Q^T & P \end{bmatrix}$ is positive definite if

$$\begin{bmatrix} v \\ w \end{bmatrix}^T \begin{bmatrix} P & Q \\ Q^T & R \end{bmatrix} \begin{bmatrix} v \\ w \end{bmatrix} > 0 \qquad (\star)$$

holds for all $(v, w) \neq 0$. Consider the cases of $v = 0$ and $w \neq 0$ separately:

(I) If $v = 0$, then the condition $(\star)$ is equivalent to $w^T R w > 0$ for all $w \neq 0$, i.e. $R \succ 0$.

(II) For non-zero v, consider the minimum of the LHS of $(\star)$ over all w. This is achieved with $w = -R^{-1}Qv$ so that

$$\min_w (v^T P v + 2v^T Q w + w^T R w) = v^T (P - Q^T R^{-1} Q) v.$$

Hence $(\star)$ holds for all $v \neq 0$ if and only if $P - Q^T R^{-1} Q \succ 0$.

From (I) and (II) we conclude that $(\star)$ is equivalent to the Schur complements $R \succ 0$ and $P - Q^T R^{-1} Q \succ 0$.

(c) Pre- and post-multiplying $P - A^T P A \succ 0$ by $S = P^{-1} > 0$ gives the equivalent inequality $S - SA^T S^{-1} AS \succ 0$. Using Schur complements (noting that $S \succ 0$), this is equivalent to

$$\begin{bmatrix} S & SA^T \\ AS & S \end{bmatrix} \succ 0$$

Using Schur complements again gives the equivalent conditions:

$$S - ASS^{-1}SA^T \succ 0, \quad S \succ 0$$

i.e. $S - ASA^T \succ 0$ whenever $P - A^T PA \succ 0$ if $S = P^{-1} \succ 0$.

2 (a) For $x \in \mathcal{X}$ we require $V_i x \leq 1$ for $i = 1, \dots, n_V$. Inserting P and $S = P^{-1}$ into these conditions gives $V_i S^{1/2} P^{1/2} x \leq 1$, so the Cauchy-Schwarz inequality implies $V_i S^{1/2} P^{1/2} x \leq (V_i S V_i^T)^{1/2} (x^T P x)^{1/2}$, which immediately shows that the conditions

$$V_i P^{-1} V_i^T \leq 1, \quad i = 1, \dots, n_V$$

are sufficient to ensure that $V_i x \leq 1, i = 1, \dots, n_V$ for all $x \in \mathcal{E}$. However these conditions are also necessary for $\mathcal{E} \subseteq \mathcal{X}$ because the maximum of $V_i x$ subject to $x^T P x \leq 1$ is equal to the quantity on the LHS since

$$\max_{\substack{x \\ x^T P x \leq 1}} V_i x = \max_{\substack{\xi \\ \|\xi\| \leq 1}} V_i P^{-1/2} \xi = V_i P^{-1} V_i^T.$$

(b) The result of part (a) implies $(F_i + G_i K)x \leq 1$ for all x such that $x^T S^{-1} x \leq 1$ if and only if $1 - (F_i + G_i K) S (F_i + G_i K)^T \geq 0$. Using Schur complements, the last condition is equivalent to $S \succ 0$ and

$$\begin{bmatrix} 1 & F_i S + G_i Y \\ (F_i S + G_i Y)^T & S \end{bmatrix} \succeq 0$$

for $i = 1, \dots, n_C$.

3 (a) The conditions given in the question ensure that:

(i) $x_k \in \mathcal{E}$, where $\mathcal{E} = \{x : x^T P x \leq 1\}$,
(ii) $-1 \leq u_{i|k} \leq 1$ for all $i \geq 0$ if $x_k \in \mathcal{E}$ and $u_{i|k} = K_k x_{i|k}$,
(iii) $\check{J}(x_k, K_k) \leq \gamma_k x_k^T P x_k \leq \gamma_k$ if $x_k \in \mathcal{E}$ and $u_{i|k} = K_k x_{i|k}$.

(b) When expressed in terms of the variables $S = P^{-1}$ and $Y = K_k P^{-1}$, the optimization becomes the semidefinite programming problem given in (5.18). Solving this using the model parameters given in the question and $x_0 = (4, -1)$ results in

$$\gamma_0 = 152.4, \quad S = \begin{bmatrix} 16.24 & -3.932 \\ -3.932 & 1.019 \end{bmatrix}, \quad Y = \begin{bmatrix} -1.170 & 0.034 \end{bmatrix}$$

and therefore $K_0 = \begin{bmatrix} -0.962 & -3.678 \end{bmatrix}$.

(c) The constraints on Θ ensure that $\|x_{i+1|k}\|_\Theta^2 \leq \|x_{i|k}\|_\Theta^2 - \|x_{i|k}\|_Q^2 - \|u_{i|k}\|_R^2$ holds along all predicted trajectories of the model under the predicted control law $u_{i|k} = K_k x_{i|k}$. Therefore the worst case predicted cost has the upper bound

$\check{J}(x_k, K_k) \leq \|x_k\|^2_{\Theta}$. The cost bound in (c) will in general be smaller than the bound in (a) because $\Theta = P\gamma$ is feasible for the optimization in (c) whenever P and γ satisfy the constraints of the optimization in (a), and because (a) includes additional constraints.

(d) The suggested optimization is likely to result in an MPC law with improved performance since the implied online MPC optimization minimizes a tighter upper bound on the worst case predicted cost. However the constraints involve products of optimization variables and hence are nonconvex, and furthermore there is no convexifying transformation of variables that can be employed in this case. Therefore it will in general be difficult to compute efficiently the global optimum for the suggested optimization, and the computational is likely to grow rapidly with problem size.

4 (a) Any point u that belongs to the projection onto the u-subspace of the set $\mathcal{E} = \{x = (u, v) : x^T P x \leq 1\}$ satisfies, by definition,

$$\min_v \left(u^T P_{uu} u + 2u^T P_{uv} v + v^T P_{vv} v\right) \leq 1,$$

where P_{uu}, P_{uv}, P_{vv} are the blocks of P:

$$P = \begin{bmatrix} P_{uu} & P_{uv} \\ P_{uv}^T & P_{vv} \end{bmatrix}.$$

Since $P \succ 0$ implies that $P_{vv} \succ 0$, the u-subspace projection of $\mathcal{E}$ is therefore given by

$$\text{proj}_u(\mathcal{E}) = \left\{u : u^T (P_{uu} - P_{uv} P_{vv}^{-1} P_{uv^T}) u \leq 1\right\}$$

Thus P_u is equal to the Schur complement $P_{uu} - P_{uv} P_{vv}^{-1} P_{uv^T}$. This can equivalently be expressed in terms of the blocks of P^{-1} as

$$P_u^{-1} = \begin{bmatrix} I_m & 0 \end{bmatrix} P^{-1} \begin{bmatrix} I_m \\ 0 \end{bmatrix}.$$

(b) The x-subspace projection of $\mathcal{E}_z$ is maximized by solving the SDP:

$$\begin{aligned} \underset{S}{\text{maximize}} \quad & \log\det(S_{xx}) \\ \text{subject to} \quad & \begin{bmatrix} S & \Psi^{(j)} S \\ S\Psi^{(j)T} & S \end{bmatrix} \succeq 0, \quad j = 1, 2 \\ & \begin{bmatrix} 1 & \begin{bmatrix} K & E \end{bmatrix} S \\ S \begin{bmatrix} K^T \\ E^T \end{bmatrix} & S \end{bmatrix} \succeq 0 \end{aligned}$$

The optimal solution gives $P_x = S_{xx}^{-1} = \begin{bmatrix} 0.839 & 3.211 \\ 3.211 & 13.23 \end{bmatrix}$ and $\det(P_x) = 0.783$.

(c) The matrix W appearing in the expression $J(x_k, \mathbf{c}_k) = z_k^T W z_k$ for the nominal cost could be determined by solving the Lyapunov equation $W - \Psi^{(0)\,T} W \Psi^{(0)} = \mathrm{diag}\{Q, 0\} + \begin{bmatrix} K & E \end{bmatrix}^T R \begin{bmatrix} K & E \end{bmatrix}$. However the question states that K is the unconstrained optimal feedback gain for the nominal cost, and by Theorem 2.10 we must therefore have

$$W = \begin{bmatrix} W_x & 0 \\ 0 & W_c \end{bmatrix}, \quad W_c = \mathrm{diag}\{B^{(0)\,T} W_x B^{(0)} + R, \ldots, B^{(0)\,T} W_x B^{(0)} + R\}$$

where W_x is the solution of the Riccati equation that is provided in the question and $B^{(0)\,T} W_x B^{(0)} + R = 4.891$.

The minimization of the nominal predicted cost is equivalent to

$$\mathbf{c}_k^* = \arg\min_{\mathbf{c}_k} \ \|\mathbf{c}_k\|^2 \ \text{ subject to } \ (x_k, \mathbf{c}_k) \in \mathcal{E}_z,$$

which is a convex quadratic programming problem with a single quadratic constraint. In applications requiring very fast online computation this can be solved using an efficient Newton-Raphson iteration as discussed in Sect. 2.8. If computational load is not important, it can alternatively (and more conveniently) be rewritten as a second order cone programming problem and solved using a generic SOCP solver. The solution for $x_0 = (4, -1)$ gives

$$J^*(x_0) = \|x_0\|_{W_x}^2 + 4.891\|\mathbf{c}_0\|^2 = 39.9.$$

5 (a) Whenever the constraint $(x_k, \mathbf{c}_k) \in \mathcal{E}_z$ is inactive in the optimization in Question 4(c) (i.e. whenever $\mathbf{c}_k^* \neq 0$), the line search defined in the question results in $z_k = (x_k, \alpha_k^* \mathbf{c}_k^*) \notin \mathcal{E}_z$. Therefore the constraints of the line search are needed in order to ensure that:

(i) The input (or more generally mixed input/state) constraints are satisfied at the current sampling instant
(ii) The optimization in Question 4(c) is feasible at time $k + 1$

The second of these conditions is imposed in the line search through a robust constraint on the one step-ahead prediction $z_{1|k}$ in order to ensure that the predicted cost decreases along closed loop system trajectories. This provides a way to guarantee closed loop stability.

(b) Minimizing the value of $\sigma \geq 0$ subject to the LMI of the question gives the optimal values:

$$\sigma^2 = 9.849, \quad \Theta = \begin{bmatrix} 2.856 & -0.009 \\ -0.009 & 15.385 \end{bmatrix}.$$

From Lemma 5.3 it follows that the state x_k under $u_k = Kx_k + c_{0|k}$ satisfies the bound

$$\sum_{k=0}^{\infty} \|x_k\|^2 \leq \|x_0\|_\Theta^2 + 9.85 \sum_{k=0}^{\infty} c_{0|k}^2.$$

(c) The constraints of the line search ensure that the optimization is recursively feasible, since $z_k = (x_k, \mathbf{c}_k^*)$ satisfies $\Psi^{(j)} z_k \in \mathcal{E}_z$, $j = 1, 2$, and hence by convexity we have $z_{k+1} = \Psi_k z_k \in \mathcal{E}_z$ whenever $z_k \in \mathcal{E}_z$.
The feasibility of $z_{k+1} = \Psi_k z_k$ and the definition of W_c implies that the solution of the online optimization, $\alpha_k^* \mathbf{c}_k^*$ satisfies the bound

$$\|\alpha_{k+1}^* \mathbf{c}_{k+1}\|^2 - \|\alpha_k^* \mathbf{c}_k\|^2 \leq -\alpha_k^* c_{0|k}^2$$

and hence

$$\sum_{k=0}^{\infty} \alpha_k^* c_{0|k}^2 \leq \|\alpha_0^* \mathbf{c}_0\|^2.$$

From the answer to part (b) the state of the closed loop system therefore satisfies the quadratic bound

$$\sum_{k=0}^{\infty} \|x_k\|^2 \leq \|x_0\|_\Theta^2 + 9.85 \|\alpha_0^2 \mathbf{c}_0\|^2,$$

implying asymptotic convergence: $x_k \to 0$ as $k \to \infty$. Since the origin of the closed loop system state space is necessarily Lyapunov stable (because $u_k = Kx_k$ is feasible at all points in some region that contains $x = 0$), it follows that the closed loop system is asymptotically stable. The region of attraction is the feasible set for the optimization in Question 4(c), namely the projection of $\mathcal{E}_z$ onto the x-subspace.

6 (a) Performing the optimization

$$\underset{\Xi^{(1)}, \Xi^{(2)}, \Gamma, X, Y}{\text{maximize}} \quad \log\det(Y) \ \text{ subject to (5.47) and (5.44b)}$$

and using the inverse transformation (5.45), we get the values of $A_c^{(1)}$, $A_c^{(2)}$ and C_c given in the question and

$$P_z = S^{-1} = \begin{bmatrix} 19.11 & -4.68 & -19.11 & 0.00 \\ -4.68 & 1.24 & 4.68 & -0.10 \\ -19.11 & 4.68 & 19.11 & 0.00 \\ 0.00 & -0.10 & 0.00 & 0.10 \end{bmatrix}^{-1}.$$

(b) Minimizing $\mathrm{tr}(W_c)$ subject to the LMI in the question gives

$$W_c = \begin{bmatrix} 0.561 & -0.175 \\ -0.175 & 3.715 \end{bmatrix}$$

Hence the minimum value of $J(x_k, \mathbf{c}_k)$ over $\mathbf{c}_k$ subject to $(x_k, \mathbf{c}_k) \in \mathcal{E}_z$ is $J^*(x_0) = 44.66$.

(c) The LMI satisfied by W_c implies that the optimal $\mathbf{c}_k^*$ sequence for $k = 0, 1, \ldots$ satisfies

$$\|\mathbf{c}_{k+1}^*\|_{W_c}^2 \leq \|\mathbf{c}_k^*\|_{W_c}^2 - 4.891(C_c\mathbf{c}_k^*)^2$$

and hence

$$\sum_{k=0}^{\infty}(C_c\mathbf{c}_k^*)^2 \leq \frac{1}{4.891}\|\mathbf{c}_0^*\|_{W_c}^2.$$

Therefore, from the quadratic bound on the l^2-norm of the closed loop state sequence in Question 4(c) we get

$$\sum_{k=0}^{\infty}\|x_k\|^2 \leq \|x_0\|_\Theta^2 + 9.85\sum_{k=0}^{\infty}(C_c\mathbf{c}_k)^2 \leq \|x_0\|_\Theta^2 + 2.01\|\mathbf{c}_0^*\|_{W_c}^2$$

which implies $x_k \to 0$ as $k \to \infty$, and hence by the argument that was used in Question 5(c), the origin of the closed loop system is asymptotically stable with region of attraction equal to the projection of $\mathcal{E}_z$ onto the x-subspace.

(d) For $x_0 = (4, -1)$, the maximum scaling σ such that σx_0 is feasible in each case is given in the following table.

Optimization:	Question 3(b) (Algorithm 5.1)	Question 4(c) (Algorithm 5.3)	Question 6(b) (Algorithm 5.4)
σ	1.17	1.02	1.09

This is consistent with the expectation that Algorithm 5.1 has the largest feasible set of these three algorithms, since it computes (where possible) a robustly invariant ellipsoidal set online that contains the current state, whereas the robustly invariant ellipsoidal sets in the other two algorithms are determined offline, and hence without reference to the current state, so as to maximize the volume of their x-subspace projections. Similarly the feasible set of Algorithm 5.4 is expected to be at least as large as that of Algorithm 5.2 since it coincides with the maximal volume robustly invariant ellipsoidal set under any linear feedback law. These observations are confirmed by the feasible sets plotted in Fig. A.4.

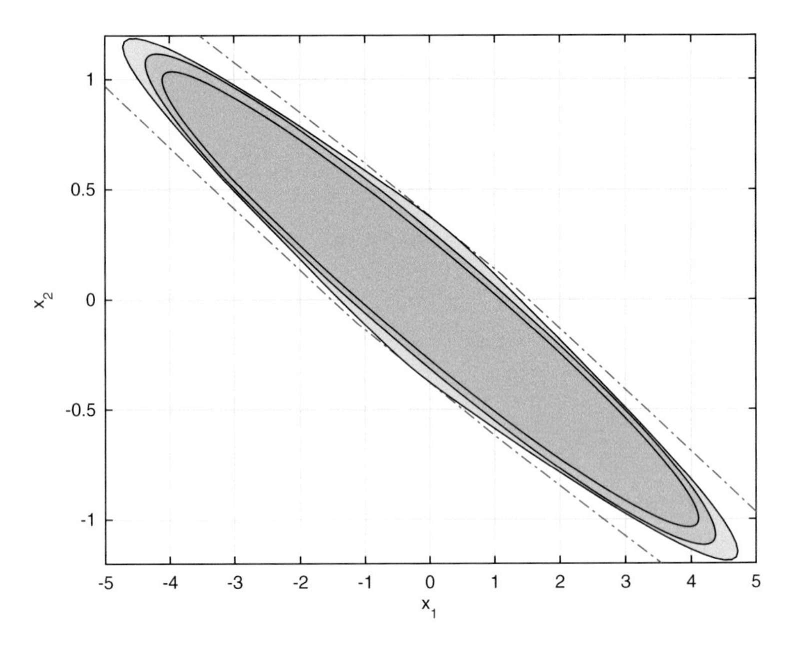

Fig. A.4 The feasible sets for Algorithm 5.2 (*inner shaded set*), Algorithm 5.4 and Algorithm 5.1 (*outer shaded set*) in Questions 6(b), 4(c) and 3(b). The *dashed line* shows the boundary of the maximal controllable set for this system and constraints

7 The volume of the low-complexity polytope $\Pi(V, \alpha) = \{x : |Vx| \leq \alpha\}$ in $\mathbb{R}^{n_x}$ can be evaluated by considering it to be a linear transformation of the hypercube $\{x : |x| \leq \mathbf{1}\}$. This gives

$$\text{volume}\big(\Pi(V, \alpha)\big) = C_{n_x} \, |\det(W)| \prod_{i=1}^{n_x} \alpha_i$$

where $W = V^{-1}, \alpha = (\alpha_1, \ldots, \alpha_{n_x})$, and C_{n_x} is a constant. Although the maximization of the determinant of a symmetric positive definite matrix P can be expressed in terms of the maximization of a concave function of its elements, e.g. $\log\big(\det(P)\big)$, which can therefore form the objective of a convex optimization, the matrix W is here neither symmetric nor positive definite in general. However for fixed V, maximizing the product of the elements of then non-negative vector α is equivalent to maximizing the determinant of the symmetric positive definite matrix $P = \text{diag}\{\alpha_1, \ldots, \alpha_{n_x}\}$, which can be expressed as a concave function of α.

8 (a) For the matrix V given in the question, the matrix $\bar{\Phi}$ whose (i, j)th element is equal to the larger of the (i, j)th element of $V\Phi^{(1)}W$ and the (i, j)th element of $V\Phi^{(2)}W$, is given by

$$\bar{\Phi} = \begin{bmatrix} 0.719 & 0.229 \\ 0.031 & 0.583 \end{bmatrix}.$$

The maximum eigenvalue of $\bar{\Phi}$ is equal to 0.760, and since this is less than unity, it follows that with α equal to the corresponding eigenvector we necessarily obtain

$$V\Phi^{(j)}W\alpha \leq \alpha, \quad j = 1, 2$$

which is the condition for robust invariance of the set

$$\Pi(V, \alpha) = \{x : |Vx| \leq \alpha\}$$

under the dynamics $x_{k+1} \in \text{Co}\{\Phi^{(1)}x_k, \Phi^{(2)}x_k\}$ (see Lemma 5.4 for the proof of this result). Here we also require that $-1 \leq Kx \leq 1$ holds for all $x \in \Pi(V, \alpha)$, and this can be ensured by scaling α.
Checking this result numerically, we have

$$\alpha = \begin{bmatrix} 0.985 \\ 0.175 \end{bmatrix}, \quad |V\Phi^{(1)}W|\alpha = \begin{bmatrix} 0.748 \\ 0.098 \end{bmatrix}, \quad |V\Phi^{(2)}W|\alpha = \begin{bmatrix} 0.615 \\ 0.133 \end{bmatrix}$$

which confirms that $|V\Phi^{(j)}W|\alpha \leq \alpha$ for $j = 1, 2$.

(b) The volume of $\Pi(V, \alpha)$ is maximized by the optimization

$$\begin{aligned} \underset{\alpha=(\alpha_1,\alpha_2)}{\text{maximize}} \ \log(\alpha_1\alpha_2) \ \text{subject to} \ & \alpha > 0 \\ & |V\Phi^{(j)}W|\alpha \leq \alpha, \quad j = 1, 2 \\ & |KW|\alpha \leq 1 \end{aligned}$$

which is convex and can be solved using, for example, any method for solving determinant maximization problems subject to linear constraints. For the problem data in the question, the optimal solution is

$$\alpha = \begin{bmatrix} 1.247 \\ 0.181 \end{bmatrix}$$

for which the conditions for robust invariance are satisfied since

$$|V\Phi^{(1)}W|\alpha = \begin{bmatrix} 0.938 \\ 0.108 \end{bmatrix}, \quad |V\Phi^{(2)}W|\alpha = \begin{bmatrix} 0.770 \\ 0.144 \end{bmatrix}, \quad |KW|\alpha = 1.$$

9 (a) Although this question uses the general complexity polytopic tube framework of Sect. 5.5, the set $\{x : Vx \leq \mathbf{1}\}$ is a low-complexity polytope. Hence the linear programs (5.94) and (5.95) that define the matrices $H^{(1)}$, $H^{(2)}$ and H_c with minimum row-sums have closed form solutions (described on p. 214). Using these solutions (or alternatively by solving an LP to determine each row) we get

$$H_c = \begin{bmatrix} 1 & 0 & 0 & 0 \\ 0 & 1 & 0 & 0 \end{bmatrix},$$

$$H^{(1)} = \begin{bmatrix} 0 & 0.041 & 0.678 & 0 \\ 0.041 & 0 & 0 & 0.678 \\ 0.391 & 0 & 0.379 & 0 \\ 0 & 0.391 & 0 & 0.379 \end{bmatrix}, \quad H^{(2)} = \begin{bmatrix} 0 & 0 & 0.999 & 0 \\ 0 & 0 & 0 & 0.999 \\ 0.348 & 0 & 0.002 & 0 \\ 0 & 0.347 & 0 & 0.002 \end{bmatrix}$$

(b) Assume that $\mathbf{c}_k = (c_{0|k}, \ldots, c_{N-1|k})$ and $\boldsymbol{\alpha}_k = (\alpha_{0|k}, \ldots, \alpha_{N|k})$ satisfy the constraints given in the question at time k. If $u_k = Kx_k + c_{0|k}$, then a feasible solution at time $k+1$ is given by

$$\mathbf{c}_{k+1} = (c_{1|k}, c_{2|k}, \ldots, c_{N-1|k}, 0)$$
$$\boldsymbol{\alpha}_{k+1} = (\alpha_{1|k}, \ldots, \alpha_{N|k}, \alpha_{N|k})$$

Therefore the constraint set is recursively feasible (Fig. A.5).

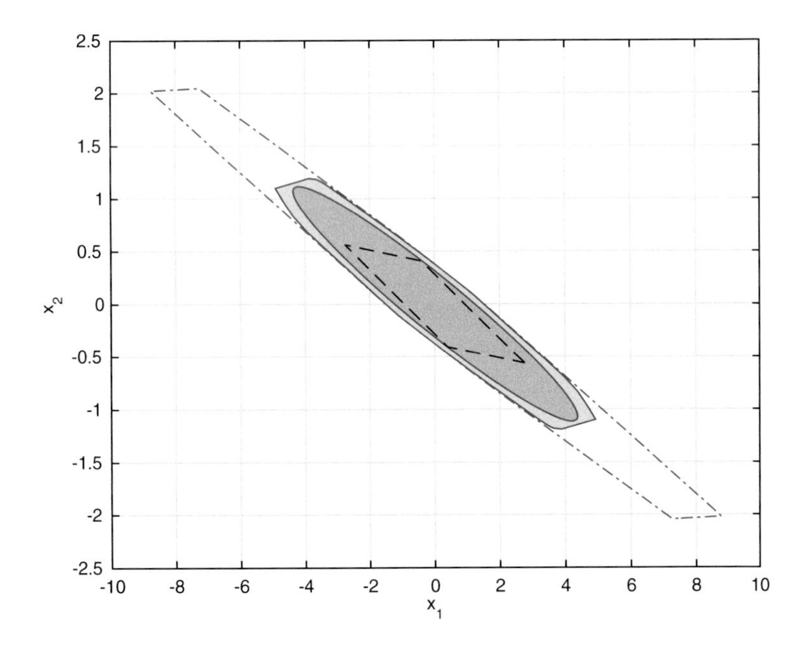

Fig. A.5 The feasible sets for the constraints in Question 9(b) (*outer shaded set*) and for Algorithm 5.4 (*inner shaded ellipsoidal set*). The boundary of the maximal controllable set (*dash-dotted line*) and the boundary of the maximal robustly invariant set under $u = Kx$ (*dotted line*) are also shown

(c) For a horizon of $N = 8$, the maximum scaling σ such that σx_0 is feasible, where $x_0 = (4, -1)$, is $\sigma = 1.134$.

(d) Solving the MPC optimization with $x_0 = (4, -1)$ gives the optimal solution for $\mathbf{c}_0$ as

$$\mathbf{c}_0^* = (0.482, 0.306, 0.251, 0.195, 0.128, 0.070, -0.020, -0.062)$$

and $s_{0|k}^* = (4, -1)$. Hence the optimal value of the nominal predicted cost is $\|s_{0|k}^*\|_{W_x}^2 + \|\mathbf{c}_0^*\|_{W_c}^2 = 37.88$.

References

1. D. Avis, *User's Guide for Irs Version 5.0* (McGill University, Montreal, 2014)
2. K. Fukuda, *cddlib Reference Manual, cddlib Version 094b* (Swiss Federal Institute of Technology, Lausanne, 2005)

Index

B. Kouvaritakis and M. Cannon, *Model Predictive Control*,
Advanced Textbooks in Control and Signal Processing,
DOI 10.1007/978-3-319-24853-0

D

R

S